S0-AKJ-186

MICROCHEMISTRY

SPECTROSCOPY AND CHEMISTRY IN SMALL DOMAINS

North-Holland
Delta Series

NORTH-HOLLAND
AMSTERDAM • LONDON • NEW YORK • TOKYO

Microchemistry

Spectroscopy and Chemistry in Small Domains

Proceedings of the JRDC-KUL Joint International Symposium on 'Spectroscopy and Chemistry in Small Domains', Brussels, Belgium, August 11–14, 1993

Edited by

Hiroshi Masuhara (Editor-in-Chief)
Osaka University
Research Development Corporation of Japan
Japan

Frans C. De Schryver
Katholieke Universiteit Leuven
Belgium

Noboru Kitamura
Hokkaido University
Japan

Naoto Tamai
Kwansei Gakuin University
PRESTO, Research Development Corporation of Japan
Japan

1994

NORTH-HOLLAND
AMSTERDAM • LONDON • NEW YORK • TOKYO

North-Holland
ELSEVIER SCIENCE B.V.
Sara Burgerhartstraat 25
P.O. Box 211, 1000 AE Amsterdam
The Netherlands

ISBN: 0-444-81513-9

This book is printed on acid-free paper.

Printed in The Netherlands.

Preface

Since the invention of laser in 1960, it has been introduced extensively in chemistry. Time-resolved laser spectroscopy using pulsed lasers was proposed and developed by chemists, and nanosecond laser flash photolysis was reported in 1967. Its time resolution was soon improved to the picosecond time domains, and nowadays femtosecond laser spectroscopy is quite common. In 1987, direct observation of the dissociation process of triatomic molecule was reported, which is one of examples showing how chemistry of ultrafast phenomena is indeed an active field of chemistry. On the other hand, monochromaticity of laser light has made the blossoming of high resolution spectroscopy possible. It is also well known that state-to-state chemistry based on the spectroscopy is another active research field in chemistry.

The other advantage of laser light is its spatial characteristics such as interference properties and focusability of the light into a small spot. When the target system is homogeneous, the spatial resolution has no chemical meaning. However, most of the chemical systems in nature are not homogeneous with respect to chemical composition, structure, and properties. One of the representative examples is the biological cell where functional organs are spatially arranged for conducting sequential reaction processes. A laser has also a high potential to interrogate such reaction dynamics in small domains. It is indispensable to understand chemical reactions as a function of the position. Furthermore, we consider that sequential reactions from site to site can be completely controlled by laser. This was the initial idea at the start of our research concerning microchemistry in 1988.

For microchemistry studies small reaction sites are indispensable where reactions are much affected by surrounding conditions such as pH, polarity, and viscosity, catalyzed by the functional surfaces, and controlled by an electrode. Partly these can be produced on materials surface by laser ablation, and can be designed and prepared by microfabrication techniques. Laser and microfabrication, both representative techniques of the 20th and 21st centuries, are thus combined with each other for chemistry studies. We considered that fabrication and spatial arrangement of minute chemical reaction sites on materials surface, measurement and elucidation of relaxation dynamics and chemical reactions in the minute sites, and reaction control in such sites would be probable by both techniques.

Indeed, novel methodologies for manipulation and creation of small materials, fabrication and functionalization of materials surface, and dynamic spectroscopies have been innovated and developed by use of laser and microfabrication. New concepts on relaxation dynamics and chemical reactions in small domains are being proposed by applying new methods. On the basis of these advances artificial reaction systems comparable to biological cells will be constructed. To simulate and develop reactions occurring at the arranged small sites is important at the present stage of investigation. We named such a study Microchemistry.

I realized sometime ago that reports of such studies were still limited and presented a possible novel area in chemistry. Microchemistry is an interdisciplinary area and relevant results are presented and published rather dispersedly (as photochemistry; spectroscopy; optics; applied physics; electrochemistry; and polymer science), and as a result hereof efficient exchanges of opinions and discussions were made difficult.

Hence, we planned to hold an International Symposium on Microchemistry and subsequently wrote to many scientists asking their opinion about possible participation in an event of this kind. All answers were positive and expressed that such a symposium would be both timely and very exciting. Thus, this volume contains the Proceedings of the Symposium on "Spectroscopy and Chemistry in Small Domains" which was held in the summer of 1993. Many excellent and interesting lectures as well as posters were presented, followed by active discussions which were indeed exciting and conducive to future research. To convey the atmosphere and also the scientific content of the Symposium we have edited the manuscripts of the lectures and classified them into five main parts. Parts of optical manipulation and creation, microfabrication and functionalization, and dynamic microspectroscopy are novel methodologies for microchemistry, where exploratory ideas and future perspectives are included. Parts of microphotochemistry, and microelectrochemistry and microphotoconversion are concerned with the relaxation dynamics and chemical reactions in small domains. I hope readers will find that these results are indeed interesting and show a high potential for future development.

Finally, on behalf of the Organizing Committee I would like to express my sincere thanks to lecturers and all participants who contributed to posters and active discussion. Also the graduate and undergraduate students of the Spectroscopy and Molecular Dynamics Laboratory of KUL are acknowledged for their efficient organization.

Hiroshi Masuhara

Hiroshi MASUHARA
Chairman of the Symposium

Organization of
The JRDC-KUL Joint International Symposium

Masuhara Microphotoconversion Project started research on microchemistry in October 1988 as part of the Exploratory Research for Advanced Technology (ERATO) program which fosters the creation of advanced technologies and advances future interdisciplinary scientific activities. This program is organized and financed by the Research and Development Corporation of Japan (JRDC), which is part of the Science and Technology Agency, an administration of the Japanese government that promotes programs a wide field of research and development. At the occasion of the end of the Masuhara Microphotoconversion Project 5-year term, JRDC and the Katholieke Universiteit Leuven (KUL) came to the conclusion to hold the symposium on "Spectroscopy and Chemistry in Small Domains" in view of the importance and future potential of this new exploratory research field. The Symposium was held from August 11 to August 14, 1993, at the Sheraton Brussels Hotel, Belgium, where 91 scientists from 11 countries convened.

Organizing Committee

Chairman: H. Masuhara (Osaka University; JRDC)
Co-Chairman: F. C. De Schryver (KUL)
K. Honda (JRDC; University of Tokyo)
N. Kitamura (JRDC, Hokkaido University)
N. Tamai (JRDC)
M. Van der Auweraer (KUL)
R. Mewis (KUL)

Program Committee

Chairman: H. Masuhara (Osaka University; JRDC)
F. C. De Schryver (KUL)
K. Honda (JRDC; University of Tokyo)
N. Kitamura (JRDC; Hokkaido University)
N. Tamai (JRDC)
M. Van der Auweraer (KUL)

Local Committee

Chairman: F. C. De Schryver (KUL)
M. Van der Auweraer (KUL)
R. Mewis (KUL)
R. Kimura (JRDC European Office)

August 11-14, 1993
Sheraton Brussels Hotel, Belgium

Participants List

Prof. J. C. André	CNRS
Dr. T. Asahi	Osaka Univ.; Microphotoconversion Project, JRDC
Ms. P. Ballet	Katholieke Univ. Leuven
Prof. A. J. Bard	Univ. of Texas
Mr. V. Bollaert	Katholieke Univ. Leuven
Prof. A. M. Braun	Umweltmesstedmik
Mr. C. Catry	Katholieke Univ. Leuven
Mr. G. Chiba	Vice President, JRDC
Dr. B. Crystall	Imperial College
Prof. F. C. De Schryver	Katholieke Univ. Leuven
Mr. D. Declercq	Katholieke Univ. Leuven
Dr. F. W. Deeg	Univ. München
Dr. R. Dekeyser	Agfa-Gevaert N.V.
Dr. G. Delzenne	Katholieke Univ. Leuven
Mr. S. Depaemelaere	Katholieke Univ. Leuven
Prof. D. D. Dlott	Univ. of Illinois
Dr. A. K. Engel	International Science & Technology Associates, Inc. ERATO Overseas Representative
Mr. H. Faes	Katholieke Univ. Leuven
Prof. J. Faure	Ecole Normale Superieure de Cachan
Prof. C. W. Frank	Stanford Univ.
Mr. R. Fujisawa	Microphotoconversion Project, JRDC
Mr. S. Funakura	Microphotoconversion Project, JRDC
Prof. K. P. Ghiggino	Univ. of Melbourne
Dr. M. C. Gower	Exitech Ltd.
Dr. L. Haeussling	BASF A.G. ZKS/H
Dr. S. Hamai	Microphotoconversion Project, JRDC
Mr. B. Hermans	Katholieke Univ. Leuven
Dr. T. Hiraga	Electrotechnical Lab.; PRESTO, JRDC
Prof. H. Hiraoka	The Hong Kong Univ. of Sci. and Tech.
Mr. J. Hofkens	Katholieke Univ. Leuven
Prof. K. Honda	PRESTO, JRDC; Univ. of Tokyo
Prof. K. Horie	Univ. of Tokyo
Dr. N. Ichinose	Microphotoconversion Project, JRDC
Prof. T. Ikeda	Tokyo Inst. Tech.; PRESTO, JRDC
Prof. M. Irie	Kyushu Univ.
Dr. M. Ishikawa	Hamamatsu Photonics Co.; PRESTO, JRDC

Dr.	M. Ishikawa	Nissan Motor Co.,Ltd.; Microphotoconversion Project, JRDC
Mr.	T. Ito	Microphotoconversion Project, JRDC
Mr.	K. Kamada	Osaka National Research Institute, AIST; Microphotoconversion Project, JRDC
Dr.	Y. Kawanishi	Nat'l Inst. of Mat. and Chem. Res.
Mr.	R. Kimura	JRDC European Office
Dr.	A. Kirsch-De Mesmaeker	Free Univ. of Brussels
Prof.	N. Kitamura	Hokkaido Univ.; Microphotoconversion Project, JRDC
Prof.	J. Klafter	Tel Aviv. Univ.
Prof.	M. Kotani	Gakushuin Univ.
Prof.	K. Kurihara	Nagoya Univ.; PRESTO, JRDC
Dr.	R. Lazzaroni	Univ. De Mons-Hainaut
Dr.	J.-P. Lecomte	Univ. Libre de Bruxelles
Prof.	H. Masuhara	Osaka Univ.; Microphotoconversion Project, JRDC
Prof.	S. R. Meech	Heriot-Watt Univ.
Ms.	R. Mewis-Suy	Katholieke Univ. Leuven
Dr.	H. Misawa	Univ. of Tokushima; Microphotoconversion Project, JRDC
Prof.	D. Möbius	Max Planck Inst.
Prof.	K. Nagayama	Univ. of Tokyo; Protein Array Project, JRDC
Dr.	K. Nakatani	Microphotoconversion Project, JRDC
Dr.	R. M. Negri	Katholieke Univ. Leuven
Dr.	D. Neher	MPI Für Polymer Forschung
Mr.	N. Noma	Osaka Univ.
Dr.	D. Noukakis	Katholieke Univ. Leuven
Ms.	A. Onkelinx	Katholieke Univ. Leuven
Dr.	C. Pac	Kawamura Inst. of Chem. Research
Dr.	R. Pansu	Ecole Normale Superieure de Cachan
Prof.	A. Persoons	Katholieke Univ. Leuven
Prof.	M. Pluta	Institute of Appl. Optics, Poland
Dr.	C. Porter	Univ. of East Anglia; Microphotoconversion Project, JRDC
Prof.	A. Reiser	Polytechnic Univ.
Dr.	G. Rumbles	Imperial College
Dr.	K. Sasaki	Osaka Univ.; Microphotoconversion Project, JRDC
Dr.	R. Schoonheydt	Katholieke Univ. Leuven
Mr.	G. Schwalb	Univ. München
Mr.	T. Shimidzu	Kyoto Univ.
Mr.	N. Shimo	Microphotoconversion Project, JRDC
Dr.	T. A. Smith	Univ. of Melbourne
Dr.	R. Srinivasan	UV Tech
Mr.	H. Sugimura	Microphotoconversion Project, JRDC

Dr.	N. Tamai	Microphotoconversion Project, JRDC
Prof.	I. Tanaka	National Inst. for Academic Degrees
Dr.	D. R. Terrell	Agfa-Gevaert N.V.
Mr.	T. Uchida	Microphotoconversion Project, JRDC
Mr.	H. Uchino	JRDC
Dr.	H. Uytterhoeven	Agfa-Gevaert N.V.
Dr.	M. Van der Auweraer	Katholieke Univ. Leuven
Dr.	J. Van Stam	Katholieke Univ. Leuven
Mr.	P. Viville	Univ. De Mons-Hainaut
Prof.	G. M. Whitesides	Harvard Univ.
Prof.	F. Willig	Max-Planck-Gesellschaft
Prof.	M. A. Winnik	Univ. Toronto
Prof.	M. S. Wrighton	Massachusetts Inst. Tech.
Ms.	T. Yamazaki	Hokkaido Univ.
Prof.	I. Yamazaki	Hokkaido Univ.
Mr.	M. Yanagimachi	Mitsui Toatsu Chemicals, Inc.; Microphotoconversion Project, JRDC

TABLE OF CONTENTS

Part V: MICROPHOTOCHEMISTRY

Part I: Introduction

MICROCHEMISTRY
Spectroscopy and Chemistry in Small Domains
Edited by H. Masuhara et al.

Microchemistry by laser and microfabrication techniques

H. Masuhara[#]

Microphotoconversion Project,[†] ERATO, Research Development Corporation of Japan, 15 Morimoto-cho, Shimogamo, Sakyo-ku, Kyoto 606, Japan

Exploratory microchemistry study has been started in the past few years by utilizing laser and microfabrication techniques, whose aim, concept, methodology, and an ultimate goal are explained. Optical manipulation and creation, microfabrication and microfunctionalization, and dynamic microspectroscopy were developed as potential methodologies for microchemistry. 1) Slow proton transfer, solvent reorientation, and excimer formation dynamics, 2) viscosity change in a microdroplet, 3) rapid completion of mass transfer in a droplet, 4) fast response of photoinduced swelling of microgels, and 5) lasing dynamics in a microdroplet have been found with these methodologies as new chemical phenomena characteristic of small domains. These are interpreted on the bases of structural, diffusional, and optical origins which are due to molecular interaction and association/orientation, rapid completion of diffusion, and interactions between confined light and molecular systems in μm small domains, respectively. As an advantage of microchemistry spatial control of chemical reactions is demonstrated, and the present stage and future development of microchemistry is discussed.

1. INTRODUCTION

Chemical reactions have been studied in general for homogeneous systems such as bulk solution and gas phase. Laboratory experiments and chemical processes in factory plants always need vigorous stirring of solutions, while powder catalysts are sometimes added to enhance chemical selectivity and reaction efficiency. The surface of catalysts provides inhomogeneous reaction fields, however, the whole reaction system is still homogeneous in the sense that added particles, droplets, films are homogeneously dispersed. Contrary to these chemical reaction fields, one of the representative inhomogeneous chemical systems is biological cells of a few ~ a few tens of μm, where functional organs are spatially arranged corresponding to proceeding chemical reactions. Minute sites, where energy is generated, reactants are transported and catalyzed, and products are extracted, are set to process

Permanent Address: Department of Applied Physics, Osaka University, Suita, Osaka 565, Japan.

† Five-year term project: October 1988 ~ September 1993.

sequential reactions in the cascade way. Namely, spatial arrangements have a special meaning in conducting reactions efficiently and selectively. It has never been believed that such organized reaction sites are prepared and reactions on them are well controlled in artificial manners. However, recent advances in microfabrication in electronics industry remind us this dream. Arbitrary patterning of various metals and semiconductors in μm dimension is now possible, and some chemists and materials scientists have already used the fabricated microstructures as sensors and functionalized microelectrodes. Indeed, their advantages such as compactness, high sensitivity, convenience, and so on have contributed to new developments in the relevant fields.

We considered that some of the fabricated and functionalized electrodes would be suitable as chemical reaction sites and various kinds of such sites would be designed and prepared. To conduct chemical reactions from step to step on the series of sites, we should manipulate small materials, energize the specified sites, interrogate chemical reactions, and control reaction paths arbitrarily. These were considered to be achieved only by introducing lasers, since spatial characteristics such as interference properties and focusability are available in addition to monochromaticity and short pulse.

The idea to conduct chemical reactions on spatially arranged minute sites with chemical functionalities provide many challenging subjects in chemistry [1]. Various preparation methods should be innovated to construct sequential micro reaction sites. Reactants, products, and some materials should be freely manipulated in small volumes. How to measure and control chemical reaction in minute sites are also important topics. After these methodological problems are solved, questions about the nature of chemical reaction in small domains are raised. Are structures and properties of solution the same to those of bulk? What are characteristics of chemical

Table 1
Comparisons between microchemistry and microelectronics

	Microchemistry	Microelectronics
Function	Chemical Functionality	Electronic Properties
Materials	Organic, Polymer	Metal, Semiconductor
State	Solution, Wet	Solid, Dry
Leading Role	Photon	Electron
Reversibility	Irreversible	Reversible
Condition	Steady State	Dynamic
Domain	Interface & Surface Layers	Interface & Surface
Dimension	μm → Sub μm	Subμm → nm

reactions in μm-sized small volumes? Every relaxation process and chemical reaction probably reflect dynamic nature of solution in small volume. We have considered that a new field of "microchemistry" is being created [1,2].

To make clear the target where we have been involved, characteristics of microchemistry discussed here is compared in Table 1 to those of microelectronics which is well recognized in general science and technology. In microelectronics, electronic properties and functionalitics are of course the purpose, electrons have a leading role in controlling the system, and metals and semiconductors are major materials. Interface and surface have received much attention as domains exhibiting enough electronic functionality, and their size is being reduced to nm. On the other hand, function, materials, condition, leading role, and so on of microchemistry are in distinct contrast with those of microelectronics. One can understand that studies on chemical reactions in μm-sized volumes cover wide range of physics, chemistry, and materials science, and need various kinds of ideas, methodologies, and concepts. It is considered that microchemistry is one of important and indispensable research fields in modern science and technology.

Here we have to make a small comment that the word of "microchemistry" has already been used in analytical and semiconductor-related chemistry. The former is concerned with extremely diluted systems, and only the detection techniques are related to the present topic because both need high sensitivity. The latter chemistry consists of preparation and fabrication of μm pattern of silicon, GaAs, InO_2, and so on and have received much attention in view of chemical reactions [3-6]. Namely, space-resolved chemical processes in electronics industry are sometimes called microchemistry. As described above, however, we consider that microchemistry proposed and reviewed here is quite different from analytical and semiconductor-related chemistry, and our trials are novel and more general.

In the past few years, we have devoted ourselves in developing new methodologies for microchemistry on the basis of laser and microfabrication techniques. Both are representative techniques of the 20th and 21st centuries and enough to overcome the challengings subjects. Using the methodologies we have elucidated new chemical phenomena characteristic of μm-sized small volumes. Although the reactions occurring in μm dimension have been considered to be similar to those of the bulk, analysis of chemical dynamics have provided new insights. To sum up the results and to demonstrate a future high potential of microchemistry, we have presented some new control ways of chemical reactions; spatial control of reactions by light and electrodes. These advances are summarized and the concept of microchemistry is described in this review.

2. NEW METHODOLOGIES FOR MICROCHEMISTRY

To open new chemistry it is indispensable to develop new methodologies. Some tools to choose materials as well as reagents, to transfer them to a certain point, and to induced chemical reaction are required. In order to conduct chemical reactions in μm-sized small volumes, we need minute reaction sites where environmental conditions such as pH, solute concentration, viscosity, and polarity can be controlled well. Spectroscopic measurement with spatial and temporal resolutions are also indispensable. Utilizing laser and microfabrication techniques, we have developed and proposed novel methodologies, which are described in this section.

2.1. Optical micromanipulation and creation

It was considered that laser is a unique technique like hands to conduct chemical reactions in small domains. By focusing the laser beam to a small particle under a microscope, radiation force is exerted on the particle. Arranging optical set up, it is possible to catch, transfer, and fix a single microparticle at a certain position in solution against its Brownian motion; laser trapping. The particle can be decomposed, evaporated, fused, and fabricated photochemically by focusing an additional UV laser pulse into the particle. Combining the trapping technique with dynamic microspectroscopy and microelectrochemistry systems which we have developed separately, various versions of laser trapping methods have been proposed as summarized in Table 2. As reviewed in detail in this volume, spectroscopic and electrochemical analysis, photochemistry, and fabrication of individual particles in solution have been made possible just as in the bulk. Furthermore, we have succeeded in developing the laser trapping method for manipulating individual particles arbitrarily in three dimensional space.

Table 2
Various versions of laser trapping methods

combination	properties / processes
laser trapping / fluorescence spectroscopy	concentration determination
	depth profile
	excimer dynamics
	solvation dynamics
	lasing dynamics
laser trapping / absorption spectroscopy	T-T annihilation
laser trapping / photochemistry	ionization
	polymerization
	adhesion
laser trapping / electrochemistry	mass transfer
	electron transfer
	spatial control of chemical reactions
laser trapping / fabrication	cutting
	ablation
	gelation

These methods are quite unique and not comparable with any other techniques, and now can be used even to create microstructures and gels in solution. Some polymeric microspheres were dispersed in solution containing monomer cross-linking agent, catalysist, and light-absorbing molecule and trapped together with a near IR laser. The contacting area of two particles was irradiated with a UV pulse for inducing local photochemical reactions. In few seconds photopolymerization was completed, and the two particles were adhered firmly. By repeating this procedure, various microstructures were constructed.

Poly(n-isopropylacrylamide) in aqueous solution shows thermal phase transition between microgel-dispersed and transparent solutions. When the temperature was elevated by focusing the intense near IR laser, gel formation was observed only at the focal point of the microscope, which was interpreted in terms of local heating of water. The prepared particle was of course simultaneously trapped by the beam. By switching off the laser beam the particle disappeared quickly, because the temperature was lowered again. Thus, creation and deletion of a single gel microparticle was successfully controlled. It is worth noting, furthermore, that the shape of the optically prepared microparticle was large and really spherical compared to that of thermally produced gels.

2.2. Microfabrication and microfunctionalization

Since chemical reactions are always sensitive to the surrounding environmental conditions, triggered by light and electron transfer to/from electrodes, and assisted by catalysts, microchemistry needs small chemical reaction sites. Microfabrication producing small structures is already established in the fields of microelectronics, which is a powerful method for preparing minute chemical reaction sites. Materials used in electronics industry are usually metals and semiconductors, and their electronic properties have been characterized and utilized. On the other hand their chemical properties have received not so much attention, although fabrication, namely, chemical processing of materials is concerned with chemical reactions such as adsorption, dissolution, bond scission and bond formation. We consider chemical aspects of microfabrication should be more emphasized and applied in microchemistry compared to those in electronics industry. Here microfabrication and micro-functionalization methods newly developed in our project are summarized.

Firstly, we describe photochemical modification of polymer surface contacted with solution of aromatic derivatives. Excited molecules near the surface attack the reactive groups of the polymer. Since the introduced substituent and the chemical bond newly formed on the surface can be characterized well, it is easy to design further reactions for planting functional groups on the surface. One of representative examples is photoinduced electron transfer reaction between aromatic alkene/aromatic cyclopropane and electron acceptor leading to production of the former cation and the latter anion. Produced cation of aromatic derivatives under-went addition to OH group of poly(2-hydroxyethyl methacrylate) (PHEMA). The surface layer with thickness of 7 nm was modified by pyrenyl group, and lateral resolution was obtained to be a few tens μm by using contact photomask. It is noticeable that the spatial resolution could be determined only by microspectroscopy, since no surface morphology was changed. Usually NMR, IR, and XPS measurements cannot be applied, while fluorescence spectroscopy is effective because of its high sensitivity. The modification was due to solution phase photochemistry near the surface, hence diffusion of ion radicals and optical conditions in alignment determine the spatial resolution.

Secondly, excitation of polymer surface with an intense laser pulse causes decomposition, melting, vaporization, and ejection of materials, leaving small holes. This is called laser ablation and has been used for microfabrication [7,8]. Since photochemical reactions are densely induced on the surface layers, active radicals are left on the fabricated areas. These radicals can react with small molecules in the gas phase and, aromatic molecules and dyes dissolved in solution. Namely, the materials were fabricated in the gas, or the ablated materials were immersed in solution immediately after irradiation. The spatial resolution of laser ablation is usually μm, so that simultaneous microfabrication and microfunctionalization were attained. Repeating the process, it is possible to modify the fabricated surface with plural functional molecules.

Thirdly, hole and electron are formed on the semiconductor surface upon photoexcitation, whose charge separation can be used for its microfabrication and microfunctionalization. Spatially patterned deposition of metals from the corresponding ions in solution was conducted by irradiating the semiconductor surface through a photomask. Reversibly, metal film on the semiconductor surface contacting with etching solution was photochemically dissolved, and its patterning was completely controlled by using a photomask. Since metals and semiconductors contacting solution are proved to be electrochemically and photochemically active, the present method is quite useful for preparing small reaction sites with 2 μm resolution.

Fourthly, scanning tunneling microscopy (STM) is another important and powerful method for microfabrication and functionalization, although it has been used as an observation method. Its geometrical arrangement constitutes an electrolyte cell between the tip and the substrate, if the electrolyte solution is inserted, and electrochemical reactions are controlled in the small volumes. It is worth noting that the current is not tunneling but faradaic in this case. The method is called scanning electrochemical microscopy (SECM) and receives much attention as microfabrication and microfunctionalization. The electrochemical reactions are conducted in the small domains, and oxidation-reduction reactions may form the products. By scanning the tip, we can make arbitrarily micropatterns of chemical products with high spatial resolution. Most of the applications of SECM to materials surfaces have been limited to semiconductors and metals, however, we considered that organic molecules and polymers are also nice targets. It was demonstrated that sub μm patterns of electrochemical products were formed in wet ion-conducting polymer films. This means that any kind of functional molecules can be distributed in the polymer film with high space resolution by applying SECM.

To develop SECM as microfabrication and microfunctionalization method for the surface, the following points can be carefully examined. The resolution is determined by lateral diffusion in electrochemical reactions. Confining the electrolyte solution, the space resolution should be improved. Adsorbed water on the surface cannot be ignored in SECM fabrication and has an important role for determining electrochemical reactions. Electrolysis of water in the small space or electrolysis in aqueous solution induces etching of substrate, deposits metals ions on the surface, and so forth. The volume of adsorbed water was adjusted to be very small, then space resolution was confirmed to be reduced to 25 nm. It is an irony that micropattern of organic molecules is hardly confirmed, because STM and SEM observation of planted organic molecules are impossible. We have demonstrated that fluorescence observation of functionalized pattern is rather easy and promising, although its resolution is limited to sub μm-order.

Fifthly, chemical vapor deposition (CVD) is another method to pattern organic molecules on the substrate, of which idea is as follows. CVD processes of organic molecules are very much affected by surface material, hence spatial patterning of deposited materials can be controlled well by using photolithographically fabricated surface as a substrate. Demonstrations were given to form micropatterns of phthalocyanine derivatives from 1, 2, 4, 5-tetracyanobenzene or 1, 2-dicyanobenzene on copper patterns. The substrate was silicon wafers and sapphire plates where copper was patterned. This area-selective CVD was extended to direction-selective CVD where whiskers and debrises showed directional growth. Both CVD methods will be representative microfabrication and microfunctionalization method because of its high potential due to the dry process.

2.3. Dynamic microspectroscopy

One of extremely useful characteristics of laser is a short pulse duration. Nowadays pulse width of a few fs is available, so that laser phtolysis can now probe relaxation and reaction processes ranging fs to ms. Molecular vibration and rotation take place in the time range of sub ps, and electronic relaxation and excitation energy as well as electron transfer occur in the late stage. Elementary processes at the various stages after excitation are now directly measured in the time domains. However, most of time-resolved spectroscopy methods have been applied to bulk systems of solution and gas phase until now. Quite recently, fabricated and inhomogeneous systems such as biological cells, powders, colloid solutions, polymer solutions, quantum wires, ultrafine particles, have received much attention in view of dynamics in small domains. Furthermore, molecular dynamics in zeolite, polymer film, LB film, and so forth is well known as important processes in restricted geometries. All these inhomogeneous systems have been measured and evaluated as a sum of each elements (particle, microcrystal, microdomain,...).

Simple application of conventional time-resolved spectroscopy to these inhomogeneous systems gives the dynamic information, however, the data are ascribed to an average and/or a sum of properties of each elements. In some cases real nature of the small element could not be clarified or completely misunderstood. If yields, concentrations, rate constants, and so on scatters between elements, the measured value for the bulk may not reflect those of each elements. It is indispensable to analyze the dynamics by referring position, size, shape, and chemical composition of each element, which requires developments of various space- and time-resolved spectroscopic methods.

Space- and time-resolved spectroscopy is classified by how to define space-resolution. Using confocal microscope, fluorescence and transient absorption spectra are measured from point to point in three-dimensional space. In the case of fluorescence spectroscopy, sub μm space, ps time, and, nm spectral resolutions were simultaneously achieved. Using μs flash lamp or fs continuum as a monitoring light pulse, transient absorption spectra of ten μm spot were obtained in μs ~ ms or fs ~ ps time regions, respectively.

In solid materials, solutions, films, and so on, structures and dynamics of the surface are believed to be different from those of the bulk. The thickness of surface is quite thin, and bulk characteristics are considered to extend just to the inside of the surface. This is true for crystals and simple solution where interactions between solvent molecules are weak. However, in polymers, amorphous solids, biological systems, and hydrogen-bonding solvents, molecular association and orientation, and ordering changes gradually from the interface and surface, forming particular interface/surface layers. Total internal reflection spectroscopy is useful to interrogate structures and dynamics in such interface/surface layers with thickness of a few tens nm. Light passing from the materials with higher refractive index to the sample with lower refractive one with larger angle than the critical one is completely reflected at the interface, while the light penetrates into the sample to some extent. This is called evanescent wave and is used for exciting molecules, forming interference pattern, and probing dynamics. Total internal reflection fluorescence spectroscopy and attenuated total reflection UV-visible spectroscopy are representative ones and have been used for clarifying the properties of their interface/surface layers. Ps dynamics occurring in the small domains can be analyzed just in the bulk phase.

Thin films are very potential material systems for devices, sensors, and so forth, and the physical and chemical nature should be elucidated in relation to their functionality. Fluorescence spectroscopy is sensitive, so that it is very useful for studies on thin films, however, it gives nothing when thin films are nonfluorescent. In that case absorption spectroscopy is expected to be efficient, but absorbance is usually weak and the signal is behind the noise as optical path length of thin film is extremely short. Absorbance is given as a relative value of incident and transmitted light, and the signal-to-noise ratio (S/N) is really worse compared to fluorescence spectroscopy. On the other hand fluorescence is based on absolute counting of photons giving a wide dynamic range. From this viewpoint we considered that transient grating measurement would be an alternative to transient absorption spectroscopy, since diffraction measurement is also based on absolute counting of diffracted light and has a high sensitivity. The S/N value in transient grating experiment is as good as in fluorescence spectroscopy, hence thin films, where the number of molecules is extremely small, can be interrogated well. The transient grating phenomenon is based on the spatial pattern of the laser-induced refractive index change, from which inhomogeneous distributions of excited states, chemical intermediates, and released thermal energy are analyzed. Relaxation dynamics, photoacoustic processes, thermal conduction, and mass transfer can be elucidated with high time-resolution in addition to the thickness resolution.

Thus, transient grating experiment has both space- and time-resolutions, but did not give spectral information, because the probe laser beam was usually minor part of the excitation pulse. Relaxation processes, which we can assume from photophysical data of the bulk system, can be interpreted, while the experiment is not effective if the chemical reactions are involved. We proposed for the first time to introduce fs continuum as a probe pulse of transient grating experiments. Since the beam diameter of the fs continuum ranging visible and near infrared wavelength regions does not expand even after passing several meters (a small divergence), it is the best pulse for getting the diffraction spectra. Indeed it was made possible to estimate the transient absorption spectra by analyzing the diffraction spectra. Fs and ps dynamics in nonfluorescent thin film of 10 nm thickness was elucidated in detail.

Thus, defining minute space by adjusting total internal reflection and interference conditions or by adopting confocal microscope, sub μm space resolutions were attained. Fs and ps time-resolution is actived by introducing fast response detector, analysis method, and so forth which are just the same as those of conventional (spatially unresolved) spectroscopy for the bulk systems. It is now possible to monitor fast processes occurring in the small space, which will contribute to future development of microchemistry.

Finally, we make two comments concerning dynamic microspectroscopy, one of which is further improvement of space resolution. It was demonstrated in our project that the photoexcitation of the surface triggers the tunneling current in STM measurement [9]. This means that the current in STM can be synchronized to laser induced processes, leading to time resolved STM observation. Quite recently, fs time-resolved STM has been proposed and applied to measurement of dynamic processes on the substrate [10,11]. The other is related to fluorescence measurements of a single molecule [12-14]. Single molecular fluorescence detection is considered to be an ultimate goal of space- and time-resolved spectroscopy, but the space resolution is not directly given by optics. These two kinds of recent methodologies will be fruitful in the studies to analyze the dynamics on a single atom/molecule, however, no information on dynamic processes based on mesoscopic structures, molecular diffusion, and optical resonance is given. As discussed later, these origins result in characteristic behavior in μm dimension and are elucidated only by the present space- and time-resolved spectroscopy.

3. NEW CHEMICAL PHENOMENA IN SMALL DOMAINS

Combining and applying the above new methodologies for microchemistry, novel and interesting phenomena have been elucidated. Some of them are proved for thin surface/interface layers, and others are observed in individual droplets and small microgels. Furthermore, some optical phenomena characteristic of a single particle were found to have a high potential in application to photochemistry. All these constitute new aspects of microchemistry, which are here summarized.

3.1. Slow proton transfer, solvent reorientation, and excimer formation dynamics

Proton transfer of 1-naphthol upon excitation results in a formation of naptholate anion, which is well known as a representative primary photoprocess. Since dipole moment in the excited coumarin dye is larger than that of its ground state, it shows a large time-dependent fluorescence spectral shift in polar solvents reflecting solvation dynamics. Excimer formation is a fundamental process of bimolecular association and its rate is determined by molecular concentration and diffusion coefficient. At an early delay time, initial distribution and mutual orientation of two molecules affect the process. All the dynamics has been studied systematically for bulk solution and polymer films, assuming that the molecules are distributed homogeneously. With total internal reflection fluorescence spectroscopy it is possible to analyze the dynamics by changing the thickness of the surface/interface layers and monitoring variable depth of solution or film. It was found that proton transfer of 1-naphthol in aqueous solution, solvation reorientation process of coumarin in 1-butanol, and excimer formation in dilute polymer solution in the interface layer of a few tens nm thickness was slower than those of the corresponding bulk solutions. On the other hand, no abnormal behavior was observed in solvation dynamics in aprotic solvent and excimer formation in organic solvents containing no polymer. Physical basis of the interesting behavior is molecular interactions through hydrogen-bonding interaction, dipole-dipole interaction, or rather stiff polymer chain. Cluster formation may also have an important role in water. Thus, it is considered that the physical and chemical properties of the surface is transmitted to the bulk solution, and its gradient ranges a few tens nm.

Similar studies have been conducted in viewpoints of molecular dynamics in restricted geometries [15]. Small gaps between flat mica surfaces, zeolites, polymer films, and so on provide such a small space, and the structures are mainly elucidated by NMR, small angle X-ray scattering, and force measurement. The meaningful dimension is usually less than 10 nm. The difference of the dimension between these and our studies may be ascribed to physical quantities on which the considerations are based. We have always used space- and time-resolved spectroscopy and elucidated dynamics as a position of small volume, and dynamic information obtained may be more sensitive to weak molecular association and ordering compared to NMR, small angle X-ray scattering, and direct force measurements. Therefore, our approach is expected to be more excellent to understand mesoscopic properties of molecular systems.

Such inhomogeneous properties in solution lead to inhomogeneous distribution of the dopant concentration and micropolarity of polymer films. The films are in general prepared by spin-coating and casting of polymer solutions, hence the special characteristics are remained through evaporation process of solvent from the polymer solution. Using total internal reflection fluorescence spectroscopy, photophysical and photochemical processes can be elucidated for interface layers of polymers with different thickness as described above. Pyrene was doped to probe micropolarity and aggregation degree with fluorescence spectra, intensity, and dynamic behavior. Poly(methyl methacrylate), poly(p-hydroxystryrene), and poly(urethaneurea), actually, showed various inhomogeneity depending upon their chemical structure. On the other hand, fluorescence behavior of pyrene was common to substrate/polymer interface, bulk, and air/polymer surface in polystyrene. The homogeneity and inhomogeneity of polymer films seem to be consistent with those of solutions as described above.

3.2. Viscosity change in a microdroplet

Utilizing various versions of laser trapping methods, photophysical and photochemical processes in a single oil droplet dispersed in aqueous solution can be elucidated. Molecular diffusion is considered by analyzing triplet-triplet annihilation process and excimer dynamics, which is eventually ascribed to viscosity of the inner solution. In the case of microdroplets, exit and entry of solute molecule to and from the surrounding solution are rapidly completed and penetration of the outer liquid into the droplet is appreciable because of large surface-to-volume ratio; rapid re-equilibrium. We have introduce two examples. In a tri-n butyl phosphate droplet (TBP), effective viscosity was found to be different from that of the bulk and became higher with decreasing the diameter. This was consistent with high solubility of water in tri-n-butyl phosphate and interpreted by assuming efficient formation of reverse micelles in the droplet. Secondly, the re-equilibrium of solute, TBP, and water molecules is suppressed in the case of microcapsules dispersed in solution, since polymer resin wall completely prohibits their efficient exit and entry. Then, solute and inner solvent is not mixed with outer aqueous solution. This was clearly confirmed by measuring excimer formation/dissociation rate constants of individual microcapsules with different sizes. Namely, the rate constants were common to all the microcapsules measured and also to the bulk solution, indicating no change of viscosity in microcapsules.

3.3. Rapid completion of mass transfer in a microdroplet

Rapid attainment of re-equilium is due to the fact that transfer of molecules and ions across liquid-liquid interface between droplet and outer solution is completed soon. This beautiful demonstration was done by laser trapping-electrochemistry technique. A single oil droplet was trapped and contacted on a microelectrode and its oxidation-reduction reactions of molecules in the droplet were conducted by adjusting the electrode potential. The oxidized molecules were repelled to water phase owing to the distribution equilibrium, and interestingly the oxidation was completed during the first potential sweep in cyclic voltanmetry. Since the size of the microdroplet was in the order of ten μm, the time for ions to cross the interface was shorter by 4 orders of magnitude compared to the bulk (1 mm order). After stopping the electrolysis neutral molecules dissolved dilutely in water came back to the droplet, taking a few hundreds seconds. This is because neutral molecules should diffuse over a mm in the bulk. Thus, exit and entry processes of molecules across the liquid-liquid interface between droplets and surrounding solvents were directly confirmed in the time domains by measuring electrochemical behavior of a single droplet.

3.4. Fast response of photoinduced swelling of microgels

Polyacrylamide gel containing triphenylmethane leuco derivatives in water shows volume expansion (swelling) upon excitation, which is due to ionic dissociation of the pendant dye. Hydrophilicity is increased and water penetrates into the gel, leading to the swelling. This is well known as one of representative photoresponsive behavior of gels. Our idea is that the response time becomes fast as the size of the microgel is reduced, because the diffusion of polymer network in water, in other words, the diffusional flow of water into the gel is the rate-determing step. It was demonstrated that the response time of the photoinduced swelling was improved by two orders of magnitudes by reducing the size from sub cm to 10 μm. Although the swelling mechanism in microgels was confirmed to be the same to that of the bulk one, diffusion motion of water and polymer network is completed in a short time. Microgels are expected as micromachine, microactuater, sensor, and so on, hence the present result is extremely important not only from fundamental viewpoints but also from future applications.

3.5. Lasing dynamics in a microdorplet

An oil droplet, a polymeric microsphere, and a glass bead dispersed in water are deemed to be a wonderful spherical microcavity, since their refractive index is larger than that of surrounding aqueous solution. Fluorescence from the interface layer is totally reflected at the interface and comes back to the original position. Satisfying the conditions of the whispering gallery mode, optical resonance peaks are observed in the fluorescence spectrum which is replaced by lasing emission by increasing the excitation intensity. The oscillation wavelength is a function of the diameter, hence emission spectral analysis gives a correct information on the diameter. Lasing efficiency is also dependent on the diameter and, when the diameter is smaller than 10 μm and larger than a few tens μm lasing is difficult because of the low Q value.

Photophysical processes occurring in individual μm particles are well elucidated by analyzing lasing dynamics. For example, energy transfer between two dyes incorporated in the particle is enhanced compared to the bulk system, which is ascribed to the fact that the light can propagates so many times during its lifetime in the present microcavity. Furthermore, lasing emission is used as a monitoring light for transient absorption measurement in the microparticle. Dye and aromatic molecules were incorporated in a single particle and excited with ps visible and ultraviolet laser pulses, respectively. Absorption rise and decay dynamics of the excited singlet state of the aromatic molecule was obtained by measuring lasing emission intensity of the dye as a function of the delay times of two pulses. The phenomenon can be alternatively considered that molecules in the microparticle are interacted with each other through lasing emission.

3.6. Origins of μm size effects

Novel chemical phenomena summarized here have been revealed only by our new methodologies, and the physical origins are considered to be classified into structural, diffusional, and optical factors. The structural viewpoint comes from molecular interactions, association/orientation of molecules, cluster formation, and so forth which range from 10 nm to sub μm. Those in the small domains are very much affected by the interface/surface, leading to various deviation from the characteristics of the bulk. The relaxation processes and chemical dynamics are very sensitive to such structural factors and rather easily detected by time-resolved spectroscopy, which should be deemed as chemistry in mesoscopic regions. Solvent reorientation dynamics in alcohols, proton transfer in aqueous solution, excimer formation in polymer solution, and viscosity change in a single droplet can be interpreted from this structural viewpoint.

In μm volume, on the other hand, bulk solution of any kind of solvent is deemed to be homogeneous except interface/surface layers described above. Even in such case the diffusional motion results in μm size effects, which is based on the fact that molecular diffusion is completed soon in small domains. According to the equation

Table 3
A List of Microparticles

Physics	
Aerosols, hydrosols	air pollution problems
Interstellar dust	astrophysics
Liquid spray	diesel engine etc.
Optical	
Microlens	
Artificial jewelry	
Display elements	movie screen, road sign, decoration based on high light scattering power
Chemical	
Pigments	ink, paints, printing
Tonors	zerographic copying machine
Capsules	teletype, laser printer, personal computer, fax machine
Catalysts	
Gas absorbing resin particles	CO_2, O_2 for therapy of patient
Silica gel	affinity chromatography
Glass ball	for nuclear fusion
Mechanical	
Standards	for measuring number and size of dusts in clean rooms in electronics industry
Glass beads	Spacer for pricision instruments, liquid crystal display
Microbaloons	for light structures and parts in aircraft for suppressing thermal conductivity in building
Shot-brust beads	for clearing car parts, dicast, camera frame
Life and Biological	
Cosmetics	make-up substances comprised of nylon 12, face powder made of TiO_2, light scattering materials
Flavable microparticles	Japanese pepper,beefsteak plant,peppermint
Biological cells	blood, macrophage, plankton, microorganism
Microcapsules	missile drug delivery system, slowly releasing drug
Medicine	immunolatex (<µm) / immunobeads (>µm)

by the random walk problem, $r = \sqrt{6D\tau}$,where r is diffusion length, D is diffusion coefficient, and τ is the time for molecule to diffuse over r, diffusion motion in 10 μm dimension is finished faster by 6 orders of magnitude compared to 1 cm. Short time of diffusion gives transient species a chance for reacting with other molecules at the interface. This opens a new possibility to design chemical reaction systems and is indispensable to understand reactions in biological cells.

The third factor is based on enhancement of optical field in the microcavity. Lasing is one of such characteristic behavior, however, the more interesting behavior will be due to interactions between optically resonating field and molecular system. We could expect various photophysical and photochemical processes in the μm cavity, which will be a completely new area.

Micrometer dimension is characteristic of thin films and microparticles which are interesting and important materials morphology in nature, covering astrophysics, high technology industry, cosmetics, medicine, and biology. Representative examples of microparticles are listed in Table 3. It is expected that new chemical phenomena in small domains summarized here are observed and analyzed in those particles. Our methodologies and scientific understandings of structures, diffusion motions, and optical phenomena in μm domains will constitute wide research fields, all of which will be understand as microchemistry.

4. SPATIAL CONTROL OF CHEMICAL REACTIONS

The developments of new methodologies and understanding of relaxation and reaction dynamics in small domains are opening a novel approach for controlling chemical reactions. In general chemical reactions in bulk solution are conducted by adjusting temperature, pH, and polarity, adding catalysts and reactants, and removing products. These procedures are processed from step to step of chemical reactions. In microchemistry studies we can prepare small reaction sites where photoresponsive and electrochemically functional molecules are planted, manipulate various kinds of particles and powders, introduce and remove necessary substances, interrogate chemical reactions, and control reaction processes. These have been partly achieved, however, we should explore how to spatially control chemical reactions. Such trials will have a key role in microchemistry studies, on which some experimentals have been done and are explained here briefly.

4.1. Reaction control by microarray electrodes

Interdigitated μm electrodes of Pt and TiO_2, and transparent microelectrodes of Sn were fabricated on the substrate, and electric potential of each electrode was adjusted independently for elucidating effects of spatial arrangement of μm electrodes upon electrochemical reaction rate and efficiency. It was demonstrated that reactions on a certain electrode were influenced very much by the potential of neighboring electrodes. Diffusion time and collection efficiency of products from one microelectrode to the other were elucidated well, which will be useful for understanding nature of spatial arrangements of μm reaction sites.

4.2. Electrochemical and spatial control of photoinduced electron transfer

Electron donor and acceptor molecules were incorporated in a single oil droplet which was manipulated freely in aqueous solution. The photoinduced electron transfer was directly probed by measuring fluorescence decay curves in ns time region by space- and time-resolved fluorescence spectroscopy. The electron donor in the microdroplet was easily oxidized by contacting the droplet on a microelectrode with an adjusted potential, then oxidized molecule (cation) was transferred to water phase across the liquid-liquid interface. The mass transfer proceeded quickly and efficiently since diffusion length in the droplet was short. Consequently, no electron donor was left during electrolysis and fluorescence quenching was completely suppressed. When the applied potential was cut off or the particle was took away from the electrode by laser manipulation, donor and acceptor molecules were again dissolved enough in the droplet by rapid re-equilibrium between the droplet and the outer phase. Thus, photoinduced electron transfer was again observed, indicating that electrochemical and spatial control is an extremely sophisticated experiment.

4.3. Spatial control of dye formation reaction

A precursor of dye was contained in an oil droplet and the reactant was produced electrochemically on the microelectrode, which was designed for demonstrating spatial control of chemical reactions. The reactant molecules migrated from the electrode to the bulk, and some of them encountered with the droplet. At the interface of the droplet the reactant molecule and the precursor reacted with each other, leading to the dye formation. Consequently the droplet was colored and the degree of the formation reaction was easily monitored by space-resolved absorption spectroscopy. The dye formation yield in the droplet could be well controlled by adjusting the distance between the electrode and the microdroplet with our scanning laser manipulation method.

5. CONCLUDING REMARKS

As one of minute chemical reaction sites we have used microelectrodes whose potential is arbitrarily adjusted. Utilization of microelectrode are expected to be a simulation of biological cell, since functional sites can be arranged spatially in order to conduct a series of reaction processes in the cascade way. In principle electrochemical reactions can be controlled well, hence such studies seemed very fruitful. Actually the relevant report was already given by Bard in 1983, proposing integrated chemical systems [16,17]. Multicomponent and multiphase systems were considered, synergistic effects between reactions were expected, and photochemical and photocatalytic driving was concluded to have key roles in conducting reactions in integrated chemical systems. The idea of integrated chemical systems started from electrochemical studies and might be named in analogy to integrated circuit. On the other hand, our approach originates from ultrafast laser spectroscopy and photochemistry, and includes creation of materials, fabrication of materials surface, construction of microstructures, and manipulation of various substances, interrogation of chemical reactions, reaction control by light, and so forth. All of them

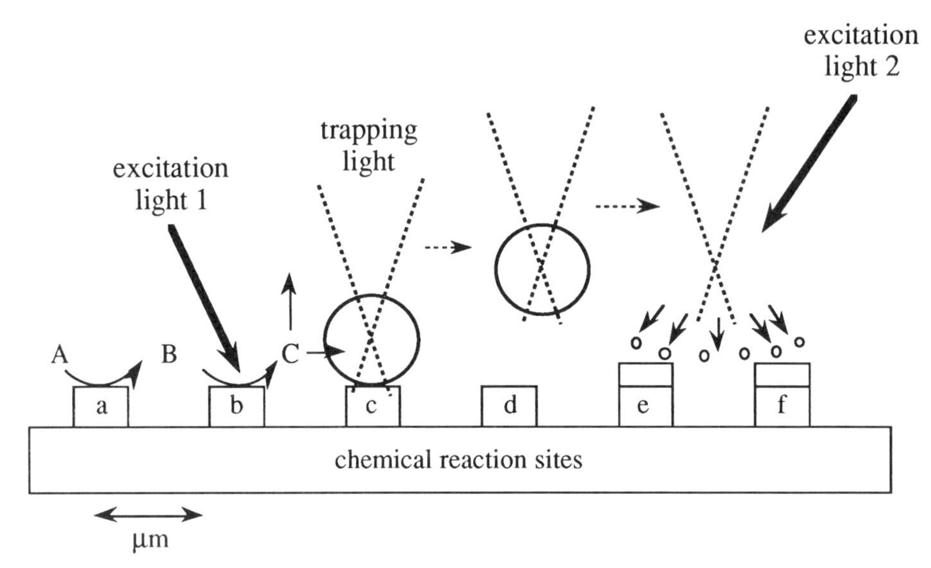

Figure 1. A shcematic diagram of Microphotoconversion system where chemical reactions are sequentially controlled by lasers on minute chemical reaction sites which are prepared by laser and microfabrication techniques.

are indispensable for chemistry in small domains and are achieved only by making the most of the laser. We understand that full utilization of lasers will make microchemistry dynamic and active.

To show our idea schematically, a prototype of microphotoconversion system is depicted in Figure 1. This has been considered to be our ultimate goal in microchemistry studies [1]. One can see that microfabrication of the materials are important and lasers are indispensable for conducting chemical reactions. Many reaction sites are already prepared and some of processes are demonstrated to be controlled by us.

Microchemistry is realized only when laser and microfabrication techniques are utilized freely. Microfabrication and microfunctionalization of materials surface are achieved by laser ablation, in addition to STM, SECM, and CVD. Furthermore, manipulation, creation, and construction of small materials and interrogation of chemical reactions in small domains are conducted exclusively by lasers. Microchemistry is made possible by lasers; in other words, microchemistry is a novel laser chemistry. Laser has superior properties such as coherency, monochromaticity, focusability, and short pulse, hence their utilization have opened new research areas on methodologies, instrumentations, reaction dynamics, and designing chemical reactions. One representative example of such laser chemistry is state-to-state chemistry. This is closely related to high resolution spectroscopy due to monochromaticity of laser. Reactions can be initiated from particular electronic, vibrational, as well as rotational energy levels of reactants and led to products with specific energy levels. Reaction path can be elucidated in detail for isolated molecules and clusters in gas phase. Multiphoton absorption, vibrational energy redistribution, and photodecomposition are practical subjects in the laser chemistry. High-time resolution in spectroscopy was attained only by introducing pulse lasers. Now fs laser is available, and studies on primary photoprocesses such as electron transfer, proton transfer, energy transfer, isomerization, dissociation, and so forth have constituted an important field called chemistry of ultrafast phenomena. This is another laser chemistry. We believe the present microchemistry will be the third laser chemistry in the sense that microchemistry is opened and made possible only by lasers [18-20].

Microchemistry is interdisciplinary between photochemistry, physical chemistry, colloid and interface science, polymer science, and materials chemistry [21-25]. Extensions to analytical chemistry are also promising. Compared to microelectronics where electron conduction should be controlled and materials are silicon and GaAs, microchemistry covers quite wide fields. Microchemistry is diverse in nature and needs detailed clarification of chemical reactions in μm dimension. Microchemistry has been started just in the past few years, and will contribute to future microscience and microtechnology via materials analysis and creation.

REFERENCES

1. H. Masuhara, Microphotoconversion: Exploratory Chemistry by Laser and Microfabrication, *In* Photochemical Processes of Organized Molecular Systems ed. by K. Honda et al., Elsevier, Amsterdam, 1991, p. 491
2. N. Kitamura and H. Masuhara, ibid., p. 509
3. R. M. Osgood Jr. Ann. Rev. Phys. Chem., 34 (1983) 77.
4. T. T. Kodas and P. B. Comita, Acc. Chem. Res., 23 (1990) 188.
5. D. J. Ehrlich and J. Y. Tsao, J. Vac. Sci. Tech. B, 1 (1983) 969.
6. A. L. Ruoff and K. -S. Chan, Materials Science, Chemistry, and Physics at Small Dimensions, *In* VLSI Electronics: Microstructure Science, 5 (1983) 329.
7. H, Masuhara, A. Itaya, and H. Fukumura, ACS Symposium Ser. 412 (1989) 400.
8. H. Masuhara and M. Fukumura, Polymer News, 17 (1992) 5.
9. H. Sugimura, N. Kitamura, and H. Masuhara, Jpn. J. Appl. Phys., 31 (1992) L1506.
10. G. Nunes Jr. and M. R. Freeman, Science, 262 (1993) 1029.
11. S. Weiss, D. F. Ogletree, D. Botkin, M. Salmeron, and D. S. Chemia, Appl. Phys. Lett., 63 (1993) 2567.
12. W. E. Moerner and L. Kador, Phys. Rev. Lett., 62 (1989) 2535.
13. M. Orrit and J. Bernard, Phys. Rev. Lett., 65 (1990) 2716.
14. E. Betig and R. J. Chichester, Science, 262 (1993) 1422.
15. J. Klafter and J. M. Drake (ed.), Molecular dynamics in restricted geometries, Wiley Interscience, New York, 1989.
16. A. J. Bard, F. F. Fan, G. A. Hope, and R. G. Keil, ACS Symposium Ser. 211 (1983) 93.
17. M. Krishnan, J. R. White, M. A. Fox, and A. J. Bard, J. Am. Chem. Soc., 105 (1983) 7002.
18. H. Masuhara, Pure and Appl. Chem., 64 (1992) 1279.
19. H. Masuhara, J. Photochem. Phtobiol. A. Chem., 62 (1992) 397.

20. H. Masuhara, N. Kitamura, H. Misawa, K. Sasaki, and M. Koshioka, ibid., 65 (1992) 235.
21. M. Toriumi, M. Yanagimachi, and H. Masuhara, SPIE Advances in Resist Technologh and Processing VIII, 1466 (1991) 458.
22. M. Toriumi and H. Masuhara, Spectrochimica Acta Rev., 14 (1991) 353.
23. M. Toriumi and H. Masuhara, ACS Symposium Ser. 527 (1993) 167.
24. N. Kitamura, Proc. of the 1st Int'l Conf. on Intelligent Material, (1992) 47.
25. H. Masuhara, N. Kitamura, and H. Misawa, Proc. of Laser Adv. Materials Processing, (1992) 1023.

Part II: Optical Micromanipulation and Creation

MICROCHEMISTRY
Spectroscopy and Chemistry in Small Domains
Edited by H. Masuhara et al.

Laser trapping and scanning micromanipulation of fine particles

Keiji Sasaki[*,#] and Hiroaki Misawa[†]

Microphotoconversion Project,[‡] ERATO Program, Research Development Corporation of Japan, 15 Morimoto-cho, Shimogamo, Sakyo-ku, Kyoto 606, Japan

Laser micromanipulation system has been developed for realizing noncontact and nondestructive positioning control of individual particles in a micrometer three-dimensional space. The microparticle was trapped against the thermal Brownian motion and freely manipulated by the radiation pressure of an infrared laser beam. Simultaneous trapping of plural particles was made possible by a scanning micromanipulation technique in which a focused laser beam was repetitively scanned by computer-controlled galvano mirrors. An arbitrary spatial pattern consisting of microparticles could be created along the locus of the laser beam. Furthermore, optical trapping and manipulation of a metal particle or a low refractive index droplet in solution, which cannot be achieved by the conventional trapping technique, was successfully accomplished using an optical caging method with the present system.

1. INTRODUCTION

Refraction, reflection, and absorption of light in/on matters cause the force on the matters, which is called radiation pressure. This force can be theoretically derived from Maxwell's equations, and the result shows that the radiation pressure is extremely weak, i.e., in pN (10^{-12} N) order, which is comparable to attractive forces between single atoms or molecules. Although macroscopic objects cannot be moved by such the weak force, the motion of micrometer-sized particles is appreciably influenced by the radiation pressure. The gravity force exerted on a 1 μm water droplet is ~5 x 10^{-15} N, and the viscous resistance on a 1 μm particle moving at the velocity of 1 μm/s in water is ~9 x 10^{-15} N, that are much smaller than the radiation pressure. Hence, we can observe a small object being levitated and transferred by light under a microscope.

---

* To whom correspondence should be addressed.
# Present address: Department of Applied Physics, Faculty of Engineering, Osaka University, Suita, Osaka 565, Japan.
† Present address: Department of Mechanical Engineering, Faculty of Engineering, The University of Tokushima, 2-1 Minami-josanjima, Tokushima 770, Japan.
‡ Five-year term project: October 1988~September 1993.

Optical trapping of a microparticle, based on the radiation pressure, was demonstrated by Ashkin in 1970 for the first time [1]. In his experiment, a polystyrene latex particle was put between opposing two Gaussian laser beams. The Gaussian beam attracts a particle toward the high intensity region at the center of the beam and pushes it in the direction of beam propagation, so that the particle is trapped in stable equilibrium at the symmetry point of the opposing two beams. In the next year, his group succeeded in levitating a particle by a single vertically directed laser beam, like a ball lifted by a fountain [2]. Since the upward radiation pressure exerted on the levitated particle should be balanced with the gravity force on the particle, the radiation pressure can be precisely estimated based on this experiment. Indeed, Ashkin et al. observed the wavelength dependence of the levitation force and clarified the relationship between the radiation pressure and optical resonances in a particle [3]. In 1986, they proposed a single-beam gradient force trapping method, in which a laser beam was focused on a particle [4]. The radiation pressure is directed to the focal point of a focused laser beam so that the particle is three-dimensionally trapped in the vicinity of the focused beam spot. This method is conceptually and practically one of the simplest and most flexible optical trapping techniques. Hence, the gradient force trapping has been widely applied in the fields of biology, chemistry, and physics [5-7].

The Ashkin's trapping technique is, however, essentially limited to a single-particle manipulation. Although the number of manipulated particles can be increased by installing multiple laser beams, the instrumental restriction determines the maximum number of the beams. In addition, a particle to be trapped by the Ashkin's method has to be transparent at the wavelength of laser light, and its refractive index must be higher than that of the surrounding medium. These limitations make a difficulty in applying the trapping technique to physical and chemical studies. In this paper, we describe a new micromanipulation technique which enables us to simultaneously trap plural particles and align them on arbitrary spatial patterns. Optical trapping and manipulation of a low refractive index droplet or a metal particle is also made possible by this technique.

2. PRINCIPLE OF LASER TRAPPING

2.1. Theory of radiation pressure

Although light has no mass, a single photon possesses the momentum given as Planck's constant divided by the wavelength of light (h/λ). When a photon flux is reflected by a mirror, the momentum of each photon is changed in its direction. Based on conservation law of momentum, the photon momentum transfers from the incident light to the mirror, which causes the force on the mirror surface. This force is called radiation pressure. The net force of a photon flux (F) can be quantitatively represented as the change of the photon momentum multiplied by the number of photon colliding on the mirror, which is expressed as

$$F = \frac{2P}{c} \cos \theta, \tag{1}$$

where P and c are the laser power and the velocity of light, respectively, and θ is an incident angle. For example, a 1 mW laser beam perpendicularly incident on a mirror exerts the force of 7 x 10^{-12} N.

The radiation pressure exerted on a microparticle under the irradiation of a focused laser beam is schematically illustrated in Figure 1. The laser light incident on a particle is refracted twice on the boundary when it goes in and out the particle. As in the case of reflection, refraction of light induces the photon momentum change, since the propagation direction and the wavelength are changed in the different refractive index media. The momentum change, $\Delta\mathbf{P} = \mathbf{P}_1 - \mathbf{P}_0$, where $\mathbf{P}_1$ and $\mathbf{P}_0$ are momentums before and after refraction, respectively, is transferred to the particle, so that the radiation pressure is exerted in the direction opposite to that of $\Delta\mathbf{P}$ (i.e., $-\Delta\mathbf{P}$), as shown in Figure 1. If the refractive index of the particle (n_p) has no imaginary part (no absorption) and a larger real part than that of the surrounding medium (n_m), the sum of force at each point of the particle irradiated by the laser beam is directed to the high intensity region at the focal point. Hence, the particle is attracted to the focused beam and three-dimensionally trapped in the vicinity of the focal point against the thermal Brownian motion, gravity, and convection.

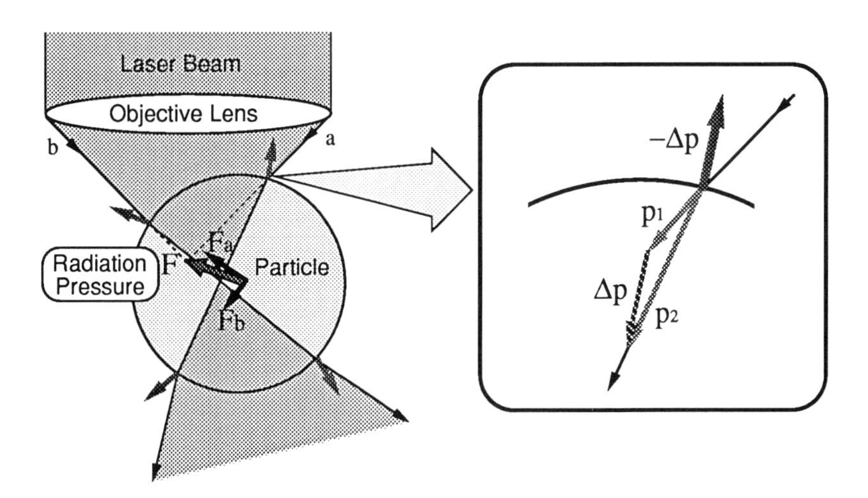

Figure 1. Principle of laser trapping.

This explanation of the optical trapping is based on geometrical optics, which can be easily understood by the photon momentum change. However, the theory cannot be applied to the particle whose size is small compared to the wavelength. For such a particle, the path of beam propagation in the particle cannot be represented as a ray, and the focal spot is not a point but a wavelength-sized spot. Hence, wave optics is indispensable for understanding the radiation pressure exerted on the small particle.

According to the Layleigh scattering theory, a particle whose diameter is much smaller than the wavelength works as a single electric dipole. The dipole experiences the Lorenz's force exerted by the optical electromagnetic field. This force corresponds to the radiation pressure, which can be theoretically expressed as

$$F = \frac{1}{2}\alpha \nabla E^2 + \alpha \frac{\partial}{\partial t}(E \times B), \tag{2}$$

where E and B are electric field and magnetic flux density, respectively, and Δ represents a gradient with respect to the spatial coordinates. α is a polarizability of a particle, which is given by

$$\alpha = r^3 \frac{(n_p/n_m)^2 - 1}{(n_p/n_m)^2 + 2}, \tag{3}$$

where r is a radius of a particle. The first term of Eq. (2) is an electrostatic force acting on the dipole in the inhomogeneous electric field, which is called gradient force. When $n_p > n_m$, the polarizability α is a positive value so that the gradient force is directed to the high electric field intensity region. The second term is derived from the change in the direction of a pointing vector, which is called scattering force. Since the gradient force is usually much stronger than the scattering force, the radiation pressure attracts the particle to the high intensity region, which is the same phenomenon as for the case of the large particle. Hence, arbitrarily-sized particles can be trapped at a focal spot of a laser beam.

2.2. Single particle trapping

Figure 2 shows laser trapping of a polystyrene (PSt) latex particle with a diameter of 5 μm in water [8]. A CW Nd:YAG laser (54 mW) was focused on a particle (indicated by an arrow in Figure 2) perpendicularly to the plane of the photograph. Untrapped particles were transferred with moving a microscope stage along (a) x and (b) y directions, while the irradiated particle was fixed at the same position. In addition, the particle was always in focus even if the stage was scanned in the perpendicular (z) direction, which shows the three-dimensional trapping.

Laser trapping of a single microparticle such as a poly(methyl methacryrate) (PMMA) particle, a toluene droplet, a liquid paraffin droplet, or a melamine-resin microcapsule containing a toluene solution of pyrene was also successful in water. PSt and PMMA latex particles could be captured in ethylene glycol and diethylene glycol. Furthermore, nonspherical particles such as a titanium dioxide (needle-like), salmonella typhymurium (ellipsoid), and calf thymus DNA (rod-like) were also optically trapped in water. Laser trapping of higher refractive index particles was always successful in relatively low refractive index media. Ashkin et al. theoretically showed that the minimum particle size that can be trapped with a laser power of 1.5 W is 20 nm at room temperature. Indeed, trapping of a 26 nm PSt latex particle in water was reported[4].

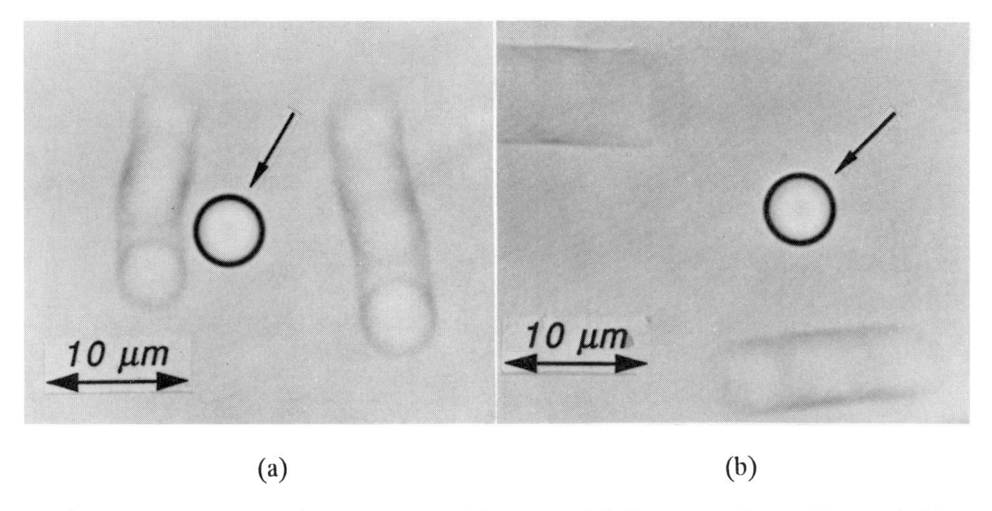

(a) (b)

Figure 2. Laser trapping of a PMMA latex particle in water along (a) x and (b) y directions.

2.3. Measurement of radiation pressure

An optically-trapped particle experiences viscous flow in a medium (i.e., resistance) when the sample stage is driven at a constant velocity. With increasing the velocity, the particle is slightly shifted so that the trapping force increases to keep the balance with the resistance. When the resistance overcomes the radiation pressure, the particle is released from the trapping. Thus, we could quantitatively determined the radiation pressure by measuring a maximum velocity (v_0) at which the particle is de-trapped. The maximum resistance (F) is given by the Stokes's law, as follows,

$$F = 6\pi\eta r v_0, \qquad (4)$$

where η is a viscosity of the medium, and r is a radius of the particle.

Figure 3 shows the laser power dependence of the trapping force along the z axis. The laser beam irradiated on a PMMA particle (6.8 μm) in ethylene glycol (η = 17.3 cP), and the sample stage was scanned by a piezo actuator under a microscope. The maximum radiation pressure was in the order of pN and proportional to the laser power, which can be explained by the fact that the number of photons interacting with a particle determines the trapping force. Figure 4 is a plot of the trapping force as a function of the particle size. The radiation pressure is stronger as the diameter increases, so that the larger particle is trapped in a deeper potential well. Besides the laser power and the particle size, refractive indices of particles and media, optical conditions such as numerical aperture and beam profile are important factors for determining the radiation pressure.

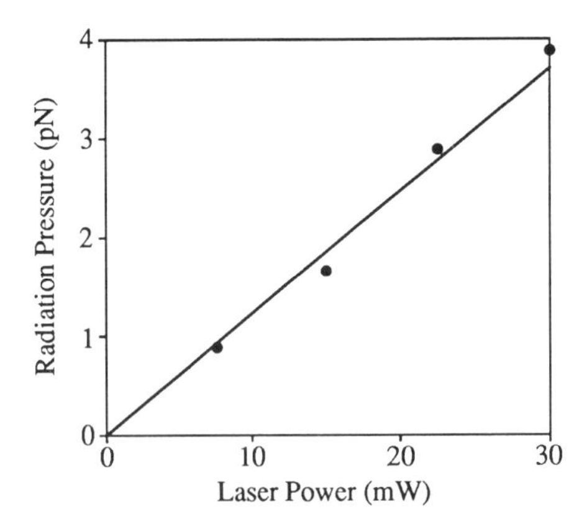

Figure 3. Trapping force along the z direction as a function of the laser power.

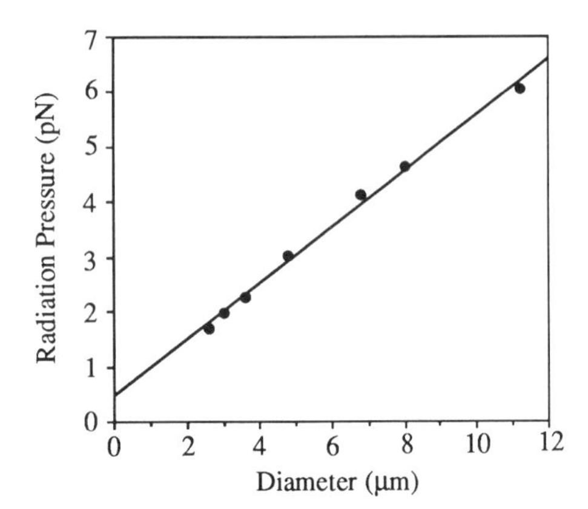

Figure 4. Particle-size dependence of the trapping force.

The calculation of the Boltzman factor with the trapping potential indicates that a 10 μm particle irradiated with a 30 mW laser beam, which experiences the radiation pressure of 5.5×10^{-12} N (see Figure 3), can be trapped against the thermal motion when the temperature is lower than 10^6 K. Hence, the pN force is sufficiently strong to trap a micrometer-sized particle at the temperature below the damage threshold of a particle. At room temperature, the walking distance of a thermally-moved particle in the trapping potential well is ~3.5 nm, which can be reduced in inverse proportional to the root of the laser power. Thus the laser trapping has the capability of nanometer positioning control of a microparticle.

3. SCANNING LASER MICROMANIPULATION

3.1. Pattern formation by scanning micromanipulation

The single focused laser beam method as mentioned above has been widely used as the noncontact and nondestructive positioning technique for a single particle manipulation. On the other hand, we demonstrated that plural particles were simultaneously trapped at the positions of intensity maxima of a standing wave field formed by a laser beam [9]. The particles could be aligned on concentric circles or a line pattern by adjusting an interference optical system. The potential of this technique is the ability to organize new functional materials and systems composed of various reactive particles. The spatial patterns produced by the interference method are, however, essentially limited to simple fringe patterns. Another possible approach is the use of photomask as is widely applied in photolithographic technology. Unfortunately, the projected image is usually degraded by speckle and/or unexpected interference fringes due to the high coherence of laser.

Furthermore, since the laser beam will be greatly attenuated by a photomask, a high-power laser will be required to achieve spatial patterning, which is likely to damage the microscope optics and the mask.

As a new approach to plural particle manipulation, we have proposed a scanning laser micromanipulation technique, which makes it possible to align microparticles on arbitrary patterns without coherent noises and loss of laser power [10]. Figure 5 shows a schematic diagram of the scanning micromanipulation system. The principal key to this technique is the repetitive scanning of a focused laser beam in a sample space. If the repetition rate of the scanning is much higher than the cutoff frequency for the mechanical response of particles in the medium, the particles cannot follow the scanning beam so that the trapping potential is given by the time-averaged intensity distribution. Hence, plural particles can be simultaneously trapped and aligned along a path defined by the scanning of a laser beam. To achieve this, we employed galvano mirrors for spatially modulating a trapping beam of a CW Nd:YAG laser. The galvano mirrors were operated by a driver and controlled by a microcomputer. The modulated laser beam was introduced into a microscope through lenses L1 and L2, which worked for matching the beam diameter to the numerical aperture of the microscope and for imaging the galvano mirror surfaces in the plane of an aperture diaphragm. In the microscope, the laser beam was focused into ~1 μm spot on a sample by an oil-immersion objective lens (NA=1.30). The micromanipulation process was monitored and recorded by a CCD camera and a video recording system.

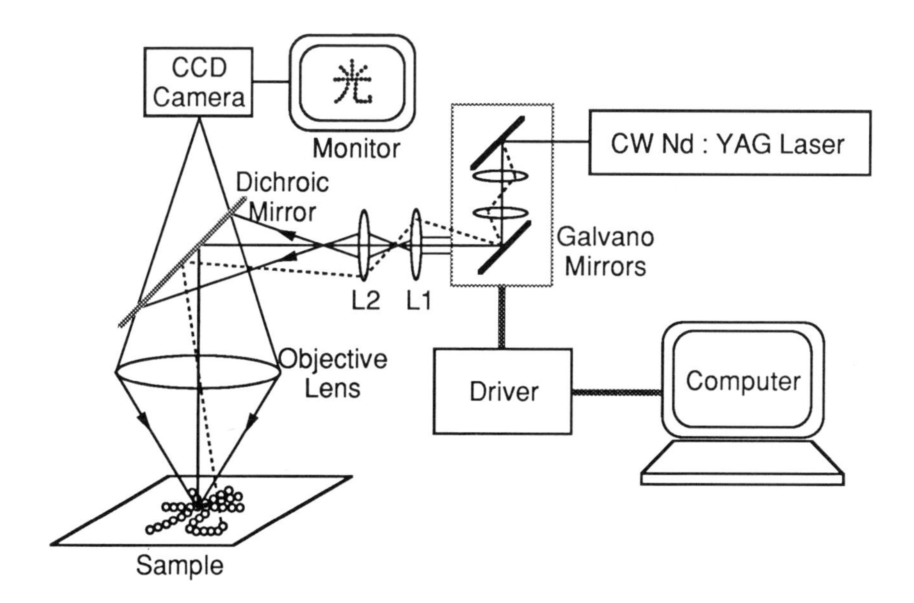

Figure 5. A schematic diagram of a scanning laser micromanipulation system.

Figure 6 demonstrates spatial patterning of 1 μm PSt latex particles in ethylene glycol along an italiclike letter of "μm". The sample solution containing the particles was placed between two quartz plates separated by a spacer of 100 μm and mounted on a microscope stage. A focal spot was scanned along the pattern at the bottom of the liquid layer with the repetition rate of 13 Hz. The spatial pattern of "μm" was formed by ~100 particles under the beam irradiation of 145 mW power, i.e., ~1.5 mW/particle. After the laser was switched on, the particles were rarely observed in the ocular field. After several tens of seconds, the radiation pressure successfully attracted the dispersed particles and created the spatial pattern on the lower quartz plate. The formed pattern could be transported in the lateral and longitudinal directions without deformation of the pattern. After the laser was switched off, the particles immediately disappeared from the pattern.

Figure 7 shows spatial alignment of 2 μm PSt particles in ethylene glycol along a Chinese character for "light". The laser power and the repetition rate of the beam scanning were 290 mW and 12 Hz, respectively. Similarly, microparticles such as PMMA latexes, titanium dioxide particles could be aligned on the various geometrical figures, letters, and so forth. Since the thermal motion is dependent on the size of the particles and the viscosity of the medium, the repetition rate were optimized for a given particle and medium as well as the complexity of the pattern to be produced.

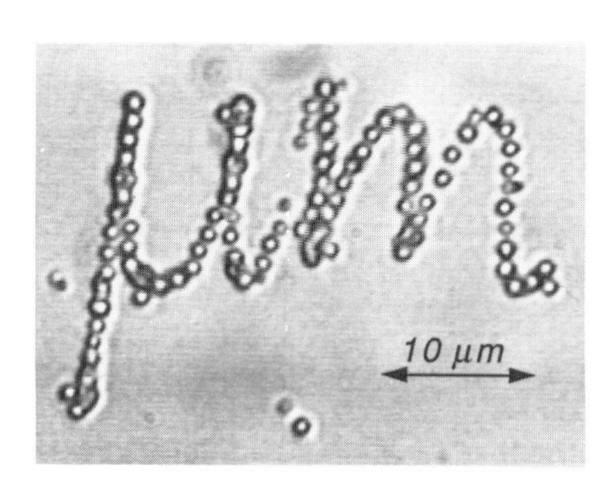

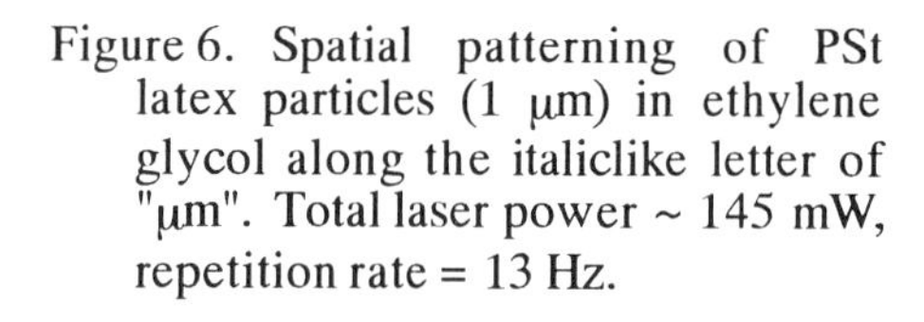
Figure 6. Spatial patterning of PSt latex particles (1 μm) in ethylene glycol along the italiclike letter of "μm". Total laser power ~ 145 mW, repetition rate = 13 Hz.

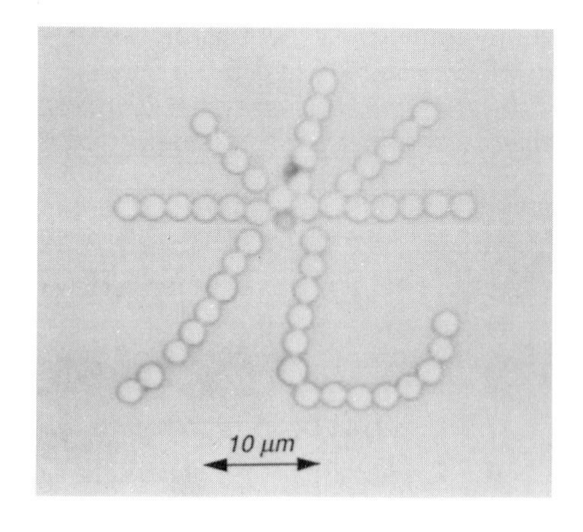

Figure 7. Scanning micromanipulation of PSt particles (2 μm) in ethylene glycol along a Chinese character for "light". Total laser power ~ 290 mW, repetition rate = 12 Hz.

The scanning laser micromanipulation is highly novel since arbitrary spatial pattern of particles can be produced by a single laser beam. The spatial pattern of particles can be easily constructed and distracted by switching the laser on or off, respectively, and is continuously varied by programming the patterns to be produced. Furthermore, the pattern formation is based on the incoherent imaging as in the case of a confocal scanning microscope so that the present technique can be further extended to the three-dimensional patterning of particles. Indeed, we succeeded in creating the spatial pattern of particles with some distance from the quartz plate, that is, the levitation of the aligned particles.

3.2. Optical manipulation of metal and low refractive index particles

The relationship between the refractive indices of a particle (n_p) and the medium (n_m) is quite important in the laser manipulation. As mentioned in Section 2.1, the gradient force trapping method is based on the attractive radiation pressure, which is obtained only for the case of a transparent particle with $n_p > n_m$. When $n_p < n_m$, contrarily, the situation is reversed so that the direction of the radiation pressure is opposite to that of the laser beam as shown in Figure 8(a). Hence, the particle experiences the repulsive force. Wave optics also indicates that the polarizability α given by Eq. (3) is negative when $n_p < n_m$ so that the particle is repelled by the laser beam toward the lower intensity region. For simplicity, we shall call such a force *repulsive radiation pressure*. For example, a water droplet (n_p=1.33) in liquid paraffin (n_m=1.46~1.47) cannot be optically trapped by a single focused laser beam. Another class of particles which cannot be trapped is those with a high reflection coefficient at the wavelength of an incident laser beam. Simple geometrical optics in Figure 8(b) indicates that such particles experience the repulsive radiation pressure analogous to those with $n_p < n_m$. Hence, metal or semiconductor particles are pushed out to the outside of the laser beam. Although the laser trapping of low refractive index droplets and highly reflective particles is required to the chemical applications, a single beam trapping technique is not applicable to these particles.

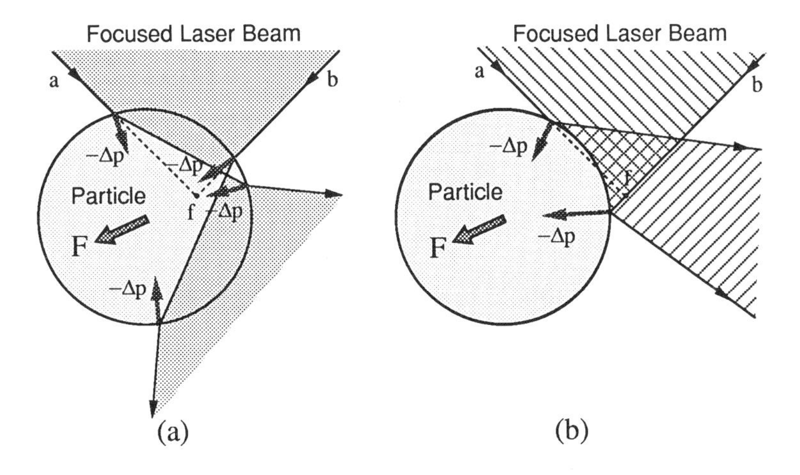

Figure 8. Repulsive radiation pressure exerted on (a) a low refractive index particle and (b) a highly reflective particle.

The scanning laser micromanipulation technique has the ability to trap such particles [11]. The principle is explained by the trapping potential shown in Figure 9. Figure 9(c) schematically shows the potential of the radiation pressure exerted on a microparticle with $n_p<n_m$, which has potential well at the focal spot. This indicates that the particle is attracted and trapped in this well. In the case of $n_p<n_m$ or a high reflective particle (Figure 9(d)), however, there is no potential well for trapping so that the particle is pushed out by the repulsive force. When a focused laser beam is repetitively scanned circularly around the particle with $n_p<n_m$ or with a high reflection coefficient, as shown in Figure 9(b), on the other hand, the potential of the radiation pressure given by the time-averaged spatial intensity distribution can be drawn in Figure 9(e). This demonstrates that circular scanning of a focused laser beam produces the potential well surrounded by relatively high potential regions. A particle which experiences the repulsive radiation pressure is expected to be *caged* inside of this particular potential well if the repetition rate of the beam scanning is faster than the mechanical motion of the particle. The focused laser beam produces high intensity region over and under the particles so that the potential well is created in the longitudinal (z) direction in addition to the x-y plane. Thus, the particle will be three-dimensionally trapped at the position where the repulsive radiation pressure are balanced with each other or with other forces.

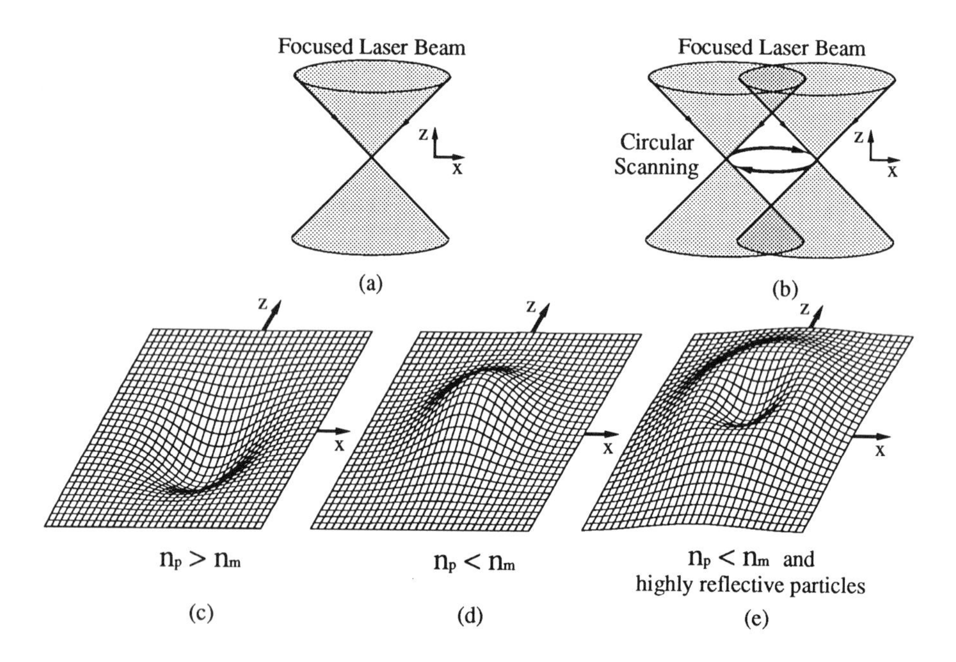

Figure 9. Schematic representations of the spatial distributions of a focused laser beam intensities (a) and (b) and the relevant potentials of the radiation pressure (c)-(e).

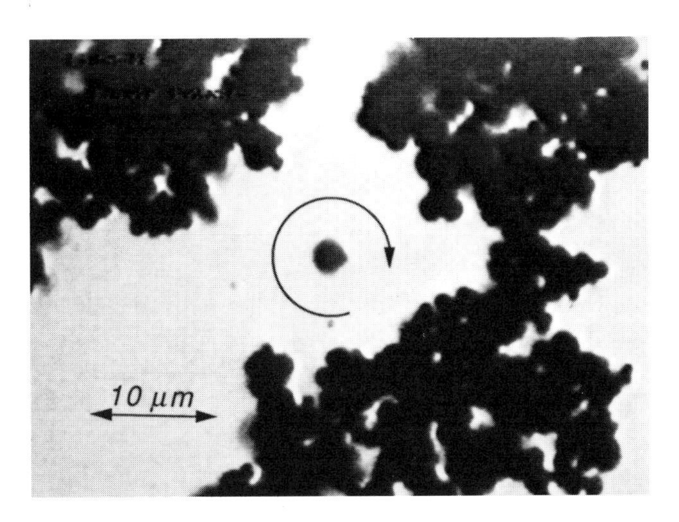

Figure 10. Optical trapping of an iron particle (3 μm) in water. The arrow indicated in the figure represents the locus of the scanning laser beam, and the particle inside of the locus is optically caged.

Figure 10 shows optical trapping of an iron particles (3 μm) in water. A focused beam of CW Nd:YAG laser (145 mW) was scanned along the circle shown in Figure 10. The particle inside the *laser cage* was optically trapped, and it could be manipulated by moving the position of the cage. Since the repulsive radiation pressure is also exerted on the particles outside of the cage, the particles neither enter nor come close to the cage, that is, the stable trapping of a single particle is achieved. For the conventional trapping of a particle with $n_p > n_m$, the prolonged irradiation of the focused beam frequently induces the aggregation of trapped particles.

Besides the scanning manipulation, a TEM_{01}* mode laser beam, which has the intensity minimum on the beam axis, also provides the potential well for a particle with $n_p < n_m$ or with a high reflection coefficient, similarly to Figure 9(e). Indeed, Ashkin et al. reported the levitation of a hollow dielectric sphere by the TEM_{01}* beam [12]. However, the potential well given by the TEM_{01}* beam is only two-dimensional, that is, there is no well in the z direction. Hence, the three-dimensional trapping cannot be performed with the TEM_{01}* beam. Furthermore, the TEM_{01}* beam trapping is restricted by the size and/or sharp of particles. On the other hand, the laser caging technique can be applied to nonspherical particles by adapting the scanning pattern and its size, which can be easily controlled by a computer.

By the present laser caging technique, iron, aluminum and carbon black particles in any kind of solvent, and water and ethylene glycol droplets dispersed in liquid paraffin were added to the list of trapped particles in Table 1. Optical trapping based on the attractive and repulsive radiation pressure is complementary with each other, and any microparticles can be optically manipulated either by scanning laser trapping or conventional trapping techniques, depending on the nature of the particle and the surrounding medium.

4. CONCLUSION

We have described the scanning laser micromanipulation, which made it possible to simultaneously trap plural particles and to manipulate a metal and a low refractive index particles. In addition to the pattern formation, this technique has the ability to move all the trapped particles along the produced pattern, that is, the continuous flow of the aligned particles can be induced by the radiation pressure [13]. Furthermore, this technique has been applied to the assembling of polymer particles where the two scanning laser beams are used for constructing an integrated latex structure like hands [14]. These applications of the scanning laser micromanipulation are expected to play a major role for the advances in physical and chemical studies of fine particles, microoptical and microelectronic devices, and micromachines.

ACKNOWLEDGMENTS

The authors express their sincere thanks to Prof. H. Masuhara and Dr. N. Kitamura, and Mr. M. Koshioka of our project for their collaboration.

REFERENCES

1. A. Ashkin, Phys. Rev. Lett., 24 (1970) 156.
2. A. Ashkin and J. M. Dziedzic, Appl. Phys.. Lett., 19 (1972) 283.
3. A. Ashkin and J. M. Dziedzic, Phys. Rev. Lett., 38 (1977) 1351.
4. A. Ashkin, J. M. Dziedzic, J. E. Bjorkholm, and S. Chu, Opt. Lett., 11 (1986) 288.
5. R. W. Steubing, S. Cheng, W. H. Wright, Y. Numajiri, and M. W. Berns, Proc. SPIE, 1202 (1990) 272.
6. A. Ashkin, J. M. Dziedzic, and T. Yamane, Nature, 330 (1987) 769.
7. S. M. Block, D. F. Blair, and H. C. Berg, Nature, 338, 514 (1989).
8. H. Misawa, M. Koshioka, K. Sasaki, N. Kitamura, and H. Masuhara, J. Appl. Phys., 70 (1991) 3829.
9. H. Misawa, M. Koshioka, K. Sasaki, N. Kitamura, and H. Masuhara, Chem. Lett., (1991) 469.
10. K. Sasaki, M. Koshioka, H. Misawa, N. Kitamura, and H. Masuhara, Jpn. J. Appl. Phys., 30 (1991) L907.
11. K. Sasaki, H. Misawa, M. Koshioka, N. Kitamura, and H. Masuhara, Appl. Phys. Lett., 60 (1992) 807.
12. A. Ashkin and J. M. Dziedzic, Appl. Phys.. Lett., 24 (1974) 586.
13. K. Sasaki, M. Koshioka, H. Misawa, N. Kitamura, and H. Masuhara, Opt. Lett., 16 (1991) 1463.
14. H. Misawa, K. Sasaki, M. Koshioka, N. Kitamura, and H. Masuhara, Appl. Phys. Lett., 60 (1992) 310.

MICROCHEMISTRY
Spectroscopy and Chemistry in Small Domains
Edited by H. Masuhara et al.

Optical harmony of microparticles in solution

N. Kitamura,*,# K. Sasaki,† H. Misawa,‡ and H. Masuhara*,†

Microphotoconversion Project,§ ERATO Program, Research Development Corporation of Japan, 15 Morimoto-cho, Shimogamo, Sakyo-ku, Kyoto 606, Japan

An optical manipulation technique has been applied to control Brownian motion of microparticles in solution and to produce dynamic optical patterns of the particles like rhythms and harmony in their movement. Microparticles were arbitrarily aligned in any letters or geometrical figures by a focused infrared laser beam and, the patterns made of the particles were shown to be manipulated in three-dimensional space as well. Besides optical alignment, active control of particles flow in any geometrical patterns and continuous changes of the patterns were also successful by a potential application of a scanning laser micromanipulation method. Possible roles of optical manipulation of microparticles in solution and future perspective of the method in science and technology are discussed.

1. INTRODUCTION

Microparticles such as polymer beads, oil droplets, capsules, semiconductors, and so forth undergo vigorous thermal Brownian motion in solution at ambient temperature. Brownian motion of particles is in a random fashion, so that we cannot control the movement of microparticles under ordinary conditions. This limits advances in chemistry and physics of microparticles. Furthermore, although studies on microparticles have a long history, any kind of measurements has never been performed for individual particles owing to a lack of a manipulation method in solution. The experimental results so far obtained are therefore always the sum or

---

* To whom correspondence should be addressed.
# Present address ; Department of Chemistry, Faculty of Science, Hokkaido University, Sapporo 060, Japan.
† Present address ; Department of Applied Physics, Osaka University, Suita 565, Japan.
‡ Present address ; Department of Mechanical Engineering, Faculty of Engineering, The University of Tokushima, 2-1 Minamijosanjima, Tokushima 770, Japan.
§ Five-year term project ; October 1988 - September 1993.

average of those for a number of particles and, a clear picture of the chemistry and physics of microparticles in solution will never be obtained. An analytical method called as "flow cytometry" enables one to observe spectroscopic characteristics of a single particle in solution [1]. However, the particle to be studied cannot be monitored during the experiments so that single particle detection or measurement is not necessarily warranted. If one can control Brownian motion of microparticles in solution, we will be able to select a particle(s) from their mixture and to investigate individual particles. It is easily expected, therefore, that the chemistry and physics of microparticles will be greatly advanced by development of a manipulation method for individual microparticles. Furthermore, arbitrary control of the Brownian motion of microparticles in solution will lead to production and control of particle motion like optical rhythms and harmony of microparticles.

A principal key to control the Brownian motion and manipulate microparticles in solution is "*radiation force of light*", which is generated by refraction of light through a microparticle [2, 3]. Refraction of light by a particle results in a change in light momentum and, the amount of the momentum change upon the light refraction should be exerted to the particle according to conservation law. This particular force is called as "radiation force" or "radiation pressure" and, can be applied to non-contact and non-destructive "*optical trapping*" or "*optical manipulation*" of various microparticles as has been described in details by Sasaki and Misawa in this volume [4]. Although optical trapping/manipulation of a microparticle was firstly demonstrated by Ashkin in 1970 [3], the method has never been applied in the field of chemistry until our first report in 1990 [5]. Interestingly, about 100 years ago in Japan, "radiation pressure" was introduced in one of the famous novels, "Sanshiro", written by a Japanese novelist, Souseki Natsume, in 1908 ;

"He set up an apparatus in midday and started experiments in night after all traffics outside of his room stopped. In the dark room, he was struggling with a number of strange lenses to test radiation pressure." ...... "Sanshiro was very surprised for his friend. Radiation pressure? How does light produce pressure? Sanshiro couldn't understand reality and any importance of radiation pressure of light."

If Sanshiro and Souseki Natsume were alive until 1970, he or they would be very much surprised for "reality" and potential applicabilities of "*radiation pressure of light*" in a variety of scientific fields.

Actually, optical (laser) trapping/manipulation has been growingly interested in a various research fields such as chemistry, physics, biology, micromachining, and so forth, and its potential applications are expected to open further new research fields. The principle, apparatus, and experimental results on optical manipulation of microparticles in solution, including a currently developed technique of scanning laser micromanipulation, have been reviewed in this volume [4, 6]. In this article, therefore, we demonstrate very elegant applications of the optical manipulation methods to produce static and dynamic optical patterns and to control the motion of microparticles in solution (optical rhythms and harmony). We show a new possibility of microparticle science.

2. OPTICAL HARMONY IN MOVEMENT OF MICROPARTICLES

2.1. Optical Control of Brownian Motion of Microparticles in Laser Interference Patterns

A single microparticle in solution can be arbitrary trapped and manipulated in three-dimensional space when a particle is irradiated by a focused, 1064 nm laser beam under the condition of $n_1 > n_2$, where n_1 and n_2 are the refractive indices of the particle and the surrounding medium, respectively [4 - 9]. Besides single-particle trap, simultaneous optical trapping of a number of microparticles can be also achieved. Namely, when the laser beam intensity is high enough, a number of particles can be optically patterned in laser beam images. Quite recently, indeed, Burns et al. reported that optical trapping could create three-dimensional arrays of polymer microparticles in interference patterns produced by several trapping laser beams and they called such the particle patterns as "*optical matters*" [10]. However, the patterns produced are static similar to crystal structures of materials. Independently, we demonstrated that both *static and dynamic* spatial patterns of microparticles in solution could be created in interference patterns produced by *a single laser beam* [11].

Figure 1 shows spatially-patterned optical trapping of polystyrene particles with the diameter (d) of ~1 μm in water. Upon irradiation of a 1064 nm laser beam from a CW Nd^{3+}:YAG laser to a sample solution under an optical microscope with interference optics, the particles flowing near the laser beam were attracted and trapped in the high laser-intensity regions of the interference pattern (Figure 1a). Further prolonged irradiation led to optical trapping of a number of the polymer particles and, after several minutes, spatially-patterned optical trapping of the particles along the concentric circles was achieved as shown in Figure 1b. This is one example for optical control of the Brownian motion of microparticles in interference patterns by a laser beam. The optical pattern of the particles disappeared immediately after switching-off the laser owing to thermal Brownian motion (Figure 1c). As characteristic features, the spatial pattern of the laser beam in the sample solution almost coincides with that of the polymer particles. Modulation of the laser beam diameter therefore renders the change in the diameter of the spatial pattern. We succeeded in controlling the diameter of the pattern in Figure 1b between 20 and 40 μm by modulating optical alignment of the laser - optics system. The diameter of the concentric circles made of the microparticles was dynamically controlled, producing like rhythms and harmony in the movement of microparticles in solution. Besides the concentric circles, we also succeeded in producing line images of the particles by interference of the laser beam.

Control of the interference pattern of the laser beam leads to separation of microparticles through the particle diameter. Namely, when we use a mixture of particles with d ~ 1 and 0.25 μm as a sample solution, spatially-patterned optical trapping of the particles results in particle size-selection. Although the particles with both d ~ 1 and 0.25 μm in the vicinity of the laser beam image are trapped to form a spatial pattern, optical trapping is more favorable for the larger-sized particle (~ 1 μm) at given laser power [11]. Upon prolonged irradiation of the laser, therefore, the number of the ~ 1 μm particle increases in the concentric circles (Figure 2a). When

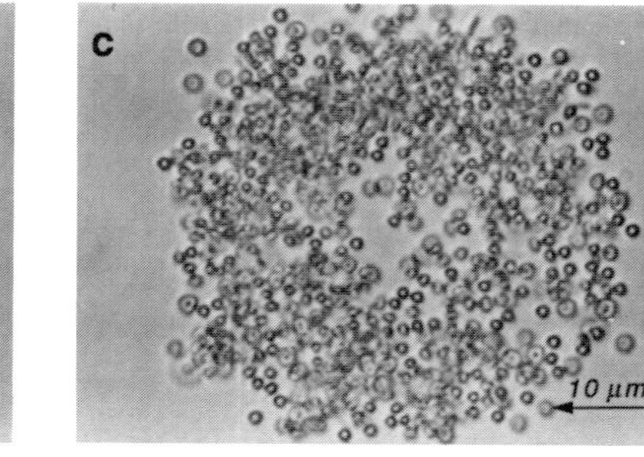
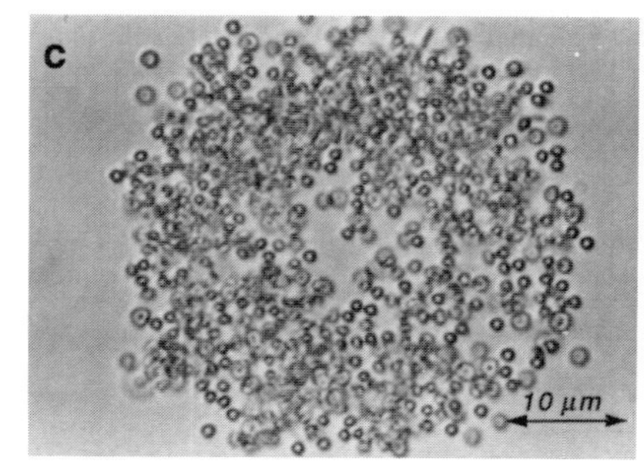

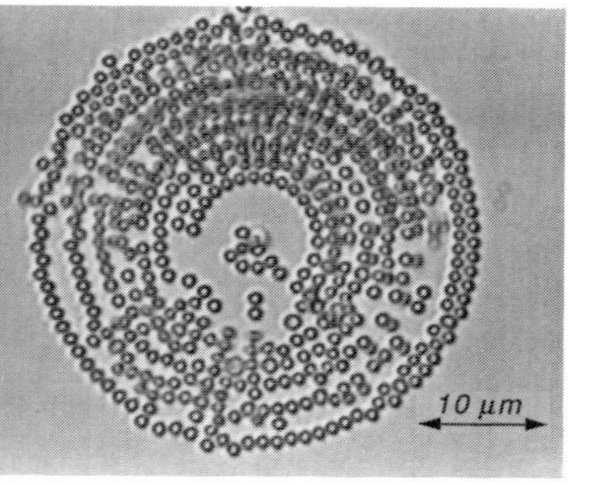

Figure 1. Spatially-patterned laser trapping of polystyrene particles (d = 1 μm) in water (a and b). 2 second after switching-off the laser beam. (c).

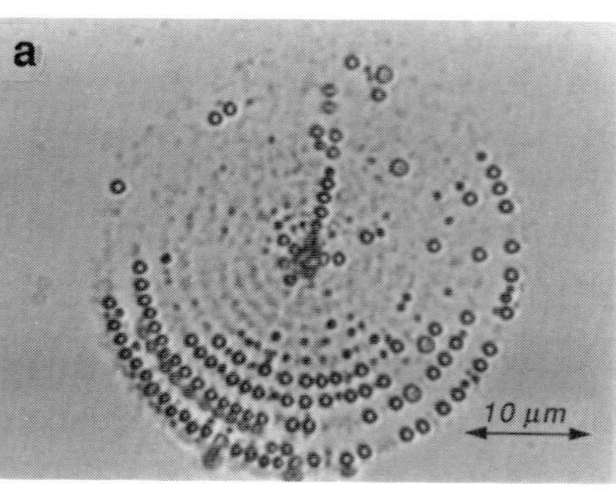
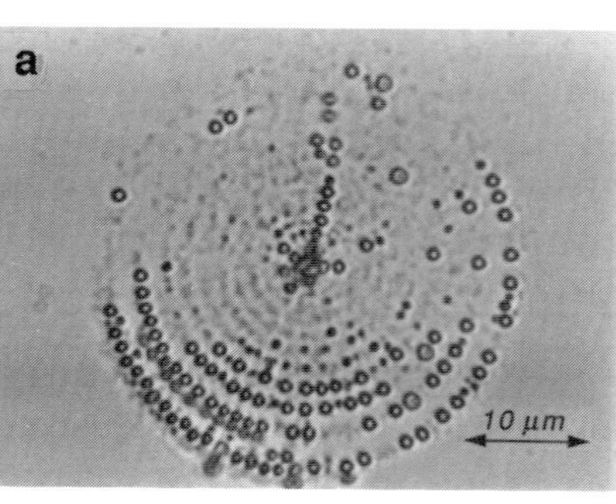

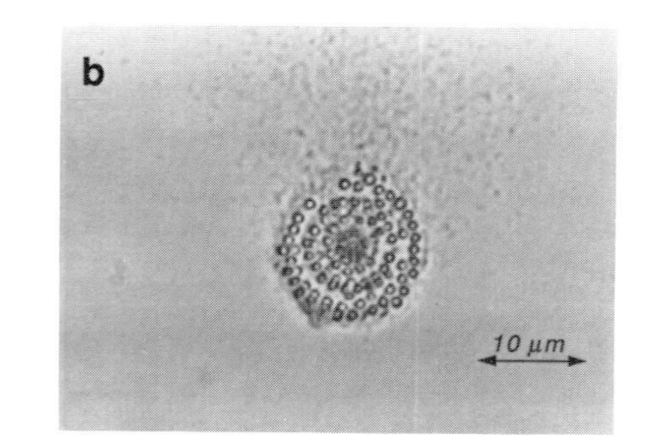

Figure 2. Size-selective laser trapping of polystyrene particles (d = 1 and 0.25 μm) in water. The particles with d = 0.25 μm are excluded from the pattern (a) by reducing the diameter of the concentric circles (b).

the diameter of the spatial pattern is reduced as mentioned above, the smaller particles are pushed out from the pattern into the water phase as seen in Figure 2b. Size-selection of the particles in solution by optical trapping is demonstrated for the first time. For a given particle diameter, optical trapping is more feasible for the particle with a higher refractive index under the condition of $n_1 > n_2$ [2, 7]. The results in Figure 2 thus imply a future possibility of selection and concentration of higher refractive index particles from a mixture of particles with various refractive indices by a laser beam. Such approaches will contribute to advances in various microparticle science and industries.

2.2. Optical Control of Brownian Motion of Microparticles by Scanning Laser Micromanipulation

Optical patterning of microparticles described above is essentially based on interference of an incident laser beam(s), so that there is severe limitation in producing arbitrary spatial patterns. An another possible approach to produce spatial patterns of light is the use of a photomask. In this case, however, interference fringes and/or a speckle noise generated by a laser beam will disturb the clear pattern formation of trapped particles. Furthermore, since laser power will be greatly reduced by passing the laser beam through a photomask, a high-power laser will be required to achieve spatially-patterned optical trapping. An elegant and sophisticated method to produce patterns of microparticles is to scan a focused laser beam in a sample solution. We thus developed a new method of "*scanning laser micromanipulation*" as described in detail elsewhere [4, 6, 12 - 15]. Briefly, a focused (~ 1 μm) 1064 nm laser beam was repetitively scanned in a sample solution by modulating computer-controlled mirrors inserted between the laser and the optical microscope. If the rate of repetitive scans of the laser beam is faster than that of flow and/or Brownian motion of particles, the particles will be optically trapped in a locus of the laser beam in a sample solution. Actually, the scanning laser micromanipulation technique enables one, i) patterning of an arbitrary number of particles in three-dimensional space [9, 12, 13], ii) manipulation of various microparticles irrespective of the refractive index condition for trapping as well as of the shape of a particle or substance [15], and also, iii) producing rhythms and harmony in the movement of microparticles in solution [13]. Several examples of our achievements are as follow.

Figure 3 shows spatial alignment of titanium dioxide particles (d ~ 0.25 μm) in ethylene glycol along the locus of the laser beam in a pattern of "star". The spatial pattern of "star" is written with laser power of 145 mW [12]. In this experiment at room temperature, the repetition rate of the beam scan along the pattern was 24 Hz and the rate was high enough to retain particles in the pattern against the Brownian motion of particles. Since the thermal motion of a particle is dependent on the size of a particle and the viscosity of a medium, the repetition rate should be optimized for a given particle size and a medium. The repetition rate should be also changed with the complexity of a pattern to be produced. More complexed patterns or characters made of microparticles can be also drawn by scanning laser micromanipulation such as an italic-like letter of "μm" and a Chinese character for "light" as demonstrated in the article by Sasaki and Misawa [4, 9, 12, 13]. Since we confirmed optical trapping of various microparticles in an appropriate medium [7], the technique will be applicable to produce various spatial patterns with any microparticles. The spatial pattern of the particles was also manipulated along the XY directions without deformation of the pattern. Although microparticles in solution undergo random thermal motion, a laser beam can certainly produce optical patterns and harmony of particles in any desired fashion.

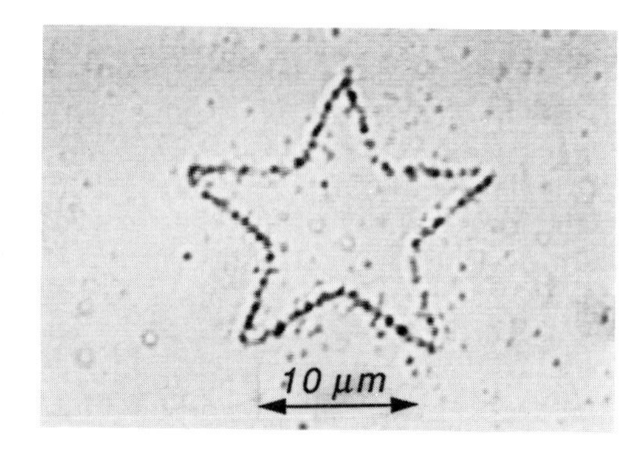

Figure 3. Scanning laser micromanipulation of titanium dioxide particles (d ~ 0.25 μm) in ethylene glycol. Total laser power ~ 145 mW and repetition rate = 24 Hz.

Computer programming of the movement of the focused laser beam in a sample solution leads to creation of any spatial patterns of particles as well as to continuous changes of the patterns like an animation movie made of microparticles. A typical example is demonstrated in Figure 4, in which polystyrene particles (d = 1 μm) are patterned continuously in the letters of "L", "V", "M" and "H" in ethylene glycol with the repetition rate of 10 Hz. This is an elegant application of the scanning laser micromanipulation technique to control the Brownian motion of particles in arbitrary patterns as well as to produce static and dynamic optical harmony and rhythms of microparticles in solution.

2.3. Optical Control of Movement of Microparticles by Scanning Laser Micromanipulation

Further potential applications of the scanning laser micromanipulation method have been demonstrated for optical control of dynamic movement of microparticles in solution [13]. In such experiments, particles are not trapped at a fixed position, but those flow in a spatial pattern as shown in Figure 5. In the actual experiments, firstly, polystyrene particles (d = 1 μm) were trapped in a circular pattern with the diameter of 13.5 μm in 1-pentanol (n_2 = 1.41 ; Figure 5a). Then, secondly, the laser beam was scanned repetitively along the circle with the rate of 15 Hz in a right-handed rotation. As demonstrated in Figure 5 as the sequential images of the pattern observed at intervals of 0.6 s, the slightly larger particle, indicated by the arrow in the photograph, followed the same direction. The particles moved together in an ordered fashion with a flow velocity of 12.2 μm/s. The scan speed of the laser beam was calculated to be 642 μm/s under applied laser power of 120 mW. An anti-clockwise rotation of the particles in the circular pattern was also successful by scanning the laser beam in a left-handed rotation.

The physical nature of the driving force for particle flow and optical control of the movement of particles is now to be explained [13]. As described in elsewhere [4], the radiation force exerted on a particle is directed to the focal spot of a laser beam

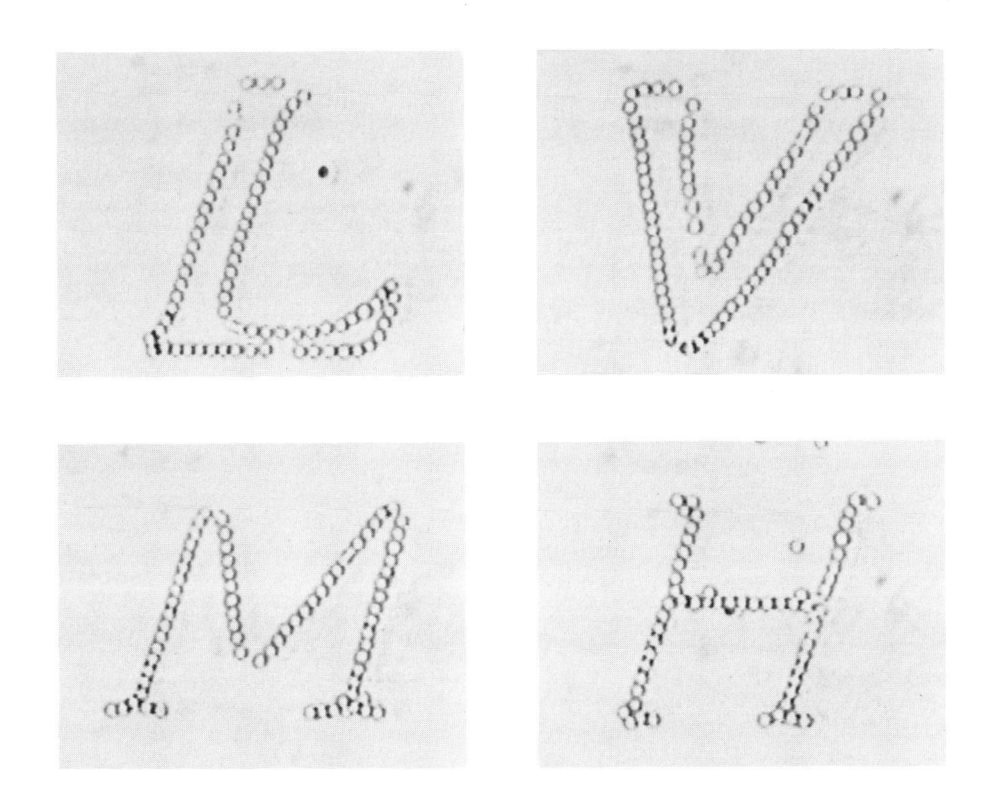

Figure 4. Continuous spatial pattern formation by scanning laser micromanipulation. Polystyrene particles (d = 1 μm) are aligned successively in the letters of "L", "V", "M", and "H" in ethylene glycol.

when $n_1 > n_2$. In such the case, the magnitude of the radiation force is dependent on the distance between the focal point of the laser beam and the particle. When the focal point of the laser beam is at the center of the particle, no force is exerted to the particle as a sum. Hence, the particle should sit in an equilibrium position. If the focused laser beam is shifted from the equilibrium position, the attractive force is experienced by the particle, which implies generation of the transaxial component of the radiation force. With an increase in the distance between the equilibrium position of the particle and the laser beam position, the magnitude of the attractive force should initially rise and then decay.

When the focused laser beam is scanned over a fixed particle, the radiation force exerted to the particle varies with following the curve in Figure 6a. The attractive force to the right-side is given as a positive value at each position of the scanning beam. The force to the left-side is vise versa. The net force, given by an integral of the curve, is zero, since the curve is symmetrical with respect to the origin. However, in the actual conditions, the position of the particle also shifts sequentially due to the

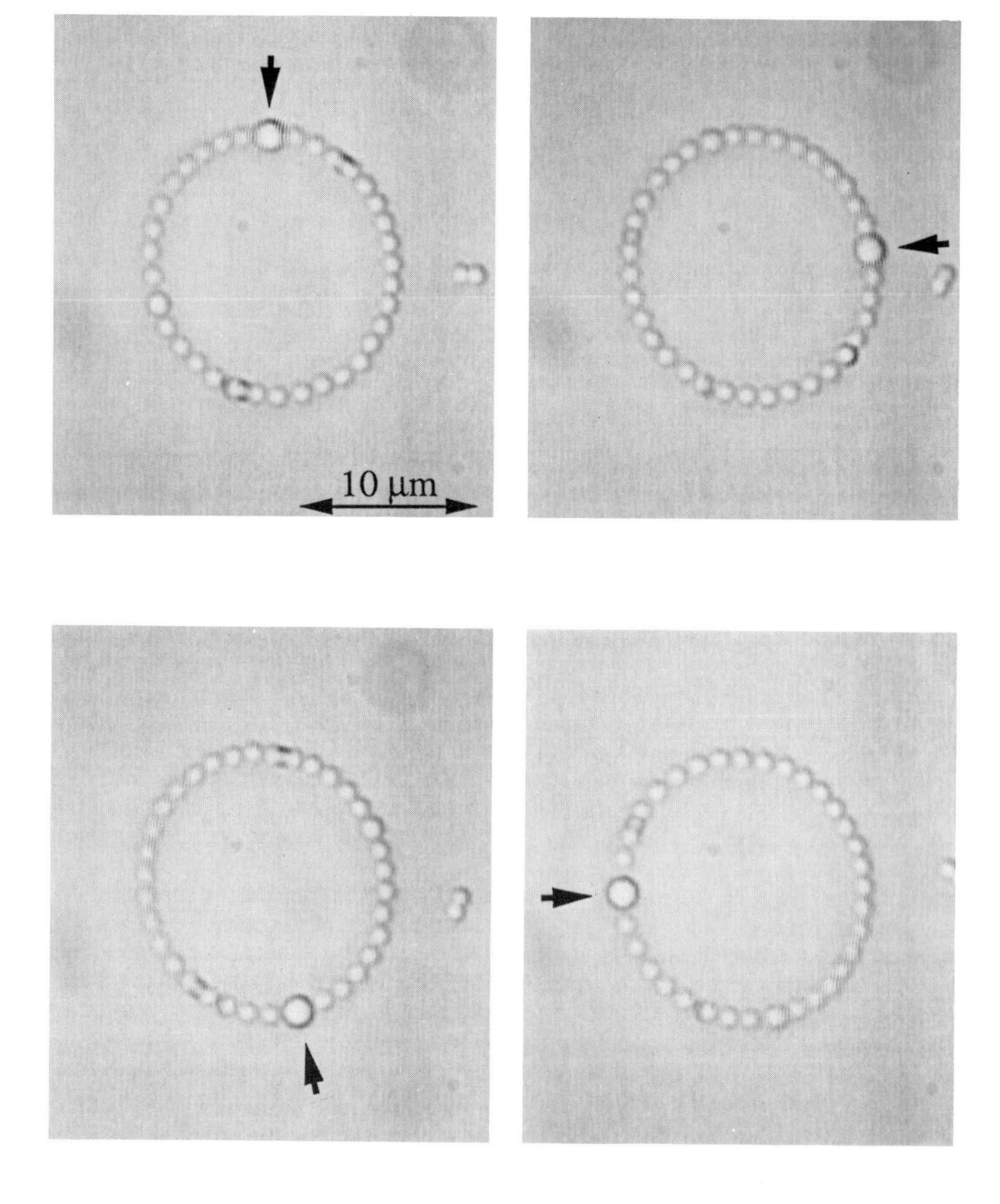

Figure 5. Sequential images of the particle flow observed at intervals of 0.6 s. The sample is polystyrene particle (d = 1 μm) in 1-pentanol.

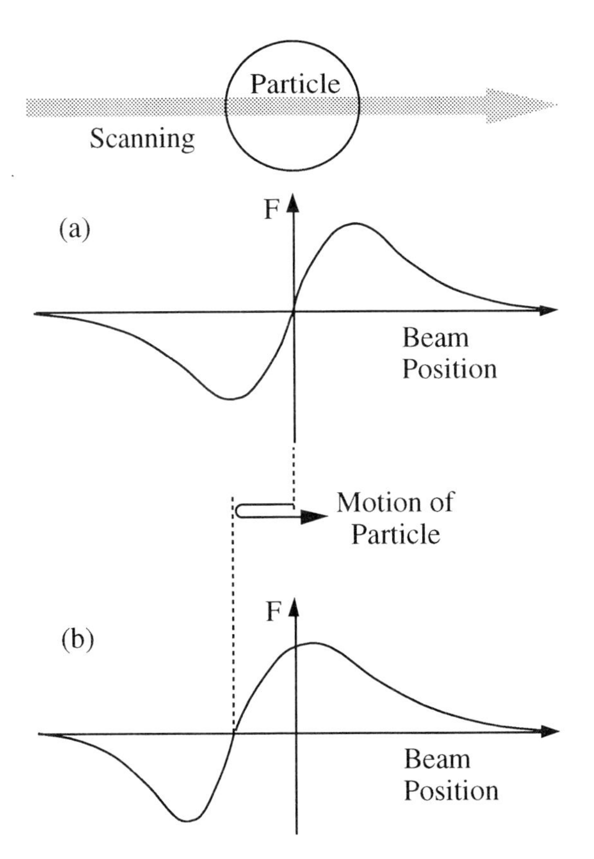

Figure 6. Schematic representation of the driving force for optical rhythms. (a) The radiation force exerted on a fixed particle as a function of the relative position of a focused beam. (b) The radiation force exerted on a moving particle.

radiation force exerted, which leads to deformation of the curve in Figure 6a. Initially, the particle moves to the left-side of the origin by the attractive force until the laser beam comes to the center of the particle. The left-half of the curve is therefore compressed as shown in Figure 6b. As the beam moves to the other side of the origin, the particle is attracted to the right such that the right-half of the curve is expanded. The asymmetrical curve in Figure 6b causes the driving force, which moves the particle in the direction of the scanning laser beam. Actually, the particles in the pattern experience such the force repetitively and the particles flow in one direction due to the continuous driving force.

The flow velocity of the particle is determined by the driving force, frictional forces between particles and the sample cell, and viscous resistance by the surrounding medium. In order to control the movement of particles, the most important factor governing the magnitude of the driving force for the particle flow

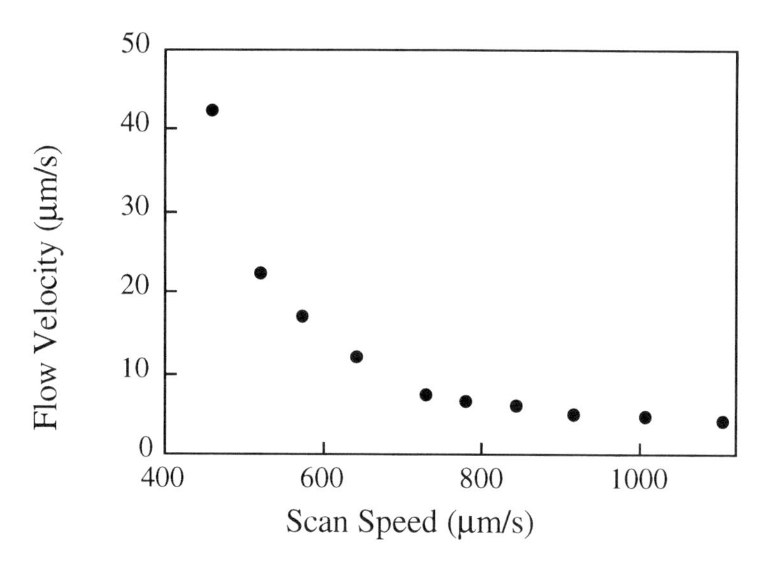

Figure 7. Velocity of the particle flow as a function of the scan speed.

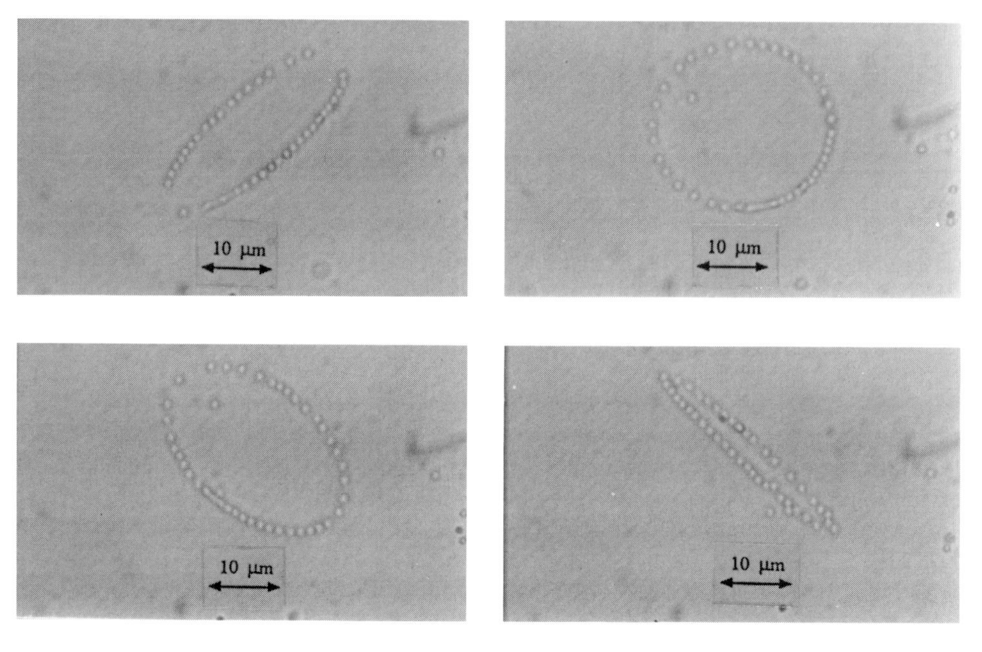

Figure 8. Flow and pattern controls of polystyrene particles (d = 1 μm) in ethylene glycol by scanning laser micromanipulation. Laser power at 1064 nm = 430 mW. Scan rate = 5 Hz.

should be elucidated. Figure 7 shows the flow velocity of particles as a function of the scan speed of the laser beam. The flow velocity of the particle becomes slower with increasing the scan speed of the laser beam. Since the mechanical response function of the particle in a viscous solution is slower in a high-frequency region, the driving force caused by asymmetricity of the force curve in Figure 6 also decreases in the case of high-speed scanning. Therefore, the motion of the particles slows down as the scanning rate is increased. This is the most important factor for the control of particle flow. In other words, the rhythms and harmony in the movement of microparticles can be controlled through the scan rate of the laser beam.

As a more sophisticated example, Figure 8 shows sequential images of the pattern of polystyrene particles (d = 1 μm) undergoing continuous flow of the particles with simultaneous changes of the pattern. The patterns were created continuously by repetitive scanning of the trapping beam at a repetition rate of 5 Hz in a right- or left-handed rotation. All the particles moved together in an ordered fashion with a flow velocity at 0 ~ 20 μm/s under laser power of 430 mW.

2.4. Optical Control of Fusion of Microdroplets

A further elaborated experiment based on the scanning laser micromanipulation method was demonstrated for optical trapping/manipulation of dye-contained water and ethylene glycol (EG) droplets in liquid paraffin [15]. It is noteworthy that the refractive indices of water (n_1 = 1.33) and EG (n_1 = 1.43) are lower than that of the surrounding medium (liquid paraffin ; n_2 = 1.46 ~ 1.47), so that these droplets cannot be optically trapped by the conventional method with a focused laser beam owing to repulsive force exerted to lower refractive index particles under the condition of $n_1 < n_2$ [4]. This drawback of the conventional method has been also overcome by scanning laser micromanipulation, by which a water or EG droplet is manipulated in a "photon cage" produced by repetitive circular scans of a focused laser beam around the droplet. Since the droplet experiences the repulsive force from all the directions, the droplet is trapped at the center of the "photon cage" similar to optical manipulation of high-reflective (iron) and high-absorbing (carbon black) particles [4, 15].

In the actual experiments, two focused laser beams [14] were scanned independently in a sample solution (25 Hz, 320 mW for each beam) to trap dye-water and EG droplets. As clearly seen in Figure 9a, scanning double laser-beam trapping was successful to manipulate the dye-water (right) and EG (left) droplets independently in the two photon cages. Three-dimensional manipulation of each droplet was also attained by controlling scanning laser beams. The dye-water and EG droplets were manipulated in the lateral (XY) direction in the sample solution and fused into one droplet by coinciding the position of the two laser cages (Figure 9b ; 640 mW in total power, 25 Hz). The dye-water droplet was diluted by the EG droplet and the color of the droplet turned to pale red. The experiment is very simple and very important for future researches of microparticles and microdroplets, since the method provides a potential means to induce chemical reactions arbitrarily in micrometer dimension. If one combines the technique with spectroscopic and/or electrochemical methods, the research on chemistry and physics of individual microparticles will be greatly advanced [16 - 19].

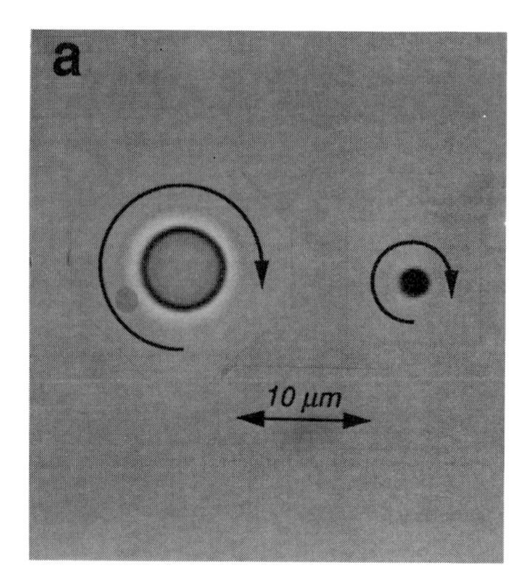

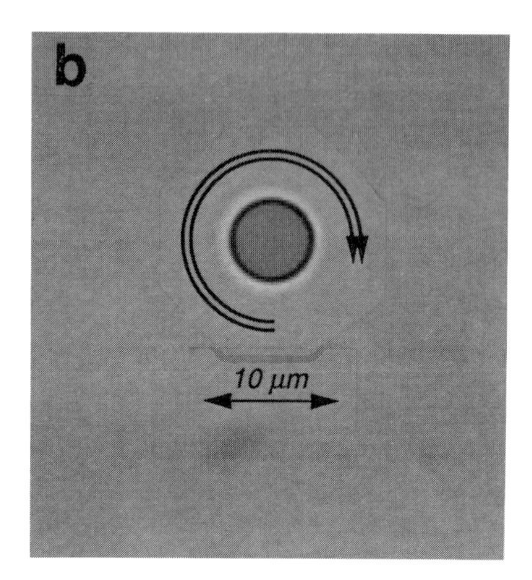

Figure 9. Double laser-beam manipulation of dye-water and ethylene glycol (EG) droplets in liquid paraffin. The arrows represent the loci of the scanning laser beams. a) EG (left) and dye-water (right) droplets are optically manipulated independently. Laser power at 1064 nm = 320 mW. Scan rate = 25 Hz. b) Two droplets were fused into one droplet by coinciding the loci of the two laser beams. Laser power = 2 x 320 mW. Scan rate = 25 Hz.

3. FUTURE PERSPECTIVE

On the basis of the optical manipulation methods, we have opened new interdisciplinary research fields. Microparticles made of organic, inorganic, and metallic materials are deeply concerned with various basic science and industries, so that their manipulation, spatial alignment, and active flow in three-dimensional space by a laser beam(s) will play essential roles for advances in microparticle science and industries. Control of thermal Brownian motion of microparticles in solution is certainly promising for development of various researches. By combining the optical manipulation method with other chemical tools, indeed, photochemical and redox reactions have been shown to be arbitrarily induced in a single microparticle undergoing Brownian motion in solution and, simultaneous spectroscopic and electrochemical characterizations of the particle are also now possible [16 - 23]. Chemical and physical properties of individual particles can be made clearer, which is the most important for advances in basic researches on microparticles. Formation and dissolution of a single polymer microparticle in solution have been also shown to be arbitrary controlled by a focused laser beam and the particle produced is manipulated simultaneously by the incident laser beam [24, 25]. Furthermore, microfabrication of individual particles and construction of microstructures composed of the desired number of particles have been demonstrated [6, 9, 14], which indicates that scanning laser micromanipulation has high potential in microparticle science.

Various novel research fields will be opened in addition to great advances in the present microparticle-related science. For example, interactions between two particles will be directly measured by manipulating them and analyzing radiation forces operating on the particles. The size, shape, and chemical structures will correlate with intra-particle forces, so that measurements of individual microparticles will contribute to basic understandings of microparticles themselves. Similarly, interaction forces between a particle and a substrate such as glasses, metals, electrodes, semiconductors, and polymer films will be evaluated and interaction mechanisms will be made clearer. One promising study will be optical manipulation and characterization of a single microparticle consisting of one macromolecule. DNA, RNA, and large molecular weight proteins are such examples. Also, studies on an artificially-synthesized macromolecule like star/burst polymers will be very fruitful. It is very easy to expect that such approaches can be applied to various research fields of biology.

The present optical manipulation techniques with other sophisticated methods have high potentials in future technologies as well. All the results guarantee new and important advances in the relevant microtechnologies : microcapsules for drug delivery systems ; polymer particles for coating films, paintings, and cosmetics ; various kinds of catalysts ; dye droplets and inorganic particles in photographs ; detergent micelles in emulsion systems. These are now the essential basis of the important industries and, their studies and technologies will be greatly advanced by introducing the optical manipulation methods. Optical driving of microstructures will provide a new technique in micromachining. Optical creation of new materials based on radiation force will give novel materials important for electronics and optics.

Beside chemistry, physics, and other science and industry, microparticles also play important roles in "art", since, for example, paints, cosmetics, and so forth indispensable for art are made of microparticles. Optical manipulation and control of the movement of microparticles therefore greatly contribute to art as well. It is noted, interestingly, that our present approaches of "*Optical Harmony of Microparticles in Solution*" has been awarded by the Moët Hennessy Louis Vuitton Science Foundation (LVMH) for "*science for art*" in 1993. The optical methods have certainly opened new fields of both science, industry, and art.

ACKNOWLEDGMENT

The authors would like to express their sincere thanks to Mr. M. Koshioka for his intimate collaborations and discussion.

REFERENCES

1. J. W. Hofstraat, C. Gooijer and N. H. Velthorst, Molecular Luminescence Spectroscopy. Methods and Applications ; Part 3. S. G. Schulman (ed.), Wiley-Interscience, New York ,1993, p.323.
2. A. Ashkin, Science, 210 (1980) 1081.
3. A. Ashkin, Phys. Rev. Lett., 24 (1970) 156.
4. K. Sasaki and H. Misawa, in this volume.
5. H. Misawa, M. Koshioka, K. Sasaki, N. Kitamura and H. Masuhara, Chem. Lett., (1990) 1479.
6. H. Misawa and K. Sasaki, in this volume.
7. H. Misawa, M. Koshioka, K. Sasaki, N. Kitamura and H. Masuhara, J. Appl. Phys., 70 (1991) 3829.
8. H. Misawa, N. Kitamura and H. Masuhara, J. Am. Chem. Soc., 113(1991) 7859.
9. H. Misawa, K. Sasaki, M. Koshioka, N. Kitamura and H. Masuhara, Macromolecules, 26 (1993) 282.
10. M. Burns, J. M. Fournier and J. A. Golvchenko, Science, 249 (1990) 749.
11. H. Misawa, M. Koshioka, K. Sasaki, N. Kitamura and H. Masuhara, Chem. Lett., (1991) 469.
12. K. Sasaki, M. Koshioka, H. Misawa, N. Kitamura and H. Masuhara, Jpn. J. Appl. Phys., 30 (1991) L907.
13. K. Sasaki, M. Koshioka, H. Misawa, N. Kitamura and H. Masuhara, Opt. Lett., 16 (1991) 1463.
14. H. Misawa, K. Sasaki, M. Koshioka, N. Kitamura and H. Masuhara, Appl. Phys. Lett., 60 (1992) 310.
15. K. Sasaki, M. Koshioka, H. Misawa, N. Kitamura and H. Masuhara, Appl. Phys. Lett., 60 (1992) 807.
16. K. Nakatani, T. Uchida, H. Misawa, N. Kitamura and H. Masuhara, J. Phys. Chem., 97 (1993) 5197.
17. K. Nakatani, T. Uchida, S. Funakura, A. Sekiguchi, H. Misawa, N. Kitamura and H. Masuhara, Chem. Lett., (1993) 717.
18. K. Sasaki and M. Koshioka, in this volume.
19. K. Nakatani, T. Uchida and N. Kitamura, in this volume.
20. K. Nakatani, H. Misawa, K. Sasaki, N. Kitamura and H. Masuhara, J. Phys. Chem., 97 (1993) 1701.
21. K. Sasaki, M. Koshioka and H. Masuhara, Appl. Spectroscopy, 45 (1991) 1041.
22. M. Koshioka, H. Misawa, K. Sasaki, N. Kitamura and H. Masuhara, J. Phys. Chem., 96 (1992) 2909.
23. M. Ishikawa, H. Misawa, N. Kitamura, and H. Masuhara, Chem. Lett., 481 (1993).
24. M. Ishikawa, H. Misawa, N. Kitamura and H. Masuhara, Chem. Lett., (1993) 481.
25. N. Kitamura, M. Ishikawa, H. Misawa and R. Fujisawa, in this volume.

MICROCHEMISTRY
Spectroscopy and Chemistry in Small Domains
Edited by H. Masuhara et al.

Photochemical microfabrication and machining of individual polymer particles in solution

Hiroaki Misawa[#,*] and Keiji Sasaki[‡]

Microphotoconversion Project,[+] ERATO Program, Research Development Corporation of Japan, 15 Morimoto-cho, Shimogamo, Sakyo-ku, Kyoto 606, Japan

Laser ablation of an optically-trapped polymer latex particle in water was demonstrated. A minute hole with its diameter of ~ subμm was fabricated on the latex particle (~ 6 μm diameter). The hole size produced was much smaller than the effective diameter of the excitation laser pulse, suggesting nonlinear optical and photochemical mechanisms for the present laser trapping-ablation. Similar laser ablation of a single microcapsule containing pyrene/toluene solution was also performed. Deformation or ablative decomposition of the capsule was observed depending on the pulsed-laser intensity. Furthermore, a multi-beam laser trapping-reaction system was developed to demonstrate independent manipulation of plural microparticles and to induce photochemical reactions of laser-trapped polymer latex particles. Integrated microstructures created by the successive manipulation/polymerization procedures were shown to be freely manipulated by laser beams. Possible roles of the optical trapping/manipulation techniques in studying chemistry and physics of polymeric microspheres are also discussed.

1. INTRODUCTION

Among various fine particles, polymer latex particles and microcapsules receive broad interests in wide fields of research. In addition to their important roles in the painting and printing industries, much efforts have been currently devoted to develop functional microparticles applicable to medical science, chromatography, catalysts, and so forth. However, these microparticles have been generally studied and utilized

---

* To whom correspondence should be addressed.
# Present address; Department of Mechanical Engineering, Faculty of Engineering, The University of Tokushima, 2-1 Minamijosanjima, Tokushima 770, Japan.
‡ Present address; Department of Applied Physics, Faculty of Engineering, Osaka University, Suita, Osaka 565, Japan.
+ Five-year term project: October 1988 ~ September 1933.

as bulk materials, and chemical/physical properties of a single particle have been assumed to be the same with those of the aggregates. This will be simply due to the lack of a manipulation method for individual particles.

Noncontact and nondestructive manipulation of individual microparticles is quite fascinating and could contribute to further advances in both basic and applied research on polymeric microparticles and colloids. If one can manipulate individual particles undergoing Brownian motion in solution, for example, chemical modification and/or fabrication of particles could be arbitrarily performed and, therefore, a new class of materials composed of various microparticles will be developed. Studies on chemical/physical interactions between particles as well as on properties of colloids will be also advanced on the basis of manipulation of polymer microparticles.

A key method to manipulate individual particles in Brownian motion was reported by Ashkin over twenty years ago [1]. However, the method have never been so far applied to chemistry of polymeric particles. The principle of the method known as "*optical trapping*" or "*laser trapping*" is described in detail by Sasaki and Misawa in this volume. Briefly, when the refractive index of a particle (n_1) is higher than that of the surrounding medium (n_2), the amount of light momentum change caused by refraction of light through the particle is transferred to the particle. This is the driving force (i.e. radiation force) of laser trapping and the particle is optically manipulated or trapped in the vicinity of the focal point of the laser beam. Steering of the laser beam leads to three-dimensional manipulation of a particle. Clearly, the optical trapping/manipulation method is highly potential to perform various studies on polymeric particles. Indeed, optical trapping of polymer latexes, microcapsules, oil droplets, and so forth have been experimentally proved to be tweezered by a single focused laser beam [2-5].

As a possible extension of the optical trapping/manipulation method, we currently developed a multi-beam scanning laser micromanipulation technique, by which plural microparticles can be arbitrarily manipulated [6-8] and excited by CW and pulsed lasers, respectively [9, 10]. In this article, we report laser ablation of optically-trapped individual polymer latex particles and microcapsules dispersed in solution [2, 4, 5]. Furthermore, we demonstrate photochemical fixation of polymer latex particles to create three-dimensional micrometer-sized structures by the multi-beam technique [9, 10]. A possible role of laser trapping/manipulation in chemistry and physics of polymeric microparticles is also discussed.

2. EXPERIMENTAL

2.1. Experimental setup of laser micromanipulation-reaction system

A block diagram of a laser system employed in this study is shown in Figure 1 [9, 10]. Briefly, a 1064 nm TEM_{00} mode Gausian beam from a CW Nd:YAG laser (Spectron, SL903U or SL902T) was used as a trapping laser source and was focused (~ 1 μm) into a sample solution through an objective lens (x100, NA = 1.30) of an optical microscope (Nikon, Optiphoto XF or 2). A sample solution was placed between two quartz plates and was set on the stage of the microscope. The laser beam was split into two beams by a polarizing beam splitter and each beam was modulated by two sets of galvano mirrors (GSI, G325 DT). The two trapping laser beams were scanned independently in the plane of the sample solution if necessary. The galvano mirrors were controlled by a controller (Marubun, TI-325) and a computer (NEC, PC9801RA). Photopolymerization in sample solutions was performed with irradiation of 355 nm laser pulses from a Nd:YAG laser (Quantel, YG501-10, pulse width ~ 30 ps or Quanta-Ray, DCR-II, pulse width ~ 7 ns). Since both trapping (1064 nm) and excitation (355 nm) laser beams were introduced coaxially into the microscope, an optically-trapped microsphere(s) or the vicinity of a trapped particle(s) was correctly irradiated by the pulsed laser beam. Both trapping and excitation laser power irradiated to individual microparticles were determined by the reported methods [4]. All the behavior occurring in the sample solution was monitored by a CCD camera-video-monitor set (Sony, DXC-750, BVU-950, and PVM-1442Q) equipped to the microscope.

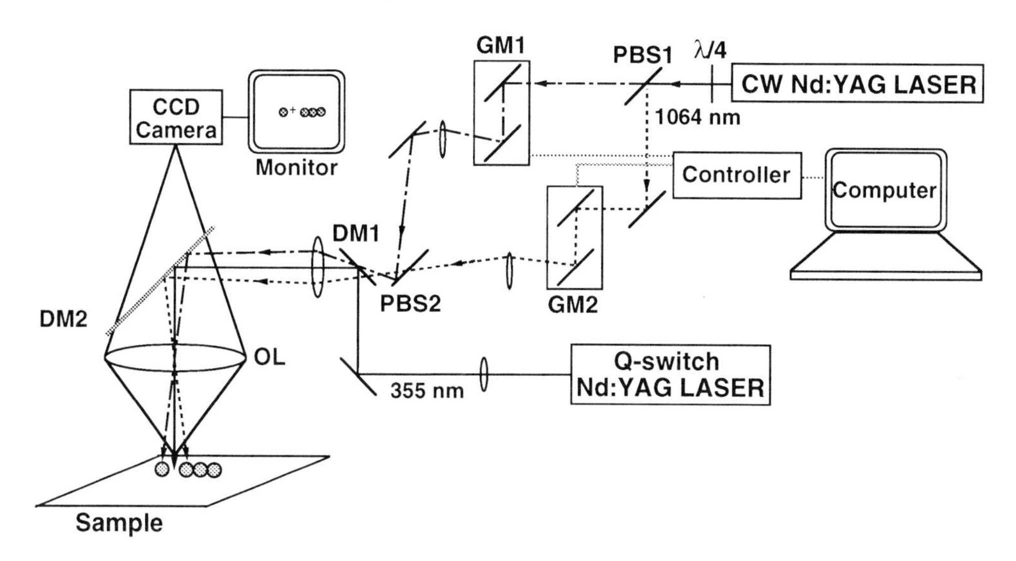

Figure 1. Block diagram of multi-beam laser manipulation-reaction system. PBS, polarizing beam splitter; GM, galvano mirrors; DM, dichroic mirror; OL, objective lens [9, 10].

2.2. Sample preparations and chemicals

Pyrene-doped poly(methyl methacrylate), PMMA, latex particles were prepared as follows. 350 mg of detergent-free PMMA latex particles (Mw = 5 ~ 10 x 10^5) with a diameter distribution of 5 ~ 16 μm were immersed in 5 ml of a methanol solution of pyrene (6.0 x 10^{-3} mol dm^{-3}) for 1 hr. The latex particles were collected by filtration with a Nuclepore membrane (pore size 1 ~ 1.2 μm) and washed with enough water. The resulting particles were dispersed in water and sonicated for few seconds before measurements.

For preparation of microcapsules containing pyrene in toluene as an inner solution, a mixture of acacia (3.5 g in 70 ml water), pyrene in toluene (8.1 x 10^{-3}, 1.0 x

10^{-2}, or 0.11 M ; 5 ml), and an aqueous solution (30 ml) of melamine (4.7 x 10^{-3} mol) and formaldehyde (1.4 x 10^{-2} mol) was vigorously stirred by a homogenizer for 15 min at room temperature after adjusting the solution pH to 4 ~ 5 with phosphoric acid. The mixture was transferred to a 100 ml round-bottom flask and ammonium sulfate (~1 x 10^{-2} mol) was added into the mixture. The reaction was allowed to stand for further 2 h at 55 °C under mild stirring. Microcapsules containing toluene alone were also prepared as a reference sample.

1-(4-Isopropylphenyl)-2-hydroxy-2-methylpropan-1-one (Darocur 1116, Merk) and other chemicals (all Nakalai Tesque Inc.) were used as received. Polystyrene (PSt, Polyscience Inc.) latex particles were used without further purification.

3. RESULTS AND DISCUSSION

3.1. Laser ablation of a polymer latex particle

As a characteristic laser-driven chemical reaction, ablative photodecomposition of solid materials has been extensively studied for surface modification/fabrication of polymers [11, 12] as well as for microelectronics applications [13-14]. The laser ablation phenomena can be understood as the results of both high-energy state photochemical and local-heated thermal reaction processes of solid surface by a laser beam with high peak power and short pulse width. Further details are described by Shimo and Uchida in this volume.

Laser ablation of a single PMMA latex particle in water was explored [2, 4]. A pyrene-doped PMMA latex particle was optically-trapped in water by the 1064 nm laser beam and the particle was irradiated simultaneously by the 355 nm pulsed laser. Morphological changes of the particle upon 355 nm laser ablation are shown in Figure 2 [4]. When laser ablation of the particle was performed at 30 Hz, complete (Figure 2a) or partial decomposition (Figure 2b) of the particle was observed depending on the number of the laser pulse. On the other hand, single-shot laser ablation of the optically-trapped particle resulted in fabrication of a minute hole on the particle (less than 1 μm diameter, Figure 2c). To our knowledge, this is the first observation of laser ablation of organic polymers in water. Controlled fabrication of individual micrometer-order particles in solution is possible by optimizing i) the laser fluence, ii) the pulse repetition rate, and iii) the number of the laser pulse. It is worth emphasizing that, when an untrapped particle was irradiated by the pulsed-laser, we could not confirm laser ablation since the untrapped particle disappeared from the ocular field of the microscope. Laser trapping is necessary for precise microfabrication of the particle dispersed in solution.

Since PMMA latex particles do not absorb the 355 nm laser pulse at all, pyrene doped in the latex should be responsible for the present laser ablation; dye-sensitized laser ablation. In order to elucidate the laser ablation phenomena, a relationship between the pulsed-laser intensity (P_{355}) and the ablative penetration depth (i.e., the length of the minute hole on the particle produced by laser ablation) was examined under the fixed optical trapping conditions (P_{1064} = 72 mW). Each experimental point in Figure 3 represents the limiting particle size on which complete ablative penetration by a minute hole is successful by the corresponding value of P_{355}. Under

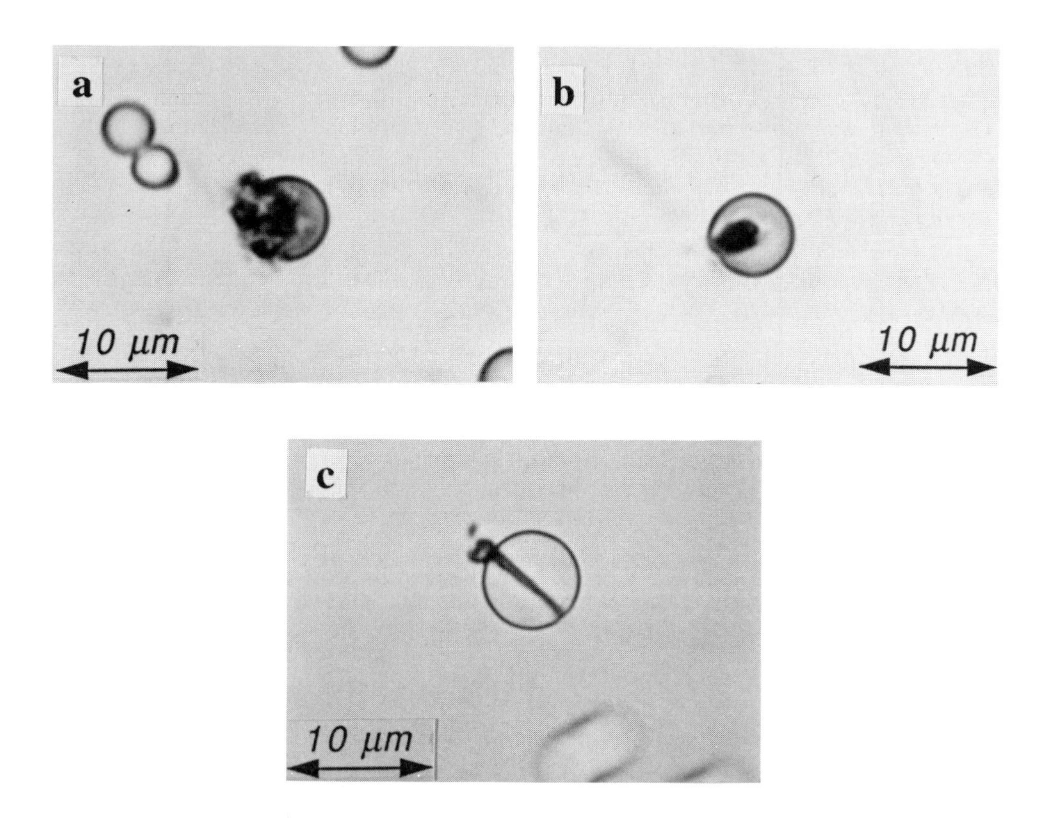

Figure. 2 Laser trapping-ablation of a pyrene-doped PMMA latex particle in water. (a) and (b); multi-shot laser ablation at 30-Hz pulsed-laser operation. P_{355} was not determined. (c); shingle-shot laser ablation, P_{355} was 23 J cm^{-2} (P_{1064} = 190 mW) [4].

the present experimental conditions, the diameter of the minute hole (sub-μm) produced in the particle was almost independent of P_{355}. According to Figure 3, the threshold energy of the trapping-ablation was estimated to be ~ 3 J cm^{-2} and ablative penetration of the 3 or 6 μm latex particle by a subμm hole could be attained by the laser fluence of 7 or 15 J cm^{-2}, respectively [4]. These values are much higher than the threshold energy for conventional laser ablation of polymer films in air or in vacuum [11, 12, 15, 16]. For example, it has been reported that the threshold energy for excimer laser (248 nm) ablation of a PMMA film is ~ 500 mJ cm^{-2} in air [17] and the etch rate of the film in vacuum is strongly dependent on the static pressure of foreign gas applied [15]. Static pressure being applied to the irradiated part of the latex particle by the surrounding water phase will be one of the possible reasons of the high threshold of the present ablation. It will be worth emphasizing, however, that the 355 nm laser pulse is focused into a μm-order spot, so that the actual laser power, P_{355}, is the order of nJ μm^{-2}. Photon density per unit area is of primary importance rather than the total laser energy.

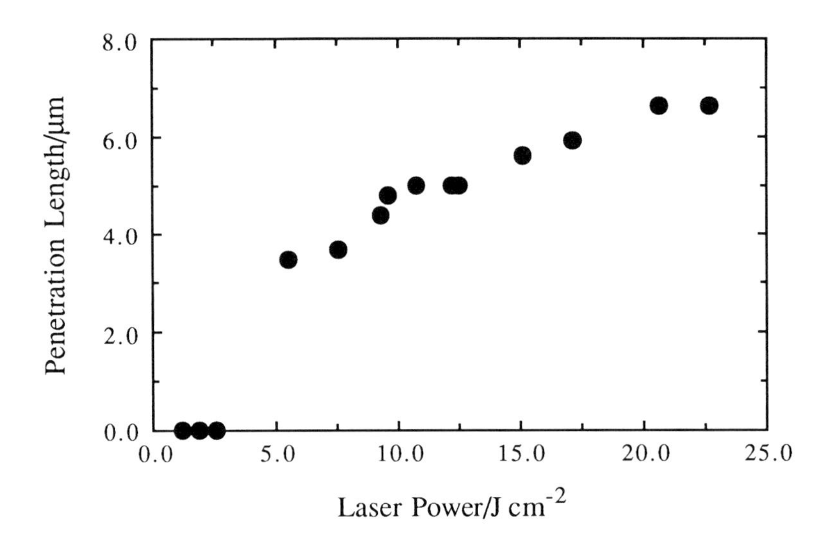

Figure 3. Ablative penetration of an laser-trapped, pyrene-doped PMMA latex particle by a minute hole in water [4].

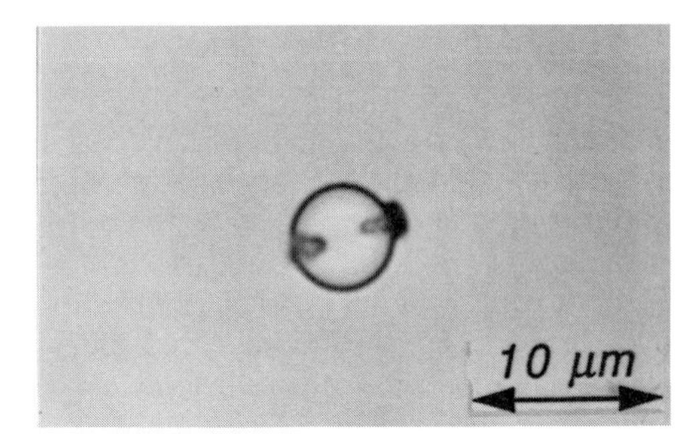

Figure 4. Laser trapping-ablation of a pyrene-doped PMMA particle in water. Single-shot laser ablation was conducted at P_{355} = 12 J cm^{-2} (P_{1064} = 190 mW) [4].

When P_{355} is not strong enough, the ablation phenomena were observed only at the entrance and exit edges of the pulsed-laser beam in the particle as depicted in Figure 4 [4]. The results suggest that, although bond scission of PMMA takes place immediately after pulsed-laser irradiation, ablative decomposition of PMMA follows with time delay. Indeed, Srinivasan et al. recently reported that ejection of PMMA fragments to air started from 1 μs after pulsed-laser irradiation [18]. Ablated products in the entrance and exit edges of the pulsed-laser in the particle possess enough internal energy, so that the products are easily ejected into the bulk water phase. However, ejection of the ablated products into water seems to be suppressed in the inner part of the particle by the surrounding PMMA and/or ablated products. Since a minute hole can be fabricated on the particle upon intense pulsed-laser irradiation, ejection of ablated products strongly depends on the laser intensity and probably, on the internal energy of the ablated products gained from the laser pulse.

It is important to note that the size of the minute hole produced by ablation is much smaller than that expected from the present experimental conditions of an aperture angle of the objective lens and the refractive indices of the particle and water. The results were explained by both nonlinear photochemical and optical effects. Namely, since laser ablation is induced by multi-photon absorption of the laser pulse by the material, the efficiency of ablation is dependent nonlinearly on the incident photon number [19]. Therefore, ablation of the polymer takes place at the central part of the pulsed-laser beam, giving a minute hole on the particle. This multi-photon absorption will be accelerated by a self-focusing phenomenon of the laser beams [20, 21]. It is known that an intense laser beam induces an increase in the refractive index of the material. Since the average laser intensity reaches very huge (few MW and GW for the trapping and excitation beams, respectively), the self-focusing effect of the laser beams is considered to be responsible for minute-hole fabrication. We suppose that both nonlinear photochemical and optical effects determine the size of the ablated minute hole on the polymer particle.

3.2. Laser ablation of a microcapsule

Analogous experiments were performed for melamine-resin wall microcapsules containing a pyrene/toluene solution [5]. When the laser-trapped microcapsule in water (Figure 5a) was irradiated by an intense laser pulse (355 nm, 15 J/cm^2), deformation of the spherical capsule was observed as clearly seen in Figure 5b. Further increase in pulsed-laser energy led to ablation of the capsule as shown in Figure 5c. The melamine resin capsule wall was decomposed by the one-shot laser pulse (490 J/pulse/cm^2), and pyrene/toluene droplets shown by the arrow in Figure 5c were ejected into the water phase. Upon laser pulse irradiation around 25 J/pulse/cm^2, on the other hand, a small bubble was confirmed to be produced inside of the microcapsule as shown in Figure 6.

The threshold energy for deformation, bubble formation, or ablative decomposition analogous to the result in Figures 5b, 6, or 5c, respectively, was estimated as summarized in Table 1. Since the capsule without pyrene is transparent at 355 nm, ablative photodecomposition of the capsule should proceed via simultaneous multi-photon absorption of the 355 nm laser pulse by the melamine-resin capsule wall and/or toluene. Simultaneous multi-photon absorption of a 355 or 266 nm laser pulse has been reported for neat toluene [22-24], solid polystyrene films [25], and so forth [26]. The two-photon absorption energy of a 355 nm laser pulse (7.0 eV) is close to the ionization energy of toluene or related compounds (6.3 ~ 7.0 eV), so that photoionized species are likely to be produced during simultaneous multi-photon absorption processes of these compounds as proved by picosecond transient

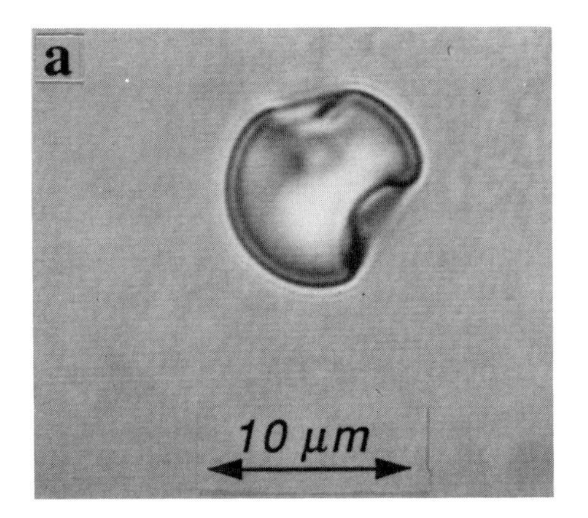

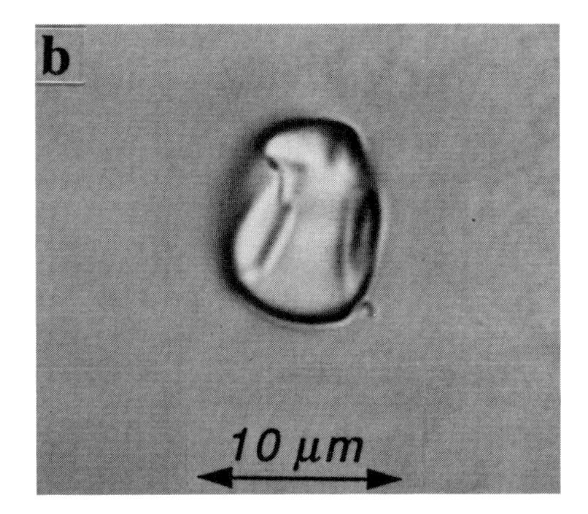

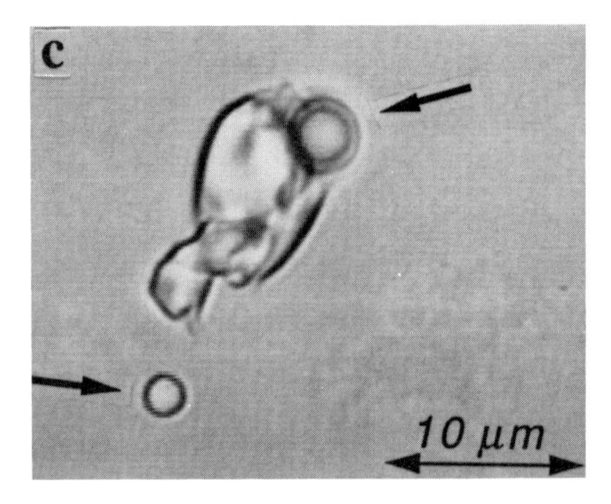

Figure 5. Laser-induced deformation and ablation of the laser-trapped pyrene/toluene microcapsule ([Py] = 1.0 x 10^{-2} M) in water (P_{1064} = 72 mW): (a) before pulse irradiation and (b) laser-induced deformation (pulse energy 15 J/pulse/cm^2) of the microcapsule. The spherical particles indicated by the arrows in Figure 5c are the pyrene/toluene droplets ejected from the capsule upon laser ablation [5].

absorption spectroscopy [22-25]. The trialkylamino-s-triazine structure of the capsule resin is also expected to undergo simultaneous multi-photon absorption of the 355 nm laser pulse similar to toluene and polystyrene. In the absence of pyrene, therefore, multi-photon ionization and subsequent bond scission of the melamine-resin wall will be one of the possible origins for ablative decomposition of the microcapsule.

On the other hand, multi-photon absorption by toluene producing ionic species and high excited states does not directly lead to bond scission of the capsule-wall, but contributes to ablative decomposition via thermal effects. Namely, recombination of the ionic species and/or nonradiative decay of the high excited states will result in local heating and subsequent vaporization of the inner toluene solution. Since local heating/vaporization of toluene accompanies an increase in the vapor pressure inside of the capsule, the melamine-resin wall will be decomposed by expansion of the inner volume via such the thermal effects. Vaporization of the inner toluene solution upon laser pulse irradiation is experimentally supported by the observation of a small bubble inside of the capsule (Figure 6).

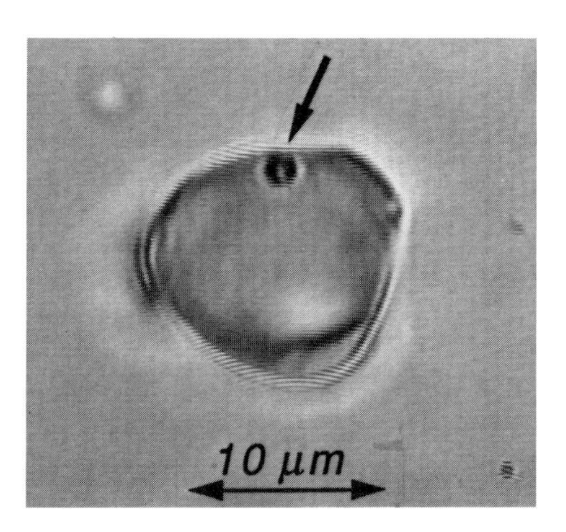

Figure 6. Laser-induced bubble formation in the laser-trapped pyrene/toluene microcapsule ([Py] = 1.0×10^{-2}M) in water (P_{1064} = 72 mW). The particle indicated by the arrow is the bubble produced by pulsed-laser irradiation (P_{355} =25 J/pulse/cm^2) [5].

For the pyrene/toluene microcapsules, pyrene in an inner toluene solution absorbs the 355 nm laser pulse. Therefore, the threshold energy for deformation, bubble formation, or ablation should be dependent on the concentration of pyrene, [Py], in the capsule. Indeed, the ablation threshold was determined to be 40 or 13 J/pulse/cm^2 for the capsule with [Py] = 1.0×10^{-2} or 0.11 M, respectively. Similarly, the pulse energy for deformation or bubble formation of the capsule was depended on [Py] as summarized in Table I [5]. A decrease in the threshold energy with increasing[Py] in the capsule clearly indicates an important role of pyrene for the results in Figures 5 and 6. Local heating/vaporization of the inner toluene solution will be responsible for deformation, bubble formation, or ablation of the capsule analogous to the results of multi-photon absorption by toluene in the capsule. Local heat generated in the inner solution should be higher for the capsule with higher [Py], so that a decrease in the threshold energy with increasing [Py] is a reasonable consequence as expected from the thermal effects mentioned above. For the capsule

Table 1
Pyrene concentration dependence of the threshold energies for deformation, bubble formation, and ablation of the laser-trapped microcapsule in water [5].

	threshold energy,[a] J/pulse/cm^2		
[Py], M^{-1}	deformation[b]	bubble formation[b]	ablation[b]
0	42	42[c]	100
1.0 x 10^{-2}	7.8	25	40
0.11	5.0	13	13

[a] Threshold energy upon single-shot pulse irradiation (355 nm) otherwise noted. [b] Typical examples of deformation, bubble formation, and ablation of the capsule are in Figures 5b, 6, and 5c, respectively. [c] Threshold energy upon multi-shots pulse irradiation (355 nm).

with [Py] = 0.11 M, the bubble formation accompanies simultaneous decomposition of the capsule-wall as revealed from the same threshold energy for both ablation and bubble formation; 13 J/pulse/cm^2 (Table 1). The results support vaporization of toluene upon laser irradiation and subsequent vapor-pressure induced decomposition of the capsule.

Another characteristic feature of the present laser ablation is the variation of the ejection mode of the pyrene/toluene droplets with the pulsed-laser energy. Namely, when laser irradiation on the trapped capsule ([Py] = 1.0 x 10^{-2} M) was performed around the ablation threshold (40 J/pulse/cm^2), small pyrene/toluene droplets were ejected or permeated from the capsule as clearly seen in Figure 7a. An increase in the laser pulse energy (95 J/pulse cm^2) resulted in ejection of a number of small droplets from the capsule (Figure 7b). Upon pulse irradiation far above the ablation threshold (490 J/pulse/cm^2), on the other hand, ejection of relatively large droplets was observed (Figure 5c). In Figure 7a, the spherical melamine-resin wall of the original capsule can be seen while the capsule-wall is completely decomposed to eject the droplets in Figures 5c and 7b. With increasing the laser pulse energy, local heat or vapor-pressure generated in the capsule increases and, therefore, the number of the ejection or permeation site of the droplet increases as demonstrated in Figure 7. Irradiation of the capsule at 490 J/pulse/cm^2, however, leads to simultaneous decomposition of the capsule-wall like *popcorn* which renders ejection of relatively large droplets (Figure 5c). The results indicate that the ejection or release mode of the inner solution from the microcapsule can be controlled by the laser pulse energy. Besides laser microfabrication of various particles in an arbitrary fashion, laser manipulation-ablation of individual microcapsules containing chemical reagents will play an important role for future drug delivery and controlled drug-release systems in body.

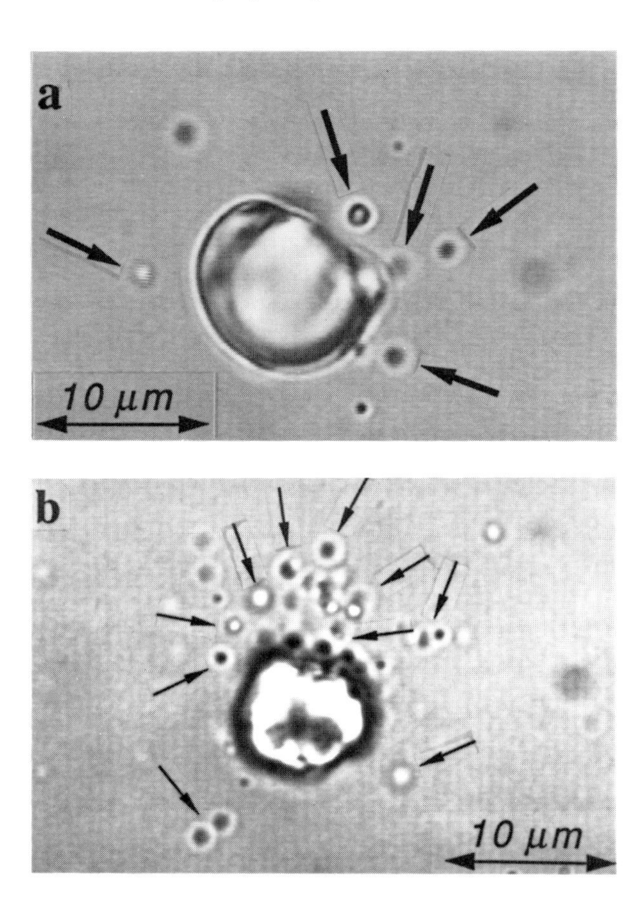

Figure 7. Laser ablation of the laser-trapped, pyrene/toluene microcapsule ([Py] = 1.0 x 10^{-2} M) in water at P_{355} of (a) 40 and (b) 95 J/pulse/cm^2 (P_{1064} = 72 mW). The small particles indicated by the arrows are the pyrene/toluene droplets ejected or permeated from the capsule. The small droplets in the vicinity of the capsules were also ejected upon pulse irradiation [5].

3.3. Photochemical assembling of microparticles

The multi-beam scanning laser micromanipulation system in Figure 1 and a photopolymerization technique can be applied to construct a microstructure composed of a desired number of a polymer particle [9, 10]. In these experiments, PSt particles (~ 3 μm diameter) dispersed in ethylene glycol solution of acrylic acid (AA, 1.9 M), N, N'-methylenebisacrylamide (MBA, 2.6 x 10^{-2} M), and Darocur 1116 (5.0 x 10^{-2} M), were used as a sample solution. The procedures are shown in Figure 8. Firstly, two polymer particles were optically manipulated independently by the two 1064 nm laser beams in a solution containing the vinyl monomers and the radical photoinitiator (Figure 8a). The two particles were forced to contact with each other and ultraviolet (355 nm) laser pulses were irradiated to the contacting area between the particles to initiate photopolymerization of the vinyl monomers. Although we could not see any morphological change on the particles, the two particles were stuck together as confirmed by the fact that the two particles were never diffused apart even after switching off the trapping laser beams (Figure 8b). Gelation of the monomers in the interfacial layer of the particles led to permanent fixation of the two polymer particles. Even if the focal points of the trapping laser beams were separated with each other by more than 3 μm, the particles were not divided at all. Since radiation force generated by the laser beam is in the order of several pN [4], adhesive forces between the particles will be larger than several pN.

Analogous procedures of laser trapping-photopolymerization were performed successively to assemble an arbitrary number of the particles with any desired geometry [9, 10]. An important key technique to assemble the particles is as follows. Since two 1064 nm laser beams are used for trapping, the two particles stuck together (Figure 8b) should be manipulated by one laser beam to manipulate the third particle by the other beam (Figure 8c). One trapping laser beam was repetitively scanned (12.5 Hz) between the two ends of the structure to keep and manipulate the structure in the plane of the sample solution. With the structure being trapped by the scanning laser beam, the other trapping beam was used to manipulate the third particle (Figure 8c). The particle was then fixed to the one end of the two-polymer-particle structure as demonstrated in Figure 8d. Similar procedures were repeated to construct microstructures comprised of an arbitrary number of polymer particles (Figure 8e and f). In these procedures, the particles were optically lifted-up in the sample solution to avoid adhesion of the particles with the sample glass plate.

Besides polymer particles, laser manipulation of micrometer-sized iron particles, carbon black particles, and so forth is possible by the scanning laser manipulation technique [8]. Therefore, the laser manipulation-photopolymerization method will be further extended to create various microstructures comprised of polymer particles, metal particles, and so on. We expect that such approaches will lead to new composite materials with various physical and chemical functions as well as to development of micromachines.

The integrated microstructure comprised of three-polymer particles was lifted-up by irradiating the two trapping-laser beams independently to both ends of the structure. With the focal spot of one trapping beam being fixed at the same position, the other laser beam was scanned circularly in the plane of the sample solution with the rate of 120 °/s. The integrated structure rotated circularly around the one end of the structure as seen in an overlaid photograph in Figure 9. Active movement of the structure with clockwise or counter clockwise circular rotation, and a swing motion like a pendulum was successfully achieved [9]. The rate of the clockwise rotation of the structure (120°/s) in Figure 9 agreed completely with that of the circular scan rate of the laser beam.

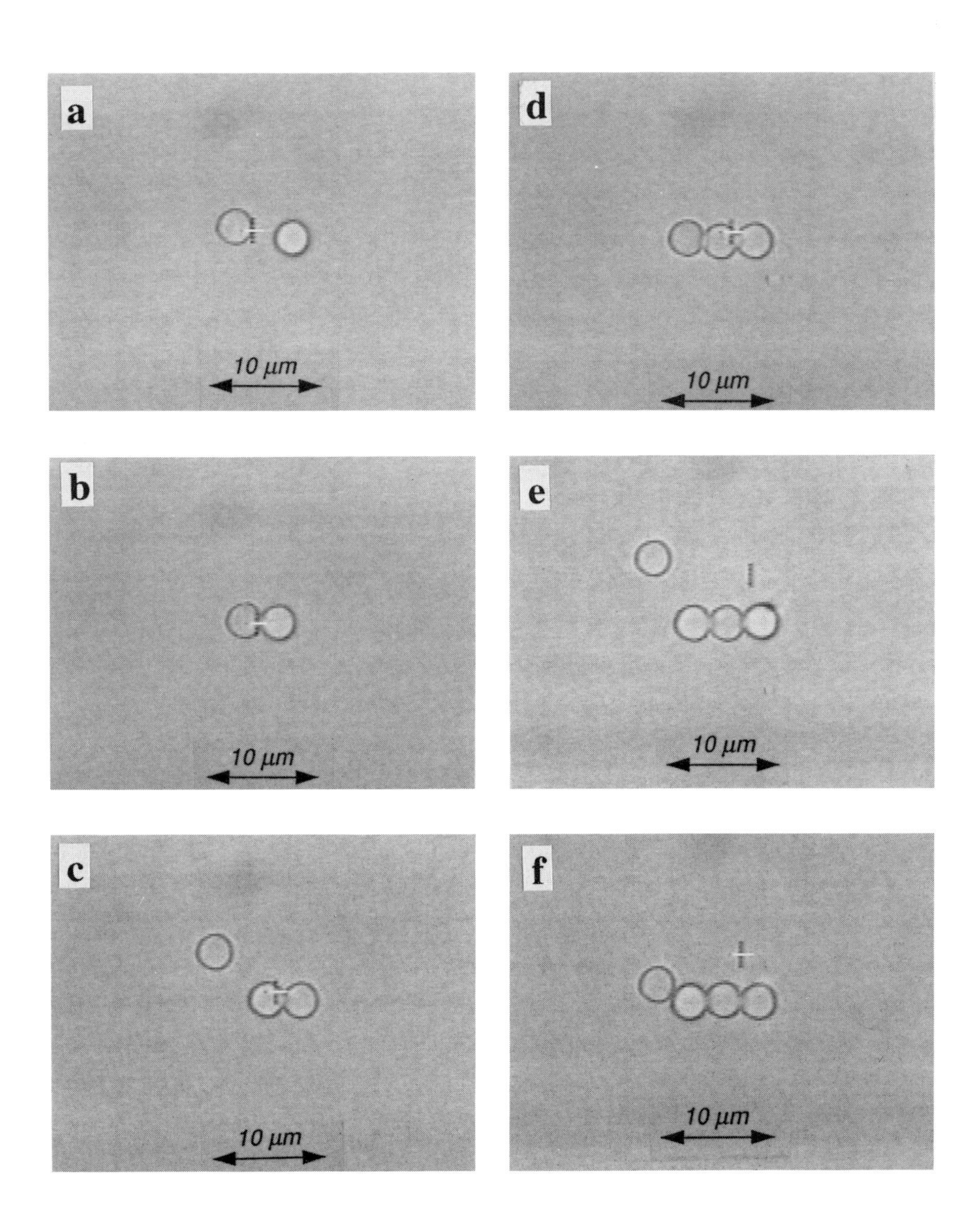

Figure 8. Photochemical assembling of PSt particles in ethylene glycol [10].

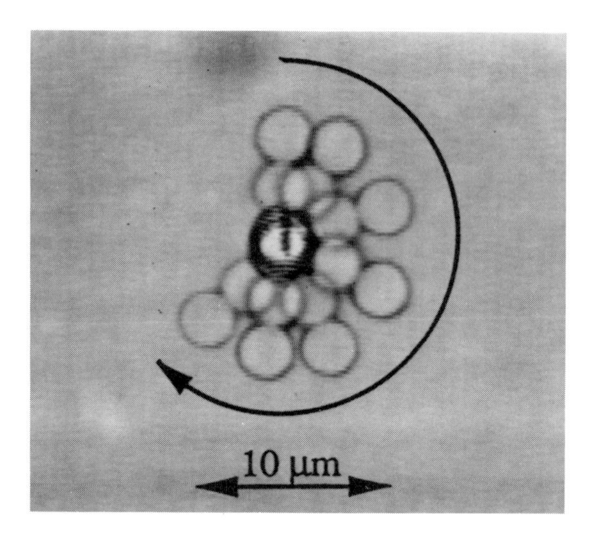

Figure 9. Optical clockwise rotation of an integrated PSt structure by scanning laser beams (overlaid photograph). One trapping beam was fixed at one end of the structure (P_{1064} = 290 mW) while the other beam was scanned circularly as shown by the arrow (P_{1064} = 290 mW) [9].

4. CONCLUSION

The optical trapping/manipulation and related techniques are highly potential to study chemistry and physics of polymeric microparticles randomly dispersed in solution. Micrometer-order particles are freely manipulated in three-dimensional space, including spatial alignment in the locus of a scanning laser beam. In addition to optical trapping itself, introduction of an excitation laser beam can induce various reactions on an optically trapped particle(s). Beside photochemical processing of particles described here, actually, ns ~ ps time-resolved fluorescence spectroscopy has been experimentally demonstrated. Elucidation of chemical reactions or various molecular interactions proceeding in or on an individual microparticle by spectroscopic methods will greatly contribute to understand chemical/physical properties of polymeric microspheres. Although experimental studies on laser trapping and related phenomena have just started in the fields of chemistry, its application is expected to be very broad. In particular, we believe that the techniques will open new research fields in polymeric microspheres.

ACKNOWLEDGMENT

The authors are greatly indebted to Prof. H. Masuhara (Osaka Univ.), Prof. N. Kitamura (Hokkaido Univ.), Mr. M. Koshioka (Kaneka Corp.), and Mr. R. Fujisawa (Mita Industrial Corp.) for intimate collaborations and discussions during the research in the Microphotoconversion Project.

REFERENCES

1. A. Ashkin, Phys. Rev. Lett., 24 (1970) 156; A. Ashkin and J. M. Dziedzic, Ber. Bunsenges. Phys. Chem., 93 (1989) 254.
2. H. Misawa, M. Koshioka, K. Sasaki, N. Kitamura and H. Masuhara, Chem. Lett., (1990) 1479.
3. H. Misawa, M. Koshioka, K. Sasaki, N. Kitamura and H. Masuhara, Chem. Lett., (1991) 469.
4. H. Misawa, M. Koshioka, K. Sasaki, N. Kitamura and H. Masuhara, J. Appl. Phys., 70 (1991) 3829.
5. H. Misawa, N. Kitamura and H. Masuhara, J. Am. Chem. Soc., 113 (1991) 7859.
6. K. Sasaki, M. Koshioka, H. Misawa, N. Kitamura and H. Masuhara, Jpn. J. Appl. Phys., 30 (1991) L907.
7. K. Sasaki, M. Koshioka, H. Misawa, N. Kitamura and H. Masuhara, Opt. Lett., 16 (1991) 1463.
8. K. Sasaki, M. Koshioka, H. Misawa, N. Kitamura and H. Masuhara, Appl. Phys. Lett., 60 (1992) 807.
9. H. Misawa, K. Sasaki, M. Koshioka, N. Kitamura and H. Masuhara, Appl. Phys. Lett., 60 (1992) 310.
10. H. Misawa, K. Sasaki, M. Koshioka, N. Kitamura and H. Masuhara, Macromolecules, 26 (1993) 282.
11. R. Srinivasan, J. Vac. Sci. Tech., B1 (1983) 923.
12. R. Srinivasan and B. Braren, Chem. Rev., 89 (1989) 1303.
13. D. J. Ehrlich and J. Y. Tsao, J. Vac. Sci. Tech., B1 (1983) 969.
14. C. G. Dupuy, D. B. Beach, J. E. Hurst Jr., and J. M. Jasinski, Chem. Materials, 1 (1989) 16.
15. N. Shimo, T. Uchida and H. Masuhara, Laser Ablation for Materials Synthesis, in Proceedings of Materials Research Society Symposium, Vol.191, edited by D.C. Paine and J.C. Bravman (Materials Research Society, Pittsburgh, 1990), p. 91.
16. T. Uchida, H. Sugimura, K. Kemnitz, N. Shimo and H. Masuhara, Appl. Phys. Lett., 59 (1991) 3189.
17. R. Srinivasan, B. Braren, R.W. Drefus, L. Hadel, and D.E. Seeger, J. Opt. Soc. Am., B3 (1986) 785.
18. R. Srinivasan, B. Braren, K.G. Casey, and M. Yeh, Appl. Phys. Lett., 55 (1989) 2790.
19. H. Masuhara, H. Hiraoka and E.E. Marinero, Chem. Phys. Lett., 135 (1987) 103.
20. Y.R. Shen, The Principles of Nonlinear Optics (Wiley-Interscience, New York,1984), Chap. 17, p.303.

21. W. Koechner, Solid-State Laser Engineering, Springer Series in Optical Science Vol. 1, edited by D.L. MacAdam (Springer-Verlag, New York, 1976), Chap. 12, p.585.
22. H. Masuhara, N. Ikeda, H. Miyasaka, and N. Mataga, Chem. Phys. Lett., 82 (1981) 59.
23. K. Hamanoue, T. Hidaka, T. Nakayama, and H. Teranishi, Chem. Phys. Lett., 82 (1981) 55.
24. H. Miyasaka, H. Masuhara, and N. Mataga, J. Phys. Chem., 89 (1985) 1631.
25. H. Miyasaka, F. Ikejiri, and N. Mataga, J. Phys. Chem., 92 (1988) 249.
26. H. Miyasaka, H. Masuhara, and N. Mataga, Laser Chem., 7 (1987) 119.

MICROCHEMISTRY
Spectroscopy and Chemistry in Small Domains
Edited by H. Masuhara et al.

Microstereophotolithography : a reality or a dream for tomorrow?

S. Zissi[a], S. Corbel[a], J.Y. Jézéquel[a], S. Ballandras[b] and J.C. André[a]

GdR "Optical processes ; applications to microtechniques", CNRS

[a]ENSIC - INPL, B.P. 451, 54001 NANCY Cedex, France

[b]LPMO - CNRS - IMFC, 32, avenue de l'Observatoire, 25000 BESANCON, France

The laser stereophotolithography allows to manufacture 3D objets made of polymers from the data computed by a CAD (Computer Aided Design) software. The basic principle is the space resolved polymerization of a multi-functional resin. The technology tends to be developed to its limits to manufacture small objects or optical parts. Nevertheless, it seems important to broaden this principle to manufacture elements immediately usable in microtechnology.

1. INTRODUCTION

The stereophotolithography (SPL) has been developed to manufacture screen copies in 3D [1 - 4]. This means that from the dimension figures of an object stored in a computer, a space-resolved material-conversion process has been defined. If several processes were in competition at the outset of this technology, the laser stereophotolithography technique tends to become standardized and can be described as follows

- one makes a laser-induced photochemical polymerization. The use of this principle induces a chemical amplification resulting from the chain reaction. Under these conditions, one can realize space-resolved material transformations with low energy lasers (several tens mW).
- an object is manufactured voxel (volume element) by voxel till a layer is completed and then layer by layer, as depicted in Figure 1, by adding new layers, "e" thick, of a photopolymerizable resin on the open surface of the reactive medium.

On these bases, it is possible to manufacture 3D objects having a complex structure, as it can be seen in Figure 2. The used resins are mixed with at least one photochemical initiator which limits the polymerization depth to a value roughly equal to the thickness "e" of a layer. Figure 3 depicts a so-called "1st generation" machine for laser stereophotolithography.

Several hundreds of these machines have already been sold in the world [5]. They can manufacture true 3D objects for various uses : moulding prototype pieces, scale models, etc.

Most of the work in this new domain of researches deals with the choice of the photopolymerizable material which must have a set of specific characteristics

- mechanical characteristics : hardness, elasticity, thermal resistance,
- chemical characteristics : small shrinkage, low toxicity, fast polymerization rate, slow aging effects, etc.

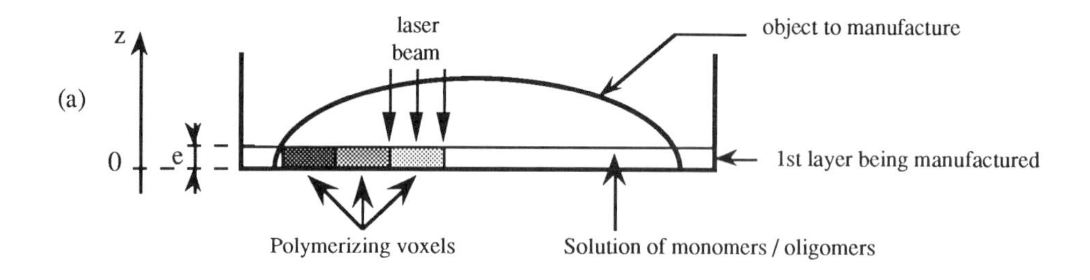

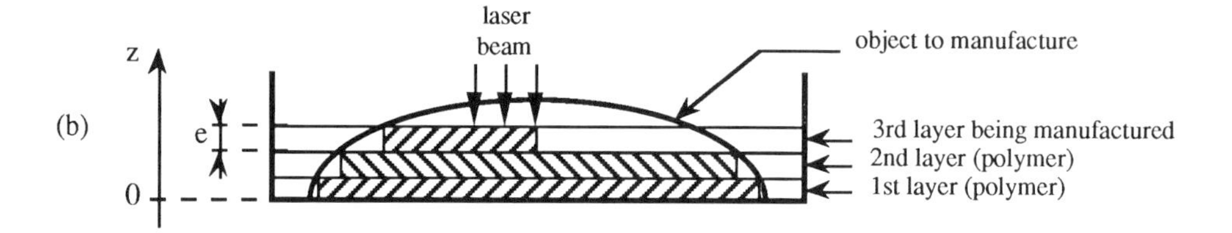

Figure 1. Manufacture of an object by polymerizing (a) successive adjacent voxels, (b) successive layers.

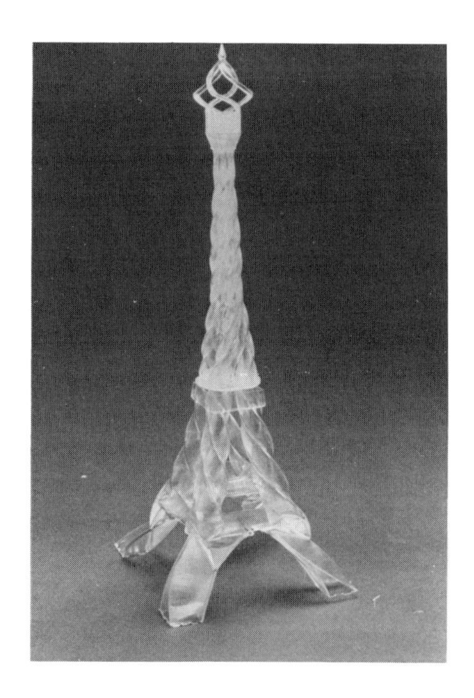

Figure 2. 3D objet having a complex structure.

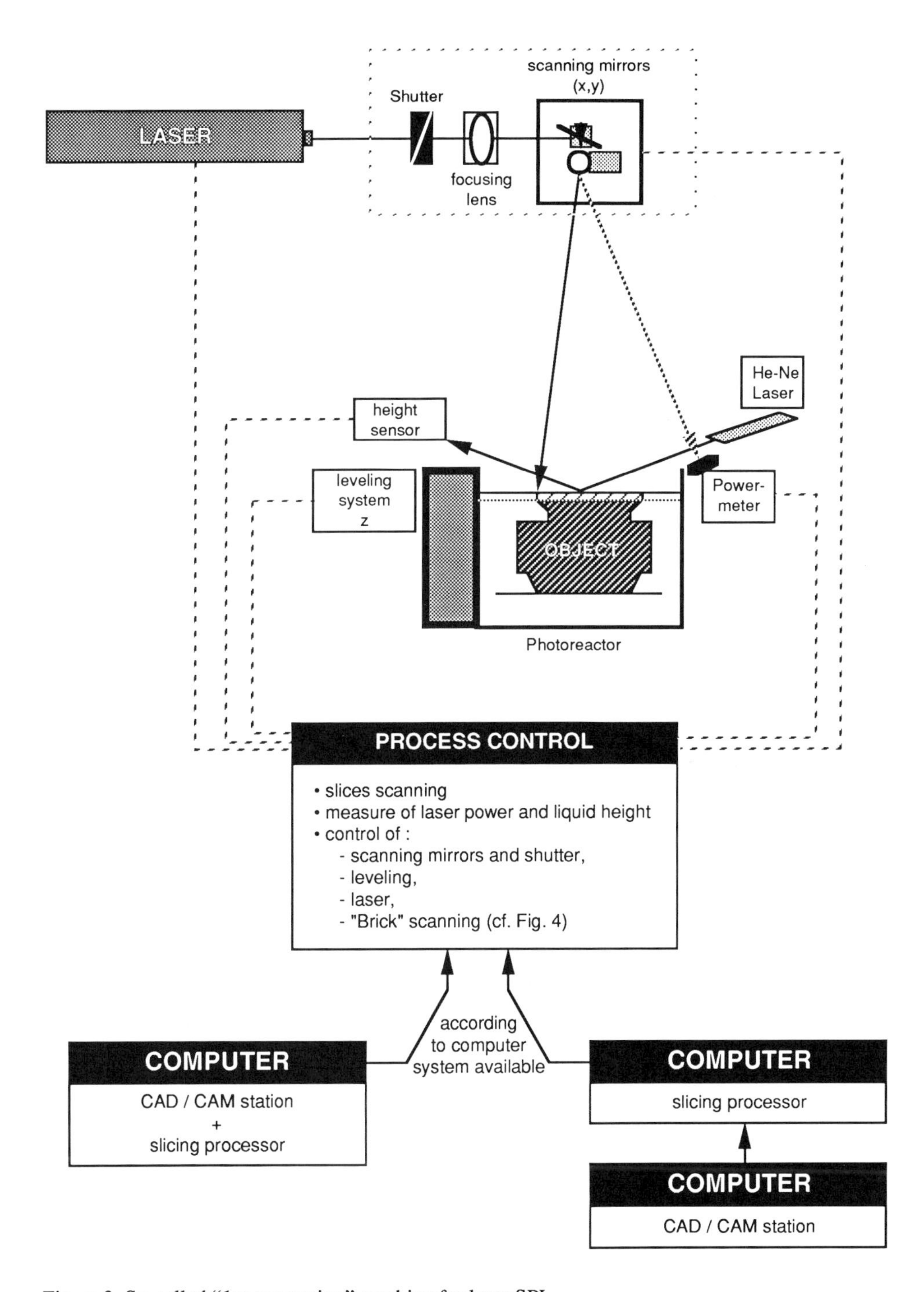

Figure 3. So-called "1st generation" machine for laser SPL.

Although the commercialized resins still lead to an important shrinkage (5 to 7 %), one can use the skill of the engineer to manufacture objects with a small shrinkage [6 - 8]. Figure 4 depicts a way to obtain this by polymerizing disjoined voxels. One waits for the end of the shrinkage of every voxel prior to polymerize them together in a second step. The global shrinkage is then much smaller and the real dimensions of the manufactured objects are very close to the expected ones.

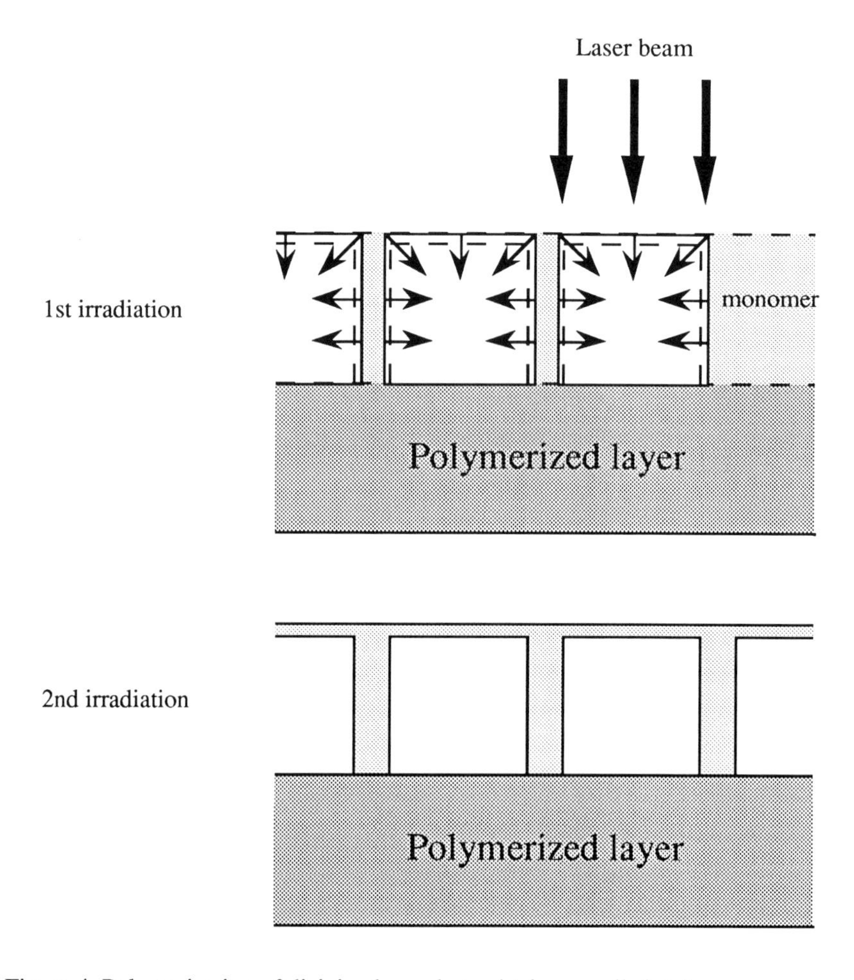

Figure 4. Polymerization of disjoined voxels to obtain a small shrinkage.

Since we have said almost everything about the "1st generation" laser stereophotolithography from a general point of view, we shall try to find the way to associate the micro-technical engineering to this manufacture technology.

The fields in which the microtechnical activity takes its full place are mainly : sensors and microsensors, micromotors, microactuators, microrobots and energy microsources [9]. In every case an energy transfer occurs (electrical, optical, thermal, etc.) inducing a modification of the information or a spatial displacement. If the optical properties of the polymers can be satisfactory, their actual mechanical and electrical properties are definitively inadequate to manufacture directly usable microtechnical elements.

We can point out a second problem due to the size of the objects. For the time being, one can use piezoelectrical motors only because only them have a high enough torque. Their sizes are several mm typically. There are indeed papers describing electrostatic micromotors made of polysilicium [10] even molecular nano-motors [11] but none of them have yet found any practical application, as far as we know.

For this main reason, the size of the microtechnical objects must be about several tenths mm even if for certain applications, one makes treatments on very thin thicknesses (molecular grafting by covalent bounding, molecular lubrification, etc.). But in these cases, the treatment is performed along *one* space-direction only.

The today know-how of the laboratories involved in this type of researches mainly deals with the "1st generation" stereophotolithography using photopolymerizable resins. With such tools it is just possible to manufacture micro-scale models or micro-prototypes but their line manufacture is impossible, with the exception of optical fibers manufacturing systems. But these objects can later be used in moulding for example.

This paper describes the present state of the know-how in these different fields. It also anticipates several ways to extend the application areas of this new technology.

2. MANUFACTURE OF PASSIVE STRUCTURES BY STEREOPHOTOLITHOGRAPHY

2.1. Smallest size of a voxel

The highest reachable precision dimension figures of an object depends on several optical and chemical factors that we describe hereafter.

2.1.1. Accuracy of the optical parts

In the classical processes one uses computer controlled galvometric mirrors to deflect the laser beam. The quality of the current systems [General Scanning Co. for example] and the numeric apertures of the laser sources allows to produce a spot about 150 μm diameter [12] on the open surface of the resin.

To increase the accuracy, it is necessary to "clean up" the laser beam by passing it through a spatial filter and to use one focusing lens at least (such as a microscope objective) which increases the numerical aperture and in consequence reduces the size of the focused light spot to several μm. On the other hand it is no longer possible to use convenient systems such as computer controlled galvometric mirrors with this technique. It is then necessary to use precision X-Y-Z motorized translators as depicted in Figure 5. These simple apparatuses are set up to prove the feasibility of this process because fast deflections of the laser beam are no longer possible. We shall in fact have to use different irradiation techniques but we can't describe them in this paper because they are not yet patented.

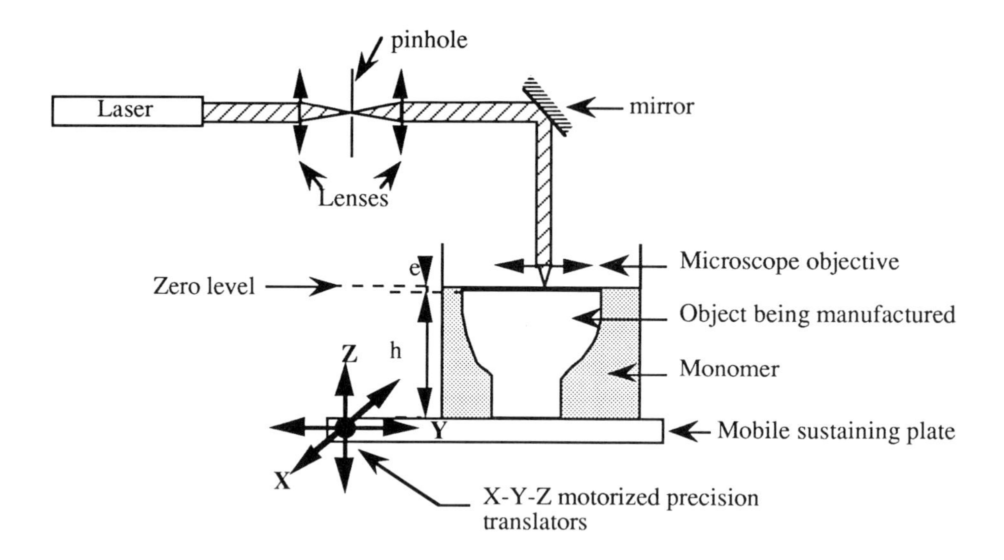

Figure 5. Schematic block diagram of a micro-SPL set-up with a microscope objective and motorized precision X-Y-Z translators.

2.1.2. Photochemical initiator

In the "monophotonic" processes which are presently used, an important notion is the optical thickness "μ" of the photo-transformable material which is

$$\mu = 1 / \sum_i \varepsilon_i c_i$$

where ε_i and c_i are respectively the molecular extinction coefficient and the concentration of the ith product absorbing the light at the wavelength λ.

If only one product absorbs the light at the wavenlength λ, it is possible to prove that the polymerized thickness x is a logarithmic function of the voxel irradiation time t. Figure 6 is an example of the spatial evolution of the polymerization front *vs.* time when one uses a Gaussian light source with a standard deviation σ. The equation of this front is

$$Ln \frac{fo \varepsilon ct}{S} = \frac{y^2}{\sigma^2} + \varepsilon cx$$

where fo is the photon flux density at $y = 0$
y is the radial coordinate
t is the irradiation time
S is the minimum energy required to initiate the photopolymerization.

Figure 7 depicts the variation of x_o (at $y = 0$) *vs.* time. After a time partly required by the reaction of the inhibitors usually existing in the resins, the evolution of x_o (at $y = 0$) *vs.* time is fairly logarithmic.

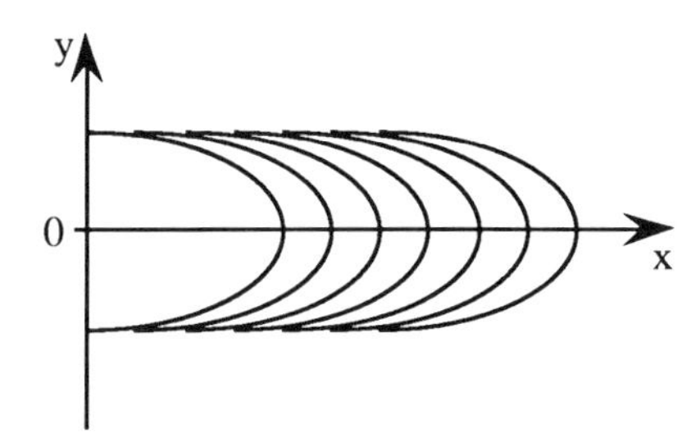

Figure 6. Translation of the parabolic front of a polymerization initiated by a Gaussian profile laser beam *vs.* time.

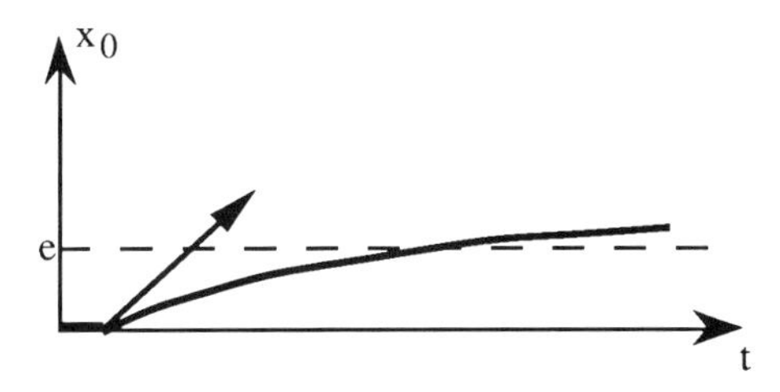

Figure 7. Slow logarithmic variation of the polymerized depth x_0 (at y = 0) *vs.* the characteristic time of hardening.

Under these conditions and if the translations are fast enough, it should be possible to polymerize thin layers of monomer. Nevertheless the quality of the phototransformation is better when μ and e are about equal. This implies to use photoinitiators the molecular extinction coefficients ε of which are high enough to lessen the optical thickness μ. In practice one can use ketones such as the 4,4'-bis (dimethylamino)benzophenone as photoinitiators and chemicals belonging to the family of the hydroxyphenylbenzotriazole as unreactive UV absorbing substances. It is also possible to increase the photoinitiator concentration.

Up to now, the thinnest layer of polymer that we manufactured was about 50 µm thick [12].

Note : Polychromatic sources.

Using UV lasers lasing at several slightly different wavelengths (for example Ar^+ lasers lasing mainly at λ_1 = 351.1 and λ_2 = 363.8 nm in the UV region) induces different effects :

- there are several focal points if the optical lenses are not perfectly achromatic,
- several optical depths μ_i exist. If the μ_i are very different one from the others, it is usually the longest which determines the accuracy along the z axis. In the case of two wavelengths one can counterbalance this effect by mixing different photoinitiators or unreactive absorbing substances so that $\mu_{\lambda_1} = \mu_{\lambda_2}$. It is also possible to select one of these wavelengths only if the remaining power of the beam is high enough.

2.1.3. Diffusion induced by the chemical reaction

An important point is the relationship - when it exists - between the impact point of one photon on the surface of the reactive medium and the precision of the resulting voxel.

The first phenomenon which may occurs proceeds from the possible solubility of the polymer in the monomer (no relationship). If this happened, the 3D part would disappear as it is manufactured. It is easy to avoid this problem by using multi-functional monomers or oligomers which reticulate.

The second one proceeds from the matter transport resulting from the long chain photopolymerization reaction. For example if the average number of monomer / oligomer molecules reacting per photon is 10^4 and if their sizes are about 1 nm, the longest path will be about several µm. This last value has the same order of magnitude as the optical resolution. To prove that the spatial migration of the reactive center (cf. Figure 8) is not a basic phenomenon, we simulated a photochemical polymerization by the MONTE-CARLO simulation method. A typical result appears in Figure 9 and shows that the polymerized volume is only several molecular units long.

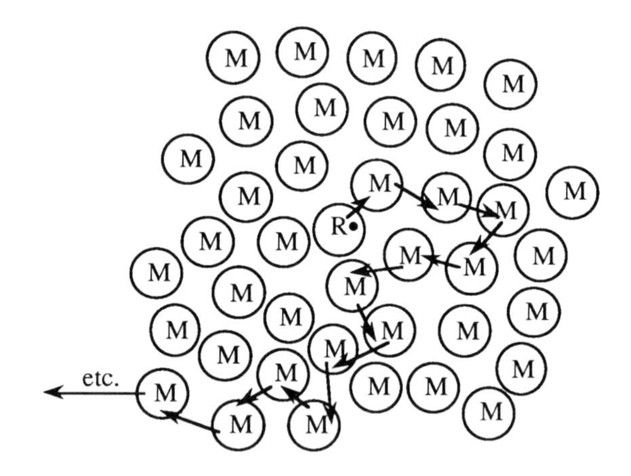

Figure 8. Spatial migration of a reactive center.

This conclusion is most important in the case of microtechnical applications because the precision of the dimensions of a part depends on the local photon absorption rate by the resin. To illustrate this result, we manufactured a pyramid (cf. Figure 10) composed of layers of disjoined voxels. This enlargement shows that the precision of the edges is about several µm effectively. In this case we think that the precision results more from the optical quality of the excitation laser than from the resin itself. Two other examples appear in Figure 11. These objects have been manufactured by using computer controlled galvanometric mirrors and their precision is about 50 µm [12].

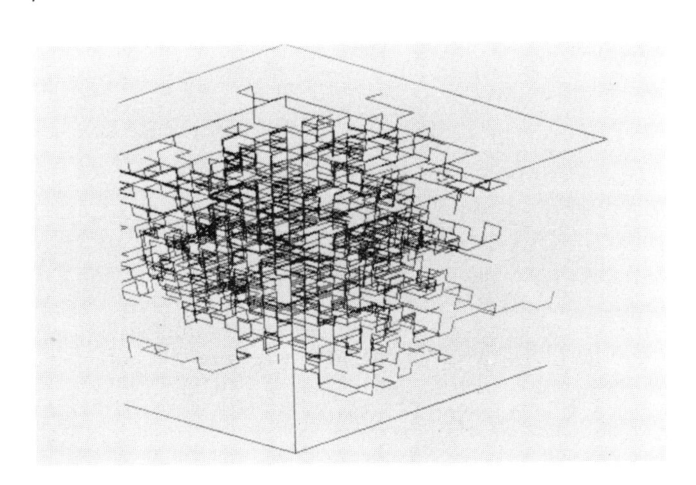

Figure 9. Typical result of the computer simulation of a 3D photochemical reticulation by the MONTE CARLO simulation method.

Figure 10. Enlargement of a part of a 3D pyramid manufactured voxel by voxel (base dimensions : 2 mm by 2 mm).

(a)

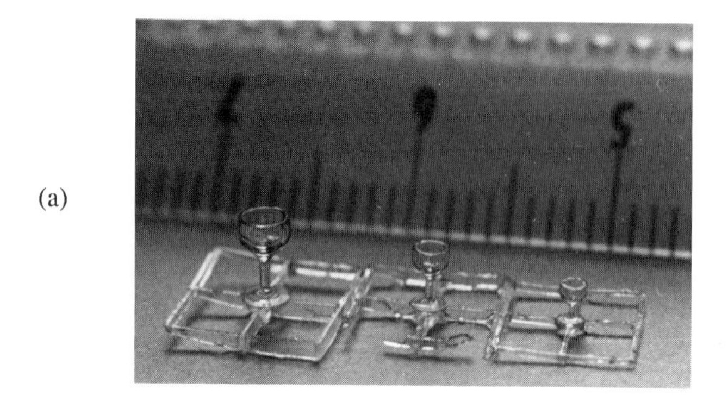

(b)

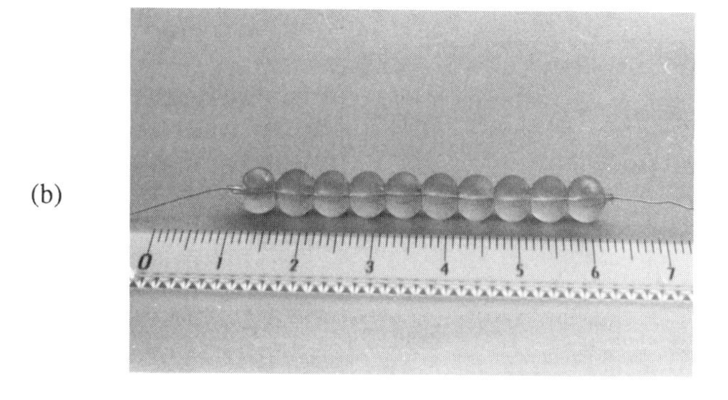

Figure 11. Two objects manufactured using computer-controlled galvanometric mirrors :
- (a) mini-glasses (4.1, 3.2 and 2.1 mm high respectively),
- (b) 250 μm diameter platinum wire inserted into a 300 μm diameter empty cylinder after the manufacture of the polymerized part.

2.2. Applications of the stereophotolitography in optics

The classical stereophotolithography processes can't obviously allow the direct manufacture of good quality optical parts. They need to be polished.

The usable processes then result from the confinement of the phototransformation as it appears in Figure 12. It is possible to manufacture lenses having different shapes by fixing the rotation speed of a cylindrical reactor. In it, a photopolymerizable resin floats on the surface of a liquid carrier non-miscible with it. The shape of the manufactured optical parts depends on their diameters, on their densities relatively to the one(s) of the liquid carrier(s) but mainly on the surface force coefficients. We tested other similar techniques and we manufactured lenses and micro-lenses (cf. Figure 13).

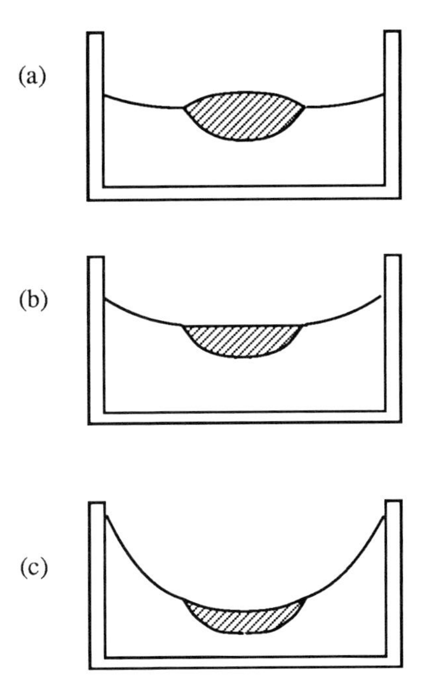

Figure 12. Manufacture of optical parts in a vortex: - (a) convex lens,
- (b) plano-convex lens,
- (c) meniscus lens.

This reaction confinement can also be used to manufacture optical fibers. This technique is based on the REYNOLDS's experiment (cf. Figure 14) : two isokinetical co-axial laminar flows of monomer flow out of their own tanks, but only the inner one contains a photoinitiator. The irradiation of these flows results in the manufacture of an optical fiber cladded by a weakly reticulated material (diffusion transfer zone of the photoinitiator betwen the two flows).

Other optical parts can be manufactured on these bases. Nevertheless we must admit that the optical quality of the parts manufactured by these techniques is not yet as good as those manufactured by the classical techniques. But these processes that we have briefly described present important advantages from the viewpoint of flexibility and mainly in the choice of the reactive materials. This last advantage is crucial to obtain a material with a refractive index fixed in advance or with an ability for molecular grafting to develop optodes or optical sensors.

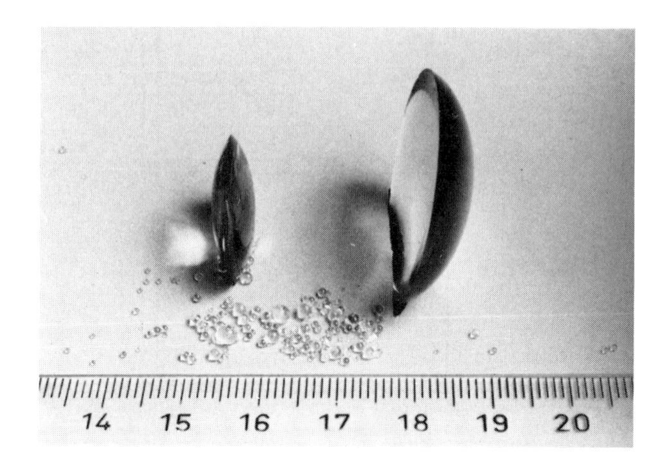

Figure 13. Lenses and micro-lenses manufactured by other similar techniques.

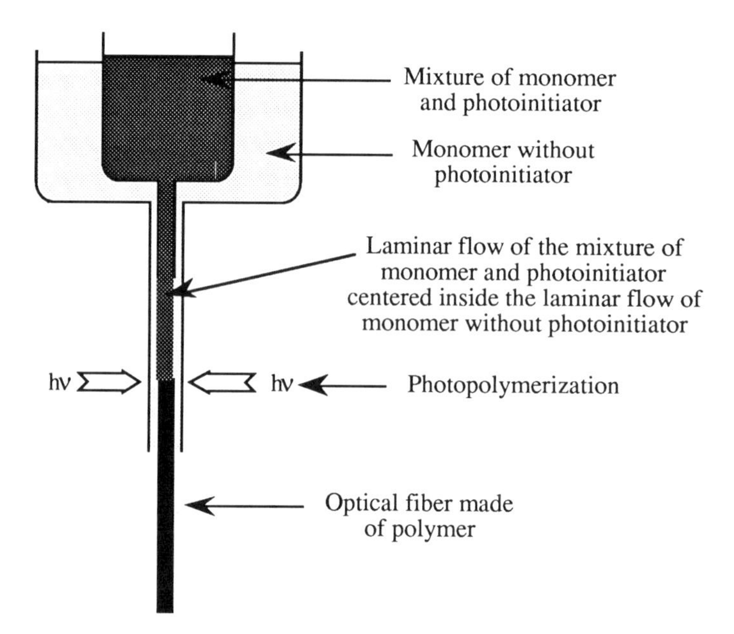

Figure 14. Manufacture of optical fibers by the photopolymerization a laminar flow of monomer.

3. WHAT ARE THE ACTIONS TO UNDERTAKE TO MANUFACTURE ACTIVE STRUCTURES ?

Following the line of this general paper, we can follow two directions : either we modify the principles of SPL to manufacture parts which could be used by the microtechniques or we revisit the processes which can be used in this field.

3.1. Towards new SPL processes

We have rapidly shown in the first part of this paper that it was possible to manufacture passive parts made of polymers. The main problems of the "1st generation" SPL processes results from the shrinkage. In microtechnical applications the obstacles are different and result from the size of the voxels and from the ability to obtain very thin layers. If this domain must become the subject of researches on the processes and on the materials, as far as we can judge, it should be particularly useful to conceive new laser CADM (Computer Aided Design and Manufacturing) processes. They should allow the manufacture of active structures made of ceramics, metals, and in a more general way the manufacture of composite systems including conducting elements, insulating elements, etc.

This field is essential because it allows industrial collective productions for the next future.

Another subject of researches is the information storage in a compacted manner. The approach can be a multiwavelength process such as the one developed by U.P. WILD in Switzerland (hole burning and interference systems) [13]. Another way to store compacted information could be a 3D approach.

The application areas of these directions of researches are already wide and will be completely developed only by interdisciplinary researches involving engineers, photonicians, chemists, mechanicians, specialists in robotics, etc. But it should be observed that the bases of these researches are already mastered well enough.

3.2. Systems revisited

On can indeed try to reach the limits of the already existing systems to manufacture very small voxels. If one follows this track, what will be the size of the smallest ball-bearing ? In fact, it is easy to understand that a collective production only can give reasonable production costs and will impose to revisit the technological principles. Will the magnetism still be the basic principal of electrical micro-motors for example ? etc. Since the classical systems will have to be questioned again, this could be a good opening to develop new technologies such as the molecular lubrification, photochemical muscles, molecular assemblies, etc.

4. CONCLUSIONS

The fields for the future that we have briefly described are already studied in many laboratories. For the time being the 3D manufacturing techniques have not yet led to industrial applications because they are in a development step which is still a research stage.

The manufacture of prototype objects made of polymer is the first step towards industrial applications. It should be followed by other realizations made of metals, ceramics, etc. This is a difficult but very stimulating activity, and we expect to persist in doing it.

REFERENCES

1. Laser fabrication of 3D-plastic objects studied, Laser Focus, 26 (1983).
2. Replication manufacture : lasers carve true 3D master patterns, Iron Coste (1982) 14.
3. A.J. Herbert, J. Appl. Photogr. Eng., 8-4 (1982) 185.
4. M. Cabrera, J.Y. Jézéquel and J.C. André, in "Application of Laser in Polymer Science and Technology", Vol. 3, C.R.C. Press, J.F. Rabek and J.P. Fouassier (eds.) (1990) 73.
5. Premières Assises Européennes du Prototypage Rapide, Paris (France), (1992).
6. P. Karrer, S. Corbel, J.C. André and D.J. Lougnot, J. Polym. Sci., 30 (1992) 2715.
7. Y. Brullé, "Procédés photoniques pour l'optique", Thèse INPL, Nancy (France) (1992).
8. J.C. André, S. Corbel and J.Y. Jézéquel, 2nd International Conference on Solar Energy Storage and Applied Photochemistry, Cairo (Egypt) (1993).
9. J.C. André, "Microstéréophotolithographie, Recherche en microtechniques : réalités et perspectives", Collection du Livre Vert, Institut de Microtechniques and CETEHOR (eds.), (1992) 73.
10. E. Bernand, L. Passatte, G.A. Racine and N. de Rooij, 4ème Congrès Européen de Chronométrie, (1992).
11. C. Fuhr, R. Hagedorn, T. Müller, B. Wagner and W. Benecke, Proceedings of the M.E.M.S. Symposium, Nara (Japan), (1991) 259.
12. S. Zissi, "Premières étapes vers la micro-stéréophotolithographie", D.E.A. INPL Nancy (France) (1992).
13. U.P. Wild, 2nd International Conference on Solar Energy Storage and Applied Photochemistry, Cairo (Egypt) (1993).

MICROCHEMISTRY
Spectroscopy and Chemistry in Small Domains
Edited by H. Masuhara et al.

Laser-controlled phase transition of aqueous poly(N-isopropyl-acrylamide) solution in micrometer domain

N. Kitamura,* M. Ishikawa,# , H. Misawa,† and R. Fujisawa‡

Microphotoconversion Project,§ ERATO Program, Research Development Corporation of Japan, 15 Morimoto-cho, Shimogamo, Sakyo-ku, Kyoto 606, Japan

Infrared (1064 nm) laser-induced, photo-thermal phase transition of non-labeled and pyrene-labeled poly(N-isopropylacrylamide) aqueous solutions (PNIPAM or Py-PNIPAM, wt % = 1.8 ~ 3.6 %) is described. The present infrared laser-induced method enabled one reversible control of formation and dissolution of a PNIPAM microparticle in water and the diameter of the particle was shown to be controlled in micrometer. Phase transition behavior was also studied on the basis of fluorescence spectroscopy on Py-PNIPAM. A possible role of the radiation force of the laser beam for the PNIPAM microparticle formation and simultaneous laser manipulation of the particle are discussed.

1. INTRODUCTION

Thermal phase transition of an aqueous poly(N-isopropylacrylamide) (PNIPAM) solution has received considerable attention due to its important roles as stimulus-responsive materials. At room temperature, an aqueous PNIPAM solution is clear and transparent, while submicrometer PNIPAM particles precipitate above 31 °C and the solution becomes turbid (phase transition temperature ; T_c = 31 ~ 32 °C) [1]. It has been well known that the phase transition of the solution is primarily responsible for large increase in hydrophobicity of the polymer chain above T_c and, the coil structure of the polymer turns to the globule structure upon the phase transition [1, 2]. For

---

* To whom correspondence should be sent to the present address at Department of Chemistry, Faculty of Science, Hokkaido University, Sapporo 060, Japan.

Present address ; Research Center, Nissan Motor Co. Ltd., 1 Natsujima-cho, Yokosuka 237, Japan.

† Present address ; Department of Mechanical Engineering, Faculty of Engineering, The University of Tokushima, 2-1 Minamijosanjima, Tokushima 770, Japan.

‡ Present address ; Mita Corp., 1-2-28 Tamatsukuri, Chuo-ku, Osaka 540, Japan.

§ Five-year term project : October 1988 ~ September 1993.

further basic understandings of the properties of the polymer solution, elucidation of molecular mechanisms of the polymer conformation change and subsequent submicrometer particle formation by temperature is quite necessary. Indeed, much efforts have been devoted to understand the phenomenon on the basis of fluorescence [2 - 7], Raman, ESR, NMR spectroscopy, and other techniques [8 - 11]. Nevertheless, a clear picture of the phase transition in water has not been given yet. Besides, the molecular mechanisms, furthermore, if one can control arbitrarily the particle formation and dissolution, applications of stimulus-responsive materials based on PNIPAM will be greatly advanced.

An infrared laser beam, in particular a 1064-nm laser beam from a CW Nd^{3+}:YAG laser, is expected to have a high potential as a heat source for the phase transition of an aqueous PNIPAM solution. Namely, since a water molecule absorbs 1064-nm light through the overtone band of the OH stretching [12], water will be heated through photo-thermal effects, by which absorbed photon energy is converted to thermal energy. When a 1064-nm laser beam is focused to ~ 1 μm and switched on/off in the solution, moreover, the phase transition will be controlled in micrometer dimension. In addition to the role of the 1064-nm laser beam as a heat source, a focused 1064-nm laser beam can also act as a light source for non-contact and non-destructive manipulation of various microparticles in solution via the radiation force generated by refraction of the laser beam through a microparticle (*laser manipulation*) as the principle [13 - 15] and various experimental results are described in detail elsewhere [14 - 17]. Infrared laser-induced chemistry of an aqueous PNIPAM solution is therefore very interested and the phase transition of the solution is expected to be arbitrarily controlled in micrometer dimension by a focused 1064-nm laser beam.

In this article, we describe characteristic features of an infrared laser-induced phase transition of PNIPAM [18, 19] and pyrene-labeled PNIPAM (Py-PNIPAM) [20] aqueous solutions (Scheme I) in detail, including fluorescence spectroscopic studies on the PNIPAM microparticle formation processes. Also, we discuss laser manipulation of a PNIPAM microparticle produced by the phase transition and an important role of the radiation force of the laser beam for the particle formation.

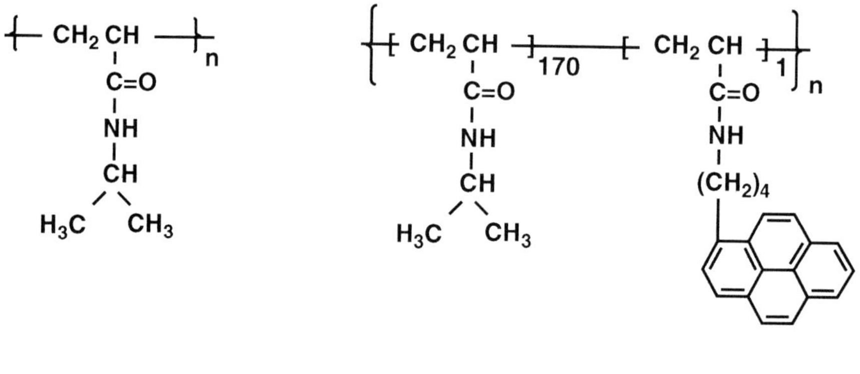

Scheme I. Structures of PNIPAM (left) and Py-PNIPAM (right).

2. CHARACTERISTIC FEATURES OF LASER-INDUCED PHOTO-THERMAL PHASE TRANSITION

2.1. Single PNIPAM Microparticle Formation by Infrared Laser Beam

Thermal phase transition of an aqueous PNIPAM solution *in the dark* produces a number of submicrometer PNIPAM particles in the entire region of the solution as has been reported by Heskins and Guillet [1]. On the other hand, when the PNIPAM solution (3.6 wt %) is irradiated with a focused (~ 1 μm) 1064-nm laser beam (CW Nd^{3+}:YAG laser, laser power ; P_{1064} = 1.2 W) through an objective lens (X 100, NA = 1.30) of an optical microscope, a single PNIPAM microparticle with the diameter of several micrometer is produced exclusively in the vicinity of the focal spot of the laser beam *even at room temperature* below T_c. A typical example of the morphological changes of the solution with laser irradiation time (t) observed under the optical microscope is shown in Figure 1.

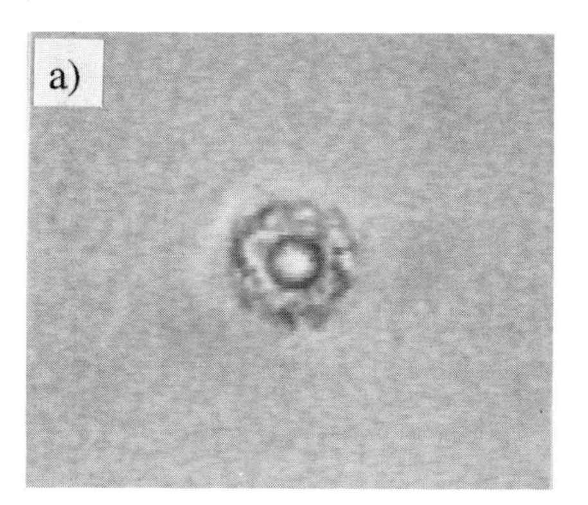

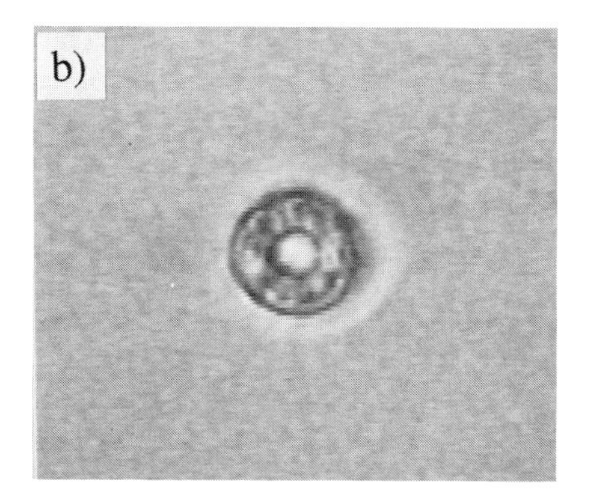

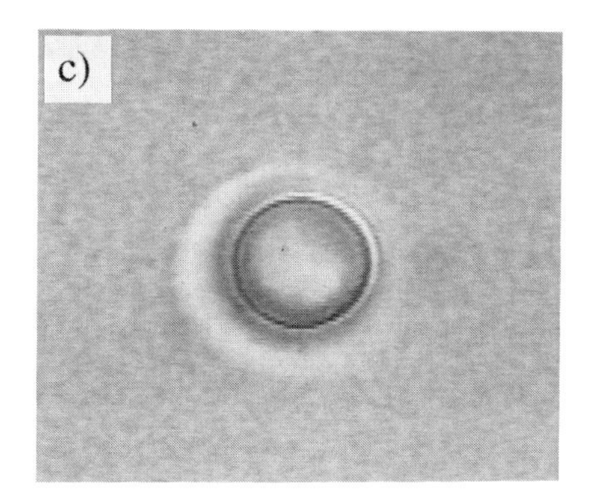

Figure 1. Single microparticle formation in an aqueous PNIPAM solution (3.6 wt %) by focused (~ 1 μm) 1064-nm laser beam irradiation (P_{1064} = 1.2 W, 20 °C). Irradiation time (t) ; a) t = 0.5 s, b) t = 5.0 s, and c) t = 50 s.

The phase transition takes place immediately after 1064-nm laser beam irradiation as recognized from the change in the solution morphology at the center of the optical micrograph. At $t = 0.5$ and 5.0 s (Figure 1a and 1b, respectively), namely, the diameter of the PNIPAM particle (d) remains ~ 4 μm while further prolonged irradiation ($t = 50$ s) renders growing-in of the PNIPAM particle to d ~ 8 μm (Figure 1c). It is noteworthy that, in the initial stage of irradiation ($t = 0.5$ and 5.0 s), the particle seemed to be surrounded by a number of submicrometer PNIPAM particles. However, the morphology around the periphery of the particle changes during laser irradiation and, at $t = 50$ s, the surface of the particle becomes smooth and the vicinity of the particle in the solution phase becomes clear as long as observation under the optical microscope. Detailed analyses of the PNIPAM particle formation indicate that growing-in of the particle to d ~ 4 μm is very fast and almost finishes within 100 ms. Further laser irradiation brings about a gradual increase in d and the diameter of the particle reaches to an equilibrium value (~ 8 μm) at $t > 40$ s (Figure 2). When the laser is switched off, the microparticle disappeared with a time constant of ~ 150 ms. The formation and dissolution of the PNIPAM particle was highly reversible for several cycles without any appreciable change in the equilibrium diameter. This is the first demonstration for laser-controlled formation and dissolution of a PNIPAM particle in solution.

2.2. Photo-Thermal Phase Transition of Aqueous PNIPAM Solution

PNIPAM itself does not absorb the 1064-nm laser beam, so that the present laser-induced phase transition and the microparticle formation are not ascribed to direct photoresponse of PNIPAM. As described earlier, since a water molecule (H_2O) possesses an absorption band at 1064 nm, the PNIPAM particle formation is induced by absorption of the laser beam by H_2O and subsequent phase transition of the

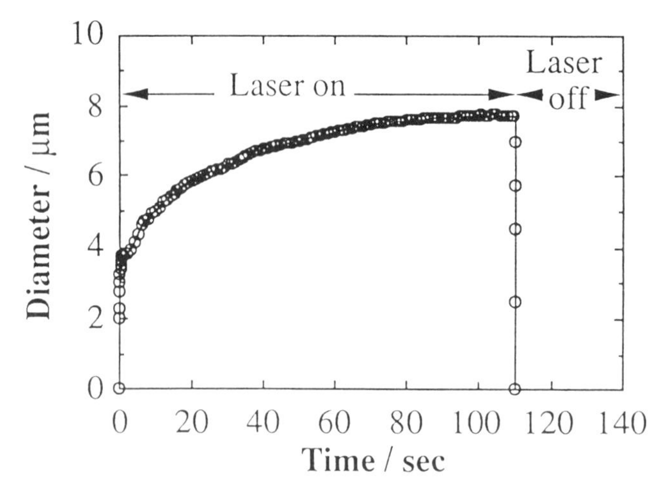

Figure 2. Time-response profile of the diameter of the PNIPAM microparticle. Polymer concentration = 3.6 wt %, P_{1064} = 1.2 W, temperature = 20 °C.

polymer solution via photo-thermal local heating of H_2O in the vicinity of the focal spot of the laser beam produces the PNIPAM microparticle. This is supported by analogous experiment in D_2O. Namely, since D_2O is almost transparent at 1064 nm [19], the laser-induced phase transition of the polymer solution is not expected in this solvent. Indeed, we could not observe the PNIPAM particle formation in D_2O under analogous experimental conditions with those in Figure 1 (polymer concentration = 3.6 wt %, P_{1064} = 1.43 W, 20 °C) [19].

An increase in the solution temperature through absorption of the 1064-nm laser beam by H_2O was roughly estimated to confirm the photo-thermal effect. For calculation, we assumed that a water droplet with a radius of 1 μm (R) was heated by the laser beam (heat energy ; Q) and relevant local heat generated in the droplet was equilibrated with the surrounding water phase. In such the case, a temperature increase in the droplet (ΔT) can be calculated by the equation [21] ; $\Delta T = Q/4\pi\kappa R$, where κ is the thermal conductivity of H_2O. Q was estimated from the absorbance of H_2O at 1064 nm (5.6 x 10^{-6} for 1 μm optical path) and P_{1064}. For P_{1064} = 1.2 W, ΔT was calculated to be ~ 4 °C for 1 μm^3. In the actual experiments, the laser beam is irradiated to the solution with a large cone angle (~ 120 °) due to the use of a large NA (1.30) objective lens, so that water is heated in a larger volume than 1 μm^3. The larger the volume heated, the lower a cooling rate by the surrounding water phase since a surface area/volume ratio of the droplet decreases with increasing in R. Therefore, we suppose that ΔT will be much higher than 4 °C under the present experimental conditions. The experimental results in Figure 1 indicate that ΔT produced by 1064-nm laser beam irradiation is sufficiently high enough to induce the phase transition of an aqueous PNIPAM solution *even at room temperature.*

Further evidence of the laser-induced effects on the phase transition was shown by the fact that the PNIPAM particle formation strongly depended on P_{1064}, a solution temperature before laser irradiation, and a polymer concentration as summarized in Figure 3. The particle formation takes place only when P_{1064} exceeds a certain threshold value (P_{th}). For a given a polymer concentration (3.6 wt %), for example, P_{th} increases with decreasing the solution temperature. Indeed, P_{th} at 20 °C is > 0.7 W (curve **b**) while that at 23 °C is > 0.2 W (curve **a**). A slight change in the solution temperature (3 °C) brings about a large change in P_{th} (~ 0.5 W). At a fixed polymer concentration, the diameter of the particle becomes larger with increasing P_{1064}, since the heat generated by the laser beam is much higher for higher P_{1064} as expected from the above equation. When a solution temperature is identical (20 °C), a decrease in the PNIPAM concentration leads to an increase in P_{th} and the formation of a smaller size of the particle for given P_{1064} (curves **b** and **c**). It is easily expected that a higher polymer concentration is favorable for the formation of a larger size of the particle, since the number of the polymer chain being taken the phase transition per a unit volume is larger for a higher concentration. Furthermore, local heating and subsequent phase transition of the solution compete with cooling by the surrounding water phase, so that P_{1064} necessary to produce the same size of the particle decreases with increasing the polymer concentration. All the results including the absence of the phase transition in D_2O prove that the primary origin of the PNIPAM particle formation is photo-thermal local heating of water by the 1064-nm laser beam irradiation.

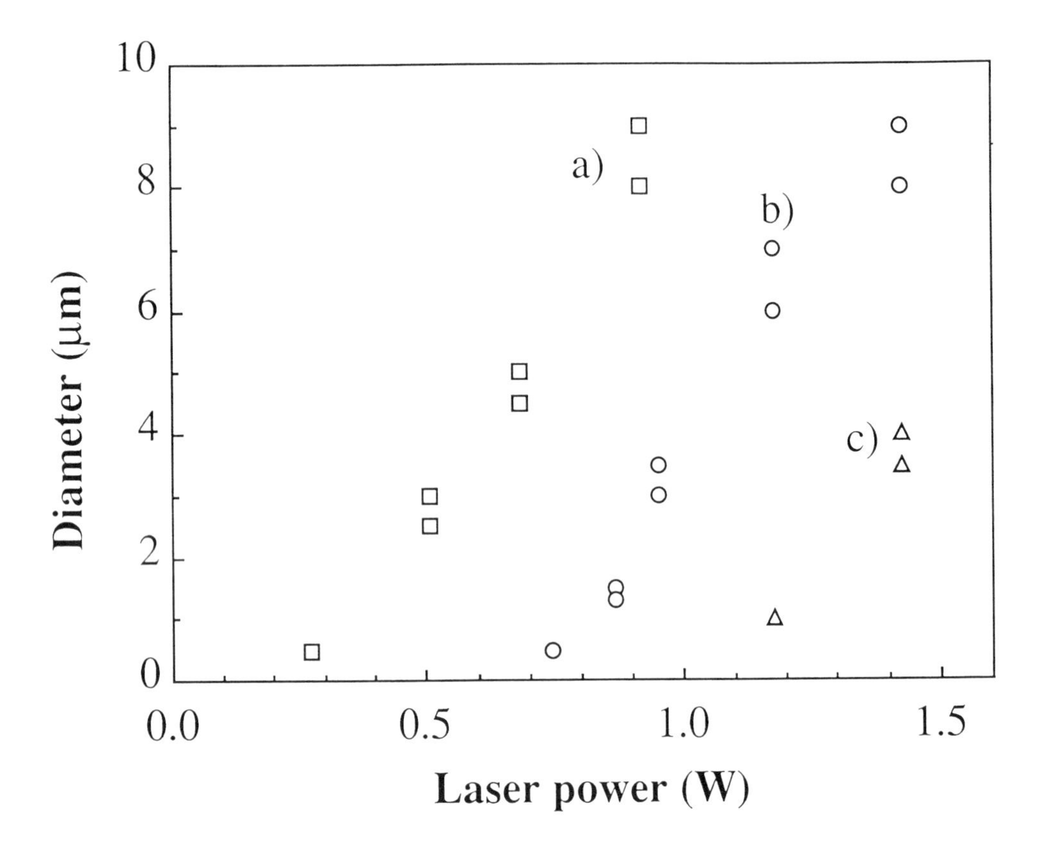

Figure 3. Photo-thermal phase transition of aqueous PNIPAM solutions under various conditions. The polymer concentration and the solution temperature before laser irradiation are ; a) 3.6 wt %, 23 °C, b) 3.6 wt %, 20 °C, and c) 1.8 wt %, 20 °C.

It is very important to note, furthermore, that the diameter of the PNIPAM particle produced can be controlled in 2 ~ 9 μm through P_{1064}, the solution temperature before laser irradiation, and the polymer concentration as demonstrated in Figure 3. This is particularly unique since a single PNIPAM particle with arbitrary diameter can be prepared, which is not attained by the thermal phase transition in the dark.

3. FLUORESCENCE SPECTROSCOPIC STUDIES ON PNIPAM MICROPARTICLE FORMATION

3.1. Fluorescence Probe Study by 1, 8 - ANS

The phase transition of the polymer solution will lead to a change in micropolarity near the polymer chains owing to hydrophilic and hydrophobic nature of the polymer below and above T_c, respectively. Since a fluorescence probe method is quite potential to monitor microenvironments in polymers, biological samples, and so forth [22], we studied the phase transition processes of an aqueous PNIPAM solution on the basis of fluorescence measurements in the presence of ANS (1-anilino-naphthalene-8-sulfonate) as a micropolarity probe.

Figure 4 shows fluorescence spectra from a bulk ANS - PNIPAM solution obtained by excitation at 355 nm without 1064-nm laser irradiation. The maximum wavelength of the fluorescence from ANS is strongly dependent on temperature as clearly seen in Figure 4. Namely, the fluorescence spectrum exhibits the peak (λ_{max}) around 520 nm at 25 °C, while that above T_c (34 °C) is shifted to the shorter wavelength by 54 nm (λ_{max} = 466 nm). Although the spectrum above T_c is supposed to be the sum or average of the ANS fluorescence from both the submicrometer particles and the aqueous phase, λ_{max} of 466 nm indicates that polarity (ε) *in* or *in the vicinity of* the PNIPAM particles is close to that in acetone ~ ethanol (ε = 20.7 ~ 24.6) as judged from the ε dependence of λ_{max} for ANS [23, 24]. On the other hand, λ_{max} of 520 nm below T_c corresponds to micropolarity in ethanol/H_2O = 1/4 (v/v, ε ~ 69). As recognized from Figure 4, furthermore, the fluorescence intensity increases considerably upon the phase transition. It has been reported that the decrease in ε from 69 to 20.7 ~ 24.6 results in an increase in the ANS fluorescence quantum yield from 0.02 to 0.31 ~ 0.37 [23, 24]. The results are in good agreement with the

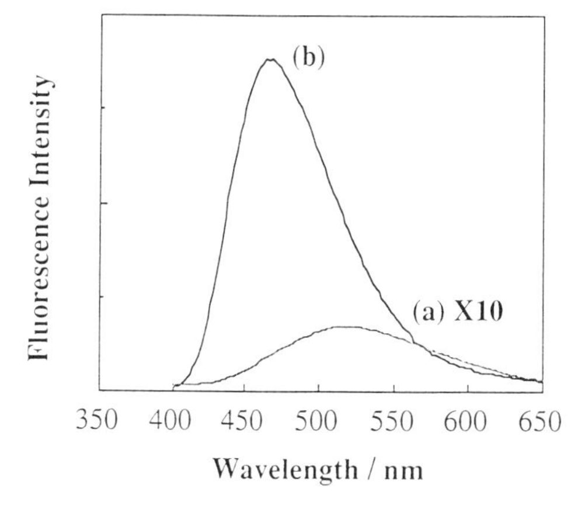

Figure 4. Fluorescence spectra of an aqueous ANS-PNIPAM solution below (25 °C, a)) and above (34 °C , b)) T_c ([ANS] = 2 x 10^{-3} M, PNIPAM = 3.6 wt %, λ_{exc} = 355 nm, without 1064-nm laser irradiation).

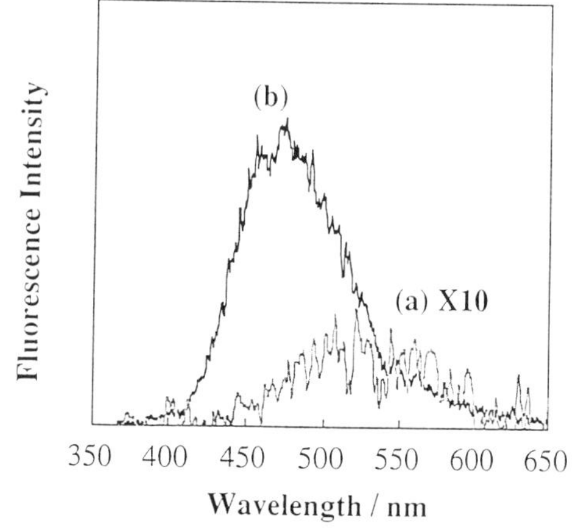

Figure 5. Fluorescence spectra of an aqueous ANS-PNIPAM solution observed before (a) and after (b) photo-thermal phase transition ([ANS] = 2 x 10^{-3} M, PNIPAM = 3.6 wt %, λ_{exc} = 355 nm, P_{1064} = 1.2 W, 20 °C).

changes in the fluorescence characteristics of ANS with solvent polarity and, the micropolarity *in* or *in the vicinity of* the PNIPAM particles above T_c is very low as compared with that in the solution below T_c.

When the single PNIPAM microparticle formation was induced by the 1064-nm laser beam under the microscope (Figure 5, P_{1064} = 1.4 W, 20 °C), the fluorescence from ANS became highly intense as compared with that before laser irradiation and exhibited the maximum wavelength around 475 nm. Although λ_{max} of 475 nm is slightly longer than that above T_c without 1064-nm laser irradiation (λ_{max} = 466 nm, Figure 4), the value corresponds to micropolarity in the single PNIPAM particle to be as low as that in ethanol/H_2O = 9/1 (v/v) ~ N,N-dimethylformamide (ε = 30.1 ~ 36.7) [23, 24]. Qualitatively, the changes in the fluorescence characteristics of the solution upon the photo-thermal phase transition in Figure 5 are quite similar to those observed by the phase transition in the dark (Figure 4). A considerable amount of water molecules is dehydrated from the polymer chains upon the photo-thermal phase transition of the solution. We conclude that the 1064-nm laser-induced PNIPAM particle formation is certainly originated from the large increase in hydrophobicity of the polymer chains analogous to the thermal phase transition in the dark.

3.2. Fluorescence Probe Study by Pyrene-Labeled PNIPAM

A fluorescence spectroscopic study on the phase transition has been also performed for an aqueous solution of pyrene-labeled PNIPAM (Py-PNIPAM, 0.75 wt %) and the results are summarized in Figure 6. Upon excitation at 355 nm, the polymer solution below T_c (20 °C) exhibits both monomer and excimer fluorescence from the pyrenyl chromophore around 380 ~ 420 and 490 nm, respectively (Figure 6a). Above T_c (35 °C ; without 1064-nm laser irradiation), on the other hand, the pyrene excimer formation is largely suppressed owing to the thermal phase transition of the solution, by which the segment mobility of the polymer chains decreases and, therefore, the bimolecular encounter efficiency of the pyrenyl chromophores attached to the polymer backbone is greatly lowered (Figure 6b). Such results are in good agreement with those reported by Winnik [2].

When the solution is irradiated with the focused 1064-nm laser beam below T_c (20 °C, P_{1064} = 1.0 W), we can confirm the formation of a single micrometer Py-PNIPAM particle analogous to the results in Figure 1. Figure 6c shows the fluorescence spectrum of the pyrenyl chromophore thus obtained after the photo-thermal phase transition of the polymer solution. Although the fluorescence spectrum from the Py-PNIPAM particle showed the intense pyrene monomer fluorescence similar to the spectrum in Figure 6b, no clear peak corresponding to the excimer fluorescence was observed, but the spectrum exhibited a broad structureless tail extending to the longer wavelength. Clearly, the fluorescence spectrum from the single particle produced by the photo-thermal phase transition (Figure 6c) is different from that obtained at 35 °C without 1064-nm laser irradiation (Figure 6b), indicating that 1064-nm laser irradiation influences the conformation and/or the aggregation state of the polymers. To confirm further this point, the temperature of the Py-PNIPAM solution was once elevated above T_c (38 °C), and then irradiated with the 1064-nm laser beam (P_{1064} = 1.0 W). The fluorescence spectrum thus obtained is shown in Figure 6d. The

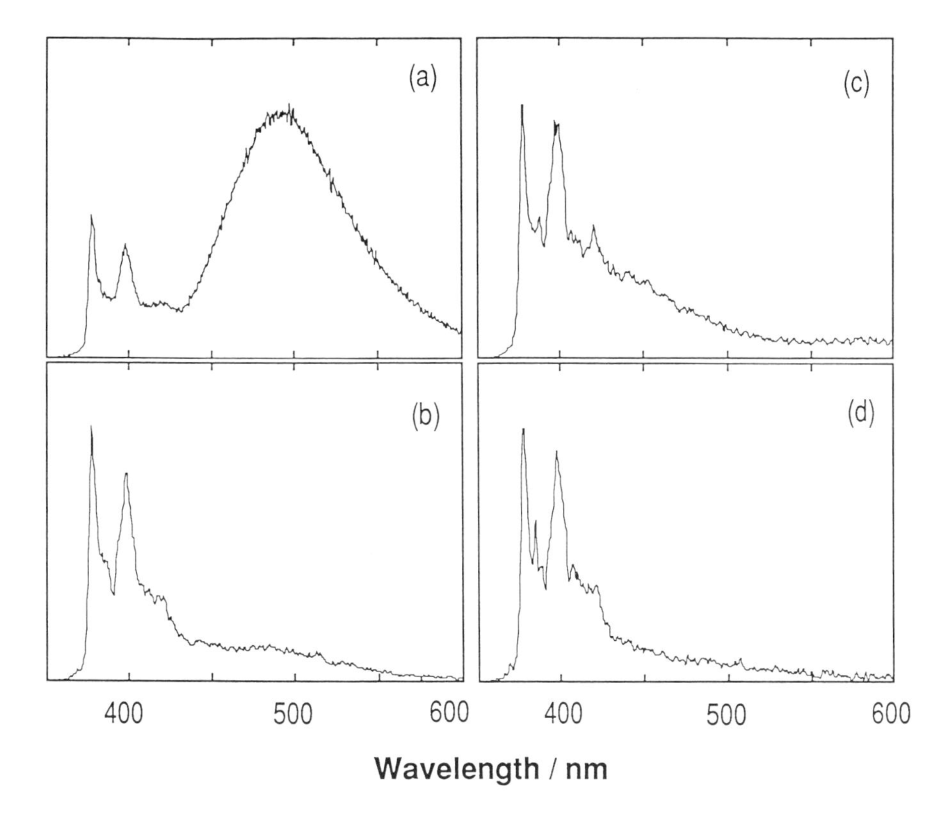

Figure 6. Fluorescence spectra of an aqueous Py-PNIPAM solution (0.75 wt %) under various temperatures (excitation wavelength = 355 nm). Without 1064-nm laser irradiation ; a) 20 °C, b) 35 °C. With 1064-nm laser irradiation (P_{1064} = 1.0 W) ; c) 20 °C, d) 38 °C.

spectrum is characterized by the monomer fluorescence at 380 - 420 nm with a minor contribution of the broad fluorescence band extending to the longer wavelength. The result agrees with neither that in Figure 6b nor Figure 6c. Although temperature for the experiments in Figures 6b and 6d is slightly different, the 1064-nm laser irradiation brings about changes in the fluorescence spectrum from the pyrenyl chromophore.

In order to assign the fluorescence spectrum in Figure 6c, we performed nanosecond time-resolved fluorescence spectroscopy. Figures 7a and 7b show early- (0 - 30 ns) and late- (61 - 404 ns) gated fluorescence spectra of the Py-PNIPAM solution, respectively, recorded at 20 °C under 1064-nm laser irradiation (excitation wavelength = 355 nm, ~ 6 ns pulse width, P_{1064} = 1.0 W). The lifetime of the broad fluorescence around 400 - 500 nm was very short and we could not confirm the broad fluorescence band in the late-gated spectrum. A difference spectrum obtained

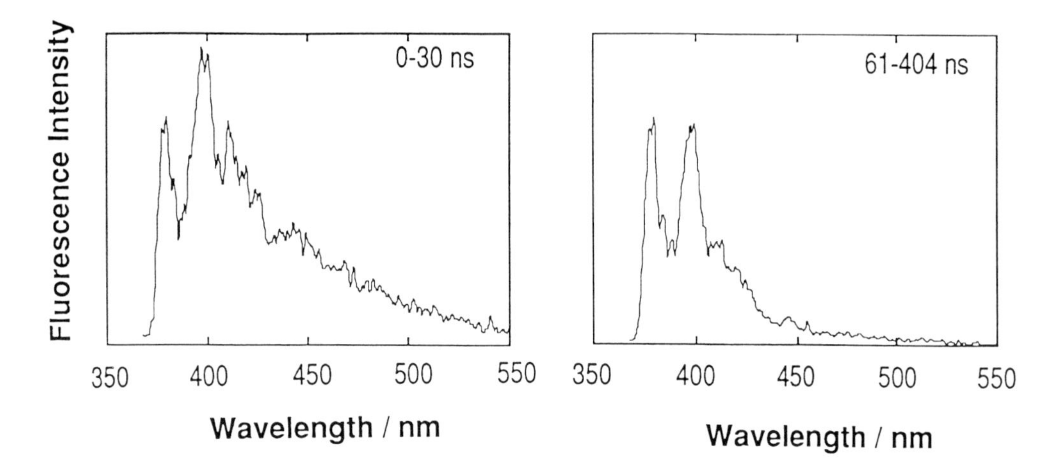

Figure 7. Time-resolved fluorescence spectra of an aqueous Py-PNIPAM solution (0.75 wt %) at 20 °C under 1064-nm laser irradiation (P_{1064} = 1.0 W). a) early - gated (0 - 30 ns after laser excitation) spectrum, b) late - gated (61 - 404 ns after laser excitation) spectrum.

by subtracting the late-gated spectrum from the early-gated one exhibits a broad fluorescence band with a peak around 410 nm. These results indicate that the spectrum in Figure 6c can be assigned to the monomer (380 - 420 nm) and one-center-type (or partial overlap) excimer fluorescence (peaking around 410 nm), while the broad structureless bands around 490 nm in Figure 6a and 6b are well characterized by the sandwich-type excimer fluorescence [20]. The one-center-type excimer of pyrene has been so far reported for the molecular aggregated systems represented by Langmuir-Blodgett films [25]. Therefore, the observation of the one-center-type excimer under 1064-nm laser irradiation (Figure 6c) suggests that the infrared laser beam can change the conformation and/or the aggregation state of the polymer chains in the PNIPAM particle. A comparison of the spectra between Figures 6b and 6d also indicates that 1064-nm laser irradiation to the solution above T_c decreases the efficiency of the pyrene sandwich-type excimer formation. Such effects of the infrared laser beam on the pyrene excimer formation in polymer systems and, therefore, on the polymer conformations in solution have never been reported. The 1064-nm laser beam directly influences the polymer conformation and the PNIPAM particle formation as demonstrated by fluorescence spectroscopy on pyrene-labeled polymer solution. We suppose that the radiation force generated by refraction of the laser beam through a particle play an important role in the present system as discussed separately in the following section.

4. LASER MANIPULATION OF PNIPAM MICROPARTICLE AND ROLE OF RADIATION FORCE

Irradiation of the focused 1064-nm laser beam in the solution leads to simultaneous non-contact and non-destructive manipulation of the PNIPAM microparticle produced by the photo-thermal phase transition. As reported previously, refraction of a light beam through a particle is the primary origin of laser manipulation of the particle [13 - 17]. Namely, since light momentum of a laser beam changes upon refraction at the particle/solution boundary, the amount of the momentum change (ΔP) is exerted to the particle with the direction opposite to that of ΔP ($-\Delta P$) owing to conservation law. The sum of $-\Delta P$ generated at all points of the particle irradiated by the laser beam (F) is the driving force of laser manipulation and, the force is called as *radiation force* [13]. When the refractive index of the particle (n_1) is higher than that of the surrounding medium (n_2), F directs to the focal spot of the laser beam. Therefore, the particle is trapped and manipulated in the vicinity of the focal spot of the laser beam as various examples have been reported in elsewhere [14 - 17].

As shown in Figure 8, actually, the PNIPAM particle produced by the photo-thermal phase transition is manipulated in water by the incident 1064-nm laser beam (P_{1064} = 1.2 W, 20 °C) [18]. Brownian motion and viscous flow of the particle are suppressed by the laser beam, while a polystyrene particle as a reference sample (not irradiated by the 1064-nm beam) moves along the flow of the sample solution. As discussed in the previous section, a PNIPAM polymer chain is dehydrated upon the phase transition and the refractive index of PNIPAM (n_1 = 1.508) [26] is higher than

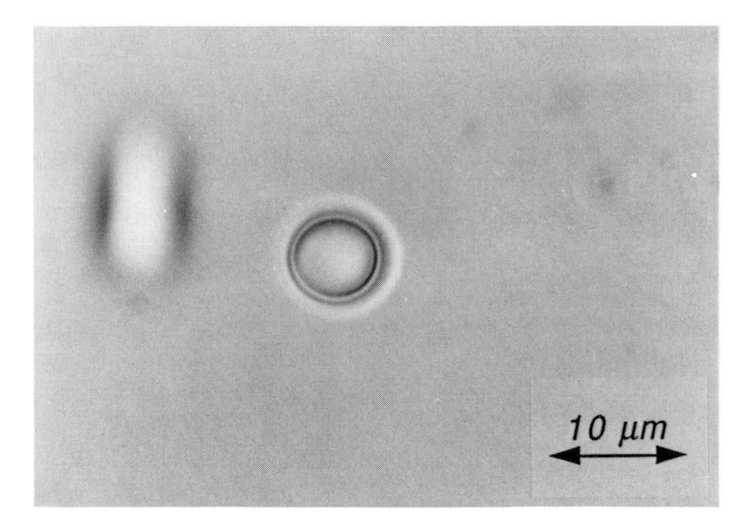

Figure 8. Laser manipulation of a single PNIPAM microparticle by the focused 1064-nm laser beam (P_{1064} = 1.2 W, 20 °C). The particle is laser trapped while that transfers along Y axis is a polystyrene particle (not irradiated) as a reference sample.

that of water (n_2 = 1.33) [27]. Therefore, the refractive index condition for laser manipulation (i.e., $n_1 > n_2$) is satisfied upon the particle formation. This is the first demonstration of simultaneous PNIPAM microparticle formation and manipulation by the focused 1064-nm laser beam.

Besides laser manipulation of the PNIPAM microparticle in solution, we suppose that the radiation force plays an important role for the PNIPAM particle formation (Figure 9) as suggested from the spectroscopic studies on the Py-PNIPAM solutions. It is noteworthy, furthermore, that the morphologies of the particle and its surroundings change with irradiation time and, finally, the surface of the particle becomes smooth as seen in Figure 1. The local heat generated by the laser irradiation is not expected to exceed T_g of the polymer (145 °C) [26], so that the changes in the morphologies of the PNIPAM particle will not ascribed to the thermal effects alone. In the case of the particle formation, the aggregated PNIPAM polymer chains above T_c whose volume and density are enough for refraction of the laser beam would be attracted towards the focal spot of the laser beam through small but finite radiation force exerted to the aggregated polymer chains. Indeed, we confirmed experimentally that submicrometer PNIPAM particles produced above T_c in D_2O were attracted and concentrated by the 1064-nm laser beam [20]. Since the photo-thermal effect is neglected in D_2O as described above, the driving force for attraction and concentration of the particles would be the radiation force of the laser beam. This is also proved by the changes in the fluorescence spectra from the Py-PNIPAM solution with and without 1064-nm laser irradiation, by which the conformation and/or the aggregation state of the polymers are shown to be influenced by the 1064-nm laser beam (Figure 6). Again, PNIPAM itself does not absorb at 1064 nm, so that the force responsible for such effects should be the radiation force of the laser beam. For the particle dissolution, on the other hand, switching off the laser beam leads to disappearance of the force attracted the polymer chains and rapid cooling of the solution by the water phase. The very fast dissolution may be therefore accounted partly by the radiation force of the laser beam as well.

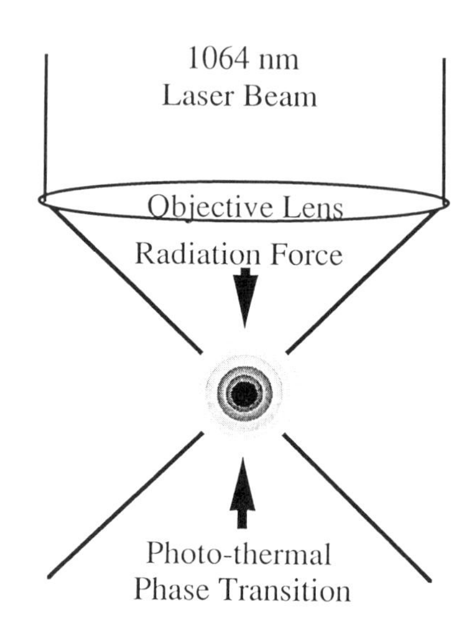

Figure 9. Role of radiation force on the photo-thermal phase transition of aqueous PNIPAM solution.

5. CONCLUSIONS

The photo-thermal phase transition of PNIPAM by an infrared 1064-nm laser beam is quite unique since, i) fast and reversible microparticle formation and dissolution are attained, and ii) a relatively large PNIPAM microparticle which cannot be obtained thermally in the dark is arbitrary prepared and manipulated in three-dimensional space. The diameter of the particle was also shown to be controlled in 2 ~ 9 μm through P_{1064}, the polymer concentration, and the solution temperature before laser irradiation. Since the phase transition of the solution can be induced through absorption of the laser beam by H_2O as a solvent and not by PNIPAM, reversibility of the phase transition is quite high without decomposition of the polymer. One of the most important findings of the present study is, furthermore, that the radiation force of the 1064-nm laser beam contributes to the particle formation and the conformation change of the polymer solution as demonstrated by the fluorescence spectroscopic studies. Although a laser beam has been frequently used as a light source for various studies, experimental evidence on control of polymer conformations in solution by the radiation force of a laser beam has not been so far reported. The present results suggest a future possibility for controlling polymer conformations and the aggregation state of polymer solutions, which may provides a potential means to study physical properties of polymers in solution. As a possible application of the present achievements, we suppose that simultaneous PNIPAM microparticle formation and laser manipulation will lead to laser-controlled microactuators, microdevices, drug delivery systems, and so forth.

ACKNOWLEDGMENT

The authors are greatly indebted to Dr. F. M. Winnik (McMaster Univ.) for generous gift of the Py-PNIPAM sample.

REFERENCES

1. M. Heskins and J. E. Guillet, J. Macromol. Sci., Chem., (1968) 1441.
2. F. M. Winnik, Macromolecules, 23 (1990) 233.
3. F. M. Winnik, Polymer (1990) 2125.
4. H. G. Schild and D. A. Tirrell, Macromolecules 25 (1992) 4553.
5. F. M. Winnik, M. F. Ottaviani, S. H. Bossman, M. Garcia-Garibay and N. J. Turro, Macromolecules 25 (1992) 6007.
6. F. M. Winnik, Macromolecules 23 (1990) 1647.
7. F. M. Winnik, M. F. Ottaviani, S. H. Bossman, W. Pan, M. Garcia-Garibay and N. J. Turro, J. Phys. Chem., 97 (1993) 12998.
8. H. Ohta, I. Ando, S. Fujishige and K. Kubota, J. Mol. Struct., 245 (1991) 391.
9. K. Kubota, S. Fujishige and I. Ando, J. Phys. Chem., 94 (1990) 5154.
10. H. Inomata, Y. Yagi, K. Otake, M. Konno and S. Saito, Macromolecules 22 (1989) 3494.

11. I. Yamamoto, K. Iwasaki and S. Hirotsu, J. Phys. Soc., Jpn., 58 (1989) 210.
12. M. R. Thomas and H. A. Scheraga, J. Phys. Chem., 69 (1965) 3722.
13. A. Ashkin, Science, 210 (1980) 1081.
14. H. Misawa, M. Koshioka, K. Sasaki, N. Kitamura and H. Masuhara, J. Appl. Phys., 70 (1991) 3829.
15. K. Sasaki and H. Misawa, in this volume.
16. H. Misawa and K. Sasaki, in this volume.
17. N. Kitamura, K. Sasaki, H. Misawa and H. Masuhara, in this volume.
18. M. Ishikawa, H. Misawa, N. Kitamura and H. Masuhara, Chem. Lett., (1993) 481.
19. M. Ishikawa, H. Misawa, N. Kitamura, R. Fujisawa and H. Masuhara, in preparation.
20. R. Fujisawa, H. Misawa, M. Ishikawa, N. Kitamura and H. Masuhara, in preparation.
21. J. Takimoto, personal communication.
22. Photophysical and Photochemical Tools in Polymer Science, M. A. Winnik, (ed), Reidel, Dordrecht, The Netherlands (1986).
23. D. C. Turner and L. Brand, Biochem., 7 (1968) 3381.
24. L. Stryer, J. Mol. Biol., 13 (1965) 482.
25. I. Yamazaki, N. Tamai and T. Yamazaki, J. Phys. Chem., 91(1987)3575.
26. F. M. Winnik, personal communication.
27. J. A. Riddik and W. B. Bunger, Organic Solvent, Technique of Chemistry, Vol. II, Wiley-Interscience, New York, (1970).

Part III: Microfabrication and Functionalization

MICROCHEMISTRY
Spectroscopy and Chemistry in Small Domains
Edited by H. Masuhara et al.

Photochemical micromodification of organic surfaces with functional molecules

Nobuyuki Ichinose*

Microphotoconversion Project,† ERATO, Research Development Corporation of Japan, 1280 Kami-izumi, Sodegaura, Chiba 299-01, Japan

Selective photochemical processes toward organic surfaces such as polymer films and self-assembled monolayers have been demonstrated as an approach for μm-order chemical modification, where some photochemically active molecules were covalently immobilized at the surface. The processes are based on photochemical reactions of surface functional groups, namely i) regioselective polar addition of hydroxy group to radical cation of arylalkene or arylcyclopropane and ii) selective photofragmentation of S-S or S-C bond of sulfur-containing monolayers upon KrF excimer laser irradiation followed by chemical immobilization of functional molecules onto the monolayers.

1. INTRODUCTION

A correlation of surface to chemical/physical properties of a material is expected to become more important when the size of a material is reduced from cm or mm (as a bulk material) to μm or subμm, since a surface/volume ratio increases with decreasing its size [1]. It is, therefore, necessary to consider the effect of surface or interface in order to study chemical processes in microdomains. One of our interests is preparation of functionalized microreaction sites on material surfaces, especially on organic surfaces by means of microfabrication techniques.

Reactivity of organic surfaces such as those of polymer films and self-assembled monolayers has been less studied because of difficulties in characterization of small quantity of products in spite of their importance in broad applications such as printing, adhesive, biocompatible materials. Studies on their physical properties, which seem closely related to the reactivity, also face problems such as their complexity and instability in structure to analyze [2]. In order to avoid or minimize these problems, we intended to prepare chemically well-defined surfaces by the use of selective photoreactions to various functional groups at the surface to which functional molecules are to be immobilized.

In this article, selective photochemical processes for μm-order chemical

* Present address: Osaka Laboratory for Radiation Chemistry, Japan Atomic Energy Research Institute, 25-1 Mii-minamimachi, Neyagawa, Osaka 572, Japan
† Five-year term project: October 1988 - September 1993.

modification of polymer films and self-assembled monolayers as organic surfaces are described in two sections, where the origin of the reactivity and selectivity in photochemical and immobilization processes is also discussed.

2. SURFACE MODIFICATION BY AROMATIC COMPOUNDS VIA PHOTOINDUCED ELECTRON TRANSFER

In 1973, a report made by Arnold on methyl-p-cyanobenzoate-sensitized photochemical polar addition of methanol to 1,1-diphenylethene initiated a fuge amount of studies on photoreactions via radical ions formed by photoinduced electron transfer [3]. Several years later, similar reaction of 1-methyl-2-phenylcyclopropane was reported [4]. Besides their historical value, important feature of these reactions is their regioselectivity in the products. The photoreaction of aromatic alkenes and cyclopropanes (S) with methanol in the presence of an electron-accepting sensitizer (Sens.) takes place in a regioselective manner as

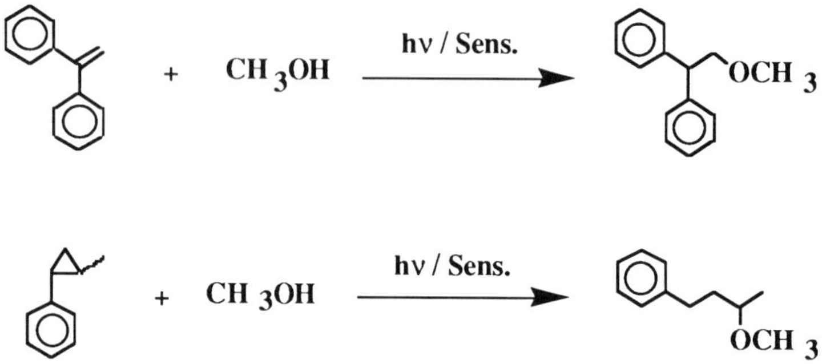

+ CH_3OH hν / Sens. → OCH_3

+ CH_3OH hν / Sens. → OCH_3

Scheme 1

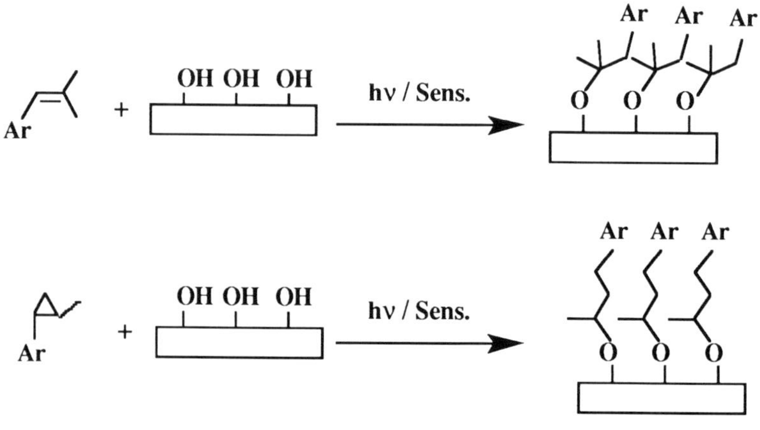

Ar + OH OH OH hν / Sens. → Ar Ar Ar O O O

Ar + OH OH OH hν / Sens. → Ar Ar Ar O O O

Scheme 2

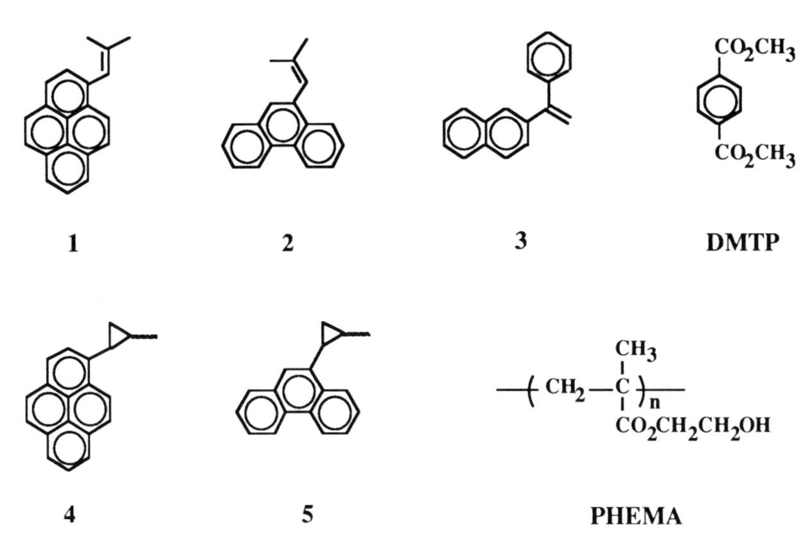

Figure 1. Structures and abbreviations of the materials.

depicted in Scheme 1. Polar addition of alcohol to radical cations of these molecules, which are generated by one-electron transfer from the excited state of S to Sens. followed by back-electron transfer, gives ether derivatives. In addition to the regioselectivity, polar addition of alcohol to cyclopropane radical cation proceeds with complete inversion of carbon center [5]. We therefore expected that the photoreaction of this class of compounds with a material bearing -OH group would bring about functionalized surfaces with chemically well-defined bonding of aromatic group to the materiall (Scheme 2).

In the present study, poly(2-hydroxyethyl methacrylate) (PHEMA) film was used as a material with -OH surface. For aromatic functionality of arylalkenes and arylcyclopropanes, a series derivatives of pyrene, phenanthrene, and naphthalene were synthesized (Figure 1). These aromatic chromophores are all fluorescent and known to act as an electron or energy transfer sensitizer for a variety of photoreactions. We show general features of the heterogeneous photoreaction between S and PHEMA film including the mechanisms and micro-patterning of the PHEMA film.

A PHEMA film (20-nm thickness) was spin-coated onto a quartz disk (29.5-mm diameter and 1-mm thickness) from a methanolic solution of the polymer (10 mg/ mL) and dried for 2-3 days at room temperature. The film was washed with boiling benzene and dried prior to the experiments. The PHEMA film on the quartz disk was fixed with a Teflon cell using a rubber O-ring. For patterning experiments, a photomask was placed in contact with the disk. A 2-mL aliquot of acetonitrile solution containing appropriate amounts of S and dimethyl terephthalate (DMTP) as an electron acceptor was poured into the cell and was purged with a N_2 stream for 5 min. The cell was then sealed with another set of a quartz disk and an O-ring. The film was irradiated from the film side to facilitate photoreactions at the film/solution boundary. A light source was a 500-W super high-pressure Hg lamp, and 313-nm line was isolated by a combination of UV-D33S and aqueous $K_2Cr_2O_7$ solution filters. After the photoreaction, the sample quratz disk was immersed in acetonitrile for several hours to remove unreacted compounds and dried in air prior to spectroscopic measurements. Characterization of the film was performed by means of X-ray photoelectron spectroscopy (XPS) and static and dynamic secondary ion mass spectroscopy (SIMS). Other experimental detail is described elsewhere [6].

A photoreaction of a PHEMA film with an acetonitrile solution of the alkene (**1** - **3**: (1 - 50) x 10^{-3} M) or cyclopropane (**4**, **5**: (5-7) x 10^{-2} M) was performed in the presence of DMTP ((1-5) x 10^{-3} M) . Introduction of the aromatic chromophore to the film was verified spectroscopically. A typical example of the absorption spectrum of the film reacted with **1** or **2** is shown in Figure 2, together with that of **1** or **2** in acetonitrile. In these figures, absorption by PHEMA film itself was subtracted from the observed spectrum, so that absorption is essentially ascribed to the chromophore introduced onto the film. The spectrum is well characterized as the pyrenyl or phenanthryl chromophore introduced to the film owing to close similarities between the film and alkene.

Fluorescence spectra corresponding to the data in Figure 2 are shown in Figure 3. Appearance of the vibrational structure around 370 - 400 nm will be explained by the reaction of **1** or **2** with PHEMA film, but not by simple adsorption. Generally, arylalkene shows a broad fluorescence owing to the presence of rotational isomers concerning to the vinyl group around the carbon(aromatic)-carbon(vinyl) bond, whereas arylalkane exhibits structured fluorescence. In addition to these monomeric fluorescence, excimer fluorescence of pyrenyl or phenanthryl chromophore observed in the long wavelength region indicates dense introduction of the chromophore to the film. Analogous results were obtained for the photoreactions of the film with arylcyclopropanes **4** and **5**, as confirmed by characteristic absorption and fluorescence spectra of the modified films.

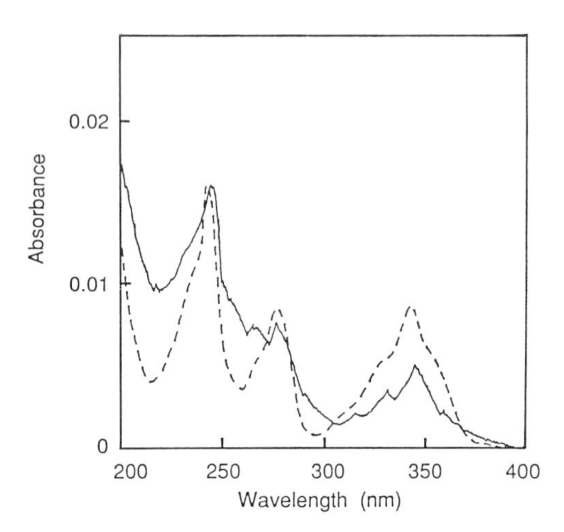

Figure 2. Absorption spectra of the **1**-modified film (solid line; [**1**] = 2 x 10^{-2} M, [DMTP] = 5 x 10^{-3} M, t = 40 min) and **1** (dashed line; in acetonitrile). The absorbance of **1** is normalized to that of the film at 245 nm.

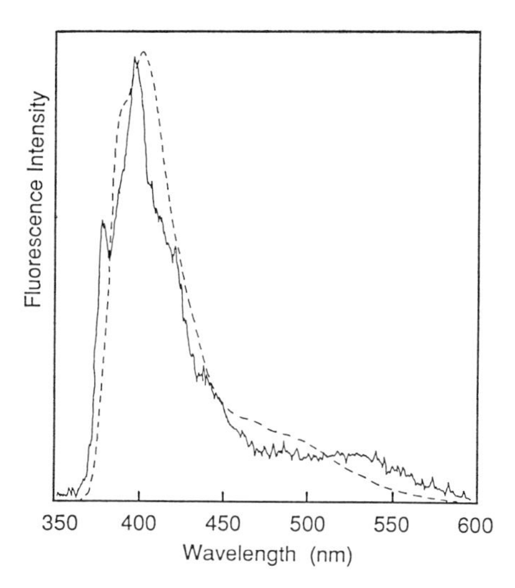

Figure 3. Fluorescence spectra of the **1**-modified film (solid line) and **1** in acetonitrile (dashed line). Fluorescence intensities are normalized to those at the maximum wavelength (λ_{ex} = 330 nm).

Characterization of the modified films by XPS and static SIMS indicated that the chromophores was introduced to the PHEMA film with structures in Scheme 3. From depth profiles of C/O and H/O atomic ratios in the film obtained by dynamic SIMS, it was concluded that the photoreaction may proceed from the surface to the depth of 7 nm of the film. The present heterogeneous photoreaction is localized to the surface layer of the film. Assuming the functionalized depth (7 nm) and $A_{346} = 5 \times 10^{-3}$, the concentration of pyrene in the film was estimated to be 0.26 M [6].

The dense introduction of the chromophores were in good agreement with the fluorescence properties of the modified films as seen in Figure 3. We have observed excimer emission of phenanthrene for the 2-modified films around 450 nm. Generally, phenanthrene does not show excimer emission in solution owing to the low stabilization energy for the excimer formation. The excimer emission of phenanthrene has been observed in amorphous solids containing dense phenanthrene chromophores such as poly(9-vinylphenanthrene) [9] or vapor deposited film of phenanthrene at low temperature [10]. The excimer emission observed in the modified films can be attributed to ground state aggregation of the chromophores.

The mechanism of the photoreaction was studied through control experiments as summarized below.

(1) Fluorescence of **1** or **2** is efficiently quenched by DMTP in acetonitrile. Since photoinduced electron transfer from **1** to DMTP is highly exothermic (free energy change < -10 kcal mol^{-1}), this fluorecence quenching is ascribed to the electron transfer mechanism.

(2) Functionalization of a PHEMA film by **1** or **2** proceeded when DMTP was replaced by 1-cyanonaphthalene (1-CN). The photoreaction in the absence of electron acceptor (DMTP or 1-CN), however, did not give the modified film.

(3) Addition of a strong electron donor, triethylamine (8×10^{-3} M) suppressed the heterogeneous photoreaction.

All these results clearly prove that the present functionalization reaction proceeds via photoinduced electron transfer from the excited singlet state of the alkene or cyclopropane ($^1S^*$) to DMTP to give their radical cations ($S^{+\cdot}$) and radical anion of DMTP ($DMTP^{-\cdot}$). The formation of pyrene radical cation and its quenching by the secondary electron transfer to triethylamine have been demonstrated experimentally by means of laser flash photolysis [7]. Furthermore, the radical cations of arylalkenes and arylcyclopropanes are known to react with alcohol to give relevant ethers [3,4]. It is supposed that the radical cations of **1**,**2**,**4**, and **5** reacts with the -OH group in the PHEMA film (Scheme 3). In fact, model reaction of phenanthryl drivatives of S, **2** and **5** with methanol gave the corresponding ethers in a regioselective manner as reported previously [3,4]. In the case of **3**, though the formation of its radical cation was expected by the fluorescence quenching by DMTP in acetonitrile, the naphthyl group was not introduced to the PHEMA film. A possible reason for this will be discussed later.

As a characteristic feature of the present functionalization reaction, the amount of S introduced to the film can be controlled by several factors: photoirradiation time (t), and the concentration of S ([S]) or an electron acceptor ([DMTP]). Time dependence of the photoreaction was examined under the condition of constant [S]/[DMTP]. The amount of S introduced to the film increased with the irradiation time, and then leveled off gradually. A final amount of S was ca. one-ninth of the amount expected from complete conversion of the hydroxy group to the ether bond (PHEMA film 20 nm thickness). The XPS and dynamic SIMS measurements of the modified film suggest that S is introduced to the film surface ($\approx$ 7 nm) in every three monomer units of the polymer. Net one-ninth of the hydroxy group will react with S. On the other hand, the amount of S introduced to the film showed almost linear relationship with the concentration of DMTP under the condition of constant irradiation time.

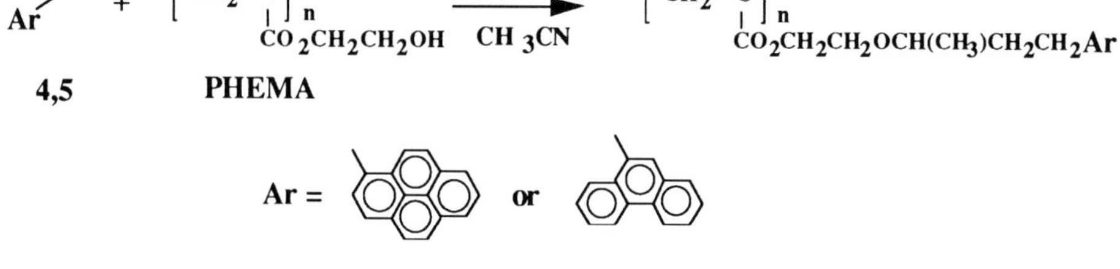

1,2 + **PHEMA** ($-[CH_2-C(CH_3)(CO_2CH_2CH_2OH)]_n-$) —DMTP, hν (313 nm), CH_3CN→ $-[CH_2-C(CH_3)(CO_2CH_2CH_2OC(CH_3)_2CH_2Ar)]_n-$

4,5 + **PHEMA** ($-[CH_2-C(CH_3)(CO_2CH_2CH_2OH)]_n-$) —DMTP, hν (313 nm), CH_3CN→ $-[CH_2-C(CH_3)(CO_2CH_2CH_2OCH(CH_3)CH_2CH_2Ar)]_n-$

Ar = (pyrenyl) or (phenanthryl)

Scheme 3

The DMTP-concentration dependence was further analyzed by a double reciprocal plot of (quantum yield)$^{-1}$ vs. [DMTP]$^{-1}$, giving the Stern-Volmer constant for fluorescence quenching involved in the photoreaction (K_{SV}) to be 300 - 400 M^{-1} for **2**. The K_{SV} obtained from fluorescence quenching experiment with **2** was 340 M^{-1}, so that the results are in good accordance with each other. This strongly suggests that the primary process of the heterogeneous photoreaction takes place in the solution phase, since the rate constant for fluorescence quenching (k_q) corresponding to that of electron transfer is strongly dependent on viscosity and polarity of the medium. The agreement of K_{SV} (= $k_q\tau$) will indicate that the rate constant for fluorescence quenching and the fluorescence lifetime (τ) in the reaction field may be considered to be the same values as those in acetonitrile. The photoinduced electron transfer gives active species in the solution, but not in swollen polymer phase. Actually, acetonitrile is a poor solvent for PHEMA in the present system.

Quantum yield measurements also supported this coclusion. We estimated the quantum yield of the heterogeneous photoreaction (Φ) on the basis of surface coverage of the film by S (α), which corresponds to the concentration of the chromophore in a unit area. α was calculated by $\alpha = A/1000\varepsilon$, where A and ε are absorbance and a molar extinction coefficient at the wavelength measured, respectively. Φ was thus calculated by $\Phi = \alpha/I = A/1000\varepsilon I$. Φ was in the order of 10^{-7} -10^{-8}, which was 10^{-2} -10^{-3} times as large as that for the homogeneous photoreaction of S with methanol. This can be the evidence for the heterogeneity of the photoreaction. The reactivity of the radical cation toward the -OH group was also estimated from Φ. The radical cations of the alkenes are more reactive as compared with those of the cyclopropanes and, the phenanthryl derivatives are more reactive than the pyrenyl derivatives. These can be explained in terms of positive charge density at the reactive carbon center of the radical cations. Namely, the positive charge is delocalized over the double bond via direct vinylic conjugation in alkenes while it is less delocalized over the cyclopropane ring via hyperconjugation. By the same reason, the charge density at the reaction center was lowered by the larger aromatic ring to make the compounds less reactive toward nucleophilic attack of alcohol in the comparison of pyrenyl and phenanthryl derivatives.

The absence of the introduction of the naphthyl chromophore to the PHEMA film by the use of **3** can be explained by the primry electron transfer in the solution phase and heterogeneity of the reaction. The photoreaction of **3** with PHEMA film gave exclusively a dimeric product of **3** in the solution. This is markedly contrasting to the photoreaction of **3** and DMTP with methanol, which gives its methanol adduct in a quantitative yield. Radical cation of 1,1-diphenylethene, an analogue of **3**, reacts with methanol and dimerizes competitively at a diffusion-controlled rate. The yields of the reactions are dependent on the relative concentration of alkene to methanol [8]. In the case of **3** and PHEMA film, the nucleophilic addition was restricted by the heterogeneous condition to give the dimer of **3** dominantly. The dimerization of the radical cation as well as its formation will not be restricted in solution phase. The results also rule out the incorporation of the swollen polymer in the present photoreaction. The dimerization or oligomerization of **1** and **2** was inhibited by the two methyl groups at β-position of the double bond in **1** and **2**.

We performed the photoreaction of the film through a photomask. Photoirradiation of a PHEMA film and **4** through the photomask gave a fluorescent pattern on the film. A typical example of the fluorescent pattern on the film is shown in Figure 4, which is comparable to the pattern of the original photomask used in the experiments. Spatial resolution of the present patterning is ≈ 50 μm, as judged from the observed fluorescence and photomask patterns in Figure 4. The resolution seems to be lowered by the divergence and scattering of the incident light source because of the 1-mm gap between the film and photomask. The effect of the diffusion of reactive intermediates will be negligibly small owing to their short lifetime. Direct drawing with a focused laser beam will improve the spatial resolution.

The present photochemical surface functionalization method is unique and novel, since introduction of the chromophore proceeds in a regioselective manner. It may be applicable to immobilization of molecules for nonlinear optical or ferroelectric materials which are designed in the form of arylalkene or arylcyclopropane as appeared here. In particular, arylcyclopropanes like **4** or **5** with an asymmetric carbon will afford optical activity to the film surface, if optically resolved arylcyclopropane are used as starting materials. We have also confirmed that the present photoreaction is applicable to functionalization of self-assembled monolayers bearing -OH or -SH functionalities such as 3-hydroxypropylsilane derivative on glass surface.

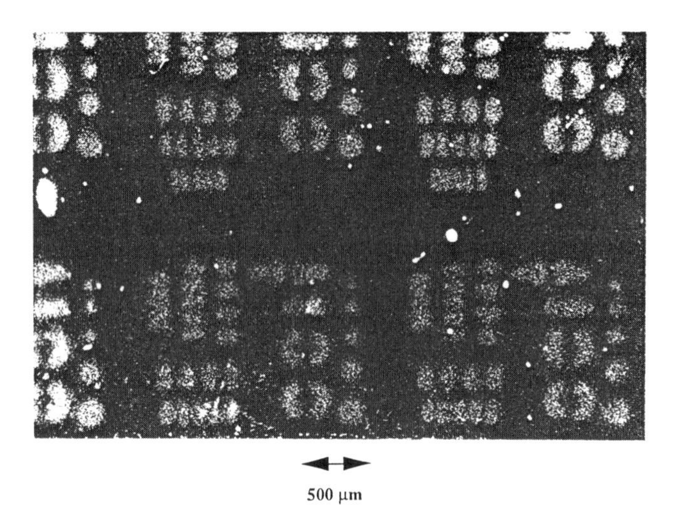

Figure 4. Micrometer fluorescent patterning of a PHEMA film: fluorescence micrograph of the film ([**4**] = 5 x 10^{-2} M, [DMTP] = 5 x 10^{-2} M, t = 32 min).

3. PHOTOCHEMICAL MICROPATTERNING OF SELF-ASSEMBLED MONOLAYERS FOR CHEMICAL FUNCTIONALIZATION OF MATERIAL SURFACE

Chemical modification of self-assembled monolayers has attracted much interest concerning microfabrication for electronics or biochemical devices [11-13]. Physical and chemical properties of material surface are sensitively influenced by the presence of a monolayer. For instance, introduction of a monolayer on silicon [11] or gold [12] surfaces prevents them from chemical etching. Photochemical micromodification of the monolayer may serve a use of as ultra thin photoresist. On the other hand, spatial arrangement of surface functionality such as hydroxy, amino, or mercapto groups, which provide a key linkage between the surface and biochemically active molecules, has also emerged [13-15]. This section describes designing of several self-assembled monolayers sensitive to 248 nm KrF laser light, their photochemical micropattering to give spatially arranged surface functionalities, and chemical modification of the pattern with polyaminopolymers including protein as shown in Figure 5.

All the self-assembled monolayers shown in Figure 6 were synthesized from 3-mercaptopropyltrimethoxysilane (MPS) *in situ* before silylation of a qurartz substrate. The mercapto group in MPS was converted into disulfides, acylalkyl groups for patterning of mercapto and acyl functionalities, respectively. The silylation (silane coupling) of the qurtz substrate (9.5 x 15 mm^2, 1mm thickness) was performed with a methanol solution containing the MPS derivative (1% v/v), water (4% v/v), and acetic acid (4% v/v) for 15 min.

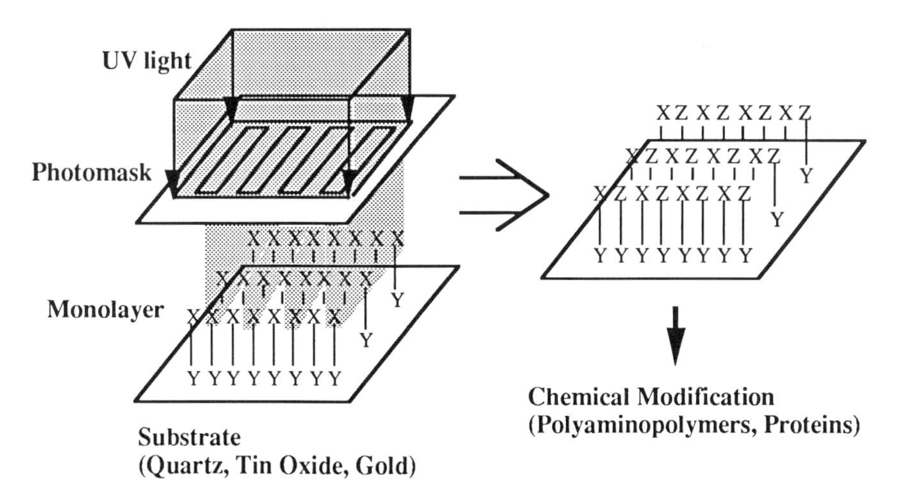

Figure 5. Schematic illustration of photochemical micropatterning of self-assembled monolayer.

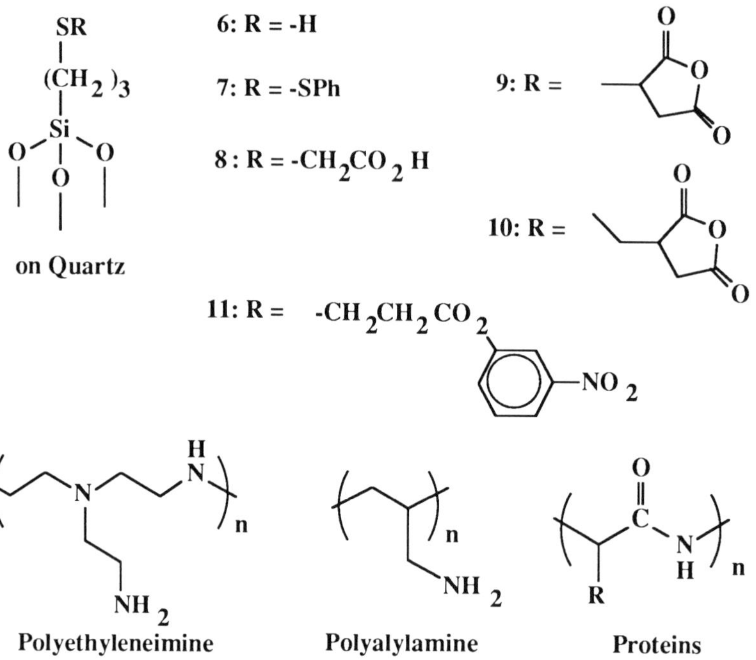

6: R = -H

7: R = -SPh

8: R = $-CH_2CO_2H$

9: R =

10: R =

11: R = $-CH_2CH_2CO_2$–(C_6H_4)–NO_2

on Quartz

Polyethyleneimine Polyalylamine Proteins

Figure 6. Structures of the monolyers and polyaminopolymers.

The monolayer of MPS on quartz (**6**) was irradiated in air at 248 nm with a KrF excimer laser (Lambda Physik LPX 200) through a mesh (40 mm width) as a photomask. Upon irradiation of 600-shots laser pulses with a fluence of 80 mJ cm^{-2} $pulse^{-1}$, the irradiated area became hydrophilic. After treatment with dithiothreitol (DTT) in methanol, the sample was reacted with an acetonitrile solution of 3-maleimidobenzoic acid N-hydroxysuccinimide ester (MBS), a cross-linking reagent for mercapto and amino groups. The substrate was immersed in a phosphate buffer solution (pH 7.4) of protein (ovalbumin, bovine serum albumin, etc.) to immobilize the protein onto the photoimage. The protein was labeled with a fluorescent dye such as fluorescein isothiocyanate (FITC) or tetramethylrhodamine isothiocyanate (TRITC) for observation. The fluorescence micrograph of the substrate indicates the formation of the image of the photomask, which means that thiol is reacted at the irradiated area. The selectivity of the reaction of **6** with MBS/protein was confirmed by a control experiment: oxidation of the unirradiated surface of **6** to sulfonic acid with H_2O_2 [16] followed by the cross-linking procedures prevented the pattern formation.

The monolayer of 3-phenyldithiopropylsilane (**7**) was prepared by *in situ* reaction of MPS with diphenyldisulfide. The advantage of this conversion will be as follows: the introduction of the phenyldithio group will afford benzene-like absorption ($\varepsilon \approx 10^4$ M^{-1} cm^{-1}) around 248 nm, which is larger than those of alkyldisulfides ($\varepsilon = 10^2$ - 10^3 M^{-1} cm^{-1}) and thiols ($\varepsilon < 100$ M^{-1} cm^{-1}); photoreactions of alkyldisulfides are known to proceed via excitation to n,σ^* state of S-S bond to give products mainly from homolytic cleavage of the S-S bond and fewer C-S bond [17]. In the case of diaryldisulfide, the cleavage of S-S bond is more selective. This group can be removed to give thiol easily by the reduction with DTT. It can be said as "photoreactive masked thiol". Similar irradiation of the monolayer **7** with 600-shots pulses of 50 mJ cm^{-2} $pulse^{-1}$ followed by the same treatment as done for **6** resulted in the formation of the fluorescent micropattern on the unirradiated area as shown in Figure 7 [18]. The disulfide in irradiated area was converted to inactive oxides toward DTT by reaction with molecular oxygen in air. When the fluence was lowered to 30 mJ cm^{-2} $pulse^{-1}$, the micropattern was formed only for the monolayer **7** but not **6** which required more than 3000 shots of the pulses. This result indicates that **7** is more sensitive than **6** toward KrF laser light.

In situ alkylation of MPS with iodoacetic acid, maleic anhydride, and itaconic anhydride afforded 3-(acylalkyl)thiopropylsilane monolayers **8** - **10**. The monolayer **8** was similarly irradiated with KrF laser pulses of 30 mJ cm^{-2} $pulse^{-1}$. After the irradiation, the sample was treated with an aqueous solution of ethyl(3-dimethylaminopropyl)carbodiimide hydrochloride (EDC) as a condensation reagent and polyethyleneimine (PEI) to immobilize the polymer on the pattern. The micropattern of the polymer was also labeled with FITC. In the case of **9** and **10**, the photopatterning required a larger fluence of 60 mJ cm^{-2} $pulse^{-1}$ as compared with **8**. The advantage of the use of **9** will be the high reactivity of the succinic anhydride moiety toward primary amines, which allows one to omit the use of condensation reagent such as EDC. In a similar manner, other polyaminopolymers such as polyallylamine and proteins were immobilized on the pattern [19].

Analysis of the irradiated surface of **7** by XPS indicated growth of a peak at 169 eV in S_{2p} region corresponding to oxidation of divalent sulfur to hexavalent sulfur

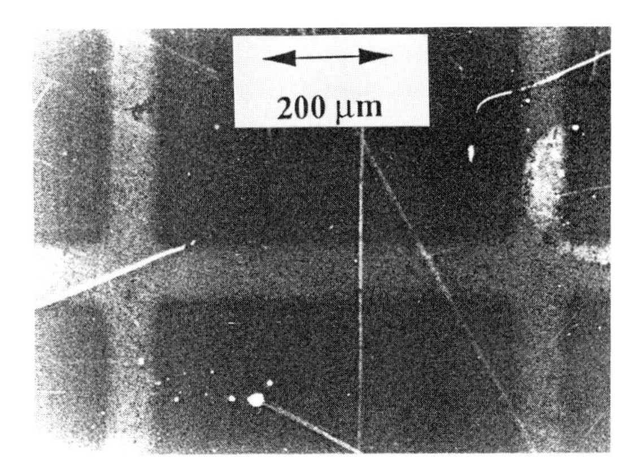

Figure 7. Fluorescence micrograph of a pattern of TRITC-labeled ovalbumin immobilized on a quratz by the use of monolayer **7**.

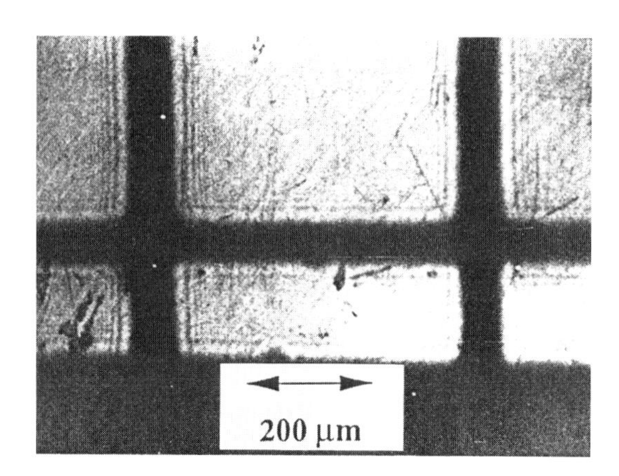

Figure 8. Fluorescence micrograph of a pattern of rhodamine 6G perchlorate adsorbed on the irradiated area of monolayer **7**.

oxide and dissapearance of the satelite peak of C_{1s} corresponding to the loss of the phenyl group. We did not obtain any evidence by XPS for other bond cleavage leading to loss of sulfur atom such as C-S, C-C, or C-Si bonds even with high fluence as observed in the irradiation of aromatic and aliphatic organosilane monolayers with ArF laser (193 nm). The irradiated area was hydrophilic for both **6** and **7** to suggest oxygenation of sulfur atom to sulfonic acid is involved in their photoreactions and this suppresses the adhesion of protein. Furthermore, adsorption of a cationic dye, rhodamine 6G perchlorate from its aqueous solution onto the irradiated area was observed as shown in Figure 8, where the fluorescent pattern contrasted with the pattern of the fluorescor-labeled protein in Figure 7 [18]. This can be attributed to ionic exchange [20] of perchlorate anion to sulfonate anion of the surface. This also supports the formation of sulfonic acid at the irradiated surface.

XPS analysis of the irradiated area of **9** revealed that the sulfur atom in the film was converted almost quantitatively to a hexavalent oxide and the carbonyl carbon was removed. The formation of sulfonic acid was also expected as a sulfur oxide from the results of hydrophilicity and adsorption of cationic dye of the irradiated area. In this case, homolytic cleavage of C-S bond was involved within the photoreaction.

In all cases, the introduction of the substituents on the sulfur atom of MPS made the monolayers photoreactive. The key factors are: 1) increase of the extinction coefficients (ε) of the compounds at 248 nm as compared to those of simple alkanethiol and dialkylsulfide ($\varepsilon \approx 10^2$ M^{-1} cm^{-1}), which is based on the absorption of the benzene-like chromophore ($\varepsilon \approx 10^4$ M^{-1} cm^{-1}) and α-acylalkylthio group ($\varepsilon \approx 10^4$ M^{-1} cm^{-1}) [21], and 2) stability of the primary photoproducts, phenylthio and α-acylalkyl radicals. However, all the monolyers does react upon KrF laser irradiation, but not upon irradiation with a high-pressure Hg lamp (> 280 nm) or low-pressure Hg lamp (254 nm), though excitation of the chromophores was expected. This nonlinear photochemical behavior with respect to light intensity is in contrast to the photoreactions of disulfides and sulfides in solution. One explanation for the present photochemical behavior is a fast recombination process between the thiyl ($SiCH_2CH_2CH_2S\cdot$) and fragment radicals ($R\cdot$) in the case of low fluence because of an extremely low diffusion condition of solid. On the other hand, when the radicals are formed densely with a high fluence of laser light, coupling between the radical fragments will retard the recombination to leave the thiyl radical as an intermediate. Incorporation of a multiphotonic process, where some excess energy isused for elimination of the fragment radicals from the film, however, cannot be ruled out as an explanation for the light intensity dependence of the present reaction.

Present method for photopatterning of self-assembled monolayers lead to the spatial control of surface mercapto and acyl functionalities, which allows us to modify the pattern with several functional molecules. For instance, we have demonstrated the immobilization of polyaminopolymers including proteins which can be also modified through reactions of the amino group with functional molecules bearing acyl, isocyanate, isothiocyanate moieties (so-called affinity labeling). We have also succeeded in the photochemical patterning of polyaminopolymers by the use of a monolayer with a photoreactive activated ester moiety **11**, which is hydrolyzed upon irradiation in contact with water [22].

In the present review, we have demonstrated the surface functionalization through some selective reactions of surface functionalities of polymer films and self-assembled monolayers. Recent advances in microfabrication of material surfaces has entered in the region of nanometer to submicrometer. However, it has not provided surfaces with definite chemical strucutres of organic molecules. Selectivities in chemical processes for surface modification will be of great importance to control chemical structures in minute spaces toward organic molecular devices in future.

ACKNOWLEDGMENT

The author thanks Dr. Kitamura (Hokkaido University) and Mr. Shimo (Idemitsu Kosan Central Research Lab.) for helpful discussions.

REFERENCES

1. H. Masuhara, Pure Appl Chem., 64 (1992) 1278.
2. G. M. Whitesides and G. S. Ferguson, Chemtracts, Org. Chem., 1 (1988) 171; G. M. Whitesides and P. E. Laibnis, Langmuir, 6 (1990) 87.
3. R. A. Neunteufel and D. R. Arnold, J. Am. Chem. Soc., 95 (1973) 4080.
4. V. R. Rao and S. S. Hixson, J. Am. Chem. Soc., 101 (1979) 6458.
5. J. P. Dinnocenzo, W. P. Todd, T. R. Simpson, and I. R. Gould, J. Am. Chem. Soc., 112 (1990) 2462; J. P. Dinnocenzo, D. R. Lieberman, and T. R. Simpson,. Ibid., 115 (1993) 366.
6. a) N. Ichinose, N. Kitamura, and H. Masuhara, Chem. Lett., (1990) 1945; b) J. Chem. Soc., Chem. Commun., (1991) 985; c) Macromolecules, 26 (1993) 2331.
7. H. Masuhara, S. Ohwada, Y. Seki, N, Mataga, K. Sato, and S. Tazuke, Photochem. Photobiol., 32 (1980) 9.
8. S. L. Mattes and S. Farid, J. Am. Chem. Soc., 108 (1986) 7356.

9. A. Itaya, K. Okamoto, and S. Kusabayashi, Polym. Prepr. Jpn., 1973, G7C02; Bull. Chem. Soc. Jpn., 50 (1977) 52.
10. N. J. Tro, A. M. Nishimura, and S. M. George, J. Phys. Chem. 93 (1989) 3276.
11. C. S. Dulcey, J. H. Georger, V. Krauthamer, D. A. Stenger, T. L. Fare, and J. M. Calvert, Science, 252 (1991) 551; J. M. Calvert, J. H.Georger, M. C. Peckerar, P. E. Pehrsson, J. M. Scnur, P. E. Schoen, Thin Solid Films, 210/211 (1992) 359; and references cited therein.
12. A. Kumar, H. A. Biebuyck, N. L. Abbott, and G. M. Whitesides, J. Am. Chem. Soc., 114 (1992) 9188.
13. S. K. Bhatia, J. J. Hickman, and F. S. Ligler, J. Am. Chem. Soc., 114 (1992) 4432. The photoreaction of monolayer of MPS **6** was reported to give a hydrophilic surface upon UV irradiation with low-pressure mercury lamp, which was based on the oxygenation of sulfur atom to sulfonic acid. On consideration of light absorption, the 185-nm line not 254-nm should essentially participate in the photoreaction.
14 S. P. A. Fodor, J. L. Read, M. C. Pirrung, L. Stryer, A. T. Lu, and D. Solas, Science 251 (1991) 767; L. F. Rozsnyai, D. R. Benson, S. P. A. Fodor, and P. G. Schults, Angew. Chem. Int. Ed. Engl., 31 (1992) 759.
15. S. Britland, E. Perez-Arnaud, P. Clark, B. McGinn, P. Connolly, and G. Moores, Biotecnol. Prog., 8 (1992) 155.
16. N. Balachander and C. N. Sukenik, Langmuir, 6 (1990) 1621.
17. E. Block, Q. Rep. Sulfur Chem., 4 (1969) 283.
18. N. Ichinose, H. Sugimura, T. Uchida, N. Shimo, and H. Masuhara, Chem. Lett., (1993) 1961.
19. N. Ichinose, N. Shimo, and H. Masuhara, to be submitted.
20. T. Uchida, H. Sugimura, K. Kemnitz, N. Shimo, and H. Masuhara, Appl. Phys. Lett., 59 (1991) 3189.
21. E. A. Fehnel and M. Carmack, J. Am. Chem. Soc., 71 (1949) 84.
22. N. Ichinose, N. Shimo, and H. Masuhara, in preparation.

MICROCHEMISTRY
Spectroscopy and Chemistry in Small Domains
Edited by H. Masuhara et al.

Simultaneous microfabrication and functionalization of polymeric materials by laser ablation

N. Shimo*, # and T. Uchida‡

Microphotoconversion Project†, ERATO, JRDC, c/o Idemitsu Kosan Cent. Res. Lab., 1280, Kami-izumi, Sodegaura, Chiba 299-02, Japan

Laser ablation was applied for patterning functional molecules on polymer surface. Oxygen, silicon, and nitrogen compounds were introduced on the ablated area from gas phase. Dye molecules were also introduced on the polymer surface by immersing the ablated sample into their aqueous solution. Positive or negative pattern can be obtained just by changing the irradiating laser fluence. The incorporation mechanism is summarized and discussed on the basis of morphological observations, measurements of XPS, ESR, and contact angle of water, and chemical structure of dye molecules used for incorporation.

1. INTRODUCTION

One of the purposes in Microphotoconversion Project is to establish a "microphotoconversion system" where efficient chemical reactions are controlled along spatially well-arranged reaction sites with high selectivity. In order to fabricate these minute reaction sites with chemical functionality, microfabrication and functionalization of material surfaces are indispensable. Many methods are well known to change properties of material surfaces. For example, polymer surface can be modified by chemical treatments with solution, treatment with coupling agent, and so on. Dry processes such as plasma etching, irradiation with ultraviolet light sources, and so on are important because of the process ability.

Interaction of intense laser pulse with various kinds of solid materials, i.e., polymers, metals, semiconductors, and inorganic compounds, causes efficient removal

---

* To whom correspondence should be addressed.
# Present Address: Idemitsu Kosan Co., Ltd. Central Research Laboratories, Sodegaura 299-02, Japan
‡ Present Address: Itaya Electrochemiscopy Project, JRDC, Sendai 982, Japan
† Five-year term project: October 1988 ~ September 1993.

of surface layers, which is called laser ablation [1-4]. Ablation behaviors have been extensively studied from fundamental and technological viewpoints. In the ablation process, after surface layers of polymer absorb the laser photons, photochemical / photothermal bond scissions take place and various kinds of fragments are ejected from the surface. Most of the works on laser ablation of polymeric materials are concerned with ablated surface properties, etching phenomena, chemical species of fragments, and so on. In spite of the ablated fragments in the gas phase being well investigated with various kinds of analytical methods, the active species on the ablated surface were not so clarified.

Physical and/or chemical changes of surface properties are recognized on the ablated polymeric materials. Hence, microfabrication and functionalization of polymeric materials can be achieved by laser ablation, which is schematically shown in Figure 1. Since bond cleavages of polymer take place during the laser ablation processes, the ablated polymer surfaces should have active species such as radicals and ionic species. Foreign reactive molecules would react with them to produce a chemically functionalized surface. For example, Lazare and Srinivasan pointed out the possibility of introducing metal compounds on the irradiated surface of polymer [5]. We consider chemical modification of polymer surface by laser ablation technique is highly possible and will be useful for the fabrication of micrometer-sized

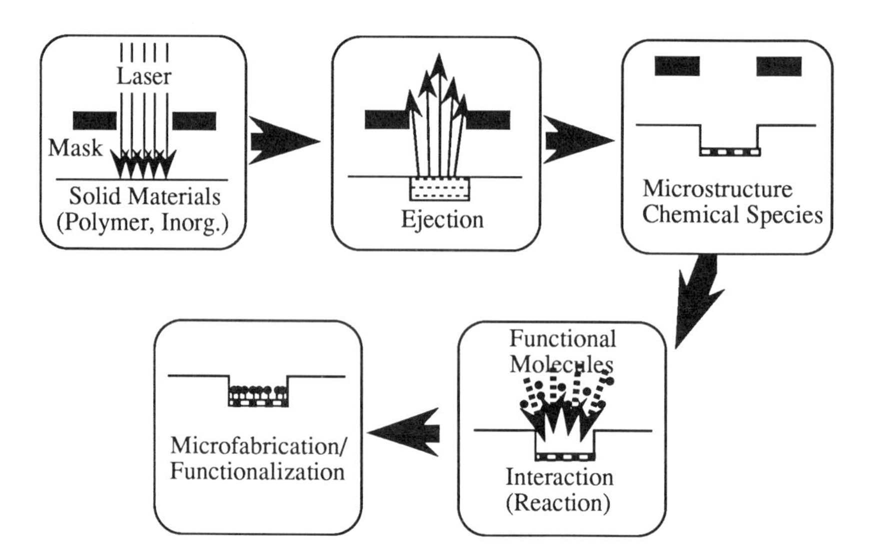

Figure 1. Microfabrication and functionalization of polymeric materials by laser ablation.

structure with chemical functionality. In this article, microstructures, morphologies, and chemical properties of the ablated polymer surface are first reviewed, and then simultaneous microfabrication and functionalization processes are discussed.

2. MICROSTRUCTURES GENERATED ON ABLATED SURFACE

Just after ablative photodecomposition was demonstrated, microstructure formation on the surface, such as conical shape, ripple-like shape, and so on, were reported. Some of the examples are listed in Table 1. The size of most microstructures are about a few micrometer, while recently Bolle et al. reported the formation of sub-micrometer sized periodic structures on polymer surface using polarized laser light at low fluences [6,7]. These microstructures were applied to improvement of polymer films, alignment of liquid crystal molecules, and so forth. Phillips et al. reported that well-defined line structures with a period of 167 nm can be produced on the surface of polyimide using an interferometric technique [8]. The period was in agreement with the calculated value that was estimated from the geometry of the interferometer. The high electric conductivity was obtained after

Table 1 Microstructures generated on the ablated surface

Microstructures / Purpose	Size	Polymer	Laser	Reference
Rough	~μm	PET	ArF	Srinivasan, 1982
Conical	~μm	Polyimide	XeCl	Dyer, 1986
Ripple-like	~μm	PET	ArF	Lazare, 1986
Periodic	~μm	PEN-2,6, PES	XeCl	Yabe, 1989
Micromachine	10 ~ 100 μm	Various Kinds of Materials	Excimer	Gower, 1990
Periodic	~sub μm	PET, PS, PC etc.	ArF, KrF	Lazare, 1992
Imaging Transfer	~μm	PMMA	YAG	Dlott, 1992
Periodic	~sub μm	Polyimide	KrF	Phillips, 1992
Periodic, Dot-like	~sub μm	PI, PET	YAG	Hiraoka, 1993

a) PET = polyethylene terephthalate, PI = polyimide, PEN-2,6 = polyethylene-2,6-naphthalate, PES = polyether sulfone, PS = polystyrene, PC = polycarbonate, PMMA = polymethyl methacrylate.

laser irradiation [9], although polyimide is widely used as an insulating material in the microelectronics industry. The sheet conductivity of unirradiated polyimide was $10^{-15}\ \Omega^{-1}\ cm^{-1}$, while the value was increased to $1\ \Omega^{-1}\ cm^{-1}$ after 6000 laser shots, The increase of electric conductivity was confirmed to be permanent. Furthermore, they reported that the submicron electrically conducting wires were formed in polyimide by laser irradiation using a holographic technique [10]. An anisotropy in the conductivity was greater than a factor of 10^7.

3. FOREIGN GAS EFFECT ON ETCHING RATE AND SURFACE MORPHOLOGY

During the laser ablation processes, foreign gas affects ablation behaviors such as etching rate, surface morphology, and surface chemical composition. Different fragments are produced by the ablation under different conditions of surrounding gas atmosphere. Koren and Yeh showed the different emission spectra observed in vacuum and under some foreign gas conditions [11]. However, the effect of foreign gas on ablated polymer surface has been scarcely reported, although excited species and radicals generated on ablated polymer surface would interact with these gases. Two kinds of effect can be expected in the laser ablation processes; one is quenching of generated active species on the ablated surface, and the second is interaction of active surface species with foreign gas such as oxygen. Here, we summarize our studies on the foreign gas effect in laser ablation processes [12].

Figure 2 shows the pressure dependence of argon and oxygen on etching rate of spin-coated polymethyl methacrylate (PMMA) film. Here, etching rate was determined by measuring the numbers of laser pulses which remove a certain thickness of PMMA film. When argon was used, the rate was suppressed with increasing the pressure probably because inert gas prevents the removal of fragments from the ablated surface. On the other hand, oxygen enhanced the etching rate above 100 Torr, although the rate decreased below the latter pressure. The acceleration effect indicates that oxygen reacts with active species on the ablated surface.

Foreign gas also affects the morphology of ablated surface of polymer. Figure 3 shows optical microscope pictures of the ablated poly (ethylene terephthalate) (PET) surface under several atmospheres. Ablated surface in vacuum (Figure 3(a)) was smooth, while that in inert gas was rather rough (Figure 3(b)). In the case of ablation with 760 Torr of oxygen, the irradiated polymer surface was more smooth than that with argon gas and many fragments were deposited under the masked area. (Figure 3(c)). The morphology observation was consistent with the oxygen effect on ablation rate.

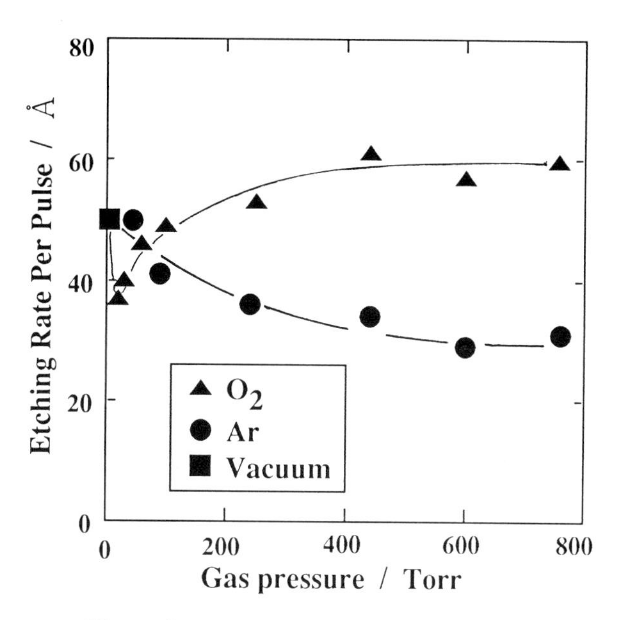

Figure 2 Pressure dependence of Ar and O_2 on etching rate of PMMA

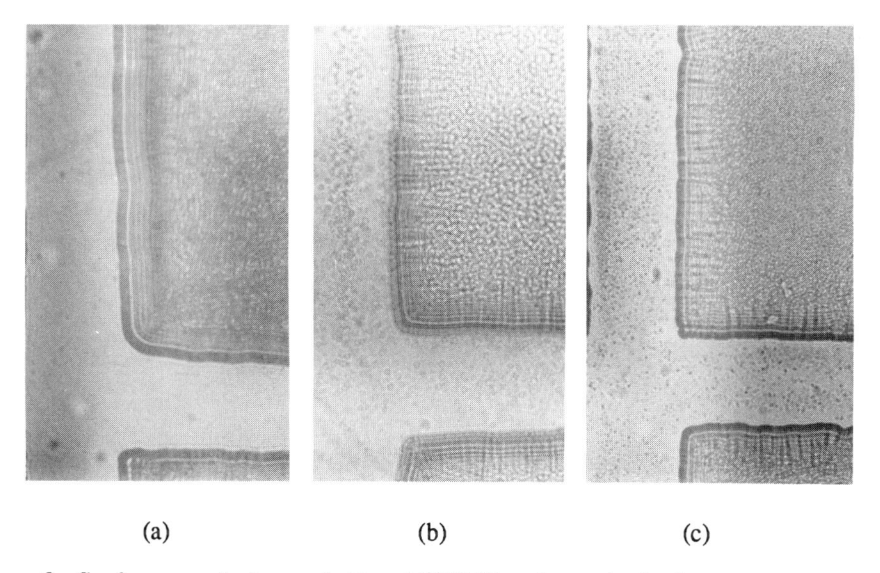

(a) (b) (c)

Figure 3. Surface morphology of ablated PET films by optical microscope. (a) Ablated in vacuum, (b) with 760 Torr of argon, and (c) with 760 Torr of oxygen. Wide area was irradiated with laser, while narrow one was prohibited from irradiation with a mesh mask (30 μm width).

4. CHEMICAL SPECIES GENERATED ON THE ABLATED AREA

For clarifying the chemical species generated on the ablated surface, XPS analysis is an effective method and gives information on the chemical composition in the first layer with a thickness of 100 Å. The study of ablated surfaces was reported by several research groups for various kinds of polymers as summarized in Table 2. As is clearly recognized from this list, the ratio of O/C decreases by laser irradiation above the ablation threshold. Therefore, oxygen containing compounds are generally removed from the ablated polymer surface.

Table 2 XPS studies on laser ablated polymer surface

Polymer	Laser	XPS Analyses	Reference
PET	ArF, 150 mJ cm^{-2}	O/C 1.00 →0.70	Lazare, 1984
polyimide	ArF, 150 mJ cm^{-2}	O/C 1.00 →0.57	Lazare, 1984
PEN-2,6	KrF, 500 mJ cm^{-2}	O/C 0.76 →0.64	Yabe, 1989
PES	XeCl, 750 mJ cm^{-2}	O/C 0.19 →0.12	Yabe, 1989
polyether etherketone	ArF, 60 mJ cm^{-2}	O/C 0.155→0.052	Occhielo, 1989
polyimide	KrF, 135 mJ cm^{-2}	O/C 0.21 →0.06	Kokai, 1989
PET	KrF, 200 mJ cm^{-2}	O/C 0.33 →0.27	Shimo, 1990

The chemical alternations on PET surface by laser irradiation under various foreign gas circumstances were summarized in Table 3 [12]. It is clearly shown that when the laser ablation was performed under vacuum, irradiated surface was almost the same as the original one. On the other hand, in the presence of foreign gases, carbon signal from C-C species is relatively increased compared to that from C=O and C-O species and the ratio O / C is decreased. These results indicate that C=O and CO_2 containing species are easily removed from PET surface. In addition, an increase of FWHM of carbon (c) means the generation of various kinds of C-C species. It is worth noting that nitrogen atom is introduced on PET surface in the ablation under gaseous ammonia atmosphere.

As laser fluence should affect the incorporation mechanism, we have investigated the chemical species generated on the surface by laser ablation at two different fluences; high fluence above the ablation threshold and low fluence below the threshold. Figure 4 shows the XPS signals of carbon atom for three PMMA samples, a) before irradiation, b) ablation at 40 mJ cm^{-2}, c) ablation at 570 mJ cm^{-2} [13]. C1s

Table 3 XPS analyses of PET surface under various conditions

Ablation Conditions	Ratio of carbon (%) [2] (a)	(b)	(c)	FWHM [3] (eV)	O/C [4]	N(%)
original [1]	14	16	70	1.40	0.33	0
under vacuum	12	16	72	1.51	0.34	0
in air	10	10	80	1.66	0.27	0
in oxygen	10	12	78	1.84	0.30	0
in ammonia	10	12	78	1.62	0.27	1.5

1) PET film which is not irradiated. 2) PET film has three kinds of carbon atoms as shown in the molecular structure; (a) carbonyl carbon, **C**=O**;** (b) ether carbon, **C**-O; (c) **C**-C carbon such as benzene ring. 3) FWHM of carbon (c) 4) Ratio of oxygen to carbon

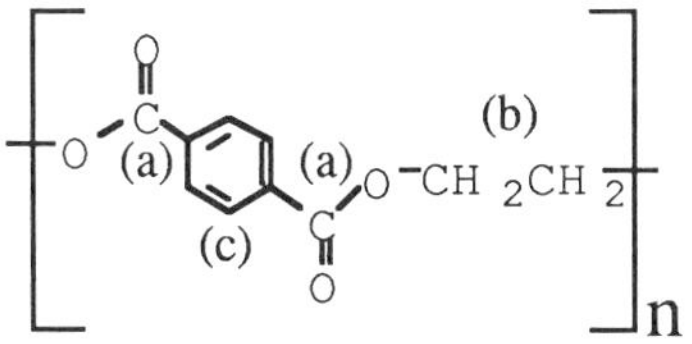

peaks have three components; (a) C-C , (b) C-O , and (c) C=O, and O1s have two components, O-C and O=C, with an equal signal intensity. The XPS signals which we observed were identical with the data previously reported [14]. As is clearly shown in this figure, the signal of C-O increased when the polymer was irradiated at 40 mJ cm^{-2}, while that of C=O decreased at 570 mJ cm^{-2}.

The value of O1s / C1s on the ablated PMMA surfaces at the two laser fluences was also investigated as a function of total input laser fluences [13]. In the case of low laser fluence (60 mJ cm^{-2} $pulse^{-1}$), the ratio increased as the total input fluences increased, which means chemical species containing oxygen component were formed on the ablated surfaces. These species on the ablated surface were probably produced by reactions of the generated active species with atmospheric oxygen. On the other hand, the ratio for the ablated surfaces at the laser fluence of 600 mJ cm^{-2} $pulse^{-1}$ reached constant above the total fluences of 80 J cm^{-2}. The lack of oxygen component on the ablated surface at 600 mJ cm^{-2} $pulse^{-1}$ irradiation can be interpreted by removal of O containing groups such as CO, OH, and OCH_3. As a result, the surfaces became carbon-rich state.

The most possible explanation for the different surface properties of the ablated surface can be given in terms of reaction probability of ambient oxygen to the surface. Lazare et al. reported modification of polymer surface with a mercury resonance lamp and an ArF excimer laser [5]. The ratio of O/C on the surface increased when irradiated by CW 185 nm mercury lamp, whereas the ratio of PET and polyimide irradiated at 193 nm laser showed a net decrease. The distinct contrast by lamp and laser irradiation was considered as follows. Since the ejected materials with

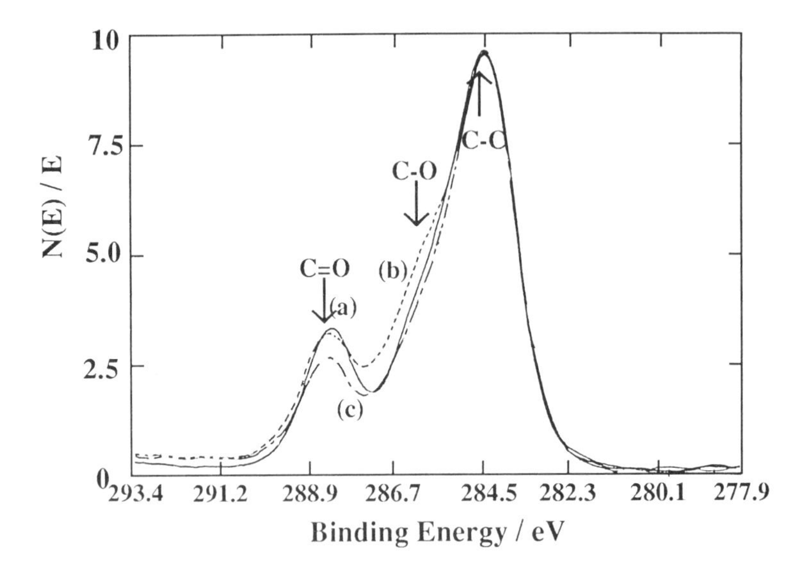

Figure 4 XPS signals for PMMA; (a) before irradiation, (b) irradiated at 40 mJ cm^{-2}, and (c) irradiated at 570 mJ cm^{-2}

supersonic velocities may prevent oxygen molecules from reaching the activated surface during the laser ablation processes, the carbon radicals produced on the ablated area are not trapped by oxygen. On the other hand, PMMA appeared to be less reactive with oxygen than PET and polyimide under their experimental conditions and showed 35 % increase in the ratio of O/C with a mercury lamp and 5 % increase with an ArF excimer laser. In our experiment, the increase of O/C at the laser fluence of 60 mJ cm^{-2} $pulse^{-1}$ is similar to the results by the mercury lamp irradiation, which is acceptable because the fluence is far below the threshold value.

When PMMA was irradiated at the fluence of 600 mJ cm^{-2} $pulse^{-1}$ in our experiment, the ratio of O/C decreased just as in the laser ablation of PET and polyimide with ArF excimer laser. However, the ratio was reported to slightly increase in the case of 193 nm laser ablation at 150 mJ cm^{-2} $pulse^{-1}$ by Lazare et al. The different results between our and their experiments are probably due to the different input fluences. The input energy of laser radiation are used for bond cleavage of polymer and excess energy could be used for the expansion of fragments, therefore the high energy causes the fragments to be expelled at supersonic velocities. Danielzik et al. studied velocity distributions on the stable polyatomic product MMA by time-of-flight mass spectroscopic measurement [15]. The measured velocity distributions corresponded well to Maxwell-Boltzmann distribution at a temperature of 1200 K at a laser fluence of 80 mJ cm^{-2} which is under the threshold value. Above the threshold, 200 mJ cm^{-2}, the distributions were broadened because a fast component generated by multi-photon processes are additionally overlapped. Thus excitation at higher laser fluence can give higher translational energy to the ablated fragments. Since the energetically higher fragments are expelled at supersonic velocity from the surface, they are less reactive with oxygen molecules in the gas phase and the fragments or those reacted with oxygen easily fly away from the ablated area. As a result, the radicals generated on the ablated surface densely remained. The concentration of the generated radicals is also higher at a high laser fluence than that at a low fluence. Accordingly, the radicals advantageously recombine to form new carbon-carbon bonds which result in carbon-rich state on the ablated area.

Niino et al. also reported surface chemical alternation of polyethylene 2,6-naphthalate (PEN 2,6) upon ablation [16]. They irradiated PEN-2,6 (the threshold value at 248 nm is ~ 50 mJ cm^{-2}) with a KrF excimer laser in air at two different fluences. After 200 shots irradiation of 3 mJ cm^{-2} $pulse^{-1}$, the ratio of O/C measured by XPS increased from 0.76 to 0.86, while the value decreased to 0.64 by 100 shots irradiation at 500 mJ cm^{-2}. Namely, the different chemical processes below and above the threshold have also been demonstrated. As described above, the different surface character was obtained by changing the laser fluence. Probably, different chemical processes are involved in fabrication processes at these two laser fluences.

As one of active species produced by the bond scission of PMMA, so-called " propagating radical " was detected by ESR as shown in Figure 5. This radical is well known to be produced in the photodegradation process of PMMA [17], and it is noteworthy that the radical is very stable. The ESR measurement was possible even at one hour after laser irradiation. Active species formed during the ablation processes are probably remained inside of the irradiated polymer, since the surface part of the polymer should react with surrounding oxygen molecules. The propagating radicals once formed on the ablated polymer probably decompose to

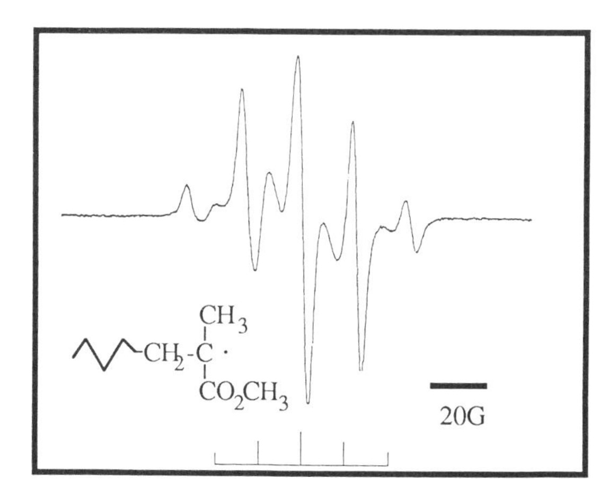

Figure 5 ESR signal of propagating radical detected in the ablation sample of PMMA

produce CO, CO_2, MMA monomer, and so on, which are coincident with the reported results on the main fragmented products in the gas phase [18]. The decomposed radicals, for instance, propyl type radicals react with oxygen to form polar components at a low laser fluence or recombine with each other to form hydrophobic surface at a high laser fluence. As a result, the surface irradiated at low and high laser fluences have oxygen containing compounds and carbon rich species, respectively.

5. SURFACE PROPERTIES OF ABLATED POLYMER SURFACE

In order to estimate the surface properties induced by laser ablation, contact angle measurements are useful because a surface free energy change arised from both an electrostatic force and a dispersion force can be evaluated. The changes of the contact angle of water can be discussed in terms of chemical and physical interactions between the ablated surface and water. We have measured the contact angle of water to the surface at two different laser fluences of 60 mJ cm^{-2} $pulse^{-1}$ and 600 mJ cm^{-2} $pulse^{-1}$ as a function of total input fluences. The surface properties are apparently different at two laser fluences. When PMMA plate was irradiated at 60 mJ cm^{-2} $pulse^{-1}$, the contact angle decreased within the range of our measurements as the accumulated laser fluences increased. Therefore, with an increase of subsequent laser shot, the wettability increased by the strong interaction with an aqueous solution. The surface properties are constantly changed as the incident total fluence increases, probably because of the morphological and chemical change of the ablated surface. In the case of laser fluence at 600 mJ cm^{-2} $pulse^{-1}$, the angle decreased and reached constant after several tens of laser shots, which means the properties of the surface did not change although the new surfaces were generated at subsequent laser pulse irradiation. In initial shots of laser irradiation, the ablated area was effectively etched accompanying with photochemical reactions (incubation) inside of the polymer film. The accumulated incubation effect alters the polymer properties probably by the recombination of radicals. The effect reached constant after several hundreds shot, then the surface properties did not change anymore.

6. SELECTIVE INCORPORATION OF FUNCTIONAL MOLECULES ON THE FABRICATED SURFACE

As summarized and discussed in the previous chapter, morphological, and chemical and physical properties are greatly altered, and reactive species are formed on the ablated area. Therefore, it is easy to induce chemical reaction of the species with functional molecules. In order to introduce chemical species from gas phase, polymer film was irradiated in the presence of reactive gaseous molecules. As an example, we summarize here ablation of PET and its functionalization with hexamethyldisilane (HMDS) as a reactive foreign molecule. The XPS analyses of both irradiated and masked areas of PET surfaces are shown in Table 4. As HMDS has no absorption at the laser wavelength (248nm), HMDS does not decompose in the gas phase. Since the peak position of Si(2p) is different from that of free silicon atom and HMDS, new chemical bonds between silicon and polymer must be formed, which is also confirmed from wide FWHM of the silicon signal. Wide FWHM could be interpreted as follows. The fragment ejected from the polymer surface reacts with HMDS to form new silicon-containing compounds in the gas phase, and these are deposited on every area of polymer surface. HMDS molecules also react with active polymer surface to form other silicon-containing species which show the different XPS signals. Accordingly, two types of silicon species, one formed in the gas phase and the other formed on the ablated polymer surface, were measured by XPS, while only the former was detected on the masked area.

When laser ablation was performed in the presence of HMDS and 892 Torr of helium, the ratio of silicon signal from irradiated area (A) to that from masked area (B) is 3.6. This value was reduced to be 0.9 for small number of laser pulse irradiation (8 pulses) and to be 1.3 for low helium pressure (89 Torr). These results mean that in the presence of inert gas, the silicon containing compounds formed in the gas phase are suppressed to fly away from the ablated area. Actually the fragments were observed as shown in Figure 3 (b). These fragments also could not leave from the surface at the small numbers of laser pulse irradiation. Therefore, the intensity of Si(2p) signal attributed to the fragments is relatively weak compared to that from the irradiated surface, which causes the decrease of the ratio (A) / (B). We can safely conclude that HMDS molecules react with the ablated polymer surface and form a new C-Si bond.

Since the chemical and physical properties of polymer surface are changed by laser irradiation, patterning of the surface can be possible by utilizing the difference.

Table 4 XPS analyses of PET surface ablated in the presence of hexamethyldisilane and helium

Ablation Conditions		Signal Intensity of Si(2P)			FWHM of Si(2P) / eV	
Helium / Torr	Number of laser pulses	Irradiated area (A)	Masked area (B)	Ratio of (B)/(A)	Irradiated area (A)	Masked area (B)
892	40	0.59	2.11	3.6	2.63	1.79
890	8	0.53	0.48	0.9	2.5	-
89	40	0.31	0.39	1.3	2.5	1.71

Hexamethyldisilane, 20.2 Torr; Laser fluence, 89 mJ cm^{-2}

Dye molecules, one of the functional molecules, can be selectively incorporated on PMMA surface. PMMA plates were irradiated with an excimer laser in air through a mesh mask which was contacted on the plate. In order to change the properties of the ablated surface, different irradiation conditions were investigated. Immediately after the laser irradiation, the samples were immersed in a saturated aqueous solution of dye molecules at room temperature for 15 minutes, then washed with ion-exchanged water under an ultrasonic operation for 20 minutes. Incorporation of dye molecules was confirmed by observing a fluorescent pattern under a fluorescent microscope. Details of the experiment were presented in elsewhere [13].

When PMMA surface was irradiated at the laser fluence of 40 mJ cm^{-2} $pulse^{-1}$, dye molecules were selectively incorporated on the ablated area as shown in Figure 6 (a). On the contrary, in the case of laser irradiation at 570 mJ cm^{-2} $pulse^{-1}$, the dyemolecules were not introduced on the ablated area, but on the unirradiated part hidden with a mesh mask. It is worth emphasizing that positive (a) and negative (b) fluorescent micropattern were controllably obtained only by changing the laser fluence. Thus, simultaneous microfabrication and chemical modification of polymer surface are successfully achieved. The technique is highly useful for introduction of chemical functionality on a minute area of polymers.

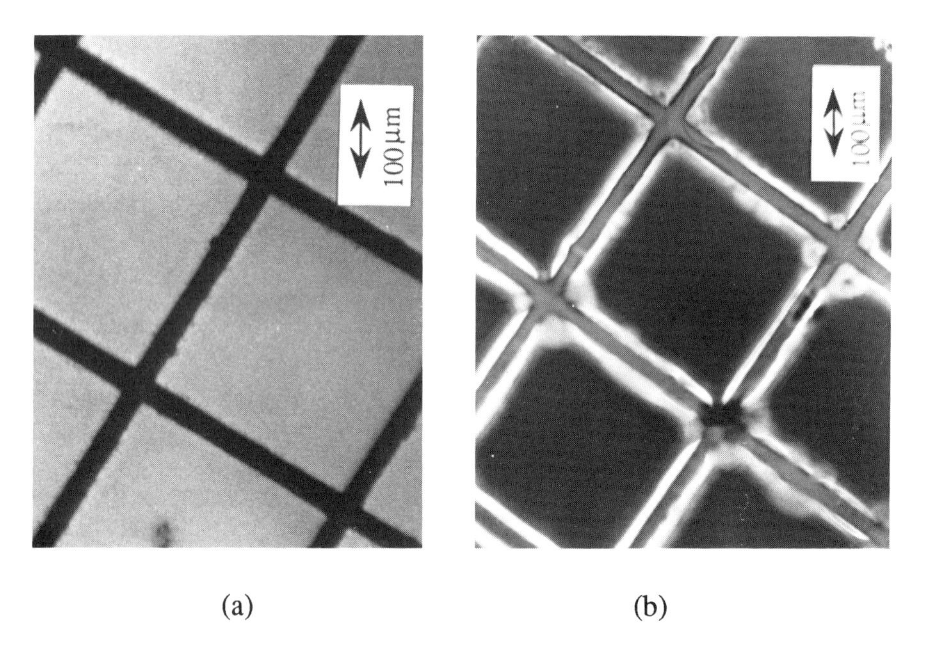

Figure 6 Selective incorporation of dye molecules on irradiated PMMA surface. (a) 40 mJ cm^{-2} x 4000 pulses, (b) 570 mJ cm^{-2} x 700 pulses

Since the threshold fluence for PMMA ablation with KrF excimer laser is known to be 500 mJ cm^{-2} $pulse^{-1}$ [19], polymer surface is rapidly ablated at the fluence of 570 mJ cm^{-2} $pulse^{-1}$. It is also well known that polymer surface is actually etched even below the threshold fluence and decomposition of polymer apparently took place even at the laser fluence of 40 mJ cm^{-2} $pulse^{-1}$. The chemically active species formed on the surface and its concentration are different between the processes at two laser fluences.

In the dye molecules used, only the basic type molecules having a cationic chromophore with a counter anion such as Rhodamine B, Rhodamine 6G, and so on, could be incorporated on the ablated surface. The other types of dye, anionic, and neutral ones, did not react with the ablated surfaces. These results indicate that the acidic chemical species such as COOH generated on the ablated area play an important role in the incorporation mechanism of dye molecules.

Recently, Niino and Yabe reported the micro-patterning of metallic materials by an area-selective electrodeless plating on the ablated films of PET, polyimide, and polyethersulfone [20]. Since the surface potential of the ablated polymer was positive, electrodeless plating could be selectively realized on the ablated sample after dipping it in palladium colloid solution.

In conclusion, oxygen, nitrogen, and silicon compounds are introduced into the fabricated polymer surface by laser ablation. This result opens a possibility to implant desirable molecules only in the ablated area. We have demonstrated the incorporation of silicon compounds and dye molecule, which is one of the representative functional molecule. New microfabrication technology by excimer laser will be achieved in near future by using special optics, so that the present study is a successful demonstration of simultaneous microfabrication and chemical modification for preparing chemically functional materials.

REFERENCES

1. Y. Kawamura, K. Toyoda, and S. Namba, Appl. Phys. Lett., 40 (1982) 374 .
2. R. Srinivasan and V. Mayne-Banton, Appl. Phys. Lett., 41 (1982) 576 .
3. R. Srinivasan and W. J. Leigh, J. Am. Chem. Soc., 104 (1982) 6784.
4. R. Srinivasan and B. Braren, Chem. Rev., 89 (1989) 1303 .
5. S. Lazare and R. Srinivasan, J. Phys. Chem., 90 (1986) 2124 .
6. M. Bolle and S. Lazare, J. Appl. Phys., 73 (1993) 3516.
7. M. Bolle, S. Lazare, M. Leblanc, and A. Wilmes, Appl. Phys. Lett., 60 (1992) 674 .
8. H. M. Phillips, D. L. Callahan, R. Sauerbrey, G. Szabó, and Z. Bor, Appl. Phys. Lett., 58 (1991) 2761.
9. M. Schumann, R. Sauerbrey, and M. C. Smayling, Appl. Phys. Lett., 58 (1991) 428 .
10. H. M. Phillips, S. Wahl, and R. Sauerbrey, Appl. Phys. Lett., 62 (1993) 2572.
11. G. Koren and J. T. C. Yeh , J. Appl. Phys., 56 (1984) 2120.

12. N. Shimo, T. Uchida, and H. Masuhara, Mater. Res. Soc., Symp. Proc., 191 (1990) 91 .
13. T. Uchida, N. Shimo, H. Sugimura, and H. Masuhara, J. Appl. Phys., submitted.
14. F. A. Houle, Laser Chem., 9 (1988) 107.
15. B. Danielzik, N. Fabricius, M. Rowekamp, and D. von der Linde, Appl. Phys. Lett., 48 (1986) 212.
16. H. Niino, A. Yabe, S. Nagano, and T. Miki, Appl. Phys. Lett., 54 (1989) 2159.
17. A. Gupta, R. Liang, F. D. Tsay, and J. Moacanin, Macromolecules, 13 (1980) 1696.
18. W. Kesting, T. Bahners, and E. Schollmeyer, Appl. Surf. Sci., 46 (1990) 326 .
19. R. Srinivasan, B. Braren, R. W. Dreyfus, L Hedel, and D. E. Seeger, J. Opt. Soc. Am., B3 (1986) 785.
20. H. Niino and A. Yabe, Appl. Phys. Lett., 60 (1992) 2697.

MICROCHEMISTRY
Spectroscopy and Chemistry in Small Domains
Edited by H. Masuhara et al.

Highly time- and space- resolved studies of superfast image production using laser ablation transfer*

David E. Hare, I-Yin Sandy Lee, and Dana D. Dlott+

School of Chemical Sciences, University of Illinois at Urbana-Champaign, Box 37-1 Noyes Lab, 505 S. Mathews Ave., Urbana, IL 61801, USA

In this paper, we describe a photothermal laser polymer surface ablation process which can be used to explosively transfer material from a film to a receiver sheet, thereby producing a high resolution color image at high speed. The creation of a single pixel of an image involves many interesting and important phenomena occurring in a small volume of a polymer thin film, including polymer superheating, polymer thermochemistry at ultra high heating rates (e.g. 10^{12} deg/s), high speed mass transfer and shock waves. New experimental methods to study these phenomena in detail have been developed, which provide high time resolution (picoseconds) and space resolution (micrometers).

1. INTRODUCTION

Laser ablation is a well known method for microlithographic alteration of a polymer surface. This paper is about highly time and space resolved studies of the fundamental mechanisms of a novel method for producing high resolution color images by laser ablation. The process [1], termed *laser ablation transfer* (LAT), involves both the explosive removal of polymer from a substrate, and the nearly intact transfer of this material to a receiver, thereby simultaneously producing a positive and a negative image. In order for this technique to become practical and economical, it will be necessary to use a relatively inexpensive source of laser photons. This requirement seems to eliminate the possibility of using ultraviolet (excimer) laser ablation [2], so instead we have concentrated on photothermal ablation using near-infrared (near-IR) photons from an efficient solid-state laser. The essential elements of the LAT process are diagrammed schematically in Fig. 1, adapted from ref. 1.

The ablatable coating on the film, typically 0.5-1 μm thick, contains polymers, colorant agents, and a near-IR absorbing sensitizer dye [1]. When the coating is irradiated with an intense pulse from a near-IR laser, the sensitizer dye becomes excited. The dye most commonly used, IR-165, efficiently absorbs light near 1.0 μm, which coincides with the emission of the Nd:YAG or Nd:YLF lasers, among others. Upon excitation, an efficient nonradiative

*This research was supported by National Science Foundation grant DMR-91-04130, the US Army Research Office grant DAAH04-93-G-0016, and a gift from Graphics Technology International (GTI).
+Author to whom correspondence should be addressed.

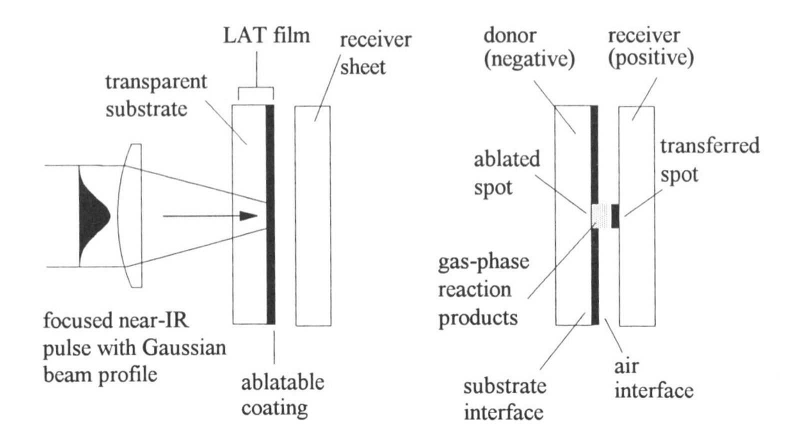

Figure 1. Essential elements of the laser ablation transfer (LAT) process used for high speed image production. Adapted from ref. 1.

relaxation process converts the optical excitation into heat. The sudden temperature jump in the polymer induces a photothermal ablation process, which propels a spot of the coating off the film, and onto the receiver. Figure 2 shows digitized images of spots with two different radii, ablated in LAT films by picosecond pulses [3] (1 ps = 10^{-12}s). The production of high resolution images by LAT has been discussed in several recent publications [1,3,4].

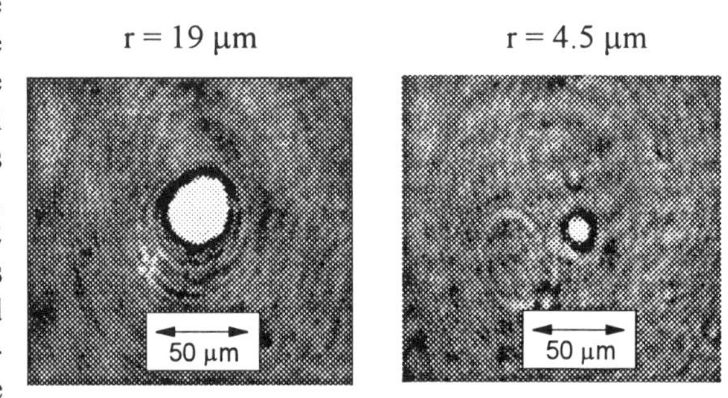

Figure 2. Computer digitized images of spots ablated in LAT film, taken from ref. 3.

Experimental investigations of the dynamics of LAT must overcome significant obstacles. The volume studied is quite small. The smaller spot in Fig. 2 has radius 4.5 μm and thickness 0.5 μm, so the volume is 3 x 10^{-11} cm^3, and the mass of polymer removed is about 40 picograms. The LAT process can occur on a very short time scale, and indeed we have shown it is possible to produce a spot on a film with a pulse as short as a few picoseconds [3]. High spatial resolution is required because the behavior of the LAT process is quite different [1] at the spot center, at the spot edges, just above the plane of the film, at the surface of the receiver sheet, etc. Due to the explosive nature of LAT, the environment near the ablated spot is very harsh. Due to the rather complicated materials used in practical applications, the chemistry and physics of this process are correspondingly complex.

The LAT project is the result of an exceptionally fruitful collaboration between scientists at Graphics Technology International, Inc., S. Hadley MA. (GTI) and our group at the University of Illinois. The group at GTI is primarily concerned with practical problems and applications, e.g. how to produce images with high color fidelity, how to decrease the ablation threshold, etc. The group at Illinois is primarily concerned with fundamental problems, e.g. the mechanisms of LAT, the role of the near-IR dye, and the chemical and physical response of polymers to sudden large-amplitude jumps in temperature. In the experiments discussed here, the rate of polymer heating is so great that a variety of polymer states which are very far from equilibrium are produced and studied. Besides their practical importance, understanding such phenomena is a central issue in chemical physics.

2. IR DYE SENSITIZER

The IR-165 dye is characterized by several highly desirable properties: the ability to deposit a large amount of heat into the polymer in a very short time [5-7], high chemical and thermal stability, chemical compatibility with a wide variety of liquid and polymeric media, and the ability to act as an ultrafast molecular thermometer [5-8]. Figure 3 shows the chemical structure of IR-165, and the optical absorption spectrum. The strong absorbance at $\lambda \approx 1\ \mu m$ is denoted the "heater" transition, because optical pumping there is used to heat the polymer film. The weaker absorbance in the visible region (0.4-0.8 μm) is denoted the "thermometer" transition because the absorption there increases with increasing temperature [6-8], as shown in Fig. 3. The thermometer calibration factor depends somewhat on the probing wavelength. As shown, when probed [5] at 0.532 μm the relative change in optical absorbance, dA/A_0, where A_0 denotes ambient temperature absorbance, increases *linearly* with increasing temperature above 40°C. Using conventional heating techniques, the thermometer data in Fig. 3 were obtained in the range 25-150°C. Using pulsed laser heating, it was verified that the linear temperature dependence is maintained [8] up to 600°C. Although most dyes can be used as molecular thermometers [9], a linear temperature dependence is a valuable attribute since it can greatly simplify the interpretation of experimental data [8].

When an absorbing dye is irradiated by intense laser pulses, the absorption becomes saturated. Optical saturation places a photophysical limit on the number of photons a dye molecule absorbs per unit time [6]. The optical saturation intensity I_{sat} was measured using the arrangement shown in Fig. 4, with 23 ps duration optical pulses [6,7]. The sample, a dilute solution of IR-165 in *poly*-methyl methacrylate (PMMA), has optical absorbance A(I), measured as a function of input intensity I. For a sample with small-signal absorbance A_0,

$$A(I) = \frac{A_0}{1 + \frac{I}{I_{sat}}}. \tag{1}$$

Equation (1) shows when $I = I_{sat}$, the absorbance of the sample is reduced by one-half. Figure 4 shows an experimental measurement of $I_{sat} = 5(\pm 0.2)\ GW/cm^2$. When this data is corrected for effects such as optical pulse shape, Gaussian beam profile, and finite sample

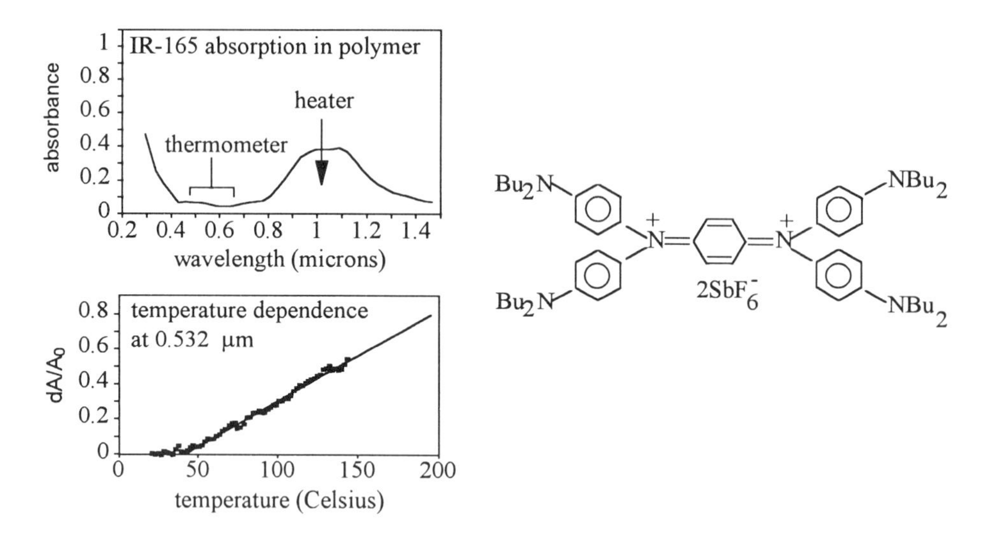

Figure 3. (right) Chemical structure of IR-165 sensitizer dye. (top left) Optical absorption of IR-165 in a polymer thin film showing absorption maximum at $\lambda \approx 1$ μm. (bottom left) Relative increase in optical absorption at 0.532 μm, dA/A_0, versus temperature.

absorbance [7], the corrected value is $I_{sat} = 1.6(\pm 0.06)$ GW/cm^2. We can then compute the ground-state recovery lifetime τ, using the relation $I_{sat} = h\nu/2\sigma\tau$, where $h\nu$ is the photon energy, and σ the absorption cross-section, $\sigma = 6.1 \times 10^{-17}$ cm^2, giving $\tau \approx 1$ ps. This result [7] implies that once an IR-165 molecule has absorbed a photon, an ultrafast nonradiative relaxation process induces a recovery of the ground-state absorption in one ps. It further implies that during a suitably intense 25 ps duration optical pulse, *a single molecule can absorb ≈25 photons*, each photon here having about one eV of energy (with longer duration pulses, even more photons can be absorbed [5]). The ultrafast ground state recovery of IR-165 is the principal attribute which allows a large amplitude temperature jump to be produced with a short duration optical pulse.

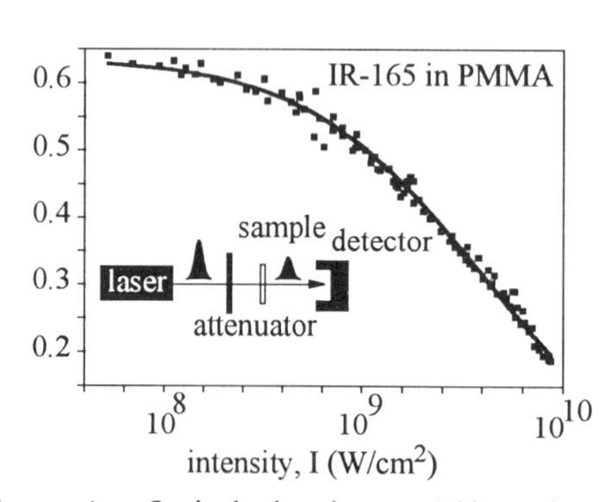

Figure 4. Optical absorbance A(I) at $\lambda = 1.064$ μm, versus intensity I for IR-165 dye in a polymer film. The saturation intensity I_{sat} is the value of I where $A(I)/A(0) = 0.5$ (from ref. 6).

We have also determined the rate at which the IR-165 dye molecules release their excess mechanical energy to the surrounding polymer [7]. In this experiment, the sample consisted of a large area (200 x 300 cm^2) polymer film sample on a

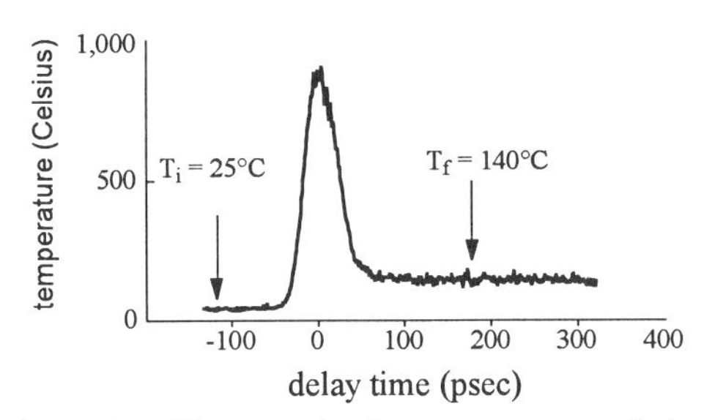

Figure 5. Time-resolved measurement of the temperature of the IR-dye in a polymer thin film, adapted from ref. 7.

float glass substrate mounted on a motorized positioner, so that a fresh sample volume was present each time the laser was pulsed. The dye was pumped by an intense 23 ps duration optical pulse at λ = 1.064 μm. The thermometer transition of the dye was probed by a somewhat shorter pulse at λ = 0.532 μm, to determine the time-dependent temperature of the dye in the polymer. The result is shown in Fig. 5. During the pump pulse the dye, initially at 25°C, becomes quite hot. It then cools down, and at ≈100 ps, the dye and the polymer have equilibrated at the final temperature, T_f = 140°C. The time constant for cooling, t_{VC}, was determined through a detailed analysis [7] of similar temperature curves at different values of T_f. Knowing the rate energy is input to the dye and the peak temperature, we found that t_{VC} = 3.8 ps.

Near-IR dyes are becoming increasingly important in technological applications such as compact disk fabrication, and various imaging schemes involving near-IR solid-state lasers [10]. The techniques used here are broadly applicable to near-IR dyes, and have specifically revealed several important attributes of IR-165, namely an ultrafast rate of energy uptake via optical absorption, an ultrafast rate of energy dissipation into the surrounding medium, and the unique property of being an optical thermometer with a nearly linear temperature response.

3. DIRECT MEASUREMENT OF THE TEMPERATURE DURING LASER ABLATION

We have used the IR-165 optical thermometer to measure the *in situ* temperature of ablating polymers, and also to determine how the heat capacity changes near the ablation threshold [8]. In these experiments, the ablation optical pulse duration was 150 ns, which is representative of that used in commercial LAT applications [1]. Presumably the mechanism of ablation with such pulses involves photothermal decomposition of the polymer into gas-phase products, propelling the coating off the substrate as shown in Fig. 1. A great deal of effort has gone into understanding the thermal decomposition of polymers [11], but almost all this effort involves experiments where the polymer is heated relatively slowly. Our experiments were designed to investigate the possibility that enormously rapid heating of polymers can produce unique chemical behavior. A typical heating rate in this experiment is 5×10^9 deg/s.

The samples studied here consisted of a 1 μm thick film of PMMA doped with about 10% IR-165 dye, supported on a float glass substrate [8]. Conventional thermal analysis was used to characterize this polymer at a heating rate of 10 deg/s, as shown in Fig. 6. Differential scanning calorimetry (DSC) shows the heat capacity suddenly increases, and thermogravimetric analysis (TGA) shows the onset of mass-loss occurs at about 250°C, indicating that thermal decomposition at 10 deg/s occurs at $T_d \approx 250°C$. The DSC also shows that in the temperature range below decomposition, 25-250°C, the average value of the heat capacity is C = 2.0 J/(g·deg).

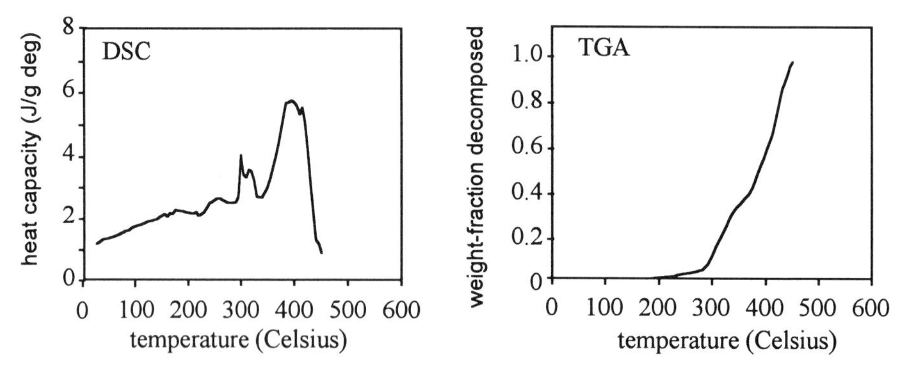

Figure 6. Conventional thermal analysis of PMMA film with IR-dye. DSC = differential scanning calorimetry. TGA = thermogravimetric analysis. At 10 deg/s, the polymer decomposes at about 250°C. Figure adapted from ref. 8.

Some typical results of instantaneous temperature measurements [8] on the 100 ns time scale of the ablation pulse are shown in Fig. 7. The data show the ns-duration ablation pulse at several energies E, the experimentally determined increase in absorption due to the molecular thermometer, and a smooth fitted curve. On the left of the figure, the pulses used were below ablation threshold. The temperature increases during the pulse, and then remains level on the time scale of observation. On the right of the figure, the pulses are above threshold. The temperature climbs, and near the end of the pulse there is a sudden increase in the apparent absorbance change. This increase does not denote a further increase in temperature, instead it denotes the onset of ablation [8]. It is caused by scattering of the probe laser beam by the ablation plume (see Section 4 and Fig. 9, below).

In these experiments, important parameters such as the optical properties of the sample and the rate of heat input are accurately known. The only parameter required to generate the fitted curves in Fig. 7 is the heat capacity C. Below ablation threshold, the value obtained, C = 2.0 J/(g·deg), is *identical* with that determined by DSC. Above threshold, an abrupt increase in C is observed, as indicated in Fig. 7. This heat capacity increase is attributed to the onset of significant thermal decomposition processes at ablation threshold [8].

Figure 8 is obtained by converting the absorption increase data (e.g. Fig. 7) into temperature. There is a temperature gradient [8] in the polymer film; the substrate interface facing the laser is hotter than the air interface away from the laser (see Fig. 1). The experimental observable is the temperature *averaged* over this gradient. Ablation begins first at the hotter surface, and since the gradient is accurately known, we can determine the peak surface temperature at ablation threshold, as shown in Fig. 8. Ablation occurs at a temperature of T_{abl} = 600 (±50)°C, when the heating rate is 5 x 10^9 deg/s. This T_{abl} is significantly higher than the ordinary thermal decomposition temperature of about 250°C, and it indicates a radical change in the mechanisms of thermal decomposition at the enormous heating rates characteristic of laser ablation [8]. Experiments designed to develop a molecular-level understanding of polymer ablation chemistry are discussed in Section 5.

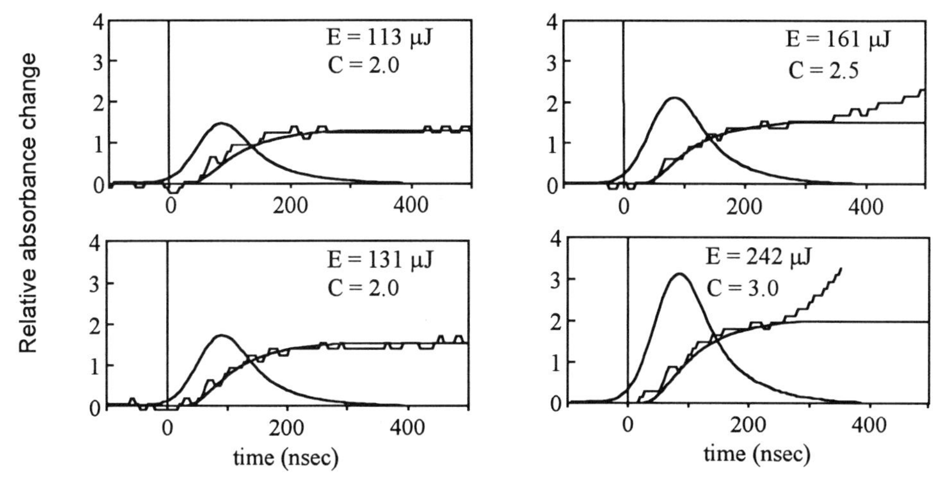

Figure 7. Relative absorbance change of the molecular thermometer in films below (left) and above (right) threshold, at pulse energies E. The optical pulse is also shown. The fitted curves are obtained by varying the heat capacity C (J/g·deg). Figure adapted from ref. 8.

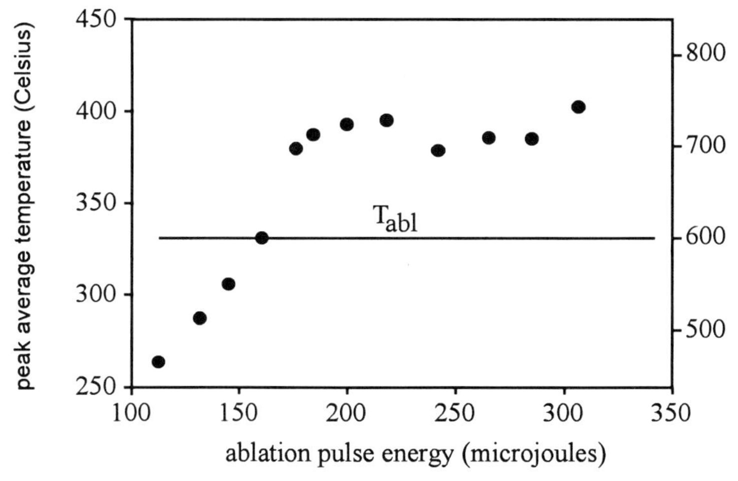

Figure 8. The peak temperature of the polymer film [8] as a function of pulse energy. The left ordinate is the average in the film. The right ordinate is the temperature at the surface facing the laser.

4. DYNAMICS OF IMAGE FORMATION USING LAT

Two important practical questions studied jointly by GTI and at Illinois are how to improve the image quality of LAT films, and how to lower the ablation threshold [3,12]. Ultrafast optical microscopy [13] was used to study the material ablated from the surface of the LAT film by a 150 ns duration optical pulse [1,4], as shown in Fig. 9. The ablated material appears in the air as a shower of tiny particles or droplets, whose velocities [1] are on the order of Mach 1. The appearance of these particles causes the sudden upturn in the data in Fig. 7. The tiny particles are not desirable because they do not produce uniform coverage of the receiver sheet [4]. Many of the properties of this film were understood by modeling the distribution of heat in the film at the time of ablation, using a finite-element thermal conduction calculation [4]. The results of this calculation, for a film pumped at the experimentally determined ablation threshold fluence J_{th} = 150 mJ/cm^2 are shown in Fig. 9. The laser pulse propagates through the substrate onto the coating (see Fig. 1). In the coating, absorption of this pulse by the IR-dye obeys Beer's law. During the 150 ns pulse, there is time for the heat within about 0.1 μm of the interface to diffuse into the substrate. Just before ablation occurs, the entire coating is above its melting point of ≈100°C, and the location of the temperature maximum is about 0.15 μm into the coating. The position of this maximum is determined by the competition between heat input via Beer's law absorption, and thermal conduction into the substrate [4]. Thermal decomposition products are created near the maximum in the coating interior, causing it to explode into a shower of small fragments.

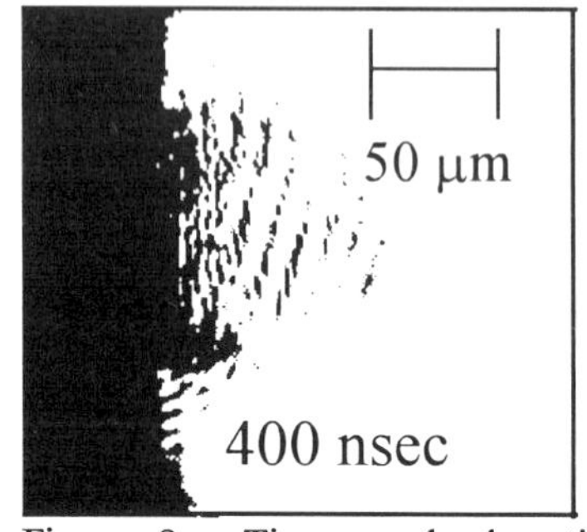

Figure 9. Time resolved optical micrograph of a film a few hundred ns after the 150 ns ablation pulse, which propagates from the left. The dark region indicates the film surface; the light region is air. (From ref. 4.)

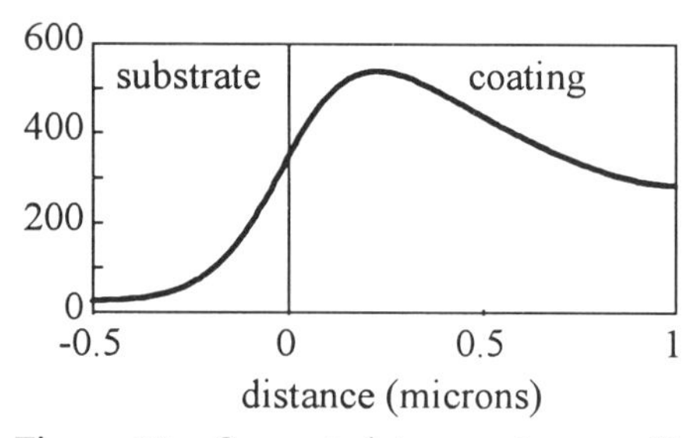

Figure 10. Computed temperature profile in the coating just prior to ablation (ref. 4). The ordinate is Celsius temperature).

Recently an improved film [4], denoted type II, was developed by GTI. The type II film, diagrammed in Fig. 11, has a dynamic release layer (DRL) consisting of 30Å thick evaporated aluminum. The optical properties [4] of this layer at 1.064 μm were: reflection R = 0.27, transmission T = 0.33, absorption fraction A_f = 0.40. Since this interlayer can absorb much of the incident pulse in a very thin region, its presence helps localize heating at the substrate interface (see Fig. 1). The DRL substantially reduced the ablation threshold and improved the coverage on the substrate [4]. An ultrafast micrograph [4] of type II film ablation is shown in Fig. 11. In contrast to Fig. 9, the ablated material appears in large flaps, like the skin of a bursting balloon. The type II films have a lower ablation threshold and a higher covering power, which allowed GTI to reduce the thickness of the coating to about 0.5 μm. Although IR-dye is not required to induce ablation in the presence of a DRL, some IR-dye is used to heat the coating to help it adhere to the receiver sheet. In optimized type II films, the concentration of IR-dye was a factor of four smaller than in type I films [4].

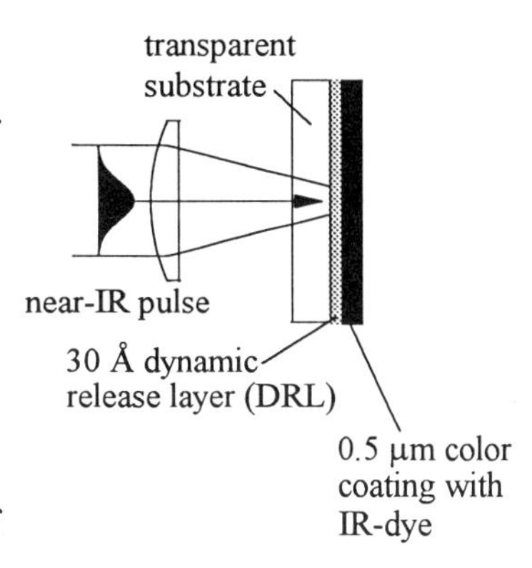

Figure 11. Diagram of a laser ablation transfer (LAT) film with a dynamic release layer (DRL).

In the type II films, most of the ablation pulse is absorbed in the DRL. The most important source of heat in the coating is thermal conduction from the DRL, with a lesser contribution due to absorption by IR-dye in the coating [14]. Figure 12 shows the computed temperature distribution in type II films irradiated at the threshold of J_{th} = 80 mJ/cm^2. The DRL localizes most of the heating at the substrate-coating interface. Except for a narrow region near this interface, the rest of the coating is below its melting point. The build up of decomposition products near the interface ablates the coating almost intact onto the receiver [4].

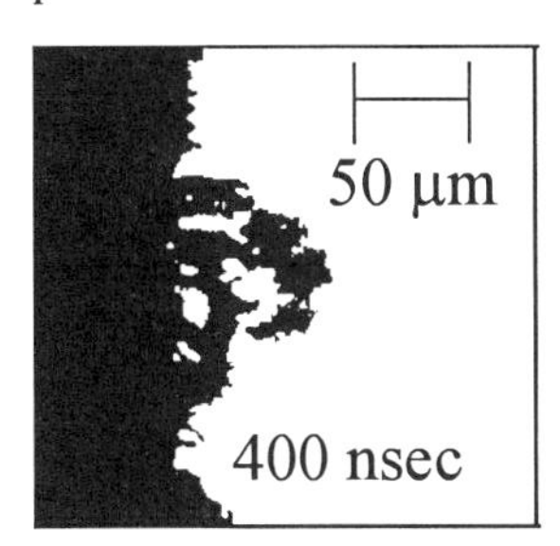

Figure 12. Time resolved optical micrograph of a film with a dynamic release layer at 400 nsec (from ref. 4).

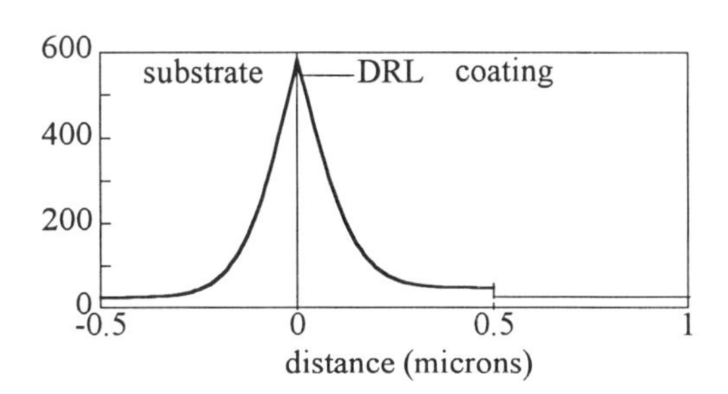

Figure 13. Computed temperature distribution in a film with a dynamic release layer, irradiated by a threshold pulse, at the time of maximum temperature. (From ref. 4).

Figures 10 and 13 indicate that the ablation threshold can be reduced by optimizing, in time and space, the heat added by the optical pulse [12]. The type II films have a threshold about one-half of the type I films, despite the optical losses incurred by the 30% reflection of the incident pulse on the DRL, because the DRL localizes more of the energy at the interface between coating and substrate [4]. It seemed likely that the use of shorter duration (ps) optical pulses could further reduce the threshold, by overcoming the effects of thermal conduction of heat into the substrate. Table I shows this is indeed the case. The type I films do not ablate well with ps pulses because the IR-dye undergoes optical saturation. The metallic absorber of the DRL does not saturate at these intensities, and a tenfold reduction of threshold was achieved. The 6 mJ/cm^2 threshold is *about 150 times smaller* than that required to remove a 1 μm thick layer of PMMA using ultraviolet (excimer) ablation [12,14].

Table 1
Ablation Fluence thresholds, J_{th} for LAT films

Film designation	150 ns pulses	23 ps pulses
Type I (no DRL)	J_{th} = 150 mJ/cm^2	----------
Type II (with DRL)	J_{th} = 150 mJ/cm^2	J_{th} = 150 mJ/cm^2

5. PICOSECOND CARS STUDIES OF LASER ABLATION

The experiments discussed so far do not provide much information about the microscopic processes which underlie the mechanisms of laser ablation. Recently we have devoted a great deal of effort to develop a new technique capable of providing such information [15]. We reasoned that time-resolved vibrational spectroscopy could be used to provide details of physical processes such as shock wave generation and polymer melting, and chemical processes involved in polymer decomposition. Although there are many infrared and Raman techniques which can be used on polymer thin films, most of the conventional methods (e.g. waveguides, frustrated internal reflection, etc.) either do not deliver the necessary time resolution, or cannot be used on thin films which are exploding during the time of observation. Coherent Raman scattering (CARS) with ps pulses overcomes most of the problems caused by ablation, and particularly important is that CARS is not adversely affected by optical emissions which arise from irradiating the sample with intense ablation pulses.

The apparatus we use [15] for ps CARS studies of ablating thin films is shown in Fig. 14. The Nd:YAG laser produces a train of ps pulses which pump a pair of tunable dye lasers. In addition, a cavity dumper in the laser produces a giant near-IR pulse used for photothermal ablation. Optical harmonic generators are available to shift the near-IR ablation pulse into the visible and UV if desired. A multiplex CARS configuration is used. That means one dye laser has a narrow output spectrum and the second a broad spectrum. In our system, the entire spectrum of a selected 100-200 cm^{-1} wide region is obtained on a single shot. High quality spectra are obtained by signal averaging at a repetition rate of up to 625 pulses per second. A stepper-motor translator moves the ablated sample through the laser beams, so a fresh volume is probed on every shot. The ablation pulse duration is about 100 ps, and the dye laser pulse duration is about 30 ps.

Figures 15 and 16 show some data we have obtained on a 3 μm thick film of PMMA doped with IR-dye. In our initial experiments, we have concentrated on a Raman transition at ca. 810 cm^{-1}, which is a stretching mode of the ester group. In Fig. 15, the 100 ps duration near-IR pulse, incident on the sample at t = 0, is below ablation threshold, and we estimate a temperature jump of about 200°C. Therefore the heating rate is 2 x 10^{12} deg/s. With the delay set at -400 ps (the CARS pulses *precede* the ablation pulse by 400 ps), the ordinary spectrum of PMMA is obtained. At t = 0 (at the peak of the ablation pulse), a small spectral shift and broadening are observed. At t = 1,000 ps, a broader, red shifted spectrum is seen. The broadening and shifting provide information about the conformation of the PMMA chains and the pressure build up caused by the sudden temperature jump.

Figure 16 shows data obtained with pulses above the ablation threshold. Again at t = -400 ps, the ordinary spectrum is seen. At t = 1,000 ps a quite broad and shifted spectrum is observed. Although we have not yet finished a detailed interpretation of the data, we expect it to provide information about the pressure build up in the polymer film and the appearance of reaction products, probably MMA monomer [16]. A thorough survey of the ablation behavior of all Raman active vibrations in PMMA is expected to help elucidate the chemical degradation mechanisms which dominate in large heating rate polymer decomposition.

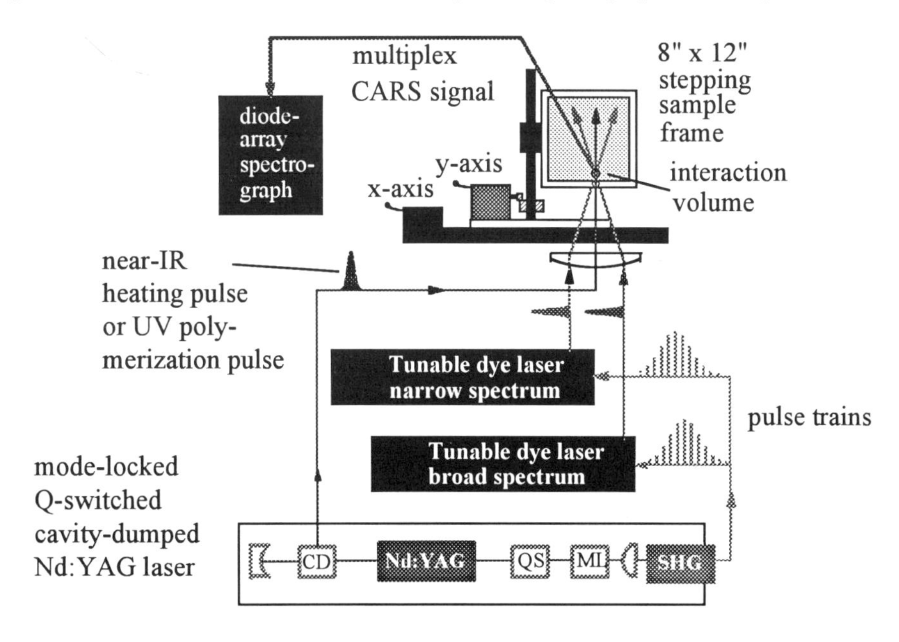

Figure 14. Apparatus used for multiplex ps CARS studies of polymer film ablation.

6. CONCLUSIONS

The combination of several highly time and space resolved optical techniques have been used to understand a rather complicated but important practical process, laser ablation transfer. Despite obstacles such as the harsh environment near the exploding polymer, the small sample volumes, and the short time scale associated with laser ablation, it is now possible to use advanced laser techniques to directly probe important quantities such as the temperature in the solid, the pressure in the solid, and the chemical composition of the solid. Of particular interest are nonlinear coherent techniques such as coherent Raman scattering, because coherent probes overcome to a large extent the problems associated with ablation debris near the sample plane and interfering optical emission from materials irradiated by high intensity optical pulses. Our ability to make a large amplitude temperature jump in a bulk polymer on the ps time scale leads to the observation of many new phenomena, and opens a window to the behavior of interesting polymer systems under extreme conditions representative of state-of-the-art materials processing techniques. Our fundamental understanding of the mechanisms of laser ablation has led to the development of systems with ablation thresholds two orders of magnitude smaller than obtained by conventional methods [3,12].

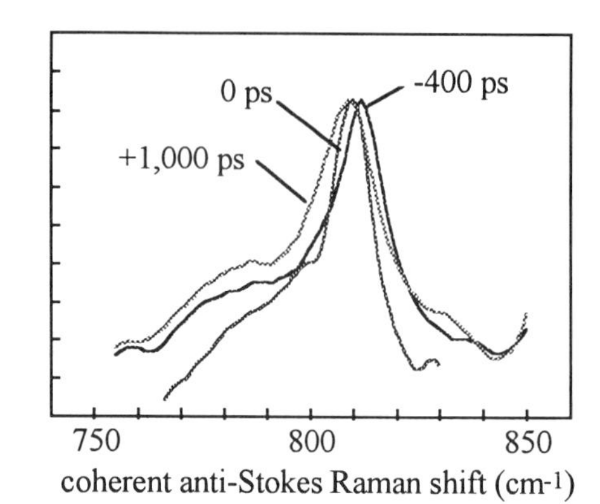

Figure 15. Coherent Raman spectrum of a PMMA film before (-400 ps), during (0 ps) and subsequent to (1,000 ps) an intense near-IR pulse which induces a 200 degree temperature increase (from ref. 15).

7. ACKNOWLEDGEMENTS

We acknowledge the contributions of our collaborators in this work, including William A. Tolbert, Xiaoning Wen, Mark M. Doxtader and Ernest R. Ellis.

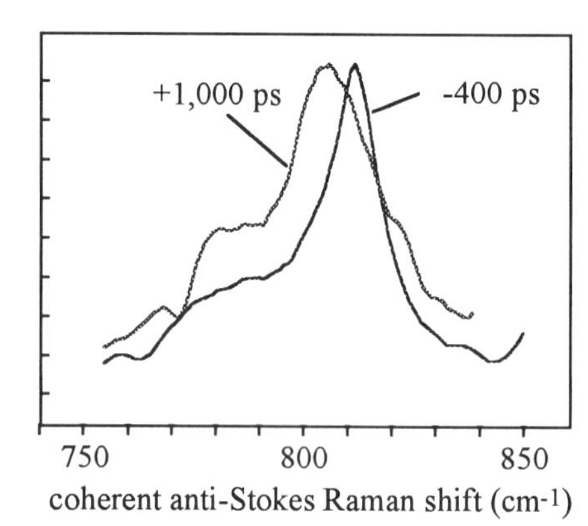

Figure 16. Coherent Raman spectrum of a PMMA film before (-400 ps) and after (1000 ps) an intense near-IR pulse which induces laser ablation (from ref. 15).

REFERENCES

1. I-Y. S. Lee, W. A. Tolbert, D. D. Dlott, M. M. Doxtader, D. M. Foley, D. R. Arnold, and E. W. Ellis, J. Imag. Sci. Tech., 36 (1992) 180.
2. J. T. C. Yeh, J. Vac. Sci. Technol. A4, (1986) 653; R. Srinivasan, Science, 234 (1986) 559; P. E. Dyer and R. Srinivasan, Appl. Phys. Lett., 48 (1986) 445.
3. W. A. Tolbert, I-Y. S. Lee, M. M. Doxtader, E. W. Ellis, and D. D. Dlott, J. Imag. Sci. Tech., in press.
4. W. A. Tolbert, I-Y. S. Lee, M. M. Doxtader, E. W. Ellis, and D. D. Dlott, J. Imag. Sci. Tech., 37 (993) 411.
5. X. Wen, W. A. Tolbert, and D. D. Dlott, Chem. Phys. Lett., 192 (1992) 315.
6. S. Chen, I-Y. S. Lee, W. A. Tolbert, X. Wen, and D. D. Dlott, J. Phys. Chem., 96 (1992) 7178.
7. X. Wen, W. A. Tolbert, and D. D. Dlott, J. Chem. Phys., 99 (1993) 4140.

8. I.-Y. S. Lee, X. Wen, W. A. Tolbert, D. D. Dlott, M. M. Doxtader, and D. R. Arnold, J. Appl. Phys., 72 (1992) 2440.
9. F. Wondrazek, A. Seilmeier, and W. Kaiser, Chem. Phys. Lett., 104 (1984) 140; A. Seilmeier, P. O. J. Scherer, and W. Kaiser, J. Phys. Chem., 90 (1986) 104.
10. M. Matsuoka, Ed., *Infrared Absorbing Dyes* (Plenum: New York, 1990).
11. For example, see J. H. Flynn, in *Thermal Analysis in Polymer Characterization*, E. A. Turi, Ed., (Philadelphia: Heyden & Son, Inc., 1981), pp. 43-59.
12. W. A. Tolbert, I-Y. S. Lee, D. E. Hare, X. Wen, and D. D. Dlott, in *Laser Ablation Mechanisms and Applications*, 1993, in press.
13. H. Kim, J. C. Postlewaite, T. Zyung, and D. D. Dlott, J. Appl. Phys., 64 (1988) 2955.
14. E. Sutcliffe and R. Srinivasan, J. Appl. Phys., 60 (1986) 3315.
15. D. E. Hare and D. D. Dlott, Appl. Phys. Lett., in press.
16. H. G. Jellinek and M. D. Luh, Makromol. Chem., 115 (1968) 89; R. Srinivasan, J. Appl. Phys., 73 (1993) 2743.

MICROCHEMISTRY
Spectroscopy and Chemistry in Small Domains
Edited by H. Masuhara et al.

Chemical dynamics of the interaction of ultraviolet laser radiation with organic polymers

R. Srinivasan

UVTech Associates, 2508 Dunning Drive, Yorktown Heights, NY 10598, U.S.A.

Ultrafast photography of a polymer surface and the plume above it during excimer laser ablation is used as a tool to study the chemical mechanism of the decomposition pathways that drive the etching process.

1. INTRODUCTION

Research interest in the ablation and etching of organic polymers by pulsed, ultraviolet laser radiation [1,2] has extended over the last ten years. Early efforts were concentrated mostly in acquiring the etch data that were necessary to develop the phenomenon into a useful technology. A rigorous examination of the chemical dynamics of the ablation process did not begin until 1986 and has been pursued actively up till now. These studies fall into two categories which are 1) analytical studies on the integrated results of the action of a single laser pulse on a given surface, and 2) studies on the physical and chemical changes caused by a single pulse in times of the order of the pulse width. A knowledge of the timing of the ablation process is fundamental to an understanding of the chemical physics of the phenomenon. The time-dependent probing methods that have been used to date include acoustic methods [3], fast photography of the ablating surface and the plume that rises from it by using a conventional camera [4], a streak camera [5], or by Schlieren photography [6], time-resolved reflectivity [7], beam deflection measurements [8], absorption spectroscopy [9] and emission spectrocopy [10]. In this brief review, the perspective will be from the chemistry of a given polymer in relation to the pathways for its ablative decomposition.

2. POLYIMIDE (KAPTON™)

The polyimide that is derived by the condensation of pyromellitic dianhydride (PMDA) with *p,p'*-oxydianiline (ODA) (Fig. 1) which is marketed under the trade name "Kapton" has been the subject of numerous studies on its ablative decomposition by ultraviolet laser pulses from an excimer laser. A detailed analysis of the UV absorption spectrum of Kapton has been published [11] and the intense absorption in the 300-330 nm region is attributed to a π - π^* transition which is associated with the central PMDA group. The weaker absorption at the longer wavelengths may be due to a n - π^* transition which is associated with the four >C=O groups. The extended, low-intensity absorption which gives the material its intense color is identified as being due to the substitution of the nitrogen atoms by a phenyl group. This knowledge is important because it makes it

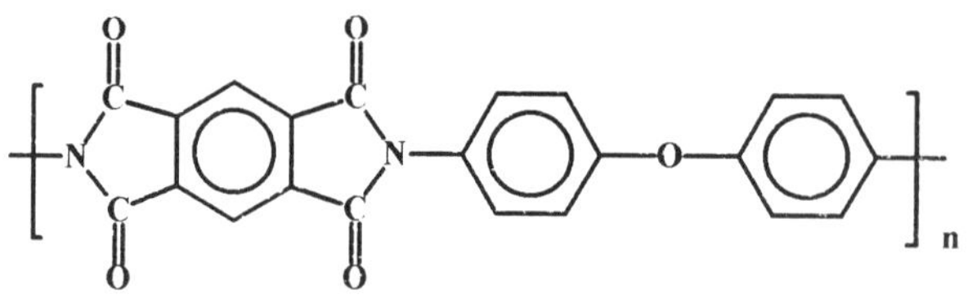

Figure 1. Formula of the polyimide, PMDA-ODA (Kapton™)

likely that the excitation of the PMDA group will result in the energy being channeled to the remaining ODA moiety so that the entire unit will be able to undergo reaction. In contrast, in the ground electronic state, the π systems of the PMDA and ODA parts are **not** expected to overlap significantly because of the torsional rotation of the phenyl ring about the nitrogen-carbon bond by ~ 60° with respect to the plane of the PMDA [12]. The simplest aromatic π-system which is present in benzene is incorporated in PMDA. Benzene in the condensed phase has been shown [13] to undergo not only two-photon excitation through a virtual state but even three-photon excitation *via* an excimer state at power densities as low as 10^7 W/cm^2. Decomposition from these upper electronic states which is accompanied by ablation has been reported [14]. A direct measurement of the lifetimes of the excited states that are reached on excitation of Kapton with 193 nm, 248 nm or 308 nm photons is yet to be made.

Discussions about the mechanism of the UV laser ablation of Kapton have leaned heavily on one set of data which is the etch depth *vs.* log fluence (or fluence) curve from its threshold to some arbitrary fluence value. The products that are formed in this process have been analyzed in many publications. Of particular importance are those studies [15-17] in which the dynamics of the product stream has been followed as a function of the ablation conditions. These investigations point to the numerous intermediates that exist in this system and the complex chemistry that may lead to them.

In a recent publication [18], the dynamics of the ablation of this material by pulsed (ns) UV laser radiation and pulsed (ns) infrared laser radiation have been compared. The contrast between the reaction pathways in the two cases has been brought out by this approach. These results will be discussed in some detail here as the fast photographic method that was used in this study gives a visual picture that is easy to understand.

The experimental arrangement is shown schematically in Figure 2 [19]. UV pulses from an excimer laser (248 nm, 20 ns FWHM) were used to ablate the polymer film. A second UV excimer laser which was connected to the first one by an electronic trigger, pumped a dye laser which produced visible pulses of < 1 ns duration. A single one of these pulses was timed to illuminate the polymer surface after a preset time delay from the ablation pulse. A standard photographic camera recorded the ablation. When the UV source for producing the ablation pulse was replaced by a TEA CO_2 laser, the ablation could be caused by a single infrared laser pulse with a wavelength of 9.17 μm. This pulse had a width of ~170 ns (FWHM) in the TEM_{00} mode.

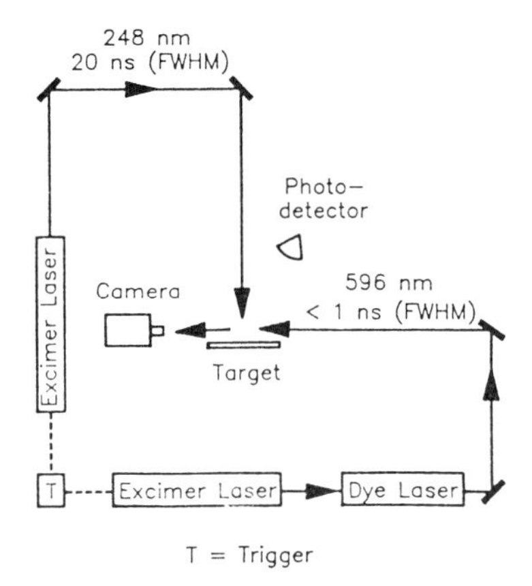

Figure 2. Schematic representation of the arrangement to photograph the blast wave and the ablation products rising from a polymer surface during UV laser ablation in an air medium.

The imaging of a polymer surface which is undergoing ablation by a single UV laser pulse gives information about the chronology of the blast wave, the spacing of the contact front behind it, the presence of heavy organic vapors in the plume and the ejection of opaque material from the surface. Typical examples of the photographs of the

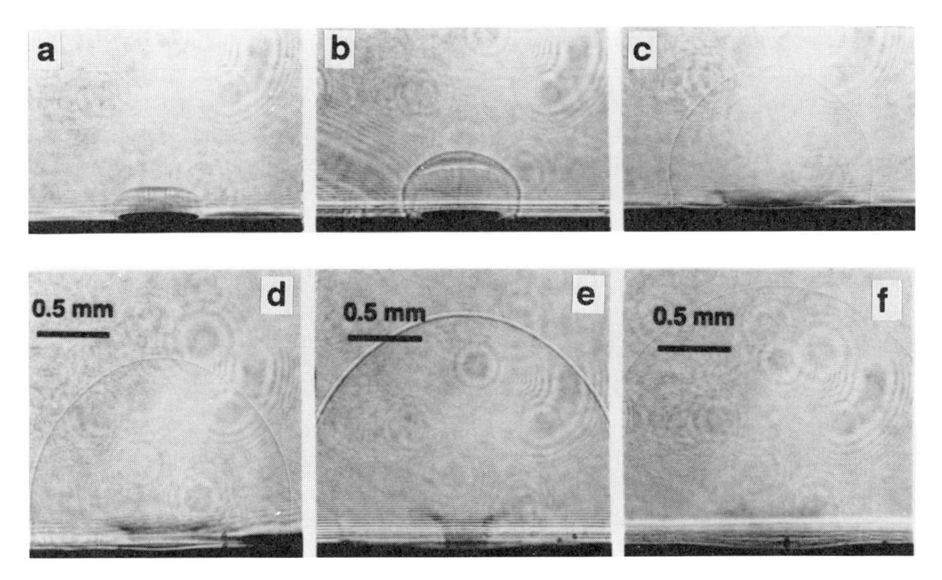

Figure 3. Photographs of blast wave rising above Kapton surface during ablation by a single pulse of 248 nm laser radiation. Laser fluence: 4.9 J/cm^2; a) 50 ns; b) 100 ns; c) 300 ns; d) 500 ns; e) 750 ns; f) 1000 ns

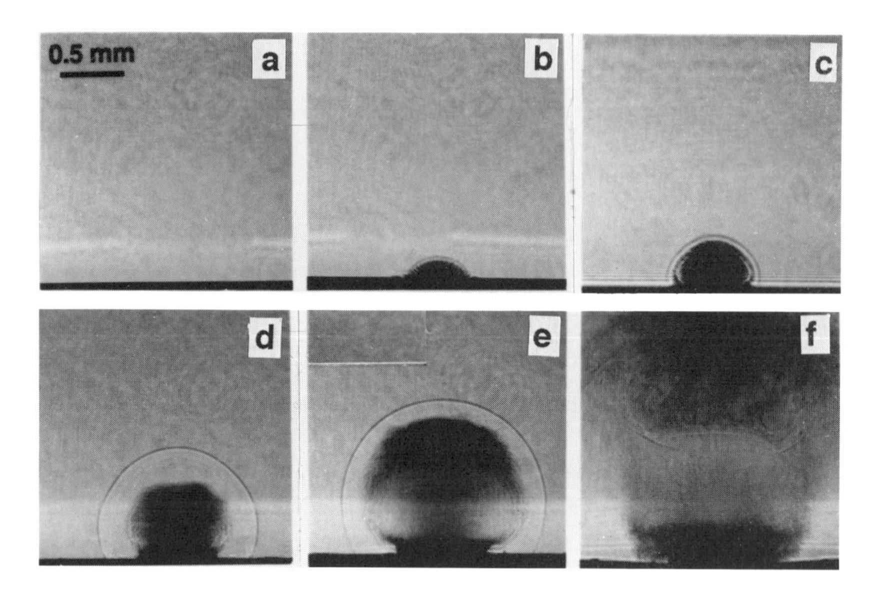

Figure 4. Photographs of the blast wave rising above a Kapton surface during ablation by pulsed, CO_2 laser radiation at 9.17 μm. Laser fluence: 4.7 J/cm^2; a) 0 ns; b) 150 ns; c) 300 ns; d) 500 ns; e) 800 ns; f) 2600 ns.

blast wave and the ablation plume from the etching of Kapton by single pulses of 248 nm laser radiation are shown in Fig. 3. Behind the blast front which is seen very distinctly in this instance, the contact front is barely visible. There is only a faint trace of opaque material leaving the surface. In similar photographs which have been publuished elsewhere [18], it is shown that more opaque material is visible during ablation by 308 nm laser pulses. Also, excited C_2 is one of the ablation products and its presence can be seen in the emission of the Swan bands [10].

The photographs in Fig. 3 should be compared to those in Fig. 4 which were obtained when the Kapton surface was ablated by a single pulse of infrared laser radiation. In this case, opaque solid material is seen to be an important product. The exposed surface is also blackened by the pulse in contrast to the action of the UV laser pulse.

A detailed discussion of the chemical evidence for the reaction pathways under these two ablation conditions is not possible here. The reader is should consult Ref.18 where the known facts are discussed. It is sufficient to point out that the ablation of a Kapton surface by an infrared laser pulse bears a striking resemblance to the pyrolysis of this material in the absence of a laser. The surface is blackened and a product which is a polymeric material of a different composition from the starting material is left behind. In contrast, the UV laser ablation process gives small (C_1, C_2 or C_3) fragments and carbon clusters ($>C_{18}$). It gives CN and HCN which are minor products in the pyrolysis [20], and there are no detectable amounts of products in which a benzene ring is present.

On the basis of the product composition, a thermochemical analysis of the decomposition has led to the deduction [18] that the UV laser break-up of Kapton cannot be achieved with less than six photons per monomer group. This has to be taken into account in any proposal for the mechanism of the laser ablation process. The chemical complexity of the product mixture suggests that it will not be easy to derive a model for the UV laser ablation of this system.

3. POLYMETHYL METHACRYLATE (PMMA)

PMMA has attracted a significant share of the published work on the ultraviolet laser ablation of polymers. It is viewed as a typical polymer with a moderate (at 193 nm) to weak (at 248 nm) ultraviolet absorption. Its etching by UV laser pulses shows the phenomenon of "incubation" [21-24]. It is used as a model and a control material for ablation in the practice of photorefractive keratectomy.

There has been considerable speculation concerning the mechanism by which organic polymers in general, and PMMA in particular, are broken up by UV laser pulses. It was first proposed [25] that the reaction is mostly photochemical when 193 nm laser pulses were used. The argument was based on the analogy to the efficient (quantum yield ~1) photodecomposition in the condensed phase of small organic molecules by low-intensity, cw, far-UV radiation [26]. Subsequently, the potential contribution of two-photon excited states to the decomposition process, especially at 248 nm, was pointed out [27]. The proposal that the UV laser decomposition of organic polymers proceeded by a photothermal mechanism originated [2] in the observation that ablation by ns UV laser pulses of all polymers has a threshold fluence which has to be exceeded. At all fluences (according to this explanation) the photon energy goes to heat the sample and the threshold corresponds to the energy that is required to raise the temperature to the point at which the rate of decomposition is fast enough to cause ablation.

As was already pointed out Section 1 in the case of Kapton, a way to sort out these mechanisms is to look for significant differences in the composition of the products that are formed in the laser ablation process when photons of widely different wavelengths are used. An additional condition would be that only those wavelengths at which the polymer has a direct absorption should be used. Such a study has been carried out recently [28] in which the plumes from the ablation and etching of a PMMA sample by UV (193 nm and 248 nm; ns pulses) and infrared (9.17 μm) laser radiation were photographed by the high-speed method that was described earlier and compared.

Before the results of that study are presented, it is useful to remember that the decomposition of PMMA by a cw, infrared laser has been carried out [29] and the sole product that was detected was the monomer, methyl methacrylate (MMA). There was no char or solid material left behind. The decomposition of PMMA by UV laser pulses under ablative conditions [27] was reported to give not only light gases (CO, CO_2) but also a solid polymeric material. Starting from a sample of PMMA of $M_n > 10^6$, the polymeric product had $M_n = 2500$ when the laser wavelength was 248 nm and an even lower mol. wt. when the wavelength was 193 nm. The monomer, MMA amounted to 18 % of the decomposed material at 193nm but was only ~1 % at 248 nm.

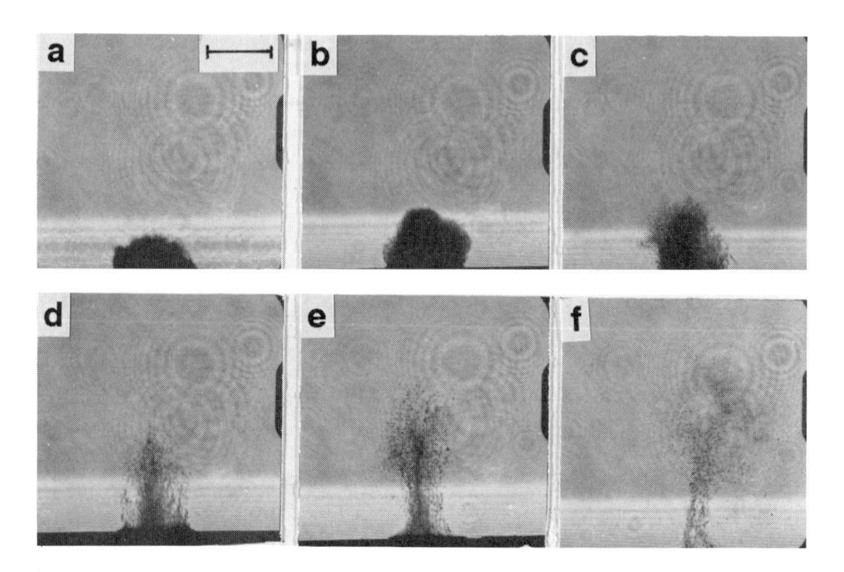

Figure 5. Photographs of the blast wave and the ejected material from the surface of a PMMA film on the impact of a single UV (248 nm) laser pulse. Scale: bar=0.5 mm. The time intervals following the start of the laser pulse are: a) 1.0 μs; b) 3.0 μs; c) 6.1 μs; d) 9.7 μs; e) 15.0 μs; f) 20.7 μs.

The series of photographs shown in Fig. 5 were taken at various time-delays during a single laser laser pulse. They show that while the blast front leaves the surface at supersonic velocity, the plume of opaque material follows at only half the velocity of sound. The contact front is barely visible and the region between the blast front and the opaque material is clear. Two features of the stream of opaque material are of interest. The opacity of the material clears up gradually which can be identified as being due to the replacement of solid particles in the stream by droplets of liquid. The stream is also seen to narrow in its width with time. The enlarged views of Fig. 5b and 5e which are shown in Figure 6 illustrates these features more clearly.

PMMA of mol. wt. 10^6 which was the target material does not really have a melting point because the molecular entanglements are so extensive as to make it incapable of flow. As the molecular weight of the material drops to 2500 during ablation by 248 nm laser pulses, its melting point must be $< 50°C$. It is remarkable that the plume material that first comes off the surface is solid because it shows how meager the heating effects are in the ablation. The subsequent melting of this solid stream and its narrowing can be attributed to the evolution of sub-surface gases which melt the residue that is left on the surface as well as the solid in the plume. This point is discussed in greater detail in Ref. 28.

There is general agreement that infrared laser pulses of MW/ cm^2 of power can give multiphoton excitation along the vibrational manifold of an absorbing group. When 9.17 μm laser pulses were used to ablate PMMA [28], information was obtained for the

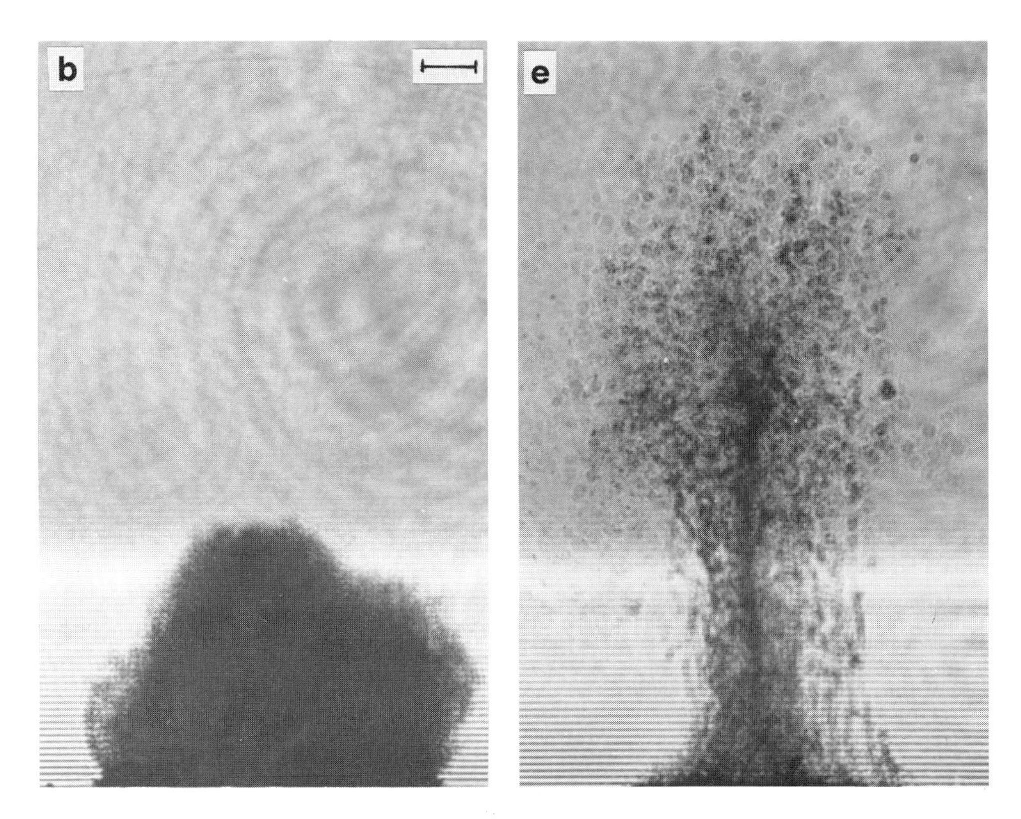

Figure 6. Enlarged views of Figs. 5(b) and (e). Scale: bar = 100μm; The narrowing of the stream of particulate matter and the abundance of droplets is seen here.

first time on the products of the ablation by purely thermal radiation of a pulsed nature. Figure 7 shows a series of photographs of the blast wave and the plume from the ablation of PMMA by a single pulse of infrared laser radiation. It is seen that 350 ns after the start of the laser pulse of ~170 ns (FWHM), the blast front and the contact front which is separated from it by a distance of < 100 μm are driven by a refractive vapor which is seen to fill the hemispherical volume. This was identified as the monomer, MMA, which is the only significant product when the decomposition of PMMA is brought about by a cw, CO_2 laser. The only other product that is seen in the plume is a faint, dark material which spreads slowly from 1 μs onward and is dissipitated by ~10 μs. This dark material was identified as a minor decomposition path [28].

This brief review was mainly intended to emphasize the complex decomposition pathways which the ablation of organic polymers by UV laser pulses may follow. It is essential to accumulate a variety of data on a single system before any speculation about the decomposition mechanism can be undertaken. Somewhat surprisingly, this is the state of affairs in the understanding of this phenomenon even after a decade of study by many

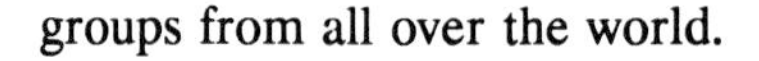
groups from all over the world.

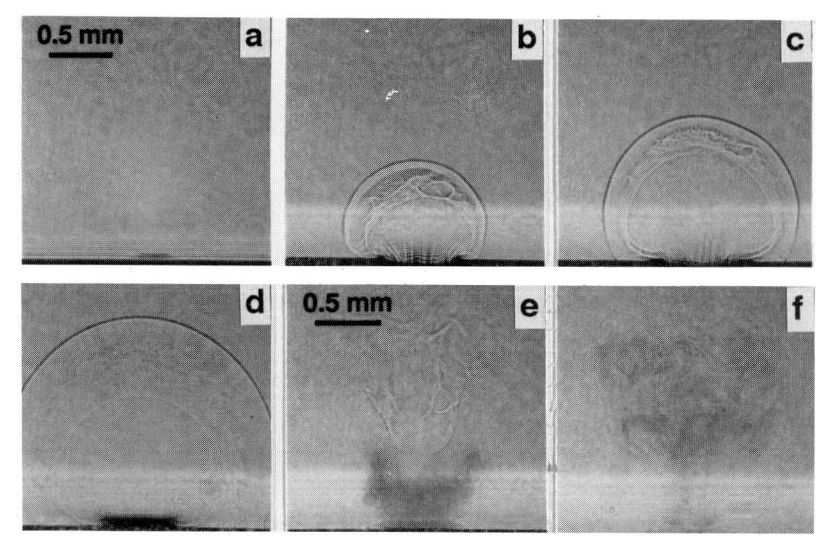

Figure 7. Photographs of the blast wave and the ejected material from the surface of a PMMA film on the impact of a single infrared (9.17 μm) laser pulse. The time intervals following the start of the laser pulse are: a) 60 ns; b) 350 ns; c) 700 ns; d) 1100 ns; e) 3 μs; f) 10 μs.

REFERENCES

1. R. Srinivasan and B. Braren, Chem. Rev., 89 (1989) 1303.
2. P.E. Dyer in Photochemical Processing of Electronic Materials, I.A. Boyd (ed.), Academic, London, 1992, pp. 359 - 385.
3. P.E. Dyer and R. Srinivasan, Appl. Phys. Lett., 48 (1986) 445.
4. P. Simon, Appl. Phys., B 48 (1989) 253.
5. P.E. Dyer and J. Sidhu, J. Appl. Phys., 64 (1988) 4657.
6. P.L.G. Ventzek, R.M. Gilgenbach, J.A. Sell and D.M. Heffelfinger, J. Appl. Phys., 68 (1990) 965.
7. G. Paraskevopoulos, D.L. Singleton, R.S. Irwin and R.S. Taylor, J. Appl. Phys., 70 (1991) 1938.
8. M. Dienstbier, R. Benes, P. Refir and P. Sladky, Appl. Phys., B 51 (1990) 137.
9. M. Golombok, M. Gower, S.J. Kirby and P.T. Rumsby, J. Appl. Phys., 61 (1987) 1222.
10. G. Koren and J.T.C. Yeh, J. Appl. Phys., 56 (1984) 2120.
11. H. Ishida, S.T. Wellinghof, E. Baer and J.L. Koenig, Macromolecule, 13 (1980) 826.
12. R. Haight, R.C. White, B.D. Silverman and P.S. Ho, J. Vac. Sci. Tech. A6 (1988) 2188.

13. A.C. Albrecht in Photochemistry and Photophysics above 6 eV, F. Lahmani (ed.), Elsevier, Amsterdam 1985, pp 227 - 241.
14. R. Srinivasan and A. Ghosh, Chem. Phys. Lett., 143, (1988) 546.
15. J.T. Brenna, W.R. Creasy and W. Volksen, Chem. Phys. Lett., 163 (1989) 499.
16. G. Ulmer, B. Hasselberger, H.-G. Busmann and E.E.B. Campbell, Appl. Surf. Sci., 46 (1990) 272.
17. E.E.B. Campbell, G. Ulmer, B. Hasselberger, H.-G. Busmann and I.V. Hertel, J. Chem. Phys., 93 (1990) 6900.
18. R. Srinivasan, Appl. Phys., A 56 (1993) 417.
19. R. Srinivasan, K.G. Casey and B. Braren, Chemtronics, 4 (1989) 153.
20. H-J. Dussel, H. Rosen and D.O. Hummel, Makromol. Chem., 177 (1976) 2343.
21. E. Sutcliffe and R. Srinivasan, J. Appl. Phys., 60 (1986) 3315.
22. S. Kuper and M. Stuke, Appl. Phys., B 44 (1988) 199.
23. S. Kuper and M. Stuke, Appl. Phys., A 49 (1989) 211.
24. R. Srinivasan, B. Braren and K.G. Casey, J. Appl. Phys., 68 (1990) 377.
25. R. Srinivasan, J. Vac. Sci. Tech., B 4 (1983) 923.
26. W.J. Leigh and R. Srinivasan, Acc. Chem. Res., 20 (1987) 107.
27. R. Srinivasan, B. Braren, D.E. Seeger and R.W. Dreyfus, Macromol., 19 (1986) 916.
28. R. Srinivasan, J. Appl. Phys., 73 (1993) 2743.
29. M. Hertzberg and I.A. Zlochower, Combustion and Flame, 84 (1991) 15.

MICROCHEMISTRY
Spectroscopy and Chemistry in Small Domains
Edited by H. Masuhara et al.

Micrometer patterning of organic materials by selective chemical vapor deposition

A. Sekiguchi[‡] and H. Masuhara[*,#]

Microphotoconversion Project,[†] ERATO, Research Development Corporation of Japan, 15 Morimoto-cho, Shimogamo, Sakyo-ku, Kyoto 606, Japan

Micrometer-sized patterns of copper phthalocyanines were fabricated from 1, 2, 4, 5-tetracyanobenzene by selective chemical vapor deposition. The deposition was achieved to produce phthalocyanine thin films only on copper micropatterns prepared on silicon wafers and sapphire plates. By thermal annealing in vacuum, fabricated films were converted to polymer of copper phthalocyanine. Chemical vapor deposition of 1, 2-dicyanobenzene on copper micropatterns under controlled conditions resulted in directional growth of copper phthalocyanine whiskers and debrises. By examining effects of deposition rate and geometrical structure of micropatterns, the deposition mechanism and an important role of micrometer reaction volume in chemical vapor deposition are discussed.

1. INTRODUCTION

Microfabrication of materials surface is an important and indispensable technique for preparing functionalized surfaces and films. Simple evaporation, molecular beam epitaxy, chemical vapor deposition (CVD), and so on have been applied to prepare various kinds of thin film. Among them CVD is the most advantageous and has been used by several research groups. Indeed, CVD has been applied to various inorganic compounds for producing and patterning their films as useful electronic devices. Although organic thin films have received much attention, their micropatterning by CVD is still limited. One representative method is to prepare thin films followed by photoetching, which is a simple extension of conventional etching technique. Physical and chemical properties of the bulk film are remained after etching, and orientation and association of organic molecules characteristic of micropatterns are not easily provided. As an alternative method, we have developed new methods of

* To whom correspondence should be addressed.
‡ Present Address:Anelva Corporation Research and Development Div., 5-8-1, Yotsuya, Fuchu, Tokyo 183, Japan.
# Present Address:Department of Applied Physics, Osaka University, Suita, Osaka 565, Japan.
† Five-year term project: October 1988 ~ September 1993.

area- and direction-selective CVD for micrometer patterning of phthalocyanines.

Our strategy for preparing micropatterns is very simple; copper micropatterns were first prepared on appropriate substrates, and CVD of organic molecules was conducted [1-3]. Namely, the process order is opposite to the conventional approach which consists of preparation of homogeneous films followed by patterning. The study has led to demonstrations of interesting phenomena and mechanistic understandings.

Phathalocyanines are one of representative organic materials which are known as electronic elements, sensing materials, photoactive compounds, and catalysts, hence they were adopted as target materials. Phthalocyanine derivatives and their polymers can be synthesized by reaction of 1, 2, 4, 5-tetracyanobenzene (TCNB) with metal salts or metal powders [4,5]. The reaction of evaporated TCNB with a copper plate in vacuum resulted in homogeneous films of copper phthalocyanine on the copper plate. Using an X-ray photoelectron spectrometer and a scanning electron microscope (SEM), Ueda and co-workers characterized the films [6]. Similarly, Yudasaka et al. reported preparation of copper phthalocyanine polymer films by double source evaporation of copper and TCNB [7]. Furthermore, in-situ analysis of CVD process of the relevant materials was conducted by Raman scattering spectroscopy [8]. Thus, phthalocyanines are considered to be the most fruitful target for our demonstrations of micrometer patterning by selective CVD methods, which are summarized here.

2. AREA-SELECTIVE CHEMICAL VAPOR DEPOSITION

Our procedures are schematically given in Figure 1 [1]. Copper film with thickness of 600 nm was deposited on silicon wafers by using a conventional magnetron sputter, and its micropatterns were fabricated by photolithography and subsequent wet etching. The patterned substrate and crystalline TCNB were set apart from each other in a glass tube and flushed with argon gas. The tube was sealed at 3 x 10^{-4} Pa and treated at different temperature. An average deposition rate was set to 10 nm/h. The treatment temperature is a key factor for area-selective CVD, which is clearly demonstrated in Figure 2. SEM pictures show the deposition in the borderline between silicon substrate and patterned copper. The films were prepared on both surfaces when treated above 220 °C, and at 260 °C the borderline between two substrates was not clearly identified. Since TCNB is sublimated effectively at 180 °C, the deposition was not observed on both areas below the temperature. At 180 °C deposition was started only on the copper line, and beautiful crystalline needles were identified on the copper at 200 °C. Two different morphologies were observed on the copper substrate; fine crystalline needles and small spherical particles. The former and the latter were dominant at 200 and 180 °C, respectively. This was the first demonstration of area-selective CVD of organic materials, which was achieved by controlling the treatment temperature.

The films deposited at 200 °C was then annealed at 500 °C for 3h in vacuum. The temperature was slowly raised from room temperature to 500 °C. After the annealing a few hillocks and voids were observed in the deposited film and its color was turned from green to dark green. It is worth noting that selectively prepared

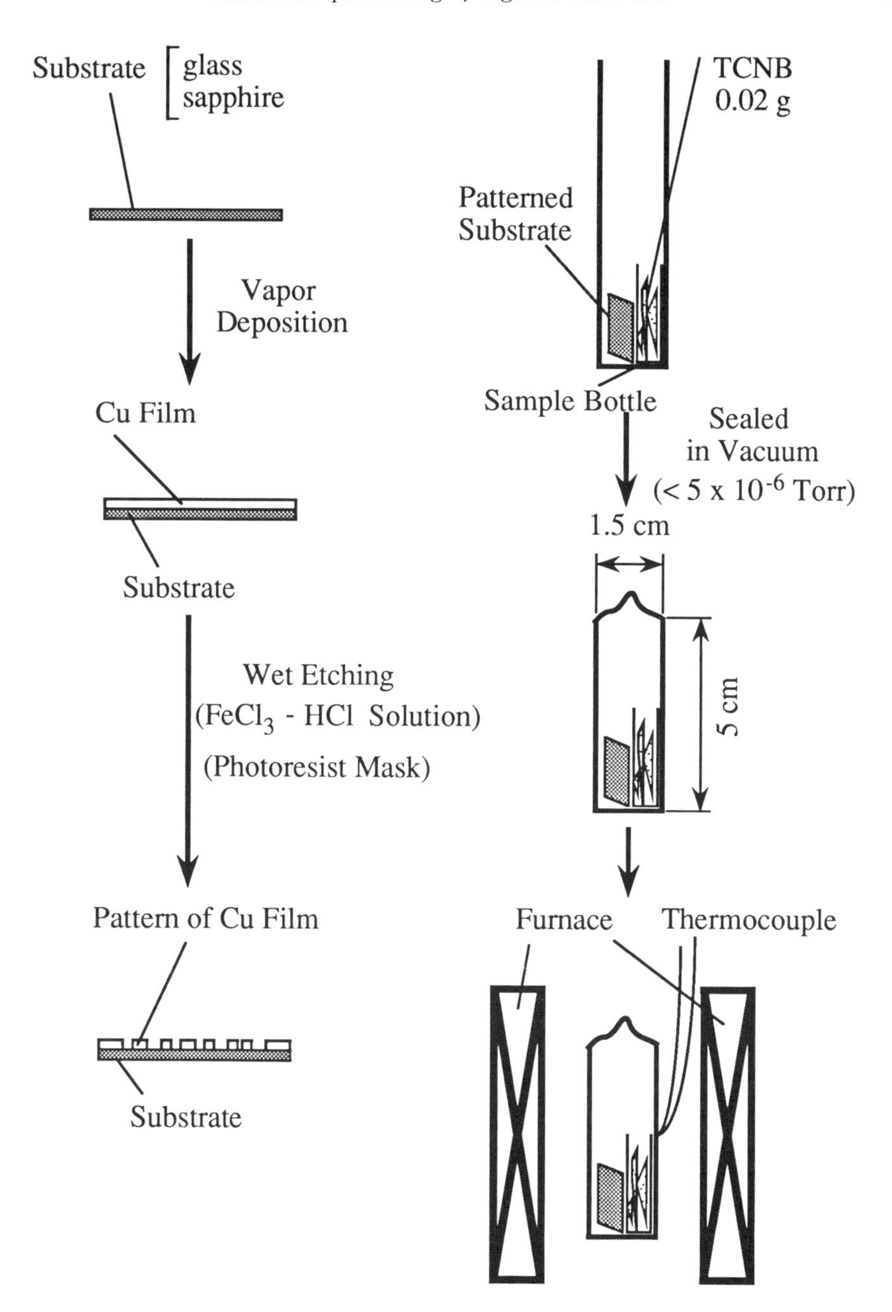

Figure 1. Schematic representation of area- and direction-selective chemical vapor deposition.

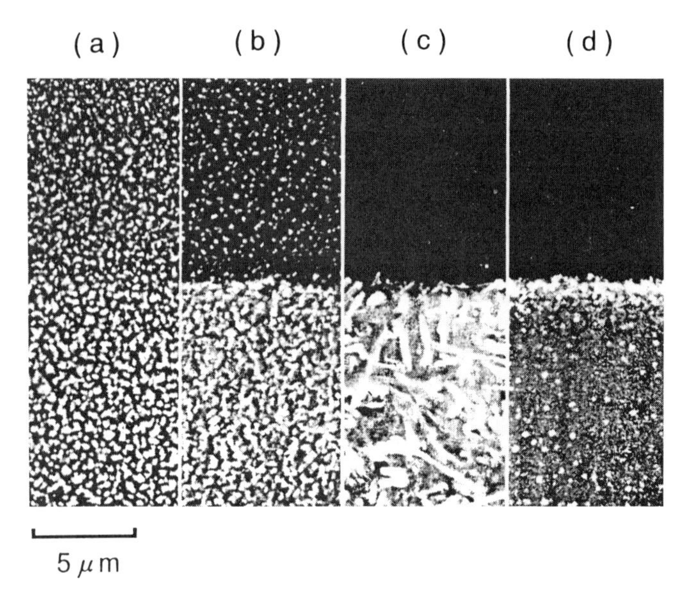

Figure 2. SEM pictures of the films, prepared by depositing TCNB on copper (lower) and silicon (upper) under the conditions; (a) 260 °C for 10h, (b) 220 °C for 48h,(c) 200 °C for 48h, and (d) 180 °C for 48h.

Figure 3. SEM pictures of the films, prepared by depositing TCNB on copper (lower) and silicon (upper) at 200 °C. (a) Before and (b) after annealing at 500 °C for 3h.

films on the copper were not transferred to the silicon area and the borderline was still clear.

To characterize chemical composition of the films, they were dissolved in concentrated sulfuric acid and their UV-visible absorption spectra were measured. The spectrum of the materials prepared at 200 °C was a little different from that of metal-free octacyanophthalocyanine, but showed the absorption maximum corresponding to Q and Soret bands. The spectral similarity suggests that phthalocyanine derivatives were produced by the ring formation of four TCNB molecules, and copper atoms from the plate were included into the phthalocyanine rings. Indeed, absorption spectrum of metal-free octacyanophthalocyanine was observed for the deposited film on SiO_2 (as the reference substrate with no copper) at 300 °C.

For annealed film on the copper of Figure 3, the absorption spectrum similar to Cu-octacyanophthalocyanine was observed. As a characteristic feature of the film, however, the absorption intensity of the Q band decreased upon annealing while that of the Soret band was almost unchanged. This result is in good accordance with those by Berezin and Shormanova [9], and by Yudasaka et al. [7] who reported that polymerization of Cu-octacyanophthalocyanine decreases the absorption intensity of the Q band. In the present spectrum of the annealed film, furthermore, the peaks shifted to the shorter wavelength by 30 nm relative to that of the deposited film prepared at 300 °C. Ashida et al. already reported that the similar spectral shift was observed for the films produced at 400 °C and interpreted that the shift is due to polymerization [10]. On the basis of the results, reaction schemes were summarized as in Figure 4. Therefore, we concluded that area-selective CVD can successfully prepare the micrometer-sized patterns of phthalocyanine derivative and its polymer.

3. MICROMETER PATTERNING OF PHTHALOCYANINE DERIVATIVES

The use of complex patterns of copper on substrate materials opens a new way to fabricate functional devices of phthalocyanine derivatives with micrometer resolution. Using the standard photomask we have prepared positive and negative copper patterns on silicon wafers, and the selective CVD was conducted on them. The SEM pictures in Figure 5 again demonstrate that deposition materials were prepared only on the copper areas. Positive and negative micrometer-sized patterns of octacyanophthalocyanine were prepared by the treatment of TCNB at 200 °C [1]. The films were observed as gathering of crystalline fine needles, and the surface of the film was rough. As far as we know this is the first example of micrometer-sized pattern of organic materials by area-selective CVD.

For in-situ spectroscopic and conductivity characterizations, fabrication of phthalocyanine should be conducted on glass and sapphire plates which are transparent and insulating substrates. The deposition procedures and conditions were the same to those for silicon wafers. Figure 6 shows that the optical microscope pictures of copper microelectrode and deposited films [2]. A gap size of the two parallel electrodes, with 10 μm width and 255 μm length, was about 5 μm. When TCNB and glass substrate with microelectrode patterns were treated at 195 - 200 °C for 120 h, the deposition was perfect on the electrodes, however, some small particles

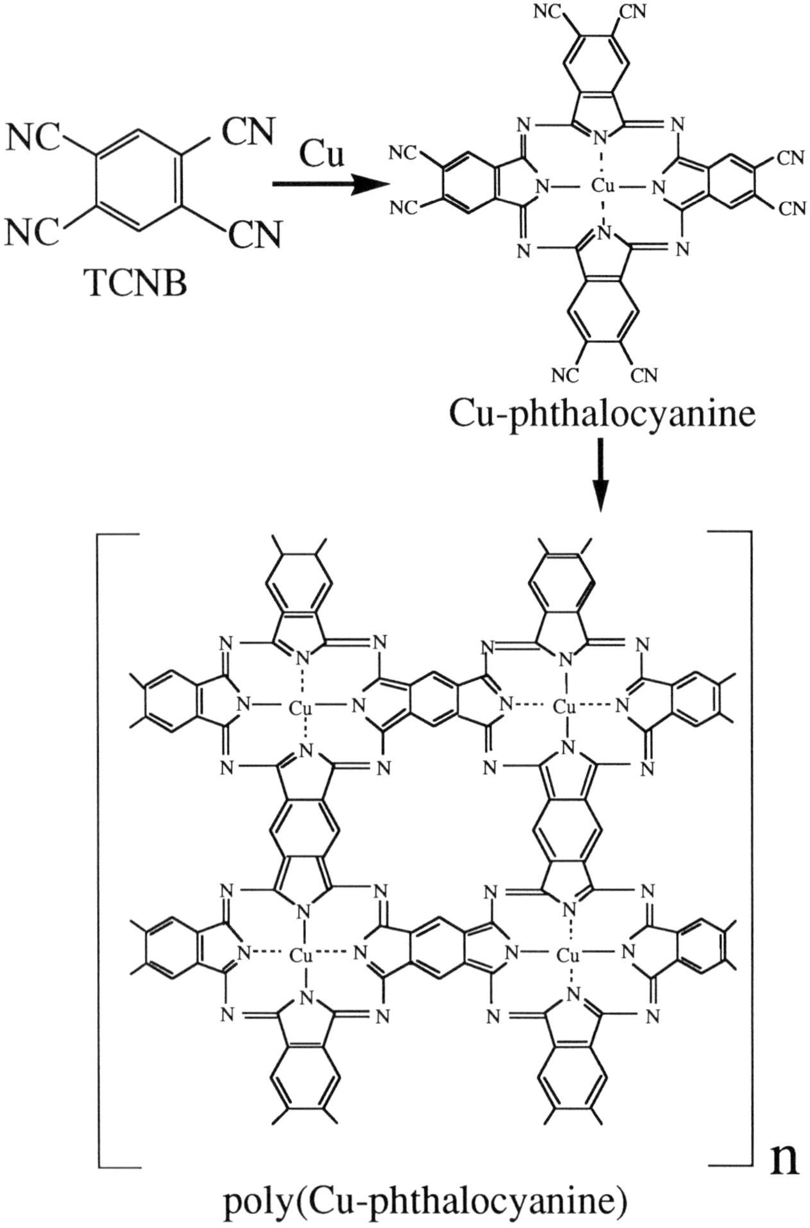

TCNB

Cu

Cu-phthalocyanine

poly(Cu-phthalocyanine)

Figure 4. Reaction scheme of chemical vapor deposition of phthalocyanines.

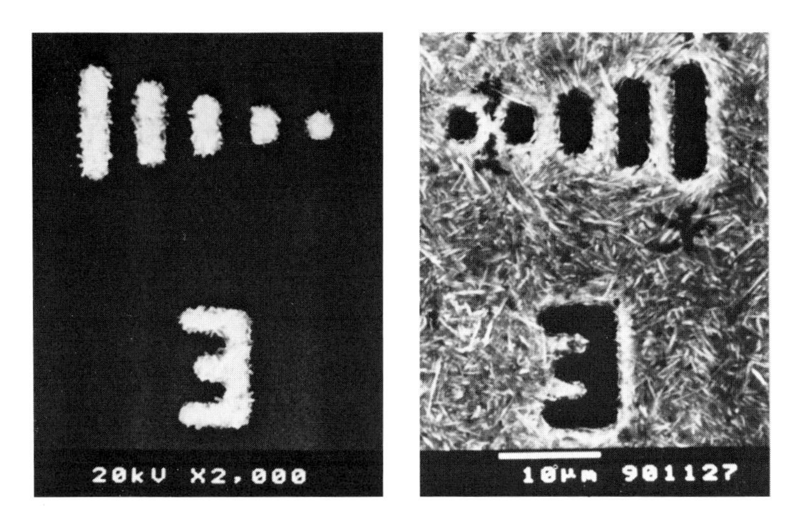

Figure 5. Positive (left) and negative (right) micrometer-sized patterns of the deposited films at 200 °C.

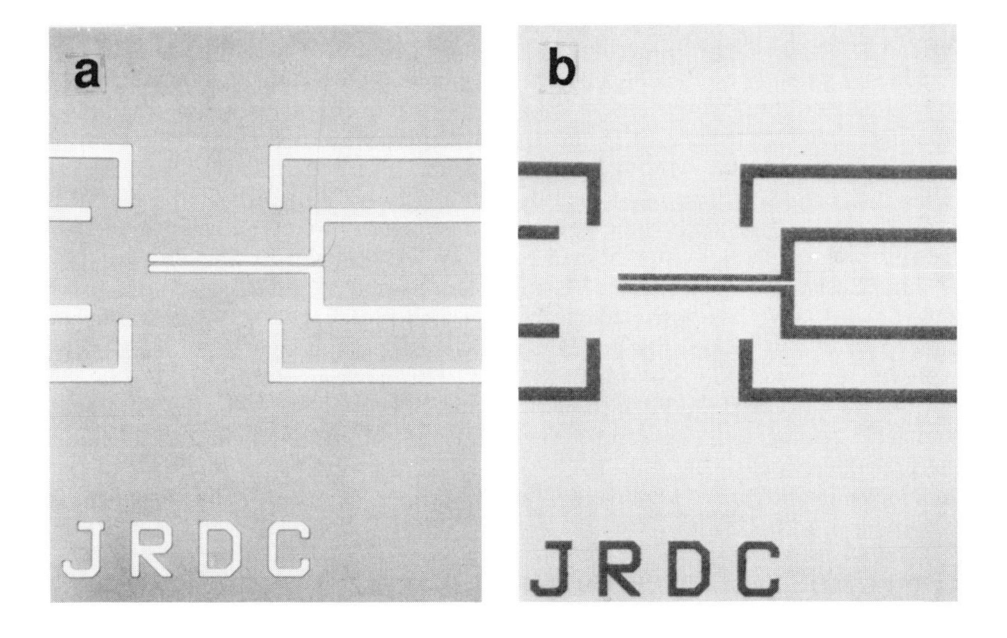

Figure 6. Optical microscope pictures of micrometer-sized electrodes and deposited films; (a) a copper microelectrode fabricated on the sapphire substrate, (b) a pattern prepared by depositing the materials on the microelectrode pattern on the sapphire at 195 - 200 °C for 120h.

were observed on the glass substrate. One possible origin of the particle formation is unexpected residues of the copper dot which could not be removed during a wet etching process of copper. We adjusted the condition to result in a litter over-etching and cleared up the metal completely on the glass surface. As a result of this over-etching, the copper microelectrodes became a little learn. Actually, the width of the electrode was narrower than that of the original electrode. Even under the condition small particles were observed. Thus, we consider that the deposited particles on the glass were related to the nature of the glass substrate, namely, metal ions in the glass may activate the ring formation reaction. It means that the area-selective CVD on the glass was not perfect, but it may be useful for some practical applications.

The selective deposition of phthalocyanine derivatives was perfect on the sapphire substrate by the same CVD procedures, and nothing was left on the sapphire. Then, the leakage current between two electrodes was measured and it was confirmed to be less than 1 pA at the applied voltage of 10 V. This value was in the same order as that of the background leakage current level, which guarantees no contact between two electrodes. It was concluded that area-selective CVD was clearly achieved on the sapphire substrate and chemically modified copper microelectrodes were fabricated. The materials were copper phthalocyanine derivatives when treated at 195 - 200 °C for 120 h. By annealing the films at 500 °C for 3 h, the morphology and color of the materials were changed, which was ascribed to a formation of polymerized copper phthalocyanine. Deposited films on copper patterns were easily removed from the copper surface, while annealed ones were very difficult to be scratched furthermore. Annealed patterns of phthalocyanine polymer are adhered strongly on the copper lines and can be used for practical applications.

Conductivity measurement is one of effective characterization methods of deposited phthalocyanines. In Figure 7, an arrangement of resistance measurement is shown. To connect the wire to the copper pattern, some parts of the deposited film on the copper patterns were stripped off mechanically. Copper wire was contacted firmly to the stripped area, while gold wire was attached to the phthalocyanine films. The film was prepared by the thermal treatment for 48 h and its thickness was about 500 nm. Resistance was calculated from the slope of I ~ V curve measured in the bias range of +1 ~ -1 V. The resistance of the deposited film was 360 ± 120 kΩ which was reduced to 2.7 ± 1.1 kΩ by annealing. It is surprising that the conductivity is improved by two orders of magnitude by thermal annealing, which should be ascribed to polymerization of phthalocyanine derivatives. The polymerization extended a conjugation degree of π-electrons, resulting in low resistance. Thus, the interpretation is consistent with UV/visible spectral assignment. Microelectrodes of phthalocyanine polymers were well prepared by area-selective CVD and confirmed to be enough useful.

4. DIRECTION-SELECTIVE CHEMICAL VAPOR DEPOSITION

CVD of phthalocyanine is also possible for 1, 2-dicyanobenzene (DCNB), although its polymer can not be prepared in principle. DCNB and silicon wafers patterned with copper were set in an evacuated glass tube and treated at controlled temperature [3], which is similar to the method described above. SEM pictures of the

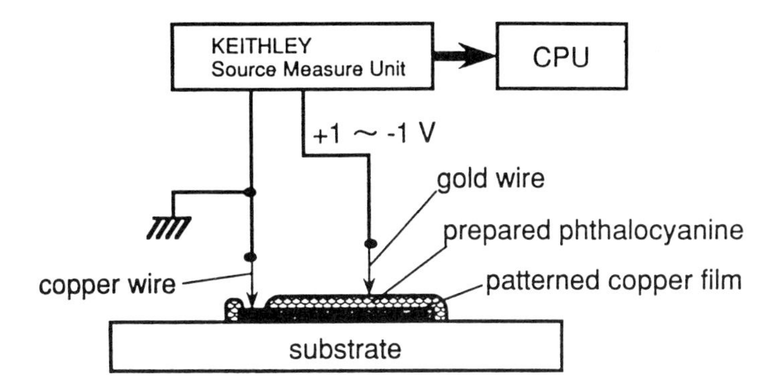

Figure 7. An arrangement of electric resistance measurement.

Figure 8. SEM pictures of the deposited materials on copper micropatterns (upper) and copper plate (lower).

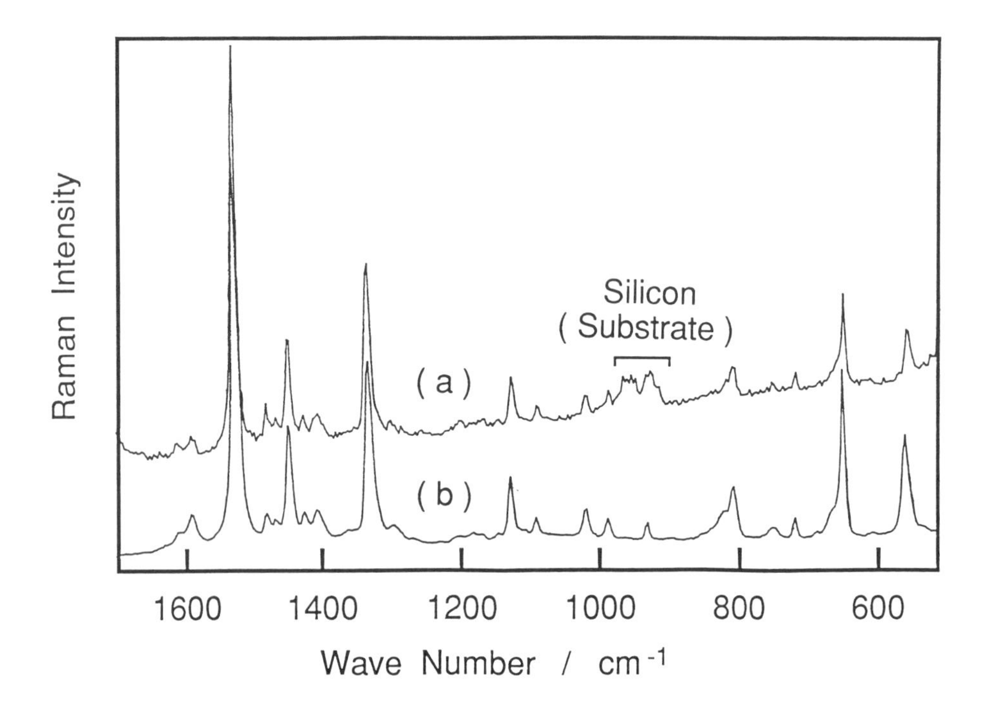

Figure 9. Raman spectra of (a) the deposited materials and (b) the reference Cu-phthalocyanine.

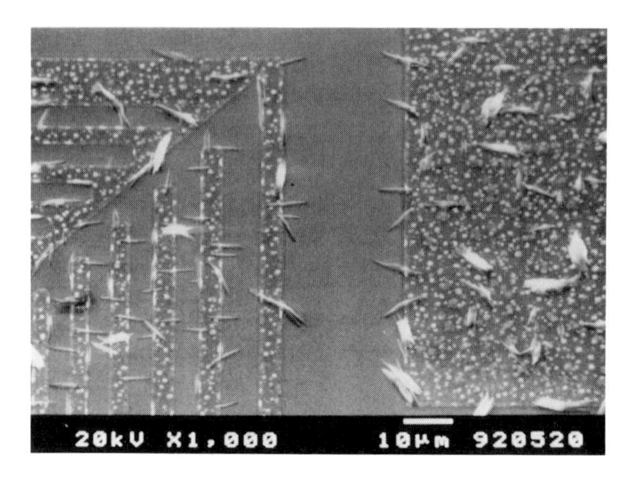

Figure 10. SEM pictures of the deposited materials on copper micropatterns with different sizes.

deposited materials on micrometer-sized patterns and homogeneous plate of copper are shown in Figure 8, where the width and the gap of the patterned copper lines are 2 and 2.5 μm, respectively. On both surfaces, μm-order deposition was observed under the preparation condition, but shape and distribution of deposited materials depended on the substrate. On the flat copper surface materials were distributed randomly, and their size was classified into three ; sub μm granules, a few μm whiskers, and large debrises. On the copper patterns the number of granules was reduced, and whiskers and debrises of similar size were formed. Furthermore, it is worth noting that whiskers and debrises were preferentially oriented along or perpendicular to the copper lines. It was confirmed that the directional growth had nothing to do with crystal axis of the substrate silicon, since the present alignment of whiskers and debrises was achieved for copper patterns on other silicon wafers with different crystal axes. Chemical structure of produced molecules was spectroscopically confirmed by Raman microscope measurement. As shown in Figure 9, Raman spectrum of the deposited crystals on the copper micropattern was similar to that of the reference Cu-phthalocyanine crystal. The whiskers oriented along and perpendicular to the copper lines showed the same spectra, furthermore, comparing X-ray fluorescence analysis data and SEM pictures, copper atoms were identified through the whisker and even at its top position. Thus, the whiskers formed can be safely assigned to Cu-phthalocyanine independent on the orientational direction.

An interesting relation between deposition behavior and size of copper patterns is presented in Figure 10. Copper lines are partially covered with phthalocyanine granules, whiskers, and debrises, and most of whiskers and debrises are parallel or perpendicular to the lines as shown in Figure 10(a). At the corner where Cu patterns crossed with each other, the directional growth was still confirmed. At copper lines being inclined to other lines, phthalocyanine again did not obey the special geometry of the line. In Figure 10(b), the results for widely spaced copper patterns are shown. On the copper area of a few tens μm dimension the deposition was rather random, although the directional growth was observed for the gap of a few μm as described above. Furthermore, on the copper line facing to wide gaps larger than ten μm, appreciably oriented deposition was not observed. It is concluded that the present characteristic deposition behavior is due to periodical structures of the micropatterns and the meaningful size should be a few μm.

5. DEPOSITION MECHANISM AND AN IMPORTANT ROLE OF MICROMETER VOLUME IN CHEMICAL VAPOR DEPOSITION

In the direction-selective CVD process phthalocyanine is produced by a ring formation of four DCNB molecules and copper atoms are supplied from the copper pattern. One can consider that copper surface enhances ring closure reaction, and phthalocyanine is selectively deposited on the copper patterns and their neighbor. To examine this idea we prepared the deposited materials from DCNB under the different conditions. Copper plates patterned with SiO_2 were prepared as a nega-pattern of the substrate used above, sealed in a glass tube with DCNB, and kept at 200 °C for 96 h. Large whiskers with length of about 100 μm were formed on the copper, which were characterized by dissolving them in 1-chloronapthalene and measuring the UV-visible absorption spectra. In the case Cu-phthalocyanine and its free form were both identified. Copper has still some role in phthalocyanine formation, but inclusion of copper atom into the phthalocyanine precursor may not be a necessary condition. Hence, it is considered that phthalocyanine is formed on copper more easily than on silicon, migrates on the surface, and attached firmly on the patterns, then copper atoms diffuse through phthalocyanine crystal, giving Cu-phthalocyanine. When phthalocyanine or/and its precursor are formed with high rate on one site of the copper lines, they may be attached on the Cu-phthalocyanine already produced on other sites. Depending on the deposition rate, copper atoms are enough supplied or not. The deposition of phthalocyanines on Cu-phthalocyanine is more efficient than that on copper, hence large whiskers and debris are formed, leaving uncovered copper area. This migration is considered to be effective for a few μm, since the periodical structure has a special meaning in the present directional growth. The important distance of a few micrometer may be determined by lifetime of involved transient or the effective volume of the diffusion. Thus, the results indicate that a micrometer volume has an important role in CVD process of phtalocyanine on copper substrate.

Area-selective and direction-selective CVD of phthalocyanines on copper micropatterns was achieved on silicon and sapphire substrates. By thermal annealing of the deposited films from TCNB, the deposited materials changed to its polymer. This was confirmed by spectroscopic and conductive measurements. Since copper phthalocyanine and its polymers are expected to be photoactive materials, sensing materials, and catalysts, the present method will largely contribute to develop new phthalocyanine-based microdevices. Further applications of the selective CVD method to other organic compounds and fabrication of their functional devices will be very fruitful.

REFERENCES

1. A. Sekiguchi, K. Pasztor, N. Shimo, and H. Masuhara, Appl. Phys. Lett., 59 (1991) 2466.
2. A. Sekiguchi, K. Pasztor, N. Shimo, and H. Masuhara, J. Vac. Sci. Tech. A., 10 (1992) 1508.
3. A. Sekiguchi, T. Uchida, H. Sugimura, N. Shimo, and H. Masuhara, Appl. Phys. Lett., (1994) submitted.
4. R. Bannehr, G. Meyer, and D. Wöhrle, Polym. Bull., 2 (1980) 841.
5. D. Wöhrle and U. Hündorf, Makromol. Chem., 186 (1985) 2177.
6. Y. Ueda, H. Yanagi, S. Hayashi, and M. Ashida, J. Electron Microsc., 38 (1989) 101.
7. M. Yudasaka, K. Nakanishi, T. Hara, M. Tanaka, and S. Kurita, Jpn. J. Appl. Phys., 24 (1985) L887.
8. K. Ishii, S. Mitsumura, Y. Hibino, R. Hagiwara, and H. Nakayama, Appl. Surf. Sci., 33/34 (1988) 1324.
9. B. D. Berezin and L. P. Shormanava, Vysokomol. Soyed. A., 10 (1968) 384.
10. M. Ashida, Y. Ueda, H. Yanagi, and K. Sayo, J. Polym. Sci. A. Polym. Chem. Ed., 27 (1989) 3883.

MICROCHEMISTRY
Spectroscopy and Chemistry in Small Domains
Edited by H. Masuhara et al.

Scanning tunneling microscope tip-induced anodization for nanofabrication of metals and semiconductors

Hiroyuki Sugimura* and Noboru Kitamura**

Microphotoconversion Project, ERATO, Research Development Corporation of Japan, 15 Morimoto-cho, Shimogamo, Sakyo-ku, Kyoto, 606, Japan

The fabrication of nanoscale oxide patterns on titanium (Ti) and silicon (Si) were accomplished by means of scanning tunneling microscope (STM) tip-induced anodization under air or nitrogen gas atmosphere. The oxide patterns were fabricated arbitrary along tip traces as a result of electrochemical reactions of the material with adsorbed water beneath the tip and the material surface. The factors governing STM tip-induced anodization were elucidated and, under optimized conditions, nanofabrication of the material surfaces with the resolution of ~ 20 nm was successfully achieved by STM.

1. INTRODUCTION

Scanning probe microscopes (SPM) such as a scanning tunneling microscope (STM), an atomic force microscope (AFM), a near-field scanning optical microscope (NSOM), and a scanning electrochemical microscope (SECM) have become powerful means to investigate material surfaces since invention of STM by Binnig et al [1]. Besides topographical imaging, many attempts have been also reported to use an SPM for nanometer ~ atomic scale surface modifications [2,3]. Such nanofabrication based on SPM has potential applications to various technologies such as high-density data storage, high-resolution lithography, and the production of novel nano-structured materials. There are various strategies towards the nanofabrication; for instance, mechanical scratching [4], van der Waals

---

*present address; Tsukuba Research Laboratory, Nikon Co., 5-9-1 Tohkodai, Tsukuba 300-26, Japan
**present address; Department of Chemistry, Faculty of Science, Hokkaido Univ., Sapporo 060, Japan

force driven atom manipulation [5], field-enhanced diffusion [6] or evaporation [7,8], and highly concentrated electron flow [9,10].

Among them, the nanofabrication through electrochemical reactions induced by an SPM probe tip is a very important field, since chemical modifications, deposition, etching, polymerization, anodization, and so forth are expected to be conducted arbitrary in space if an SPM tip is used as a working or counter electrode for electrochemical reactions. Actually, several demonstrations of the SPM-based electrochemical modification have been so far reported. As an example, Penner and co-workers reported deposition of copper or silver pillars with 10 ~ 30 nm in diameter on a graphite surface to construct a nanoscale galvanic cell [11]. Electrochemical modifications of material surfaces by SECM has been reported for the first time by Bard and co-workers [12,13], demonstrating fabrication of noble metal lines of 200 nm in width or etched holes on semiconductors in μm dimension. Recently we have extended this method to fabrication of organic polymer film surfaces with fluorescent micropatterns as well [14,15].

A very simple but interesting example along the line is electrochemical nanofabrication of material surfaces with adsorbed water on the sample substrate by STM, in which water molecules act as a chemical reagent for the fabrication. Indeed, fabrication of monolayer deep pits of highly oriented pyrolitic graphite (HOPG) with adsorbed water has been demonstrated by Albercht et al. [16] or MacCarley et al. [17], respectively. Since any external reagents are not necessary for electrochemical reactions, this idea could be applied to nanofabrication of various materials other than HOPG. Actually, we have recently demonstrated that modifications of titanium (Ti) and silicon (Si) surfaces with nanometer-scale oxide patterns by STM [18,19]. The STM-fabrication of Ti or Si was induced only when the sample substrate was biased anodically versus the STM tip and, therefore, was essentially based on electrochemical oxidation of the materials as known as *anodization*. In order to extend the STM tip-induced anodization to nanofabrication and related technologies, the electrochemical origin and the factors governing the phenomena should be elucidated in detail. In this review, we describe recent progress in the STM tip-induced anodization of several materials and chemical approaches to nanofabrication.

2. PRINCIPLE OF STM TIP-INDUCED ANODIZATION

The principle of STM tip-induced anodization is schematically illustrated in Figure 1. A sample surface placed in air or another humid atmosphere is covered with a very thin adsorbed water layer, and a water column is formed beneath an STM tip and the sample surface owing to capillarity when the tip is brought to the sample surface [17]. If an appropriate positive bias is applied to the sample with respect to the tip, anodization will be proceeded at the sample surface just under the tip, while reduction reactions will take place at the tip as a counter electrode. In such the case, since a part of the probing current for the tip control during the anodization is consumed as a faradaic current accompanied by these electrochemical reactions, the microscope works as SECM [21].

When Ti is used as a sample material, for example, oxidation of Ti to titanium dioxide (TiO_2) and hydrogen generation take place at the Ti and tip surfaces, respectively. Thus, an anodic oxide film can be produced arbitrarily on the sample along a trace of the tip in nm scale lateral resolution. The spatial resolution of STM tip-induced anodization is primarily determined by a tip shape, and by a thickness of the adsorbed water layer and, hence, is strongly affected by humidity of the atmosphere as described later.

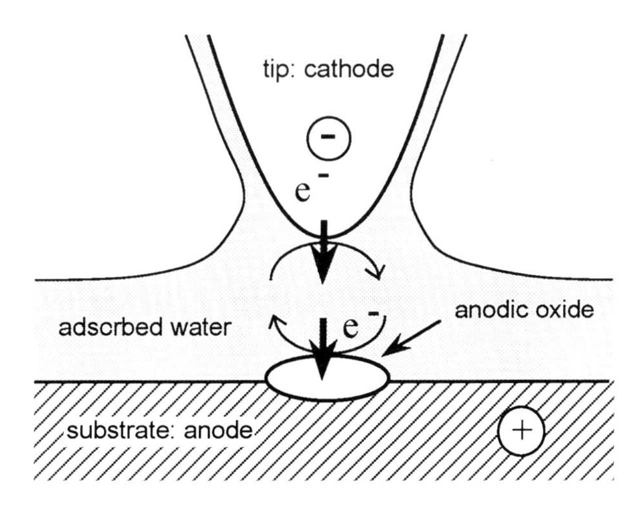

Figure 1. Principle of STM tip-induced anodization.
tip-reaction; $nH_2O + ne^- \rightarrow n/2H_2 + nOH^-$
substrate reaction; $M + nH_2O \rightarrow MO_n + 2nH^+ + ne^-$

3. FABRICATION OF NANOSCALE OXIDE PATTERNS

3.1. STM tip-induced anodization of Ti

Ti films with 10 ~ 20 nm in thickness were evaporated onto mica or HOPG substrates freshly cleaved prior to the evaporation at a deposition rate of 0.1 nm/s in vacuum (< 2 × 10^{-6} Torr). A conductive diamond tip was brought to the sample surface to allow the electron tunneling from the native oxide to the tip by setting a sample bias (V_S) to -3.0 V [22]. Piezo feedback along the Z-axis being kept active, V_S was then changed to a positive bias (+3.0 ~ +8.0 V) to anodize the sample surface. The tip was scanned along an arbitrary pattern with an appropriate scan speed (v) and constant V_S and a reference current (i).

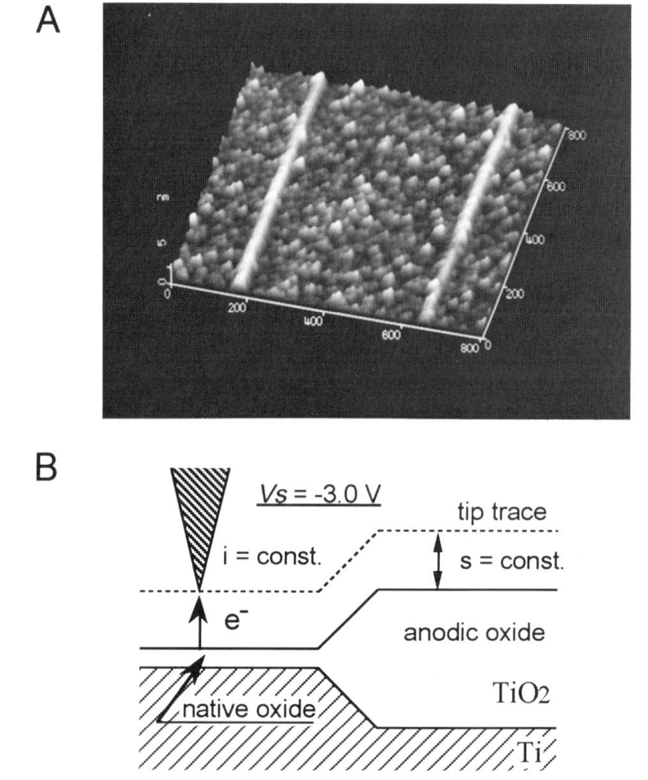

Figure 2. (**A**) A 800 × 800 nm constant current STM image acquired at V_S = -3.0 V and i = 0.1 nA. The anodic oxide lines of 30 nm in width were fabricated at V_S = +8.0 V, i = 0.2 nA, and v = 5 μm/s. (**B**) Schematic representation for STM image observation.

After the anodization, V_S was set again to -3.0 V to observe STM images of the anodized patterns. STM images were acquired in the constant current mode. We used conductive diamond tips to avoid unfavorable damages of both the tips and the samples which were frequently induced by the use of metal tips during the anodization. The mechanical and electrochemical stabilities of the diamond tips were very effective to attain reproducible nanofabrication of the materials and observation of the patterns before and after the anodization.

The molar volume of anodic TiO_2 is calculated to be 26.6 cm^3/mol through the relevant density (ρ = 3.0 g/cm^3) [23] and molecular weight (MW = 79.9), while the corresponding value for Ti is 10.6 cm^3/mol (ρ = 4.5 g/cm^3 and MW = 47.9). These values indicate that the volume increases in ~ 150% upon anodization of Ti to TiO_2. Therefore, the patterns fabricated on Ti by STM tip-induced anodization can be recognized as topographical changes in the STM image, as typically shown in Figure 2**A**. One might have a question that the STM image in Figure 2**A** represents the electronic or chemical information of the modified areas rather than the real topographic information. However, V_S was set at -3.0 V during the STM observation, so that the Fermi level of the tip is lower than the level of the valence band (E_V) edge of TiO_2 in energy [22,24]. The electron tunneling from E_V of TiO_2 to the tip is thus expected to proceed at this bias voltage, and the tip can trace on the anodic TiO_2 surface as well as the native TiO_2 surface (Figure 2**B**). We conclude that the STM image in Figure 2**A** is a real topographic image of the surface. This was also confirmed through Nomarski optical microscopy of the fabricated pattern, by which topographical contrast between the anodized and native sample surfaces were clearly resolved as demonstrated in Figure 3.

Scanning tunneling spectroscopy (STS) and Auger electron spectroscopy (AES) were conducted to analyze the anodized pattern [25]. The tunneling current - sample bias characteristics of the fabricated surface were almost identical with those of a native oxide surface on Ti, indicating that the surface electronic properties of the anodized area are common to that of the native oxide. On the other hand, AES analysis of the samples revealed that the peak height ratios of oxygen to titanium (O/Ti) were 1.40 and 1.38 for the anodized and native oxide surfaces, respectively. Therefore chemical structures of both surfaces are identical with each other, in consistent with the conclusion from the STS experiments. It is noteworthy, however, the depth profiles of O/Ti measured by Ar^+ ion beam etching are quite different between the anodized and native oxide surfaces. The O/Ti ratio for the native oxide decreased rapidly with the etching depth, whereas that for the fabricated surface slowly decreased. The anodic oxide layer on the fabricated pattern is obviously thicker than the native oxide layer. We conclude that the nanofabrication presented here is essentially based on anodization of Ti by the STM tip.

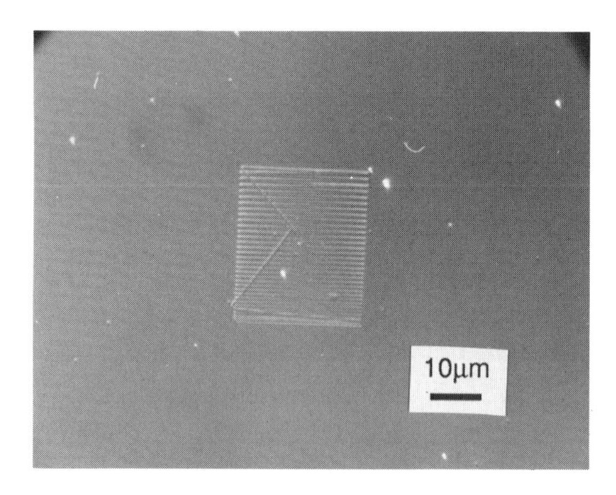

Figure 3. An optical micrograph (Nomarski) of a large scale anodized pattern of 30 × 36 μm. The pattern was fabricated at V_S = 8.0 V, i = 0.2 nA and v = 5 μm/s with repeating a 30 × 36 μm scan for 15 times.

3.2. STM tip-induced anodization of Si

The oxide pattern formation on Si or gallium-arsenide (GaAs) has been reported for the first time by Dagata et al. [26,27] through *field-enhanced oxidation* of the material surface in which oxide is considered to be produced by an intense electric field between a tip and the surface. In field-enhanced oxidation, a sample is biased *cathodically* and *oxygen* in an atmosphere plays a main role for the oxidation, while the oxidation proceeds when a sample is polarized *anodically* in the presence of moisture for tip-induced anodization. In this section, we describe nanofabrication of Si by means of STM tip-induced anodization and, differences in the characteristic features between the tip-induced anodization and the field-enhanced oxidation are also discussed [19].

n-type Si(100) substrates with the resistivity of 0.01 ~ 0.02 Ω•cm were etched in an aqueous HF solution (50 %) for 30 s and rinsed subsequently in pure water to passivate these surfaces by hydrogen prior to experiments. The experimental procedures are similar to those on Ti except for the V_S value for observing STM images (V_S = -2.0 V). The atmosphere near the sample substrates was purged with a dry N_2 stream.

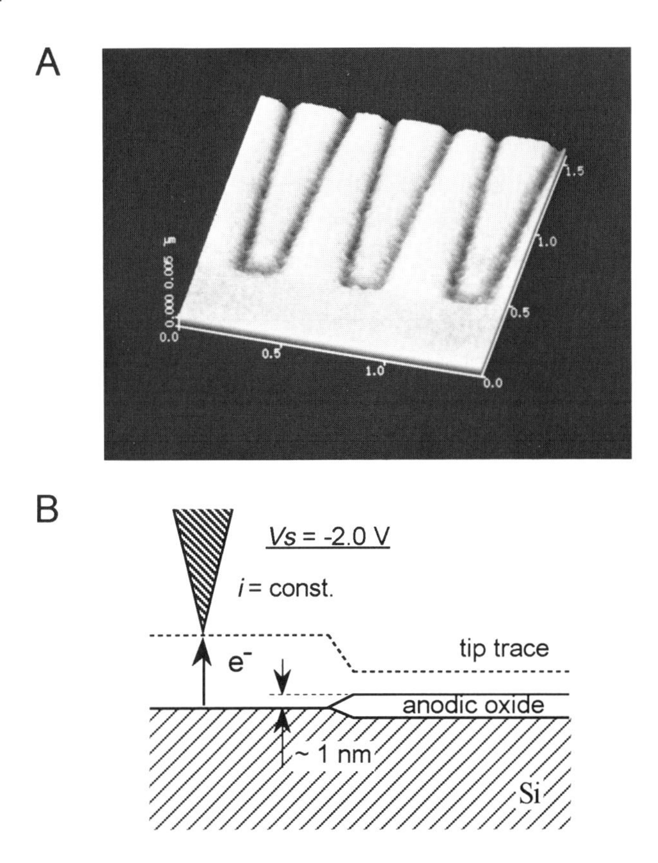

Figure 4. (**A**) A 1.5 × 1.5 μm constant current STM image acquired at V_S = -2.0 V and i = 0.1 nA. The anodic oxide lines of 60 nm in width were fabricated at V_S = +5.0 V, i = 0.1 nA and v = 0.04 μm/s. (**B**) Schematic representation for STM image observation.

As shown in Figure 4**A**, the Si surface was anodized successfully in nm scale analogous to the results on Ti. When Si is anodized to SiO_2, a mound is expected to be formed owing to an increase in molar volume from 12.0 (Si) to 19.8 (SiO_2) cm^3/mol. However, the anodized Si patterns were observed as apparent depressions 5 nm deep in the unmodified substrate analogous to the STM images of oxidized patterns by field-enhanced oxidation [26,27]. The interpretation of the STM image (Figure 4**A**) is schematically illustrated in Figure 4**B**. Once insulating

SiO_X is produced, an STM tip is lowered closer to the surface until a tunneling current reaches a reference value when the tip is scanned across the boundary of the original and the oxidized areas. At the present stage of the investigation, the highest lateral resolution of 20 nm is obtained. We consider that the thickness of the oxide produced is very thin, probably around 1 nm, since when the pattern was imaged by Nomarski optical microscopy, we could barely find its slight topographic feature. Therefore, we conclude that the pattern on Si seen in the STM image is due to the changes in electronic or chemical nature of the surface in contrasting to the STM images of nanostructures fabricated on Ti by the tip-induced anodization.

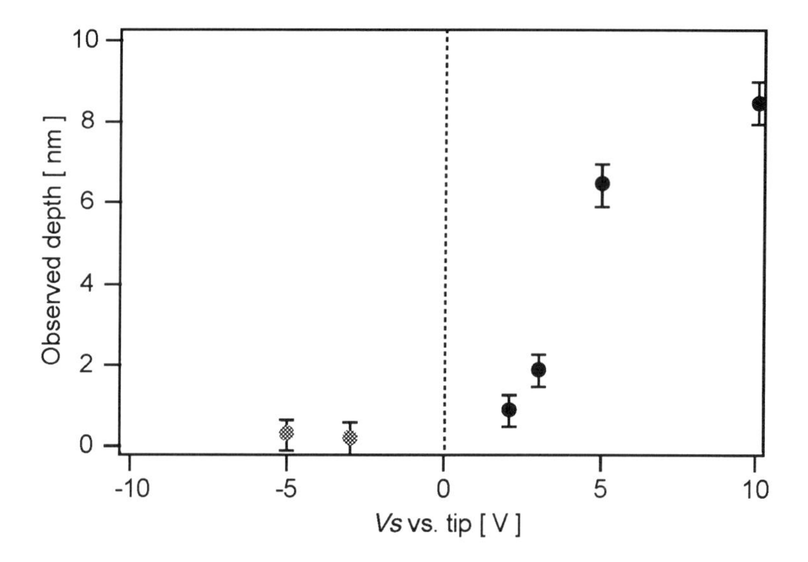

Figure 5. Bias dependence of the observed depth of the oxide pattern. $V_S > 0$: tip-induced anodization $V_S < 0$: field-enhanced oxidation. The patterns were fabricated at $i = 0.1$ nA and $v = 0.009$ μm/s.

A relationship between the observed depth of the fabricated pattern in the STM image and V_S (-5 ~ +10 V) was studied in detail and the results are summarized in Figure 5. In the positive V_S range, the observed depth of a pattern increases from 2.0 to 8.5 nm with the increase in V_S from +2.0 to +10.0 V, probably owing to an increase in the oxide thickness on the pattern. It is known that the growth of anodic oxide is governed by the drift of ionic species through the oxide layer. The drifting rate, therefore the oxide growth rate, is determined by the strength of the electric field formed in the oxide layer [28], so that the anodic oxide becomes thicker with an increase in V_S,. The thickness of the anodic oxide can be controlled by V_S. in STM-anodization as well as in macro-scale anodization. When fabrication of Si is performed with $V_S < 0$, on the other hand, field-enhanced oxidation of Si is promoted with adsorbed water in spite of the absence of oxygen in the atmosphere. However, the patterns fabricated in the negative V_S range showed very shallow depressions less than 0.5 nm in depth as shown in the left half of Figure 5. The patterns fabricated by tip-induced anodization are 6 ~ 28 times deeper than the patterns fabricated by field-enhanced oxidation in the negative Vs range. We conclude that the oxidation of Si with adsorbed water can be promoted by tip-induced anodization ($V_S > 0$) much more efficiently than by field-enhanced oxidation ($V_S < 0$).

4. HUMIDITY EFFECTS ON STM TIP-INDUCED ANODIZATION

Among various factors governing tip-induced anodization, a rate of oxide growth is expected to be affected by humidity through adsorbed water on the sample and tip surfaces and, therefore, humidity should play a key role in determining the spatial resolution of the fabricated patterns. In order to establish the method for nanolithography of metals and semiconductors, we study here humidity effects on tip-induced anodization of Ti. STM experiments were performed under a dry N_2 or water-saturated N_2 gas stream to control humidity near the sample substrate. Relative humidity of the atmosphere was set at < 25 % (below the measurable range of a humidity meter) or 90 %, respectively. Experiments were performed at ambient temperature (23 °C).

When an STM tip is held at a certain position for an appropriate period with positive V_S, Ti surface under the tip is anodized and a mound structure is produced. The mound becomes larger in diameter with an anodization time (t_a). The anodized area corresponding to the diameter of the fabricated structure was remarkably affected by humidity. As shown in STM images **A** and **B** of Figure 6, the diameter slightly increased from 200 to 300 nm with the increase in t_a from 10 to 20 min under the low humidity condition with a dry N_2 gas stream. In such the case, the anodized area is almost restricted position beneath the tip. On the contrary, the mound fabricated under the high humidity condition with a water-saturated N_2 gas stream is fairly large compared with that fabricated under the low humidity condition, and increased considerably from 700 to 2200 nm with increasing t_a from 10 to 20 min as clearly seen in STM images **C** and **D** of Figure 6. In this case, the anodization clearly proceeds in a large area than the position just under the tip, owing to a considerable thickness of the adsorbed water on Ti. The water column formed between the STM tip and the sample is expected to be so thick that the anodization occur in a very large area.

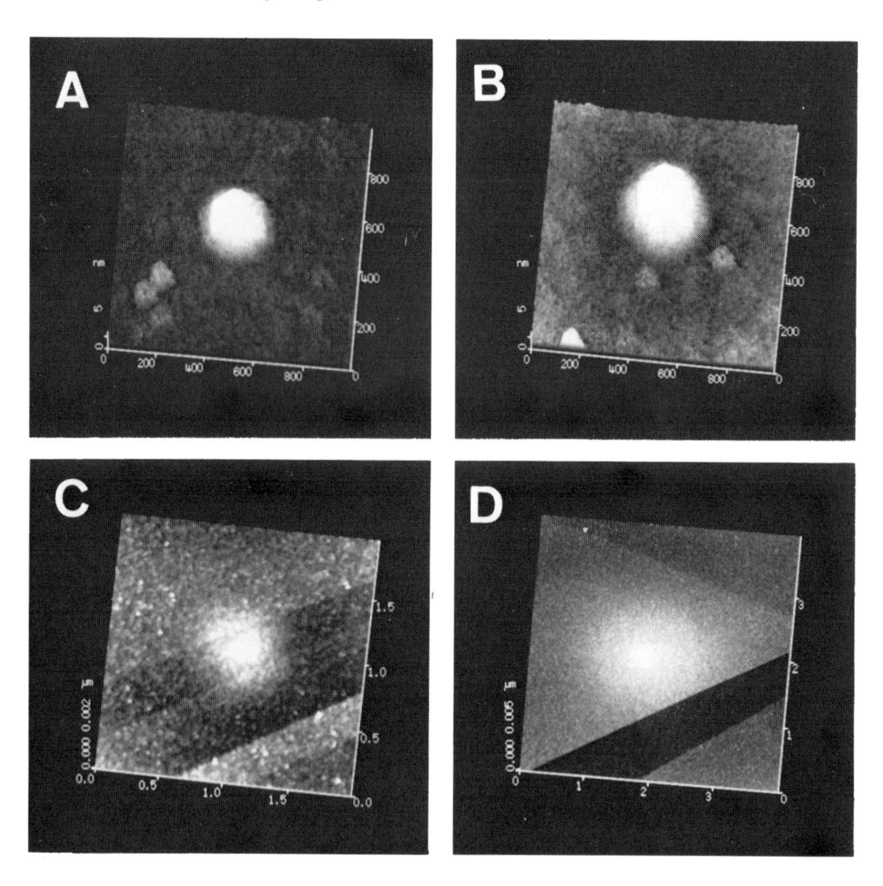

Figure 6. Mound structures fabricated at V_S = +5.0 V and i = 0.5 nA on Ti (10 nm in thickness)/HOPG. (**A**) and (**B**): 1000 × 1000 nm STM images of mounds anodized for 10 and 20 min, respectively, under the low humidity condition (RH < 25 %). (**C**) and (**D**): 2 × 2 and 4 × 4 μm STM images of mounds anodized for 10 and 20 min, respectively, under the high humidity condition (RH = 90 %).

These results are summarized in Figure 7. An identical diamond tip was used throughout the experiments to eliminate the effect of a tip shape on the spatial resolution. As shown by the curve **1**, the diameter of the mound increased rapidly with t_a and reached to ~ 4200 nm at t_a = 40 min. When Ti was fabricated under the low humidity condition, on the other hand, the increase in the diameter of the mound with t_a was very slow, and the diameter remained at ~ 300 nm even upon prolonged anodization for 40 min (curve **2** in Figure 7). It is noteworthy that, when a mound fabricated under a dry O_2 gas stream, the increasing rate of the diameter (curve **3** in Figure 7) is almost the same with that for the curve **2**. For these experiments, RH is kept very low (< 25 %) as well as under a dry N_2 gas stream, so that it will be concluded that O_2 does not contribute to the enhancement of the anodization area. The spatial resolution of STM tip-induced anodization thus becomes worse with an increase in humidity, and an anodic oxide pattern with the highest spatial resolution can be obtained under a low humidity condition.

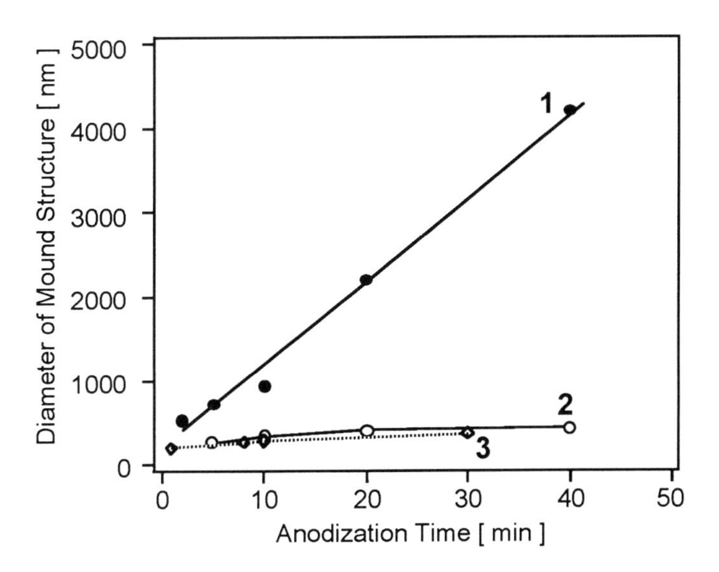

Figure 7. Changes in the diameter of the mound structure with t_a at V_S = +5.0 V and i = 0.5 nA under (**1**) high and (**2**) low humidity N_2 conditions, and (**3**) low humidity O_2 conditions.

Besides the lateral growth of the anodic oxide, the growth rate in thickness was also dependent on humidity. The height of an anodized structure is not equal to the thickness of the anodic oxide, since an anodic oxide film on Ti is supposed to grow not only to the outer direction from the surface, but also to the inside of the Ti film, as illustrated in Figure 1. In the present experiments, however, the height of the fabricated pattern was used as a measure of the thickness of the anodic oxide. 2000 × 2000 nm square-sized patterns were thus fabricated on Ti/mica substrates. Tip-induced anodization was performed repeatedly on a certain surface area for several cycles (25 s/each cycle), and changes in the height of the anodized area with the number of the writing cycle are summarized in Figure 8. The increase in the height is clearly more efficient for fabrication under high humidity (4 ~ 5 nm) than that prepared under low humidity (2 ~ 4 nm). Growth of the oxide layer in the vertical direction to the surface is enhanced by an increase in the amount of adsorbed water on Ti. However, the height of the anodized pattern tends to level off at a certain value with increasing the number of the writing cycle on the surface, and the difference in height of the pattern produced under the high and low humidity conditions becomes smaller with the writing cycle. Further increase in the writing cycle resulted in saturation of the height increase under both conditions of humidity. The saturated height was dependent on V_S, but not on *i*, *v*, and humidity, analogous to the relationship between anodic oxide thickness and V_S in the bulk anodization with constant potentials in an electrochemical cell [28]. The saturated heights at V_S = +3.0, +5.0 and +8.0 V were about 1, 3 and 6 nm, respectively.

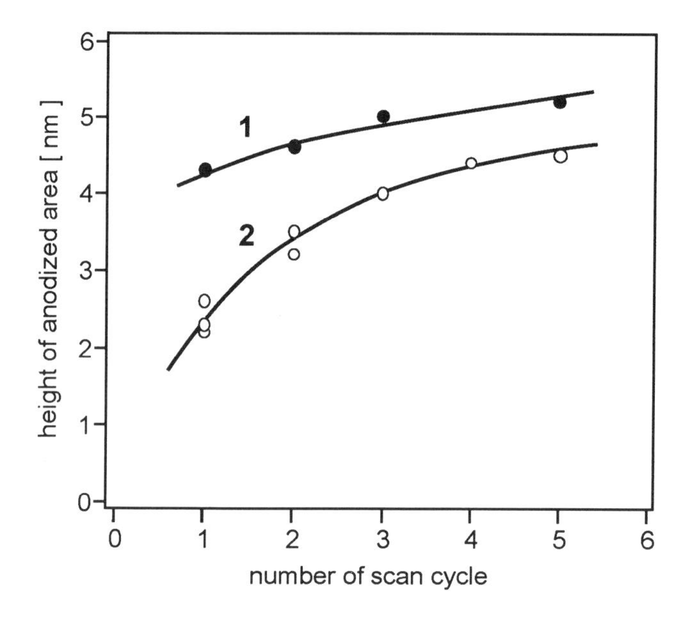

Figure 8. Increase in height of square patterns fabricated at V_S = +8.0 V, *i* = 0.1 nA and *v* = 5 μm/s under (**1**) high and (**2**) low humidity conditions as a function of the writing cycle (25 s/cycle).

5. CONCLUSION

STM tip-induced anodization was successfully applied to fabrication of Ti and Si surfaces in arbitrary spatial patterns with nm ~ μm spatial resolution as confirmed by STM images of the pattern. The spatial resolution of the method was affected by humidity and a tip shape. Thus, a humidity control in atmosphere is indispensable to achieve high spatial resolution nanolithography by STM tip-induced anodization. The best lateral resolution of 20 nm was attained for both Ti and Si under the optimized conditions. STM tip-induced anodization can be performed in N_2 atmosphere as well as in air, so that the method has a practical importance for application to nanolithography. Although the oxide thickness which can be prepared by the method is very thin, that is enough to use the oxide pattern as the mask for the chemical etching to fabricated the substrates [29].

STM tip-induced anodization is based on the electrochemical reactions of Ti or Si with adsorbed water at the surface, so that it is expected to be easily extended to nanofabrication of other materials. Since in macro scale, anodization has been widely used for the surface finishing of various metals and semiconductors, the method should be useful for nanofabrication of these materials.

REFERENCES

1. G. Binnig, H. Rohrer, C. Gerber and E. Weibel, Phys. Rev. Lett. 49 (1982) 57.
2. C. F. Quate, Scanning Tunneling Microscopy and Related Methods, eds. H. Rohrer, R. J. Behm, N. Garcia (Kluwer: Dordrecht, 1990)
3. J. A. Stroscio and D. M. Eigler, Science 254 (1991) 1319.
4. M. A. McCord and R. F. Pease, Appl. Phys. Lett. 50 (1987) 569.
5. D. M. Eigler and E. K. Schweizer, Nature 344 (1990) 524.
6. J. L. Whitman, J. A. Stroscio, R. A. Dragoset and R. J. Celotia, Science 251 (1991) 1206.
7. I.-W. Lyo and P. Avouris, Science 253 (1991) 1398.
8. H. J. Mamin, S. Chiang, H. Birk, P. H. Guethner, and D. Rugar, J. Vac. Sic. & Technol. B9 (1991) 1398.
9. M. A. McCord and R. F. W. Pease, J. Vac. Sci. & Technol. B4 (1986) 86.
10. M. A. McCord, D. P. Kern and T. H. P. Chang, J. Vac. Sci. & Technol. B6 (1988) 1877.
11. W. Li., J. A. Virtanen, R. M. Penner, J. Phys. Chem. 96 (1992) 6529.
12. A. J. Bard, G. Denualt, C. Lee, D. Mandeler and D. O. Wipf, Acc. Chem. Res. 23 (1990) 357.
13. A. J. Bard, F.-R. F. Fan, D. T. Pierce, R. R. Unwin, D. O. Wipf and F. Zhou, Science 254 (1991) 68.

14 H. Sugimura, T. Uchida, N. Shimo, N. Kitamura and H. Masuhara, Ultramicroscopy, 42-44 (1992) 468.
15. H. Sugimura, T. Uchida, N. Kitamura, N. Shimo and H. Masuhara, J. Electroanal. Chem., (in press)
16. T. R. Albercht, M. M. Dovek, M. D. Kirk, C. A. Lang, C. F. Quate and D. P. E. Smith, Appl. Phys. Lett. 55 (1989) 1727.
17. R. L. McCarley, S. A. Hendricks and A. J. Bard, J. Phys. Chem. 96 (1992) 10089.
18. H. Sugimura, T. Uchida, N. Kitamura and H. Masuhara, Jpn. J. Appl. Phys. 32 (1993) L553.
19. H. Sugimura, N. Kitamura and H. Masuhara, Jpn. J. Appl. Phys. (in press)
20. H. Sugimura, T. Uchida, N. Kitamura and H. Masuhara, Appl. Phys. Lett. 63 (1993) 1288.
21. H. Sugimura, T. Uchida, N. Kitamura and H. Masuhara, submitted to J. Phys. Chem.
22. H. Sugimura, N. Kitamura and H. Masuhara, Jpn. J. Appl. Phys. 31 (1992) L1506.
23. C. K. Dyer and J. S. L. Leach, J. Electrochem. Soc.125 (1978) 1032.
24. F.-R. Fan and A. J. Bard, J. Phys. Chem., 94 (1990) 3761.
25. H. Sugimura, T. Uchida, N. Kitamura, N. Shimo and H. Masuhara, Electrochemical Processing of Tailored Materials (2nd International Symposium), Eds. R. Alkire, N. Masuko and D. R. Sadway (Electrochemical Society, Inc., 1993)
26. J. A. Dagata, J. Schneir, H. H. Harary, C. J. Evans, M. T. Postek, and J. Bennett, Appl. Phys. Lett. 56 (1990) 2001.
27. J. A. Dagata, J. Schneir, H. H. Harary, J. Bennett, and W. Tseng, J. Vac. Sci. Technol. B9 (1991) 1384.
28. The Anodic Behavior of Metals and Semiconductors Series, Ed. J. W. Diggle (Marcel Dekker, Inc., New York, 1972)
29. H. Sugimura, T. Uchida, N. Kitamura and H. Masuhara, submitted to J. Vac. Sci. Technol.

MICROCHEMISTRY
Spectroscopy and Chemistry in Small Domains
Edited by H. Masuhara et al.

Fabrication of two-dimensional protein and colloidal arrays

Kuniaki Nagayama

Department of Pure and Applied Sciences, The University of Tokyo, Komaba, Meguro-ku, Tokyo, 153 Japan
NAGAYAMA Protein Array Project, ERATO, JRDC, 5-9-1 Tokodai, Tsukuba, 300-26 Japan

We report an effective method for the fabrication of two-dimensional (2D) arrays of submicron particles (protein and colloidal particles). The arrays are formed in a wetting film of a suspension that holds the submicron particles on flat, clean substrate surfaces such as glass, cleaved mica, and mercury. The 2D assembly process includes two steps, nucleation and growth, which are similar to that of the crystallization process for molecules in solution. The detailed steps, however, for each of these processes progress with different mechanisms. The nucleation process is initiated by a special kind of capillary force, called the lateral capillary immersion force, which is an attractive force between particles for which the tops of the particles protrude in the air. The growth process is efficiently guided by the flux of water flowing into the array area where water is continuously removed by evaporation. The key technology in this fabrication is an array making apparatus (arrayer), which can produce a contamination-free surface of mercury that leads to complete spreading of the suspension to yield a very thin liquid film.

1. INTRODUCTION

Most crystals in nature that are made of inorganic substances have three-dimensional shapes. On the other hand, crystals in living systems tend to form one-dimensional fibers like myofibril in muscle or 2D planes like biomembranes and the outer surface of bacterial cells (s-layer) [1]. Three-dimensional crystalline objects are only observable in bones or teeth where inorganic crystals dominate. Protein assemblies in life are usually not guided to yield infinitely large integration but rather small clusters, which are often called supramolecules. Here, we can recognize life's unique strategy and the important role of protein molecules. Namely, supramolecular architecture that occurs through the mutual recognition of molecular partners. The assembly from constituent protein molecules to supramolecules can be divided into two processes: diffusion-limited aggregation (random assembly) of individual protein molecules, and molecular orientation adjustment by specific interactions (mutual recognition) among the assembled proteins. Both are believed to be a consequence of the nature's robust technology, the fabrication that is guaranteed by thermally driven molecular diffusion. The assembly usually progresses in dipersions in living cells. Due to the limited cell volume (~ 1 μm length), the diffusion-limited assembly successfully avoids being a bottleneck for supramolecular construction. If the cell is 1 mm long, the process is unacceptably time-consuming because of the slow Brownian movement of such large molecules as proteins in dispersions. This biological framework of the unit assembly is, therefore, inadequate for material engineering on a larger scale where the size easily exceeds 1 mm.

In contrast with the supramolecule constructions in cells, our crystalline arrays of proteins are allowed to grow on a substrate in a 2D form. The choice of an appropriate substrate surface becomes particularly crucial for fine particles on the nanoscale to be fabricated in a well defined array form. In biosystems, protein 2D crystals are often embedded in biological membranes, such as cell membranes and liposomes [1]. To replace the membranes by artificial substrates, many kinds of interfaces such as water/air and lipid/water/air interfaces have been adopted. But, problems in the diffusion-limited assembly still exist in the use of these surfaces. Our recent innovation for the protein 2D crystallization by using a clean mercury surface has solved this problem. The breakthrough in our technology has been brought about by the use of wetting films that are stably formed on a clean mercury surface [2]. An interesting consequence of confining protein particles within such a thin liquid layer on the mercury surface is the directional particle motion that is forced by the water flow, which accelerates the crystal growth and yields a large domain of 2D crystals in a short period (Figure 1). This was first experimentally demonstrated with polystyrene latex particles of micrometer size [3, 4]. The principal aspects of the fabrication method applied for protein and colloidal particles will be reported.

2. CRYSTALLINE FILMS OF POLYSTYRENE LATEX PARTICLES ON SOLID AND MERCURY SURFACES

Crystallization generally includes three successive steps; 1) nucleation, 2) growth, and 3) molecular reorientation, in this order. In the three-dimensional (3D) crystallization of ordinary substances, the 2nd and 3rd steps can progress in a combined manner, since it is not unusual that the forces responsible for the two steps are one and the same. In the 2D crystallization of proteins on a surface, however, these two steps are identified, two different forces working during the 2nd and 3rd step. This was first suggested in a model experiment that used colloidal particles, polystyrene latexes, which allow for *in situ* observation of the dynamics with an optical microscope [3]. When proteins are largely separated, as compared to their own size, they are simply considered as classical colloidal particles that exhibit only non-specific short- or long-range interactions. This is one reason why we have been motivated to pursue the condensation process of colloids such as polystyrene particles on substrate surfaces.

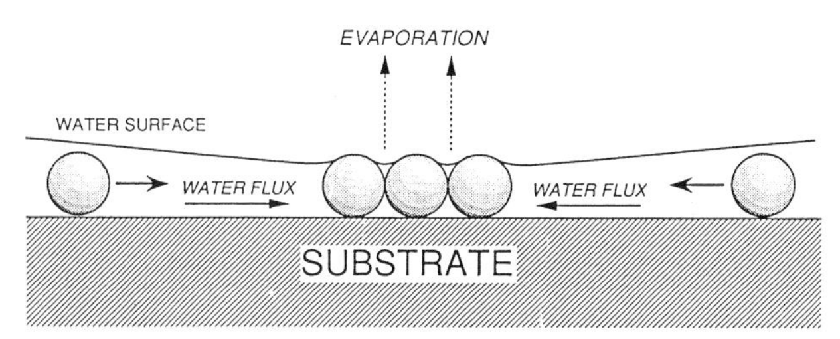

Figure 1. The growth of particles or molecules to two-dimensional crystalline arrays in a wetting film, which is mediated by the flow of water, which is driven by evaporation taking place at the array boundary (flow-mediated-assembly).

The 2D lattice formation of polystyrene particles in a thin liquid must give basic knowledge about the process of the protein crystalline array formation. With the use of various sizes of polystyrene latex particles ranging from 50 nm to 2 μm, the new surface phenomena have been studied in our project [3-6]. What is remarkable in the assembly is the sudden and rapid growth of crystalline arrays. This is in marked contrast to the 2D or 3D crystallization observed in a solution, for which growth is relatively slow due to the rate-limiting step of diffusion or activation. In contrast, the new assembly process relies on the directional flow of particles carried by water flow. With several experimental results, we can propose the assembling mechanism shown in Figure 1 [3-5]. The driving force for the crystalline array growth is not the thermal diffusion of particles but the directional flow of particles. When the evaporation rate was kept constant, the crystal grew linearly with time [5]. This explains the rapid growth of 2D arrays of colloidal or protein particles on the solid or liquid metal surface we have observed [2].

Another important factor, which has been found in this work, is the sudden appearance of a strong attractive force among particles when they protrude from the solvent surface. We call this new force the "lateral capillary immersion force," which arises from the surface tension at the deformed liquid surface around particles [7, 8]. This attractive force is long-ranged and strong even for nanometric particles [7]. This force becomes important at the final stage of the film drying and is, therefore, responsible for the packing of particles in hexagonal crystalline lattices, when the intermolecular forces are overwhelmed by the force.

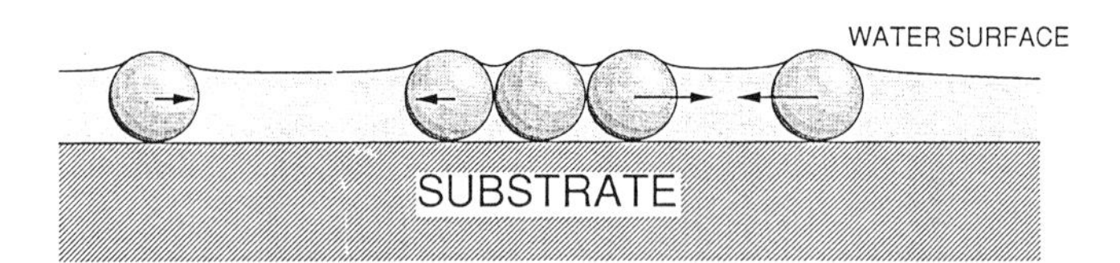

Figure 2. The lateral capillary immersion force [7, 8] effectively working at the thin liquid layer to force particles to pack and form crystalline arrays.

3. STABLE WETTING FILMS

The essence of our 2D assembly method for particles lies in the stable wetting film that is made on the substrate surface as shown in Figures 1 and 2. This wetting film plays two important roles here; 1) 2D liquid medium where particles can be driven by water flow to the crystal boundary for growth, and 2) generation of the lateral capillary force by the surface tension for particle packing at the boundary. In parallel to the usual 3D crystallization, these processes can also proceed by using the free energy difference before and after the reaction in the non-equilibrium state. This is schematically shown in Figure 3. First, particles undergo the Brownian motion in the thick film (Figure 3A). When the wetting film becomes as thin as the particle size by the removal of water, an ordered 2D domain starts to form on the surface (Figure 3B). This is driven by the very subtle energy of the water flow caused by the evaporation, which is only one explicit non-equilibrium condition in our system. The most difficult task in this fabrication is not to control the evaporation but to create stable wetting films that are appropriate for the submicron or nanometric particles because the thinning process on the solid substrate are not stable, so it is difficult to avoid raptures in the very thin films [9,10]. The film rupturing is a complicated phenomenon that depends on the roughness, wettability, and chemical stability of the substrate surface, and the suspension properties. Cleaved mica or acid-rinsed glass provides a very wettable surface and, consequently, is suitable for the array formation of colloidal particles when they are larger than 50 nm in diameter. Due to the roughness of the surface of these substrates, protein molecules or fine particles smaller than 50 nm in diameter are difficult to use to form ordered 2D arrays on the surfaces. The major reason to employ the very clean mercury surface for smaller particles such as proteins resides in this point [2]. For unknown reasons, the mercury surface becomes not only flat but also highly wettable and hydrophilic [11] under dry oxygen gas.

We adopted three different thinning methods to sustain the liquid thin layer on substrates: 1) evaporation [3-5], 2) suction [6], and 3) spreading wetting [2], For colloidal particles (polystyrene) larger than 50 nm, these three methods work well for the various substrates such as mica, glass, metal-coated mica, carbon-coated mica, metal-coated glass, and mercury. For the smaller particles such as protein molecules, the combination of spreading wetting and evaporation on the mercury surface is successful in making crystalline arrays.

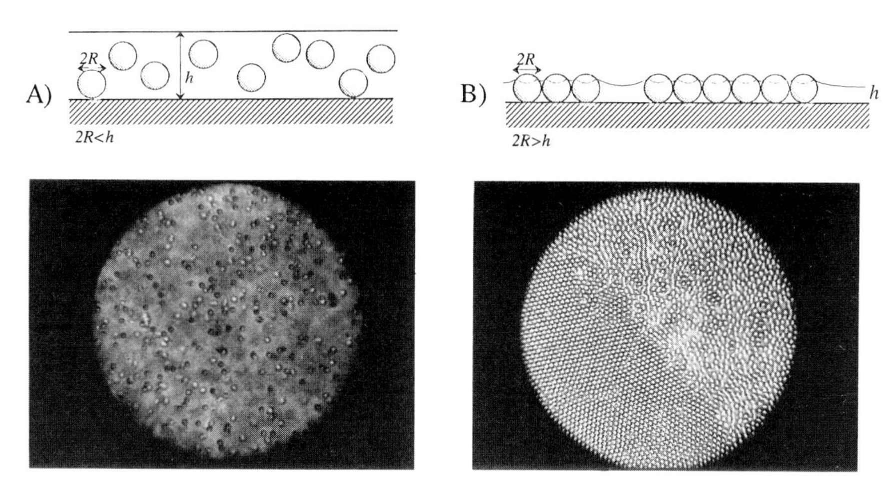

Figure 3. Two-dimensional assembly of particles in the wetting film.
A) Particles undergo the Brownian motion in the liquid layer when the thickness is much larger than the particle size.
B) Particles start to assemble in the wetting film as the thickness of the film becomes comparable to or slightly smaller than the particle size.

4. CRYSTALLINE ARRAYS OF PROTEINS ON MERCURY

As descrived above, the clean mercury, when modified to be wettable by the use of clean oxygen gas, yields stable wetting films on its surface [2, 6]. This convenient property has always been used for the formation of protein 2D arrays in our studies [2, 12-16]. The very thin wetting films of an aqueous protein solution has recently been experimentally confirmed [6, 17]. The arrayer used for preparing clean mercury surface, spreading a protein solution, and making protein crystalline arrays was a home-made machine produced by us (JEOL Ltd., the author's former affiliation) [2].

Experimental procedures are as follows. 1) Clean the inside of the trough that is isolated by a bell jar from the laboratory air, 2) chamber is filled with clean oxygen gas, 3) distilled, nitric-acid washed mercury is introduced into the trough, 4) the surface of the mercury is withdrawn by a vacuum pump, 5) the surface is swept with two barriers to clean the surface, and 6) the cleanness of the bare mercury surface is checked by two physical indices, surface tension and optical constants such as refractive index. After the surface is cleaned, 7) a small amount of protein solution (2 – 20 μl) is loaded onto the surface by using a syringe (refer to Figure 4A). 8) In a short period (8 – 10 sec), the protein solution completely spreads (spreading boundary reaches the trough wall and barriers)(refer to Figure 4B). 9) Evaporation and drying process of the wetting film immediately follows after the spreading and, hence, assembly and crystallization of protein molecules are postulated to start soon after the completion of the spreading. 10) After the water is completely removed by the evaporation for one or two minutes, the protein crystalline film formed on mercury is transferred to a carbon film covering the specimen grid for electron microscope observation (refer to Figure 4C).

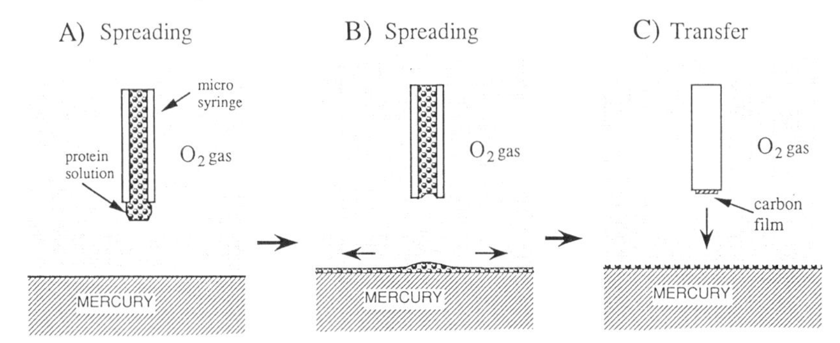

Figure 4. Schematics of spreading wetting of protein solutions and the transfer of protein 2D crystalline arrays to the second substrate surface for observation.
A) Loading on a mercury surface.
B) Spreading wetting.
C) Transfer of arrays to the carbon film surface on the specimen grid for electron microscopy.

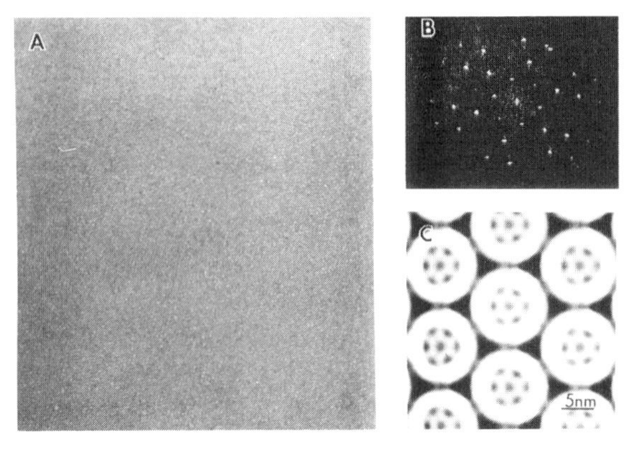

Figure 5. An electron microgram of a 2D crystal of ferritin (~480 kDa, 24 mer of subunit proteins of 18 – 20 kDa), its numerical diffraction spots, and a reconstituted image.
A) An image of ferritin 2D crystalline arrays taken with an electron microscope (JEOL, JEM1200H) at a magnification of 30,000×.
B) Numerical diffraction spots for the 2D crystalline array.
C) An image of the crystalline array reconstituted by collecting and Fourier-transforming the diffraction spots with a signal to noise ratio better than 2 (up to the fifth order).

To make the transfer successful, the carbon film should be pre-treated by ion sputtering. The electron micrograms shown later (Figure 5) were thus taken from the 2D crystals of ferritin. The packing ratio and, therefore, ordering quality of the 2D arrays are found to be directed by the balance of the growth speed and the particle-supply rate governed by the water flow driven by evaporation [5]. If both of these factors are carefully controlled, a very large, single domain of 2D crystal can be obtained. Unfortunately, such control in the spreading wetting is not easily achieved. By adjusting the protein concentration and loading sample volume, we have tried to take indirect control of the two crucial parameters.

Protein molecules are distinguishable from normal colloidal particles by their irregular, unique structures. In the 2D crystallization, therefore, the reorientation process of protein molecules should follow after the particle assembly. For spherical particles, such as polystyrene particles, this reorientation process is not required. The relative orientation of protein particles on the supporting substrate surface is important for the practical use of the crystalline arrays, because the functions in the proteins (e.g., enzymatic, oxidative/reductive, and electro-optic activities), which are restricted to the local area on the molecular surface and its orientation to the substrate surface or to the free medium space, are crucial to efficiently yield these activities. As already described, the lattice form of the 2D crystalline arrays is usually hexagonal due to the strong packing force arising from the lateral capillary immersion force (Figure 2), which is not the intrinsic intermolecular interaction but a non-specific interaction that is only emergent from the anisotropic environment at the liquid interface [18]. This, in turn, sets an external constraint on the protein orientation in 2D crystalline arrays. Examples of 2D protein crystals with hexagonal packing are shown in Figure 5A.

Table 1.
Size, Symmetry, and Ordering

* Water soluble proteins		Size (kDa)	Molecular Symmetry	Diffraction Order	References
ferritin	(horse spleen)	480	P3m1	6	7a
H^+-ATPase F1	(*Thermus thermophilus*)	340	P3	5	7a, 12, 19
chaperonin	(*Thermus thermophilus*)	420	P1	4	13
myosin S1	(rabbit skeletal)	100		1	a
metallothionein	(rabbit liver)	8		b	a
*** Membrane proteins**					
L-P ring	(Salmonella)	1,300	P6	9	11
Na^+, K^+-ATPase	(dog kidney)	120	P1	0	a
*** Protein-nucleic acid complex**					
bacteriophage		$> 10^4$	P6	11	14

a: not published
b: not transferred to the specimen grid of electron microscope

The orientations of protein molecules in the crystals could be determined with the standard 2D reconstitution technique based on the diffraction spots (Figure 5B) that were obtained by Fourier transform analysis [19]. A numerically processed picture of the crystal, including protein internal structures, that was obtained after reverse Fourier-transform is shown in Figure 5C together with an original microgram (Figure 5A). Based on the comparison between the experimentally determined projection shown in Figure 5C and the numerically obtained one based on the X-ray structure of ferritin [20], the protein molecule was found to orient with its 3-fold axis perpendicular to the substrate surface. This orientation agrees with the lattice symmetry [2]. The results of the reconstitution of various protein systems are summarized in Table 1, in which the relations between the obtained crystal quality and the molecular size and symmetry are shown. A general trend could be observed in the table: Larger (higher molecular weight) and rounder (higher symmetry) protein molecules give better crystallinity (higher diffraction order). This trend is easily understood and compatible with our postulation of the crystal packing mechanism suggested from the model experiment with polystyrene particles; 1) tight packing by the lateral capillary immersion force is favored for larger molecules due to the size-dependence of the force, which is proportional to the square of the particle radius [7, 18], and 2) tight packing appears when the molecular symmetry matches the hexagonal symmetry of the crystal lattice.

5. DISCUSSION

The 3D colloidal crystals, which appear in the desalted polystyrene latex suspensions, have been widely studied by many colloid chemists. Contrary to the 2D crystalline arrays shown here, the lattice spacing in the 3D colloidal crystals is larger than the size of the constituent colloids. This softness in the crystals is now understood to be generated by the repulsive interaction between the similarly charged colloid particles [21]. Then, a question arises whether the close-packed 2D array here presented, which naturally results from the strong attractive capillary force, is a crystal or not. Several criteria should be satisfied for this material to be called a crystal; 1) the presence of a lattice, 2) the presence of a lattice-constant, and 3) a phase-separation between different species (size, shape, and surface properties). On the first and second criteria, the strong packing force guarantees the formation of a close packed hexagonal lattice, which must have a lattice constant determined by the molecular geometry. This is clearly shown in Figure 5. The third condition has recently been confirmed by the observation of the separation of crystalline areas among two species of colloidal particles with different sizes [22]. We believe that the 2D crystalline arrays can correctly be called crystals based on the satisfaction of these three criteria.

ACKNOWLEDGEMENT

Thanks are extended to the project members who have contributed to the research work shown in this manuscript, Dr. H. Yoshimura, Dr. S. Endo, Dr. N. D. Denkov, Dr. C. D. Dushkin, Mr. A. S. Dimitrov, Mr. O. D. Velev, Mr. V. N. Paunov, Ms. K. Matsubara, Dr. T. Miwa and Ms. M. Yamaki. The author is grateful to Professor I. B. Ivanov and Dr. P. A. Kralchevsky (Sofia University), Dr. M. Matsumoto (Kyoto University), and Dr. S. Ebina (this Project) for their illuminative discussions. The author expresses his thanks to those collaborators who kindly provided valuable protein samples, Dr. N. Ishii and Dr. K. Yoshida (Tokyo Institute of Technology), Dr. Y. Kagawa (Jichi Medical School), Dr. T. Akiba and Dr. K. Namba (Hotani Project, ERATO).

REFERENCES

1. B. K. Jap, M. Zulauf, T. Scheybani, A. Hefti, W. Baumeister, U. Aebi and A. Engel, Ultramicroscopy, **46** (1992) 45.
2. H. Yoshimura, M. Matsumoto, S. Endo and K. Nagayama, Ultramicroscopy, **32** (1990) 265.
3. N. D. Denkov, O. D. Velev, P. A. Kralchevsky, I. B. Ivanov, H. Yoshimura and K. Nagayama, Langmuir, **8** (1992) 3183.
4. N. D. Denkov, O. D. Velev, P. A. Kralchevsky, I. B. Ivanov, H. Yoshimura and K. Nagayama, Nature, **361** (1993) 26.
5. C. D. Dushkin, H. Yoshimura and K. Nagayama, Chem. Phys. Lett., **204** (1993) 455
6. A. S. Dimitrov, C. D. Dushkin, I. B. Ivanov, H. Yoshimura and K. Nagayama, Langmuir, submitted.
7. P. A. Kralchevsky, V. N. Paunov, I. B. Ivanov and K. Nagayama, J. Coll. Interface Sci., **151** (1992) 79.
8. V. N. Paunov, P. A. Kralchevsky, N. D. Denkov, I. B. Ivanov and K. Nagayama, J. Colloid Interface Sci., **157** (1993) 100.
9. T. D. Blake in *Surfactants* (Ed. T. F. Tadros) (1984), 221, Academic Press, London.
10. B. V. Derjaguin, Theory of Stability of Colloids and Thin Films (translated by R. K. Johnson) (1989), Chapter 4, Consultants Bureau, New York.
11. V. S. Kulkarni, H. Yoshimura, K. Nagayama and M. Matsumoto, J. Colloid Interface Sci., **114** (1991) 586.
12. H. Yoshimura, M. Matsumoto, S. Endo, K. Nagayama and Y. Kagawa, J. Biochem., **106** (1989) 958.
13. N. Ishii, H. Taguchi, M. Yoshida, H.Yoshimura and K. Nagayama, J. Biochem., **110** (1991) 905.
14. T. Akiba, H. Yoshimura and N. Namba, Science, **252** (1991) 1544.

15. N. Ishii, H.Yoshimura, K. Nagayama, Y. Kagawa and M. Yoshida, J. Biochem., **113** (1993) 245.
16. H. Yoshimura, S. Endo, K. Nagayama and M. Matsumoto, JEOL News, **29E** (1991) 2.
17. M. Yamaki, K. Matsubara and K. Nagayama, Langmuir, in press.
18. P. A. Kralchevsky and K. Nagayama, Langmuir, in press.
19. L. A. Amos, R. H. Henderson and P. N. T. Unwin, Prog. Biophys. Mol. Biol., **39** (1982) 183.
20. a) S. H. Banyard, D. K. Stammers and P. M. Harrison, Nature, **27** (1978) 282.
    b) D. M. Lawson et al., Nature, **349** (1991) 541.
21. S. Hachisu and K. Takano, Adv. Colloid Interface Sci., **16** (1982) 233.
22. M. Yamaki and K. Nagayama, unpublished data.

Part IV: Dynamic Microspectroscopy

MICROCHEMISTRY
Spectroscopy and Chemistry in Small Domains
Edited by H. Masuhara et al.

Three-dimensional space- and time-resolved spectroscopy using a confocal microscope

Keiji Sasaki[*,#] and Masanori Koshioka[†]

Microphotoconversion Project,[‡] ERATO Program, Research Development Corporation of Japan, 15 Morimoto-cho, Shimogamo, Sakyo-ku, Kyoto 606, Japan

Time-resolved fluorescence and transient absorption spectroscopy systems have been developed for elucidating photophysical and photochemical processes occurring in micrometer-sized volumes. The fluorescence measurement is based on a confocal fluorescence microscope and a single photon timing, which enables us to observe the fluorescence dynamics with both submicrometer three-dimensional space- and picosecond time-resolutions. The time-dependent fluorescence depolarization can also be measured at each small spot. Anisotropy decays are obtained by a new analytical theory which considers the polarization change caused by the large solid angle characteristic of the microscope optics. For the transient absorption spectroscopy, excitation and monitoring laser pulses are coaxially and confocally introduced into the microscope. This optical arrangement provides the longitudinal resolution in addition to the lateral resolution, which has never been achieved by any other absorption microscope. To analyze rise and decay curves as a function of three-dimensional position and wavelength, we have proposed a new fast and accurate analyzing method using a convolved autoregressive model. The computation time is much shorter than that of a conventional nonlinear least-squares method.

1. INTRODUCTION

Chemical reactions proceed as a function of not only time but also position, so that the reaction processes are influenced by spatial structures of the reaction systems. Indeed, microstructures in various molecular assemblies such as polymer films, Langmuir-Blodgett films, composite materials, and biological cells, often play an important role for their physical and chemical functions. Polymer latexes, liquid

---

* To whom correspondence should be addressed.
# Present address: Department of Applied Physics, Faculty of Engineering, Osaka University, Suita, Osaka 565, Japan.
† Present address: Kaneka Corporation, Elmech Business Development Section R&D Group, 5-1-1, Torigai-Nishi, Settsu, Osaka 566, Japan.
‡ Five-year term project: October 1988~September 1993.

droplets, microcapsules, and so forth can also be expected to provide characteristic properties different from those of bulk materials and different between individual microparticles. In order to elucidate the dynamic structure of the inhomogeneous reaction systems and to control their processes, spectroscopic tools with both high temporal and spatial resolutions are indispensable.

The time-resolved spectroscopy has been greatly advanced with the recent development of short-pulsed lasers and high-speed detectors. The temporal resolution has been improved considerably from microsecond to femtosecond. The time-resolved spectroscopy has been used as a variable tool to clarify dynamic processes such as excitation energy relaxation, electron as well as proton transfer, molecular vibrational relaxation, and isomerization occurring in various photophysical and photochemical materials. However, most of the studies have been performed with spatially-unresolved spectroscopy systems, so that the experimental results provided the information on the average or/and sum of various components with different structures.

On the other hand, the most popular tool for the space-resolved measurement with a μm scale is an optical microscope, which has been widely applied to various sciences and industries, especially to biological studies and recent semiconductor technology. The optical microscope, however, has some difficulties in combining with the time-resolved laser spectroscopy. One is the influence of speckle and interference fringes due to the high spatial coherence of laser. Invisibly small dusts and slight distortion of optical systems cause irregular patches and/or periodic intensity distribution. These coherent noises disturb the image formation with laser light, so that the quantitative information on the spatial distribution is lost in the observed image. The other problem concerns the longitudinal (depth) resolution. The conventional optical microscopes provide only two-dimensional images. The observed image does not correspond to the structure only on the focal plane, but the distribution accumulated along the depth direction. The contamination of out-of-focus contributions makes it difficult to accurately analyze the spectroscopic information on the inhomogeneous systems.

These problems can be solved by introducing a confocal microscope which has recently received much attention in the field of optics [1]. This microscope is based on point-excitation (illumination) and point-detection systems combined with a scanning mechanism. The sample or laser beam is mechanically or optically scanned so that fluorescence, reflection, and absorption are measured point by point. The detected signal is sequentially processed on a digital or analog computer to form the images. In contrast with the conventional optical microscopes, the confocal fluorescence and reflection microscopes have the capability of three-dimensional imaging. While the confocal absorption (transmission) microscope essentially has a difficulty in the depth-resolved measurement, the transient-absorption-mode confocal microscope, which we originally developed, possesses the longitudinal resolution as well as the lateral resolution. Furthermore, the confocal microscopes provide clear images free from the speckle and interference fringe. Flare and scattered light are also negligibly weak due to the use of the point-detector.

In this paper, we describe micrometer three-dimensional space- and picosecond time-resolved fluorescence and absorption spectroscopy systems based on the confocal microscope and laser spectroscopy. In addition, a fast and accurate mathematical method is introduced, which is indispensable for the analyses of rise and decay curves as a function of position and wavelength.

2. FLUORESCENCE SPECTROSCOPY

The techniques used widely for the time-resolved fluorescence spectroscopy are frequency up-conversion [2] and single photon timing [3], and the streak camera is a representative detector instrument [4]. The frequency up-conversion method is based on nonlinear optics and possesses the highest time-resolution of less than 1 ps. The streak camera can perform the simultaneous measurement of temporal and spectral information. Compared with these techniques, the single photon timing has the advantages of high sensitivity and wide dynamic range due to the photon counting detection. Therefore, this method is appropriate for weak fluorescence measurements in the application of microscopic spectroscopy.

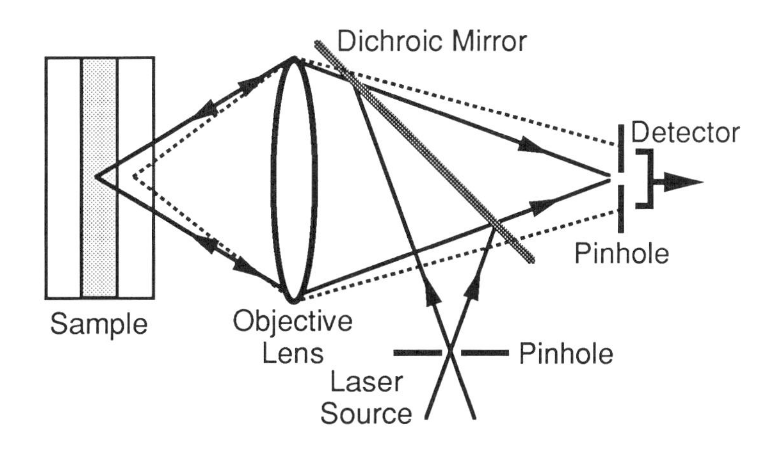

Figure 1. Optical system of the confocal fluorescence microscope.

The space- and time-resolved fluorescence spectroscopy system we developed is based on the single photon timing and the confocal fluorescence microscope [5]. Figure 1 shows the optical diagram of this microscope. The difference between this system and a conventional fluorescence microscope is the use of two pinholes. The first pinhole is set in front of a light source (usually a laser) to limit it to a small spot. Excitation light from this minute source is reflected by a dichroic mirror and focused onto a sample by an objective lens. Fluorescence emitted from the sample is collected by the same objective lens and imaged on the second pinhole, and only the fluorescence passing through the pinhole is detected. This optical arrangement is a key to providing three-dimensional space resolution. Since fluorescence from the focal spot is condensed on the second pinhole, most of its energy goes through. On the other hand, as fluorescence from out-of-focus positions is defocused on the pinhole plate as shown by dotted lines in Figure 1, most of its energy is cut off. Therefore, the observed fluorescence is ascribed to the minute volume so that the longitudinal resolution is obtained besides lateral resolutions. The sample is set on a XYZ scanning stage and moved for measuring the three-dimensional structure.

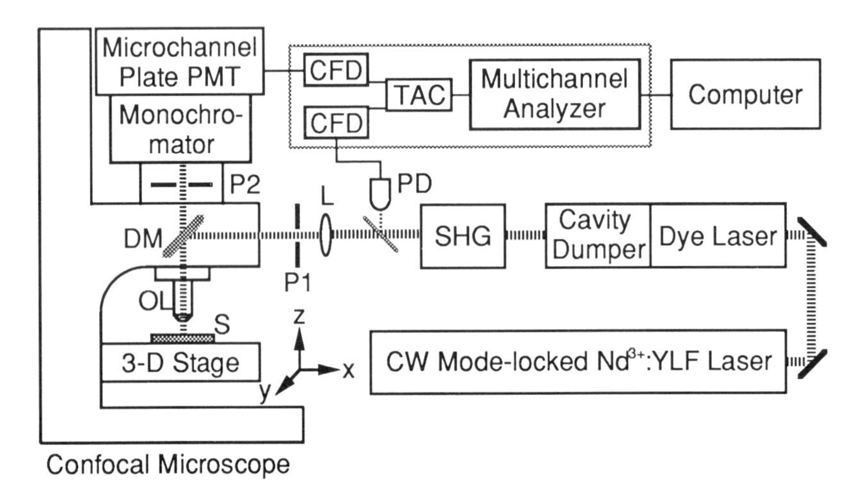

Figure 2. A schematic diagram of a three-dimensional space- and time-resolved fluorescence spectroscopy system. **L**, lens; **P1** and **P2**, pinhole; **DM**, dichroic mirror; **OL**, objective lens; **S**, sample; **PD**, photodiode; **CFD**, constant fraction discriminator; **TAC**, time-to-amplitude converter.

Figure 2 shows a schematic diagram of the three-dimensional space- and time-resolved fluorescence spectroscopy system. A cavity-dumped dye laser, synchronously pumped by the second harmonics of a CW mode-locked Nd:YLF laser, was used as a light source. Its wavelength was tunable from 560 nm to 620 nm by using a rhodamine 6G dye. The pulse width and the repetition rate were 2 ps (FWHM) and 3.8 MHz, respectively. An ultraviolet pulse (280~310 nm) produced by a second harmonic generator was condensed by a lens and introduced into a pinhole **P1** of a microscope. A zoom lens was used for matching the beam diameter with the numerical aperture of the microscope. In the microscope, the laser light was reflected by a dichroic mirror and focused onto a sample by an oil-immersion objective lens. This objective was made of quartz for the ultraviolet excitation, and its magnification and numerical aperture were 100 and 1.25, respectively. Fluorescence from the sample was imaged on a pinhole **P2**. Its diameter of 40 μm corresponds to 0.25 μm on the sample, as the magnification was 160. The three-dimensional scanning stage was driven with steps of 0.25 μm in the lateral directions and 0.1 μm in the depth direction. Fluorescence passing the pinhole **P2** was spectrally resolved by a monochromator and detected by a microchannel-plate photomultiplier. The output signals of the photomultiplier and a PIN photodiode detecting the excitation pulse were processed by constant fraction discriminators. Their outputs were time-correlated by a time-to-amplitude converter and processed by a multichannel pulse-height analyzer. A microcomputer controlled the sample stage, the monochromator, and the single photon timing apparatus, and the data analyses were performed with a workstation. The performance of the system, which was specified by measuring thin liquid layers of pyrene and rhodamine B, are the three-dimensional space resolutions of 0.3 μm (lateral) and 0.5 μm (depth) and the temporal resolution of 2 ps (FWHM of the instrumental response function, 33 ps), in addition to the spectral resolution of 1 nm (the observable wavelength range, 300~1000 nm).

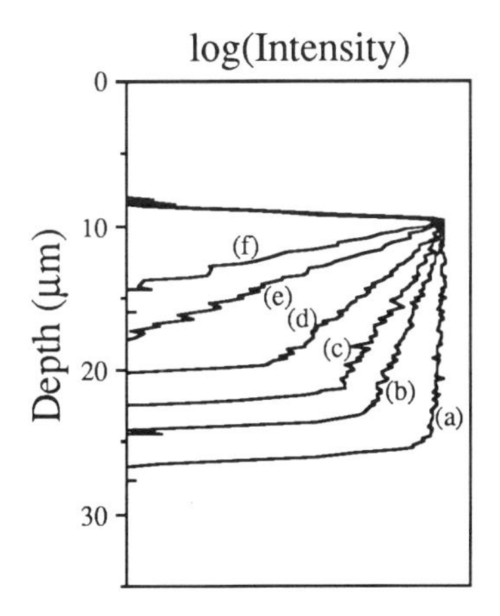

Figure 3. Concentration dependence of depth profiles of pyrene-doped PMMA films. The thickness of the films was adjusted to decrease the concentration increases.

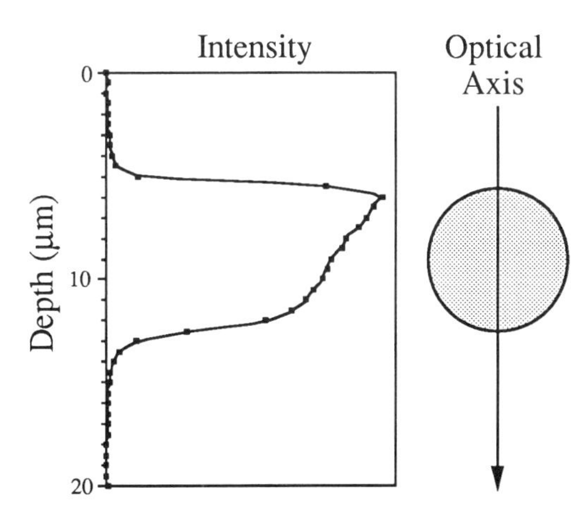

Figure 4. Longitudinal fluorescence intensity distribution of pyrene-doped PMMA latex particle (7.2 μm).

As one of the applications of the system, the quantitative concentration estimation of dye molecules in small volumes is described [5]. Figure 3 shows depth-resolved fluorescence curves of pyrene-doped poly(methyl methacrylate) (PMMA) films. Pyrene fluorescence at 385 nm was observed along the longitudinal axis perpendicular to the films. The fluorescence intensities vary inside the films, which can be explained in terms of attenuation of the excitation light in the films. The intensity of the excitation light decreases as the light goes deeply, because it is absorbed by pyrene. The logarithmic plots of the fluorescence intensities are well fitted to linear lines, which indicates that pyrene was homogeneously distributed, and the Lambert-Beer law held for the attenuation in the films. The pyrene concentration determined based on the homogeneous distribution has a good relation to the gradient of the logarithmic curves, so that the molar extinction coefficient at 293 nm (excitation wavelength) can be calculated.

Figure 4 shows a depth profile of a PMMA latex particle with a diameter of 7.2 μm. The PMMA latex was soaked in a methanol solution of pyrene, washed with cold water, and then dispersed in pure water. The use of the molar extinction coefficient given by the curves in Figure 3 makes it possible to estimate the pyrene concentration in the latex particle. The obtained concentration was 4.9×10^{-2} M. It is noteworthy that this concentration estimation is very powerful since it is not affected by surrounding materials, even if the surroundings absorb or scatter the light.

Here, some problems with the present system are considered and commented. The most serious problem with microspectroscopy is chromatic aberration. Since the focal length of an objective lens depends on the wavelength, the depth coordinates on fluorescence images observed at different wavelengths are shifted away from each other. Therefore, the chromatic aberration causes distortion of spectra obtained at small volumes. In the developed system, the chromatic aberration is automatically compensated by varying the position of both the sample stage and the zoom lens as a function of wavelength [5]. The other problem is damage of a sample caused by the high-intensity excitation. The laser beam is condensed onto the submicrometer spot, so that the excitation pulse energy often rise to over tens mJ/cm^2. Such an intense light may damage most of the molecular materials. Therefore, the intensity of the excitation laser has to be attenuated and adjusted to be below the damage threshold of the sample. Under the latter condition, the observed fluorescence usually becomes weak. Fortunately, the developed system has the advantage of high detection efficiency due to the high numerical aperture of the objective lens, which overcomes the weakness of the fluorescence intensity.

3. FLUORESCENCE DEPOLARIZATION SPECTROSCOPY

The present fluorescence spectroscopy system can be extended to the space- and time-resolved fluorescence depolarization measurement in a selected small volume [6]. For the polarization spectroscopy, the excitation laser beam is passed through a Babinet-Soleil compensator to adjust the polarization direction. The polarized fluorescence is selected by an analyzer and passed through a depolarizer, which are set in front of the monochromator. By setting the excitation polarization parallel and perpendicular to that of the analyzer, two respective decay curves are measured in each small volume. Figure 5 shows polarized fluorescence decays of liquid paraffin solution of p-terphenyl. The solution was sandwiched between two quartz plates to form a thin liquid layer with a thickness of 170 μm. The decay curves were measured at the middle of the layer. The rotational relaxation process of p-terphenyl can be analyzed from these curves.

Unfortunately, the conventional method for calculating the anisotropy decay is not applicable to the present system, because the excitation light is far from plane wave due to strong condensation by the high-numerical-aperture objective lens. The fluorescence from the sample is also collected at a large solid angle, so that various directions of polarization are mixed with each other on the detector plane. We therefore derived a practical theory of the fluorescence depolarization analysis for the microscopic measurement [6]. In this theory, changes in polarization caused by the

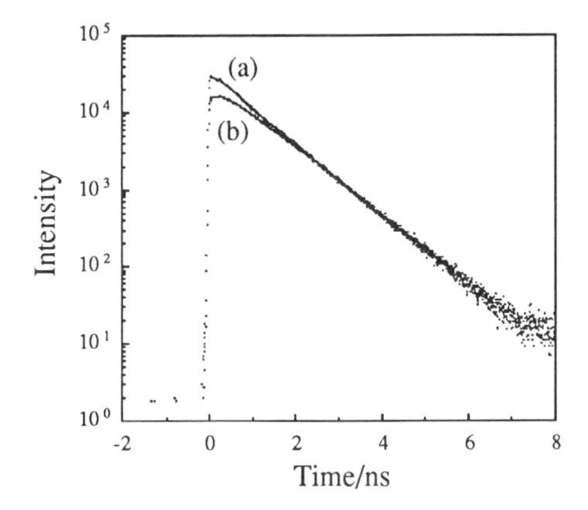

Figure 5. Polarized fluorescence decay curves of a liquid-paraffin layer containing p-terphenyl (1×10^{-3} M). The polarization of the excitation laser was set to be (a) parallel and (b) perpendicular to that of the analyzer.

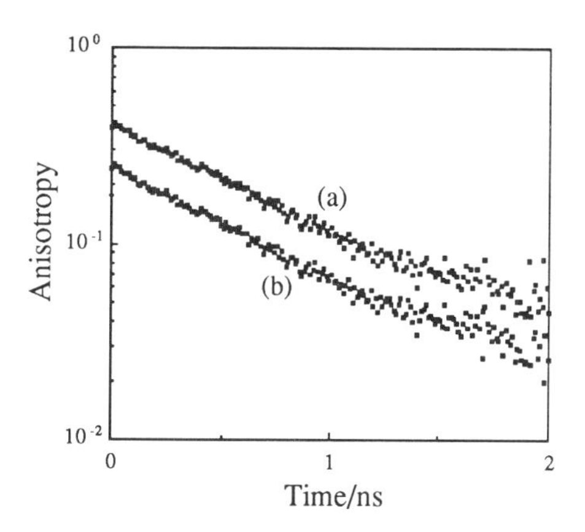

Figure 6. Anisotropy decay curves calculated from the data of Figure 5, using Eq. (7), (a) with the estimated parameters k_1 and k_2, and (b) by setting $k_1=k_2=0$.

microscope optics are represented as linear equations with two parameters k_1 and k_2, as follows,

$$D_{//}(t) = (1-k_1)\, I_{//}(t) + k_1\, I_{\perp}(t), \tag{1}$$

$$D_{\perp}(t) = k_2\, I_{//}(t) + (1-k_2)\, I_{\perp}(t), \tag{2}$$

where $D_{//}(t)$ and $D_{\perp}(t)$ are the observed fluorescence intensities of parallel and perpendicular polarization components, respectively, and $I_{//}(t)$ and $I_{\perp}(t)$ are unknown intensities of the undistorted polarization which would be obtained under the plane-wave excitation and detection systems. The parameters k_1 and k_2 can be theoretically evaluated from

$$k_1 = 1 - \frac{\int \{ E_x(\mathbf{r})W_x(\mathbf{r}) + E_y(\mathbf{r})W_y(\mathbf{r}) + E_z(\mathbf{r})W_z(\mathbf{r}) \}\, d\mathbf{r}}{\int \{ E_x(\mathbf{r}) + E_y(\mathbf{r}) + E_z(\mathbf{r}) \} \{ E_x(\mathbf{r}) + E_y(\mathbf{r}) + E_z(\mathbf{r}) \}\, d\mathbf{r}}, \tag{3}$$

$$k_2 = \frac{\int \{ E_x(\mathbf{r})W_y(\mathbf{r}) + E_y(\mathbf{r})W_x(\mathbf{r}) + E_z(\mathbf{r})W_z(\mathbf{r}) \}\, d\mathbf{r}}{\int \{ E_x(\mathbf{r}) + E_y(\mathbf{r}) + E_z(\mathbf{r}) \} \{ E_x(\mathbf{r}) + E_y(\mathbf{r}) + E_z(\mathbf{r}) \}\, d\mathbf{r}}, \tag{4}$$

where $E_x(\mathbf{r})$, $E_y(\mathbf{r})$, and $E_z(\mathbf{r})$ represent spatial intensity distributions of three polarization components in the excitation system, and $W_x(\mathbf{r})$, $W_y(\mathbf{r})$, and $W_z(\mathbf{r})$ are x, y, and z components of the vector point spread function for the observation optics. These parameters can also be determined experimentally by measuring the reference molecule, whose structure is rotationally symmetric with respect to the emission dipole and which is randomly oriented in the medium. The polarized fluorescence decay curves for such a sample are given as

$$D_{//}(t) = \frac{1}{3} D_o \exp\left(-\frac{t}{\tau_f}\right) \left\{ 1 + \frac{2}{5}(2-3k_1) \exp\left(-\frac{t}{\tau_r}\right) \right\}, \tag{5}$$

$$D_{\perp}(t) = \frac{1}{3} D_o \exp\left(-\frac{t}{\tau_f}\right) \left\{ 1 - \frac{2}{5}(1-3k_2) \exp\left(-\frac{t}{\tau_r}\right) \right\}, \tag{6}$$

where D_0 is the initial intensity, and τ_f and τ_r are fluorescence decay and rotational relaxation times, respectively. Fitting these equations to the observed decays of the reference, k_1 and k_2 are evaluated from the amplitude ratios of two decay components. Since k_1 and k_2 are the parameters characteristic of the developed system, their estimates can be used in the anisotropy analyses for any samples, if the experimental conditions are the same as that of the reference. Substituting these parameters into the following equation, an anisotropy decay $\gamma(t)$ for the microscopic depolarization measurement can be calculated from the observed curves $D_{//}(t)$ and $D_{\perp}(t)$.

$$\gamma(t) = \frac{D_{//}(t) - D_{\perp}(t)}{(1-3k_2)D_{//}(t) + (2-3k_1)D_{\perp}(t)}, \tag{7}$$

The curves shown in Figure 5 were analyzed based on the derived theory, yielding the estimated values of 3.6×10^{-1} and 9.3×10^{-2} for k_1 and k_2, respectively, and 970 ps and 710 ps for τ_f and τ_r, respectively. Figure 6 shows anisotropy decay curves calculated from Eq. (7) (a) with the obtained parameters and (b) with setting $k_1=k_2=0$. The latter corresponds to the conventional depolarization analysis. The anisotropy of curve (b) is always smaller than that of curve (a) at any delay time. The initial anisotropy of curve (b) is 0.26, that is quite different from the theoretical value of 0.4, while that of curve (a) is 0.41, which is successfully compensated. This result demonstrated that the present analyzing method is indispensable to the space- and time-resolved fluorescence depolarization spectroscopy.

4. TRANSIENT ABSORPTION SPECTROSCOPY

Absorption spectroscopy is indispensable for identifying non-luminescent excited-states, chemical intermediates, and hot molecules, and for analyzing their behaviors. Unfortunately, the absorption (transmission)-mode confocal microscope is not applicable to the three-dimensional measurement, since an optical transfer function of this microscope is angularly band-limited so that the depth resolution is degraded, especially for laterally structureless samples such as films and stacking layers. This difficulty is common to conventional absorption microscopes. In order to obtain the three-dimensional resolution, we developed a new confocal microscope [7].

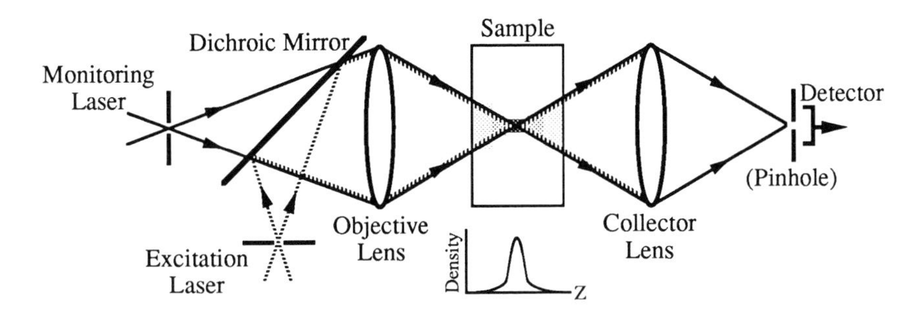

Figure 7. An optical diagram of a confocal transient absorption microscope.

Figure 7 shows an optical diagram of the microscope. Excitation light from a point source is focused on a sample by an objective lens, as in the case of the fluorescence excitation system. This light excites molecules so that their concentration is proportional to the photon density. An absorption-monitoring laser is introduced coaxially and condensed confocally with the excitation system. The transient absorption in the object is imaged by a collector lens and observed by a incoherent detector. Here, let us consider the imaging of a thin film doped with absorptive molecules. When the film is set on the focal plane shown as A in Figure 7, the excitation light is condensed to a diffraction-limited spot on the film so that the molecules should be excited densely. Since the monitoring light is transmitted through the same region as the excited area, the film works as a dense attenuation filter. On the other hand, the film shifted to an out-of-focus plane, as shown as B in Figure 7, is irradiated in a large area but with much lower intensity, so that the density in the excited state is reduced. Since the transient absorption is proportional to the concentration of the excited molecules, the absorbance is much smaller than that of the film at position A. Therefore, the observed transient absorption is mainly contributed by the focal spot. This shows that the present confocal transient absorption microscope has the three-dimensional space resolution. In contrast with this microscope, the conventional confocal absorption microscope, that measures ground-state absorption without the excitation system, has no depth resolution, because the ground-state absorption is independent of the position.

We developed a space- and time-resolved transient absorption measurement system based on the present microscope and a pump-probe method. The system utilized the same laser source and microscope as those of the fluorescence system shown in Figure 2. Second harmonic and fundamental pulses of the dye laser were used as the excitation and the monitoring lights, respectively. Since the laser power is highly intensified by condensing a beam at the focal spot, a nJ excitation pulse is enough for the transient absorption measurement under the microscope, which has been normally performed by mJ pulse lasers such as Q-switched YAG and amplified dye lasers. The monitoring pulse was optically delayed to the excitation pulse, and the delay time was scanned so that the temporal variation of the transient absorption could be observed. The detected signal was processed by a lock-in amplifier with a chopping signal of 1 kHz to precisely measure the transient absorption.

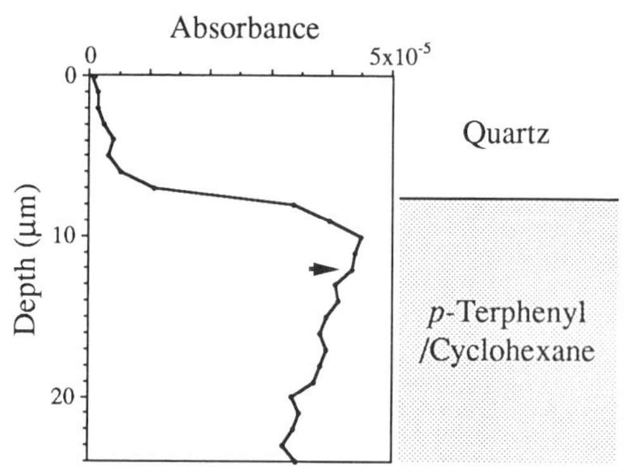

Figure 8. Depth profile of transient absorption of p-terphenyl (1 x 10^{-3} M) in a cyclohexane liquid layer.

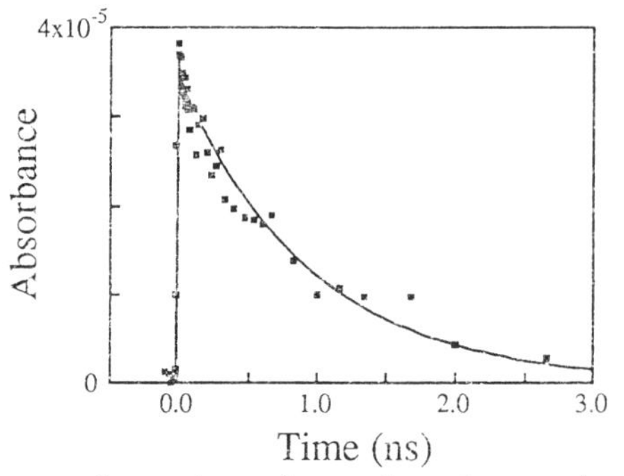

Figure 9. Decay curve of transient absorption observed at the middle of the liquid layer. The sample was the same as in Figure 8.

Figure 8 shows the result of a depth-resolved measurement of a cyclohexane solution layer containing p-terphenyl with a thickness of ~100 μm. The $S_n \leftarrow S_1$ transition of p-terphenyl molecules was measured along the longitudinal axis perpendicular to the layer. The depth resolution of this system could be determined from the profile to be 1.5 μm, which was defined as the FWHM of a differential curve at the interface between the quartz plate and the solution. Figure 9 is a decay curve of the transient absorption observed at the middle of the solution layer. The time resolution was determined from this curve to be 10 ps. To the best of our knowledge, this is the first demonstration of the depth-resolved absorption measurement under a microscope.

5. ANALYSIS OF MULTI-DIMENSIONAL DECAY CURVES

The time- and space-resolved fluorescence and absorption spectroscopy systems provide a large number of rise and decay curves as a function of position and wavelength. The mathematical method for analyzing these data is required to use a fast and, of course, accurate processing. The nonlinear least-squares method has been usually applied for decay analyses because of the advantage of the high accuracy in the estimated parameters [8]. Unfortunately, since this method uses a nonlinear iterative algorithm, it is complicated in programming, and it takes an extremely long computation time, so that the time- and space-resolved spectroscopic data cannot be analyzed for the tolerable time. In addition, this method may fail to converge if the initial parameters are set to inappropriate values. To avoid these difficulties, we proposed a new mathematical method based on a convolved autoregressive model [9].

When the dynamic process can be represented by a linear differential equation, the observed decay curve x(t) is given by the convolution of a multiexponential curve with an instrumental response function y(t) as follows:

$$x(t) = \int \sum_{i=1}^{N} A_i \exp\left(-\frac{t'}{\tau_i}\right) y(t-t')\, dt', \tag{8}$$

where τ_i and A_i are the decay constant and the amplitude of the i-th component decay, respectively. We derived that x(t) can also be expressed as a convolved autoregressive model:

$$x(n\Delta t) = \sum_{i=1}^{N} B_i\, x((n-i)\Delta t) + \sum_{i=1}^{N} C_i\, y((n-i+1)\Delta t), \tag{9}$$

where Δt is the sampling interval, and B_i and C_i are model parameters, which are directly related with τ_i and A_i. Since Eq. (9) is the linear combination of x(t) and y(t) with coefficients of B_i and C_i, a linear least-squares method can be used for fitting Eq. (2) to the observed decay curve and determining B_i and C_i. Then, τ_i and A_i are given from B_i and C_i through characteristic and recursive equations.

This method is based on the linear algorithm so that the computation time is much shorter than that of the conventional nonlinear least-squares method. To confirm this advantage, we analyzed fluorescence decay curves of cryptocyanine in ethanol by the present and conventional methods and compared their results. The computation time of the present method was 0.11 s with a workstation (SUN4-370), which was more than 100 times shorter than that of the nonlinear method (12.08 s). The decay constants estimated by the present and conventional methods were 82.1±3.3 ps and 81.3±3.2 ps, respectively. There is no difference in accuracy between the two methods. In addition, the present method does not require the program parameters such as initial values, acceleration coefficients, and dumping factors.

The convolved autoregressive model fitting method can be extended to the blind deconvolution where two decay curves are analyzed instead of the instrumental response function. The global analysis based on the present model was also proposed and applied to various reaction systems.

6. CONCLUSION

The micrometer three-dimensional space- and picosecond time-resolved fluorescence and absorption spectroscopy systems combined with a fast and accurate analyzing method have been described. Characterization of molecules and detailed analysis of chemical reactions in the inhomogeneous samples can be performed [10-12], and the information given by the systems can be used in controlling photophysical and photochemical processes in micrometer small volumes.

ACKNOWLEDGMENTS

The authors express their sincere thanks to Prof. H. Masuhara and Drs. N. Kitamura, H. Misawa, and N. Tamai of our project for their collaboration.

REFERENCES

1. T. Wilson, Ed., *Confocal Microscopy* (Academic Press, London, 1990).
2. M. A. Kahlow, W. Jarzeba, T. P. DuBruil, and P. F. Barbara, Rev. Sci. Instrum. 59 (1988) 1098.
3. D. V. O'Connor and D. Phillips, *Time-Correlated Single-Photon Counting* (Academic Press, London, 1985).
4. H. Masuhara, S. Eura, H. Fukumura, and A. Itaya, Chem. Phys. Lett., 156 (1989) 446.
5. K. Sasaki, M. Koshioka, and H. Masuhara, Appl. Spectrosc., 45 (1991) 1041.
6. M. Koshioka, K. Sasaki, and H. Masuhara, Appl. Spectrosc., submitted.
7. K. Sasaki, M. Koshioka, and H. Masuhara, J. Opt. Soc. Am. A, 9 (1992) 932.
8. D. V. O'Connor, W. R. Ware, and J. C. Andre, J. Phys. Chem., 83 (1979) 1333.
9. K. Sasaki and H. Masuhara, Appl. Opt., 30 (1991) 977.
10. M. Koshioka, H. Misawa, K. Sasaki, N. Kitamura, and H. Masuhara, J. Phys. Chem., 96 (1992)2909.
11. K. Nakatani, H. Misawa, K. Sasaki, N. Kitamura, and H. Masuhara, J. Phys. Chem., 32 (1993) 788.
12. H. Masuhara, N. Kitamura, H. Misawa, K. Sasaki, and M. Koshioka, J. Photochem. Photobiol. Chem. A, 65 (1992) 235.

MICROCHEMISTRY
Spectroscopy and Chemistry in Small Domains
Edited by H. Masuhara et al.

Time-resolved confocal microspectroscopic imaging

K.P. Ghiggino, P.G. Spizzirri and T.A. Smith

Department of Chemistry, University of Melbourne, Parkville, Vic., Australia, 3052.

The design of an instrument to undertake time-resolved scanning confocal fluorescence imaging is described. The instrumentation utilizes a picosecond laser excitation source, a fibre-optic based scanning confocal microscope and time-correlated single photon counting detection. Fluorescence decay profiles and time-gated fluorescence images can be recorded with a time resolution of ~30 picoseconds and a spatial resolution of ~1 μm. The application of this technique to image species with different fluorescence decay characteristics within cellular environments is outlined.

1. INTRODUCTION

Optical microscopy is widely used in the biological and chemical disciplines and has recently gained increased interest due to enhanced spatial resolution and improved sensitivity. In particular, the high spatial resolution (both axial and lateral) conferred by scanning confocal fluorescence imaging (SCFI) has revolutionised the field of three dimensional image reconstruction through inherent optical sectioning capabilities and simplified sample preparation techniques. We report here the development of a new technique of time-resolved SCFI to complement other forms of microscopy applicable to a wide variety of systems of chemical and biological interest.

In many experimental situations where fluorescence imaging is employed, high spatial resolution alone is inadequate to enable the extraction of all the information required about a given microscopic system under investigation. One example of this is the porphyrin and phthalocyanine families of compounds currently under investigation for use in photodynamic cancer therapy [1]. The intracellular location and environment, and the extent of aggregation of these compounds, are of fundamental interest since these factors are critical to the activity of the drug as a photosensitiser of singlet oxygen production, upon preferential uptake by neoplastic cells [1-4]. The high degree of spectral resolution necessary to perform such detailed analysis is often difficult to achieve experimentally due to the spectral properties of the compounds in question and to photon losses incurred when using many common wavelength resolving devices (e.g. monochromators). For example, porphyrins are able to aggregate under certain conditions, with the monomer and aggregate displaying very similar absorption and fluorescence spectral properties [5,6]. They do, however, have quite different fluorescence decay characteristics, potentially enabling the determination of the intracellular distributions of monomer and aggregate on the basis of their differing fluorescence lifetimes. Similarly, mixed fluorophores with similar emission spectra but differing fluorescence lifetimes may be spatially resolved using this technique. For this reason, the general term "fluorescence lifetime imaging" is sometimes used to denote time-resolved fluorescence imaging.

Since the fluorescence lifetimes of many dyes are highly sensitive to their local environment, time-resolved fluorescence imaging could also be used to reveal important information regarding the local chemical composition and details of the spatial distribution of different molecular environments surrounding a fluorophore. In particular, conditions such as pH, metal ion concentration (e.g. [Ca^{2+}]), the presence of oxygen and other quenchers of excited states, viscosity and polarity can have major effects on the temporal behaviour of a fluorophore. Additional information regarding the rotational motion of a fluorophore referenced to its location in the sample may also be extracted from the measurement and analysis of fluorescence anisotropy decays. Temporal resolution offers additional advantages over conventional fluorescence imaging by facilitating the discrimination between fluorescence from a fluorophore, and natural background fluorescence (auto-fluorescence) and scattered (Rayleigh or Raman) light. Conversely, resonance Raman scatter signals, which may be overwhelmed by fluorescence, can be temporally isolated by rejection of these emissions [7,8]. Micro-Raman imaging techniques, which have recently been the subject of growing interest, would benefit greatly from this ability to decrease the fluorescence signal contribution in a Raman image.

Techniques for time-resolved fluorescence microscopy (TRFM) have been reviewed recently [9-11, and references therein] and the efforts into developing TRFM can be categorized broadly into two general approaches. Firstly, several techniques have been reported by which fluorescence decay profiles may be collected through a microscope [9,12-26]. These methods differ in their choice of excitation sources and detection techniques depending upon the particular experimental arrangement employed. There are, of course, advantages and disadvantages associated with each of the various methods in which fluorescence decay profiles are collected at particular points on the sample. For example, "impulse response" techniques, in which a high intensity, low repetition rate, pulsed laser is used in conjunction with a fast photodetector can produce fluorescence decays very quickly. Depending on the laser pulse width and type of detector (e.g. fast photomultiplier tube, CCD or streak camera), time resolution as short as hundreds of femtoseconds [27] and, through signal averaging, decays of reasonable signal-to-noise (S/N) can be obtained. The use of high intensity pulses to achieve the required fluorescence intensity can, however, due to the high photon density at the focal point of a microscope objective (of even low magnification), cause damage to the sample if it is easily bleached, or photo- or thermally-degraded. Another problem with high intensity excitation pulses can be equipment damage such as lens failure due to intra-element focussing within complex objective lens assemblies [28].

Photon counting detection methods offer high sensitivity, thus permitting the use of low intensity light sources while conferring excellent S/N characteristics. The time-correlated single photon counting (TCSPC) technique for decay measurements [12-18,20-26], whilst offering very good time resolution (approaching ten's of picoseconds), can, however, be time consuming to achieve the desired S/N, especially with low repetition rate laser or flash lamp excitation. Multiphoton counting techniques [19,29] have been developed to facilitate the use of low repetition rate excitation sources but these techniques currently have far inferior time resolution to TCSPC. One example of a method employed to enhance the measurement of fluorescence lifetime information uses a combination of single photon counting with phase shift and demodulation techniques [30,31] in order to obtain the cumulative advantages of each technique.

Such methods can, in principal, provide important information regarding fluorophore location, environment and heterogeneity on a microscopic scale. They do, however, suffer from the problem that the spatial resolution is in practice, limited by having to collect a complete set of data at each location on the sample. Furthermore each set of data must be analysed subsequently by fitting a trial function of an assumed mathematical form. It has been shown in other cases of photophysical interest, that the choice of the most appropriate decay trial function is often difficult, and interpretation of data therefore open to debate. In the case of simple homo- or hetero-geneous systems, the appropriate choice of function is clear (in the case of data collected in the time domain, fluorescence decay data are usually analysed using single- or multi-exponential decay functions), so long as phenomena such as energy transfer or fluorescence quenching can be neglected. If energy transfer or diffusion-controlled processes are operative, more complex, non-exponential functions may be more appropriate. The time taken to analyse each data set can vary depending on the accuracy of the adopted fitting technique (c.f. simple linear least squares with iterative, non-linear least squares techniques) and the particular trial function.

The second type of TRFM method under discussion here is time resolved fluorescence imaging (or "fluorescence lifetime imaging") where, instead of collecting complete decays at particular (widely spaced) points on a sample, the fluorescence intensity as a function of time is monitored as the excitation spot is scanned at high resolution across the sample. Methods of achieving time resolution in fluorescence imaging using various detection techniques have been discussed elsewhere [11,19,32-35] but the concept is still in its infancy and these techniques have not yet been applied extensively to real chemical or biological systems. In principle such techniques should enable imaging of a variety of heterogeneous fluorescent systems, however, there are numerous problems to be overcome. As discussed elsewhere [34], pulsed image intensifier gating techniques are limited in their overall time-resolution and their inability to discriminate between fluorophores with fluorescence lifetimes which do not differ by at least an order of magnitude. In order to overcome some of these problems Lakowicz and Berndt [34] have described an imaging apparatus based upon phase sensitive detection in which 100% contrast was achieved between fluorophores with similar fluorescence lifetimes.

In this paper, we describe the development of a scanning laser, confocal fluorescence microscope capable of picosecond time resolution and micron spatial resolution in three dimensions. The method is based on gated time-correlated single photon counting methods [36] and appears quite promising with regards to time resolution, data analysis complexity and potential instrument miniaturisation.

2. EXPERIMENTAL SECTION

A single layer of onion skin was isolated and treated with a dilute aqueous solution of ethidium bromide (EB) followed by treatment with a dilute ethanolic solution of nile red (NR).

The scanning confocal microscope used in this work was based on the fibre-optic confocal microscope system reported recently [25,37] and uses TCSPC detection electronics. The experimental arrangement is illustrated schematically in Figure 1. In brief, the output from either a synchronously mode-locked and cavity dumped dye laser (Spectra Physics 3500) or a mode-locked Argon ion laser (Spectra Physics 2030) after extra-cavity pulse selection (Spectra Physics 344/345)) is launched into a single mode optical fibre (Newport, F-SA). This fibre (of ~4 μm core diameter) is used to both deliver the excitation beam to the scanning head and act as the confocal pinhole in the collection of the emission (Figure 2). We have determined that there is negligible temporal dispersion in passing the picosecond pulses through the short length of fibre used in the confocal arrangement (~1 m), as measured using background-free autocorrelation. Any dispersion is certainly far less than the time-resolution of the TCSPC technique.

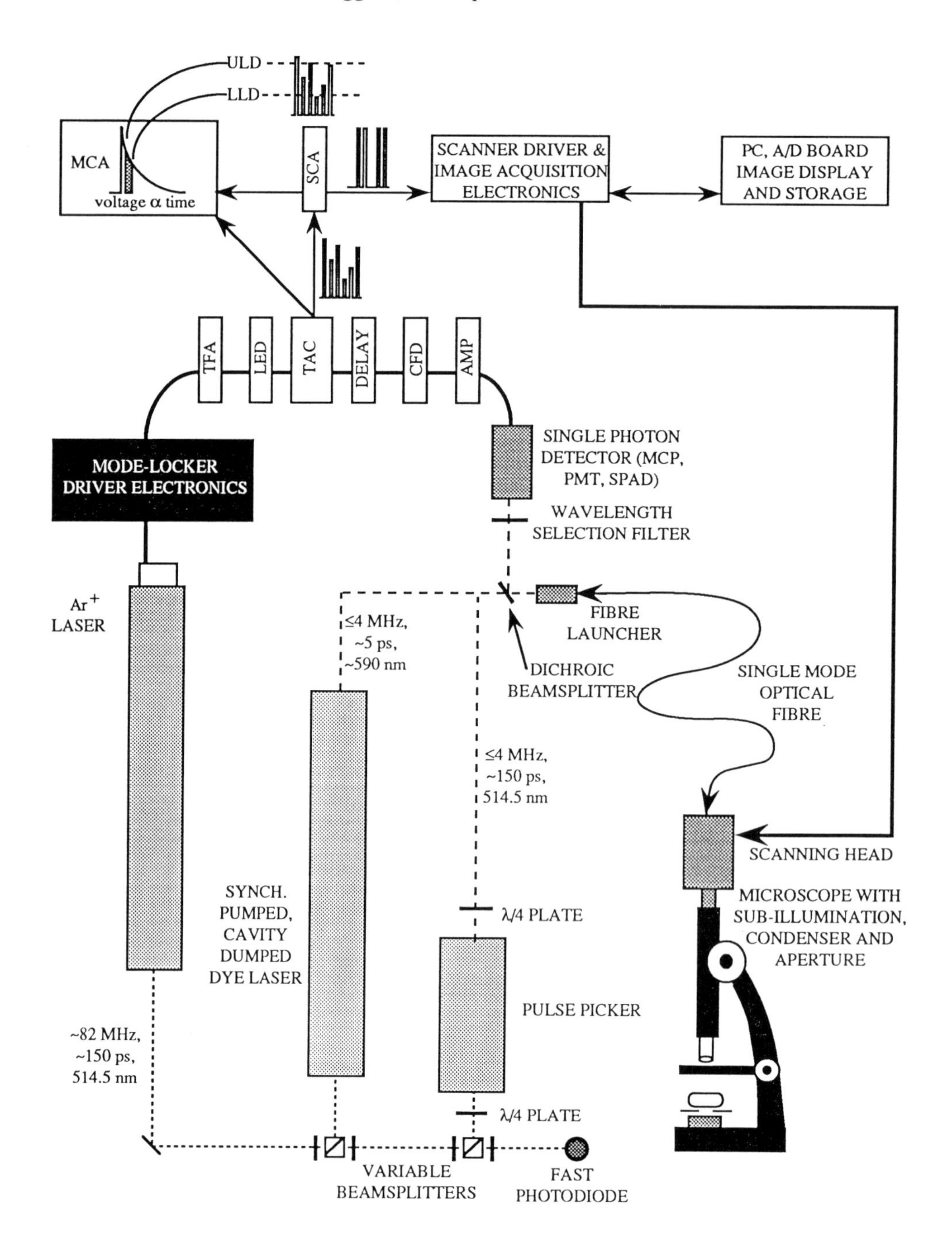

Figure 1. Schematic diagram of the experimental arrangement used for time-resolved fluorescence imaging and decay measurements.

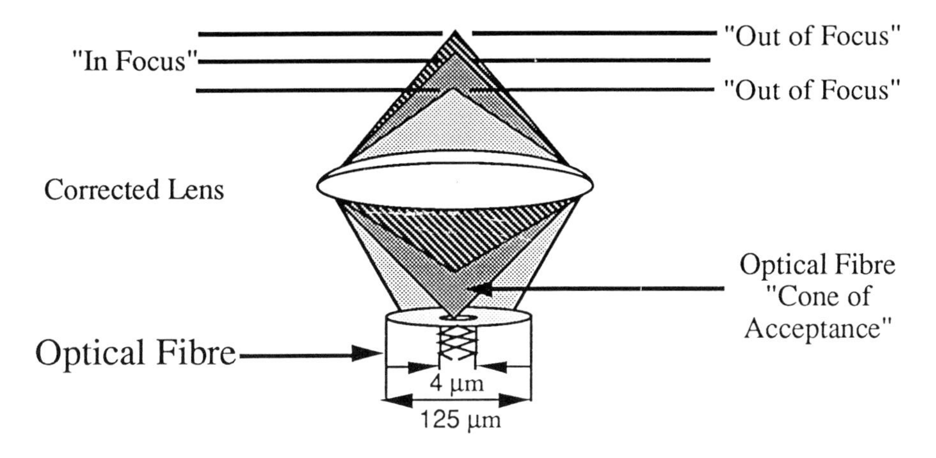

Figure 2. Diagrammatic representation of the fibre-optic, confocal configuration used.

The scanning head fits directly into the camera port of an Olympus BH-2 fluorescence microscope. The focussed (UVFL 10x, 0.4 NA) picosecond pulse laser beam is raster scanned across the sample and the resultant fluorescence is collected (through the same optical path) and isolated from scattered excitation light using a dichroic beamsplitter and an appropriate blocking filter. The emission is detected in the usual TCSPC way using one of a variety of single photon detectors. During this work various single photon counting detectors were employed including: a photomultiplier tube (PMT) (Philips XP2020Q), a microchannel plate (MCP) PMT (Hamamatsu R1564U-01), and a commercially available, passively quenched, avalanche photodiode (APD) (RCA (EG&G) C30902S). An alternative detector that can provide superior detection characteristics is the single photon avalanche photodiode (SPAD) device reported by Cova et al. [38] and sold commercially by Silena, Italy. Pulses from the detector were amplified using wideband preamplifiers (AMP, Research Communications type 9009) prior to being processed by a constant fraction discriminator (CFD, Tennelec TC454) en route to the start channel of a biased time-to-amplitude converter (TAC, Ortec 457). The delayed stop pulses for the TAC were supplied either from a fast photodiode detecting the laser output directed through another CFD, or the synchronous electronic output of the cavity dumper driver after preshaping using a timing filter amplifier (TFA, Ortec 454) and leading edge discriminator (LED, Ortec 436) combination.

The fluorescence lifetime imaging technique employed in this work is analogous to the method of recording time-resolved emission spectra (TRES) using time-correlated single photon counting [36]. In TRES, the fluorescence resulting from narrow pulse excitation is gated so that only those photons emitted at certain times after excitation are recorded. In practice this is achieved by selecting the upper and lower voltage limits of the output from a TAC module by adjusting the voltage discriminators on a multi- or single-channel analyser (MCA or SCA) input. This selects the time window (time delay and gating width) in the decay profile from which emission is collected. A monochromator can then be scanned synchronously with the MCA operating in multichannel scaling mode. In the time-resolved fluorescence imaging method under discussion here, the output pulses from the TAC, which vary in intensity proportionally with time after excitation, are passed through a SCA (Ortec 550) which outputs a well defined pulse upon each occurrence of a TAC pulse meeting the criteria set on the upper and lower level voltage discriminators on the SCA. These pulses are then directed to a fast digital frequency counter (frequency-to-voltage (F/V) converter) of our own design, which is synchronised to each pixel position on the mirror scanning current ramp. The F/V output voltage is further modified using an analogue amplifier circuit with linear offset adjustment. This facilitates a linear "back-off" control for photon counting backgrounds (particularly useful for APD devices with high residual background count rates) to ensure that the output voltage, which is sampled by a fast analogue-to-digital converter (A/D) board (Axiom 5411) in a personal computer, is acquired with the full dynamic range of the A/D converter. By adjusting the upper and lower level discriminators on the SCA while monitoring the fluorescence decay profile on the MCA, fluorescence images corresponding to early and late times after excitation can be recorded and later manipulated.

Gain control provided at this stage does not increase the signal dynamic range, however, when combined with the linear offset adjustment, the S/N and image contrast characteristics can be improved. As the computer supplies the clock signal for the synchronised X-Y scan generator and F/V converter, software adjustable "pixel dwell" times in the range ~1 msec to 100 msec are provided. This results in a signal level enhancement which eliminates the need to signal average multiple image frames since a full 8 bit dynamic A/D converter range can be utilised with a high contrast level. This eliminates the problem of over-exposing thermally- or photo-sensitive samples to a focussed laser beam during multiple image scans for signal averaging purposes. Additionally, reduced excitation intensity may be used with longer "pixel dwell" times to achieve the same effect providing sufficient sample fluorescence intensity is present and can be detected using a sensitive photon counting detector. Unfortunately, the higher dwell times lead to long image collection times for 65536 (256 x 256) image points (pixels) making a 12 bit A/D dynamic range or very low illumination level image unrealistic using this method. Detection count rate limitations imposed by the TCSPC technique and its components also determine the image acquisition time. Typical photon detection count rates were kept to <1% of the laser excitation repetition rate (4 MHz) during the image collection resulting in a total scan duration from ~2 minutes to ~120 minutes per frame depending upon the "pixel dwell" time used. Typical image collection times ranged from 4-18 minutes and the illumination intensity for image collection was ~8 μW average power (as measured after the optical fibre). Fluorescence decay profiles were collected with photon detection count rates kept to <0.5% of the laser excitation repetition rate (4 MHz) in accordance with standard TCSPC procedures, and using an illumination intensity of ~1 μW in order to reduce bleaching effects.

As the PC controls the instrument and displays the acquired image, various "off the shelf" components may be used and adapted (with suitable software) to perform the same task. For example, a suitable frequency counting board could be used to directly sample the output of the SCA without the need for the F-V converter/A-D converter combination. In our case, we developed a specialised system of software and hardware which could be easily configured for the collection of total (ungated) and time-gated fluorescence imaging, fluorescence decay curves from particular micro-regions of the sample, as well as reflection and transmission micrographs. In the following section, we report some examples of data obtained recently using this apparatus.

3. RESULTS AND DISCUSSION

Ethidium bromide and Nile Red are both common fluorescent staining agents used in fluorescence microscopy. EB, a phenanthridinium intercalator, has a high membrane permeability in non-viable cells and binds to DNA with little or no base pair preference. The intercalated form is therefore located in cell nucleii while the unbound form may be found distributed throughout the cell depending on staining procedures used and the chemical environment of the cell. Once bound (intercalated), the fluorescence intensity is enhanced approximately 20 times and the emission maximum blue-shifts from 610 to 595 nm [39]. These small wavelength shifts make spectral discrimination of the bound and unbound species difficult. The intercalated form of EB is known to have a fluorescence lifetime of ~23 ns compared to the unbound form which exhibits a lifetime of ~1.8 ns [40].

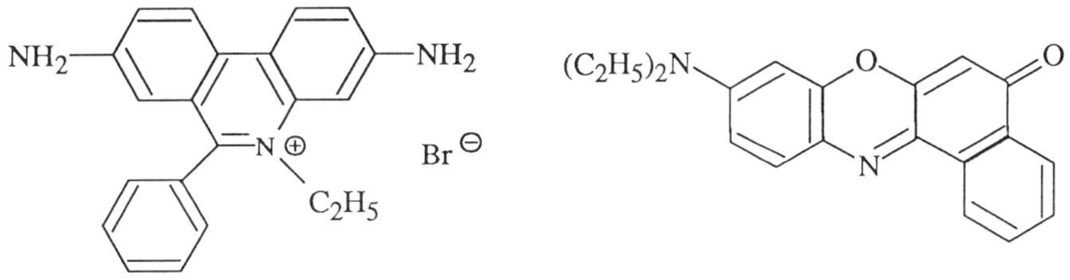

NH_2 NH_2 $N^{\oplus}$ C_2H_5 $Br^{\ominus}$ $(C_2H_5)_2N$ O O N

Ethidium Bromide **Nile Red**

Figure 3. Structures of the fluorescence probes used.

The phenoxazine dye NR, on the other hand, has been used to probe the lipid environments within and around cells as its fluorescence is enhanced in these low polarity environments compared to its almost nonfluorescent character in polar solvents. This fluorescence quantum yield dependence, with a concomitant variation in fluorescence lifetime (~1.8 to ~4.2 ns depending on the solvent (polarity)) [41], means that NR can be used as a polarity probe. The emission maximum occurs at 635 nm which is in close proximity to that of EB.

Figure 4(a) shows the total (ungated) fluorescence image collected from a thin section of onion skin stained with these two dyes. Regions of strong fluorescence intensity are evident corresponding to localisation of the various species of the dyes in the nucleus and around the cell membrane. Discrimination of emission from the intercalated EB, unbound EB, and NR is not easily achieved on a purely spectral basis. Without prior knowledge as to the location of the various cellular features (i.e. cell nucleus and membrane), one could not assign emission from a specific region as resulting from one dye species or another. We have collected fluorescence decay profiles from specific locations in the sample as indicated in Figure 4(a). Figure 4(b) shows the fluorescence decays collected by monitoring the emission from (i) the nucleus and (ii) a membrane region of the image. One problem with analysing fluorescence decays is that an assumed functional form must be fitted to the decay data. The decays illustrated in Figure 4(b) as well as decays collected from other less well defined regions required a function consisting of the sum of four exponential terms to obtain adequate fits. Whilst the recovered lifetimes could be assigned to the various emitting species (bound and unbound EB, NR and an unknown component in the EB with a lifetime of ~9 ns), the assumption that the dyes exist in distinct environments and exhibit a discrete lifetime for each environment is questionable. Other possibilities for analysis of these decays may be (i) the use of globally fitted lifetime distribution functions, or (ii) complex analysis on the basis of interactions between the dyes and other molecules in close proximity.

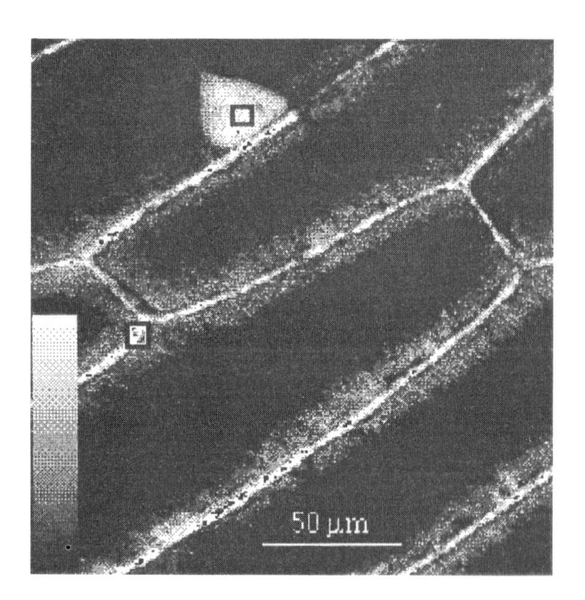

Figure 4(a). Ungated fluorescence image from a section of onion skin stained with EB & NR. The dark outlines correspond to the the nucleus and membrane regions respectively.

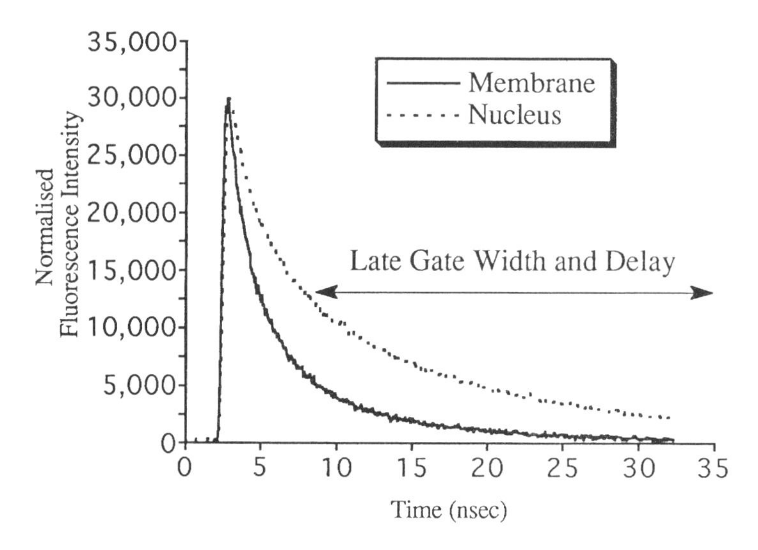

Figure 4(b). Fluorescence decay curves collected from onion skin stained with EB & NR corresponding to the areas depicted by small boxes in Figure 4(a).

Another consideration when recording fluorescence decays through a microscope is the collection of the instrument response function (IRF). To record an IRF in conventional TCSPC measurements the sample must be replaced with a scattering solution or a reflective surface such as a mirror. This change-over is particularly inconvenient when recording fluorescence decay data through a microscope since the region of interest of the sample may not be easily relocatable and wavelength selection devices must be changed. The switch between sample and scatterer can be particularly problematical in confocal measurements since this usually requires refocussing of the instrument which results in differences associated with the detection of sample fluorescence compared with that of scattered light (e.g. photon energies, optical path, coherent versus incoherent detection etc.). An analysis method coined "auto-reconvolution" has been reported recently [42,43] whereby fluorescence decay curves, collected either from two spectral regions or at the two emission polarization extremes in anisotropy measurements, are analysed simultaneously with one of the decays replacing the IRF in the iterative reconvolution analysis. The auto-reconvolution method requires that the two decays can be fitted by sums-of-exponential functions with the same number of exponential terms and with the same lifetimes i.e. only the contribution of each term differs between the decays. As discussed elsewhere [42,43], this method is particularly advantageous in situations where a suitable instrument response function (IRF) is difficult to record, as is the case when recording fluorescence decay data through a microscope. The auto-reconvolution method has the additional advantage that the whole of the decay curve (including the leading edge) can usually be analysed with no degradation of the fitting criteria.

We have also succeeded in recording fluorescence decays as a function of the emission polarization with a polarization analyser placed after the optical fibre, in front of the photon detector. We have found that there is quite a deal of polarization information retained even after both the excitation and emitted radiation have passed through the optical fibre. The maintenance of any polarization information may appear somewhat surprising since we do not, at this stage, use polarization preserving fibre. The analysis of these anisotropy decays is currently under investigation but at this time, quantitative results are difficult to obtain since the accurate determination of the instrument "G-factor" is problematical. Preliminary results, however, indicate that the auto-reconvolution method is again extremely useful for this application. The maintenance of polarization information within our instrument also makes viable the idea of polarization imaging proposed by Seidlitz et al. [44] in which separate fluorescence images are collected with an analysing polarizer set parallel and perpendicular to the polarization of the exciting light.

As discussed above, the collection of fluorescence decay curves through microscopes via one method or another is now reasonably common and is providing much useful information. The collection of time-resolved fluorescence images is, however, much rarer. As with time-resolved emission spectra, which have been shown to provide much useful information without the need to resort to an assumed functional form [45], time-gated fluorescence images can also simplify the analysis of complex fluorescence behaviour. An example of the use of the technique we have developed is illustrated in Figure 5 showing the same thin section of onion skin stained with the two fluorescent dyes discussed above. The gating delay and width were optimised to temporally isolate the intercalated EB emission from the various other species (late time gate). This allows the location of the nucleus to be clearly isolated from the other cellular components. Some residual emission can be observed in the gated image (Figure 5) around the cell boundaries. This can be attributed to the contribution of the 9 ns species which, from the decay analysis, is present in both nucleus and membrane locations with varying contributions to the total emission and can be observed as a residual component in the gating window used.

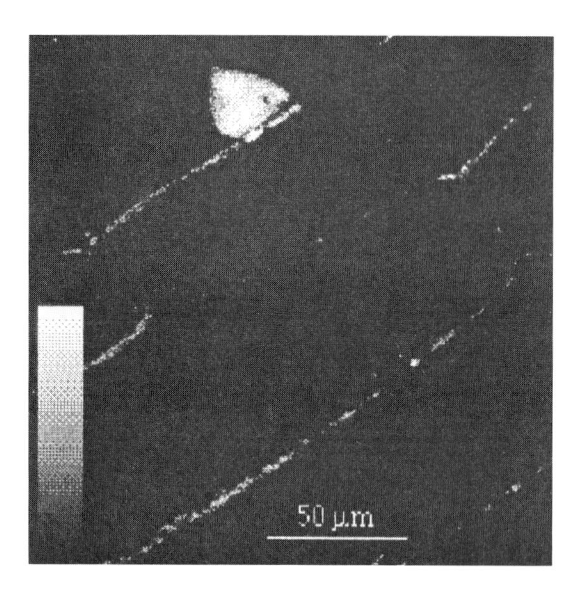

Figure 5. Late-gated fluorescence image from a section of onion skin stained with EB & NR. Gating width as indicated in Figure 4(b)

To obtain the image shown in Figure 5, the pixel dwell time had to be increased over the dwell time used to collect the ungated image (Figure 4(a)) because of the decreased number of pulses conforming to the SCA voltage criteria. This was done in order to maintain the same integrated pixel intensity in the nuclear region without resorting to increasing the laser power or gain settings.

An additional utility for the optical fibre-based microscope system described here is the facile incorporation of a scanning type 1a[†] [46] imaging mode (sensitive to transmission, refractive index, polarisation etc.). This can be achieved simply by turning off the laser and using the sub-stage illuminator/condenser facility incorporated in most fluorescence microscopes. The incoherent light transmitted through the sample is collected through the same scanning confocal fibre attachment and detected using (ungated) photon counting. An example of a transmission micrograph is illustrated in Figure 6. These images can be readily superimposed with the gated and ungated fluorescence images to obtain information about the cell shape, boundaries, size and overlapping of other cells. This permits distributions of fluorescence intensity, which are often highly localised, to be related back to the overall cell structure (particularly in cell monolayers).

[†]This mode of scanning light microscopy is non-confocal using a large area, sub-stage source for illumination and pin-hole detector, and is sensitive to transmission, refractive index and polarization properties of the sample.

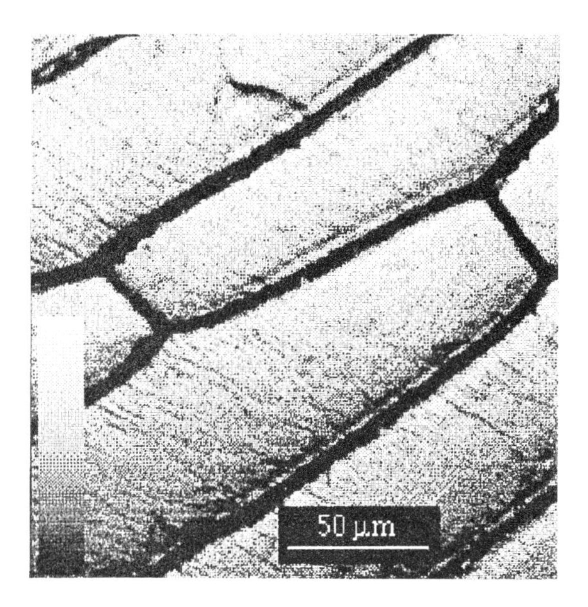

Figure 6. Scanning type 1a (transmission) image from the same region of the onion skin stained with EB & NR illustrated previously.

4. CONCLUSIONS

We have developed a simple extension of the gated TCSPC method to permit the recording of time-gated fluorescence images. The technique allows one to record (i) total (ungated) fluorescence images (ii) scanning type 1a images and (iii) time-resolved (gated) fluorescence images all with photon counting sensitivity, and with very little adjustment of the apparatus. Potentially, the technique could also be used to normalise the contribution to an image by a species which has a low quantum efficiency (or short fluorescence lifetime) in the presence of other species with different emission intensities and lifetimes. The technique also has potential application to Raman imaging where the contribution from fluorescence can swamp the Raman signal. The fluorescence contribution could be reduced by simply gating it out of the image. Conversely, Raman and Rayleigh scattering can be gated out of a fluorescence image. The experimental arrangement described here can be used readily with fibre-optic based confocal fluorescence microscope systems reported recently, offering micron spatial resolution in three dimensions with the convenience of a compact, easy to align instrument through the use of currently available fibre handling technologies. Such an arrangement can, of course, be easily adapted to any suitable excitation source such as mode-locked lasers and picosecond diode lasers, and any suitable single photon detection device such as photomultiplier tubes, microchannel plate PMT's or single photon avalanche photodiodes.

The limitations of the technique are: (i) the time-resolution (which is limited by the overall instrument response function width) and can currently be as narrow as 20-30 ps FWHM, (ii) the spatial resolution (which is limited by the excitation wavelength, objective lens used, the confocal aperture etc.), (iii) the ratio of the fluorescence lifetimes (where a factor of approximately two maybe the limit - this is currently under investigation), and (iv) the intensity of fluorescence. The latter point is perhaps the main disadvantage of this technique since many photons are wasted. Multi-photon sampling techniques may overcome this problem to some extent. Despite these limitations, we believe this technique has the potential to provide detailed information relating to the local chemical map on a microscopic scale and will complement other micro-spectroscopic techniques.

ACKNOWLEDGEMENTS

TAS is the Ernst and Grace Matthei Bequest Scholar of the University of Melbourne and support from this fund is gratefully acknowleged. PGS acknowledges the support of an Australian Postgraduate Research Award (Industry) in association with HBH Technological Industries Pty. Ltd. Financial support from the Australian Research Council is also acknowledged.

REFERENCES

1. A.J. MacRobert, S.G. Bown and D. Phillips, "What are the Ideal Photoproperties for a Sensitizer?", in "Photosensitizing Compounds: their Chemistry, Biology and Clinical Use.", (1989), John Wiley & Sons, p. 4.
2. S. Kimel, B.J. Tromberg, W.G. Roberts and M.W. Berns, Photochem. Photobiol., 50 (1989) 175.
3. J.D. Spikes, Photochem. Photobiol., 43 (1986) 681.
4. L.E. Bennett, K.P. Ghiggino and R.W. Henderson, J. Photochem. Photobiol. B: Biology, 3 (1989) 81.
5. K.P. Ghiggino, L.E. Bennett and R.W. Henderson, Photochem. Photobiol., 47 (1988) 65.
6. G.J. Smith, K.P. Ghiggino, L.E. Bennett and T.L. Nero, Photochem. Photobiol., 49 (1989) 49.
7. R.P. Van Duyne, D.L. Jeanmaire and D.F. Shriver, Anal. Chem., 46 (1974) 213.
8. J. Watanabe, S. Kinoshita and T. Kushida, Rev. Sci. Inst., 56 (1985) 1195.
9. H. Schneckenburger, H.K. Seidlitz and J. Erberz, J. Photochem. Photobiol., B: Biology, 2 (1988) 1.
10. R. Tian and M.A.J. Rodgers, "Time-Resolved Fluorescence Microscopy", in "New Techniques of Optical Microscopy and Microspectroscopy", R.J. Cherry (Ed.), (1991), MacMillan, p. 177.
11. C.G. Morgan, A.C. Mitchell and J.G. Murray, J. Microsc., 165 (1992) 49.
12. K. Kinosita, S. Mitaku, A. Ikegami, N. Ohbo and T.L. Kunii, Jap. J. Appl. Phys., 15 (1976) 2433.

13. F. Docchio, R. Ramponi, C.A. Sacchi, G. Bottiroli and I. Freitas, J. Microsc., 134 (1983) 151.
14. M.A.J. Rodgers and P.A. Firey, Photochem. Photobiol., 42 (1985) 613.
15. T. Minami, M. Kawahigashi, Y. Sakai, K. Shimamoto and S. Hirayama, J. Lumin., 35 (1986) 247.
16. R. Ramponi and M.A.J. Rodgers, Photochem. Photobiol., 45 (1987) 161.
17. G. Rumbles and A.J. MacRobert, personal communication.
18. T.A. Louis, G. Ripamonti and A. Lacaita, Rev. Sci. Inst., 61 (1990) 11.
19. X.F. Wang, S. Kitajima, T. Uchida, D.M. Coleman and S. Minami, Appl. Spect., 44 (1990) 25.
20. K. Sasaki, M. Koshioka and H. Mashuhara, Appl. Spect., 45 (1991) 1041.
21. H. Schneckenburger, W. Strauß, A. Rück, H.K. Seidlitz and J.M. Wessels, Opt. Eng., 31 (1992) 995.
22. H. Schneckenburger and K. König, Opt. Eng., 31 (1992) 1447.
23. H.K. Seidlitz, K. Stettmaier, J.M. Wessels and H. Schneckenburger, Opt. Eng., 31 (1992) 1482.
24. K. König, A. Rück and H. Schneckenburger, Opt. Eng., 31 (1992) 1470.
25. K.P. Ghiggino, M.R. Harris and P.G. Spizzirri, Rev. Sci. Inst., 63 (1992) 2999.
26. H. Masuhara, N. Kitamura, H. Misawa, K. Sasaki and M. Koshioka, Polym. Prepr., 33 (1992) 850.
27. K. Kinoshita, M. Ito and Y. Suzuki, Rev. Sci. Inst., 58 (1987) 932.
28. P.G. Spizzirri, unpublished results.
29. F. Pauker, H. Schneckenburger and E. Unsöld, J. Phys. E: Sci. Inst., 19 (1986) 240.
30. J.G. Murray, R.B. Cundall, C.G. Morgan, G.B. Evans and C. Lewis, J. Phys. E: Sci. Instrum., 19 (1986) 349.
31. C.G. Morgan and J.G. Murray, Chem. Phys. Letts., 179 (1991) 211.
32. A. Kusumi, A. Tsuji, M. Murata, Y. Sako, A.C. Yoshizawa and T. Hayakawa, Acta Histochem. Cytochem., 20 (1987) 716.
33. T. Ni and L.A. Melton, Appl. Spect., 45 (1991) 938.
34. J.R. Lakowicz and K.W. Berndt, Rev. Sci. Inst., 62 (1991) 1727.
35. R. Cubeddu, P. Taroni and G. Valentini, J. Photchem. Photobiol. (Biol.), 12 (1992) 109.
36. D.V. O'Connor and D. Phillips, "Time-Correlated Single Photon Counting", (1983), Academic Press.
37. M.R. Harris, "A Scanning Confocal Microscope", Appl. No. PCT/AU 89/00298, (1989).
38. S. Cova, A. Lacaita, M. Ghioni and G. Ripamonti, Rev. Sci. Inst., 60 (1989) 1104.
39. T. Härd and D.R. Kearns, J. Phys. Chem., 90 (1986) 3437.

40. D.P. Millar, R.J. Robbins and A.H. Zewail, J. Chem. Phys., 76 (1982) 2080.
41. G.B. Dutt, S. Doraiswamy, N. Periasamy and B. Venkataraman, J. Chem. Phys., 93 (1990) 8498.
42. A.J. Marsh, G. Rumbles, I. Soutar and L. Swanson, Chem. Phys. Letts., 195 (1992) 31.
43. G. Rumbles, T.A. Smith, J. Morgan, M. Carey and B. Crystall, in preparation
44. H.K. Seidlitz, H. Schneckenburger and K. Stettmaier, J. Photochem. Photobiol. B: Biology, 5 (1990) 391.
45. K.P. Ghiggino, S.W. Bigger, T.A. Smith, P.F. Skilton and K.L. Tan, "Kinetic Spectroscopy of Relaxation and Mobility in Synthetic Polymers", in "Photophysics of Polymers", C.E. Hoyle and J.M. Torkelson (Ed.), (1987), ACS, Washington, p. 368.
46. T. Wilson and C.J.R. Sheppard, "Theory and Practice of Scanning Optical Microscopy"", (1984), London, Academic Press.

MICROCHEMISTRY
Spectroscopy and Chemistry in Small Domains
Edited by H. Masuhara et al.

Morphology of organized monolayers by Brewster angle microscopy

R. C. Ahuja, P.-L. Caruso, D. Hönig, J. Maack, D. Möbius, and G. A. Overbeck

Max-Planck-Institut für biophysikalische Chemie, Postfach 2841, D-37018 Göttingen, F.R.G.

The potential of Brewster angle reflectometry and microscopy in characterizing phase transitions and long range order in monolayers at the air-water interface and on solid substrates is demonstrated. This recently developed method does not require fluorescent probes and yields information on tilt orientational and bond orientational order in organized monolayers. The technique is particularly useful in the investigation of dye monolayers and for following photochemical reactions by using non-actinic light.

1. INTRODUCTION

The lateral structure of monolayers and transitions between different monolayer phases have been visualized in the past using fluorescent probes (fluorescence microscopy) [1-5]. This techniques is based on different density and/or different spectroscopic properties including fluorescence quantum yield of the dye probe in the different monolayer phases. Many interesting phenomena have been observed in this way like the manipulation of domains in an inhomogeneous electric field [6] or the melting of domains upon monolayer transfer onto a solid substrate [7]. The internal structure of monolayer domains including long range tilt orientational order was observed with polarized fluorescence microscopy [8]. A fixed relation between the orientation of the probe molecule attached to the hydrophilic head group and the orientation of the hydrocarbon chain of the amphiphile had to be assumed.

The recently developed Brewster angle microscopy (BAM) does not require such probes and enables one to investigate the undisturbed morphology and different phases of monolayers at the air-water interface and on solid substrates. The first instrument was built with an elaborate optical system for achieving a high lateral resolution [9]. An independently developed Brewster angle microscope (BAM) with limited resolution provided real time video recordings of morphology changes during compression-relaxation cycles of monolayers [10]. The principle and recent developments in characterization of monolayers phases and domain morphology by Brewster angle microscopy and reflectometry are discussed in the following sections.

2. FUNDAMENTALS OF BREWSTER ANGLE MICROSCOPY

When a light beam passes an interface between two media with different refractive indices n_1 and n_2, incident from the medium with smaller refractive index n_1, a part of the light is refracted into the medium with the larger n, and a fraction of the light is reflected from the interface. The Brewster angle φ_b is defined by the condition that the reflected and the refracted beams are at an angle of 90°. The Brewster angle is therefore given by $tg(\varphi_b) = n_2/n_1$. As a consequence of the Brewster angle condition, p-polarized light is not reflected, see Fig. 1a, in contrast to s-polarized light. However, with the angle of incidence set to φ_b for the clean water surface, a monolayer formed at the air-water interface changes the optical situation, and reflected light is observed that may be registered with a photomultiplier like fluorescence emission or recorded with a video camera, (Fig. 1b). Thus, Brewster angle reflectivity and microscopy are zero background techniques with the advantage of using non-actinic light causing no photochemical damage of the monolayer as compared to fluorescence techniques.

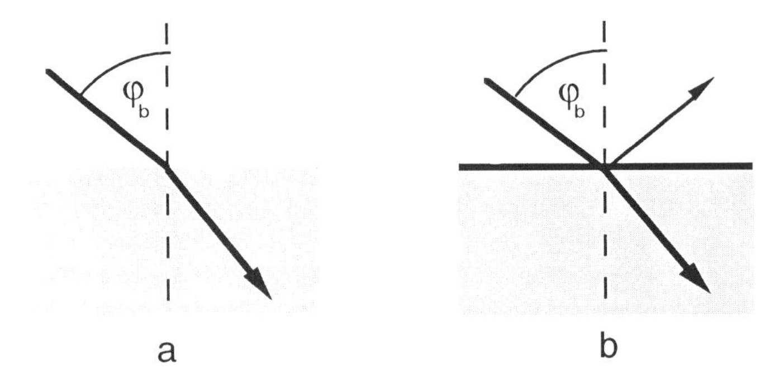

Figure 1. Principle of Brewster angle reflectometry and microscopy: (a) with light incident at the Brewster angle φ_b on an interface between two media of different refractive indices, no reflection of p-polarized light is observed; (b) the formation of a monolayer at constant angle of incidence causes reflection of p-polarized light.

The monolayer induced reflectivity of the interface depends on the refractive index of the monolayer material and the thickness of the layer. For a monolayer with isotropic refractive index, domains of different molecular density and/or thickness show up as areas of different brightness. In the more general case of refractive index anisotropy, the reflected light is not p-polarized but contains a s-polarized component. By rotating an analyzer in the reflected light beam, this anisotropy can easily be detected. As will be shown in Section 4, many monolayer phases are characterized by such an anisotropy whose origin may be an orthorhombic packing of vertically oriented hydrocarbon chains or long range tilt orientational order of the hydrocarbon chains. With dye monolayers the large polarizability of π-electron systems dominates the optical response.

Phase separation in monolayers of two or more components has been deduced from measurement of surface pressure/area isotherms [11,12] when additivity is observed. If one component is a dye, energy transfer investigations provide evidence for phase separation [12]. The phenomenon of phase separation can also be directly observed by BAM, provided the domains of the separated phases are large enough, i.e. of several μm size, and the optical properties of the component phases are sufficiently different. Such a two-component monolayer will be discussed in Section 5.

3. COUNTER ION INFLUENCE ON DOMAIN MORPHOLOGY

Many amphiphiles forming stable monolayers that are used for assembling monolayer organizates have ionic head groups like quarternary ammonium salts, carboxylates (at high pH of the subphase) or phosphates. In particular, the salts of long chain fatty acids formed with divalent cations are widely used for the construction of Langmuir-Blodgett (LB) films [13] and more sophisticated structures [14]. The divalent cations like Ca^{2+} or Cd^{2+} play an important role since they strongly facilitate stepwise monolayer transfer to solid substrates.

Large changes of the area per molecule have been observed for ionized monolayers upon interaction with components in the aqueous subphase that cannot be attributed to a penetration of the water soluble molecule into the hydrophobic moiety of the monolayer as found frequently with proteins. In contrast, it has been concluded, that the packing of the amphiphiles was modified by the counterion [15]. Such interactions have been investigated with BAM for monolayers of the cationic amphiphile dioctadecyl-dimethyl-ammonium bromide (DOMA) with different halogenides in the aqueous subphase [16]. Effects of the nature and concentration of the counterion in the aqueous subphase on the π/A-isotherm of DOMA monolayers have already been reported [17]. The halogenide ions Cl^-, Br^-, and I^- at a concentration of 1 mM have a condensating effect on DOMA monolayers as compared to the π/A-isotherm measured on water. In contrast, an expansion effect is observed with F^-. Since the smallest ion F^- is strongly hydrated, its incorporation into the head group region of the monolayer should cause an expansion.

Dramatic effects of the interactions between head groups and counter ions are seen in BAM images. DOMA monolayers on water appear homogeneous, and no features are detectable during compression to high lateral pressures. The brightness, i.e. the reflectivity, increases due to dense-packing of the molecules. With F^- or Cl^- in the subphase, the morphology of the monolayer is quite similar to that on water, and no features have been observed. However, in the presence of 1mM NaBr in the subphase, dendritic domains form and grow rapidly surrounded by coexisting liquid phase during compression at a surface pressure of 4 mN/m. An example is shown in Figure 2.

Dendritic growth of monolayer domains after surface pressure jumps has been observed in monolayers of dimyristoyl-phosphatidyl-ethanolamine (DMPE) by fluorescence microscopy, and the phenomenon has been rationalized in terms of diffusion limited aggregation [18]. The dye present in those experiments is expelled from the solid domains and accumulates in the surrounding liquid phase, causing a rise in melting pressure. This prevents further crystallization, and domain growth is limited by dye diffusion. Near the tips the dendrites, this impurity concentration is smaller, and domains grow at the tips. The morphology of domains with self-similar structures, therefore, is a non-equilibrium phenomenon. In the case of Figure 2, no dye is present in the monolayer. Nevertheless, the formation of the observed fractal structures could be caused by diffusion limited growth either due to unknown impurities of the material or due to limited mobility of the DOMA molecules. Similar domain morphology has been observed by BAM in the absence of dyes also with monolayers of DMPE [10].

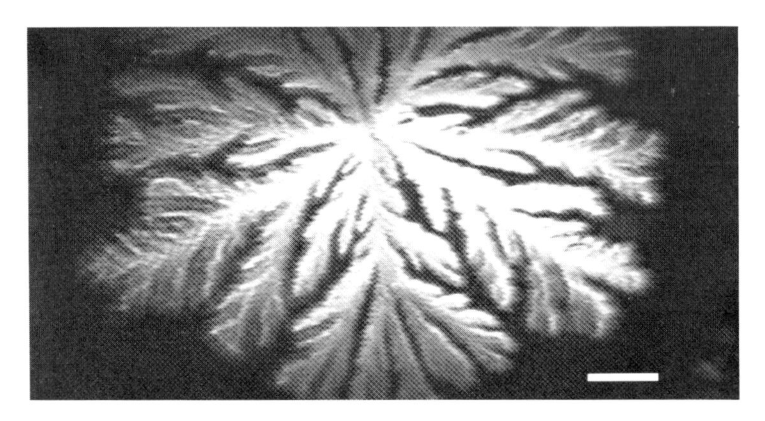

Figure 2. Brewster angle microscopy of a monolayer of dioctadecyl-dimethyl-ammonium bromide on 1mM NaBr aqueous subphase; dendritic domain of solid phase (bright) surrounded by coexisting liquid phase; bar: 100 μm; surface pressure: 4 mN/m, temperature 21°C.

The condensating effect of I^- on DOMA monolayers is even more pronounced than of Br^-, and solid phase domains of irregular shape are visible immediately after spreading.

Monolayers of J-aggregates of amphiphilic cationic cyanine dyes have been characterized by BAM [19]. Strong effects of the anion in the aqueous subphase on the morphology of such monolayers has been observed. J-aggregates show unusual spectroscopic properties [20-22] due to the particular chromophore packing [22, 23]. The excited state has been described by a model of coherent excitons of temperature dependent size [24]. The non-resonant light of the He-Ne laser used in the BAM avoids any excitation of the aggregate domains and consequently causes no photochemical damage during observation of the monolayers.

4. LONG RANGE ORDER

Monolayers of long chain fatty acids undergo phase transitions upon compression that may be detected by discontinuities in the π/A-isotherms. Since a monolayer may be considered as the smallest subunit of a smectic liquid crystal, the different phases have been identified with smectic crystal categories according to their particular type of long range order [25]. This nomenclature has the advantage of linking order phenomena known in liquid crystals to very similar observations in organized monolayers.

4.1. Types of long range order

The highest degree of order exists in crystalline systems where each molecule is located on a lattice site. This positional order is found in monolayers at the air-water interface by X-ray investigations [26]. It is, however, of short range. The types of long range order encountered in liquid crystal and in monolayer mesophases are tilt orientational order (TO) and bond orientational order (BO), see Figure 3.

A monolayer phase of a long chain fatty acid may be characterized by a particular tilt of the extended hydrocarbon chains that are often approximated by cylinders, with respect to the surface normal. The molecules are not necessarily placed on lattice sites. All molecules of a domain show the same polar and azimuthal tilt angles, φ and θ, respectively. Different domains have different azimuth θ. The nomenclature of monolayer phases derived from liquid crystal categories is associated with the type of long range order. A mesophase with long range tilt orientational order of the molecules in the direction of the nearest neighbour is S_I (smectic I), whereas a phase with tilt to the next nearest neighbour is S_F. Phases with bond orientational order of vertically oriented hydrocarbon chains may be of the hexagonal (S_{BH}) or orthorhombic type. A large variety of phases has been identified so far, and the modern experimental techniques provide us with an increasingly refined description of organized monolayer states.

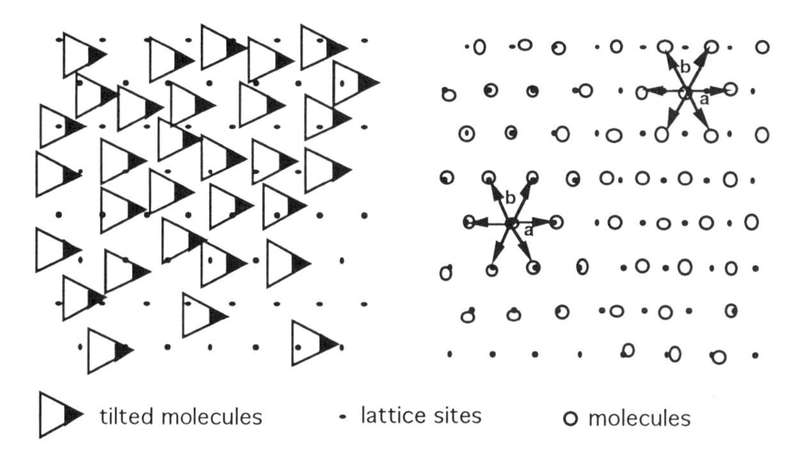

Figure 3. Types of long range order in monolayers: tilt orientational order (left), bond orientational order (right); the molecules are not necessarily located on lattice sites.

4.2. Optical anisotropy of monolayers

When the extended hydrocarbon chains of a monolayer in a liquid-condensed phase are tilted with respect to the surface normal an optical anisotropy in the monolayer plane should result, since the polarizability of the hydrocarbon chains is larger along the chain than in the cross-section of the chain. Concequently, the refractive index component along the chain is larger than perpendicular to the chain. This causes the appearance of an s-polarized component of the reflected light depending on the tilt direction which can be detected by rotating an analyzer in the reflected light beam. Alternatively, domains with different tilt azimuth angle θ can be discriminated at fixed analyzer angle by their different brightness. An example for this phenomenon is shown with a monolayer of pentadecanoic acid (PDA) on water at a surface pressure of 12 mN/m in Figure 4 [27]. The different dense-packed domains of hexagonal or pentagonal shape can clearly be seen. Further, the constant brightness within one domain of typically 100 μm size is evidence for long range tilt orientational order of the molecules. Upon further compression of the PDA monolayer, the contrast vanishes, and a solid phase with vertically oriented hydrocarbon chains in a hexagonal packing is formed. Such a phase does not have an in-plane anisotropy of the refactive index and therefore appears homogeneous.

A change in polar tilt angle is an excellent indicator for a phase transition as observed by BAM. Sudden changes and the coexistence of different phases are evidence for a first-order phase transition upon compression or expansion of the monolayer. The kinetics of a first-order phase transition, i.e. the progression of the front of a new phase, has been followed by BAM in real time upon decompression of eicosanol monolayers on water from the superliquid RotI phase to the S_I (L2) phase at about 15 mN/m and a temperature of 20.6°C [28]. The propagation of the boundary between two phases is based on a collective change of the tilt of the molecules. In the case of eicoanol monolayers, the propagation velocity depends on the rate of decompression causing a surface pressure gradient and is typically in the range of cm/min. Size and shape of the S_I phase domains formed after decompression from the RotI phase depend on temperature and decompression rate. Small structures with diameters of 20 μm may be obtained around 19°C and 24°C, whereas oblong domains with straight borders and a few mm long are formed around 21°C [28].

Figure 4. Optical anisotropy in monolayer domains due to long range tilt orientational order detected by Brewster angle microscopy: due to different azimuthal tilt angle the domains show different brightness; bar: 100 μm; monolayer of pentadecanoic acid, subphase: water, surface pressure: 12 mN/m, temperature: 20°C.

Second-order phase transitions were found upon compression of eicosanol monolayers from the S_I phase into either the S_E (solid) phase or the RotII phase [28].

Such a transition is characterized by a smooth, gradual change of the tilt angle and consequently of the contrast between domains of different azimuth angle in BAM.

A particular phase transition is the monolayer collapse. Depending on the type of monolayer, abrupt changes are observed with the Brewster angle microscope when the surface pressure exceeds the collapse pressure. In the case of behenic (docosanoic) acid, irregular bright stripes propagate very rapidly through the image [19]. However, transformation from a monolayer to three-dimensional structures may involve intermediate, very well defined multilayer systems. This has been demonstrated with monolayers of stearic (octadecanoic) acid on water and aqueous subphases like 1 mM NaCl or HCl solutions of different pH [29]. Depending on the conditions, monolayer relaxation at constant surface pressure of 30 mN/m involves formation and growth of nuclei to large domains of homogeneous brightness, i.e. well defined thickness and refactive index, also exhibiting optical anisotropy. At a subphase pH = 5, dendritic growth of the nuclei is observed.

Brewster angle microscopy has also been applied to monolayers transferred on solid substrates like glass plates. The first monolayer of cadmium stearate in contact with the glass surface appears homogeneous, whereas following bilayers exhibit anisotropic domains [30]. From X-ray investigations of transferred monolayers an orthorhombic packing of vertically oriented hydrocarbon chains has been deduced [31]. Such an arrangement representing one case of long range bond orientational order should also lead to an optical in-plane anisotropy, and it has been a particular challenge to discriminate this type of anisotropy from that caused by long range tilt orientational order as dicussed above.

In principle, the two cases, chain tilt or orthorhombic packing of vertically oriented chains, can easily be discriminated by BAM on the basis of a simple qualitative rule [32]. Rotation of the sample around the surface normal produces different variations of the reflectivity of single domains with rotation angle depending on the origin of the in-plane optical anisotropy. According to model calculations one maximum of the reflectivity is expected in the case of chain tilt, whereas two maxima should be observed in the case of orthorhombic packing upon rotation by 360°.

Such dependences have indeed been observed, see Figure 5, where the averaged reflected intensity of single domains is plotted vs the rotation angle [32]. One maximum is found in the case of a monolayer of arachidic (eicosanoic) acid on glass (Fig. 5a). The case of two maxima was found for a single domain of a system of 5 monolayers of cadmium arachidate on glass (Fig. 5b). Therefore, the anisotropy in the transferred single monolayer results from long range tilt orientational order, whereas it is due to orthorhombic packing of vertically oriented chains (long range bond orientational order) in the Langmuir-Blodgett system.

Phase transitions from a mesophase with hexagonal packing of vertically oriented chains to a phase with orthorhombic packing of vertically oriented chains have been observed now upon compression of a monolayer of arachidic acid on water although the contrast between the domains with orthorhombic chain packing is less than that observed for domains with long range tilt orientational order. These very recent developments of Brewster angle microscopy and reflectometry illustrate the amazing potential of these techniques in providing detailed information on long range order and molecular packing in monolayers and transferred layer systems that have been accessible so far only through very tedious and expensive X-ray investigations.

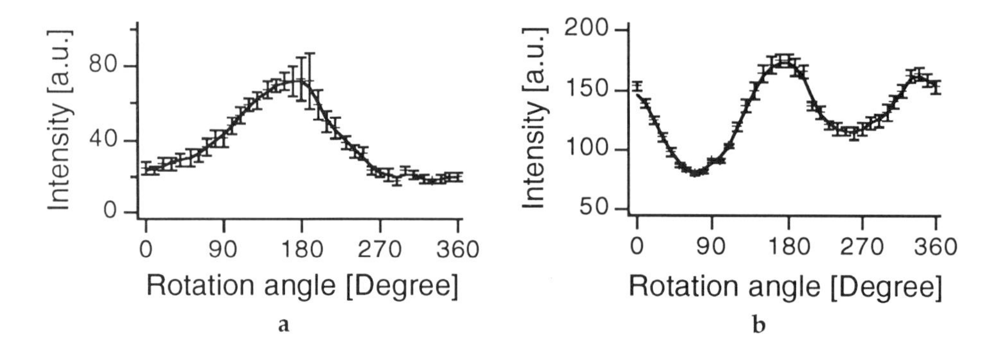

Figure 5. Discrimination of chain tilt and orthorhombic chain packing as origin of optical in-plane anisotropy by rotation of the sample around the surface normal in the BAM; reflected intensity plotted vs rotation angle: (a) monolayer of arachidic acid on glass with one maximum indicating long range tilt orientational order; (b) 5 monolayers of cadmium arachidate on glass with two maxima, characteristic for orthorhombic packing (long range bond orientational order).

4.3. Textures and their simulation

Particular textures of long range bond orientational order have been observed in freely suspended liquid crystal films of a chiral compound [33]. Similar types of textures were found in monolayers by polarized fluorescence microscopy and BAM [34] illustrating the correspondence between smectic liquid crystal phases and mesophases found in monolayers. The textural features can be described by gradual or abrupt variations of the molecular tilt inside a domain, leading to shells, stars and stripes. Possible molecular arrangements of such features are schematically shown in Figure 6. The arrows indicate the azimuthal direction of the molecular tilt. In the arrangement of Figure 6a, classified as splay type, the azimuthal orientation runs through 360° along the boundary of the domain. It is radial with respect to circles of varying radii, centered along the line OP. This molecular organization leads to a shell texture of the domain. The arrangement of Figure 6b shows a splay type azimuthal orientation typical for stripes with the tilt directed to the interior of the domain, and Figure 6c is an example for the arrangement of a star texture with the azimuthal tilt oriented to the outside of the domain.

On the basis of such arrangements, the Brewster angle reflectivity patterns have been calculated using the 4x4 matrix algorithm of Berreman [35] with reasonable parameters for the refractive indices and thickness [36]. The model domain was composed of 50x50 cells with a particular azimuth angle for each cell. Illumination with the light was the same as in the case of the experimental image. Computer simulation and BAM image are compared in Figure 7.

Upon comparison of the computer simulation of other molecular arrangements and under different illumination conditions with the experimental images, the pattern of tilted molecules as shown in Figure 6a was identified as the only one leading to coincidence of model and experiment [36]. In the same way, the possible molecular arrangements underlying stripe and star textures have also been examined. By computer simulations of the domains under different illumination conditions and comparison with the experimental result, the direction of molecular tilt as well as the change within the domains were unambiguously identified.

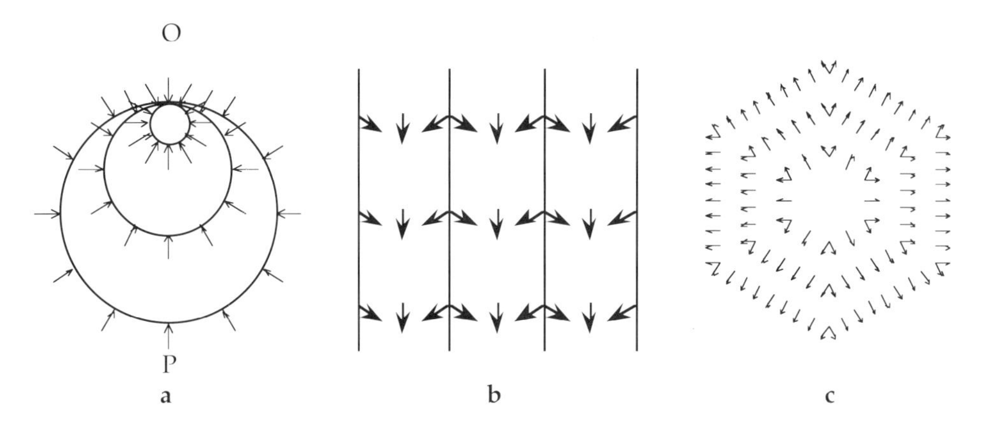

Figure 6. Possible tilt azimuthal orientation of molecules in monolayer domains with different textures; (a) shell; (b) stripe; (c) star.

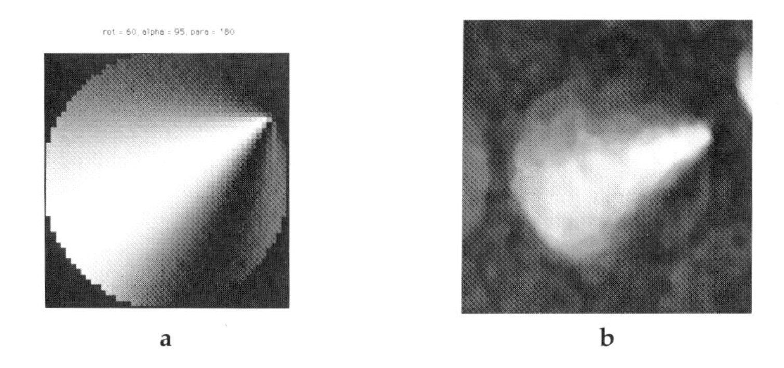

Figure 7. Shell texture in monolayers; (a) computer simulation (b) Brewster angle microscope image of a shell in a monolayer of pentadecanoic acid on water, surface pressure: 3 mN/m, temperature: 20.4°C.

These results mark a major improvement in understanding molecular arrangements in monolayer domains.

5. REVERSIBLE PHOTOCHEMICAL ISOMERIZATION

Azobenzenes are known to undergo reversible photoisomerization in solution [37]. Amphiphilic azobenzene derivatives have been synthesized for various purposes, and their behaviour in monolayers at the air-water interface including photochemical isomerization was investigated in large detail [38-42]. The reversible cis-trans isomerization of the azo double bond causes changes of the area per molecule, absorption spectrum, and dielectric properties of the monolayers. A particularly interesting observation was the modulation of the lateral conductivity of a complex layer system upon reversible isomerization of an azobenzene group in the hydrophobic region [41].

The azo dye used for this modulation has the structure shown on top of Figure 8 with n = 12 (A812P), i.e. the azobenzene chromophore is linked to the hydrophilic head group (the pyridinium ring system) via a hydrocarbon chain of 12 C-atoms. In order to demonstrate the effect of photochemical isomerization on the response of the monolayer in Brewster angle microscopy the azo dye was incorporated in a two-component monolayer with cholesterol (CHO). This biologically very important compound is known to form a separate monolayer phase in combination with many other amphiphiles. Therefore, domains of cholesterol are an excellent internal reference for the observation of photochemical cis-trans isomerization of the amphiphilic azo dye A812P.

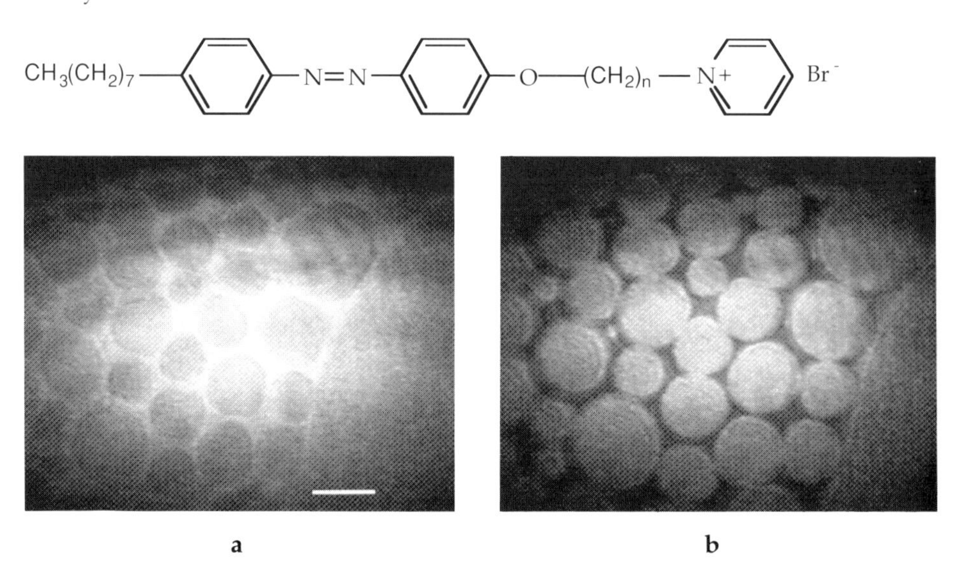

Figure 8. Phase separation in a two-component monolayer and photoisomerization of one component as recorded by Brewster angle microscopy; structure of the azo dye, n = 12 (top); (a) image of a monolayer of trans-azo dye and cholesterol, molar ratio 1:4, the dark round domains are the cholesterol phase, the bright stripes the azo dye; (b) monolayer after photochemical isomerization trans-to-cis, the contrast is inverted, and the cholesterol domains appear now bright; bar: 100 μm; surface pressure: 4 mN/m, temperature: 20°C.

Figure 8a shows a Brewster angle microscope image of the two-component monolayer with the molar ratio of trans-A812P:CHO = 1:4 on water. The system forms indeed separate phases. The large dark domains are composed of cholesterol, and the bright stripes are the azo dye. The phase separation can easily be seen by BAM since the optical properties of the two materials are sufficiently different. When the monolayer is exposed to UV radiation (λ = 366 nm) the azobenzene part isomerizes to the cis-form. Now, the cholesterol domains appear bright as compared to the dark stripes of cis-A812P, see Figure 8b.

According to the π/A-isotherms the cis-form has a larger area per molecule than the trans-form. For a potential use of this system in information storage it is desirable to organize a mixed monolayer such that the isomerization still takes place without change of the molecular packing. This sounds like a contradiction. However, one can imagine a dense-packed monolayer matrix for the azo dye with a thickness smaller than the length of the hydrocarbon chain that links the azobenzene moiety to the pyridinium head group. Then, the azobenzene lying on top of the dense-packed hydrocarbon chains would be able to isomerize although the molecule as a whole does not change its area in the mixed monolayer.

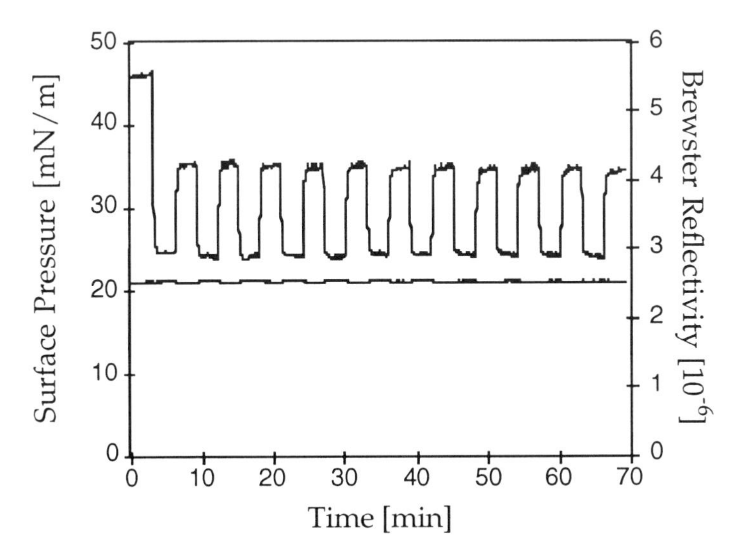

Figure 9. Reversible isomerization of a mixed monolayer of the azo dye A820P and dipalmitoyl-phosphatidic acid, molar ratio 1:1; periodic change of Brewster angle reflectivity R upon variation of the photostationary state of trans (formed by irradiation with 436 nm light) and cis (formed by irradiation with 366 nm radiation) azo dye at constant area; temperature: 20°C.

This idea has been realized using the azo dye A820P with n = 20 in the structure of Figure 8, top [43]. An appropriate matrix is dimyristoyl-phosphatidic acid (DMPA). The isomerization has been followed by measuring Brewster angle reflectivity (BAR) of the monolayer during alternating exposure to UV radiation (366 nm) and blue light (436 nm), respectively. Formation of the trans-form causes an increase, formation of the cis-form a decrease of the BAR signal, as shown in Figure 9. The structural change of the azobenzene chromophore does not induce a significant variation of the monolayer molecular packing as evidenced by the negligible variation of the surface pressure π at constant area. Therefore, the assumption, that the chromophore is located above the dense-packed hydrophobic part of the mixed monolayer is correct and the strategy of molecular organization outlined above has proven to be successful.

6. CONCLUSION

Phase transitions in monolayers and phases of different optical properties like in two-component monolayers with immiscible components are detected in real time by Brewster angle microscopy (BAM) without using fluorescent probes. Details of the molecular organization in monolayers at the air-water interface as well as on solid substrates can be revealed by BAM. The use of non-actinic light is particularly adequate in BAM of dye-containing monolayers and in Brewster angle reflectivity measurements during the investigation of photoreactions.

ACKNOWLEDGMENT

This work was funded by the BMFT (grant No. 03M4060F). Financial support by the Fonds der Chemischen Industrie, Germany, is gratefully acknowledged.

REFERENCES

1. M. Lösche, E. Sackmann, and H. Möhwald, Ber. Bunsenges. Phys. Chem., 87 (1983) 848.
2. R. Peters and K. Beck, Proc. Natl. Acad. Sci. U.S.A., 80 (1983) 7183.
3. R. M. Weis and H. M. McConnell, Nature, 310 (1984) 47.
4. H. Möhwald, Annu. Rev. Phys. Chem., 41 (1990) 441.
5. H. M. McConnell, Annu. Rev. Phys. Chem., 42 (1991) 171.
6. A. Miller and H. Möhwald, Europhys. Lett., 2 (1986) 67.
7. H. Riegler and K. Spratte, Thin Solid Films, 210/211 (1992) 9.
8. X. Qiu, J. Ruiz-Garcia, K. J. Stine, Ch. M. Knobler, and J. V. Selinger, Phys. Rev. Lett., 67 (1991) 703.
9. S. Hénon and J. Meunier, Rev. Sci. Instrum., 82 (1991) 936.
10. D. Hönig and D. Möbius, J. Chem. Phys., 95 (1991) 4590.
11. N. L. Gershfeld, Annu. Rev. Phys. Chem., 27 (1976) 349.
12. H. Bücher, O. v. Elsner, D. Möbius, P. Tillmann, and J. Wiegand, Z. phys. Chem. Neue Folge, 65 (1969) 152.
13. K. B. Blodgett, Phys. Rev., 55 (1939) 391.
14. H. Kuhn and Möbius, in B. W. Rossiter and R. C. Baetzold (eds.), *Investigations of Surfaces and Interfaces*, Physical Methods of Chemistry Series, 2nd ed., Vol. IXB, Wiley, New York, 1993, p. 375.

15. D. Möbius and H. Grüniger, Bioelectrochem. & Bioenerg., 12 (1984) 375.
16. R. C. Ahuja, P.-L. Caruso, and D. Möbius, Thin Solid Films, submitted.
17. J. Marra, J. Phys. Chem., 90 (1986) 2145.
18. A. Miller, W. Knoll, and H. Möhwald, Phys. Rev. Lett., 56 (1986) 2633.
19. D. Hönig and D. Möbius, Thin Solid Films, 210/211 (1992) 64.
20. G. Scheibe, Angew. Chem., 49 (1936) 563.
21. E. E. Jelley, Nature, 138 (1936) 1009.
22. H. Bücher and H. Kuhn, Chem. Phys. Lett., 6 (1970) 183.
23. C. Duschl, W. Frey, and W. Knoll, Thin Solid Films, 160 (1988) 251.
24. D. Möbius and H. Kuhn, Israel J. Chem. 18 (1979) 375.
25. A. M. Bibo, C. M. Knobler, and I. R. Peterson, J. Phys. Chem., 95 (1991) 5591.
26. R. M. Kenn, C. Böhm, A. M. Bibo, I. R. Peterson, H. Möhwald, J. Als-Nielsen, and K. Kjær, J. Phys. Chem., 95 (1991) 2092.
27. D. Hönig, G. A. Overbeck, and D. Möbius, Adv. Mater., 4 (1992) 419.
28. G. A. Overbeck, D. Hönig, and D. Möbius, Langmuir, 9 (1993) 555.
29. S. Siegel, D. Hönig, D. Vollhardt, and D. Möbius, J. Phys. Chem., 96 (1992) 8157.
30. D. Hönig and D. Möbius, Chem. Phys. Lett., 195 (1992) 50.
31. P. Tippmann-Krayer, R. M. Kenn, and H. Möhwald, Thin Solid Films, 210/211 (1992) 577.
32. G. A. Overbeck, D. Hönig, and D. Möbius, Nature, submitted.
33. S. B. Dierker, R. Pindak, and R. B. Meyer, Phys. Rev. Lett., 56 (1986) 1819.
34. J. Ruiz-Garcia, X. Qiu, M.-W. Tsao, G. Marshall, C. M. Knobler, G. A. Overbeck, and D. Möbius, J. Phys. Chem., 97 (1993) 6955.
35. D. W. Berreman, J. Opt. Soc. Am., 62 (1972) 502.
36. G. A. Overbeck, D. Hönig, and D. Möbius, Thin Solid Films, submitted.
37. H. Rau, in H. Dürr and H. Bouas-Laurent (eds.), *Photochromism*, Elsevier, 1990.
38. J. Heesemann, J. Am. Chem. Soc., 102 (1980) 2167.
39. H. Nakahara and K. Fukuda, J. Coll. Interface Sci., 93 (1983) 530.
40. T. Seki and K. Ichimura, Thin Solid Films, 179 (1989) 77.
41. H. Tachibana, A. Goto, T. Nakamura, M. Matsumoto, E. Manda, H. Niino, A. Yabe and Y. Kawabata, Thin Solid Films, 179 (1989) 207.
42. M. Sawodny, A. Schmidt, M. Stamm, W. Knoll, C. Urban and H. Ringsdorf, Polymers Adv. Techn., 2 (1991) 127.
43. J. Maack, R. C. Ahuja, H. Tachibana, M. Matsumoto, and D. Möbius, Thin Solid Films, in press.

Abbreviations:

A812P Amphiphilic azo dye, structure see Figure 8 with n = 12
A820P Amphiphilic azo dye, structure see Figure 8 with n = 20
BAM Brewster angle microscopy
BAR Brewster angle reflectometry
CHO Cholesterol
DMPA Dimyristoyl-phosphatidic acid
DMPE Dimyristoyl-phosphatidyl-ethanolamine
DOMA Dioctadecyl-dimethyl-ammonium salt
PDA Pentadecanoic acid

MICROCHEMISTRY
Spectroscopy and Chemistry in Small Domains
Edited by H. Masuhara et al.

Transient absorption microspectroscopy of a single optically trapped particle

N. Tamai*,#, S. Funakura‡, and C. F. Porter§

Microphotoconversion Project†, ERATO, Research Development Corporation of Japan, 15 Morimoto-cho, Shimogamo, Sakyo-ku, Kyoto 606, Japan

Transient absorption microspectroscopic systems with femtosecond ~ microsecond temporal and micrometer spatial resolutions were developed by using a microscope and a laser trapping technique. A pump pulse and a white-light probe pulse were coaxially introduced into a microscope and focused on an optically trapped microparticle by an objective lens. This method was applied to analyze photochemical and photophysical dynamics of a single dye-doped liquid droplet in water and a single perylene microcrystal. Possibilities and limitations of the transient absorption microspectroscopic system have been discussed.

1. INTRODUCTION

Recently, much attention has been paid for the chemical analysis of heterogeneous objects with high spatial resolutions in many areas of science and technology. Examples include biological cells, organic solids such as crystals and polymer particles, Langmuir-Blodgett films, catalysts, and artificial materials such as etched patterns and integrated electronic circuits, to name just a few. Absorption, fluorescence, and Raman scattering are important spectroscopic techniques and sometimes combined with a microscope for identification and understanding of the physicochemical, and biological processes of the inhomogeneous samples with high spatial resolutions. These processes include photochemical reactions such as electron and proton transfers and isomerizations which frequently occur in microsecond to femtosecond time scales [1]. Therefore, high temporal resolution in addition to spatial

---

* To whom correspondence should be addressed.
# Present Address: Light and Material Group, PRESTO, JRDC, Department of Chemistry, Faculty of Science, Kwansei Gakuin University, Uegahara, Nishinomiya 662, Japan
‡ Present Address: Chemicals Division, Dainippon Ink and Chemicals, Inc., Higashi-fukashiba-18, Kamisu-machi, Ibaraki 314-02, Japan
§ Present Address: Environmental Sciences, University of East Angria, Norwich NR47TJ, U. K.
† Five-year term project: October 1988 ~ September 1993.

resolution is indispensable for analyzing the dynamic properties of inhomogeneous samples [2].

One of the fruitful ultrafast spectroscopies is to detect fluorescence using a picosecond single-photon timing technique [3] under a microscope because of its high sensitivity and high temporal resolution. This type of fluorescence microspectroscopy has been reviewed by Ghiggino et al. and Sasaki et al. in Part IV. Time-resolved fluorescence microspectroscopy, however, is ultimately limited only for fluorescent materials.

Another important method is transient absorption spectroscopy for the identification of non-fluorescent excited species, chemical intermediates and hot molecules. Temporal resolution of this spectroscopy has been improved from microsecond to femtosecond time scale since the pioneering work by Porter and Norrish in 1949 [4]. However, very little is reported on the spatial resolution of transient absorption spectroscopy.

Recently, Sasaki et al. proposed a confocal laser-induced absorption microscope for the 3-dimensional imaging of non-fluorescent transient species of a sample, and demonstrated that the depth resolution of a microscope was $\approx$ 1.5 μm with a temporal resolution of $\approx$ 10 ps [5]. In this experiment, two-color pump-probe technique using high-repetition rate (3.8 MHz) laser pulses was used to detect very weak absorbance of the order of 10^{-5}. However, it is impossible to obtain the spectral information. The spectral information in addition to time responses are indispensable for the identification of transient species. A white-light of a xenon flash lamp or a supercontinuum generated by the interaction of an intense picosecond or femtosecond laser pulse with glasses or liquids [6] is highly desirable as a probe pulse for the measurement of transient absorption spectrum.

For the analyses of photochemical and photophysical dynamics by transient absorption spectroscopy under a microscope, a small object should be fixed in a certain position during the measurements. This is an inevitable requirement for all objects, especially for small particles in solution, where the position of a particle changes with time due to the thermal Brownian motion. Optical micromanipulation using a focused laser beam is one of the best techniques to trap and manipulate small objects such as polymer latex particles [7-9] and biological cells [10-12]. An optical micromanipulation system generally consists of a microscope and a laser beam which is focused into the objective plane. The details of micromanipulation have been described in Part II by Sasaki and Misawa. This method should be combined with transient absorption microspectroscopy.

In this article, we report transient absorption microspectroscopic systems using a microscope and an optical trapping technique with microsecond ~ femtosecond temporal and micrometer spatial resolutions. A pump pulse and a white-light probe pulse of xenon flash lamp or supercontinuum generated by focusing an intense laser pulse into water were coaxially introduced into a microscope and focused on an optically trapped microparticle by an objective lens. These systems were applied to individual microparticles dispersed in solution and perylene microcrystals. Temporal and spatial resolutions, and potential applications and limitations of transient absorption microspectroscopy are described.

2. MICROSECOND TRANSIENT ABSORPTION MICROSPECTROSCOPY

In absorption microspectroscopy, a blank measurement without a sample is very important to take a correct ground-state absorption spectrum. On the other hand, a blank measurement is not necessary in the transient absorption spectrum, since the difference of the both intensities of a probe beam with (I) and without (I_0) excitation is measured in a certain position of a sample. In transient absorption measurements, intense excitation laser pulse is irradiated on a sample to create an enough amount of transient species. High concentration of the excited species or chemical intermediates induce a rapid degradation of a sample if these species react with oxygen or contaminated water. The smaller the size of a sample is, the faster the rate of the degradation becomes. The method of sample preparation to avoid the sample damage is essential for the transient absorption microspectroscopy.

In this Section, we describe the sample preparation for microspectroscopy and a transient absorption microspectroscopic system with microsecond temporal resolution.

2.1. Preparation of a sample for microspectroscopy

A schematic illustration of a sample preparation is shown in Figure 1. By protecting a part of the surface with paraffin or polystyrene, a cover glass was partially etched by hydrofluoric acid (HF). The etching depth is 50 ~ 150 μm by adjusting the concentration of HF, temperature, and etching time. A sample should be handled in a vacuum glove box filled with argon gas, where the concentration of oxygen is less than 0.1 ppm. All the samples were degassed before use. The sample was placed on a slide glass and covered by the etched cover glass, which was then sealed by an adhesive resin. This procedure is indispensable to prevent the sample damage by the photochemical reactions with oxygen.

When the microparticles are dispersed in solution, they tend to coalesce with one another. This may interfere the absorption measurements of a single particle. The surfactant is known to reduce the interfacial tension of a droplet in solution, which is one of the important factors to avoid coalescence phenomena [13]. We have used sodium dodecyl sulfate (SDS) as a surfactant to stabilize individual microparticles of tri-n-butyl phosphate (TBP) containing zinc tetraphenylporphine (ZnTPP) dispersed in water.

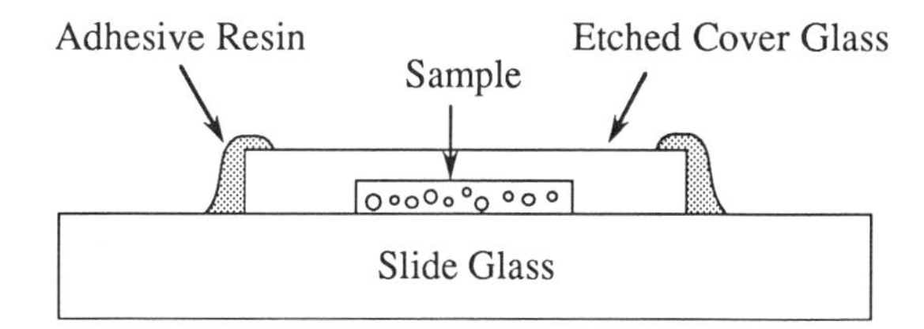

Figure 1. A sample prepared for transient absorption microspectroscopy by using a etched cover glass. All the procedures are carried out in a vacuum grove box.

2.2. Transient absorption microspectroscopic system

A block diagram of the system is illustrated in Figure 2. The excitation pulse was second (532 nm) or third (355 nm) harmonics of a Q-switched Nd:YAG laser (Quanta-

Ray, DCR-II, ≈ 10 ns pulse width). A xenon lamp (Tokyo Instruments, XF 80-60) flashed synchronously with a laser was employed as a probe pulse, the duration of which was ≈ 70 μs. It may be possible to use a steady-light of xenon lamp as a probe light when the enough intensity is available. The pump and probe pulses were coaxially introduced into a microscope and focused on a sample by an oil-immersion objective lens (Nikon CF Flour, x 100, numerical aperture (NA) = 1.3). The probe pulse passing through a sample was reflected by a beam splitter and detected by a polychromator (McPherson, 2035, 150G/mm) - a gated multichannel detector (Princeton Instruments, D/SIDA-700G) combination. The time delay between the pump pulse and the gate time was controlled by a digital delay generator (Stanford Research, DG-535). The temporal resolution of the system is limited by the gate width of the detector (50 ns).

Individual microparticles dispersed in solution should be fixed in a certain position under a microscope. A cw Nd:YAG laser (Spectron, SL-903U) was used as the optical trapping source and focused into ≈ 1-μm spot through an objective lens. The laser power at the sample was 100 ~ 250 mW. The trapping behavior was monitored by a CCD (charge-coupled device) camera - TV monitor set equipped to the microscope.

The diameter of a probe pulse was 2 ~ 3 μm in a visible wavelength region. Because of the incoherent nature of the light source, it is very difficult to focus the white-light of a xenon lamp to 1-μm diameter. In addition, the effect of the chromatic aberration of a standard objective lens cannot be neglected at the focusing point. The lateral resolution of the system with the xenon lamp was determined to be ≈ 3 μm

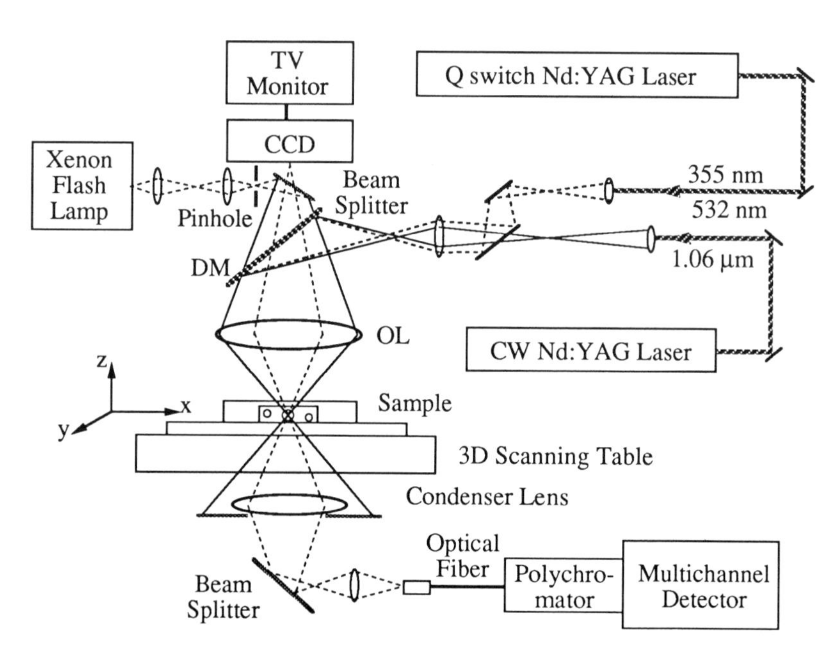

Figure 2. A block diagram of the transient absorption microspectroscopic system with microsecond and micrometer resolutions. DM, dichroic mirror; OL, objective lens.

in the wavelength region from 450 to 700 nm by a method similar to that described in Section 3.4. The excitation laser pulse, on the other hand, can be focused to a diameter less than 0.5 μm. In such a condition that the diameter of a probe pulse is larger than that of a pump pulse, the spectrum becomes broad and its absorbance is lower than the real one. The diameter of the pump pulse should be larger than the probe pulse to take a correct spectrum without distortion. The pump pulse was therefore defocused to ≈ 50 μm at the focal plane of the probe pulse.

Another important point is the conical nature of the probe pulse focused by an objective lens. Because of the longer path length than in the parallel beam, this effect make the absorbance overestimate. The error of the absorbance by the conical probe beam has been reported to be ≈ 5% [14]. An aperture angle α corresponding to an obliquity of the light is given by the following equation:

$$\alpha = \sin^{-1}(NA/n) \quad (1)$$

where n is the refractive index of the medium. α in both condenser and objective lenses were calculated to be 23° and 59°, respectively [15]. In the present system as shown in Figure 2, the probe pulse with large obliquity is cut off by the condenser lens. As a result, the effect of obliquity on the absorbance is negligibly small. This was experimentally confirmed by using liquid films (15 ~ 80-μm thickness) of ZnTPP/TBP system, in which the molar extinction coefficient was in good agreement with that in the bulk solution.

Figure 3 illustrates a typical example of ground-state absorption spectra of a single optically trapped particle (8 x 10^{-3} M ZnTPP/TBP) in water. The Q-band of ZnTPP with a maximum at 556 nm is clearly observed. We examined the linearity of the absorbance against the diameter of the droplet and found that Lambert-Beer's law held down to ≈ 10-μm diameter. Providing the molar extinction coefficient of ZnTPP in the droplet is the same as that in the bulk, the concentration is estimated to be ≈ 6.1 x 10^{-3} M. This value is smaller than that in the thin liquid film without water, indicating that the concentration of ZnTPP is diluted in the droplet. This effect is probably attributed to the partition equilibrium, which should be surely considered in dealing with the droplet in solution.

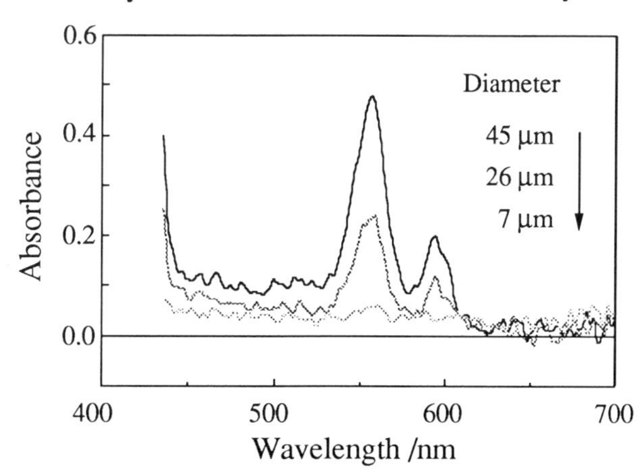

Figure 3. Ground-state absorption spectra of a single optically trapped ZnTPP/TBP droplet in water. Initial concentration of ZnTPP in TBP solution is 8 x 10^{-3} M, whereas the estimated concentration of ZnTPP in the droplet is ≈ 6.1 x 10^{-3} M.

The transient absorption spectra of an optically trapped ZnTPP/TBP droplet are shown in Figure 4. The spectrum with a peak at 465 nm is $T_n \leftarrow T_1$ absorption of ZnTPP, whereas the negative absorption at 556 nm is the ground-state bleaching. The spectrum disappeared within 10 μs. Since the decay time strongly depended on the excitation intensity, T_1 - T_1 annihilation was predominant in the relaxation process of ZnTPP in a single droplet. We found that the annihilation was a diffusion controlled reaction with a rate constant of ≈ 3 x 10^9 $M^{-1}s^{-1}$. The absorption of the bleaching was also reduced with a corresponding decay rate of $T_n \leftarrow T_1$ absorption, indicating that the ground-state was recovered by T_1 - T_1 annihilation process.

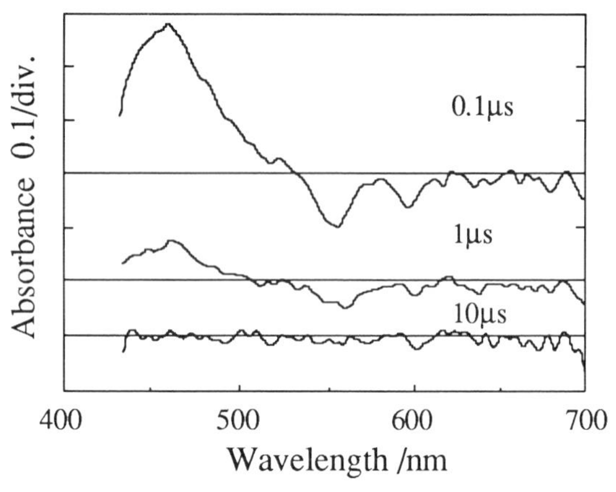

Figure 4. Transient absorption spectra of a single optically trapped ZnTPP/TBP droplet (diameter ≈ 27 μm) in water. The sample is excited at 355 nm (excitation intensity ≈ 7 mJ/cm^2). The concentration of ZnTPP is the same as that in Figure 3.

As described in this Section, physicochemical properties of a single microparticle in solution, which cannot be analyzed by fluorescence microspectroscopy, can be elucidated by transient absorption microspectroscopy with microsecond and micrometer resolutions. The method of a sample preparation as well as the optical trapping technique play a key role in such studies.

3. FEMTOSECOND TRANSIENT ABSORPTION MICROSPECTROSCOPY

Absorption spectroscopy with femtosecond resolution is required to analyze the ultrafast processes in small domains. In the pump-probe experiments such as transient absorption measurements, the temporal resolution is limited by the pulse width of the laser. The most shortest pulse width reported until now is 6 fs (pulse thickness ≈ 1.8 μm) by Fork et al. in 1987 [16]. Beside the laser in such a research level, laser systems with ≈ 100-fs pulse width are commercially available by the recent development of laser technology. So far as we know, no one has been reported on the pump-probe experiments with femtosecond temporal resolution under a microscope.

In this Section, we describe a transient absorption microspectroscopic system with femtosecond and micrometer resolutions, in which femtosecond laser pulses are introduced into a microscope [17]. The laser trapping technique is also used to fix a particle in solution during the measurements.

3.1. Femtosecond laser system

The laser system consists of femtosecond dye laser and its amplification system. A cw mode-locked Nd:YAG laser (Coherent, Antares 76S) was used for synchronously

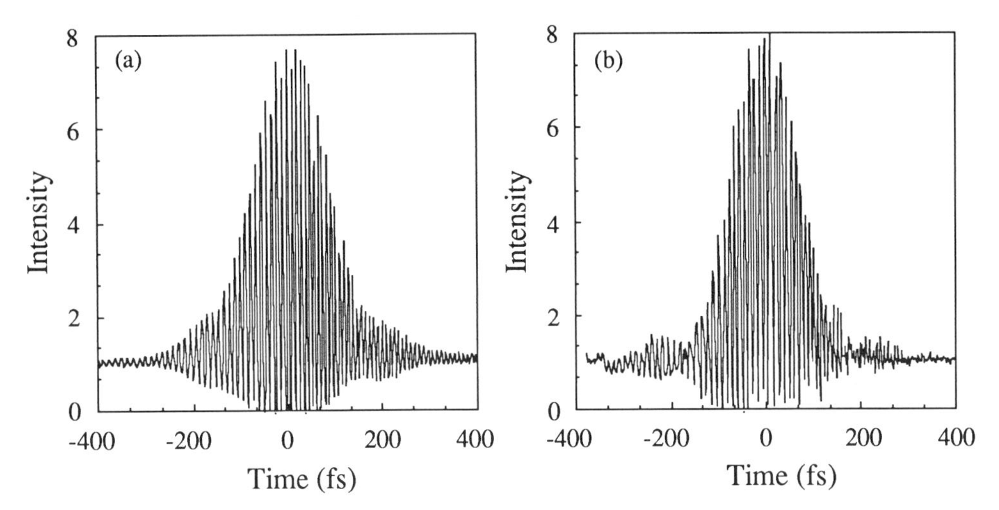

Figure 5. Interferometric autocorrelation traces of dye laser before (a) and after (b) amplification.

pumping a dye laser (Coherent, Satori 774). The hybridly mode-locked and dispersion compensated dye laser operating with a gain dye, pyridine 2, and a saturable absorber, DDI, was actively stabilized by using piezotransducers. The output power of the dye laser was over 200 mW at a 76-MHz repetition rate and a central wavelength of 720 nm. Figure 5a illustrates a typical example of the autocorrelation trace of laser pulses measured by a home-made interferometric autocorrelator, which gives approximately 200-fs autocorrelation width.

The output of the dye laser was amplified to 300 ~ 400 μJ/pulse by a three-stage dye amplifier (Continuum, PTA 60) with pyridine 1 as an amplifying dye, which was pumped by a regenerative amplifier (Continuum, RGA 60-10) at a repetition rate of 10 Hz. This regenerative amplification synchronizes the picosecond pump pulses (≈ 70 ps, 25 mJ at 532 nm) and femtosecond dye oscillator pulses, which results in the performance benefits such as lower levels of amplified spontaneous emission (ASE) and higher amplification efficiency as compared with different types of amplifications by nanosecond Nd:YAG or excimer lasers [18]. An appropriate filter (Hoya, IR-76) was inserted between the first and second stages of the amplifier to reduce the leading edge of the amplified pulses and ASE, and to minimize the effect of unamplified 76 MHz laser beam on the measured signal. Autocorrelation pulse shape of the amplified dye laser is shown in Figure 5b. The pulse width was approximately 170 fs (autocorrelation width) without side wings (110-fs pulse width by the deconvolution assuming sech^2 pulse shape). The intensity fluctuation of this system was better than ± 5 %. The pump pulse for sample excitation was obtained by frequency doubling of the amplified dye laser output using a 1-mm BBO (β-barium borate) crystal. A circular variable attenuator was used to reduce the intensity of the pump pulse in order to avoid the sample damage, which was then introduced into a microscope.

3.2. Generation of femtosecond supercontinuum

The fundamental laser beam passing through the BBO crystal was focused into a 1-cm H_2O cell to generate a supercontinuum. Figure 6 illustrates a typical example of its spectrum. An intensity of the white light at 400 nm was three orders of magnitude lower than that of the peak wavelength (≈ 720 nm). Appropriate filters should be used to reduce the intensity around the peak wavelength. An useful wavelength region for experiments is ranged from 400 nm to 1.3 μm. The spectral shapes of the supercontinuum generated through various solvents (ethylene glycol, acetonitrile, n-hexane, carbon tetrachloride, and so forth) and a pyrex glass were very similar with each other. This is in contrast with the previous supercontinuum generated by picosecond laser pulses [19], in which the spectral shape strongly depends on the used media. These results are probably interpreted in terms of various physical processes of supercontinuum generation such as parametric amplification, self-phase modulation, and stimulated Raman scattering [6]. In the current experimental condition (pulse width < 200 fs), the self-phase modulation may be a most likely process of the supercontinuum generation [20]. The spectral intensity of the continuum strongly depends on the intensity of fundamental laser pulse and consequently fluctuates from pulse to pulse. The intensity fluctuation should be corrected in absorption measurements as indicated in the next Section.

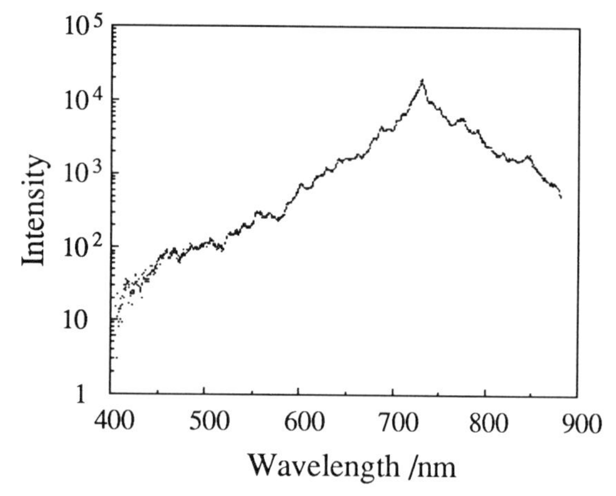

Figure 6. A typical spectrum of femtosecond supercontinuum generated by focusing amplified dye laser into 1-cm water cell.

3.3. Transient absorption microspectroscopic system with optical trapping

A block diagram of simultaneous optical trapping and transient absorption measurements with femtosecond temporal resolution under a microscope is illustrated in Figure 7. A residual 1.06-μm laser beam from the cw mode-locked Nd:YAG laser was divided into two beams, one of which was used for optical trapping of the particle and the other for seeding into the regenerative amplifier. The laser intensity for manipulation was adjusted by a combination of a half-wave (λ/2) plate and a polarizer. The laser beam was reflected by a dichroic mirror in a modified microscope (Nikon Optiphoto XF) and was focused on a sample through a reflecting (Ealing Electro-Optics, x 74, NA = 0.65) or a standard (Nikon CF Flour, x 40, NA = 0.85) objective lens. As indicated in Part II by Sasaki and Misawa, the trapping force depends on the particle diameter, laser power, NA of the objective lens, and the difference of the refractive index between the particle and the surrounding medium

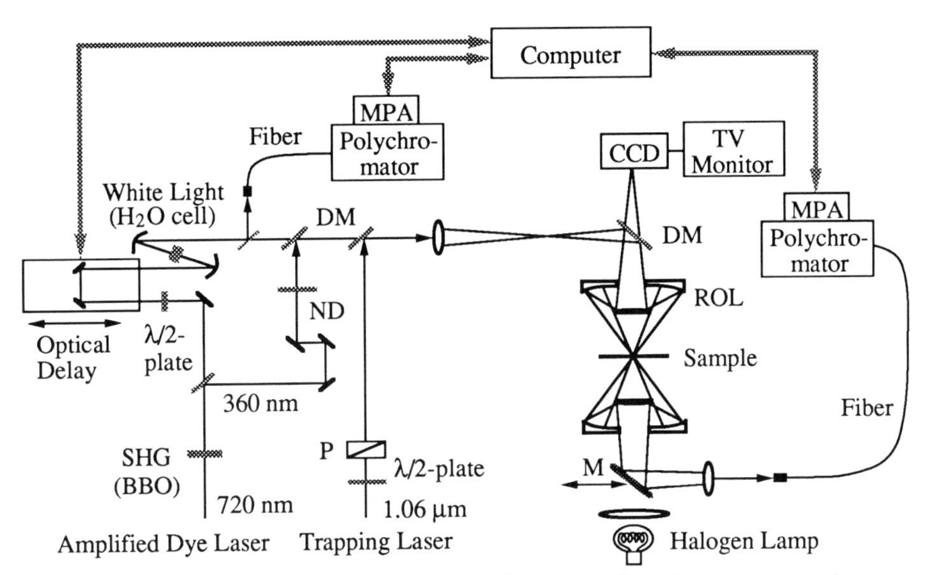

Figure 7. A block diagram of the femtosecond transient absorption microspectroscopic system with an optical trapping technique. DM, dichroic mirror; M, removable mirror; MPA, multichannel photodiode array; P, polarizer; ROL, reflecting objective lens; VA, variable attenuator.

[8, 9, 11]. In the current experimental condition, a toluene droplet of ≈ 10 μm diameter in water can be easily trapped with ≈ 50 mW input laser power into the microscope. If the sufficient power over 1 W is used to trap a particle, a nonlinear phenomenon such as second or third harmonic generation easily occurs from a particle because of the high peak intensity of the trapping laser beam (pulse width ≈ 100 ps). This is in contrast with the laser trapping by a cw Nd:YAG laser, in which such a nonlinear phenomenon hardly occurs. The laser power for manipulation was adjusted to 100 ~ 200 mW for ordinary experiments.

For the transient absorption measurements, the pump pulse (≈ 360 nm), the supercontinuum (probe pulse), and the trapping beam were coaxially introduced into a microscope and focused on the sample by a reflecting objective lens. The transmitted probe pulse through the sample was collected by the same type of reflecting collective lens with changing X and Y positions by an autoscanning stage (0.5 μm step in lateral direction) and was reflected by a removable mirror. The probe pulse was then imaged through filters on a quartz fiber coupled to an input slit of a polychromator (McPherson, 2035) and detected with a multichannel photodiode array (Princeton Instruments, PDA-1024 and ST-1000). The wavelength resolution of the present system is approximately 3 nm. The time delay between the pump and probe pulses was controlled by a translation stage (minimum step = 1 μm).

An important point to measure the transient absorption spectra is to correct the fluctuation of the probe pulse. The supercontinuum was therefore partly reflected by the surface of a cover-glass plate (≈ 120-μm thickness), and was used as a reference pulse to compensate the intensity fluctuation. The following equation was used to calculate the absorbance of the sample, ΔOD [21]:

$$\Delta OD = \log (I_R/I_S) - \log (I_R^0/I_S^0) \tag{2}$$

where R and S denote the reference and signal pulses, I_S and I_S^0 are the signal pulse intensities with and without excitation pulse, respectively.

Another significant point is to correct the time dispersion of femtosecond supercontinuum. The transient absorption spectrum obtained at each delay without the dispersion correction is not a true one. The time dispersion of the supercontinuum is originated from some optical elements and a 1-cm water cell. When the time dispersion against the wavelength is known in the current experimental condition, the spectrum can be corrected by the time-channel shift method. Namely, the absorbance of a sample at a wavelength λ_i and a delay time t_j, OD (λ_i, t_j), should be replaced by OD [λ_i, $t_j + \Delta t(\lambda_i)$], where $\Delta t(\lambda_i)$ is a time dispersion at λ_i. A plot of OD [λ_i, $t_j + \Delta t(\lambda_i)$] against λ_i gives the corrected transient absorption spectrum at a delay time t_j. The rise curve of the $S_n \leftarrow S_1$ transient absorption of ZnTPP in dimethyl formamide (DMF) liquid layer (≈ 120-μm thickness) was measured under the microscope as a function of wavelength, and the time dispersion was determined to be 3 ps in the wavelength region from 450 to 700 nm. The experimental time dispersion curve was then used for the correction of the transient absorption spectra.

3.4. Spatial resolution in transient absorption microspectroscopy

Focusability of the white-light probe pulse is essential for the measurements of transient absorption spectra in small domains. A microscope stage micrometer was used as a reference sample to examine the focusability and the chromatic aberration of the objective lens. This sample has square grids with 10-μm separation, a line of ≈ 20-μm width, and line thickness of 0.1-μm thickness deposited by pigments.

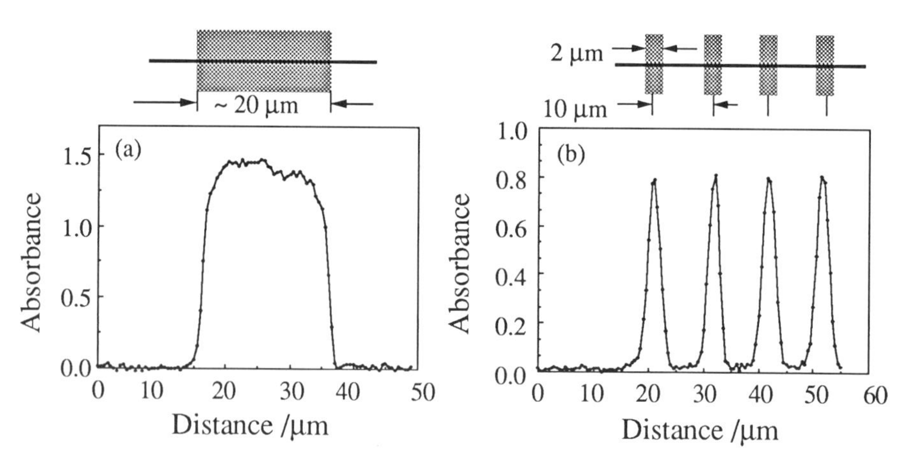

Figure 8. Ground-state absorbance of a microscope stage micrometer as a function of distance for (a) wide line and (b) comb-like lines observed at 600 nm.

Figure 8 illustrates the absorbance of the microscope stage micrometer as a function of distance for both a wide and narrow comb-like lines monitored at 600 nm by using a reflecting objective lens (x 74). A line of ≈ 20-μm wide can be clearly distinguished from a transparent region (Figure 8a), whereas a line of ≈ 2-μm wide with 10-μm separation has lower absorbance than that of the wide line as indicated in Figure 8b. This result suggests that a diameter of the supercontinuum on the focusing point is comparable with the width of a narrow line. A transmittance and its differential curves against the distance are shown in Figure 9. A lateral resolution defined by FWHM of a differential curve is approximately 1.6 μm at 600 nm. This value is almost constant irrespective of observation wavelengths from 450 to 700 nm (1.5 ~ 2.4 μm) so far as the reflecting objective lens is used. The lateral resolution of 1.6 μm at 600 nm is, however, somewhat larger than that expected from NA of the reflecting objective lens (≈ 1.1 μm at 600 nm for NA = 0.65). This may be originated from the nature of the supercontinuum. The supercontinuum cannot be regarded as a point light source because of the nonlinear generation processes, which results in lower spatial resolution than expected.

The lateral resolution becomes worse when the standard objective lens is applied: because of the chromatic aberration, the distance for 10 to 90 % rise in the transmittance curve is ranged from 2 to 9 μm over the range from 450 to 700 nm. The reflecting objective lens with aberration-free quality is clearly indispensable for absorption measurements of spatially inhomogeneous samples. If the collection lens with a small aperture angle is applied as described in Section 2.2, the lateral resolution may be improved even in the standard objective lens.

The pump pulse, on the other hand, can be focused to ≈ 1-μm diameter larger than that of the probe pulse when its diameter illuminated on the objective lens is matched to NA. One method to take an accurate transient absorption spectrum is to defocus the pump pulse at the focal plane as indicated in Section 2.2. In the reflecting objective lens, this method is useful only for thin samples but not for thick samples because of the intrinsic nature, obscuration, of the lens. Another method is to reduce the diameter of the pump pulse illuminating on the objective lens, which results in the

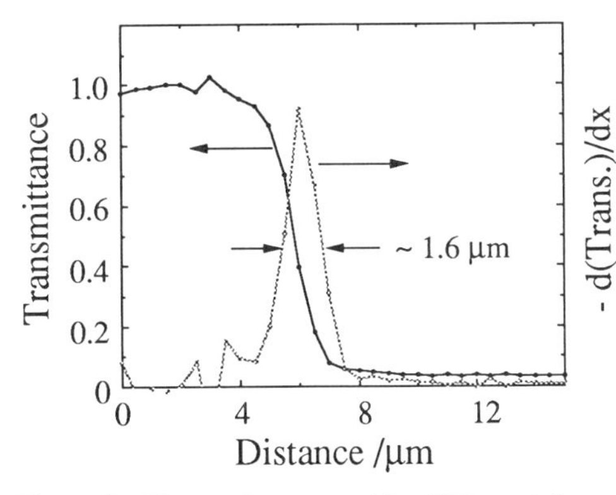

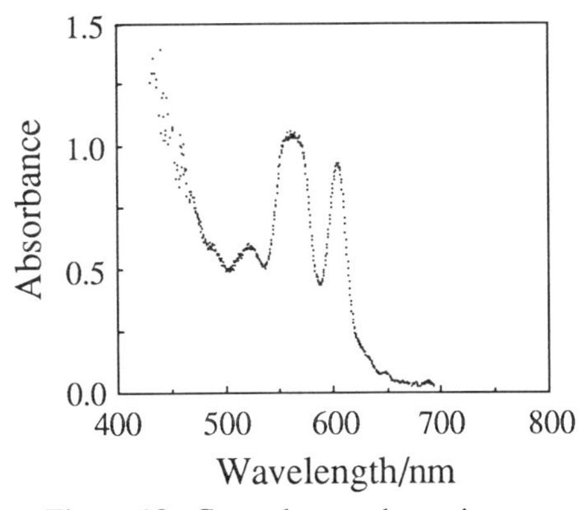

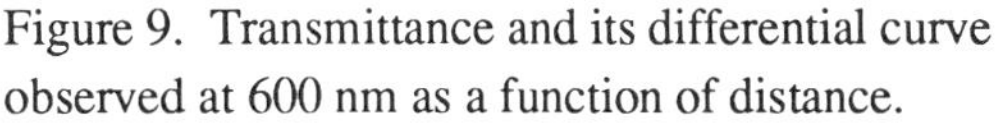

Figure 9. Transmittance and its differential curve observed at 600 nm as a function of distance.

Figure 10. Ground-state absorption spectrum of a ZnTPP microcrystal (7 x 8 μm).

larger beam diameter and deeper depth of focus of the pump pulse. This is more general method in the case of reflecting objective lens.

Using the above mentioned method, transient absorption spectra of a single Brownian particle in solution can be measured with femtosecond temporal and micrometer (< 2.4 μm) spatial resolutions.

3.5. Measurements of transient absorption spectrum in micrometer domains

The ground-state absorption spectrum in micrometer domains is easily measured by the microspectroscopic system. Figure 10 illustrates an example of a ZnTPP microcrystal with a size of 7 x 8 μm. The Q-band of ZnTPP is clearly observed in the wavelength region from 500 to 650 nm. We can also measure a polarization spectrum of micrometer-sized samples, since the polarization direction of the supercontinuum can be simply altered by changing the polarization of fundamental laser pulse with a λ/2-plate set in front of the water cell.

Figure 11 illustrates transient absorption spectra of a single optically trapped droplet in water at a delay time of 100 ps after excitation. In a ZnTPP/toluene droplet (7 x 10^{-3} M, 32 μm ϕ), the spectrum has a peak at ≈ 450 nm, which corresponds to the $S_n \leftarrow S_1$ absorption [21], and a ground-state breaching of Q-band at 556 nm (Figure 11a). The spectrum of a pyrene/paraffin droplet (5 x 10^{-2} M, 46 μm ϕ) has a peak at 472 nm and at 513 nm (Figure 11b), which is safely assigned to the $S_n \leftarrow S_1$ absorption of pyrene [22,23]. Transient absorption spectra of a single Brownian particle in solution can be obtained only if the optical trapping technique is applied to a particle under the microscope. The sensitivity of the transient absorption microspectroscopic system is ≈ 0.01 in absorbance unit as shown in Figure 11b, which is in comparable with a conventional transient absorption spectroscopy system for bulk samples [21, 23]. Using molar extinction coefficients of the $S_n \leftarrow S_1$ absorption, the excited-state concentrations of ZnTPP and pyrene were estimated to be ≈ 24 % and ≈ 1.4 % of that of the ground state, respectively. The pump energy onto the sample was estimated to 5 ~ 10 mJ/cm^2 by the excited-state concentration and the beam diameter of the pump pulse (5 ~ 6 μm ϕ). The further increase of the laser

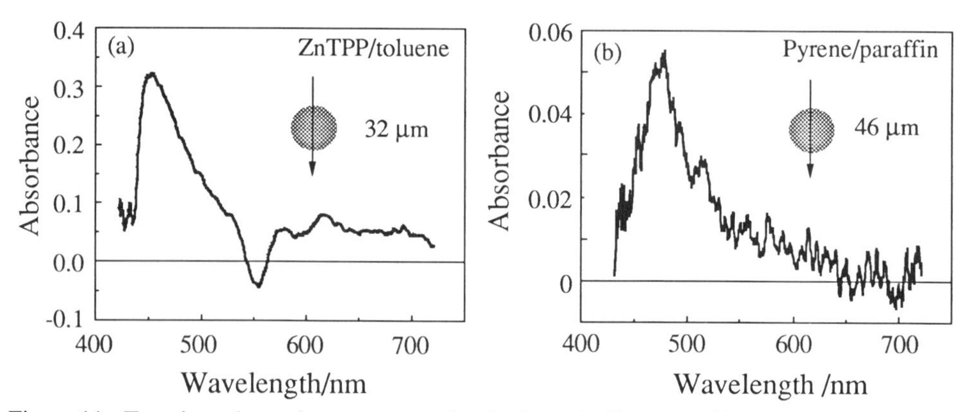

Figure 11. Transient absorption spectrum of a single optically trapped ZnTPP/toluene droplet (a) and a pyrene/paraffin droplet (b) in water obtained at 100 ps after excitation.

intensity to 20 mJ/cm^2 induces sample damage, as shown by a change of ground-state absorption of a droplet within 10 min.

Figure 12 shows a rise curve of transient absorption of an optically trapped ZnTPP/toluene droplet (30 μm ϕ) in water observed at 450 nm. This rise curve can be convoluted by a system impulse of 170 ~ 200 fs pulse width with the lifetime of ZnTPP plus rise component of approximately 1 ps. The rise component of $S_n \leftarrow S_1$ absorption is due to thermal relaxation of vibrationally hot molecules [24, 25]. As the thermal relaxation involves the energy dissipation to the surrounding medium and its kinetics strongly depends on the kinds of solvents [26], the thermal relaxation time may depend on the size of a droplet. This will be one of interesting subjects made possible by the present system.

Transient absorption microspectroscopic system is also useful to analyze the physicochemical properties of individual microcrystals by selecting a single crystal from a mixture. We have examined α- and β-perylene microcrystals, since the structures of both crystals are different from each other [27]. As a result, the luminescence spectrum of α-perylene is distinct from that of β-perylene [28]. One can easily separate each crystal by just a color of luminescence with an eye under the microscope, and then the transient absorption measurements are carried out. Rise curves of transient absorption of a single α-perylene microcrystal observed at 510 and 660 nm are illustrated in Figure 13. The absorbance at 510 nm shows an instantaneous rise and very slow decay components. On the other hand, fast decay components are observed at 660 nm. The difference in temporal behaviors is originated from the fact that transient species at two wavelengths are different with each other [29]: Cation at 510 nm and self-trapped exciton at 660 nm as shown later. The temporal behaviors at any wavelengths can be analyzed only by the present system using the supercontinuum probe pulse.

Figure 14 illustrates transient absorption spectra of both single α- and β-perylene microcrystals. The spectrum of a single α-perylene crystal shows a peak at 510 nm and a broad absorption ranging from 570 to 720 nm. On the other hand, a sharp peak was revealed at 500 nm in a single β-perylene crystal, but not a broad

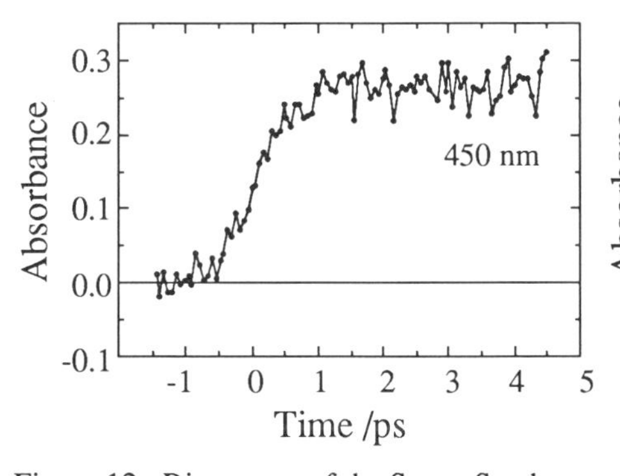

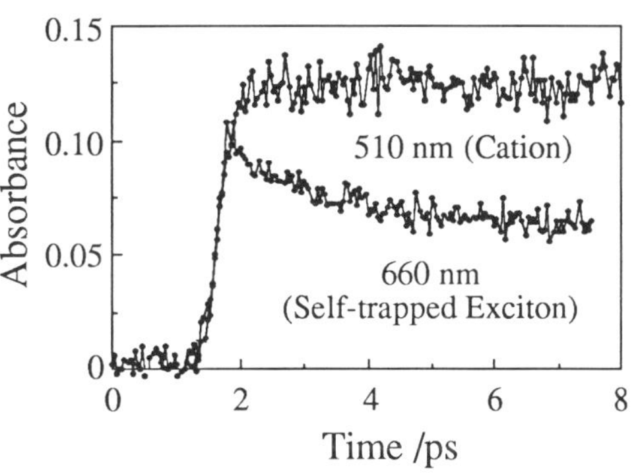

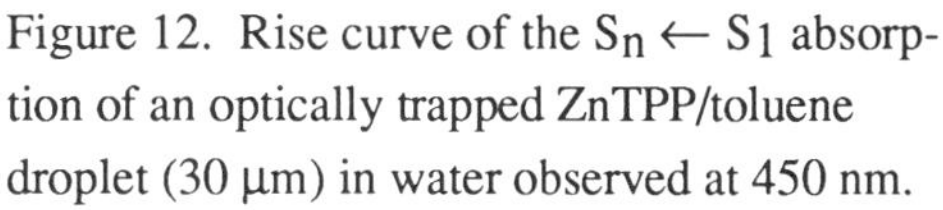
Figure 12. Rise curve of the $S_n \leftarrow S_1$ absorption of an optically trapped ZnTPP/toluene droplet (30 μm) in water observed at 450 nm.

Figure 13. Rise and decay curves of the transient absorption of a single α-perylene microcrystal observed at 510 and 660 nm.

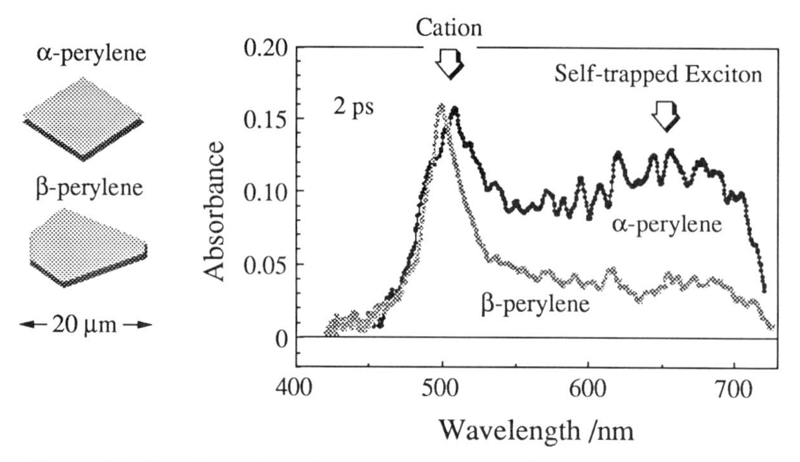

Figure 14. Transient absorption spectra of α- and β-perylene microcrystals obtained at 2 ps after excitation.

absorption in longer wavelength region. The band width of absorption at 500 nm in β-perylene crystal appears to be narrower than that at 510 nm in α-perylene crystal. These absorption bands in the shorter wavelength region is probably assigned to the partially and fully overlapped dimer cations induced by two-photon excitation because of the spectral similarity with perylene cations observed in γ-irradiated glassy solution [30]. The broad band centered at ≈ 650 nm observed only in α-perylene microcrystal is believed to be the absorption spectrum from two-center (fully overlapped) type self-trapped exciton [29]. The spectral and temporal differences between α- and β-perylene crystals are directly correlated to their structural differences in the ground and excited states.

4. POSSIBLE APPLICATIONS AND LIMITATIONS OF TRANSIENT ABSORPTION MICROSPECTROSCOPY

The transient absorption microspectroscopic systems developed here have micrometer spatial and microsecond ~ femtosecond temporal resolutions. The maximum sensitivity is comparable with that of a conventional transient absorption spectroscopic system. The present systems are very useful for the analyses of photophysical and photochemical dynamics of various kinds of nonfluorescent inhomogeneous systems such as micrometer-sized particles, polymer films, biological samples, micrometer patterns for microdevices, and so forth. The combination of an optical trapping technique is essential in transient absorption microspectroscopy. If this technique is applied to a living cell to fix it, photochemical primary processes will be revealed in a living condition without any destruction or pretreatment of the cell. This will make the complicated handling processes of the biological samples much easier, and become a new tool for analyzing the real physicochemical functions of biological organs. In addition, the growth of a large single crystal is very difficult in some cases, resulting in a mixture of microcrystals. Thin crystals are also required for absorption measurements with large molar extinction coefficients. Therefore, present systems are indispensable for studying any kind of microcrystals or microparticles. Furthermore, these systems can be easily modified with some improvements for transient reflection microspectroscopy, which is useful for analyzing non-transparent materials in micrometer domains.

In the current experiments, the input pump energy into the microscope should be reduced to avoid the sample damage by nonlinear photochemical processes even in a sample prepared in Section 2.1. When the input energy is over than 1 μJ in the femtosecond microspectroscopic system, the breakdown of water itself can be easily seen as cavitation. This phenomenon is known as a physical effect associated with optical breakdown. In this condition, the pump power with 200-fs pulse width is over than 1.2 TW/cm^2. This value is in good agreement with the reported threshold values of the optical breakdown of water using picosecond Nd:YAG laser (40-ps pulse width): 1.6 TW/cm^2 for deionized water and 0.5 TW/cm^2 for tap water [31]. It is worth noting that, for nanosecond laser pulses, threshold powers are of a few times lower than that of picosecond pulses [31]. As a result, the nanosecond laser is better for taking a transient absorption spectrum in micrometer domains without sample damages using a pulse energy of a few tens of mJ/cm^2. For the study of physicochemical properties in picosecond time scales, particular attention should be paid for the input laser intensity to avoid the sample damages during the measurements.

The sensitivity ($\Delta OD \approx 0.01$) of our systems will limit the applications for micrometer domains. Since the number of molecules included in such systems is very small, the absorbance of the transient species is expected to be low unless their molar extinction coefficients are very high. The low S/N value is mainly originated from a fluctuation of the probe pulse, which comes from a low repetition rate of the laser (10 Hz) in generating the supercontinuum probe pulse. High repetition-rate laser systems will improve the sensitivity of a femtosecond microspectroscopic system with an optical trapping technique to 10^{-5} in absorbance unit. Acousto-optically switched Nd:YAG laser regenerative amplification (~ 100 kHz) [32] or cw argon-pumped Ti:sapphire regenerative amplification (~ 250 kHz) [33] is a most likely technique for transient absorption microspectroscopy. Such systems will become a new tool in the field of microchemistry, photobiology, and any other microscience.

ACKNOWLEDGMENTS

The authors thank Dr. T. Asahi (Osaka Univ.) and Prof. H. Masuhara (Osaka Univ.) for the collaborations and discussions during the research period of the Microphotoconversion Project.

REFERENCES

1. G. R. Fleming, Chemical Applications of Ultrafast Spectroscopy, Oxford University Press, New York, 1986.
2. H. Masuhara, J. Photochem. Photobiol. A: Chem., 62 (1992) 397.
3. I. Yamazaki, N. Tamai, H. Kume, T. Tsuchiya, and K. Oba, Rev. Sci. Instrum.,

56 (1985) 1187.
4. R. G. W. Norrish and G. Porter, Nature, 164 (1949) 658; G. Porter, Proc. Roy. Soc. London, A200 (1950) 284.
5. K. Sasaki, M. Koshioka, and H. Masuhara, J. Opt. Soc. Am. A, 9 (1992) 932.
6. R. R. Alfano (ed.), The Supercontinuum Laser Source, Springer-Verlag, New York, 1989.
7. A. Ashkin, Phys. Rev. Lett., 24 (1970) 156.
8. A. Ashkin, J. M. Dziedzic, J. E. Bjorkholm, and S. Chu, Opt. Lett., 11 (1986) 288.
9. H. Misawa, M. Koshioka, K. Sasaki, N. Kitamura, and H. Masuhara, Chem. Lett., (1991) 469; J. Appl. Phys., 70 (1991) 3829.
10. A. Ashkin, J. M. Dziedzic, and T. Yamane, Nature, 330 (1987) 769.
11. A. Ashkin and J. M. Dziedzic, Ber. Bunsen. Phys. Chem., 93 (1989) 254.
12. S. M. Block, D. F. Blair, and H. C. Berg, Nature, 338 (1989) 514.
13. W. Adamson, Physical Chemistry of Surfaces, 5th ed., John Wiley, New York, 1990, Chapter 14.
14. H. Piller, Micrposcope Photometry, Springer-Verlag, Berlin, 1977, Chapter 11.
15. NA = 0.6 and n = 1.52 (slide glass) for the condenser lens, NA = 1.3 and n = 1.52 (immersion oil) for an objective lens.
16. R. L. Fork, C. H. Brito Cruz, P. C. Becker, and C. V. Shank, Opt. Lett., 12 (1987) 483.
17. N. Tamai, T. Asahi, and H. Masuhara, Rev. Sci. Instrum., 64 (1993) 2496.
18. W. H. Knox, IEEE J. Quantum Electron., QE-24 (1988) 388.
19. H. Masuhara, H. Miyasaka, A. Karen, T. Uemiya, N. Mataga, M. Koishi, A. Takeshima, and Y. Tsuchiya, Opt. Commun., 44 (1983) 426.
20. R. L. Fork, C. V. Shank, C. Hirlimann, R. Yen, and W. J. Tomlinson, Opt. Lett., 8 (1983) 1.
21. H. Masuhara, N. Ikeda, H. Miyasaka, and N. Mataga, J. Spectrosc. Soc. Jpn., 31 (1982) 19.
22. Y. Nakato, N. Yamamoto, and H. Tsubomura, Chem. Phys. Lett., 2 (1968) 57.
23. H. Miyasaka, H. Masuhara, and N. Mataga, Laser Chem., 1 (1983) 357.
24. A. Seilimeiyer and W. Kaiser, in *Ultrafast Laser Pulses and Applications*, edited by W. Kaiser (Springer, Berlin, 1988) p. 279.
25. H. Miyasaka, M. Hagihara, T. Okada, and N. Mataga, Chem. Phys. Lett., 188 (1992) 259.
26. F. Laermer, W. Israel, and T. Elsaesser, J. Opt. Soc. Am. B, 7 (1990) 1604.
27. M. Iemura and A. Matsui, The Menoirs of the Konan Univ., 27 (1981) 7.
28. H. Nishimura, T. Yamaoka, K. Mizuno, M. Iemura, and A. Matsui, J. Phys. Soc. Jpn., 53 (1984) 3999.
29. N. Tamai, C. F. Porter, and H. Masuhara, Chem. Phys. Lett., 211 (1993) 364.
30. A. Kira and M. Imamura, J. Phys. Chem., 83 (1979) 2267.
31. B. Zysset, J. G. Fujimoto, and T. F. Deutsch, Ber. Bunsenges. Phys. Chem., 93 (1989) 260.
32. A. J. Ruggiero, N. F. Scherer, G. M. Mitchell, and G. R. Fleming, J. Opt. Soc. Am. B, 8 (1991) 2061.
33. T. B. Norris, Opt. Lett., 17 (1992) 1009.

MICROCHEMISTRY
Spectroscopy and Chemistry in Small Domains
Edited by H. Masuhara et al.

Picosecond dynamics in thin films by transient grating spectroscopy

N. Tamai*,#, T. Asahi‡, and T. Ito§

Microphotoconversion Project†, ERATO, Research Development Corporation of Japan, 15 Morimoto-cho, Shimogamo, Sakyo-ku, Kyoto 606, Japan

Transient grating spectroscopy using a femtosecond supercontinuum as a probe pulse was developed and applied for the analyses of photophysical and photochemical dynamics in submicrometer organic thin films such as benzophenone-doped polystyrene and copper phthalocyanine. The correlation between time-resolved diffraction and transient absorption spectra were discussed, and the temporal profiles of first- and second-order diffraction were analyzed in terms of the excited-state interaction in thin films. Potential applications of transient grating spectroscopy to organic thin films in regular reflection and total internal reflection modes were also discussed in comparison with that in ordinary transmission mode.

1. INTRODUCTION

Transient grating experiments have been successfully applied to analyze wide variety of transient processes of physicochemical properties in various systems. For example, transient gratings were used to deduce the rates of excitation energy transport in solids and liquids [1, 2], carrier dynamics in semiconductors [3], triplet quantum yields of organic materials and their relaxation dynamics in liquids [4], rotational relaxation rates in various liquids [5-7] and liquid crystals [8, 9], thermal and mass diffusion constants in condensed phases [10, 11], and so forth.

In this technique, two coherent light pulses are crossed simultaneously at an appropriate angle to create an optical interference pattern in a sample as illustrated in Figure 1. Optical interaction of the sample with light through absorption, electrostriction [12], or any other mechanisms [13] forms a transient grating of

---

* To whom correspondence should be addressed.
# Present Address: Light and Material Group, PRESTO, JRDC, Department of Chemistry, Faculty of Science, Kwansei Gakuin University, Uegahara, Nishinomiya 662, Japan
‡ Present Address: Department of Applied Physics, Osaka University, Suita, Osaka 565, Japan
§ Present Address: Tokyo Technical Research Lab., Mita Industrial Co., Ltd. Yushima 4-chome, Bunkyo-ku, Tokyo 113, Japan
† Five-year term project: October 1988 ~ September 1993.

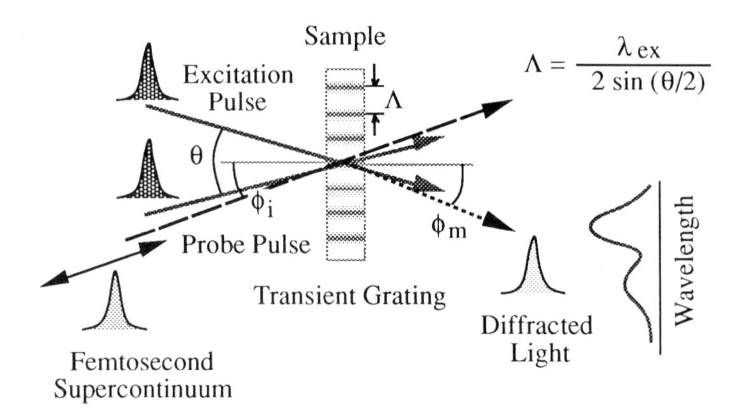

Figure 1. A schematic illustration of transient grating spectroscopy.

periodic structure of a change in complex refractive index, $\tilde{n}$. The fringe spacing of the transient grating is dependent on the excitation beam geometry:

$$\Lambda = \lambda_{ex}/[2 \sin(\theta/2)] , \tag{1}$$

where Λ is the spatial period of the grating, λ_{ex} is the excitation wavelength, and θ is the angle between the excitation pulses. When the excitation intensity is strong enough to induce a nonlinear photochemical reaction, a permanent grating is formed in micrometer-order domains. A typical example of a permanent grating formed in a polymer thin film is shown in Figure 2, in which a periodic structure is characterized by a micrometer-order period Λ. Several tens of μm in a two-dimensional domain is required for the formation of a transient grating, so that the lateral resolution of the transient grating experiments will be limited by this condition. However, picosecond dynamics of a sample with *a few nm thickness* can be measured as shown later. In the usual experiments, the excitation intensity should be reduced to avoid the formation of a permanent grating.

Rise and decay processes of a transient grating are probed by a third laser beam incident at an appropriate angle for diffraction. Since only the diffraction signal is detected, the dynamics of well-known phenomena such as excitation energy transport and electron transfer in the system can be sensitively measured as compared with the conventional transient absorption experiments. Furthermore, orientational relaxation and thermal or mass diffusion process in

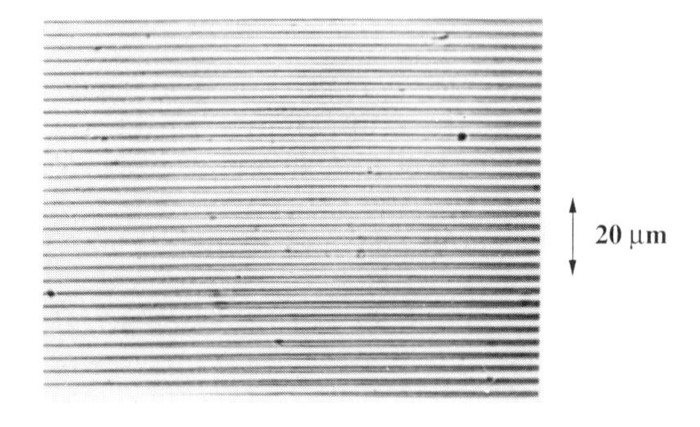

Figure 2. A permanent grating formed in polymer thin film. $\Lambda = 3.2$ μm.

a system, which are impossible to be examined by the transient absorption experiments, can be also analyzed with high sensitivity. Another advantage of transient grating experiments is to use lower excitation intensity than in the transient absorption experiments, which results in a benefit to examine the dynamic properties of molecular aggregates without any nonlinear effects.

Over the past decade, a single frequency laser beam has been widely used as a probe pulse. However, the spectral information of the excited states and/or intermediate species of the photochemical reactions in question have never been explored. Transient grating spectroscopy using a femtosecond supercontinuum as a probe pulse (see Figure 1) is extremely useful for analyzing ultrafast dynamics in various samples [14]. Using this spectroscopy, the diffraction spectrum corresponding to the transient absorption can be obtained with high sensitivity, and transient species can be identified as discussed later. In addition, it has a high potential to analyze photochemical and photophysical dynamics in organic thin films with picosecond temporal resolution, where conventional transmittance-type transient absorption spectroscopy cannot be applied.

In this article, we have examined picosecond diffraction spectra and their rise and decay dynamics in benzophenone-doped polystyrene and phthalocyanine thin films with submicrometer thickness by femtosecond transient grating spectroscopy. The diffraction spectra correlated with transient absorption spectra and the temporal profiles of first- and second-order diffraction were analyzed in terms of excited-state interaction. Potential applications of transient grating spectroscopy in regular reflection and total internal reflection modes will be also discussed with references of experimental results on a π-conjugated polymer and a photochromic compound.

2. FEMTOSECOND TRANSIENT GRATING SPECTROSCOPIC SYSTEM

A block diagram of femtosecond transient grating spectroscopy is illustrated in Figure 3. The detail of the laser system has been described in a previous paper [15]. Briefly, it consisted of a hybridly mode-locked, dispersion compensated femtosecond dye laser (Coherent, Satori 774) and a dye amplifier (Continuum, RGA 60-10 and PTA 60). The sample was excited by second harmonic ($\lambda_{ex} \approx 360$ nm) or fundamental of the amplified dye laser output (center wavelength ≈ 720 nm, pulse width ≈ 150 fs fwhm) at a repetition rate of 10 Hz. The excitation intensity was changed by a circular variable attenuator. The excitation pulse was split into two beams, which were crossed in the sample at an appropriate angle with a focusing diameter of ≈ 500 μm to produce an optical interference pattern. The induced transient grating was probed by a femtosecond supercontinuum obtained by focusing a portion of fundamental laser beam into a 1-cm H_2O cell. A computer-controlled translation stage was used to change the time delay between pump and probe pulses. By setting a $\lambda/2$-plate in front of the H_2O cell, the polarization of the probe beam was changed with respect to the pump polarization. Partial loss of polarization caused by the supercontinuum generation was compensated by a linear film polarizer.

Diffraction spectra were obtained by an intensified multichannel detector (Hamamatsu PMA-10, or Prinston Instruments ICCD-576) as a function of probe

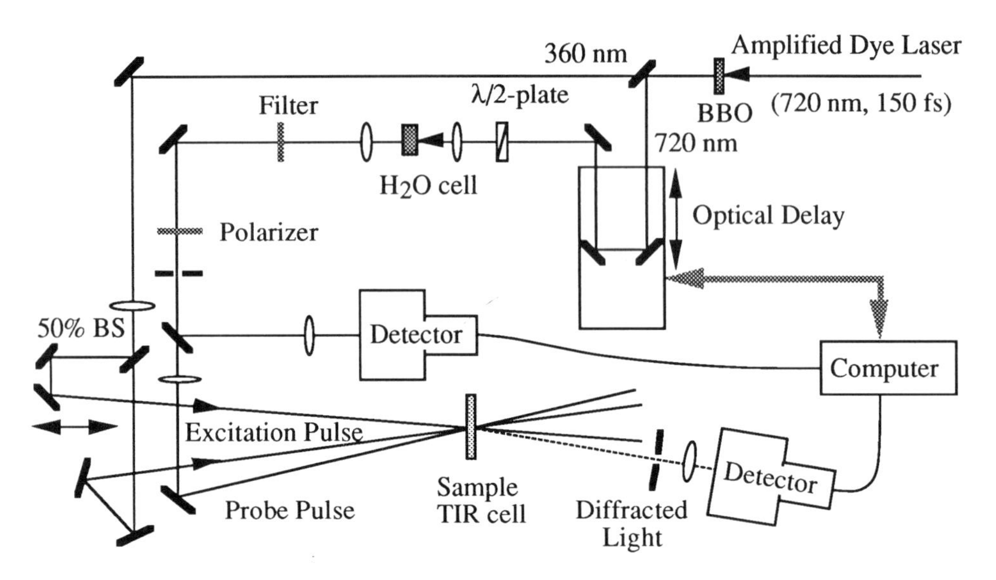

Figure 3. A block diagram of the transient grating spectroscopy. BBO, β-barium borate; BS, beam splitter; TIR, total internal reflection.

delay, and corrected for the spectral intensity variation and time dispersion of the supercontinuum. Rise and decay curves of the diffraction signals were measured with a photomultiplier (Hamamatsu, R928) - monochromator (Japan Spectroscopic, CT-10) combination. All experiments were carried out at 293 ± 1 K.

3. TIME-RESOLVED DIFFRACTION AND TRANSIENT ABSORPTION SPECTRA

According to the relative values of the sample thickness L and the grating period Λ, the diffraction phenomena can be classified into Raman-Nath and Bragg diffraction regimes. Normally, a dimensionless parameter Q defined by the following equation is used to distinguish Raman-Nath ($Q < 1$) and Bragg ($Q > 1$) diffraction regimes [16]:

$$Q = 2\pi\lambda_p L / n\Lambda^2 , \tag{2}$$

where λ_p and n are the probe wavelength and the refractive index of a sample, respectively. When the sample thickness is of the order of μm or less, Raman-Nath condition is normally satisfied. In this case, the probe pulse with an incidence angle, ϕ_i, is diffracted to a direction, ϕ_m, according to the following equation:

$$\sin \phi_m = \sin \phi_i + m \lambda_p/\Lambda , \qquad m = 0, \pm 1, \pm 2, \cdots\cdots \tag{3}$$

where m is the order of diffraction. Namely, white-light incident at an appropriate angle is diffracted. Figure 4 shows a photograph of a femtosecond super-continuum diffracted by a thermal grating of a liquid crystal thin film. As clearly shown in the Figure, a wide wavelength region is simultaneously diffracted. The spectrum should be corrected for an intensity variation of the incident supercontinuum to take a diffraction efficiency as a function of wavelength, which corresponds to the diffraction spectrum.

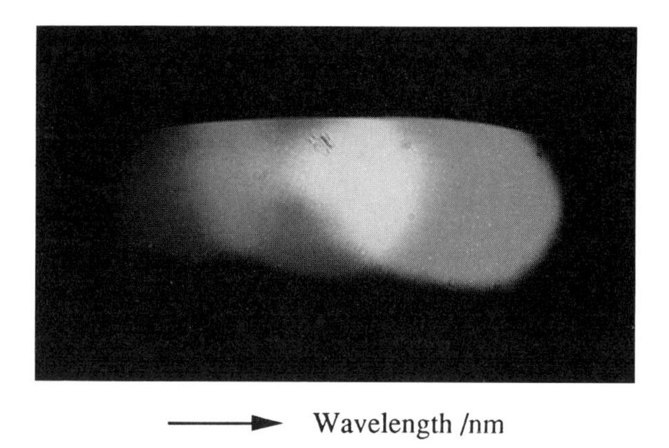

Figure 4. Diffraction of a femtosecond super-continuum by a thermal grating of a liquid crystal thin film.

In the Raman-Nath regime, m-th order diffraction efficiency, η_m, is given by the following equation [16, 17]:

$$\eta_m = \exp(-2.3OD)\, J_m^2 \left(\left| 2\pi\Delta nL/\lambda_p + i\, 2.3\Delta OD/2 \right| \right), \tag{4}$$

where OD is the absorbance of a sample at the probe wavelength, J_m is the Bessel function of the first kind, and ΔOD and Δn are the peak-null difference of absorbance and the real part of refractive index induced by pump laser, respectively. As the small signal limit, equation 4 is simplified to

$$\eta_m = \exp(-2.3OD)\, [\, (\pi\Delta nL/\lambda_p)^2 + (2.3\Delta OD/4)^2 \,]^m / m!^2 , \tag{5}$$

First-order diffraction, η_1, is normally measured, and thus the diffraction efficiency is proportional to a sum of the square of ΔOD and Δn, respectively. Δn includes terms of absorption change (Δn_{ex}), which is connected with ΔOD by the Kramers-Kronig relation, and acoustic strain (Δn_s) originated mainly from the heat release by the photoexcitation of a sample. Thus the diffraction spectrum, $\eta_1(\lambda_p)$, obviously corresponds to the transient absorption spectrum when the contribution of acoustic grating to the diffraction spectrum is negligible [14].

If the sample is rather thick and the Bragg diffraction condition is satisfied, a selected wavelength is diffracted to a certain direction ϕ_m for an incidence angle ϕ_i of the probe light. This condition is expressed by the following equation:

$$\sin|\phi_m| = \sin|\phi_i| = m\,\lambda_p/2\Lambda\,. \tag{6}$$

A small deviation from the Bragg condition is allowed. This effect was theoretically examined by Kogelnik using coupled wave theory [18]. The effect of incidence angle on the diffraction efficiency of the white-light probe pulse is calculated in Figure 5. It is clear that a wavelength rage from 400 to 700 nm is diffracted at an incidence angle of 3 ± 0.7°. Therefore, a distribution of the incidence angle (solid angle) of white-light make an observation of diffraction spectrum even in the Bragg diffraction regime [14]. Consequently, the diffraction spectrum corresponding to transient absorption spectrum is sensitively measured both for Raman-Nath and Bragg diffraction conditions.

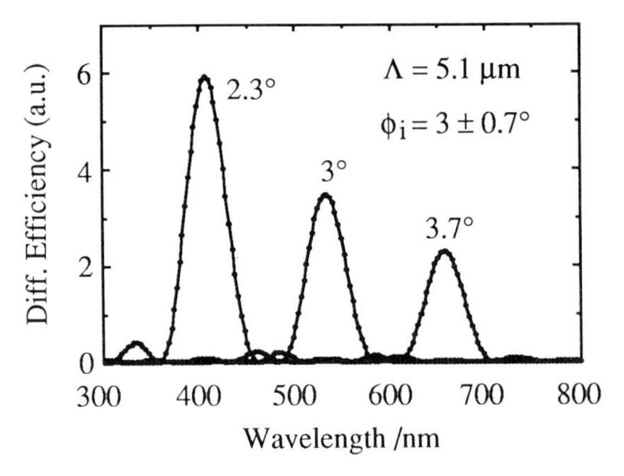

Figure 5. The wavelength dependence of $\eta_1 (\lambda_p)$ in the Bragg diffraction regime.

4. PICOSECOND DYNAMICS IN THIN FILMS BY TRANSMISSION-MODE TRANSIENT GRATING SPECTROSCOPY

4.1. Intersystem crossing of benzophenone in a polystyrene thin film

Benzophenone is known as a typical non-fluorescent organic compound (quantum yield of triplet-state formation (Φ_{ST}) = 1) to show ultrafast intersystem crossing. The dynamics of the excited states in solution has been analyzed mainly by transient absorption spectroscopy. However, the rate constants of intersystem crossing so far reported are scattered from 7 to 30 ps even in benzene solution [19, 20], not to mention in thin films.

Figure 6 illustrates time-resolved diffraction spectra of a benzophenone-doped polystyrene film (thickness ≈ 4 μm) excited at ≈ 360 nm. The spectrum with a peak at approximately 575 nm was observed at 1-ps delay time after excitation. The spectral shape was changed with increasing the time delay followed by an increase in the diffraction intensity at shorter

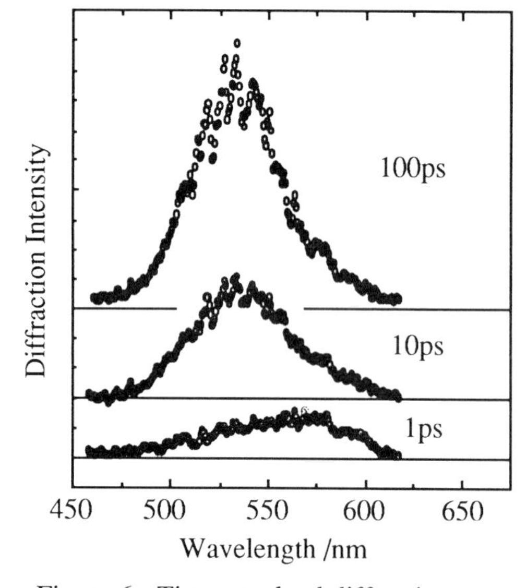

Figure 6. Time-resolved diffraction spectra of benzophenone-doped polystyrene thin film (≈ 4 μm thickness). The delay time after excitation is indicated in the Figure.

wavelength region. The spectrum at late time region (100 ps) has a peak at 530 nm, which is very similar to $T_n \leftarrow T_1$ absorption in solution [21]. A similar change in time-resolved diffraction spectra was observed in acetonitrile solution [14]. The diffraction spectrum with a peak at 530 nm is safely assigned to that of the triplet-state of benzophenone and the spectrum at 575 nm is due to the diffraction of the excited singlet-state. The spectral change clearly indicates the dynamics of intersystem crossing of benzophenone in the thin film. The rate constant of intersystem crossing, k_{ISC}, was analyzed with the temporal profile of diffraction intensity at 530 nm. Since ΔOD and Δn_{ex} show the same temporal behavior in equation 5, the population density, $\rho(t)$, of the triplet state is expressed in the form

$$\rho(t) = \sqrt{\eta_1(t)} = 1 - \exp(-k_{ISC}\, t)\,. \tag{7}$$

k_{ISC} of benzophenone in the polystyrene film was estimated to be $\approx$ 10 ps. In our recent studies, k_{ISC} was slightly dependent on the solvent polarity and in the range from 9 to 13 ps in various solvents. This result indicates that even in a rigid matrix k_{ISC} is almost the same as in the liquid phase. It was shown that transient grating spectroscopy is very useful to take a diffraction spectrum corresponding to transient absorption and to analyze ultrafast molecular processes in thin films with high sensitivity.

4.2. Nonlinear dynamics in the excited state of phthalocyanine thin film

Phthalocyanines are widely used as photoconductive materials. In addition, remarkable attention has been paid for nonlinear optical properties of phthalocyanines since a charge transfer state between a central metal and a ligand can be controlled by changing a central metal. For the fundamental understanding of the nonlinear optical processes, the analyses of photophysical and photochemical properties are indispensable. However, dynamic properties of some phthalocyanine compounds in the fluorescence state are not well known because of their low fluorescence quantum yields and of emissions in the near-infrared region. Therefore, phthalocyanines are nice target for transient grating spectroscopy. In this section, we describe the excited-state dynamics of a copper phthalocyanine thin film prepared on a glass substrate by a vacuum evaporation method. Excitation intensity dependence on the diffraction signals was discussed in terms of nonlinear interaction in the excited state.

Fig. 7 illustrates time-resolved first-order diffraction spectra of a copper phthalocyanine thin film ($\approx$ 50 nm thickness) excited at $\approx$ 720 nm. The diffraction spectrum clearly shows a peak at $\approx$ 535 nm irrespective of delay times. The intensity of this band decayed very rapidly with time, while spectral shapes at various delay times were very similar to one another. The spectrum with a peak at $\approx$ 535 nm is probably assigned to the diffraction from the excited singlet-state of copper phthalocyanine. We have examined transient absorption spectra of the film for analyzing the diffraction spectrum. The transient absorption spectrum at a delay time of 1 ps shows a peak at $\approx$ 525 nm and a ground-state bleaching at $\approx$ 620 nm, and its absorbance at $\approx$ 525 nm is very low (< 0.02). This result indicates that the conventional transient absorption spectroscopy with a low repetition-rate laser

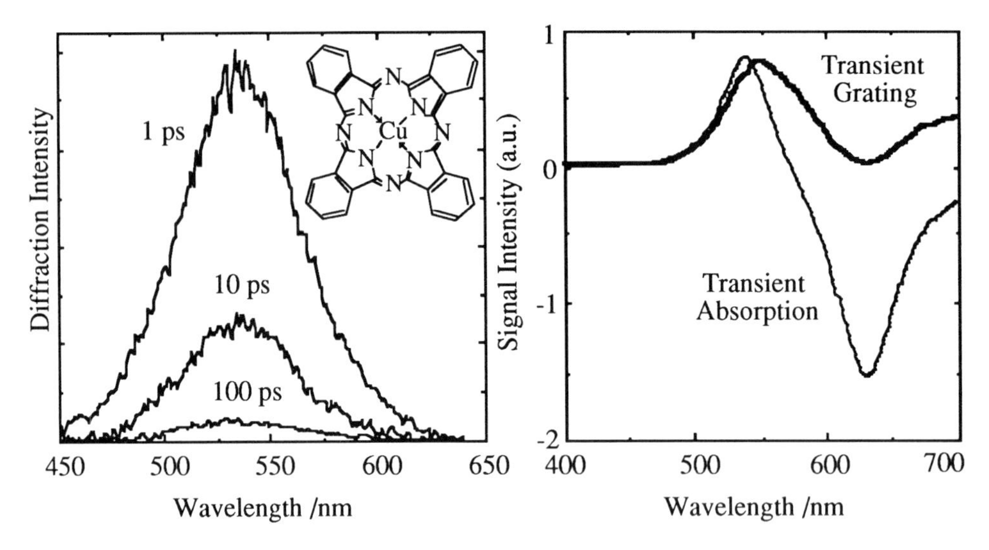

Figure 7. Time-resolved diffraction spectra of a copper phthalocyanine thin film (≈ 50 nm thickness) excited at 720 nm. Delay times after excitation are indicated in the Figure.

Figure 8. Simulation of the diffraction spectrum of copper phthalocyanine. Transient absorption spectrum is a superposition of the ground-state bleaching and $S_n \leftarrow S_1$ absorption.

system is difficult to analyze the relaxation dynamics of such a thin film. The spectrum at ≈ 525 nm can be assigned to $S_n \leftarrow S_1$ absorption. A simulation of the diffraction spectrum carried out by using ΔOD and Δn_{ex} (estimated from Kramers-Kronig relations) predicted a peak at ≈ 545 nm, the result of which is shown in Figure 8 together with the simulated transient absorption spectrum. This result is in good agreement with the experimental diffraction spectrum. In the case where the ground-state absorption is overlapped to the absorption spectrum of intermediate species, the spectral shape of the diffraction spectrum is very different from the transient absorption spectrum. A simulation calculation is helpful to estimate the transient species. This is in contrast with the result on benzophenone as discussed above (Figure 6).

Decay curves of first-order diffraction signals of copper phthalocyanine observed at 540 nm at various excitation intensities are illustrated in Figure 9. It was found that the diffraction intensity decayed nonexponentially and was dependent strongly on the excitation laser intensity. This result is probably interpreted in terms of singlet exciton-exciton (S_1 - S_1) annihilation of copper phthalocyanine. It has been reported that the dynamics of transient absorption of a β-hydrogen phthalocyanine film (submicrometer thickness) was strongly dependent on the excitation intensity in picosecond time regions, and the result was interpreted with singlet-exciton annihilation [22]. Although the film thickness is very thin in our experiment and the kind of phthalocyanines is different from the previous report, a similar annihilation mechanism is probably adapted for analyzing the excitation intensity-dependent diffraction signals.

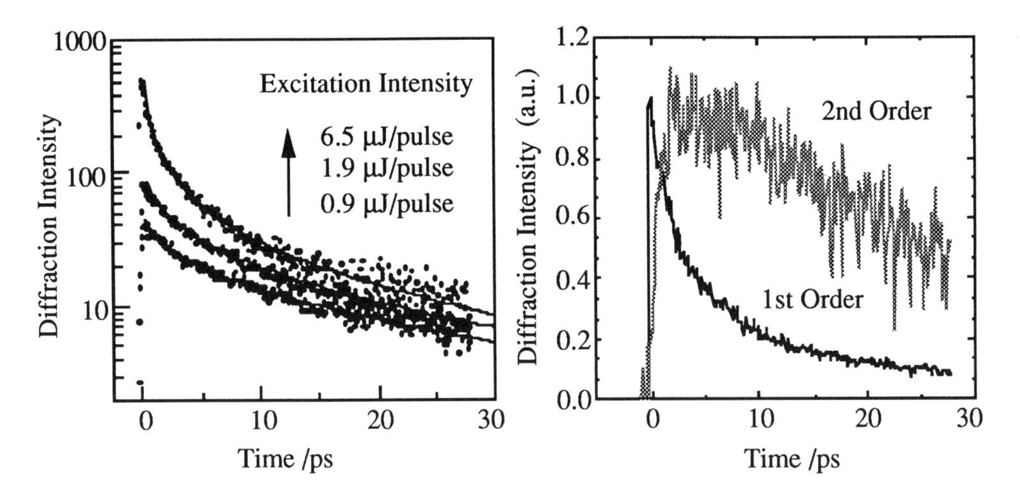

Figure 9. Time profiles of diffraction signals of a copper phthalocyanine thin film (≈ 50 nm) at various excitation intensity. Solid lines are simulation curves based on the equation 11. $\lambda_p = 540$ nm.

Figure 10. First- and second order diffraction signals of copper phthalocyanine thin film. The same excitation intensity was used for both measurements.

The exciton-exciton annihilation process can be expressed by the following rate equations:

$$-dN/dt = \tau_0^{-1} N + k_1 N^2 \tag{8}$$

$$-dN/dt = \tau_0^{-1} N + k_2 t^{-1/2} N^2 \tag{9}$$

where N is the population density of the singlet exciton, τ_0 is the intrinsic lifetime, k_1 and k_2 are the rate constants of respective models. τ_0 was estimated to be ≈ 100 ps from a decay curve at low excitation intensity. Equation 8 is based on the simple bimolecular reaction model with the time-independent rate constant k_1, whereas the time-dependent rate constant $k_2 t^{-1/2}$ is introduced in equation 9. The former model is useful for long time limits in motion-limited diffusion theory. In short time regions, the diffusion process depends on time, and thus the latter model is effective [23, 24]. Another mechanism of the annihilation is the Förster-type dipole-dipole interaction between excitons [25]. This model is described by the $t^{-1/2}$-dependent rate and is also expressed by equation 9. Solutions of both equations can be written as:

$$N(t)^{-1} = \left[N(0)^{-1} + k_1\tau_0\right] \exp(t/\tau_0) - k_1\tau_0 \,, \qquad \text{for equation 8} \tag{10}$$

$$N(t)^{-1} = \left[\left(N(0)^{-1} + k_2\sqrt{\pi\tau_0}\right) \mathrm{erf}\left(\sqrt{t/\tau_0}\right)\right] \exp(t/\tau_0), \qquad \text{for equation 9} \tag{11}$$

$$\mathrm{erf}\,(x) = (2/\sqrt{\pi}) \int_0^x \exp(-y^2)\, dy,$$

where erf(x) is the error function.

The square root of the diffraction intensity ($\eta(t)^{-1/2} \propto N(t)^{-1}$) did not give a straight line against $\exp(t/\tau_0)$, indicating that the annihilation process cannot be expressed by the time-independent rate equation. Using a time-dependent rate based on the equation 11, simulation calculations were carried out and plotted in Figure 9. Experimental data at various excitation intensities were fitted well by changing two variables, N(0) and k_2. Relative k_2 values were almost constant irrespective of the excitation intensity, whereas N(0) increased with increasing the intensity. From these results, a model described with the time-dependent rate is considered to be necessary for describing the annihilation process. In the previous paper [22], τ_0 was neglected in the decay curve analyses based on equation 9. In our experiments, however, a simulation calculation without τ_0 does not provide a satisfactory fitting.

Among two mechanisms (time-dependent diffusion vs. Förster-type interaction) the time-dependent diffusion is the most likely process as the singlet-exciton annihilation. In the Förster-type interaction without diffusion, a spectral overlap between the transient absorption and the fluorescence of copper phthalocyanine has a key role in the annihilation process. Experimental results show only weak transient absorption in long wavelength region (> 700 nm), so that the Förster-type dipole-dipole interaction is an unlikely mechanism in the exciton-exciton annihilation.

In the case where the exciton-exciton annihilation is a dominant relaxation process, the second-order diffraction ($m = \pm 2$ in equation 3) can be easily observed. The result is illustrated in Figure 10 together with a first-order diffraction signal obtained at the same excitation intensity. A rise component of a few ps was clearly observed in the second-order diffraction, which is a remarkable difference from the first-order diffraction. Rise and decay time constants became faster with increasing the excitation intensity. The intensity ratio of the second- to the first-order diffraction was experimentally estimated to be a few % at the peak intensity, although it was dependent on the excitation intensity.

An origin of the higher-order diffraction is explained as follows. The exciton-exciton annihilation induces the deformation of a spatial pattern of sinusoidal population density formed just after excitation with time as illustrated in Figure 11a. Namely, the excited-state population at high density regions decays very rapidly rather than at low density regions. The deformed population density can be expanded as a Fourier series (Figure 11b), in which a spatial pattern at a certain delay time can be written as:

$$N(x,t) = n_0(t) + n_1(t)\cos(2\pi x/\Lambda) + n_2(t)\cos(4\pi x/\Lambda) + \cdots \qquad (12)$$

where the grating vector is in x-direction, and $n_i(t)$ is the i-th Fourier coefficient. It is clear that the second-order diffraction arises from the second Fourier coefficient, $n_2(t)$, with the grating period of $\Lambda/2$, and thus the higher Fourier coefficients are

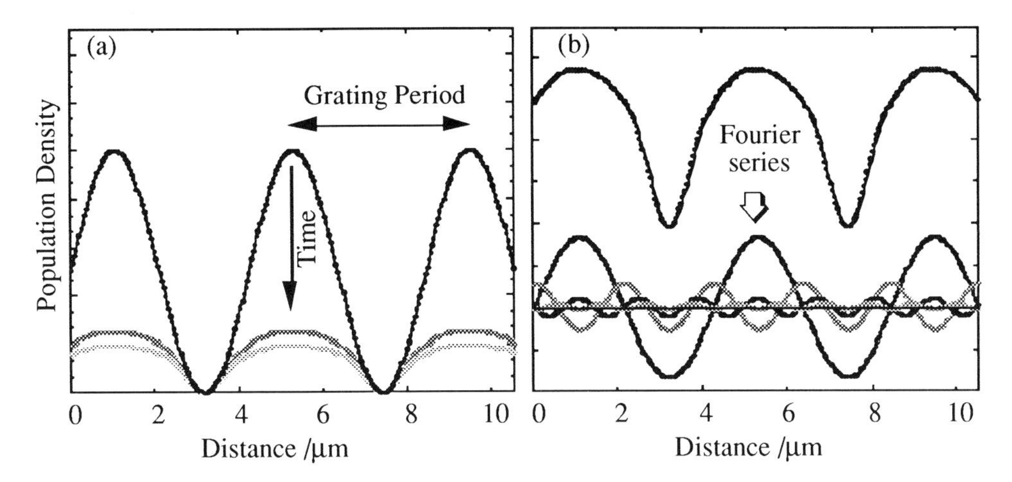

Figure 11. (a) Deformation of the spatial pattern of sinusoidal population density with time. (b) A deformed pattern is expanded as a Fourier series.

responsible for higher-order diffraction [26, 27]. The square of $n_i(t)$ is proportional to the i-th order diffraction intensity. In the initial time region, $n_2(t)$ is predicted to increase with proceeding by the annihilation, which corresponds to the observation of a rise component, and finally it decays with time. This is in good agreement with the experimental data shown in Figure 10.

The second-order diffraction should be observed under the Raman-Nath diffraction conditions even if the excited states decay without any nonlinear reactions (see equation 4). Such kind of second-order diffraction from the thin samples will be measured only if the first-order diffraction has an enough efficiency. Normally, η_1 is not so high ($\approx 10^{-4}$), which results in the undetectable value of η_2 ($\approx 10^{-9}$) as calculated from equation 5. Thus the second-order diffraction is mainly originated from the nonlinear reaction. Such examples include (a) the excited-state interaction as indicated above, (b) the saturation absorption, (c) the two-photon absorption, and so forth. Transient grating spectroscopy has potential for analyzing the nonlinear physicochemical processes mentioned above.

5. PICOSECOND DYNAMICS IN THIN FILMS BY REFLECTION-MODE TRANSIENT GRATING SPECTROSCOPY

5.1 Regular reflection-mode transient grating spectroscopy: dynamics in the surface layer of a photoconductive polymer

A transient grating technique in the regular reflection mode has been used for the generation and detection of a surface acoustic wave [28, 29]. The surface acoustic wave is believed to be characteristic of a submicrometer surface layer. On the other hand, if the transient species generated by this method has a strong absorption at a probe wavelength, photophysical and photochemical dynamics of a submicrometer layer can be analyzed in the regular reflection mode. Recently, Fishman et al.

theoretically analyzed the diffraction signals both for reflection and transmission geometries [30]. According to their results, the effective probe depth that is the average probe distance from the surface, d_p, is given by the following equation:

$$d_p = \lambda_p / \{ 4\pi n [1 + (k/n)^2]^{1/2} \} , \quad (13)$$

where n and k are the real and imaginary parts of the complex refractive index, $\tilde{n}$. It is clear that the probe depth is of the order of 10 to 100 nm, in which the characteristics features of the surface layer are expected. We report here the excited-state dynamics of a thin film of photoconductive polymer, poly(2,5-thienylene vinylene) (PTV), by transient grating spectroscopy in the reflection and transmission modes.

PTV is a one-dimensional, π-conjugated polymer and its third-order nonlinear response has been received much interest because of its potential applications. Thin films of PTV were prepared by a precursor polymer which was coated from THF solution on quartz plates followed by thermal conversion of the precursor polymer at 200 - 250 °C with an acid catalyst [31] as shown in the following scheme, and the nonlinear response was analyzed by transient grating spectroscopy,

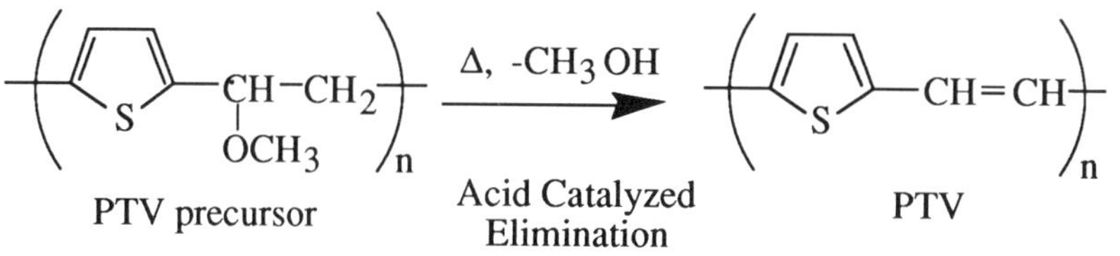

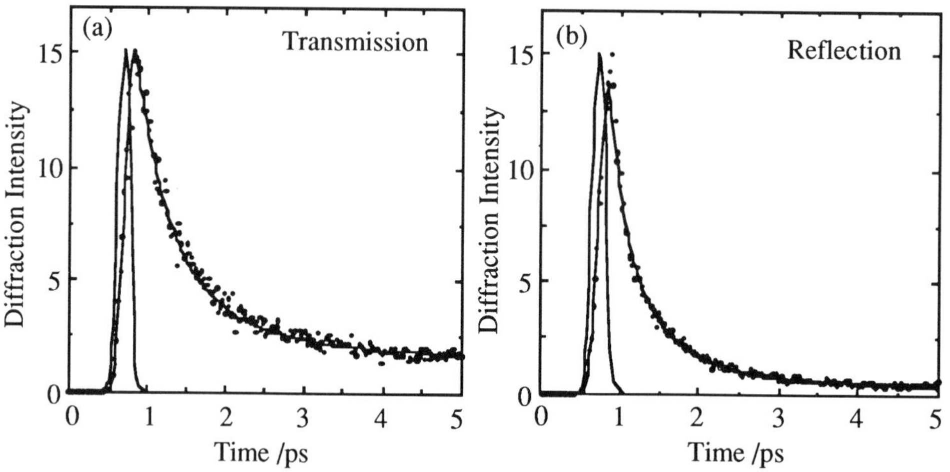

Figure 12. Time-resolved diffraction signals of a PTV thin film (≈ 90 nm thickness) observed at 660 nm in transmission (a) and reflection (b) modes. The sample was excited at 360 nm.

The time-resolved diffraction spectrum of a PTV thin film has a peak at approximately 660 nm and a shoulder at 590 nm irrespective of probe delays, which is safely assigned to the diffraction from the self-trapped exciton of PTV. The peak wavelength is in accord with that of the $\chi^{(3)}$ spectrum obtained by the third-harmonic generation method [32]. The intensity of this band was found to decay very rapidly with time. Figure 12 illustrates typical diffraction signals of a PTV thin film ($\approx$ 90 nm thickness) in both transmission and reflection modes. The decays show non-exponential behavior. It is clear that the diffraction signal in the reflection mode decays more rapidly than in the transmission mode. The fast decay component in the transmission mode was 420 fs, whereas 250 fs was observed in the reflection mode. In the current experiment, d_p was calculated to be $\approx$ 20 nm using equation 13. In addition, the fast component in the reflection mode is in good agreement with that of the ultrathin film of 5 nm thickness ($\approx$ 230 fs), so that this component is considered to be characteristic of the surface layer. Defects in the surface layer are probably responsible for the ultrafast relaxation of the self-trapped exciton.

As demonstrated here, transient grating spectroscopy in the reflection mode is very useful for analyzing the dynamics of a surface-layer with a few nm to 0.1 μm thickness.

5.2 Total internal reflection-mode transient grating spectroscopy: dynamics in the liquid/solid interface layer

Total internal reflection (TIR) confines the electromagnetic wave to the surface region with submicrometer thickness. The applications of the evanescent wave generated with TIR phenomena to time-resolved fluorescence spectroscopy have been reviewed in this Volume [33-35] and widely used to analyze the dynamic properties in the interface layers since the pioneering work utilizing a picosecond single-photon timing technique [36]. However, these applications are only limited to the fluorescent samples. In this Section, we introduce an application of the evanescent wave to transient grating spectroscopy for characterizing the dynamic properties of nonfluorescent samples.

The principle is illustrated in Figure 13. Two-excitation pulses are crossed in a sample (refractive index, n_2) with an incidence angle larger than the critical angle, θ_c, by using a sapphire prism as a TIR substrate (n_1), so that the sample is excited with an evanescent wave modulated by the interference. The penetration depth of the

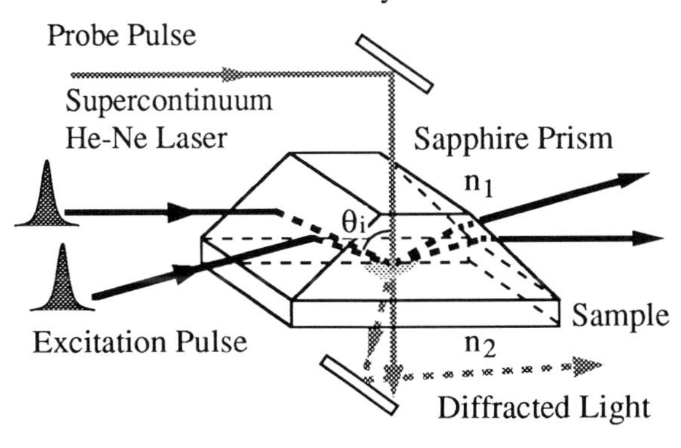

Figure 13. A schematic illustration of transient grating spectroscopy under TIR mode. The sample is excited with an evanescent wave modulated by the interference. θ_i is the incidence angle.

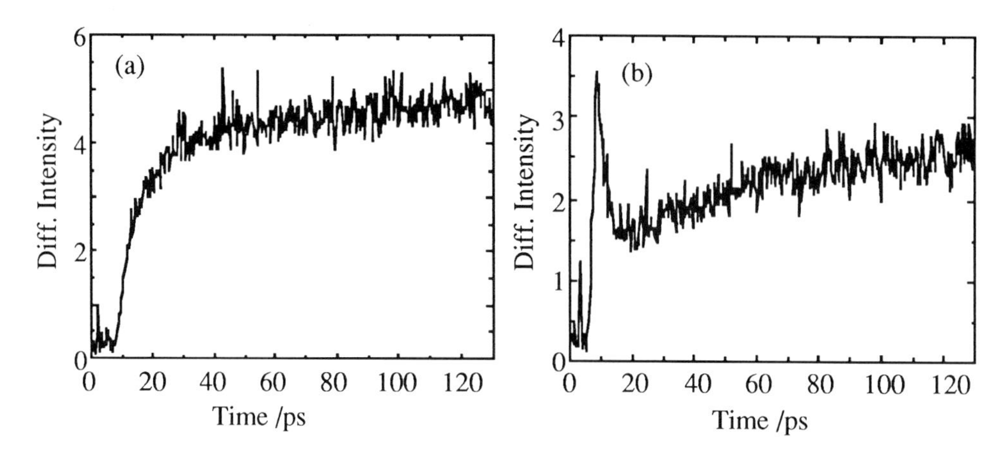

Figure 14. Diffraction signals of a spirooxazine compound in 1-butanol under normal (a) and TIR (b) excitation conditions. The sample is excited at ≈ 360 nm and monitored at 600 nm. The penetration depth is ≈ 150 nm (sapphire/1-butanol interface layer).

evanescent wave, which is defined by equation 1 in ref. [33], can be changed from μm to a few tens of nm by changing the incidence angle. The transient grating is then probed by a femtosecond supercontinuum as shown in Figure 13. In contrast to the reflection mode, transient grating spectroscopy in TIR mode has potential to analyze the photophysical/photochemical dynamics as a function of the penetration depth of the evanescent wave.

We have applied this technique to a photochromic reaction of a spirooxazine compound in the interface layer. It has been reported that spirooxazine undergo C-O bond cleavage with a rate constant of $(700\ \mathrm{fs})^{-1}$ followed by the structural change to merocyanine [37]. Furthermore, a rise time of 40 ~ 50 ps observed at a peak wavelength of merocyanine absorption was attributed to the thermal relaxation of vibrationally hot merocyanine in the ground state. It is expected that the dynamics of photochromic reaction in the sapphire/alcohol interface layers is different from that in the bulk because of the hydrogen bond between sapphire surface and alcohol [34]. Figure 14 illustrates diffraction signals of a spirooxazine photochromic reaction in 1-butanol under normal and TIR (penetration depth ≈ 150 nm) excitation conditions. In the bulk 1-butanol, a very fast and slow rise components, of ≈ 1 ps and a few tens of ps, were observed. The fast one is due to C-O bond cleavage and the slow one is probably due to the thermal relaxation from vibrationally hot merocyanine to a metastable merocyanine accompanied with some structural changes. On the other hand, fast decay in addition to very slow rise as compared with the bulk were observed in a TIR excitation condition (Figure 14b). The fast decay was confirmed to be characteristic to the surface layer by a reference experiment without spirooxazine. The presence of a slow rise component indicates that the rate of structural change from an unstable merocyanine formed just after excitation to a metastable merocyanine is slower in the sapphire/1-butanol interface layer than in the bulk. A further detailed mechanism of the photochromic reaction in the interface layer will be revealed by the measurements of time-resolved diffraction spectra.

Transient grating spectroscopy in TIR excitation mode has great potential for such an ultrafast photochemical reaction being impossible to be analyzed by fluorescence spectroscopy in the interface layer.

6. CONCLUSION

We have demonstrated here that transient grating spectroscopy using a femtosecond supercontinuum as a probe pulse is very useful for analyzing the transient processes of physicochemical properties in organic thin films and surface and/or interface layers. This spectroscopy is one of the highly sensitive method with wide dynamic range. The time-resolved diffraction spectrum corresponding to transient absorption spectrum can be obtained both for Raman-Nath and Bragg diffraction conditions, which makes it possible to identify intermediate species in various physicochemical processes. This is a great advantage as compared with the conventional transient grating experiments using a single frequency probe pulse. Detection of higher-order diffraction provides us information on the dynamics of a nonlinear reaction, which originated from the time-dependent deformation of a sinusoidal population grating. Furthermore, regular reflection and total internal reflection modes in transient grating spectroscopy have high potential for understanding the dynamics in surface and/or interface layers with submicrometer ~ a few tens of nm thickness.

ACKNOWLEDGMENTS

The authors thank Prof. H. Masuhara (Osaka Univ.) for the collaborations and discussions during the research period of the Microphotoconversion Project.

REFERENCES

1. J. R. Salcedo, A. E. Siegman, D. D. Dlott, and M. D. Fayer, Phys. Rev. Lett., 41 (1978) 131; D. R. Lutz, K. A. Nelson, C. R. Gochanour, and M. D. Fayer, Chem. Phys., 58 (1981) 325.
2. L. Gomez-Jahn, J. Kasinski, and R. J. D. Miller, Chem. Phys. Lett., 125 (1986) 500.
3. S. Komuro, Y. Aoyagi, Y. Segawa, S. Namba, A. Masuyama, H. Okamoto, and Y. Hamakawa, Appl. Phys. Lett., 43 (1983) 968.
4. M. Terazima and N. Hirota, Chem. Phys. Lett., 189 (1992) 560.
5. D. W. Phillion, D. J. Kuizenga, and A. E. Siegman, Appl. Phys. Lett., 27 (1975) 85.
6. R. S. Moog, M. D. Ediger, S. G. Boxer, and M. D. Fayer, J. Phys. Chem., 86 (1982) 4694.
7. A. B. Myers and R. M. Hochstrasser, IEEE. J. Quantum. Electron., QE-22 (1986) 1482.
8. K. Sala and M. C. Richardson, Phys. Rev. A., 12 (1976) 1036.

9. F. W. Deeg and M. D. Fayer, J. Chem. Phys., 91 (1989) 2269.
10. C. D. Marshall, I. M. Fishman, R. C. Dorfman, C. B. Eom, and M. D. Fayer, Phys. Rev. B, 45 (1992) 10009.
11. K. Thyagarayan, P. Lallemand, Opt. Commun., 26 (1978) 54.
12. K. A. Nelson, D. R. Lutz, M. D. Fayer, and L. Madison, Phys. Rev. B, 24 (1981) 3261.
13. R. J. D. Miller, T. S. Rose, M. Pierre and M. D. Fayer, J. Phys. Chem., 88 (1984) 3021.
14. N. Tamai, T. Asahi, and H. Masuhara, Chem. Phys. Lett., 198 (1992) 413.
15. N. Tamai, S. Funakura, and C. F. Porter, Part IV in this Volume.
16. H. J. Eichler, P. Gunter, and D. W. Pohl, Laser induced dynamic gratings (Springer-Verlag, Berlin, 1986).
17. D. D. Nolte, D. H. Olson, G. E. Doran, W. H. Knox, and A. M. Glass, J. Opt. Soc. Am. B, 7 (1990) 2217.
18. H. Kogelnik, Bell. Syst. Tech. J., 48 (1969) 2909.
19. R. M. Hochstrasser, H. Lutz, G. W. Scott, Chem. Phys. Lett., 24 (1974) 162.
20. D. E. Damschen, C. D. Merritt, D. L. Perry, G. W. Scott, L. D. Talley, J. Phys. Chem., 82 (1978) 2268.
21. H. Miyasaka and N. Mataga, Bull. Chem. Soc. Jpn., 63 (1990) 131.
22. B. I. Greene and R. R. Millard, Phys. Rev. Lett., 55 (1985) 1131.
23. A. Suna, Phys. Rev. B, 1 (1970) 1716.
24. V. Kenkre, Phys. Rev. B., 22 (1980) 2089.
25. R. C. Powell and Z. G. Soos, J. Lumin., 11 (1975) 1.
26. M. K. Casstevens, M. Samoc, J. P. Pfleger, and P. N. Prasad, J. Chem. Phys., 92 (1990) 2019.
27. M. Samoc and P. N. Prasad, J. Chem. Phys., 91 (1989) 6643.
28. J. J. Kasinski, L. Gomez-Jahn, K. J. Leong, S. M. Gracewski, and R. J. D. Miller, Opt. Lett., 13 (1988) 710.
29. A. Harata, H. Nishimura, and T. Sawada, Appl. Phys. Lett., 57 (1990) 132.
30. I. M. Fishman, C. D. Marshall, A. Tokmakoff, and M. D. Fayer, J. Opt. Soc. Am. B, 10 (1993) 1006.
31. T. Tsutui, H. Murata, T. Momii, K. Yoshimura, S. Tokito, and S. Saito, Synt. Metals, 41 (1991) 327.
32. H. Murata, N. Takeda, T. Tsutui, S. Saito, T. Kurihara, and T. Kaino, J. Appl. Phys., 70 (1991) 2915.
33. M. Toriumi and M. Yanagimachi, Part IV in this Volume.
34. S. Hamai, N. Tamai, and M. Yanagimachi, Part V in this Volume.
35. G. Rumbles, D. Bloor, A. J. Brown, B. Crystall, D. Philips, and T. A. Smith, Part IV in this Volume.
36. H. Masuhara, N. Mataga, S. Tazuke, T. Murao, I. Yamazaki, Chem. Phys. Lett., 100 (1983) 415.
37. N. Tamai and H. Masuhara, Chem. Phys. Lett., 191 (1992) 189.

MICROCHEMISTRY
Spectroscopy and Chemistry in Small Domains
Edited by H. Masuhara et al.

Time-resolved total-internal-reflection fluorescence spectroscopy and its applications to solid/polymer interface layers

Minoru Toriumi* and Masatoshi Yanagimachi#

Microphotoconversion Project †, ERATO,
Research Development Corporation of Japan,
15 Morimoto-cho, Shimogamo, Sakyo-ku, Kyoto 606, Japan

Time-resolved variable-angle total-internal-reflection fluorescence spectroscopy has been applied to the study of physicochemical properties of pyrene-incorporated thin films in solid/polymer interface layers. Inhomogeneities such as a concentration gradient of pyrene molecules, and a gradient of polarity (hydrophobicity) are observed in segmented poly(urethaneurea) and poly(p-hydroxystyrene) films. Those characteristics reflect the nature of polymer matrix in the interface layers.

1. INTRODUCTION

Time-resolved total-internal-reflection (TIR) fluorescence spectroscopy is well suited to studies of dynamic chemical interactions and structures in the interface layers of the sub-micrometer region [1-3]. We have developed variable-angle time-resolved TIR fluorescence spectroscopy to study thin film structures in the solid/polymer interface layers [4-9]. This TIR fluorescence technique has been applied to pyrene-incorporated segmented poly(urethaneurea) (SPUU) and poly(p-hydroxystyrene) (PHST) [6].

SPUUs are well-known biomedical materials for a surface modification of the biocompatibility because of their specific structure and properties of the surface. A PHST is a very important polymer for resist materials in the semiconductor industry [10,11]. A chemistry in micrometer size such as the inhomogeneity of structure, distribution, and reactivity of polymers and incorporated compounds has become great important in many fields like microelectronics and biotechnology.

---

* Present address: Hitachi Central Research Laboratory, Hitachi, Ltd., Kokubunji, Tokyo 185, Japan.
# Present address: Central Research Institute, Mitsui Toatsu Chemicals,Inc., Yokohama, Kanagawa Prefecture 247, Japan.
† Five-Year term project: October 1988 - September 1993

2. EXPERIMENT

2.1. Principle of TIR fluorescence spectroscopy

A polymer film is coated on the flat surface of a cylindrical TIR substrate as shown in Figure 1. The incident angle, θ_i, determines the excitation volume of the

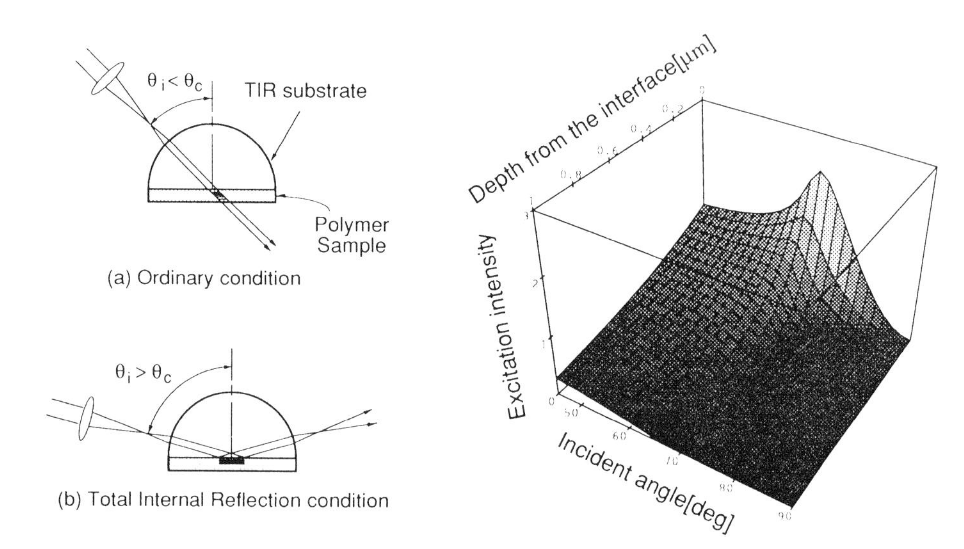

Figure 1. Principle of TIR fluorescence spectroscopy. (a) ordinary condition and (b) TIR condition.

Figure 2. Depth and incident-angle dependences of the intensity of an excitation light.

sample film. When the incident angle is smaller than the critical angle, θ_c, which is determined by the refractive index of the system, the excitation light transmits through the sample. The information of the entire film is obtained by observing the fluorescence from the excited volume under this "ordinary" condition. On the contrary, for an incident angle larger than the critical angle, the incident light is totally reflected, and there is some penetration of the excitation light by an "evanescent wave" into the interface region of the sample film. In the case of the total-internal-reflection, the information from interface layer of the sample is obtained by observing the fluorescence from the excited volume.

The penetration depth, d_p, of the evanescent wave is given by

$$d_p = \frac{\lambda_0}{2\pi}\sqrt{\frac{2}{-n_2^2(1-\kappa^2)+n_1^2\sin^2\theta_i+\sqrt{\{n_2^2(1-\kappa^2)-n_1^2\sin^2\theta_i\}^2+4n_n^4\kappa^2}}} \quad (1)$$

where λ_0 is the wavelength of the incident wave in vacuum, n is the refractive index, the suffices 1 and 2 indicate the TIR substrate and the sample, respectively, and κ is the absorption index of the sample. Figure 2 shows the depth and incident angle dependences of the intensity of an excitation light . Zero depth corresponds to the interface between the sample and the TIR substrate. The light intensity decreases exponentially with the increase of depth into the sample film. The evanescent wave decreases more rapidly with the increase of the incident angle. It should be noted that the penetration depth and the excitation volume are easily controlled by the incident angle and are limited to an interface region that is smaller than the wavelength of the excitation light. This high depth resolution is one of the great advantages of TIR spectroscopy.

The fluorescence intensity, I_f, is given by

$$I_f = k\,|F(\theta_i)|^2\,|F(\theta_o)|^2 \int_0^{th} C(z)\,\exp\left(-\frac{2z}{d_p}\right) dz \quad (2)$$

where k is the proportional factor, $F(\theta_{i,o})$ is the Fresnel factor for incident and observation angles, θ_o is the observation angle, **th** is the film thickness, and C(z) is the pyrene concentration at a distance **z** from the interface. The fluorescence intensity, I_f, as a function of the incident angle is expressed by the Laplace transformation of the concentration distribution of fluorescent molecule as given in Equation 2 [1,6], so it determines experimentally the depth profile of fluorescent molecules in a thin film [8,9]. This is another advantage of variable-angle TIR fluorescence spectroscopy.

2.2. Equipment of TIR fluorescence spectroscopy

Figure 3 shows the equipment of time-resolved TIR fluorescence spectroscopy. The frequency-doubled output of a Nd:YLF laser was used to synchronously pump a dye laser. The dye laser was cavity-dumped and frequency-doubled to give 2-ps excitation pulses at 300 nm. The laser beam was focused by a cylindrical lens onto the focal plane of the hemicylindrical TIR prism on which a sample film was coated.

The hemicylinder was firmly mounted on a rotating table. Fluorescence observed at angle θ_o was introduced by a cylindrical lens into a scanning double monochromator and detected using a microchannel-plate photomultiplier. The resolution of wavelength was approximately 0.5 nm.

Fluorescence rise and decay profiles were collected using a picosecond single-photon-timing system. The response time of the whole system was better than 30 ps.

Fluorescence spectra were also recorded by the single-photon-timing system.

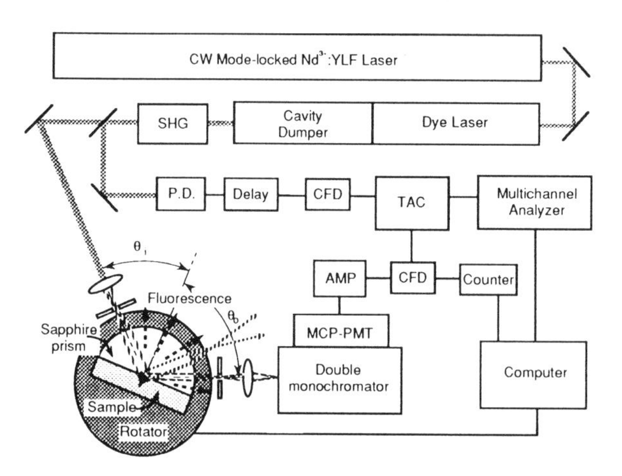

Figure 3. Experimental apparatus for TIR fluorescence spectroscopy.

2.3. Materials and film preparation

2.3.1. SPUU films

SPUU was prepared from an urethane prepolymer consisting of toluene-2,4-diisocianate (TDI) and poly(propylene oxide) (PPO), (Mitsui Toatsu Chemicals, Inc.). The molar composition of TDI to PPO was 2:3 and that of TDI unit to propylene oxide monomer (PO) unit was 1:8. The concentration of pyrene in a SPUU film is expressed in the unit of mol / 1 mol TDI unit / 8 mol PO unit. The refractive index of SPUU was determined to be 1.52 in an Abbe refractometer.

A sapphire hemicylindrical prism (the refractive index 1.82, Shinkosha) was used as an internal reflection element.

Here, we define the "interface" layer to mean the layer of a sample facing to the TIR substrate, which was measured under the TIR condition, the "bulk" layer to mean the entire film of a sample under the ordinary condition, and the "surface" layer to be the one facing the air, which is measured under the TIR condition.

The sample film of SPUU was prepared to investigate the interface and bulk layers as follows. The SPUU films doped with pyrene were spin-coated on a sapphire prism from a 60 w/v% toluene solution of SPUU prepolymer and an appropriate amount of pyrene. The films were heated at 70 °C for 10 min. and terminal TDIs of the prepolymer were condensed with water in air to liberate CO_2. Before fluorescence measurement the film was coated with poly(vinyl alcohol) (PVA) to prevent quenching of the excited pyrene by oxygen in air .

2.3.2. PHST films

PHST (Polyscience, Inc.) was purified by a repeated petroleum ether/ethyl acetate precipitation sequence. Pyrene (Aldrich) was purified by column chromatography on silica gel followed by recrystalization from ethanol. Ethylene glycol monoethyl ether acetate (ECA) (Nacalai tesque, research grade) was used as received.

The sample film of PHST was prepared to investigate the surface, interface and bulk layers as follows. PHST films doped with pyrene were coated on the TIR substrate by a spin-casting method from an ECA solution. PHST films were heated at 80 °C for ten min. The film thickness was determined to be about 2 μm by an optical interference method. The films were coated with PVA before fluorescence measurements. The properties of the surface layer were measured as follows. The sample was spin-cast onto a quartz plate and baked at 80 °C for ten min. The air-side of the sample film was then pressed to the TIR substrate mechanically.

3. APPLICATIONS

3.1. SPUU films

3.1.1. Micropolarity of SPUU films

Fluorescence spectra of pyrene-doped SPUU films have both monomer and excimer fluorescence around 380 nm and 480 nm, respectively, as shown in Figure 4. Pyrene was used as a fluorescence probe to investigate the micropolarity of a thin film structure. The solvent dependence of vibrational structure of pyrene fluorescence has been well studied [12]. In the presence of the polar solvents there is an enhancement in the intensity of the 0-0 band, I_0, whereas there are little effects on the 0-2 band, I_2. The relative intensity, I_0/I_2, of pyrene fluorescence is used as a good indicator of the micropolarity around a pyrene molecule [13,14].

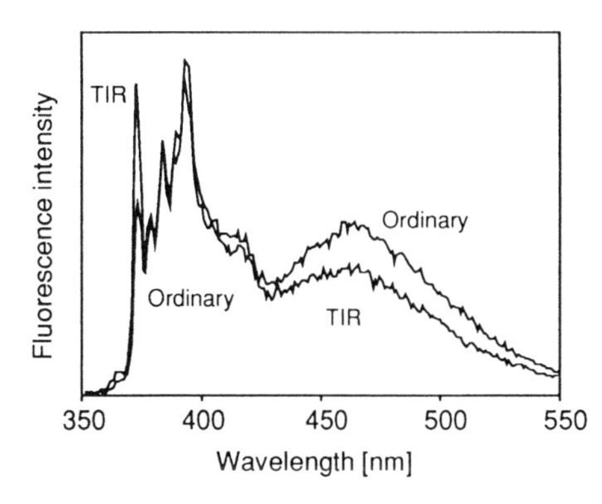

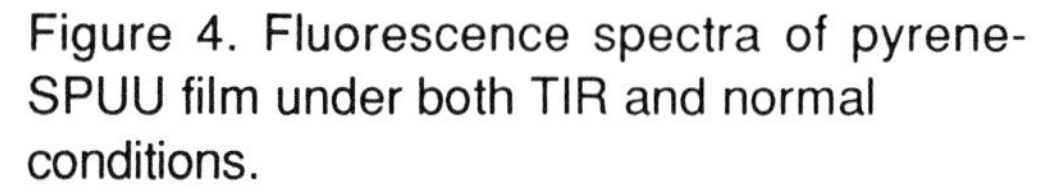
Figure 4. Fluorescence spectra of pyrene-SPUU film under both TIR and normal conditions.

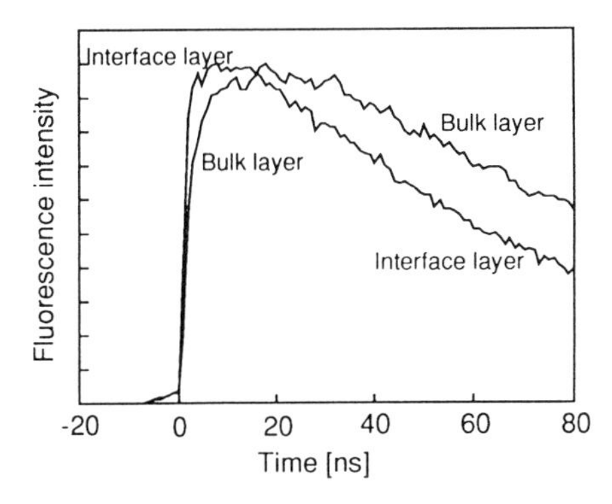

Figure 5. Rise and decay curves of pyrene excimer fluorescence under both TIR and normal conditions.

Figure 4 shows the pyrene fluorescence spectra in the interface and bulk layers of SPUU films. The fluorescence intensity of the spectra is normalized at the third vibronic band of pyrene monomer fluorescence (384 nm). The observation thickness of the interface layer under the TIR condition ($\theta_i = 75°$) is determined to be approximately 13 nm using Equation 1. The relative intensity, I_0/I_2, is larger under TIR condition than the ordinary one. The lower intensity under ordinary condition may be due to the reabsorption of the fluorescence by the ground-state pyrene. To confirm it, the dependence of the intensity ratio on the pyrene concentration was investigated. The I_0/I_2 ratio was almost constant and remained higher in the interface layer at the concentration lower than 1 mol pyrene / 1 mol TDI unit / 8 mol PO unit. Therefore, it was concluded that the microenvironmental polarity around pyrene molecules in the interface layer was higher than that in the bulk.

The origin of the environmental polarity is considered to be due to the inhomogeneous distribution of the polar TDI hard segment by the phase segregation of SPUU. According to the surface analysis by ESCA [15,16] and ATR [17] of SPUU, the interface layer showed a higher concentration of urethane hard segments than the bulk layer. In the case of the pyrene/SPUU film, it may also be shown that a higher concentration of TDI hard segment exists in the interface layer by the phase segregation. Therefore, pyrene molecules are highly incorporated in TDI segment matrices of the interface layer than the bulk layer.

3.1.2. Dynamics of pyrene monomer fluorescence in SPUU films

Since fluorescence rise and decay curves of pyrene monomer fluorescence in a SPUU film are complex and difficult to analyze as multiexponential components, we used $\tau_{1/e}$ lifetime which is the time required for the monomer fluorescence intensity to be 1/e of the initial intensity. The lifetime was 179 ns and 266 ns in the interface layer and the bulk layer at the concentration lower than 0.5 mol / mol TDI unit / 8 mol PO unit. It is suggested that the polarity enhancement by solvents may be accompanied by a decrease in the radiative lifetime [12]. For the reason, the environmental polarity is higher in the interface layer than in the bulk layer. This result agrees quite well with that of the vibronic intensity ratio of fluorescence spectra described above.

3.1.3. Dynamics of pyrene excimer fluorescence in SPUU films

The relative intensity ratio of excimer to monomer fluorescence depends also on the concentration of pyrene as shown in Figure 4. The intensity of the excimer fluorescence is lower in the interface layer than the bulk layer. It corresponds that the excimer fluorescence is more effectively quenched in the interface layer than in the bulk layer, and it agrees with the results of excimer formation dynamics mentioned below. The excimer formation is less feasible in the interface layer than in the bulk in spite of the higher concentration of pyrene in the interface. Therefore, it is considered that a nonradiative transition of the excited pyrene dimer is more effective in the interface layer than in the bulk layer. The same result is found in the case of pyrene-PMMA film [3].

Figure 5 shows rise and decay curves of pyrene excimer fluorescence in the interface and bulk layers. Fluorescence rise and decay of the pyrene excimer is faster under TIR condition than under ordinary condition. This means that pyrene excimer is more feasibly formed and effectively quenched in the interface layer than in the bulk layer. The faster rise time of excimer formation is explained by the effective excitation energy migration of pyrene in the interface layer. In the rigid film the faster rise time of excimer formation means the higher concentration of pyrene in the interface layer as compared with that in the bulk layer. As a result, the pyrene concentration in the interface layer is higher than in the bulk. The excimer fluorescence is more effectively quenched in the interface layer than in the bulk layer. In soild matrix, pyrene molecules are likely to form a ground-state dimer, and the excited pyrene dimer is known to be weakly luminescent. There may be due to the larger contributions of the pyrene ground-state dimers in the interface layer.

3.1.4. Inhomogeneous concentration distribution of pyrene in SPUU films

Depth profile of pyrene concentration is experimentally determined from the angular spectrum of TIR fluorescence spectroscopy as mentioned above. Figure 6 illustrates the incident angular dependence of fluorescence intensity of pyrene dissolved in ethylene glycol liquid film in order to verify the validity of the theoretical Equation 2. The thickness of the liquid film is approximately 10 μm. The solid curve is the simulated result with assumption that pyrene concentration is homogeneous along the depth from the interface by using Equation 2. The simulated results agree well with the experimental data.

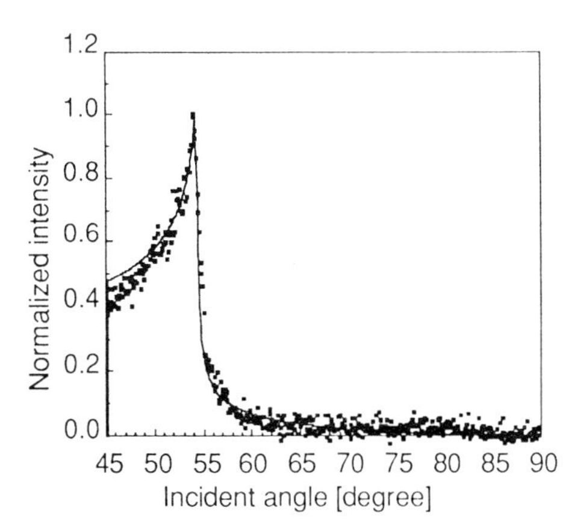

Figure 6. Incident angular spectrum of monomer fluorescence of pyrene-ethylene glycol liquid film.

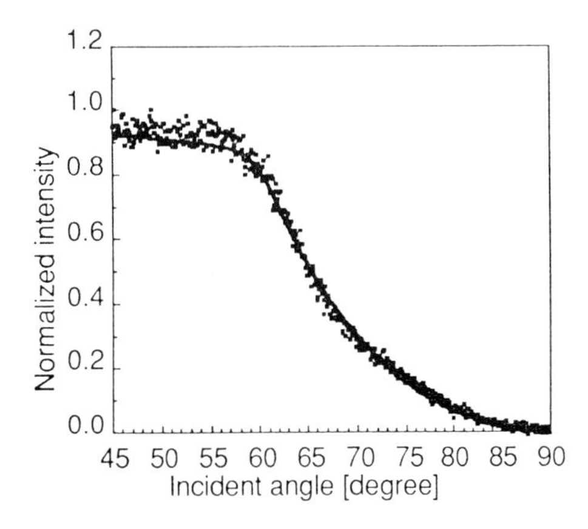

Figure 7. Incident angular spectrum of monomer fluorescence of pyrene-SPUU film.

Angular spectra of pyrene fluorescence in a SPUU film are analyzed by fitting with the step function of pyrene concentration along the depth as follows,

$$C(z)=C_i \quad \text{for} \quad 0 < z < \zeta \tag{3a}$$
$$C(z)=C_b \quad \text{for} \quad \zeta < z < \mathbf{th} \tag{3b}$$

where C_i and C_b are the relative concentrations of pyrene in the interface layer and the bulk layer, respectively, and ζ and **th** are the thickness of the interface layer and the film. Figure 7 shows the incident angular spectrum of monomer fluorescence of pyrene-SPUU film. The solid curve is the simulated results with using Equation 2 and well fitted with the experimental data. The fitting parameters, C_i, C_b and ζ, were determined to be 0.8, 0.2 and 150nm [9]. The concentration of pyrene monomer is higher in the interface layer of the thickness of 150 nm than that in the bulk layer. The result is in a good agreement with that of the time-resolved measurement mentioned above.

Judging from these experimental results obtained above, Figure 8 discribes the schematic representation of the location and the aggregation of pyrene molecules in the interface and buk layers. Pyrene molecules are located in the vicinity of the polar TDI segement in the interface layer while they are in the near less polar PPO segment in the bulk layer. Although pyrene concentration is higher in the interface layer than the bulk layer, the excimer formation is less feasible in the interface layer than the bulk layer due to the formation of the nonradiative ground-state pyrene dimer.

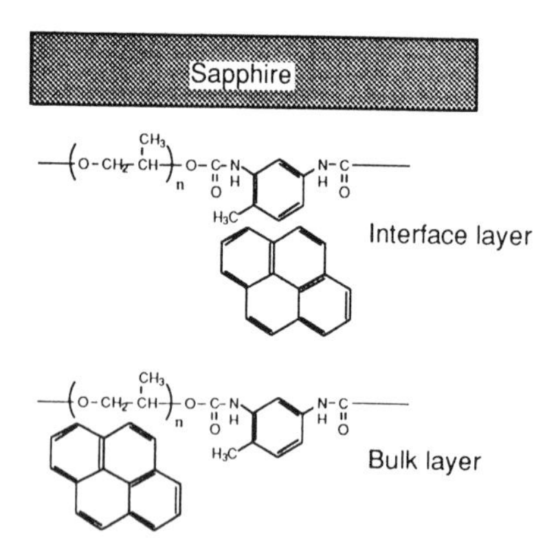

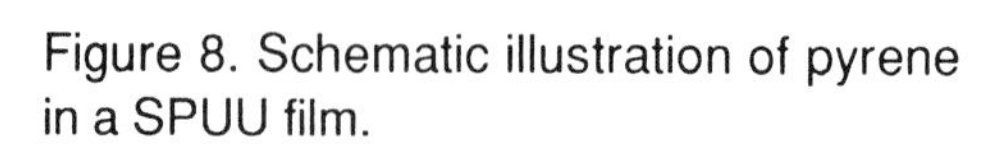

Figure 8. Schematic illustration of pyrene in a SPUU film.

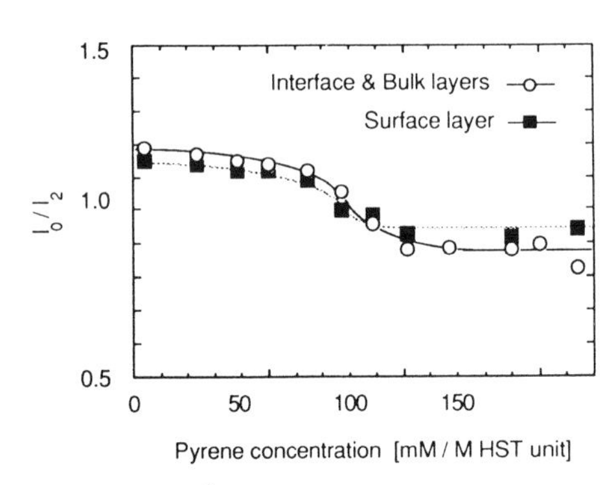

Figure 9. Concentration dependence of pyrene monomer fluorescence in the interface, bulk and surface layers.

3.2. PHST films

3.2.1. Micropolarity in PHST films

Figure 9 shows the relative intensity, I_0/I_2, of the vibronic bands of pyrene fluorescence, a good indicator of the micropolarity, as a function of pyrene concentration in the surface, interface and bulk layers. The values of I_0/I_2 in the interface layer agree with those in the bulk layer within the experimental errors. The accurate ratio should be estimated by correcting relative contributions of monomer and excimer fluorescence. At pyrene concentrations lower than 50 mM / M HSTunit, the intensity ratio is almost the vibrational intensity ratio, I_0/I_2, of independent of the pyrene concentration. In this range the value of I_0/I_2 is about 1.2. This value is reasonable, because aromatic compounds such as benzene and benzyl alcohol show the value of about 1.2. The value of I_0/I_2 determined for the surface layer is a bit smaller than that for the interface and bulk layers. This indicates that the environmental polarity around pyrene molecules in the surface layer is less polar than in the bulk and interface layers. It means that pyrene exists in the hydrophobic sites in the PHST films. The polar sites of hydroxy-groups are packed by intramolecular interaction and surrounded by non-polar sites of aromatic parts [18,19]. The lower polarity in the surface layer indicates that the effects of a hydrogen-bonding interaction are larger in the surface layer than in the bulk and interface layers.

3.2.2. Inhomogeneous concentration distribution of pyrene in PHST films

The excimer-to-monomer fluorescence intensity ratio is almost the same in the bulk and surface layers at lower pyrene concentrations such as 90 mM / HST unit M, but the excimer contribution is larger in the interface layer than in the bulk and

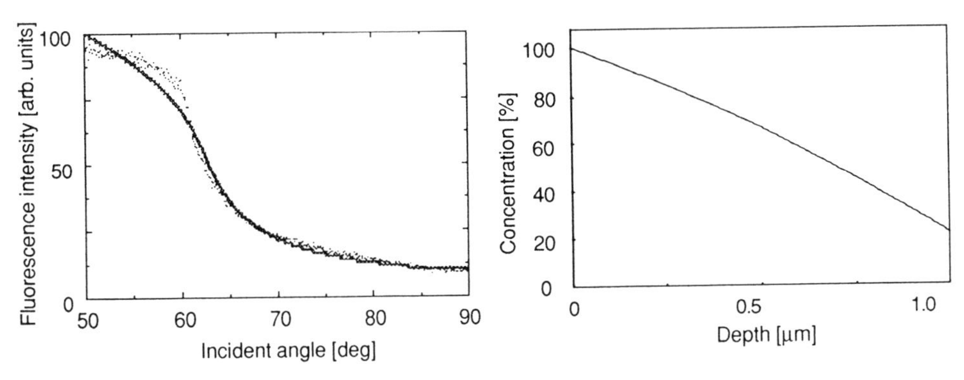

Figure 10. (a) Incident angular spectrum monomer fluorescence of pyrene-PHST film (dots) and the best fitting curve of uniform depth profile (solid curve).

Figure 10 . (b) Concentration depth profile calculated from the experimental data of Figure 10 (a) using the Equation 2.

surface layers. At higher concentration the contribution of excimer fluorescence is smaller in the surface layer and medium in bulk layer and the largest in the interface layer. The decrease of excimer contribution indicates the concentration gradient of pyrene molecules from interface layer to the surface layer.

Depth profile of pyrene concentration is experimentally determined from the angular spectrum of TIR fluorescence spectroscopy as mentioned above. Figure 10a shows the angular spectrum of pyrene monomer fluorescence in a PHST film and a solid curve is the fitting result of the uniform distribution of pyrene in a PHST film using the Equation 2. There are three fitting parameters; refractive index of the sample, background of the noise intenisty, and a proportional constant in the least-squares method of simplex minimizaiton. The fitting is bad assuming a homogeneous depth distribution. The experimental data is fitted well only if the inhomogeneous distribution of the chromophore is assumed along the depth from the interface. The depth profile determined by Equation 2 assuming a polynomial distribution function is shown in Figure 10b. It shows the concentration gradient from the interface, bulk to surface layers. It agrees with the results by analysis of the excimer formation. Note the large inhomogeneity of doped pyrene in a PHST film.

3.2.3. Rise and decay analysis of monomer fluorescence

Figure 11 shows the typical decay curves of monomer fluorescence of pyrene in the interface and bulk layers. The fluorescence decays faster in the interface layer than in the bulk layer. This is reasonable, because of the higher concentration of pyrene in the interface layer. A slow component is observed in the bulk layers as indicated by the dashed line. The contribution of the slowest component, defined as the intensity ratio of the component at time zero to that at peak intensity, is about 5 % and does not show obvious dependence on the concentration. The lifetime of this component does not depend upon the pyrene concentration and is equal to

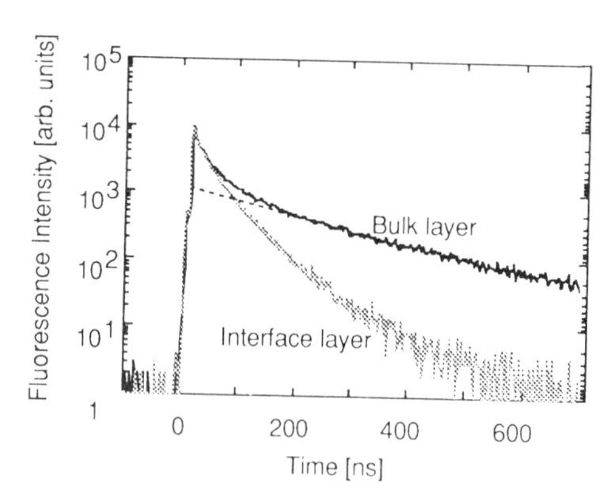

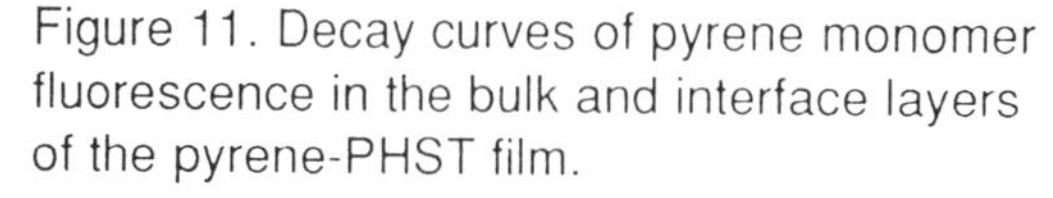
Figure 11. Decay curves of pyrene monomer fluorescence in the bulk and interface layers of the pyrene-PHST film.

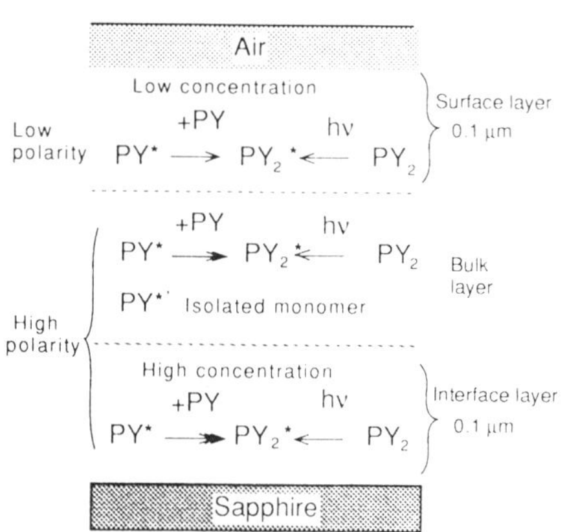

Figure 12. Schematic illustration of pyrene dynamics in a PHST film.

250 ns, which is close to the monomer fluorescence lifetime in solvents. The slowest component should be attributed to an isolated pyrene molecule in the bulk layer. The lifetime in PHST matrix is in between non-polar and polar solvents. The difference in lifetime of monomer fluorescence can be ascribed to environmental polarity as mentioned in SPUU films.

3.2.4. Dynamics of pyrene excimer fluorescence in PHST films

The rise and decay curves of pyrene excimer fluorescence in the PHST films are measured. The emission maximum intensity of excimer is observed immediately after the laser excitation pulse. This is not the case of dynamic excimer formation or the diffusional migration of two pyrene molecules. For dynamic formation the excimer fluorescence intensity should increase with time with a delay after the laser excitation as mentioned in SPUU films. The experimental results indicate static excimer formation from ground-state dimers of pyrene and no efficient movement in a PHST system. The fast rise time of the excimer fluorescence may be explained by the free volumes which provide sufficiently large voids for pyrene molecules to form excimer configurations. This static excimer formation is observed in all layers.

The experimental results obtained above can be summarized in Figure 12. A concentration gradient of pyrene-doped in the PHST film is observed. The concentration of pyrene is higher in the interface with the substrate and lower in the surface layer facing to the air. Excimers are formed from the ground-state dimers in all layers. The higher concentration of pyrene promotes the effective formation of excimers in the interface layers. There is a stable isolated pyrene in the bulk layer.

The micropolarity is lower in the surface layer than in the bulk and interface layers.

4. SUMMARY

Time-resolved total-internal-reflection (TIR) fluorescence spectroscopy has been shown to be a powerful technique for study of the structural properties of thin films, such as the inhomogeneous distribution of incorporated molecules and the hydrophobicity of the interface layer. Inhomogeneity is the intrinsic structure along the depth direction in the order of microns. This inhomogeneity reflects the variation in the static microstructure and the dynamic local motions of molecules, such as the chemical and physical structure, aggregation, interaction between polymer and substrate, and conditions of film preparation. Such information is important not only in fundamental research but also in applications to controlling and creating the desired surfaces and thin film properties in fields such as microelectronics and biotechnologies.

ACKNOWLEDGEMENTS

The authors wish to thank Professor Hiroshi Masuhara for intensitve discussions on this work and to thank Dr. Naoto Tamai for a helpful discussion on pyrene photochemistry.

REFERENCES

1. H. Masuhara, N. Mataga, S. Tazuke, T. Murao and I. Yamazaki, Chem. Phys. Lett. **100** (1983) 415.
2. H. Masuhara, S. Tazuke, N. Tamai, I. Yamazaki, J. Phys. Chem. **90** (1986) 5830.
3. A. Itaya, T. Yamada, K. Tokuda and H. Masuhara, Polymer J. **22** (1990) 697.
4. M. Toriumi, M. Yanagimachi and H. Masuhara, MRS Extended abstracts, C. Drake, J. Klafter, R. Kopelman (eds.), pp.133-136, 1990.
5. M. Yanagimachi, M. Toriumi and H. Masuhara, Chem. Mater. **3** (1991) 413.
6. M. Toriumi, M. Yanagimachi and H. Masuhara, H. Proc. SPIE Advances in Resist Technology and Processing VIII, **1466** (1991) 458.
7. M. Toriumi and H. Masuhara, Spectrochimica Acta Rev. **14** (1991) 353.
8. M. Toriumi, M. Yanagimachi and H. Masuhara, Appl. Opt., **31** (1992) 6376.
9. M. Yanagimachi, M. Toriumi and H. Masuhara, Appl.Spectrosc., **46** (1992) 832.
10. M. Toriumi, N. Hayashi, M. Hashimoto, S. Nonogaki, T. Ueno and T. Iwayanagi, Polym. Eng. Sci. **29** (1990) 868.
11. S. Nonogaki and M. Toriumi, Makromol. Chem., Macromol. Symp. **33** (1990) 233.
12. A. Nakajima, Bull. Chem. Soc. Jpn. **44** (1971) 3272.
13. D. C. Dong and M. A. Winnik, Photochem. Photobiol. **35** (1982) 17.
14. D. C. Dong and M. A. Winnik, Can. J. Chem. **62** (1984) 2560.
15. C. S. P. Sung, C. B. Hu, E. W. Merril and E. W. Salzman, J. Biomed. Mater. Res. **12** (1978) 791.
16. S. W. Graham and D. M. Hercules, J. Biomed. Mater. Res. **15** (1981) 465
17. C. S. P. Sung and C. B. Hu, J. Biomed. Mater. Res. **13** (1979) 161.
18. A. Reiser, Photoreactive polymers, John Wiley and Sons, New York, N. Y.,1989.
19. J.H. Huangm, T. Wei and A. Reiser, Macromolecules **22** (1989) 4106.

MICROCHEMISTRY
Spectroscopy and Chemistry in Small Domains
Edited by H. Masuhara et al.

Time-resolved evanescent wave induced fluorescence studies of polymer-surface interactions

G.Rumbles[a], D.Bloor[b], A.J.Brown[a], A.J.DeMello[a], B.Crystall[a], D.Phillips[a] and T.A.Smith[a]

[a]Department of Chemistry and Centre for Photomolecular Science, Imperial College of Science, Technology and Medicine, Exhibition Road, London, SW7 2AY, U.K.

[b]Department of Physics, University of Durham, South Road, Durham DH1 3LE, U.K.

Total internal reflection fluorescence spectroscopy (TIRF) or, more precisely, evanescent-wave induced fluorescence spectroscopy (EWIF) offers a sensitive, spectroscopic technique that can probe the interfacial region between two dielectric media. A common configuration uses a modified glass surface, as one medium, and a solution containing a fluorophore as the other medium. Fluorophore labelled synthetic polymers and biopolymers have been used as probes to examine the influence of the surface on the solvated polymer. Time-resolved fluorescence has extended the scope of the technique when, in some cases it has resolved issues, but in others it has created controversy. In this paper we use the techniques of steady state and time-resolved EWIF to study the photophysical properties of the soluble polydiacetylene, poly-4BCMU at a fused silica surface. The data suggests that the fluorescence quantum efficiency of the polymer is increased when in the vicinity of the glass surface, as a result of inhibiting a non-radiative decay channel. A surface-induced chromism effect is also examined and a structure for the ordered form of the polymer at the surface is proposed. Using fluorescence anisotropy measurements, the ordered phase of the polymer at the glass/solution interface appears to be aligned parallel to the glass surface. The difficulties in interpreting time-resolved fluorescence decays excited using an evanescent wave are also discussed.

1. INTRODUCTION

Soluble polydiacetylenes (e.g. Figure 1) have received considerable attention over the past few years[1,2], since they exhibit dramatic solvato-, thermo- and electro-chromic effects. The spectral shifts in both absorption and emission (where observable) is attributed to the degree of disorder in the conjugated π-electron backbone. In the disordered systems, such as those induced by the higher solubilities of high temperature solutions or good solvents, the conjugation is disrupted which reduces the effective conjugation length and results in a shift to higher energy of both the absorption and emission spectra. Conversely, a reduction in the solubility causes an increase in the effective conjugation length in the polymer backbone and results in a shift to lower energies of the absorption and emission spectra. These two cases are commonly referred to as the Y- and R- forms, respectively. In some cases the increased order in the R-form also reduces the fluorescence quantum yield to an undetectable level[3]. Figure 2 shows an example of a thermochromic shift in the absorption spectra for a solution of poly-4BCMU in toluene over the temperature range 329 K - 344 K. A similar transition can be invoked solvatochromically by adding a good solvent, for example chloroform, to the solution.

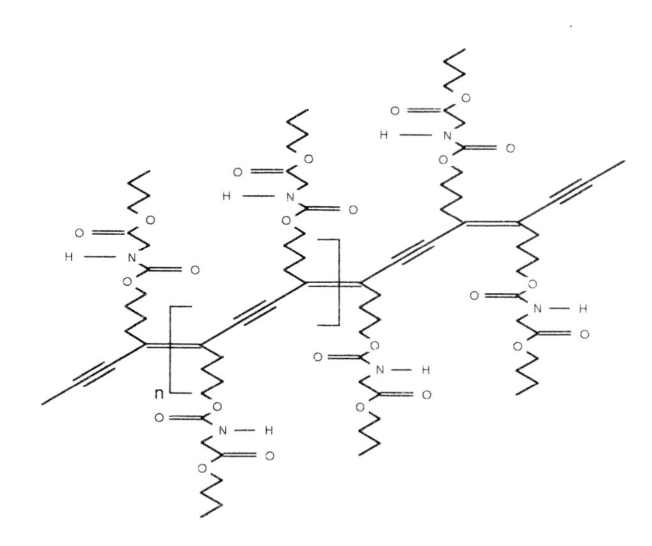

Figure 1. Structure of the soluble polydiacetylene, poly-4-butoxycarbonylmethylurethane (poly-4BCMU).

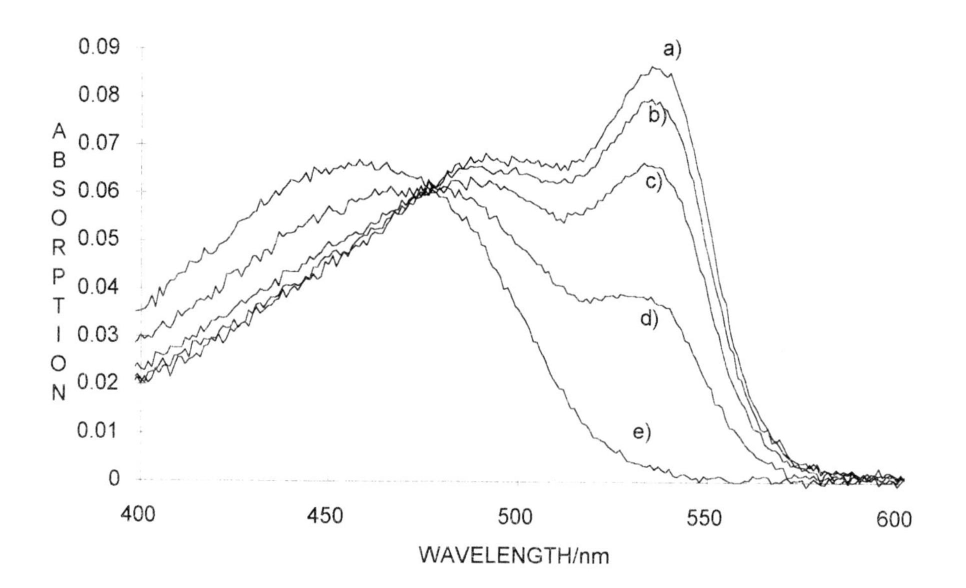

Figure 2. Absorption spectra of poly-4BCMU in toluene at a) 329 K, b) 332 K, c) 336 K, d) 341 K and e) 344 K.

The nature of the structural deformations responsible for the observed changes in spectral properties is still an area of controversy. In disordered systems it is not clear whether the deformations are regular or irregular, whilst in the ordered forms, the existence of solvated single chains is still a subject of debate[4-9]. A recent proposal[10] suggests the Y-form to R-form transition could be due to chain folding to form an intramolecular fringed micelle structure. This elegant idea, based on Raman data of very dilute solutions, combines both of the aforementioned ideas: whilst the transition is a single chain phenomenon, the product is an intramolecular aggregate, and at higher concentrations would lead to intermolecular aggregation with an almost identical type of structure. The conjugation lengths within the micellar structure are longer than within the disordered Y-form and hence absorb lower energy light. This interpretation also complies with conventional poor solvent effects on traditional polymers, where the polymer coils in order to minimise the interaction with the solvent.

The influence of surfaces on the properties of polymers in solution is currently receiving a great deal of interest[11-15]. Total internal reflection fluorescence (TIRF) spectroscopy, or more precisely evanescent wave induced fluorescence (EWIF) spectroscopy plays a major role in the investigation of processes that occur at a solid/solution interface. The aim of this present study is to determine the effect of a fused silica surface on the conformation of the soluble polydiacetylene, poly-4BCMU (poly-4-butoxycarbonylmethylurethane) (Figure 1), in good and poor solvents. Poly-4BCMU is relatively unusual in that both the Y and R forms fluoresce[3,16], although the fluorescence quantum efficiency is extremely low, the reason for which is still unclear. Internal conversion to the ground state or another non-fluorescent excited state is the most probable deactivation mechanism, although this has yet to be unambiguously confirmed. In addition to the fundamental interest in such a study, the results are also of technological importance, since it is from solution that thin films of poly-4BCMU are prepared, either by spin or dip coating, for use as planar optical waveguides in integrated non-linear optical applications.

2. THEORY

When light in a medium of refractive index, n_1, encounters a medium of lower refractive index, n_2, two processes can occur: When the angle of incidence is less than the critical angle, θ_c, defined by Snell's law as $\theta_c = \sin^{-1}(n_2/n_1)$, refraction occurs; for angles of incidence greater than θ_c total internal reflection occurs. The name total internal reflection (TIR) however, is a misnomer, as it implies that all of the light remains within the higher refractive medium. This is not the case, as predicted by Maxwell's equations of electromagnetic radiation[17,18], since a small amount of light penetrates the lower refractive medium in the form of a standing wave that decays exponentially in intensity with distance from the media interface and is known as an 'evanescent' wave. If the lower refractive index medium is a solution or even in a gas, the evanescent wave can be used to probe molecules at the interface using conventional spectroscopic techniques. The depth of penetration of the light into the solution or gas can be far less than the wavelength of the incident light, overcoming the diffraction limit. Originally developed in the IR[18], techniques utilising uv/visible absorption[18,19], Raman[20,21] and fluorescence[22,23] have become more common since the advent of the laser. EWIF uses the evanescent wave to excite molecules of interest that located in the interfacial region.

Figure 3 shows the standing wave field at the interface resulting from the interference of the incident and reflected waves. The penetration depth of the evanescent field, d_p, is related to the wavelength of the incident light and the refractive indices of the two materials at the interface.

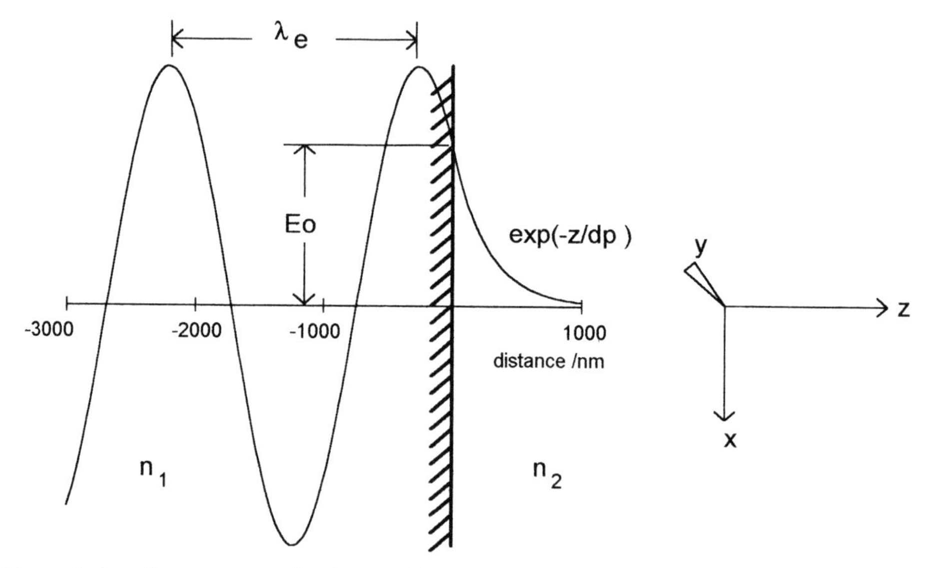

Figure 3. Standing-wave amplitudes established near a totally reflecting interface.

$$\lambda_e = \frac{\lambda_{vac}}{n_1 \cos\theta} \tag{1}$$

$$d_p = \frac{\lambda_{vac}/n_1}{2\pi\sqrt{\sin^2\theta - n_{21}^2}} \tag{2}$$

where $n_{21} = \dfrac{n_2}{n_1} = \sin\theta_c$

For the results discussed in this work, the total internal reflection medium was a hemi-cylindrical fused silica prism. Figure 4 shows a schematic of this prism along with the relative orientation of the polarisation of the exciting light and the detection system with respect to the orientation of the E-fields at the surface. Whilst the E-field penetration depth is *independent* of the polarization of the incident light, the electric field amplitudes at the interface are *dependent* upon the polarization.

The light intensity profile within the evanescent wave is related to the square of the E-field and is given by[23]

$$I(z) = U_0(\theta)\exp\left(\frac{-z}{\Lambda}\right) \tag{3}$$

where $\Lambda = \frac{d_p}{2} = \frac{\lambda_{vac}/n_1}{4\pi\sqrt{\sin^2\theta - n_{21}^2}}$ (4)

the depth of penetration at which the light *intensity* is diminished by e^{-1}. The intensity of the light at the surface U_0 (E_0^2) is dependent upon both the angle of incidence and polarization.

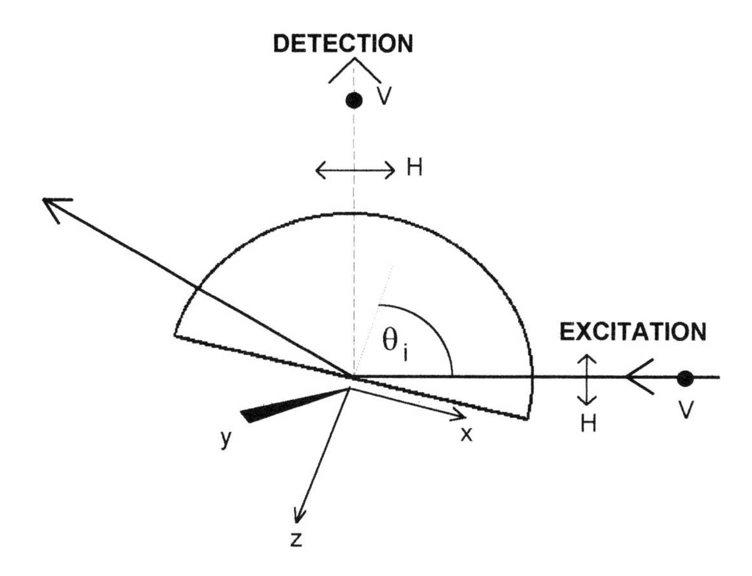

Figure 4. Hemicylindrical TIR prism used for EWIF experiments.

For unpolarized incident light, the value of U_0 for the three directions x, y and z (defined in Figure 4) are given by

$$U_{0,y} = \frac{4\cos^2\theta}{\left(1-n_{21}^2\right)} \quad (5)$$

$$U_{0,x} = \frac{4\left[\sin^2\theta - n_{21}^2\right]\cos^2\theta}{\left(1-n_{21}^2\right)\left[\left(1+n_{21}^2\right)\sin^2\theta - n_{21}^2\right]} \quad (6)$$

$$U_{0,z} = \frac{4\cos^2\theta\sin^2\theta}{\left(1-n_{21}^2\right)\left[\left(1+n_{21}^2\right)\sin^2\theta - n_{21}^2\right]} \quad (7)$$

For s or vertically polarized incident light, $U_V = U_{0,y}$. Conversely, for p or horizontal polarized incident light, $U_H = U_{0,x} + U_{0,z}$. Hence

$$U_H = \frac{4\cos^2\theta\left[2\sin^2\theta - n_{21}^2\right]}{\left(n_{21}^4\cos^2\theta + \sin^2\theta - n_{21}^2\right)} \tag{8}$$

and

$$U_V = \frac{4\cos^2\theta}{\left(1 - n_{21}^2\right)} \tag{9}$$

It is interesting to note that within the evanescent standing wave, a component of the E-field exists in each of the three directions x, y and z. This is in contrast with a conventional travelling light wave where only two components exist and has a strong influence on the interpretation of polarized emission data, discussed later.
From equation 4 it can be seen that there are two limiting cases for Λ. As θ_i deviates from the critical angle and approaches 90°, a limiting penetration depth given by

$$\Lambda_{\lim} = \frac{\lambda_{vac}/n_1}{4\pi\sqrt{1 - n_{21}^2}} \tag{10}$$

is attained and for the system described here is 850 Å, assuming that the concentration of the poly-4BCMU is sufficiently dilute so that the refractive index of the solution can be approximated by that of the pure solvent. As θ_i approaches the critical angle, θ_c, Λ approaches infinity and refraction occurs. The total intensity of the evanescent light at any angle of incidence greater than θ_c, is given by

$$I_{evan}(\Lambda) = \int_0^\infty U_0(\theta_i)\exp(-z/\Lambda)\,dz \tag{11}$$

Therefore for a fluorophore dissolved in the solvent, the intensity of fluorescence that would result from $I_{evan}(\Lambda)$, would be

$$I_{fluor}(\Lambda) = \int_0^\infty \phi(z)c(z)U_0(\theta_i)\exp(-z/\Lambda)\,dz \tag{12}$$

Where $\phi(z)$ and $c(z)$ are the fluorescence quantum yield and the concentration profile of the fluorophore in the interfacial region respectively. For homogeneous solutions that extend up to the surface, such that $\phi(z)$ and $c(z)$ are independent of z and are therefore constant, equation 12 reduces to

$$I_{Fluor,hom}(\Lambda) = U_0(\theta_i)\phi c\Lambda \tag{13}$$

If this assumption is valid, then measurements of fluorescence intensity from the evanescent wave can be used to verify the existence of the evanescent wave and prove the theory which describes it[24].

A more general treatment, without the assumptions above, can be developed as follows. The fluorescence quantum yield, ϕ, and lifetime, τ_F, are related by

$$\phi = k_r\tau_F \tag{14}$$

where k_r is the rate constant for radiative decay of the fluorophore. Assuming that k_r is independent of environment, then equation 12 can be re-written as

$$I_{fluor}(\Lambda) = k_r \int_0^{\infty} \tau_F(z) c(z) U_0(\theta_i) \exp(-z/\Lambda)\, dz \qquad (15)$$

In order to implement equation 12 and derive the concentration profile of a fluorophore within the evanescent field, the variation of the fluorescence quantum yield with distance, $\phi_F(z)$, from the surface must be known, or more ideally it must be constant. Previous studies have shown that in many cases this is not true, with fluorescence decay profiles being dependent upon penetration depth, thus causing difficulties in determining concentration profiles.

3. EXPERIMENTAL

The fluorescence spectrometer was based on a time-correlated single-photon counting apparatus using pulsed laser excitation[25]. The 10 picosecond excitation pulses at a wavelength of 414 nm and at a repetition rate of 3.8 MHz were obtained from the output of a frequency doubled (lithium iodate), cavity-dumped, synchronously-pumped, mode-locked styryl-9 dye laser (Coherent 701-3/CD), pumped by the frequency doubled output of a mode-locked, cw, Nd:YAG laser (Coherent Antares). Fluorescence was collected perpendicular to the direction of excitation by an optimised collection optic (Melles-Griot cmp 119), which focused the light on to the slits of a 1/4m monochromator and was subsequently detected by a side-window photomultiplier tube (Hamamatsu R955). Fluorescence spectra were recorded by counting fluorescence photons within a 1ns time window of the excitation pulse, operating the multichannel analyser in the multichannel scaling mode and scanning synchronously with the monochromator. The EWIF set-up used a semicircular cylindrical fused silica prism (Spanoptic. Refractive index, n_1, = 1.460) as the TIR medium, the full experimental details of which are described elsewhere[26]. Bulk spectra were recorded at an angle of incidence of 70°, corresponding to refraction conditions. All measurements were taken at a temperature of 295 K. Fluorescence decay profiles were recorded by replacing the conventional photomultiplier tube (PMT) with a microchannel plate PMT (Hamamatsu R1564-U01), which reduced the instrument response function to 50 picoseconds (FWHM).

The prism was thoroughly cleaned, using chloroform and 2-methyltetrahydrofuran (2-MeTHF), and air-dried at room temperature prior to use. Fresh solutions of poly-4BCMU were prepared in 2-MeTHF (purified by refluxing over lithium aluminium hydride) or a 7:18 chloroform:hexane (Aldrich HPLC grade, used without further purification) at an optical density of 0.13 at the excitation wavelength. All solvents were tested in the apparatus to ensure that no extraneous fluorescence could be observed. Poly-4BCMU was prepared by gamma irradiation of the monomer crystals and purified by selective solvation, as described elsewhere[27].

4. RESULTS AND DISCUSSION

4.1 Anomalous fluorescence intensities

A comparison of the spectra recorded using evanescent-wave excitation and normal refraction excitation are shown in Figure 5 for both Y-form and R-form solutions. The EWIF spectra are those measured after the system has reached equilibrium i.e. at the point at which no further spectral changes are observed (see below). Inspection of the measured fluorescence intensities recorded at equilibrium under EWIF and refraction conditions was found to be strongly dependent on the concentration of the original solution and in addition, the EWIF intensity appeared to be far more intense than would be expected from previous experiments with dye molecules[22]. This was confirmed by using the fluorescent molecule, 9,10-dicyanoanthracene in hexane solution as a reference compound. Using time-resolved EWIF this molecule has been proved not to exhibit any anomalous emission effects, and can be excited using the same excitation wavelength as the poly-4BCMU.

Table1 shows the ratio of EWIF intensity, I_{EWIF}, versus bulk fluorescence intensity, I_{BULK}, for the reference compound, the Y-form solution and the R-form solution, all taken at an optical density of 0.13. Clearly the fluorescence intensity originating from the evanescent region with both Y- and R- form solutions is far greater than would be predicted from the bulk concentration.

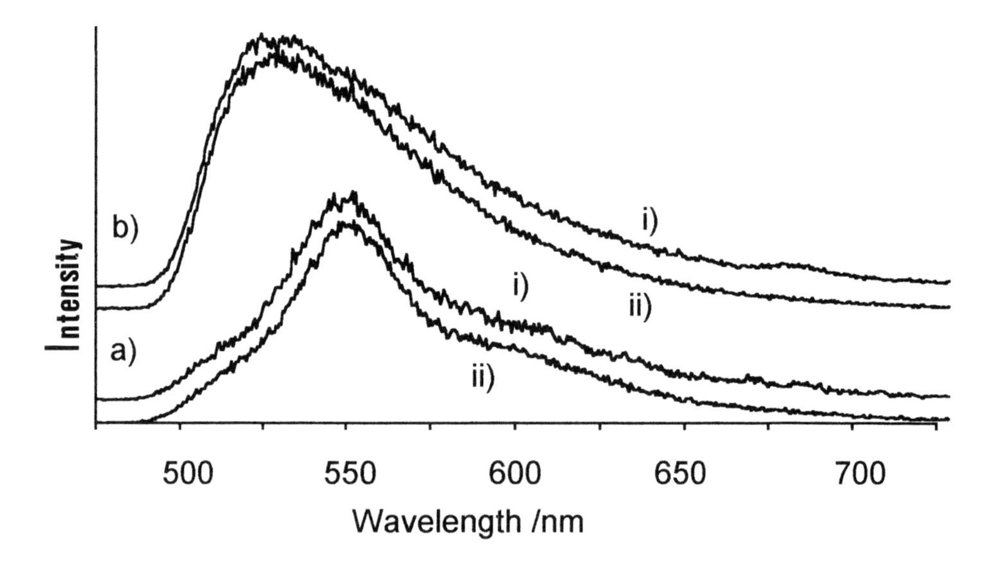

Figure 5. Y-form (b) and R-form (a) spectra recorded (i) in bulk and (ii) using EWIF.

Table 1.
Ratio of EWIF intensity:bulk fluorescence intensity recorded after equilibrium is attained.

Sample	$I_{EWIF}:I_{BULK}$
Reference	0.026
Y-form	0.56
R-form	1.67

Reduction of the sample bulk concentration has no effect on the $I_{EWIF} : I_{BULK}$ ratio for the reference sample, but causes a *decrease* in the ratio using the R-form solution and an *increase* using the Y-form solution.

In the case of the Y-form solution the EWIF intensity is, from Table 1, approximately 20 times greater than would be predicted from the bulk concentration, whereas the R-form solution has an EWIF intensity 60 times greater than would be predicted from the bulk concentration. The enhanced fluorescence intensity can be explained in terms of a contribution from two possible sources:
(i) an increase in polymer concentration in the evanescent wave region.
(ii) an increase in the fluorescence quantum efficiency, q_F, of the emitting chromophore(s), by either a decrease in the non-radiative deactivation rate constant, k_{NR}, or an increase in the radiative rate constant, k_R.
where:

$$q_F = \frac{k_R}{k_R + k_{NR}} = k_R \tau_F \tag{16}$$

In order to resolve this issue fluorescence decays were recorded for an R-form solution in both bulk and EWIF configurations.

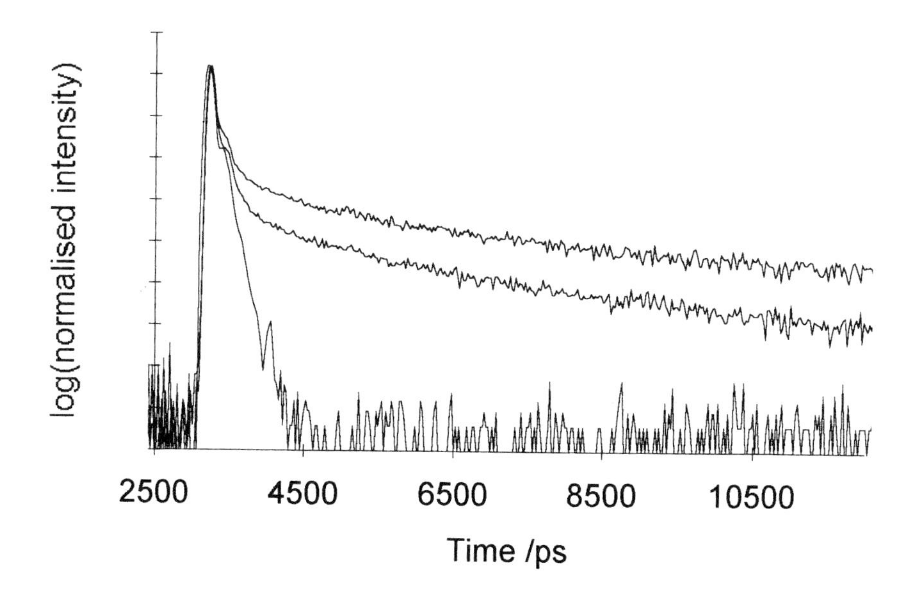

Figure 6. Fluorescence decays from an equilibrated R-form solution recorded using evanescent wave (upper trace) and bulk excitation (lower trace). A 50ps instrument response function is also shown.

Figure 6 shows a normalised plot of a bulk and EWIF decay at a penetration depth, Λ, of 1300 Å, both recorded at an emission wavelength of 560 nm. The decays are clearly non-mono-exponential in character. At long times (> 300 ps) the decay from the bulk solution is seen to be much shorter lived which, assuming that the radiative rate constant is the same in both cases, would suggest that the fluorescence quantum efficiency at the surface is higher. However, the majority of the fluorescence in both cases occurs at times less than 300 ps. These decays were analysed using an iterative re-convolution procedure in terms of a multi-exponential decay function of the form:

$$G(t) = \sum_{i=1}^{i=n} A_i \exp\left(-t/\tau_i\right) \qquad (17)$$

In both cases the quality of fitting was very poor even when using a tri-exponential function (n = 3), although the major component of the decay was clearly sub 10 ps and was the major component of the bulk decay but was less significant in the EWIF decay. Previous attempts at measuring fluorescence decays from both bulk Y-form and R-form poly-4BCMU solutions have failed, even with an instrument with a 40 ps instrument response function and it has been assumed that in both cases the major decay component is too short to measure. However, in the case of the EWIF decays the short component may lengthen to become measurable with the present apparatus. Four decays at different penetration depths were recorded on a surface equilibrated R-form solution.

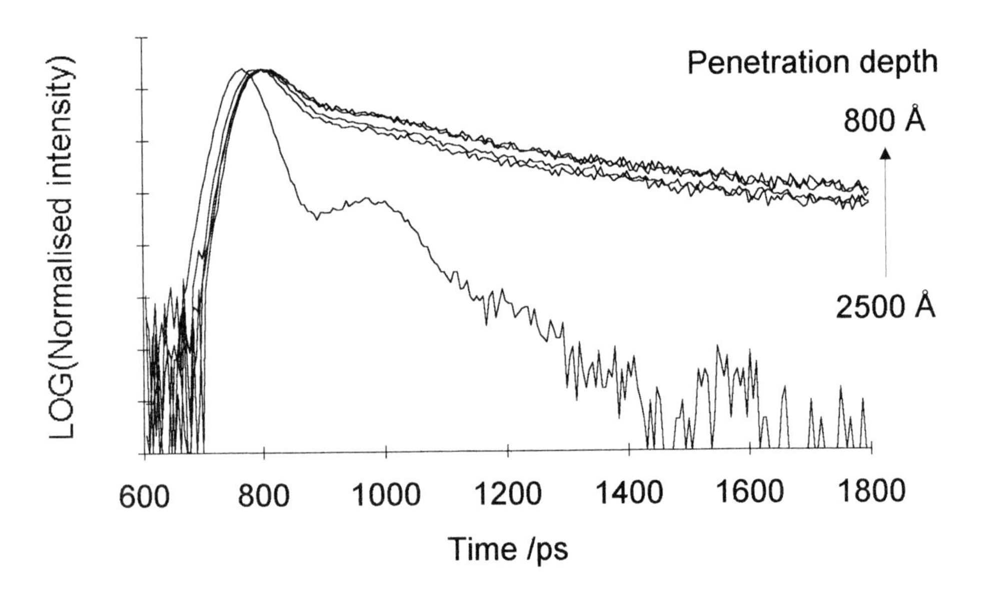

Figure 7. EWIF decays recorded at penetration depths of approximately, 2500Å, 1200Å, 900Å and 800Å

The normalised decays shown in figure 7 and the results of fitting a tri-exponential function to the data are given in table 2, the three decay times were constrained to be common to all four decays and only the pre-exponential factors were allowed to vary freely (global analysis). All fits were deemed to be good fits according standard statistical evaluations.

Table 2
Percentage yield of the decay times associated with the tri-exponential, global analysis of the fluorescence decay data shown in Figure 7.

Penetration depth /Å	$\tau_1 = 21$ ps	$\tau_2 = 160$ ps	$\tau_3 = 935$ ps
2500	31	24	45
1200	27	27	46
900	19	30	51
800	19	30	51

Inspection of the decays in figure 7 clearly shows that the decays are becoming longer as the penetration depth is reduced. This is verified by inspection of the data in Table 2 where the yield of the short 21 ps component is reducing in magnitude whilst the yield of the long 935 ps component is increasing in magnitude. At this stage it is worth noting that we do not assign the three decay times to spectroscopic species, with the tri-exponential fitting function simply providing a means of parameterising the fluorescence decay profiles. This point will be discussed at a later stage. These results confirm two important points: (i) The quantum yield of the poly-4BCMU is dependent on its proximity to the glass-solvent interface and (ii) The variation with penetration depth shows that the effect becomes more apparent as the polymer approaches the glass surface. Indeed, the 60 fold increase in intensity for the EWIF decays relative to the bulk, suggests that the detected fluorescence originates only from polymer molecules in close proximity or even slightly bound to the glass surface and therefore in this instance the technique is surface sensitive.

In conclusion, the fluorescence quantum efficiency of poly-4BCMU in the close proximity of a fused silica surface is increased relative to the bulk. Therefore, the increased EWIF intensity cannot be solely attributed to an increased concentration of the polymer within the evanescent wave, as would be required by equation 13.

4.2 Surface induced chromism

On addition of a freshly prepared solution of poly-4BCMU in 2Me-THF to a cleaned and air-dried prism, it was observed that the first EWIF spectrum did not resemble the normal Y-form spectrum. Subsequent EWIF spectra more closely resembled the expected Y-form spectrum, with a good correlation reached after 4 hours, following introduction of the solution to the prism surface. Removal of the sample followed by washing with pure solvent and then re-adding a fresh Y-form solution, did not produce the same effect, with the EWIF spectrum immediately resembling the bulk spectrum. The effect was found not to be reproducible until the prism was re-cleaned and dried.

A similar experiment was repeated using an R-form solution. Under these conditions the EWIF spectrum did closely resemble the expected R-form spectrum, although a good correlation between the EWIF and bulk spectra was not reached for up 15 minutes following the addition of the solution to the prism.

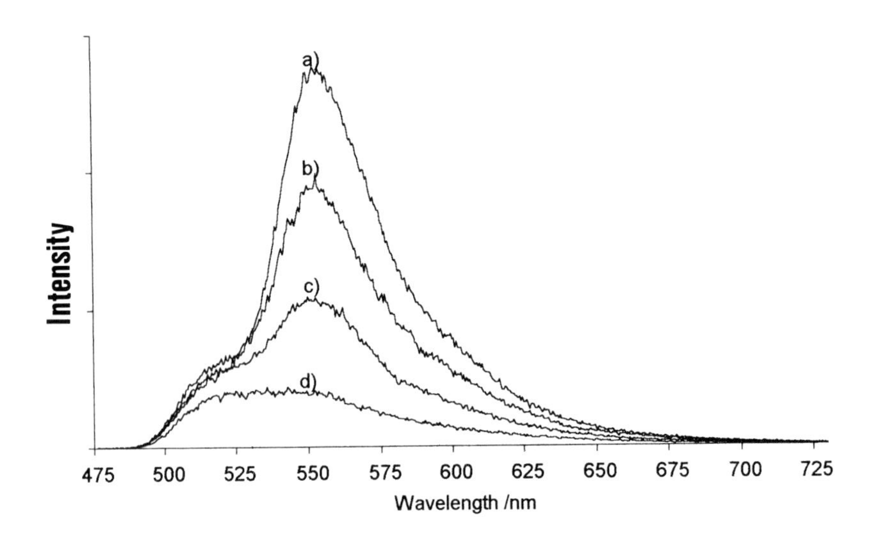

Figure 8. EWIF spectra recorded after: a) 2.5, b) 20, c) 50 and d) 195 minutes, following the introduction of a Y-form solution of poly-4BCMU solution to an air-dried fused silica surface.

The first EWIF spectrum recorded on the Y-form solution exhibited a close similarity to the R-form spectrum. Indeed, the spectrum was characteristic of a more ordered state of the polymer chain, where the Y $\leftrightarrows$ R equilibrium is even further shifted to the right, than in the case of the R-form solution. In order to investigate this phenomenon more closely, a series of EWIF spectra were recorded consecutively following the addition of the Y-form solution to the cleaned and dried prism. Four representative EWIF spectra are shown in figure 8 taken at times: 2.5, 20, 50 and 195 minutes, following the introduction of the sample to the surface. At early times the spectrum closely resembles the R-form spectrum reducing to the Y-form spectrum at long times.

Using the 2.5 minute spectrum as an 'early R-form' spectrum and the 195 minute spectrum as a 'late Y-form' spectrum, intermediate spectra were analysed as a linear combination of these two extremes. From this preliminary kinetic study the 'early R-form' spectrum decays with a rate constant of 0.03 minutes^{-1}. The time-resolution of this experiment was limited to ca. 2.5 minutes, the time taken to record a spectrum, and in order to investigate in more detail the kinetics at early times it was necessary to monitor two wavelengths, representative of the R- and Y-forms of the polymer. At 518 nm, the emission is predominantly from the Y-form, although a small contribution from the R-form cannot be ignored. Similarly, at 558 nm the emission is predominantly from the R-form, especially at early times, but in this instance the contribution from the Y-form should not be ignored. Figure 9 shows the data from this study and table 3 summarises the kinetics.

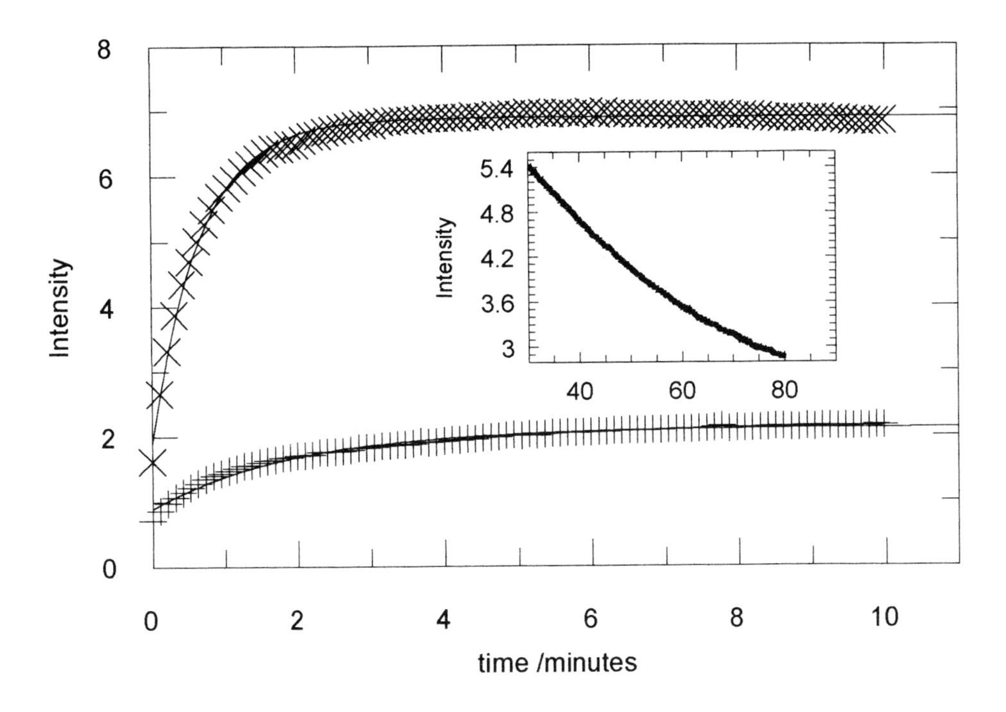

Figure 9. Time dependence of EWIF intensity at 558nm (upper trace and inset), 518nm (lower trace), following addition of a Y-form solution to the glass surface.

Table 3.
First order kinetic parameters; data in Figure 9 fitted to a function of the form: B+Aexp(−kt).

	Pre-exponential factor, A.	Rate constant, k, / minutes^{-1}	Constant, B.
0-10 minutes, Y-form solution at 518 nm.	-1.2 (±0.3)	0.50 (±0.05)	2.1 (±0.3)
0-10 minutes, Y-form solution at 558 nm.	-5 (±1)	0.7 (±0.1)	6.9 (±1.0)
20-90 minutes, Y-form solution at 558 nm.	7.7 (±1.5)	0.025 (±0.005)	1.7 (±0.5)
0-15 minutes, R-form solution at 558 nm.	-1.1 (±0.3)	0.11 (±0.01)	1.3 (±0.3)

The three most important points from this analysis are:
(i) At long times, the R-form at the surface decays with a lifetime of 0.025 minutes^{-1}, indicating that at the surface it takes in excess of two hours before equilibrium is attained.
(ii) The short time growth of the R-form at the surface occurs with a decay time of 0.7 minutes^{-1}. The process associated with this fast rate is comparable to a polymer folding process to form a fringe micellar structure as suggested by Batchelder et al.
(iii) The growth of the Y-form structure at the surface appears, to a first approximation, to be independent of the changes in the R-form, suggesting that processes other than a simple disturbance of the Y-form/R-form equilibrium is taking place.

A similar study for an R-form solution was also carried out. In this instance, the EWIF spectrum closely resembled that of the bulk but grows in with a decay time of 0.11minutes^{-1}. In both cases the introduction of the solution to a cleaned and dried glass surface produced a situation which took a number of hours before equilibrium was attained. In the case of a Y-form solution the process also produced a chromism effect, with the more ordered form of the polymer rapidly forming and then taking a number of hours to disappear. The fact that the effects were not observable when the surface had been solvent conditioned suggests that the surface is passivated when air-dried. In terms of surface science the prism is far from clean and almost certainly has adsorbates on it, one of which is probably water. The non-equilibrium situation that arises from introducing the solution to the 'air-dried' surface is probably a result of the interaction between the surface adsorbates and the solvent. By choosing a molecule that is spectroscopically very sensitive to its environment, the solvent conditioning of the surface is readily observable.

4.3 Fluorescence anisotropy

The rate of formation of the R-form at the surface from a Y-form solution of 0.7 minutes^{-1}, compares extremely well with the rate of the single chain phenomenon of fringe micelle formation, or *intra*-molecular aggregation, suggested by Batchelder et al[10]. This model for the R-form of the polymer is very attractive since it is a single chain effect and also looks like *inter*-molecular aggregation. The surface induced chromism observed at the glass surface occurs within the evanescent wave and it is suggested that this data supports the fringe micelle theory. One question which remains unanswered is: How does the polymer arrange itself at the surface? The absorption and emission dipoles lie parallel to the polymer backbone and the facile, exciton-like migration along the polymer backbone can randomise the relative orientations of the absorption and emission dipoles and thus depolarize the emission.

In order to try and understand the ordering of the polymer at the glass surface fluorescence spectra were recorded at early times (3-17 minutes following addition of the solution to the surface) using vertical and horizontally polarized excitation light, with an emission analyser either parallel or perpendicular to the excitation polarization. The definitions of the four possible configurations are ($I_{Excitation,Emission}$): I_{VV}, I_{VH}, I_{HV} and I_{HH}, where V and H are as given in Figure 4. To compensate for the changes in fluorescence intensity during the acquisition period a second set of four spectra were recorded in reverse order and added to the first set. Unlike conventional fluorescence anisotropy measurements where E-fields exist only normal to the direction of propagation, the evanescent wave experiment differs in that E-fields exist in all spatial directions.

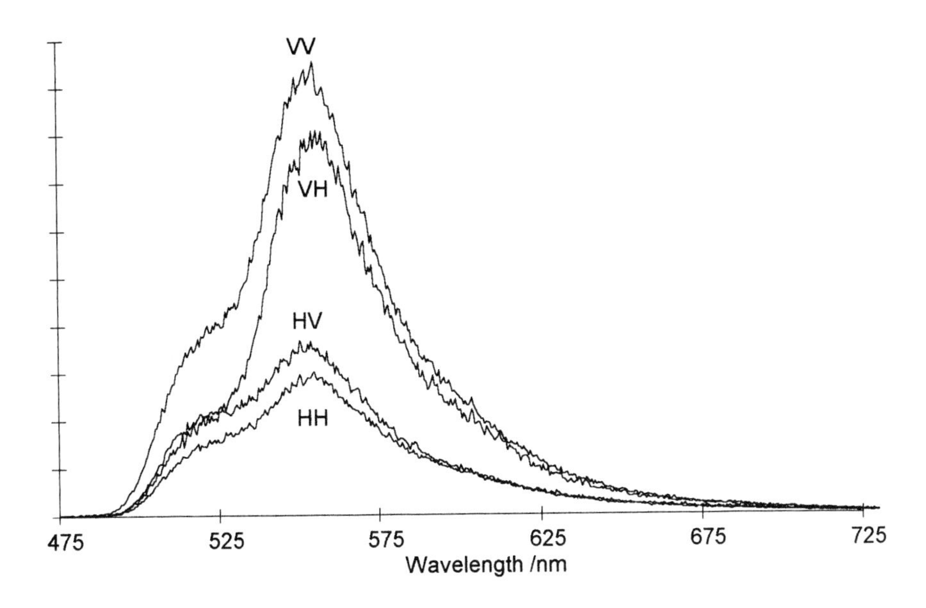

Figure 10. Anisotropy of emission from a Y-form poly-4BCMU solution at early times.

Figure 10 shows the four spectra from a poly-4BCMU Y-form solution at early times. To a first approximation vertical excitation provides maximum emission regardless of the orientation of the analyzer. However, in order to understand this data better, it is necessary to spectrally deconvolve emission into the Y and R components. By comparing the blue edge of the emission spectra with a Y-form solution bulk spectrum, the percentages of the two forms were deconvolved. Table 4 lists the relative fluorescence intensities of the two components.

Table 4.
Relative fluorescence intensities and measured anisotropies of Y- and R- forms of poly-4BCMU at a fused silica surface produced from a Y-solution after 10±6 minutes contact.

Polarization (Exc, Emis)	Y component	anisotropy, $\frac{V-H}{V+2H}$	R component	anisotropy, $\frac{V-H}{V+2H}$
HH	0.22		0.28	
		0.13		0.00
HV	0.32		0.28	
VV	0.58		0.95	
		0.26		-0.02
VH	0.29		1.00	

Table 5.
Anisotropy of fluorescence from a poly-4BCMU solution in equilibrium with a fused silica surface.

Solution	Penetration	anisotropy, $\frac{V-H}{V+2H}$
Y-form	Bulk	0.30
Y-form	EWIF	0.23
R-form	EWIF	0.29

The anisotropies may be compared with those measured at equilibrium, that is, after the solution had been in contact with the glass surface for at least two hours. These data are shown in Table 5. A comparison of the anisotropy of the Y component with vertical excitation in Table 4 is identical to that for the bulk Y component in Table 5. We may conclude that the Y component of emission excited by the evanescent wave in poly-4BCMU is due to free Y-form of the polymer in solution, close to the interface, rather than adsorbed material. For horizontal excitation some depolarization would be expected, as is observed. For the R component, significantly, total depolarization is universally observed, with vertical excitation also producing three times as much emission as horizontal excitation. Mechanisms of depolarization include diffusional motion and electronic energy migration, possibly exciton diffusion. The fluorescence decay times are, as mentioned earlier, very short and it is unlikely that diffusional motion is the depolarizing mechanism. Thus energy migration along the polymer backbone is the most likely explanation for the depolarization. The dramatic difference in intensity between vertical and horizontally polarized excitation light can be understood in terms of the difference in intensity of the light at the interface as defined by equations 5-9. Only a summary of the conclusions are given here, but a more detailed account can be found elsewhere. By substituting in the experimental parameters into equations 5-7, the following values are derived: $U_{0,y} = 2.05$, $U_{0,x} = 0.09$ and $U_{0,z} = 2.14$. Since the prism is oriented at an angle 90-θ with respect to the detection system (see figure 4), these values must be modified such that the vertical component, $U_{0,V}$, remains unchanged at 2.05, and the horizontal component, $U_{0,H}$, is almost the same as $U_{0,x}$ at 0.09. The actual ratio of the R component for vertical and horizontal excitation is 3.4:1. If all the R components were parallel to the surface then this ratio would be ca. 22:1 and if there was no preference, then it would be 1:1. The experimental ratio thus suggests that the R components are preferentially aligning themselves to the glass surface, although it must be recognised that an R component perpendicular to the surface may not experience as large an increase in quantum efficiency as a component parallel to the surface. However, combining this data with the inference of a fringe micelle type structure for the R-form, discussed earlier, suggests that the polymer folds onto the prism surface.

The large anisotropy for the Y-form at the surface and in the bulk, suggests that the diffusion process does not appear as extensive as originally thought. In the R-form, however, complete depolarization is observed. This could result from an increase in energy transfer between the parallel sections of the folded R-form, by a dipole-dipole interaction. The uncertainty in defining the true nature of the absorbing and emitting chromophores in poly-4BCMU, or indeed in any conjugated polymer, makes it very difficult to relate spectroscopic features, especially in emission, to structural changes of the polymer backbone.

5. CONCLUSIONS

The inclusion of time-resolution into an EWIF experiment provides a very reliable method for determining whether measured fluorescence intensities are a result of concentration or quantum efficiency changes, within the evanescent wave. In the case reported here, and elsewhere, quantum efficiency changes are unambiguously evident when there is a strong interaction between the solute and the surface. The goal at present is to try to use time-resolved data as a method of measuring concentration profiles within the evanescent wave[28]. In this context, the major problem arises in the interpretation of the complex functional form of the fluorescence decays. If the multi-exponential analysis could be interpreted as multiple species, then the pre-exponential factors would represent the concentration of the individual species, this however is an ideal situation and one that has yet to be confirmed. The use of stretched exponentials and other related functions has proved to be successful method of modelling the fluorescence decays, but provides little physical insight into the nature of the interfacial species. Finally, the use of distribution analysis as a method of resolving large heterogeneities shows a great deal of promise, yet it still suffers as a technique that it is open to criticism.

When EWIF is applied to the study of the influence of a solid-solution interface on the properties of the soluble polydiacetylene, poly-4BCMU, a number of interesting phenomena have been found: The nature of the surface has a strong influence on the polymer conformation at the interface with a poor solvent environment causing a structural transformation in the polymer conformation to the more ordered R-form from the Y-form found in the bulk of the solution. The kinetics for the creation of the R-form suggest that the fringed-micelle is a very likely structure for this form of the polymer The kinetics for the loss of the R-form have been attributed to the solvent conditioning of the glass surface, which takes up to three hours to complete and is almost certainly due to atmospheric species, probably water, adhering during the cleaning procedure. This observation provides an insight into the need for extreme care when both preparing the surface and in interpreting EWIF data. In the interfacial region, the quantum efficiency of both the R-form and the Y-form is far higher than similar, structural forms found in bulk solution. The complexities of the fluorescence decays used to confirm this observation reveal a greater complexity in the data, making concentration profiles difficult to determine. The interpretation of the enhanced fluorescence quantum efficiency is attributed to a restriction of molecular motion, which inhibits the internal conversion process normally responsible for the efficient non-radiative deactivation pathway of the excited states of soluble polydiacetylenes. Fluorescence anisotropy measurements suggest that the more ordered R-form preferentially folds parallel to the glass surface, but more importantly, the less ordered Y-form shows a large anisotropy both at the interface and in the bulk, suggesting that efficient energy transfer down the polymer chain does not occur, a result which is in conflict with previous thoughts.

6. ACKNOWLEDGEMENTS

We would like to thank the molecular electronics initiative of the Science and Engineering Research Council, for the generous financial support for this project.

REFERENCES

1. Polydiacetylenes, Nato ASI series E - 102, eds. D.Bloor and R.R.Chance (Martinus Nijhoff, Holland, 1985).
2. B.Cho and R.Xu, Acc.Chem.Res., 24 (1991) 384 .
3. A.J.Brown, G.Rumbles, D.Phillips and D.Bloor, Chem. Phys. Lett. 151 (1988) 247.
4. K.C.Lim., C.R.Fincher and A.J.Heeger, Phys. Rev. Lett. 50 (1983) 1934.
5. K.C.Lim, A.Kapitulnik, R.Zacher and A.J.Heeger, J.Chem. Phys. 82 (1985) 516.
6. D.G.Peiffer, T.C.Chung, D.N.Schultz, P.K.Agarwal. R.T.Garner and M.W.Kim, J. Chem. Phys. 85 (1986) 4712.
7. G.Wenz, M.A.Muller, M.Schmidt and G.Wegner, Macromols. 17 (1984) 837.
8. M.Rawiso, J.P.Aime, J.L.Fave, M.Schott, M.A.Muller, M.Schmidt, H.Baumgartl and G.Wegner, J.Physique 49 (1988) 861.
9. L.D.Coyne, C.Chang and S.L.Hsu, Makromol. Chem. 188 (1987) 2311.
10. M.A.Taylor, J.A.Odell, D.N.Batchelder and A.J.Campbell, Polymer 31 (1990) 1116.
11. C.Allain, D.Ausserre and F.Rondelez, Phys. Rev. Lett. 49 (1982) 1694.
12. B.K.Lok, Y-L Cheng and C.R.Robertson, J.Coll. and Int. Sci. 91 (1983) 87.
13. D.Ausserre, H.Hervet and F.Rondelez, Macromols. J.Physique Lett. 46 (1985) L929.
14. D.Ausserre, H.Hervet and F.Rondelez, Phys.Rev. Lett. 54 (1985) 1948.
15. D.Ausserre, H.Hervet and F.Rondelez, Macromols. 19 (1986) 85.
16. A.J.Brown, Ph.D. Thesis, University of London (1989).
17. M.Born and E.Wolf in 'Principles of Optics', McMillan, NY (1964).
18. N.J.Harrick in 'Internal Reflection Spectroscopy', Wiley interscience, NY (1967).
19. N.Ikeda, T.Kuroda and H.Masuhara, Chem.Phys.Letts., 156 (1989) 204.
20. R.Iwamoto, M.Miya, K.Ohta and S.Mima, J.Amer.Chem.Soc., 102 (1980) 1212.
21. R.Iwamoto, M.Miya, K.Ohta and S.Mima, J.Chem.Phys., 74 (1981) 4780.
22. G.Rumbles, A.J.Brown and D.Phillips, J.Chem. Soc. Faraday Trans. 87 (1991) 825. And references therein.
23. D.Axelrod, T.P. Burghhardt and N.L.Thompson, Ann. Rev. Biophys. Bioeng., 14 (1984) 247.
24. J.Edwards, D.Aussere, H.Hervert and F.Rondelez, Applied Optics, 28, 1881 (1989).
25. D.Phillips and D.V.O'Connor in 'Time-Correlated Single- Photon Counting', Academic press, London (1984).
26. A.J.Brown, Ph.D.Thesis, University of London (1989).
27. G.N.Patel, Polym. prepr. 19 (1978) 154.
28. H.Masuhara, S.Tazuke, N.Tamai and I.Yamazaki, J.Phys.Chem., 90 (1986) 5830.

MICROCHEMISTRY
Spectroscopy and Chemistry in Small Domains
Edited by H. Masuhara et al.

Picosecond lasing dynamics of an optically-trapped microparticle

Kenji Kamada,[#] Keiji Sasaki,[*,†] Ryo Fujisawa,[‡] and Hiroaki Misawa[§]

Microphotoconversion Project,[+] ERATO Program, Research Development Corporation of Japan, 15 Morimoto-cho, Simogamo, Sakyo-ku, Kyoto 606, Japan

A dye-doped polymer microparticle in water was simultaneously manipulated and pumped by 1064 nm CW and 532 nm pulsed Nd:YAG laser beams, respectively, which induced laser oscillation in the particle, based on optical resonances in a high-Q microspherical cavity. Spectral and temporal characteristics of microspherical lasing were analyzed by a picosecond time-resolved microspectroscopy system, which clarified that lasing dynamics depends on a resonance wavelength and a dye concentration, as well as the size and refractive index of the particle. Enhancements of energy transfer and transient absorption in the microspherical cavity were demonstrated, which are characteristic phenomena of micrometer-sized particles. A possibility of high-sensitive transient absorption measurements in/on a particle, based on intracavity effects, is also discussed.

1. INTRODUCTION

Optical interaction between laser light and a spherical microparticle such as a liquid droplet, a polymer latex particle, or a microcapsule leads to two interesting phenomena. One is radiation pressure caused by a photon momentum change, which has been applied to noncontact and nondestructive manipulation, spatial patterning, and assembling of microparticles, as reviewed in Part II of this volume. The other is optical resonance within a microsphere, in which light propagates in a circumferential manner to create a standing wave field just inside of the microsphere surface, that is, a spherical particle can act as an optical cavity [1]. Such a phenomenon has been called as whispering-gallery mode resonance. The characteristic behavior of light

* To whom correpondence should be addressed.
# Present address: Department of Optical Materials, Osaka National Research Institute, Agency of Industrial Science and Technology, Ikeda, Osaka 563, Japan.
† Present address: Department of Applied Physics, Osaka University, Suita, Osaka 565, Japan.
‡ Present address: Mita Industrial Co. Ltd., Tamatsukuri 1-2-28, Chuo-ku, Osaka 540, Japan.
§ Present address: Department of Mechanical Engineering, The University of Tokushima, Tokushima 770, Japan.
+ Five-year term project: October 1988~September 1993.

confined in a microsphere has been theoretically investigated by several researchers in early years of 20th century, based on the Mie-Debye light scattering theory. Experimentally, Ashkin et al. verified the microspherical resonance by precise observation of radiation pressure exerted on a particle as a function of wavelength in 1977 [2]. The theoretical calculation indicates that a micrometer-sized sphere possesses an extremely high quality factor ($Q > 10^8$), which is sufficient for inducing laser oscillation. Indeed, Tzeng et al. demonstrated lasing of a microparticle for the first time in 1984 [3]. In their experiment, a ~60 μm ethanol droplet containing rhodamine 6G was produced by a vibrating orifice aerosol generator, and the droplet falling in air was pumped by a CW argon-ion laser. After the first demonstration, various physical properties of the microspherical laser oscillation were studied in the last decade [4-6].

In order to apply the microspherical laser as a small light source for inducing photophysical and photochemical reactions in micrometer volumes and for optically probing such phenomena, we have developed a simultaneous optical trapping and lasing system which makes it possible to freely manipulate a lasing particle in three-dimensional space [7]. Furthermore, this system can be applied to precise analyses of spectral and temporal characteristics of a microspherical laser without any disturbances such as thermal Brownian motion, gravity, and convection. Three-dimensional trapping also works for avoiding optical interactions of a lasing particle with its surroundings such as glass plates and other particles, which reduce the quality factor of a microspherical cavity and affect the lasing process. By the use of the system combined with time-resolved spectroscopy, we have elucidated picosecond lasing dynamics in a microspherical cavity [8]. A microparticle acts as a short cavity (micrometer resonator length) so that a picosecond lasing pulse can be produced by a single pulsed pumping, which is one of the characteristic property of microspherical lasing. Rise and decay curves of the pulsed laser oscillation provide valuable information on the mechanism of microspherical lasing, as well as characteristic molecular dynamics in a microparticle. Some photochemical processes within a microspherical particle are influenced by the high-Q resonance effect so that the efficiency of the processes will be extremely high compared with those of bulk materials.

In this paper, we describe results of spectroscopic and temporal analyses of microspherical lasing processes and discussed on factors influencing the lasing dynamics. Characteristic phenomena in energy transfer, and transient absorption processes interacted with the microspherical resonance are also introduced, and its application to high-sensitive microspectroscopy is proposed and experimentally demonstrated.

2. LASER OSCILLATION OF A MICROPARTICLE

2.1. Optical resonance and lasing in a microsphere

The mechanism of laser oscillation in a microparticle is schematically illustrated in Figure 1. A dye-doped microparticle is irradiated by intense light so that an inverted population of dye molecules is induced in the particle. If the refractive index of the

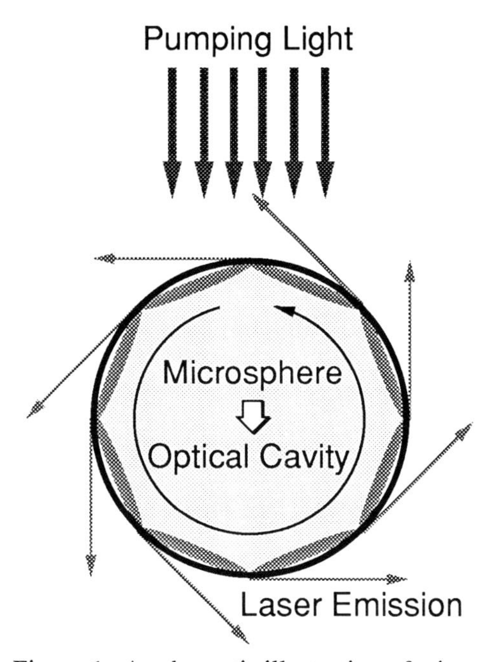

Figure 1. A schematic illustration of microspherical resonances and lasing.

particle is higher than that of the surrounding medium, emission incident at the particle-medium boundary with an angle larger than the critical angle is reflected back inside of the particle. This emission is reflected repeatedly and propagates circumferentially along the boundary, which forms an optical cavity like a Fabry-Pérot resonator. When the emission returns to the starting point with the same phase except for some integer multiple of 2π, oscillation is induced so that the emission is amplified by the population-inverted dye molecules; laser oscillation. Since the surface of the microsphere is not flat but curved, the emission is not totally reflected at the boundary, that is, the evanescent wave does not totally return to the inside of the sphere. Because of this leakage, the laser light emits from the particle boundary toward the tangents of the sphere. Thus, the laser emission is observed on the rim of the microsphere.

The analysis based on wave optics shows that the laser oscillation in a microparticle normally possesses two linearly-polarized modes, i.e., transverse electric (TE) and transverse magnetic (TM) modes in addition to spatial modes of angular and radial resonances that correspond to longitudinal and transverse modes of a Fabry-Pérot resonator, respectively [9]. Oscillation wavelengths of these modes are different from each other, and the separation between the adjacent angular resonance wavelengths is approximately given by

$$\Delta\lambda = \frac{\lambda^2}{\pi d n_2} \frac{\tan^{-1}[(n_1/n_2)^2 - 1]^{1/2}}{[(n_1/n_2)^2 - 1]^{1/2}}, \tag{1}$$

where n_1 and n_2 are the refractive indices of a particle and the surrounding medium, respectively, and d is the diameter of a particle. If absorption by a particle itself is negligibly small, the resonance with a larger angular mode number, which is obtained in a larger particle, possesses the higher quality factor.

2.2. Lasing of an optically-trapped microparticle

A schematic diagram of a simultaneous optical trapping and lasing system is shown in Figure 2 [7]. A dye-doped poly(methyl methacrylate) (PMMA) latex

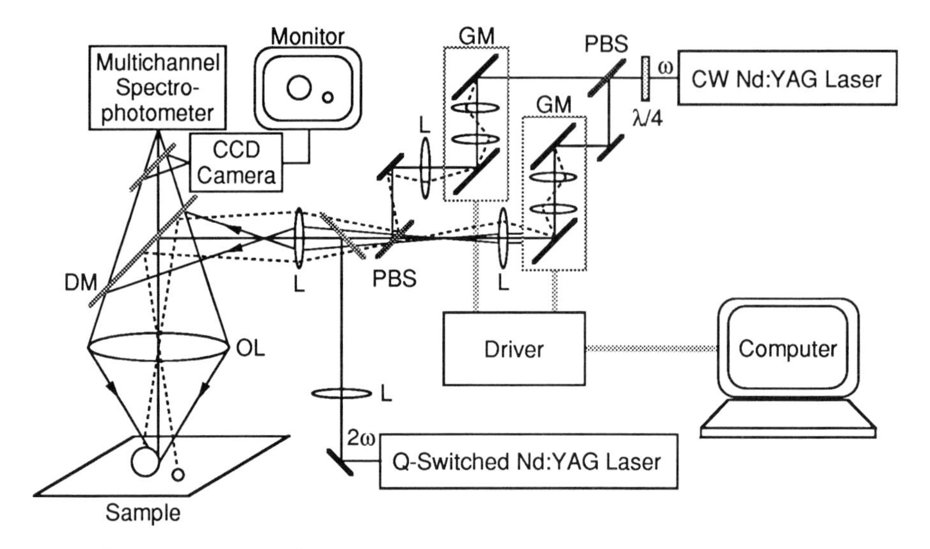

Figure 2. A schematic diagram of a simultaneous optical trapping and lasing system.

particle undergoing thermal Brownian motion in water was optically manipulated by a focused 1064 nm beam (440 mW, ~1 μm spot) from a CW Nd:YAG laser under a microscope. Dye molecules in the particle were uniformly pumped by a second harmonic pulse from a Q-switched mode-locked Nd:YAG laser (532 nm, ~40 ps, 10Hz). The pumping laser intensity was carefully adjusted to be lower than the damage threshold of the particle. Emission from the particle was collected by an objective lens and detected by a polychromator and a double-intensified multichannel photodiode array detector (spectral resolution = 0.4 nm). Time-resolved emission measurements were performed with a streak camera (temporal and spectral resolutions = 10 ps and 1.5 nm, respectively). The behavior of a lasing particle was monitored by a CCD camera attached to the microscope, and photographs were printed with a video printer.

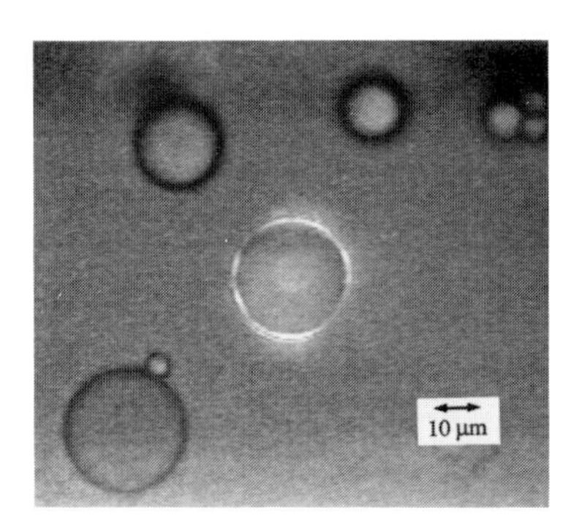

Figure 3. Lasing of an optically-trapped RhB-PMMA particle (24 μm, indicated by the arrow) in water. Pumping laser power ~ 25 mJ, trapping laser power ~ 440 mW.

Figure 3 shows a typical example for lasing of a rhodamine B (RhB) doped PMMA particle with the diameter of 24 μm [7]. Besides orange spontaneous emission from RhB molecules, more intense emission was observed near the particle-water boundary. This ring-like emission could be ascribed to laser light from the microspherical cavity. In the present experiment, the lasing particle was lifted up from a microscope glass plate by optical trapping so that coupling loss caused by leakage of laser emission to the glass plate could be avoided, which was superior conditions for inducing laser oscillation and for comparing the experimental results with the Mie-Debye scattering theory. Under the present conditions, stable lasing of RhB-PMMA particles in water was confirmed for more than 30 minutes without any degradation of the dye or damage of the particle.

2.3. Spectral characteristics of microspherical lasing

Emission spectra from a RhB-doped PMMA particle (26 μm) in water are shown in Figure 4 [7]. When the pumping laser power is low, the emission spectrum is broad and structureless (Figure 4(a)), which corresponds to a spontaneous emission spectrum of RhB. As increasing the pumping intensity, distinct resonance peaks appear, and their intensities increase (Figure 4(b) and 4(c)). Several pairs of adjacent high and low intensity peaks, that correspond to TE and TM modes, respectively, are spaced at the constant wavelength interval. This ripple structure is attributed to the whispering-gallery mode resonances. The interval of the TE mode peaks is 3.1 nm, which is the same as that of the TM modes, and well agrees with the calculated value of 3.06 nm based on Eq. (1).

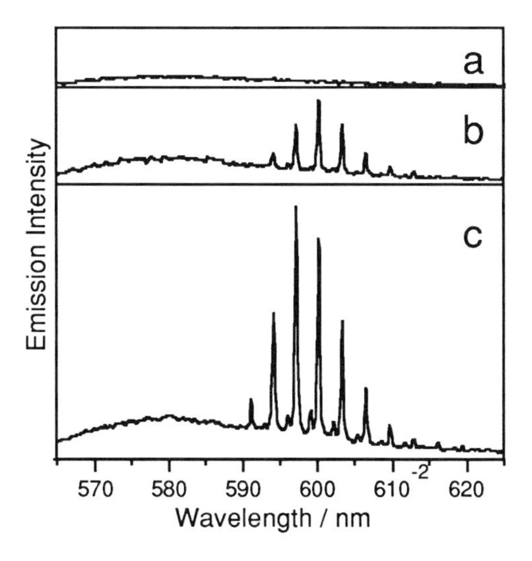

Figure 4. Emission spectra of a RhB-PMMA particle (26 μm) pumped with the laser power of (a) 1.1, (b) 3.0, and (c) 9.7 mJ cm^{-2} $pulse^{-1}$.

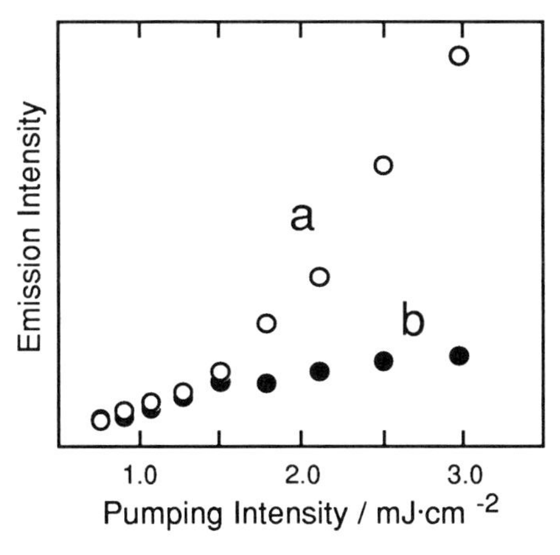

Figure 5. Pumping laser power dependence of the emission peak intensity. The sample was the same as that observed in Figure 4. The emission intensity was determined at (a) 600.4 and (b) 577.9 nm.

The emission intensities of the spectra in Figure 4 were plotted as a function of the pumping intensity, as shown in Figure 5 [7]. The intensity at 577.9 nm, where the spontaneous emission is observed, increases linearly with the pumping intensity, while the intensity of resonance mode peak at 600.4 nm exhibits a nonlinear dependence on the pumping intensity, which indicates lasing of the RhB-PMMA particle in water.

2.4. Two-photon pumped lasing

With increasing the power of the CW 1064 nm trapping laser to several watts, the laser intensity within a focal spot reaches over 100 MW cm^{-2}. Such intense light will pump rhodamine molecules in a particle through a two-photon absorption mechanism, so that trapping and lasing of a microparticle can be simultaneously performed by a single laser beam [10]. This two-photon pumping overcomes difficulty in matching optical alignments of trapping and pumping laser beams, and is favorable for the positioning controllability of the microspherical laser source.

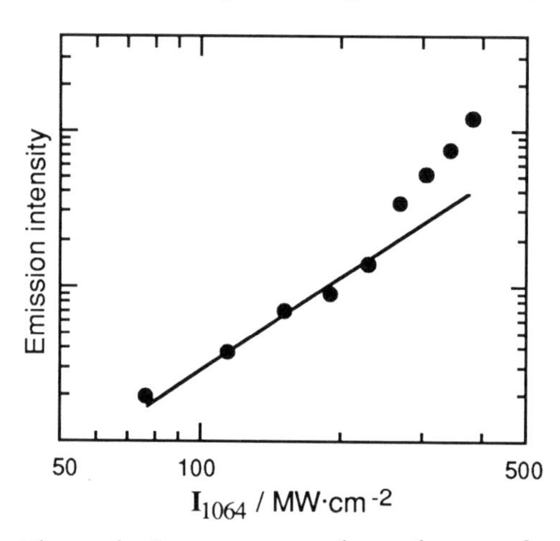

Figure 6. Laser power dependence of the emission intensity of an optically-trapped, Rh640-PSt particle (8.8 μm) in water. Also shown is a straight line with the slope equal to 2.0.

The emission intensity of a rhodamine 640 (Rh640) doped polystyrene (PSt) latex particle (8.8 μm) was observed as a function of the power of a CW 1064 nm laser beam (Figure 6). When the laser power is lower than 2.3 W, which provides the intensity of <230 MW cm^{-2} at a focal spot, the emission intensity is proportional to the square of the pumping power, which indicates that the observed emission originates from two-photon absorption of Rh640. In this power range, weak orange emission was observed at the central part of the particle, and emission spectra were broad and structureless, like that in Figure 4(a). Above the pumping power of >230 MW cm^{-2}, on the other hand, the emission intensity increases more than quadratic with increasing the laser power. At such intense pumping power, strong emission was observed at the particle-water boundary, and periodic ripple structure characteristic of microspherical resonances appeared in the observed spectra. These results clearly show that the two-photon pumped lasing occurs in the Rh640-PSt particle.

Although two-photon excitation is generally made possible by the extremely high local instantaneous intensity provided by a pulsed laser beam, the present results demonstrates that the tightly focused CW laser beam used for optical manipulation can induce two-photon absorption of dye molecules and lasing of a microparticle. If one compares between one-photon and two-photon pumping, the latter is superior in respect to the excitation wavelength region. Under a microscope, ultraviolet excitation of molecules is difficult because an objective lens does not normally transmit ultraviolet (<350 nm) light, while the two-photon process can pump the ultraviolet absorption band using visible laser light. In addition, a volume excited by the two-photon absorption is localized at a focal spot of the laser beam, so that damage of a sample and thermal effects on photochemical processes can be efficiently avoided.

3. CHARACTERISTIC DYNAMICS OF MICROSPHERICAL LASING

3.1. Picosecond pulsed laser oscillation

Emission propagating in a centimeter-sized Fabry-Pérot cavity takes nanosecond for returning the light to the starting point (velocity of light = 30 cm ns^{-1}), while emission in a microparticle can go round hundreds times during a pumping pulse width of tens picosecond, which is sufficient feedback for laser oscillation. Hence, the microspherical laser can be expected to produce a picosecond pulse. In order to elucidate physical properties of the microspherical lasing and the related phenomena, we have studied dynamics of lasing processes in the picosecond time region [8].

Temporal profiles of laser emission from a RhB-PMMA particle (21 μm) were observed with a streak camera (Figure 7). When the pumping intensity is low (Figure

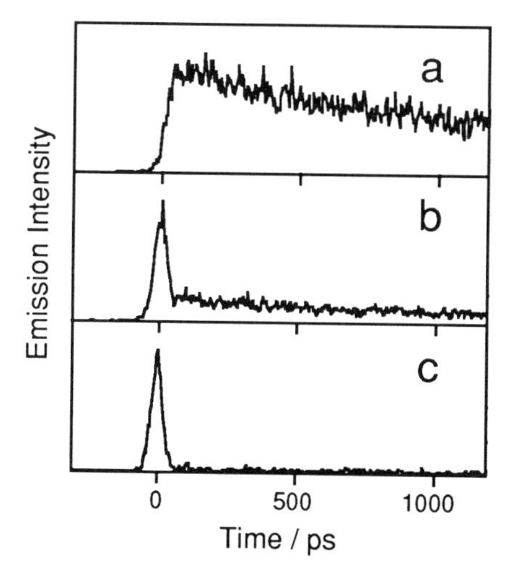

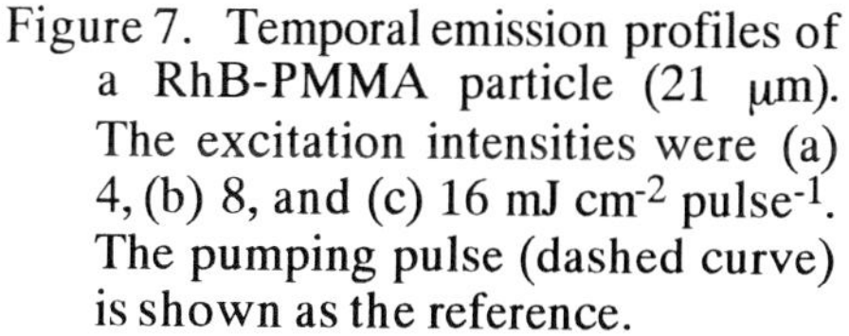

Figure 7. Temporal emission profiles of a RhB-PMMA particle (21 μm). The excitation intensities were (a) 4, (b) 8, and (c) 16 mJ cm^{-2} $pulse^{-1}$. The pumping pulse (dashed curve) is shown as the reference.

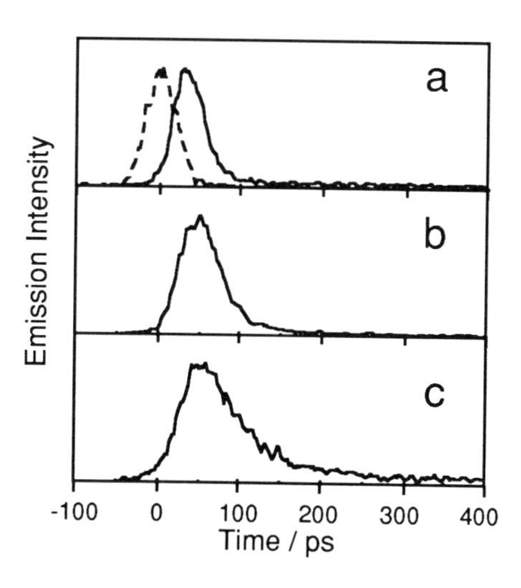

Figure 8. Temporal profiles of laser emission from a RhB-PMMA particle (38 μm), observed at wavelength regions of (a) 575-584, (b) 584-593, and (c) 593-601 nm.

7(a)), the emission slowly decays, corresponding to the fluorescence decay of RhB in PMMA (lifetime = 3.3 ns). At the pumping intensity of 8 mJ cm^{-2} $pulse^{-1}$ (Figure 7(b)), short pulsed emission appears and its relative contribution increases compared with that of fluorescence. This fast decay clearly indicates that the stimulated emission process is induced in the microparticle. The spontaneous emission after the lasing pulse can be ascribed to fluorescence from the central part of the particle. As the pumping power increases (Figure 7(c)), the intensity of the pulsed emission nonlinearly increases so that the decay curve includes no appreciable fluorescence component. The pulse width of the laser emission was determined to be ~40 ps (FWHM).

The pulse shape depends on the lasing wavelength and the dye concentration. Figure 8 shows temporal profiles of laser emission from a 38-μm RhB-PMMA particle observed at three wavelength regions [8]. Both rise and decay of the lasing emission become faster as the wavelength is shorter. In the wavelength region of 575-584 nm (Figure 8(a)), the profile is quite similar to that of a pumping pulse (dotted curve in Figure 8(a)) without an appreciable time delay. The decay curves were mathematically analyzed, and the decay time constants were determined to be <10, 10, and 50 ps for the spectral regions (a), (b) and (c), respectively.

This wavelength dependence can be explained by the cavity gain provided by stimulated emission of RhB and the cavity loss due to reabsorption of the emission by the dye in the particle. Since the intensity maximum of the emission spectrum of RhB is located at 584 nm, the cavity gain is lower in the longer wavelength region so that the rise time of curve (c) is longer than those of curves (a) and (b). On the other hand, the absorption band of RhB extends to the lasing wavelength region, which induces the loss in the microspherical cavity. This reabsorption loss causes an exponential decay of the emission intensity in the cavity. The decay time constant determined by the reabsorption loss is expressed as [11]

$$\tau_a = n_1 / [\, \varepsilon(\lambda)\, a\, c \ln 10 \,], \qquad (2)$$

where $\varepsilon(\lambda)$ is a molar extinction coefficient at the wavelength λ, and a and c are the concentration of RhB and the velocity of light in vacuum, respectively. Since $\varepsilon(\lambda)$ of RhB decreases in the longer wavelength region, the decay time constant of a microspherical lasing pulse becomes longer as the wavelength is longer. The decay times calculated from Eq. (2) are 3, 15, and 70 ps for the spectral regions (a), (b) and (c), respectively, which well agree with the experimental decay times. Although leakage from the cavity also reduces the decay time constant of the pulsed laser emission, the agreement between the decay times given by Eq. (2) and the experimental values indicates that reabsorption is a dominant factor for determining the decay process of microspherical lasing under the present conditions.

The reabsorption loss also depends on the concentration of RhB, as expected from Eq. (2). As the concentration increases, the cavity loss as well as the cavity gain increase so that the rise and decay time constants of a lasing pulse become faster, which has been experimentally confirmed.

It is concluded that the pulse shape of microspherical lasing can be controlled by adjusting a dye concentration and a lasing wavelength. The shorter pulse can be produced by the higher gain and loss of a microspherical cavity, while the low cavity loss, that is, the high quality factor is required for studies of the resonance effect on photochemical processes in a microparticle, as mentioned later.

3.2. Enhanced energy transfer in a microsphere

Energy transfer between donor-acceptor molecular pairs in a single aerosol particle, based on the Förster mechanism, has recently been reported to be enhanced by microspherical cavity effects compared with that in bulk solution [12]. The particular enhancement of energy transfer in a microparticle can be understood as the result of coupling of the transfer process with the microspherical resonances. Such enhanced energy transfer has been clarified by analyzing spontaneous emission from dye-doped microparticles. However, the enhancement is expected to be considerably efficient under the lasing condition so that lasing of acceptor molecules can be induced by the enhanced energy transfer. In order to study the energy transfer process in a lasing microparticle, we have spectroscopically analyzed lasing properties of donor and acceptor dyes-doped polymer particles [13].

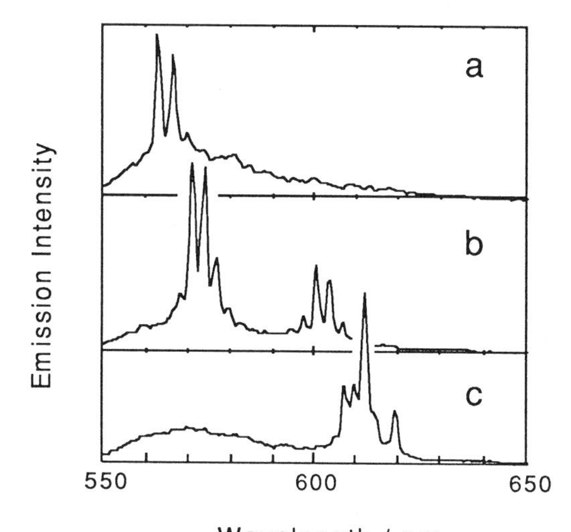

Figure 9. Lasing spectra of Rh6G-DODCI-PMMA particles with the diameters of (a) 20, (b) 25, and (c) 36 μm. Enhanced energy transfer from Rh6G* to DODCI is clearly shown in (b) and (c).

PMMA particles containing rhodamine 6G (Rh6G, 1.0 x 10^{-3} M) and 3, 3'-diethyloxadicarbocyanine (DODCI, 1.0 x 10^{-5} M) as donor and acceptor molecules, respectively, were prepared. Individual Rh6G-DODCI-PMMA particles (20, 25, and 36 μm, respectively) were optically trapped in water and pumped by a 532 nm laser pulse. Resonance emission peaks observed in the 20 μm particle (Figure 9(a)) are localized around 565 nm. This lasing wavelength region is almost identical with that of a PMMA particle containing only Rh6G with the same diameter. This indicates that the resonance emission from the 20 μm particle is ascribable to lasing of Rh6G. The lasing threshold power was several times higher than that of the Rh6G-PMMA particle, which could be explained by the energy transfer from excited state of Rh6G (Rh6G*) to DODCI. On the other hand, the spectral region of laser emission from the 36 μm particle (Figure 9(c)) does not correspond to that of a Rh6G-PMMA particle with the same size. Under the present experimental conditions, Rh6G in PMMA particles was exclusively excited by irradiation of a 532 nm laser pulse. Indeed, no appreciable laser emission was observed in any particles containing only DODCI. In addition, the emission of DODCI in a bulk solution of ethylene glycol containing the two dyes was below a detection limit. These results show that lasing of DODCI is induced in the 36 μm particle by the energy transfer from Rh6G* to DODCI, which is enhanced by the microspherical resonance effect. Furthermore, it is reasonable to assign the emission peaks around 570 and 605 nm in Figure 9(b) to lasing of Rh6G and DODCI, respectively.

In conclusion we have shown that enhanced energy transfer in individual microparticles is considerably efficient under the lasing condition. Our experimental results indicate that the efficiency of the enhanced energy transfer can be controlled by the size of a microparticle. We expect that the particular phenomenon of energy transfer in a microspherical cavity will open a new aspect of photophysics and photochemistry in microparticles.

3.3. Intracavity transient absorption effect

Besides the energy transfer between donor and acceptor molecules, the photon energy of laser emission in a microspherical cavity can be transferred to transient species; transient absorption process, which is also expected to influence the dynamics of microspherical lasing. Since the transient absorption can be adjusted by the intensity of excitation light, the quality factor of a microspherical cavity can be optically controlled; optical Q-switching. Hence, the resonance peak intensity and the pulse shape of microspherical lasing can be varied by irradiation of an excitation pulse.

Figure 10 shows emission spectra of a 46-μm PMMA particle doped with RhB and 9,10-diphenylanthracene (DPA) [14]. Lasing of RhB was induced by a second harmonic pulse (532 nm) from a Q-switched mode-locked Nd:YAG laser (Figure 10(a)). DPA absorb neither 532 nm laser light nor emission of RhB. When DPA was excited by the third harmonic pulse (355 nm), the transient absorption was induced, whose spectrum (dotted curve in Figure 10(a)) overlapped with the emission band of RhB. Hence, the intensity of laser emission from the microparticle was appreciably reduced, as shown in Figure 10(b). By cutting off the 355 nm excitation light, the laser emission intensity immediately recovered. Furthermore, it was confirmed that the intensity of microspherical lasing increased as decreasing the excitation intensity.

On the other hand, the spontaneous emission intensity exhibits no appreciable difference between the two spectra in Figures 10(a) and (b). This can be explained by the fact that the optical path length in the microparticle is no more than tens micrometer so that the transient absorbance is negligibly small under the present experimental conditions. However, the laser emission goes round the particle so that the effective path length reaches millimeter or centimeter order, which is hundreds or thousands times longer than that of the straight path. Hence, the transient absorption is enough probed not by fluorescence but by laser emission.

Based on the theory of intracavity laser absorption spectroscopy [15], the degree of the transient absorption enhancement in microspherical lasing is given by

$$\xi = (g_0 / L) / (g_0 - L), \tag{3}$$

where g_0 represents the unsaturated gain, and L is the intrinsic intracavity loss per round. ξ is called as an enhancement factor, which is ~100 for the microspherical lasing observed in the present experiment. If transient absorption per round (ΔL) is much smaller than the intrinsic loss L, the change in the laser emission intensity is approximately given by

$$(I_0 - I) / I_0 = \xi \Delta L, \tag{4}$$

where I and I_0 are the emission intensities of microspherical lasing with and without irradiation of 355-nm excitation light, respectively. We have confirmed a linear relationship between $(I_0-I)/I_0$ and transient absorption under the condition of $(I_0-I)/I_0 < 0.4$.

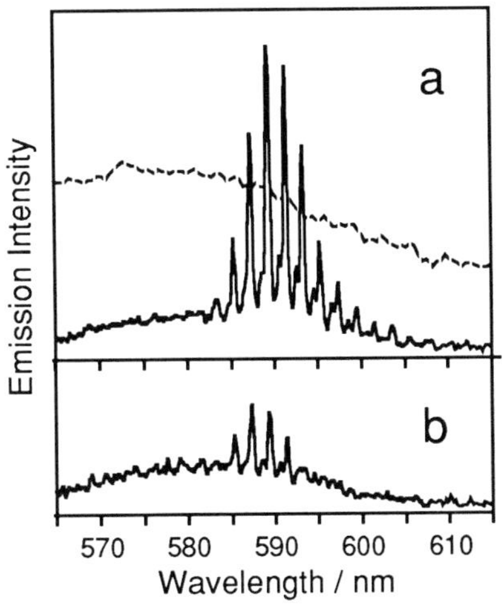

Figure 10. Emission spectra of a RhB-DPA-PMMA particle (46 μm) in water. (a) Lasing of RhB was induced by a 532 nm laser pulse, and (b) quenched by transient absorption of DPA which was induced by a 355 nm laser pulse.

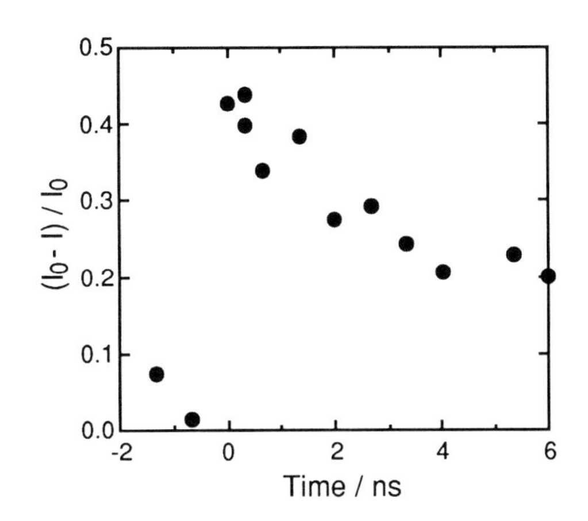

Figure 11. Decay curve of transient absorption of DPA in a PMMA particle (31 μm) observed by an intracavity spectroscopy system.

Based on this linear enhancement, high-sensitive absorption measurements can be realized, which is indispensable for studies on absorption dynamics of micrometer-sized particles. In addition, the picosecond pulse of the microspherical lasing makes it possible to extend the technique to time-resolved spectroscopy. The 532 nm pumping pulse for lasing is optically delayed to the 355 nm excitation pulse, so that the temporal variation of transient absorption can be obtained with the picosecond time-resolution.

Figure 11 shows a decay curve of transient absorption observed in a RhB-DPA-PMMA particle (31 μm). The decay time constant of this curve was estimated to be

7 ns., which almost agreed with the lifetime of DPA reported elsewhere [16]. On the other hand, a PMMA film containing RhB and DPA with the same concentrations as those of the microparticle exhibited no appreciable signal of transient absorption. These results clearly demonstrated the high sensitivity and the high temporal resolution of the present technique.

3.4. Photon tunneling between lasing and nonlasing particles

The Mie-Debye light scattering theory indicates that the electric field formed by the whispering-gallery mode resonance is localized at the vicinity of the particle-surrounding boundary, that is, surface wave. Hence, the intracavity transient absorption measurement mentioned in the previous section can be applied for analyzing photochemical dynamics at surfaces and interfaces. In addition, the microspherical resonance forms an evanescent field around a particle, so that lasing dynamics is also sensitive to changes in absorption and the refractive index of the surroundings just outside of the particle. Indeed, recent theoretical studies on light scattering properties of two-closely positioned microspheres predict that the high-Q resonances in a single particle are extremely broadened by interparticle coupling effects [17]. We have employed a multibeam laser scanning manipulation technique to elucidate the optical interaction between lasing and nonlasing particles [7].

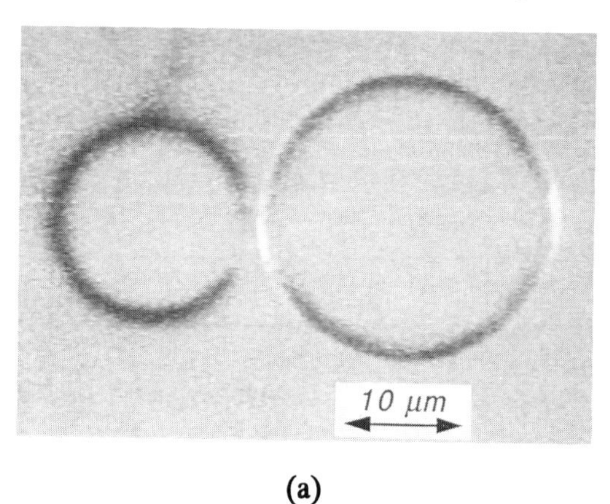

(a)

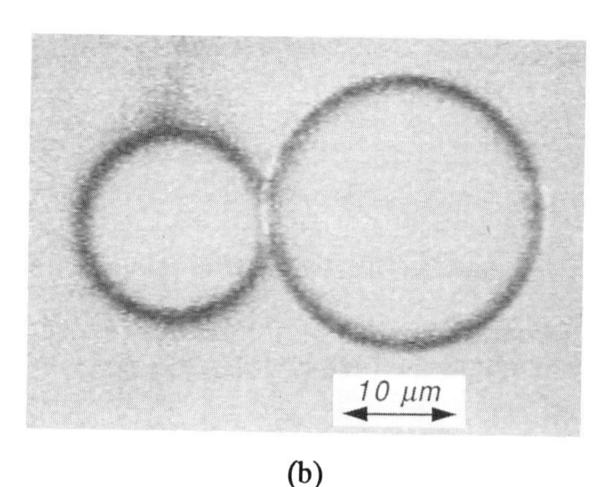

(b)

Figure 12. Optical interaction between lasing and nonlasing RhB-PMMA particles (25 and 17 μm, respectively). (a) The right particle was excited in a limited region by a picosecond pulse so that lasing was observed in the horizontal direction in the photograph. (b) Quenching of lasing by photon tunneling with a nonlasing particle (left).

Two RhB-PMMA particles (25 and 17 μm) were independently trapped and manipulated by two laser beams, as shown in Figure 12. Only a limited region of the right edge in the particle (right in Figure 12(a)) was pumped, so that laser emission could be observed in two narrow parts at the left and right sides. The cavity for lasing might be formed along a circular path through these two spots. In particular, the path perpendicular to the plane of the photograph would mainly contribute to the observed laser emission. When the distance between the two particles is large enough (> 1 μm), the nonlasing particle (left in Figure 12(a)) does not influence lasing of the other particle. If the nonlasing particle is positioned at the close vicinity (<100 nm, noncontact) to the lasing particle by optical manipulation, laser oscillation in the RhB-PMMA particle is suppressed as seen in Figure 12(b), which may be ascribed to reduction of the quality factor of the relevant modes for lasing. The results will be understood as photon tunneling from the lasing to nonlasing particles through the evanescent field.

Although we have not determined the absolute distance between the two particles in Figure 12(b), our rough estimation suggests that the distance between two particles can be controlled within 10 nm by optical trapping. Therefore, three-dimensional manipulation of a lasing particle will play an important role in studies on the mechanism of a laser oscillation process in a microspherical cavity and on the distance dependence of photon tunneling between particles. We also expect that lasing microparticles can be utilized as a probe for a photon-mode scanning tunneling microscope or a near-field scanning optical microscope. By scanning a lasing particle on a sample substrate, the variation of the emission intensity and the lasing spectrum provide information on microstructures of the sample and spectroscopic properties at each position. If a very small particle is attached to a microparticle to be lased by the reported technique, the spatial resolution of the method as a spectroscopic microscope will be further improved.

4. CONCLUSION

We have described the spectral and temporal properties of the microspherical lasing and introduced the characteristic photochemical processes occurring in a microparticle. The high-Q resonance in a microspherical cavity can be expected to induce various interesting phenomena different from bulk behaviors, which can be utilized for constructing highly efficient photoconversion system. In addition, manipulated lasing microparticles can be used as a small movable light source for observing chemical reactions occurring in the photoconversion system and for controlling their processes.

ACKNOWLEDGMENTS

The authors express their sincere thanks to Prof. H. Masuhara and Dr. N. Kitamura, and Mr. M. Koshioka of our project for their collaboration.

REFERENCES

1. P.W. Barber and R.K. Chang ed., *Optical Effects Associated With Small Particles*, World Scientific, Singapore, 1988.
2. A. Ashkin and J. M. Dziedzic, Phys. Rev. Lett., 38 (1977) 1351
3. H.M. Tzeng, K.F. Wall, M.B. Long, and R.K. Chang, Opt. Lett., 9 (1984) 499.
4. H.-B. Lin, A.L. Huston, B.L. Justus, and A.J. Campillo, Opt. Lett., 11 (1986) 614.
5. A. Biswas, H. Latiffi, and R.L. Armstrong, and R.G. Pinnik, Opt. Lett., 14 (1989) 214.
6. H.-B. Lin, J.D. Eversole, and A. Campillo, J. Opt. Soc. Am. B, 9 (1992) 43.
7. K. Sasaki, H. Misawa, N. Kitamura, R. Fujisawa, and H. Masuhara, Jpn. J. Appl. Phys., 32 (1993) L1144.
8. K. Kamada, K. Sasaki, H. Misawa, N. Kitamura, and H. Masuhara, Chem. Phys. Lett., 210 (1993) 89.
9. M. Kerker, *The scattering of Light and Other Electromagnetic Radiation*, Academic Press, San Diego, 1969.
10. H. Misawa, R. Fujisawa, K. Sasaki, N. Kitamura, and H. Masuhara, Jpn. J. Appl. Phys., 32 (1993) 788.
11. A. Yariv, *Introduction to Optical Electronics*, Holts, Rinehart and Winston, New York, 1985.
12. L.M. Folan, S. Arnold, and S.D. Druger, Chem. Phys. Lett., 11 (1985) 322.
13. H. Misawa, R. Fujisawa, K. Sasaki, N. Kitamura, and H. Masuhara, in preparation.
14. K. Kamada, K. Sasaki, H. Misawa, N. Kitamura, and H. Masuhara, in preparation.
15. S.J. Harris, Appl. Opt., 23 (1984) 1311.
16. D.J.S. Birch and R.E. Imhof, Chem. Phys. Lett., 32 (1975) 56.; S. Hirayama, H. Yasuda, M. Okamoto, and F. Tanaka, J. Phys. Chem., 95 (1991) 2791.
17. K. A. Fuller, Appl. Opt., 30 (1991) 4716.

MICROCHEMISTRY
Spectroscopy and Chemistry in Small Domains
Edited by H. Masuhara et al.
1994 Elsevier Science B.V.

Near-field optics: Chemical sensors, photon supertips and subwavelength spectroscopy

Weihong Tan[a,b], Duane Birnbaum[a], Craig Harris[b], R. Merlin [c], B. Orr[c], Zhong-You Shi[a], Steve Smith[c], Bjorn A. Thorsrud[b] and Raoul Kopelman[a,c] *

Department of Chemistry[a], and Toxicology Program, Department of Environmental & Industrial Health[b] and Department of Physics[c], The University of Michigan, Ann Arbor, Michigan 48109, USA

The promise of near-field optics for microscopy, spectroscopy, biochemical sensing, nanofabrication and monitoring of electronic devices and optical memories is of much current interest. Nanometer fiber-optic chemical sensors have been fabricated by a near-field photopolymerization technique and applied for pH measurements in microdomains, single cells and early stage embryos. These robust and durable sensors have tips that are a thousand times smaller than previous state-of-the-art devices, reduce the required sample volume by a factor of a billion and the response time by a factor of a thousand or better. The submicrometer tip diameter is as small as 100 nm, contains as little as 1000 active molecules, detects as few as 1000 hydrogen ions, has a response time of less than 50 ms and a precision of 0.01 pH units. It can be used for spatially resolved chemical measurements, with a resolution of 0.1 μm.

Near-field, spatially resolved luminescence spectra have been obtained on mixed crystals, composite polymeric materials and on multiple quantum wells. They reveal submicrometer structures with a spatial resolution of 40-100 nm. In addition, near-field optical measurements provide several image contrast effects.

We have also nanofabricated subwavelength light sources as small as 40 nm in diameter and exciton sources that are probably smaller. These include micropipette tips with nanocrystals as well as polymeric matrices. They also include fiber-optic tips with polymeric or crystalline tips. Further miniaturization is achieved by the addition of crystalline or polymeric supertips. The near-field photo-nanofabrication has also been used for stepwise nanofabrication of optical probes, growing supertips onto tips. This is an ongoing effort to produce the ultimate photon or exciton nano-tip, consisting of a single active molecule or chromophore. Furthermore, we discuss the near-field effect in photofabrication and in optical measurements.

1. INTRODUCTION

The rapidly developing fiberoptic sensors and scanning probe microscopies demand more sophisticated probe and sensor fabrication. The realization of better resolution by subwavelength light sources has led to the concept of Near-Field Optics (NFO). NFO makes it possible to by-pass the optical diffraction limit ("uncertainty principle") through the use of a small light source which effectively focuses photons through a tiny aperture that may be as small as $\lambda/50$ [1-2]. Near-field Scanning Optical Microscopy (NSOM) is another form of scanning probe microscopy that has recently been developed and is generating considerable interest [1-5]. NFO has enabled researchers to examine optically a variety of specimens without being limited in resolution to one half the wavelength of light.

There are several NFO applications: NSOM, Near-field Scanning Optical Spectroscopy (NSOS), Molecular Exciton Microscopy (MEM) and Near-field Fiber Optic Chemical and Biological Sensors (FOCS/FOCBS). The development of specialized subwavelength optical probes is critical for all of them. We have applied NFO in nanofabrication and have prepared different types of subwavelength optical probes, both passive and active, with or without specific chemical or biological sensitivity, for NSOM, MEM, NSOS and FOCS by using micropipettes and nanofabricated optical fiber tips [2-4]. Near-field optics has played key roles in the nanofabrication of subwavelength optical probes and chemical and biological sensors and their operation in biochemical analysis. Several techniques have been developed for the preparation of subwavelength optical and excitonic probes. Specifically, a new and controllable nanotechnology, photo-nanofabrication, based on nanofabricated optical fiber tips and near-field photopolymerization, has been developed to make nanometer optical fiber chemical and biological sensors with extremely fast response times. The same technique can be used to prepare smaller light and exciton probes for NSOM, NSOS and MEM. The miniaturized optical fiber biochemical sensors have shown their great potential application in *in-vivo* chemical and biomedical measurements in intact biological units and single cells. We have also demonstrated the ability of NSOS as a technique for obtaining high spatial resolution in a spectral dimension by studying tetracene and perylene doped polymer films as well as quantum wells.

In this chapter, we are going to describe some new developments in near-field optics. Photo-nanofabrication, optical nanoprobes, active light sources, supertip development, NSOS and subwavelength optical fiber biochemical sensors will be covered. We also discuss the near-field effect in photofabrication and in optical excitation. We believe that NFO promotes both processes, and there is a difference between NFO and conventional optics (far-field optics) in the molecular cross-section, which makes NFO more attractive.

2. NEAR-FIELD OPTICS AND OPTICAL NANOPROBES

2.1 Near-field optics

Conventional ("far-field") optical techniques are based on focusing elements such as a lens. This leads to the "diffraction limit" of about $\lambda/2$ (half the wavelength). The realization of near-field optics is through optical nanoprobes. The principle underlying this concept is schematically shown in Figure 1. The near-field apparatus consists of a near-field light source, sample and far-field detector. To form a subwavelength optical probe, light is directed to an opaque screen containing a small aperture. The radiation emanating through the aperture and into the region beyond the screen is first highly collimated, with dimension equal to the aperture size, which is independent of the wavelength of the light employed. The region of collimated light is known as the "near-field" region. The highly collimated emissive photons only occurs in the near-field regime. To generate a high resolution image, a sample has to be placed within the near-field region of the illuminated aperture. The aperture then acts as a subwavelength sized light probe which can be used as a scanning tip to generate an image. That is why this optical microscopy is called near-field scanning optical microscopy [1, 5].

Unlike scanning tunneling microscopies (STM) or atomic force microscopies (AFM), imaging in NSOM is via the interaction of light with the surface by either a simple refraction/reflection contrast, or by absorption and fluorescence mechanisms. The advantages of NSOM are its non-invasive nature, its ability to look at non-conducting and soft surfaces, and the addition of a spectral dimension, the latter of which does not exist in either STM (room-temperature) or AFM. This potential for extracting spectroscopic information from a nanometer-sized area makes it particularly attractive for biomedical research and materials science. The resolution of a NSOM image is limited by the size of the light probe. We have designed several techniques to miniaturize optical and excitonic probes for NSOM, MEM, NSOS and FOCS.

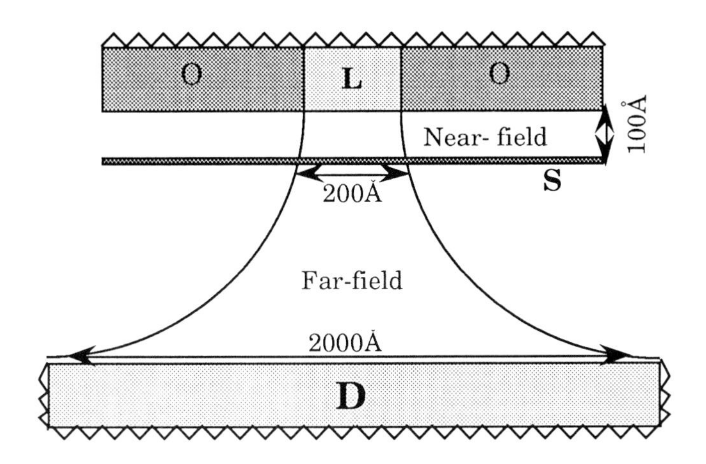

L: Active or passive light source;
O: Opaque material;
S: Sample (support not shown);
D: Far field detection system(e.g. lens).

Figure 1. Near-field optics.

2.2 Nanofabrication process

There are two major probes used in NFO: metal coated glass micropipettes and nanofabricated optical fiber tips [1, 4, 5]. Optical fiber tips and micropipettes are easily fabricated to sizes of approximately 50 nm, and the smallest nanofabricated optical fiber tip reported to date is about 20 nm [1]. The fabrication of miniaturized optical probes has keyed the development and application of NFO in a wide variety of fields.

The first step in the probe nanofabrication process is the pulling of micropipette and fiberoptic tips of appropriate size and shape. The second step is the metal-coating of such tips. This is followed by crystal (or polymer) growing if active optical or excitonic probes are desired. Here we illustrate the nanofabrication of optical fiber tips. Very similar techniques have been applied in the fabrication of micropipettes [1, 8].

Optical fiber tips have been used in many areas [1, 2, 4, 6] and can be fabricated either by heating and stretching or by chemical etching. Our apparatus for fiber tip pulling consists of a P-87 Micropipette Puller from Sutter Instrument Co. and a 25 watt CO_2 infrared laser from Synrad Co. The CO_2 infrared laser beam replaces the electric filament in the puller to heat the optical fiber for the pulling process. The laser beam is reflected by a mirror and directed to heat the optical fiber which is fixed on the puller. The details of the pulling set-up and procedures can be found in a series of references [1, 4, 7, 8]. By using appropriate program parameters and laser power, optical fibers can be tapered to subwavelength diameters. Most of the tips used in fiber sensor fabrication and nanometer light and exciton sources are from 0.05 μm to about 0.5 μm diameter [2, 4, 7]. After pulling, the optical fiber tip is coated with aluminum by vapor deposition to form a small aperture. The procedure of vacuum deposition of metals is well known but far from trivial. A specially built high vacuum chamber is employed for coating these pulled fiber tips: only the fiber tip sides are coated with aluminum, leaving the end face as a transmissive aperture [1, 2]. To make it into a light source, a visible or UV laser beam is coupled to the opposite end of the pulled tip. This probe delivers light very efficiently since most of the radiation is bound to the core until a few micrometers away from the tip. A randomly chosen 0.2 μm optical fiber probe gives 10^{12} photons per second [7].

Using the same puller, glass micropipettes have been pulled with different diameters. In order to pull short shank micropipettes, we developed a multistep pulling program [9]. In multistep pulling, the general idea is to use initial puller settings to produce a rapid taper over the desired distance, allowing them to operate over one or more pulling cycles, and to form an ultrafine tip during the final pulling cycle. In a multistep program the last pulling step is still the most crucial one. As in a one step pulling program, after inserting the capillary tube and programming the puller, the remainder of the process was controlled by the microprocessor. We constructed different multistep programs for pulling both borosilicate and aluminosilicate glass tubes. By varying the parameters in each step and the glass type and tube dimensions, a wide variety of micropipette shapes and sizes were achieved. According to the SEM micrographs, the smallest tips used in the crystal growth experiments were about 500 Å inner diameter with a usable shank. Tips pulled by multistep pulling programs showed great promise for crystal growth inside them. Micropipettes have to be metallized on their outsides before being used as subwavelength light probes. The same apparatus is used for the coating of micropipettes as described for fiber tips.

2.3 Optical nanoprobes

Near-field optical nanoprobes can be classified into three different kinds: passive optical probes, such as coated micropipettes or small holes on a screen [2, 5], semi-active light sources, such as optical fiber tips [1, 4, 6], and active light sources, such as nanometer crystal light sources [3, 8]. We have fabricated all three kinds of light and exciton sources. Both optical fiber tips and micropipettes have been used as optical nanoprobes. Micropipettes have been widely used in biology and electrochemical sensors [3, 10]. They were adapted for NSOM as subwavelength optical probes [3, 5, 8]. The problem in using micropipettes is the conflict of smallness and the light intensity (see below). Thus they gained limited usage in NSOM after their initial application in NFO. Compared to a hollow micropipette tip, a nanofabricated optical fiber tip is a "semi-active" photon tip [1, 2, 4, 6]. Generally it is orders of magnitude brighter, easily coupled to an optical source and at least as mechanically sturdy as a micropipette. It is interesting to notice that the top of a fiber tip is really very resistant to breakage. The photochemical stability for optical fiber tips is excellent and under very intense illumination it is the heat that damages the aluminum coating at the tip. Both probes have been made around 500 Å in diameter without difficulties in applications as light sources.

3. PHOTO-NANOFABRICATION AND SUBWAVELENGTH BIOCHEMICAL SENSORS

3.1 Photo-nanofabrication

NFO enables a revolution in nanofabrication techniques. We have developed a novel nanofabrication technique based on NFO: photo-nanofabrication, a new and controllable nanofabrication technology. It can produce nanometer sized optical and exciton probes with or without a specific chemical or biological sensitivity. For probes with a specific chemical sensitivity, they are automatically FOCS. Using the near-field optics principle, photo-nanofabrication controls the size of the luminescent material grown at the top of a light transmitter or conductor, such as a micropipette or optical fiber tip, by photochemical reactions. These reactions are initiated and driven by an appropriate wavelength of light. The luminescent material is formed (synthesized) only in the presence of light and is "bonded" only to the area where light is emitted. We notice that the key to photo-nanofabrication is a near-field photochemical reaction, in which the electromagnetic waves of the light sources are mapped by the photochemical process. Thus the size of the luminescent probe is defined by the light emitting aperture and is independent of the wavelength of the light used to promote the chemical reaction. The photochemical reaction only occurs in the near-field region [4, 7], where the photon flux and the absorption cross-section are the highest (see below).

To illustrate the principle of photo-nanofabrication, we here describe the near-field photopolymerization process, by which submicrometer optical fiber pH sensors have been prepared. After silanization of the metal coated fiberoptic tip, the photopolymerization is controlled by the light emanating from the near-field light source. The size of the light source and the near-field evanescent photon profile control the size and shape of the immobilized photoactive polymer [7]. The pH sensors are prepared by incorporating fluoresceinamine derivative, acryloylfluorescein (FLAC), into an acrylamide-methylenebis(acrylamide) copolymer that is attached covalently to a silanized fiber tip surface by photopolymerization, as shown in the following polymerization scheme. The polymerization process is schematically shown in Figure 2. The size of the polymer grown on the aperture of the optical fiber tip is equal to or smaller than that of the aperture. By using multi-step photochemical synthesis, we have further miniaturized optical and exciton probes to sizes much smaller than the sizes of the original light conductors. Also, by using a multi-dye system for photochemical synthesis, we have prepared multi-functional probes with extremely small size [11]. The ultimate goal of our photo-nanofabrication technique is to produce optical, exciton and sensor probes with molecular size for NSOM, NSS, MEM and FOCS by controllable molecular engineering.

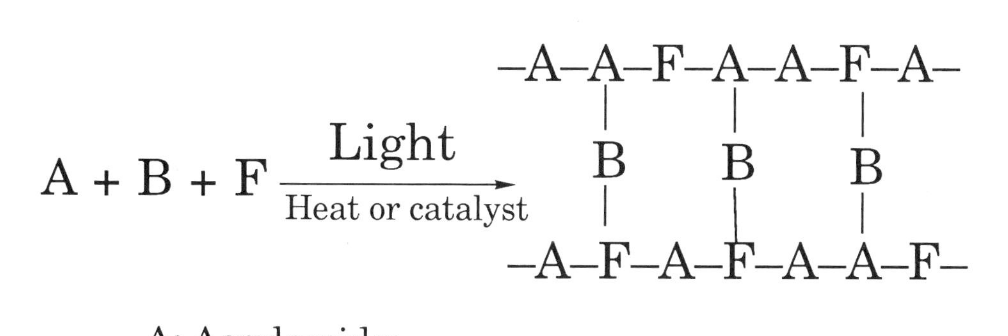

A: Acrylamide;
B: N, N-methylene (bis) (acrylamide);
F: Acryloylfluorescein.

Polymerization Scheme

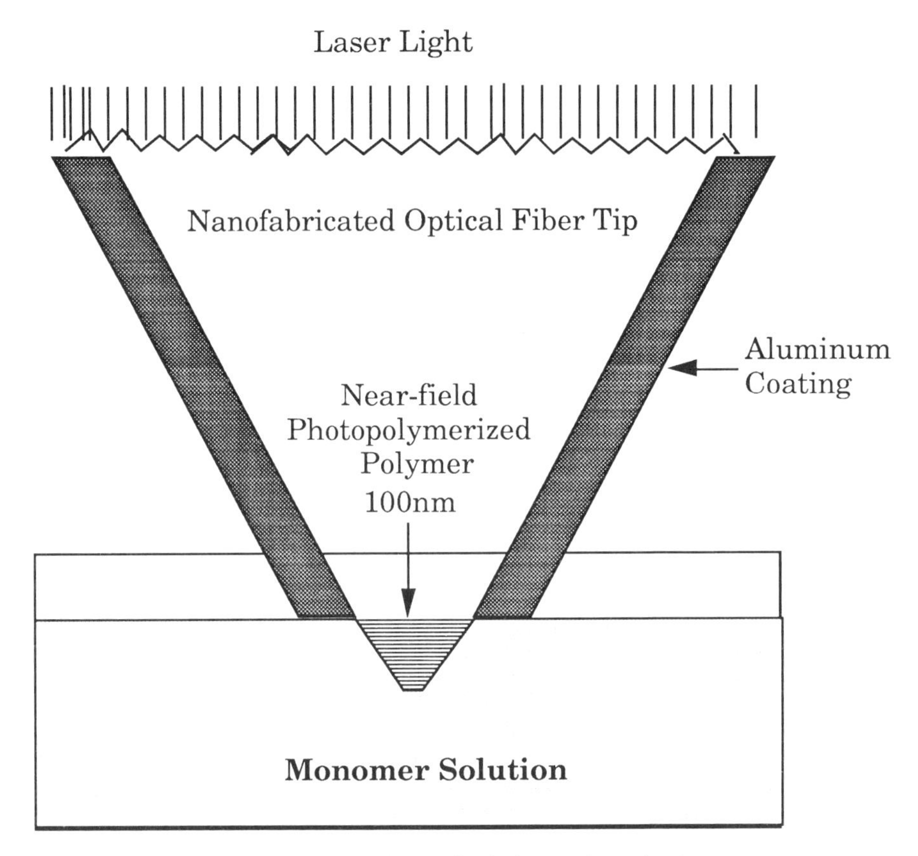

Figure 2. Schematic drawing of near-field photopolymerization.

3.2 Near-field biochemical sensors

Photo-nanofabrication has enabled us to prepare subwavelength optical probes. One of the most successful applications of photo-nanofabrication to date is the preparation of submicrometer FOCS. We have taken Scanning Electron Microscopy (SEM) micrographs of the submicrometer optical fiber pH sensors, and they clearly show that the sensor dimension is smaller than the original light conductor and appears to become even smaller as the polymer grows. For the fabrication of the submicrometer sensor, the polymerization reaction temperature, polymerization time, and laser coupling efficiency had to be optimized in order to grow submicrometer sized polymers on the activated fiber tip surfaces. We produced many different size chemical sensors, down to about 0.1 μm.

3.3 Sensor response times, sensitivity and stability

The miniaturization of the sensor results in very fast response times. The size of the sensor is between 0.1 to 1 μm, and no mechanical confinement is used. Thus the analytes have immediate access to the dye on the sensor tip. This gives our sensors the shortest response times among any reported optical fiber sensors. The response times are well below 50 ms [4].

The submicrometer pH sensor has a very low detection limit. We notice that the smallest volume that gave definite pH measurements had only a few thousands of hydrogen ions (at pH=8) [7]. The submicrometer pH sensor also has reasonable optical stability even when a high laser power is used. We established that removal of oxygen and/or efficient stirring do significantly reduce the photobleaching. Furthermore, the bleaching is practically imperceptible at the low laser power used under standard operating conditions. With the above performance values and considering the high fluorescence intensity recorded (10^5 cps) [7], we believe that it should be easy to downsize the sensor to tens of nanometers.

The feasibility of the miniaturized optical fiber sensor [4] has been tested with unknown chemical and biological samples. A few different kinds of biological samples were tested. Most notable are the *in-vivo* static and dynamic pH measurements on single and living rat conceptuses. The ultramicrofiberoptic pH sensor is shown to be able to discriminate pH changes of less than a tenth of a pH unit over the range of one pH unit above and below the physiologic pH of 7.4. The pH measurements were obtained in real time, on a single, intact, viable rat conceptus under conditions of environmental change and direct chemical exposure. The insertion of the ultrasmall sensor through the visceral yolk sac appeared to cause no damage to or leakage from the involved tissues. This demonstrates the advantage of an essentially non-invasive approach, as compared to conventional means, necessitating the disruption of large numbers of conceptuses to obtain less sensitive and indirect measurements. Conceptuses of ascending developmental age undergo an acidification of the external embryonic fluid. The ability of ultramicrofiberoptic sensors to measure pH changes, in real time, in the intact rat conceptus, demonstrates their potential applications for dynamic analysis in small multicelluar organisms. Working sensor probe dimensions and response characteristics also make this approach feasible for use in single cells. The application of this novel technology to studies of developmental regulation, pharmacokinetics, toxicology and physiology will provide valuable spatial and temporal information not heretofore available using conventional techniques.

4. NEAR-FIELD SCANNING OPTICAL SPECTROSCOPY

Near-field scanning optical spectroscopy (NSOS) [2, 13] is based on NSOM. It basically adds one more dimension, spectroscopy, to NSOM and can be used to obtain spectrum of various nanostructures, such as quantum wells. NSOS inherits all the advantages of NSOM: its non-invasive nature, its ability to look at non-conducting and soft surfaces, and the addition of a spectral dimension. The ability to obtain spectroscopic information with a nanometer-sized resolution makes NSOS very promising for a wide variety of scientific researches. Examples include the detection of fluorescent labels on biological samples and isolating local nanometer-sized heterogeneities in microscopic samples. Here we present the potential for addressing spectroscopy on a submicrometer scale. We have studied systems of tetracene and perylene doped in PMMA as well as microscopic crystals in order to demonstrate that nanoscopic inhomogeneities can be detected in what might at first appear to be a homogeneous sample. The eventual goal is to obtain spectroscopic information with a spatial resolution of nanometer or even molecular sizes.

In NSOS, an optical probe with an emissive aperture that is submicrometer in size is positioned such that the sample is within the near-field region. With piezoelectric control of the fiber tip, the tip can be accurately positioned over a fluorescing region of the sample and a spectrum recorded. Excitation of the sample can be either external with detection through the fiber tip or with the fiber tip itself and subsequent detection of the emitting photons. This means that it is not necessary for the sample to be of any particular thickness or opacity, however it should be a relatively smooth surface. The optical probes used in NSOS are the same nanometer-sized optical fiber light sources for sensor preparation [4, 7]. The experimental apparatus for measuring fluorescence spectra with high spatial resolution is shown schematically in Figure 3. The 442 nm line from a He-Cd laser or one of several lines from an argon ion laser is coupled to an optical fiber with a high precision coupler. The fiber tip is mounted in a hollow tube of piezoelectric material which is positioned by the usual STM control electronics. The sample (deposited on a glass slide) is mounted on the near-field microscope such that it is perpendicular to the exciting tip and the entire apparatus rests on the base of an inverted frame microscope with a reflected light fluorescence attachment. Excitation of the sample via the fiber tip generates fluorescence which is collected by an objective, filtered (to remove laser light) and collimated before exiting the microscope. The fluorescence is then focused onto an optical multichannel analyzer (OMA). The data is then collected and analyzed on a computer.

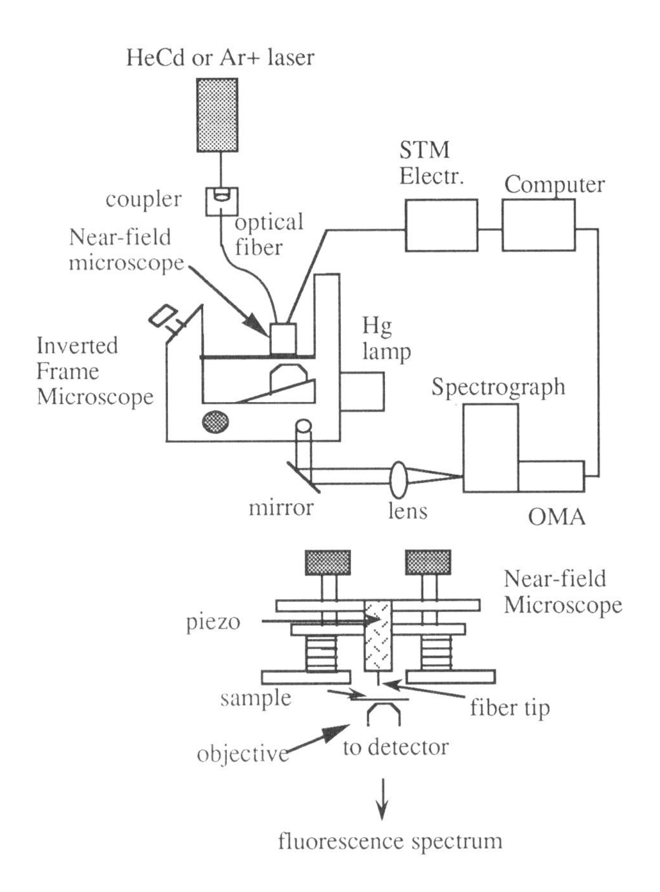

Figure 3. Schematic of Experimental Apparatus used for NSOS.

Films of a 1.0 wt.% mixture of tetracene in polymethylmethacrylate were prepared by evaporation from dichloromethane. The film thickness was approximately 200 to 300 μm and it appeared optically clear. Thin films (<10 μm) of 15 wt. % perylene (Aldrich) in PMMA were prepared by spin-coating a dichloromethane solution on a glass slide. Microcrystalline samples of perylene were grown from benzene solution on a glass slide.

4.1 Spectra of tetracene/PMMA film
Tetracene/PMMA films examined under the fluorescence microscope show microaggregates of tetracene with an average size of ~10 μm embedded in the polymer [14]. The background fluorescence from the film appears greenish-yellow and is presumed to be from either isolated molecules or crystals which are of a size smaller than what can be resolved with the conventional microscope. What is surprising about the aggregates is that these fluorescence ranges in color from green to yellow to red. Thus, the macroscopic fluorescence spectrum obtained with Hg lamp excitation is very broad, containing contributions from the background and all colors of aggregates. With NSOS it is then easy to excite a specific aggregate and record its fluorescence spectrum.

4.2 Nanoscopic perylene crystal spectra
Similar results are obtained for microcrystals of perylene which were grown on a glass slide from benzene solution. Perylene is known to crystallize in two distinct forms [15], α form and ß form crystallizes. Under the microscope, α and ß perylene are easily distinguished as bright green (ß) and bright yellow (α) crystals. To demonstrate spatial resolution by monitoring spectral features, samples of spin-coated perylene doped PMMA films were studied. After spin-coating, these films were found to have regions of sharp color contrast easily observable under the microscope. That is, aggregates which fluoresce yellow are found next to blue-green fluorescent regions of the film which are presumably due to the isolated molecule. Figure 4 shows the result of bringing a fiber optic tip within the near-field region of the sample and monitoring the fluorescence spectrum as the tip is scanned across such an interface. The spectra were obtained for a 0.5 μm tip which was scanned across a blue-green region towards a rather large aggregate (~5 μm) embedded in the film. The increments of only 400 Å were sufficient to measure a noticeable change in the spectrum. Note that the blue-green intensity does not change as significantly since it is surrounding the aggregate and therefore contributes a constant background fluorescence.

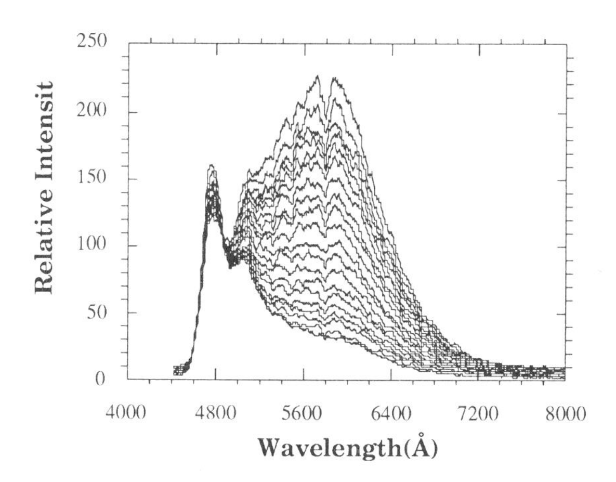

Figure 4. NSOS spectra of a perylene doped PMMA spin-coated film. The size of the fiber tip was a 0.5 μm tip positioned in increments of 400 Å.

That a measurable change in fluorescence spectra can be obtained with lateral movements of an exciting fiber tip that are one order of magnitude less than the diameter of the tip is highly encouraging in terms of our eventual goal of achieving molecular resolution. If different fluorescent species are separated by a distance of only 10 Å, then it is conceivable that a light source of only 100 Å will be sufficient to resolve this separation provided enough photons are available. Of course, interpretation of unknown heterogeneities will be difficult if the exciting tip is not of a comparable size to the fluorescing region. If the tip is much larger than this region, then it would be difficult to quantify the number of such regions compared to their size within the near-field.

4.3 Microscopic spectra of multiple quantum wells

Near-field spectroscopy of GaAs/AlGaAs multiple quantum wells in the collection mode has been obtained at low temperatures (150K). Figure 5 shows the luminescence spectra taken both by conventional optics and by collection mode near-field spectroscopy. The sample is an n-i-n structure and the luminescence spectra were taken with the sample biased at 6.0 volts. The tip size was approximately 800 Å. The excitation was at 6748 Å.

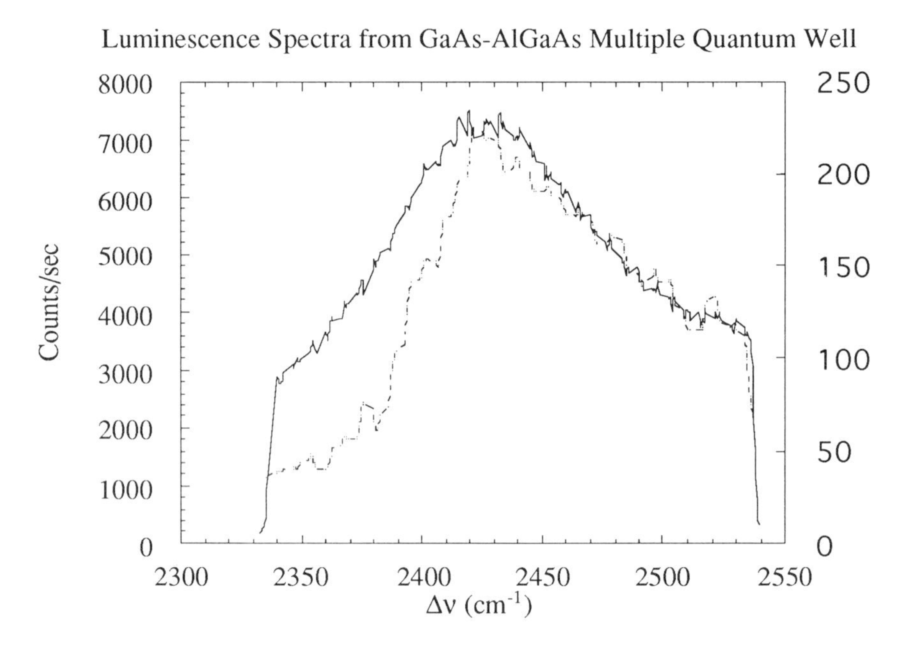

Figure 5. Luminescence of GaAs-AlGaAs multiple quantum well taken at 150K, 6748 Å excitation. Solid line spectrum taken with conventional optics, dashed line spectrum taken in collection mode by near-field optical spectroscopy.

We have demonstrated the ability of NSOS as a technique for obtaining high spatial resolution in a spectral dimension by studying different microscopic and nanoscopic samples as well as quantum wells. In the heterogeneous samples it is possible to isolate heterogeneities on a nanometer-size scale. Measurable changes in the fluorescence spectra are obtained for lateral increments of approximately 1/10 the size of the aperture of the exciting fiber tip. This implies that it may be possible to differentiate heterogeneities at molecular sizes (~10 Å), with a light source 100 Å in diameter.

5. ACTIVE LIGHT SOURCES AND MOLECULAR EXCITON MICROSCOPY

In NSOM, the size of the light source determines the resolution of the imaging, provided that it can be scanned in the near field region. At nanometer size, the number of photons emitted from coated micropipettes or nanofabricated fiber tip is so low that it becomes very difficult to transmit, in the transmission mode NSOM [1, 5], or excite a large enough number of fluorophores, in the fluorescence mode NSOM [1, 5], such that their subsequent transmission or emission can be detected. Of course, then it is impossible to detect a single molecule. Alternative light sources have been proposed and prepared for obtaining molecular resolution, most notably, the molecular exciton source [3, 8, 9]. Such a light source incorporates a nanometer size organic crystal at the end of a pulled micropipette with an inner diameter of a few nanometers. Excitation of the crystal then results in the creation of excitons whose travel through the crystal is only limited by the crystal size, which potentially could be of a molecular dimension. Excitation of a sample would then proceed through an energy transfer mechanism or other near-field operation, which may be orders of magnitude more efficient than emission and subsequent reabsorption by the sample. Therefore we could accomplish NSOM with molecular resolution and sensitivity.

5.1 Active light sources

In NSOM, the smaller the light source, the better the resolution of NSOM. However, the requirements of smallness and intensity are in direct conflict. All these passive light sources are typically apertures letting light through, and when the size of the aperture gets to be significantly below that of a wavelength, most of the light will be diffracted or reflected back, rather than transmitted [8, 16]. This resolution limitation derives from the less than ideal characteristics of the aperture. There are no propagating electromagnetic modes in a subwavelength cylindrical metallic waveguide such as a metal coated micropipette. The least attenuated mode for a round aperture has been found to be the TE_{11} mode for which the energy decays at a rate of approximately

$$E = E_o \exp(-3.62\ L/r) \tag{1}$$

where L is the length of the aperture formed by the metal coating and r is the radius [8, 16]. With a sufficiently rapid tapering of the micropipette, however, this evanescent region can be kept short enough to obtain a fairly large throughput of light. What ultimately limits the resolution is the finite conductivity of the metallic coating around the micropipette. The electromagnetic wave penetrates the coating and decays within it at a finite rate as well, given by [8]:

$$E = E_o \exp(-d/c) \tag{2}$$

where d is the depth of the penetration and c is the extinction length of the metal. When the attenuation due to the waveguide effect exceeds the attenuation in the metal the contrast between the aperture and surrounding medium becomes insufficient for super-resolution applications. The metal with the largest opacity in the visible region is aluminum for which d = 500 Å [17] when the wavelength is 5000 Å. For instance, for a hole in a metal plate, with an apertures of 500 Å or less, the intensity goes down super-exponentially with decreasing aperture size. This occurs because the metal surrounding the hole must be at least about 500 Å thick [17] in order to be opaque enough to define a hole.

To solve this problem, an active light source, based on molecular excitons, for NSOM and MEM was proposed by Lieberman et al [3]. The basic point is to convert a passive light source into an active light and exciton source. The micropipette approach has been extended in several ways. (a) The tip has been filled with a photoactive material [3, 18]. (b) The "pre-tip" region is specially bent, to double as a "force probe" [19]. These extensions have kept the micropipette tip in competition with the optical fiber tip. Very recently, we prepared active light sources with nanofabricated fiber tips [8, 20] and developed supertips for both NSOM and MEM [2, 8, 9].

The solution to the severe loss of light problem is to use the energy packaging capabilities of certain materials to circumvent the boundary problem of the edge of the aperture. According to fundamental understanding of energy propagation in materials, excitation can be confined to molecular and atomic dimensions under appropriate conditions [21, 22]. Using this property of materials, one can develop a subwavelength light or exciton source by growing a suitable crystal within the subwavelength confines of a micropipette. With this approach, energy can be guided directly to the aperture at the tip instead of being allowed to propagate freely in the form of an electromagnetic wave. Such a material can be excited through an electrical or radiative process to produce an abundance of excitons that allows light to be effectively propagated through the bottleneck created by the subwavelength dimensions of the tip near the aperture. The excitons can be generated directly at the tip or within the bulk of the material and allowed to diffuse to the tip via an excitonic (electric dipole, Förster) transfer [21, 22]. In either case, in a suitable material these excitons will then undergo a radiative decay producing a tiny source of light at the very tip of the micropipette. The excitonic throughput is basically independent of the wavelength and is a linear function of the cross-section area of the aperture [8].

5.2 Preparation of active light sources

Active light sources are those light and excitonic sources by the means of luminescent materials. There are two major techniques in the preparation of active subwavelength optical and exciton probes. The first one uses optical fiber probes and bonds luminescent materials covalently to the top surfaces of probes by photo-nanofabrication [4, 7], while the second one uses micropipettes or nanofabricated optical fiber tips to hold crystals or doped polymers at their tips [2, 3, 20]. The nanofabricated optical fiber probes have been used successfully for the preparation of active light and exciton sources. For example, we described above the submicrometer optical fiber pH sensors which are prepared by incorporating fluoresceinamine into an acrylamide-methylenebis (acrylamide) copolymer that is attached covalently to an activated fiber tip surface by near-field photopolymerization [4, 7]. The same technique can be used to prepare other subwavelength active light and exciton sources.

Nanometer crystal light and exciton sources were prepared with organic and inorganic crystals or molecularly doped polymers grown inside the very tip of a micropipette from a solution. We have used a great variety of crystals and doped polymers. The luminescent materials involved are: anthracene, perylene, 9,10-diphenyl-anthracene (DPA), 9,10-dimethyl-anthracene (DMA), pyrene, fluorescein and its derivatives, rhodamine series dyes, various aminoanthracenes, tetracene, DCM and BASF dyes [[9], rubrene, dendrimers [23], uranyl compounds, CsCl, zinc sulfide (ZnS) and cadmium sulfide (CdS) as well as dye doped polymers. These active probes can be excited by different laser lines from UV to visible. The techniques used in the crystal growth inside the micropipette are: crystal growth from solution, from melt, from vapor, from chemical reactions [9] and from other sources. The details of preparation of nanometer crystal light sources have been described previously [2, 3, 9, 20]. The physical sizes of the active light or exciton probes are as small as 0.05 μm, and mostly in the range of 0.1 to 1 μm. The effective size of these active light or exciton sources has not been exactly determined. Tiny polymer aggregates for microscopic studies have also been prepared. These light sources have subwavelength dimensions and could be used in a wide variety of fields.

5.3 Molecular exciton microscopy [24]

The concept of active light sources enables a totally new mode of NSOM, based not on the blocking or absorption of photons but rather on quenching directly the energy quanta that otherwise would have produced photons. For instance, a thin, localized gold film (or cluster) can quench an excitation (or exciton) that would have been the precursor of photons. Furthermore, a single atom or molecule on the sample could quench (i.e., by energy transfer) the excitations located at the tip of the light source. For simplicity, we assume that the active part of the light source is a single atom, molecule or crystalline site, serving as the "tip of the tip". This quenching energy transfer from the excitation source's active part (donor) to the sample's active part (acceptor) may or may not qualify technically as an NSOM technique. However, it is the best hope, currently, for single atom or molecule resolution and sensitivity. This technique basically is a quantum optics microscopy. It has been called Molecular Exciton Microscopy (MEM) [24].

MEM is conceptually quite similar to STM [8, 24]. The excitons "tunnel" from the tip to the sample. However, there is no driving voltage or field. Rather, it is the energy transfer matrix element which controls the transfer efficiency. Its unusual matrix elements allow for the highest sensitivity to distance, higher than that of STM and comparable to that of AFM. In addition, the most striking result of this direct energy transfer is its ultrahigh sensitivity to isolated or single molecular chromophores. The quantum-optics energy transfer is highly efficient within the range of the "Förster radius". Thus, a single excitation could be "absorbed" by the sample acceptor. In contrast, based on the Beer-Lambert law [22], about a billion photons are needed to excite a single acceptor in the absence of other acceptors. Furthermore, as the distance range is limited to about 10 nm for the direct energy transfer, MEM is as much a near-field technique as STM or AFM, i.e., very sensitive in the single digit nm range and much less sensitive beyond 10 nm. However, in combination with conventional NSOM, the range can be extended to about 200 nm. Thus MEM is a technique which is able to "zoom in" from macroscopic to nanoscopic distances. Obviously such "zooming in" enhances the speed of operation. It also allows for a much more universal range of samples, from metal spheres and clusters to soft, *in vivo* biological units. In addition, MEM can use fluorophores, metal-clusters, etc. to enhance contrast, sensitivity and resolution with the help of NSOM. It can also be used in conjunction with lateral force feedback, in the same way as NSOM [1].

6. THE DEVELOPMENT OF OPTICAL AND EXCITONIC SUPERTIPS

There are many methods and reasons for making supertips. For example, the optical fiber tip can be treated, in principle, chemically, so as to produce specific supertips for a variety of purposes: 1) Wavelength shifters, *e.g.*, crystallites that fluoresce to the red of the tip emission; 2) Time "extenders," same as (1), utilizing prompt or delayed fluorescence or even phosphorescence; 3) Highly sensitive opto-chemical nanosensors; 4) Energy transfer supertips; 5) Heavy-atom sensors. Supertip development is the key to MEM which relies on quantum optics mechanisms, such as Förster energy transfer [22] or Kasha effect (external heavy atom effect) [25]. These interactions occur at the interface of the tip (its active center) and the sample (which are quantum mechanically coupled). For the highest resolution, this active center would consist of a single molecule, or molecular cluster, that does the imaging. This molecule is then the energy donor site for the Förster energy-transfer, or the spin-orbit interaction site for the Kasha effect. Figure 6 shows the relation between the tip, the supertip and the active center. As much as the supertip is part of the tip, the active center is part of the supertip. We note that a completely analogous situation is found in atomic force microscopy, where the active center is the force contact site at the tip of the supertip. In MEM the active center has to be optically excited repeatedly. The design of the system of tips is thus geared towards the need of supplying the active center with plenty of excitation quanta. We describe below several approaches for the present developments of supertips. Supertips have been constructed on both micropipette tips and optical fiber tips. The following gives a few examples of supertips.

6.1 Nanocrystal designer probes

One approach employs exciton conducting crystallites that absorb the light of the fiber-optic tip, convert it into excitons, which ultimately produce again photons. We have successfully grown such crystals of perylene, diphenyl-anthracene, etc. onto the fiber tip. Alternatively, the crystallite is grown at the tip of a micropipette and the fiber-optic tip is pushed deep into the pipette, very close to the crystal tip. The crystallite acts as an antenna (compare photosynthetic antenna) that channels the excitons to an *active center* which acts as an exciton trap [21]. This trap collects excitation from as far as 500-1000 Å. The active center is the key of the *molecular engineering*.

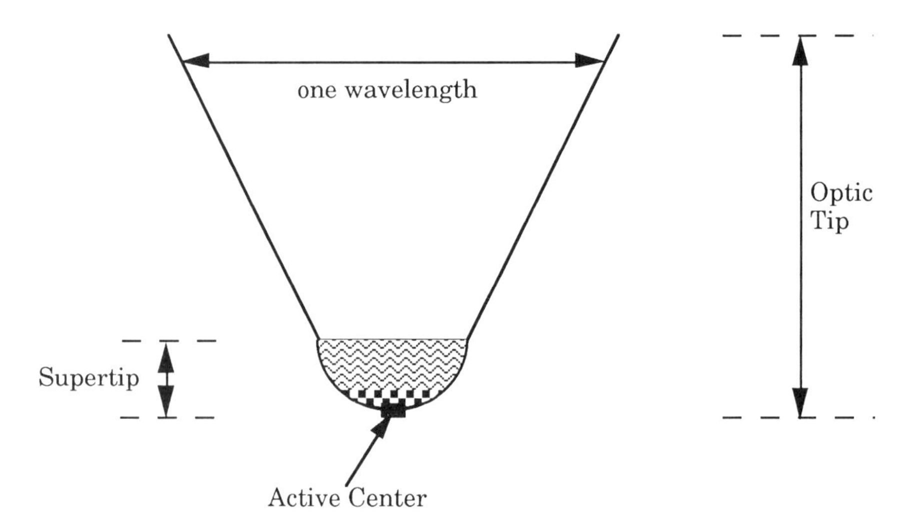

Figure 6. Schematic drawing of tip, supertip and active center.

The single impurity molecule (the "supertrap") creates a host "funnel" around it [21]. This funnel consists of host crystal molecules perturbed by the impurity ("guest") molecule. The closer the host molecule is to the trap, the lower its excitation energy. The molecules in the funnel act as exciton traps, catching the excitation from the host crystal and passing it deeper and deeper (in energy) to the deepest of them all, the supertrap. For MEM the guest molecule is deposited onto the surface of a molecular microcrystal and thus creates an energy funnel at its apex (active center). The best microcrystals are grown onto the tip of a micropipette. This tip is excited *internally* by a fiber-optic tip. It gives as much intensity as with epi-luminescence, *i.e.*, external excitation method [19]. We have observed acceptable levels of luminescence intensities from a few molecules of rhodamine-B embedded onto the surface of a DPA crystal, as shown in Figure 7. From the spectra, it is clear that even for dipping into very dilute rhodamine B solutions (10^{-7} M), the emission of a tiny rhodamine B active center is still detectable in an optical multichannel analyzer (OMA) apparatus. This preliminary work clearly demonstrates that supertips can be prepared by dipping nanometer crystal tips into dye solutions. The crystal tips absorb the incident light, create excitons and emit light again (with their typical fluorescence spectrum) or transfer energy to the active center and emit a lower energy fluorescence. These tips are exciton supertips. By optimizing the selection of host-crystal guest-supertrap pairs, we have been able to prepare supertips with high intensity and good stability [9].

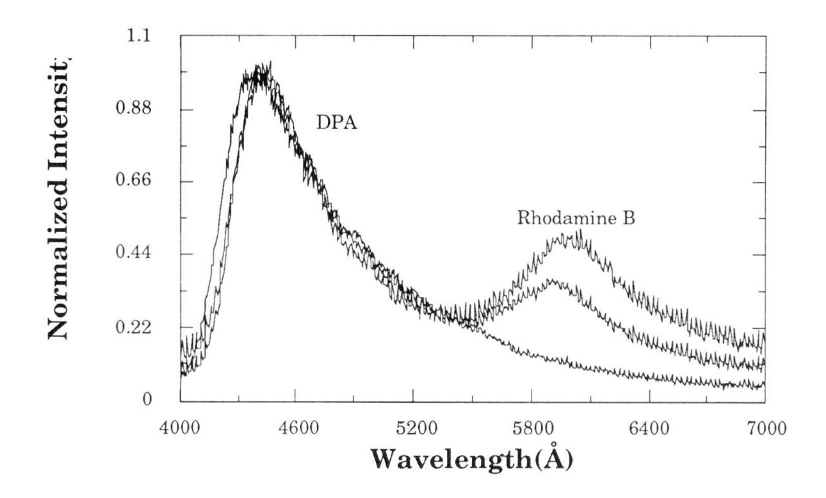

Figure 7: Fluorescence spectrum of active centers (Rhodamine-B) and supertip nanocrystal (diphenyl-anthracene). Note that an optical filter easily separates the two spectra.

6.2. Polymer matrix supertip

Another approach altogether involves a polymeric matrix attached to the optical-fiber tip by spatially controlled photopolymerization. The polymer is a copolymer, consisting of acrylamide, N,N-methylenebis (acrylamide) (BIS) and appropriate dye monomer groups [4, 11]. This polymer supertip acts as an antenna, even though a less efficient one compared to a molecular crystal. However, with the large photon flux emanating out of the tip this lower efficiency should suffice. At the very tip of this supertip the "active center" is attached by physical and chemical methods. The dye molecule on the top surface of the polymer is produced by photo-chemical reaction from a layer of precursor molecules deposited on the polymer tip by dipping into a solution. The highest probability for photo-chemical reaction is at the center of the tip, thus making it likely that the first and only active molecule is produced at this center. Several parameters can be used to control this operation: Depth and duration of dipping, precursor concentration, and the intensity and/or exposure time and the photo-chemical reaction light intensity. By using a multi-dye solution for photochemical synthesis, we have prepared multi-functional optic and excitonic probes with extremely small size. These probes emit multi-wavelength photons or produce excitons of different energy levels. By using multi-step photochemical synthesis, we have miniaturized optical and exciton probes to sizes much smaller than the sizes of the original light conductors. The multi-step near-field photopolymerization [11] has been very effective in fabricating extremely small optical and excitonic supertips [8].

We have also designed a supertip made of a single symmetric macromolecule by using newly developed dendrimer supermolecules [23]. The architecture of this "dendrimer" is controlled by synthetic methods. For example, the overall shape of a "D-127" molecule is bowl-like. Its size is about 125 Å. It can thus act as both optical and force active center. The large "rim" is bound to the tip by cumulative van der Waals bonding or covalent bonds. We have synthesized similar dendrimers containing higher aromatic rings, *e.g.*, substituting benzene rings by naphthalene or anthracene rings. A single such substituted group acts as a supertrap [8, 21], collecting most of the excitation. This intramolecular exciton supertrap also plays the role of the active center. It may act as an exciton donor, transferring excitation to an acceptor on the sample, and as the smallest possible light source.

The super-tips can double as NSOM and as MEM tips. The super-resolution imaging of biologically interesting species relies on this dual function. The MEM operation itself involves different mechanisms: 1) Active energy transfer (Förster-Dexter) [21, 22] where the supertip is the exciton (energy) donor and the sample is the acceptor. 2) Passive interaction (sensor mode), exemplified by the Kasha effect [25], where the sample either quenches the supertip's exciton or transforms it from "singlet" to "triplet."

7. NEAR-FIELD OPTICAL CROSS-SECTIONS

NFO has begun to gain its recognition in several fields, including photofabrication, biomedical research, computer memory, microscopy and biochemical sensing. A special point of both theoretical and practical interest is the increased photoexcitation cross-section of optically active molecules in the near-field range. This factor is important for both the photofabrication process and the utilization of such probes. For example, NFO has been applied in two different ways in chemical and biological sensing: 1). For the nanofabrication of optical fiber chemical and biological sensors and 2). For the operation of these subwavelength sensors. In the first case, by applying near-field optics, we successfully demonstrated a new concept of near-field photochemical synthesis, in which the dimension of a produced sensor is solely determined by the size of the light source. The synthesis maps the electromagnetic far-field or near-field profile of the light source. We further extend the miniaturization of chemical sensors by applying multi-step photo-nanofabrication. The second utilization of NFO occurs during the operation of these sensors. In NFO, the molecular absorption probability is much higher than that in far-field optics where about one billion photons are needed to excite one isolated molecule [2, 8]. This near-field effect is important for both the photo-nanofabrication process and the utilization of such optical probes in biochemical analysis.

Near-field excitation [8, 11] is a new concept introduced for the excitation of subwavelength optical and excitonic probes. There are three major reasons to introduce this new concept. First, geometrically, near-field optics encompasses a very small space, thus only those molecules inside the near-field region will be excited efficiently. The molecules to the left and right of the near-field light source will not be excited at all. This is in contrast to an evanescent wave spreading over a large region; Second, in the near-field region, the light intensity flux is much higher than that in the far-field of the same light source. Thus the excitation will be much stronger. Third, we believe that in the near-field region, the molecular absorption probability is much higher than in the far-field region (where about one billion photons are needed to excite one highly absorbing isolated molecule). Moreover, in the special case, when a molecule is within the Förster radius of an active near-resonance light source [8, 9], the molecular excitation probability is near unity. In this case, the probability in the near-field region should be between unity and that in far-field optical excitation. From a quantum mechanics point of view, the electromagnetic wavefunction of light in the near-field region has become constrained. This may imply that the overlap between the molecular wavefunction and this constrained wavefunction should be significantly better. Thus a higher excitation probability is expected. A classical analogy is the effect of the refractive index on the Einstein absorption coefficient. The absorption coefficient increases with λ^{-3}, where λ is the real wavelength. We assume that the effects of "photon squeezing" by the refractive index and by the near-field conditions are comparable. This leads to a much enhanced cross-section.

The promise of near-field photon sources for the nanofabrication and monitoring of electronic devices and optical memories is obviously of much current interest. So is the use of optical nanoprobes for microscopy, spectroscopy and biochemical sensing [1, 2, 4].

For the comparison of optical fiber tips and micropipettes, we note that the higher refractive index (n) of the fiber material reduces the photon wavelength inside it to $\lambda = \lambda_0/n$, compared to the wavelength in vacuum or air (λ_0), using a classical optics description. This reduces significantly the diffraction of the light at the orifice. In principle, as λ approaches the optical absorption of the dielectric, n increases, and eventually becomes a complex quantity [8]. Alternatively, one can use a quantum approach and consider the exciton-polariton resonance or quasi-resonance [28, 29]. Thus the optical-fiber tip exhibits a crossover with wavelength from a passive to an active photon tip. This may be one of the major reasons for the optical fiber tip's high light throughput.

8. SUMMARY

Different subwavelength light and exciton probes have been prepared by micropipettes and nanofabricated optical fiber tips. A battery of nanofabrication techniques have been developed to design and prepare these nano-optic and excitonic probes. Specifically, a new near-field nanotechnology, photo-nanofabrication, has been developed to prepare nanometer light and exciton probes. *In-vivo* dynamic and static measurements of biochemical parameters have been carried out on individual, viable, intact rat conceptuses during the period of organogenesis or external environmental changes. This is the first time such an experiment is done on a single and live rat embryo. The ability of the sensors to measure pH changes, in real time, in the intact rat conceptus, demonstrates their great potential applications for dynamic analysis in small multicelluar organisms and single cells. Also, we have demonstrated the ability of NSOS as a technique for obtaining nanometer-size spatial resolution in a spectral dimension by studying different microscopic and nanoscopic samples. Furthermore, supertips have been prepared and their development has been discussed. The large variety of nanoprobes should lead to applications in NFO as well as other microscopic analysis in biomedical and materials science. NFO has played key roles in all of these nanofabrication and analysis methods.

We expect that NFO will be further developed and that NSOM, NSOS and FOCS will become conventional tools in scientific research and development. The advances towards nanometer-resolved microscopy, spectroscopy and biochemical sensing promise to push chemical and biological analysis much closer to one of its ultimate goals—the non-invasive detection of a single molecule, radical or ion, the determination of its precise coordinates and the characterization of its structural conformation, as well as its internal dynamics and energetics, as a function of time and environmental perturbations.

ACKNOWLEDGMENT

This work is supported by NSF grant DMR-9111622.

REFERENCES

1. E. Betzig, and J.K. Troutman, Science, 257 (1992) 189.
2. R. Kopelman, S. Smith, W. Tan, R. Zenobi, K. Lieberman and A. Lewis, SPIE-Int. Soc. Opt. Eng., 1637 (1992) 33.
3. K. Lieberman, S. Harush, A. Lewis and R. Kopelman, Science, 24 (1990) 59.
4. W. Tan, Z-Y. Shi, S. Smith, D. Birnbaum and R. Kopelman, Science, 258 (1992) 778.

5. D.W. Pohl, in Advances in Optical and Electron Microscopy, C.J.R. Sheppard and T. Mulvey, (eds.) Academic Press, London, 1990, pp. 243-312.
6. T. Vogelmann, G. Martin, G. Chen and D. Buttry, in Advances in Botanical Research, Academic Press, 18 (1992) 255.
7. W. Tan, Z-Y. Shi and R. Kopelman, Anal. Chem., 64 (1992) 2985.
8. R. Kopelman and W. Tan, in Spectroscopic and Microscopic Imaging of the Chemical State, M. D. Morris (ed.), Marcel Dekker Inc., New York, 1993, pp. 227-254.
9. W. Tan, Ph. D. Thesis, University of Michigan, Ann Arbor, MI, 1993.
10. K.T. Brown and D.G. Flaming, Advanced Micropipette Techniques for Cell Physiology, John Wiley & Sons, New York, 1986.
11. W. Tan, Z-Y. Shi, B.A. Thorsrud, C. Harris and R. Kopelman, SPIE, 2068 (1993) 10.
12. C. Harris, M.R. Juchau and P.E. Mirkes, Teratology, 43 (1991) 229.
13. D. Birnbaum, S. Kook and R. Kopelman, J. Phys. Chem., 97 (1993) 3091.
14. S.K. Kook and R. Kopelman, J. Phys. Chem., 96 (1992) 10672.
15. J. Tanaka, Bull. Chem. Soc. Jpn., 36 (1963) 1237.
16 A. McDonald, IEEE Trans. Microwave Theory Tech. MTT-20 (1972) 698.
17. A. Lewis, E. Betzig, A. Harootunian, M. Isaacson and E. Kratschmer,in Spectroscopic Membrane Probes, L. M. Loew (ed.), CRC Press, Vol. II, 1988, p. 81.
18. W. Tan and R. Kopelman, in Dynamics in Small Confining Systems, J.M. Drake et al, (eds.), Materials Research Society, Pittsburgh, PA, 290 (1993) 287.
19. A. Lewis and K. Lieberman, Anal. Chem., 63 (1991) 625A.
20. W. Tan, Z-Y. Shi, S. Smith and R. Kopelman, Mol. Cry. Liq. Cry., (submitted, 1993).
21. A.H. Francis and R. Kopelman, R., Topics in Applied Physics , 49, Laser Spectroscopy of Solids, 2nd ed., W. M. Yen and P. M. Selzer (eds.), Springer, Berlin, 1986, p. 241.
22. M. Pope and E. Swenberg, E., Electronic Processes in Organic Crystals, Oxford Univ. Press, New York, 1982.
23. Z. Xu, Z-Y. Shi, W. Tan, R. Kopelman and J.S. Moore, Polymer Preprints, 33 (1993) 130.
24. R. Kopelman, A. Lewis and K. Lieberman, in X-Ray Microimaging for the Life Sciences, D. Attwood and B. Barton, (eds.), Lawrence Berkeley Laboratory, 166, 1989.
25. M. Kasha, J. Chem. Phys., 20 (1952) 71.
26. R. Kopelman, W. Tan, S.J. Smith, A. Lewis and K. Lieberman, Microbeam Analysis, D. G. Howit (ed.), 1991, 91.
27. A. Roberts, J. Appl. Phys., 65 (1989) 2896.
28. V.M. Agranovich and M.D. Galanin, Electronic Excitation Energy Transfer on Condensed Matter, North Holland, Amsterdam, 1982.
29. J. Aavikso, A. Freiberg, J. Lipmaa and T. Reinot, J. Lumin., 37 (1987) 313.

Part V: Microphotochemistry

MICROCHEMISTRY
Spectroscopy and Chemistry in Small Domains
Edited by H. Masuhara et al.

Dipolar interactions in the presence of an interface

Michael Urbakh and Joseph Klafter

School of Chemistry, Tel-Aviv University, Tel Aviv 69978, Israel

The nature of the pair interaction between dipoles embedded in a liquid in the vicinity of a non-metallic interface is investigated in the continuum limit. It is shown that the dipole-dipole interaction can be significantly modified in the presence of a boundary in liquids characterized by a nonlocal dielectric function or in liquids whose dielectric properties change due to geometrical restrictions. Different limits are studied and relationships to experimental observables are discussed.

1. INTRODUCTION

Recent experimental and theoretical studies [1-5] have demonstrated that spatial restrictions can strongly affect the dynamic and thermodynamic properties of embedded liquids and molecules. Examples include the influence of interfaces on the pair-interaction energy between charged or uncharged particles [6-7], translational and rotational diffusion of probe molecules [4,8,9] as well as chemical reactions and energy transfer properties [1,8].

The influence of boundaries on molecular properties in their vicinity is of course related to the nature of the intermolecular forces and to how they are modified near an interface. These modifications can directly influence processes such as adsorption and electron, or energy, transfer at interfaces.

Here we discuss changes in the interaction between point dipoles embedded in a liquid near a nonmetallic interface when compared to the bulk liquid. We follow Refs.[10,11] where we represented the liquid in the *continuum approximation*, in terms of a nonlocal dielectric function $\varepsilon(\mathbf{k},\omega)$. In analogy to these previous works the effect of the interface is introduced through the concept of additional boundary conditions [12]. We analyze the dependence of the dipole-dipole interaction on the distance between the dipoles and between the dipoles and the boundary as well as on the dielectric parameters that characterize the interface region. The approach used here is suitable for both static and dynamic limits. Although the continuum framework does not explicitly include molecular level details, it enables to derive closed expressions that relate microscopic quantities to measurable observables such as dielectric functions [13-15]. Continuum approximations and their modifications have been shown to be powerful in unraveling leading physical processes in complex systems.

The nonlocal nature of the liquid defines a length scale, Λ, which is a measure of spatial correlations in the liquid. This parameter can be estimated on the base of diffraction experiments [16] and on molecular dynamic simulations of liquids [17,18]. For instance, in aqueous solutions the correlation length, Λ, is of the order of the extension of local hydrogen-bonded clusters, $\Lambda \sim 4$ Å ; in solutions of dipolar polymers the correlation length may be much larger [19]. The introduction of this new length scale is of particular importance in the case of a liquid under geometrical restriction.

When the characteristic size of the system is comparable to the correlation length, Λ, fundamental differences between the response of the liquid under geometrical restriction and its bulk response are observed. Only beyond distance Λ are the limits of the macroscopic description reached.

The nonlocal description introduces a short range order within the continuum model of the liquid, at least phenomenologically, and leads to dipole-dipole and dipole-boundary distance and pore size dependencies which do not appear in the case of a local dielectric function. In order to mimic the possibility that the liquid itself changes its dielectric behavior due to interaction with the boundary [4,20,21] we assume a region of modified liquid near the boundary. The influence of a modified surface layer of a liquid on ion-ion and dipole-dipole interactions at a metal-electrolyte solution interface was previously considered and shown to provide interpretations to experimental observations of ionic adsorption in electrochemical systems which could not be understood in the framework of the traditional description [7,22].

The formalism developed in this paper can be also applied to the calculation of the van der Waals and hydration forces in atomic force microscopy (AFM) operating in liquids. Recent AFM studies in polar solvents [23] have demonstrated some new features arising in liquids. Particularly it has been shown that AFM operates more stably in water than in air or in a vacuum. Most of previous attempts to devise a theory of atomic force microscopy [24,25] neglected the solvent structure effect. However, the latter gives rise to a hydration force which together with steric repulsion are probably responsible to the strong repulsive forces between polar surfaces at small distances (less than 2 nm [26]). The first attempt to take into account the interfacial hydration in AFM has been done in ref. [27].

2. THE MODEL

We now consider the model we use for the description of electromagnetic interaction of time-dependent point dipoles embedded in a liquid near a substrate. As in [10,11] we assume that the substrate is characterized by a local dielectric function, $\varepsilon_{sub}(\omega)$, and the liquid is described by the nonlocal dielectric function

$$\frac{1}{\varepsilon(k,\ \omega)} = \frac{1}{\varepsilon_*(\omega)} - \left[\frac{1}{\varepsilon_*(\omega)} - \frac{1}{\varepsilon_b(\omega)}\right]\frac{1}{1 + k^2\Lambda^2} \tag{1}$$

Here $\varepsilon_*(\omega)$ is the short-wavelength dielectric constant of a bulk liquid, $\varepsilon_b(\omega) = \varepsilon_b(k{=}0, \omega)$ and $\mathbf{k}$ is a wave vector. The quantity ε_* is close to unity and in the low frequency range which we consider in this paper the relationship $|\varepsilon_*(\omega)| \ll |\varepsilon_b(\omega)|$ is satisfied. Generalization to a more complex nonlocal dielectric function is possible.

As already mentioned in the Introduction, the continuum approximation is valuable in obtaining analytical relationships between the dielectric properties of the participating media and the details of the dipole-dipole interaction. The nonlocal nature of the liquid, as introduced in Eq.(1), accounts phenomenologically for some structural aspects of the liquid through the length Λ.

Let us denote the coordinates of the centers of dipoles with dipole moments μ_d and μ_a by $\mathbf{r}_d = (0, 0, z_d)$ and $\mathbf{r}_a = (\mathbf{R}, z_a)$, respectively, where z_d and z_a are the distances of dipoles from the substrate and R is the distance between dipoles along the surface (see Figure 1). The substrate surface plane coincides with the plane $z = 0$.

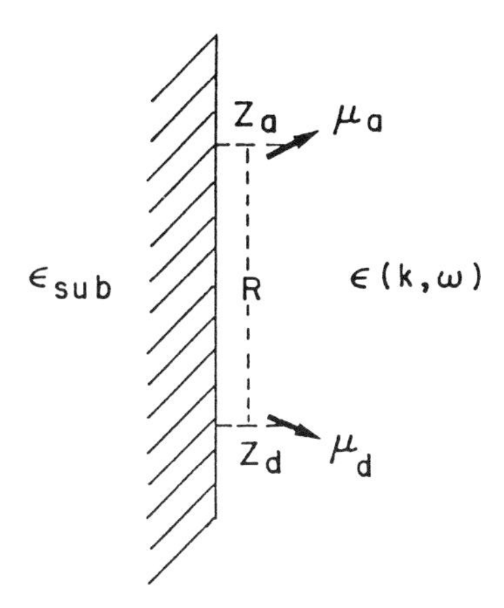

Figure 1. Dipoles at the substrate-liquid interface.

The energy of the electromagnetic interaction between dipoles with moments μ_d and μ_a in a general orientation with respect to the interface can be written in the following form

$$U = - \mu_d \frac{\partial^2 \Phi(\mathbf{r}_d,\mathbf{r}_a)}{\partial \mathbf{r}_d \partial \mathbf{r}_a} \mu_a = \mu_d \left[\mathbf{T}^{(dip)} + \mathbf{T}^{(ind)} \right] \mu_a \tag{2}$$

where $\Phi(\mathbf{r}_d,\mathbf{r}_a)$ is the potential at a point $\mathbf{r} = \mathbf{r}_a$ created by a point charge located at $\mathbf{r} = \mathbf{r}_d$ near the interface and the tensors $\mathbf{T}^{(dip)}$ and $\mathbf{T}^{(ind)}$ describe direct interaction between dipoles and interaction induced by the presence of an interface correspondingly. In a *nonlocal* medium the tensor $\mathbf{T}^{(dip)}$ has the form [13,28]

$$\mathbf{T}^{(dip)}_{ij} = - \frac{2}{\pi} \frac{\partial^2}{\partial \mathbf{r}_i \partial \mathbf{r}_j} \int_0^\infty \frac{dk}{\varepsilon(k,\omega)} \frac{\sin kr}{kr} =$$

$$= - \frac{3\, r_i r_j - r^2 \delta_{ij}}{\varepsilon_b(\omega) r^5} -$$

$$- C_2\exp(-r/\Lambda)\left[\frac{3\, r_i r_j - r^2\delta_{ij}}{r^5} + \frac{2\, r_i r_j - r^2\delta_{ij}}{\Lambda\, r^4} + \frac{r_i r_j}{\Lambda^2 r^3}\right] \quad (3)$$

Here $\mathbf{r} = \mathbf{r}_a - \mathbf{r}_d$ is the vector connecting two dipoles. The first term in Eq.(3) constitutes the classical result and the second term describes nonlocal effects. The second term, in contrast to the first, does not vanish after averaging over directions of the vector $\mathbf{r}$, and this leads to some important physical consequences, in particular to the existence of an electric field near a homogeneous dipole layer. It is evident from Eq.(3) that the effect of nonlocality does not change the energy of dipole-dipole interactions in the bulk liquid at distances larger than the characteristic correlation length, Λ, namely for $r \gg \Lambda$. The situation is however quite different at interfaces.

The interaction tensor $\mathbf{T}^{(ind)}$ which arises due to the polarization at the interfacial region has been previously calculated in [11]. Using the results obtained in Ref.[11] we will analyze here the dependence of the interaction tensor, $\mathbf{T}$, on the distance between the dipoles and on the substrate and liquid parameters. We will first consider the interaction between molecules with dipole moments μ_d and μ_a *perpendicular* to the surface. The zz-component of the interaction tensor $\mathbf{T}^{(ind)}$ has the following asymptotic behaviors of the interaction tensor T_{zz}:

1. *at large distances between dipoles,* $R \gg z_d, z_a, g\Lambda$

$$T_{zz} \approx \frac{2}{R^3}\,\frac{\varepsilon_{sub}}{\varepsilon_b\varepsilon_*^2(\varepsilon_{sub} + \varepsilon_b)}\left\{\varepsilon_*^2 + (\varepsilon_b - \varepsilon_*)^2\exp(-(z_d + z_a)/\Lambda) + \varepsilon_*(\varepsilon_b - \varepsilon_*) \times \right.$$

$$\left. \times \left[\exp(-z_d/\Lambda) + \exp(-z_a/\Lambda)\right]\right\} \quad (4)$$

where $g = \varepsilon_b\varepsilon_{sub}/\{\varepsilon_*(\varepsilon_{sub} + \varepsilon_b)\}$. As an example, at the interface between silica and water $\varepsilon_{sub} = 11.6$, $\varepsilon_b = 80$ and $\varepsilon_* = 1$ and for low frequencies the parameter $g \simeq 10$.

2. *in the intermediate region,* $z_d, z_a, \Lambda \ll R \ll g\Lambda$

$$T_{zz} \approx \frac{2}{R\Lambda^2}\,\frac{(\varepsilon_{sub} + \varepsilon_b)}{\varepsilon_b\varepsilon_{sub}(\varepsilon_b - \varepsilon_*)^2}\left\{\varepsilon_*^2 + (\varepsilon_b - \varepsilon_*)^2\exp(-(z_d + z_a)/\Lambda) + \varepsilon_*(\varepsilon_b - \varepsilon_*) \times \right.$$

$$\times \left[\exp(-z_d/\Lambda) + \exp(-z_a/\Lambda)\right]\Bigg\} \qquad (5)$$

We can extend the domain of applicability of the intermediate asymptotics in Eq. (5) to smaller values of R (but bound by the condition $R \gg z_a, z_d$) if we add to the expression (5) the term (3) which takes into account direct interaction between dipoles.

The region of intermediate asymptotics, Eq.(5), with anomalously slow (Coulomb-like) decrease of dipole-dipole interaction with distance R exists only in systems with high values of the dielectric constants of the substrate, ε_{sub}, and of the liquid, ε_b, when $\varepsilon_{sub}/\varepsilon_* \gg 1$ and $\varepsilon_b/\varepsilon_* \gg 1$, and therefore correspondingly the coefficient $g \gg 1$. For instance at silica-water interface, as discussed in case 1, and for $\Lambda \sim 4$Å [13] the limit in Eq.(5) can apply for a wide range of distances 3Å < R < 20Å. Also at an interface between two immiscible liquids (for instance water - nitrobenzene system; for nitrobenzene $\varepsilon_b = 34.8$) the slow R-dependence of Eq.(5) can play an important role.

3. *dipoles very close to the interface*, $z_a, z_d < \Lambda$

When both dipoles are placed at small distances from the surface, $z_d, z_a < \Lambda$, Eqs.(4) and (5) are simplified to

$$T_{zz} \approx \begin{cases} \dfrac{2}{R^3}\,\dfrac{\varepsilon_{sub}\,\varepsilon_b}{\varepsilon_*^2(\varepsilon_{sub}+\varepsilon_b)}, & R \gg g\Lambda \qquad (6) \\[2ex] \dfrac{2}{R\Lambda^2}\,\dfrac{(\varepsilon_{sub}+\varepsilon_b)\,\varepsilon_b}{\varepsilon_{sub}(\varepsilon_b-\varepsilon_*)^2}, & z_d, z_a, \Lambda \ll R \ll g\Lambda \qquad (7) \end{cases}$$

These equations are appropriate for the description of the interaction between dipoles located in the first few layers closest to the substrate, as well as the interaction between adsorbed molecules, $z_a = z_d \simeq r_m$, where r_m is the molecular size. In the limit of $\varepsilon_{sub} \to \infty$ Eqs.(6) and (7) reduce to the results obtained for metal-electrolyte interface [7]. It should be noted that in this case the expression of the dipole-dipole interaction energy at large R is directly proportional to the bulk dielectric constant of a liquid, ε_b, in contrast to the traditional description obtained for a local representation of the liquid in contact with a substrate, in which the energy, $\mu_a \mathbf{T}^0_{zz} \mu_d$, is inversely proportional to ε_b

$$T^0_{zz} \approx \frac{2}{R^3} \frac{\varepsilon_{sub}}{\varepsilon_b(\varepsilon_{sub} + \varepsilon_b)}, \quad R \gg z_a, z_d \tag{8}$$

When the bulk dielectric constant of a liquid, ε_b, is larger than both ε_{sub} and ε_* Eqs.(6) and (7) can be written as

$$T_{zz} \approx \begin{cases} \dfrac{2}{R^3} \dfrac{\varepsilon_{sub}}{\varepsilon_*^2} & \text{at } R \gg g\Lambda, z_d, z_a \quad (9) \\ \dfrac{2}{R\Lambda^2} \dfrac{1}{\varepsilon_{sub}} & \text{at } z_d, z_a, \Lambda \ll R \ll g\Lambda \quad (10) \end{cases}$$

The different behavior in Eqs.(4)-(5) and (9)-(10) originates solely from our nonlocal description of the liquid and is essentially insensitive to the details of the imposed boundary conditions [10,29]. At large distances between dipoles (and $\varepsilon_b > \varepsilon_{sub}$) the interaction tensor, Eq. (9), has the same form as the traditional expression, T^0_{zz}, but with a reduced effective dielectric constant, ε_*.

Our results demonstrate that nonlocal description of liquids leads to a nonuniform distribution of polarization fluctuations in the interfacial region reflected in the dependence of the dielectric response of the liquid on the distance from the substrate. A new characteristic length, the correlation length in a liquid Λ, appears in the problem. Comparison of the Eqs.(4)-(10) with the results obtained in the model of the modified liquid layer at the substrate surface [7,10] (see also the results of next section) shows that this effect corresponds to the formation of the interface layer with reduced dielectric constant, ε_*. The presence of such a layer reflects the structuring effect of a substrate [13,20]. The thickness of the layer is of the order of the characteristic liquid structure distance, Λ. Only when the dipoles are placed far beyond the interfacial layer the traditional description of Eq.(8) applies. The structuring effect (interfacial hydration) gives rise also to hydration forces which are of crucial importance in the interaction and fusion of biological membranes and macromolecules [20,30].

We see that *for all distances* between dipoles our results differ from the corresponding local behavior, Eq.(8). The effect of nonlocality may lead to an enhancement in the interaction between dipoles. At large distances, $R \gg g\Lambda$, the ratio of interaction tensors is $T_{zz}/T^0_{zz} \approx (\varepsilon_b/\varepsilon_*)^2$ which for water is of the order of 10^2 - 10^3. Similar nonlocal enhancement of dipole dipole interaction at large distances R was predicted in ref [31]. The interaction at the substrate-liquid interface can be larger than the interaction near a free substrate , as described by the following ratio, $(T_{zz}/T^0_{zz}\ (\varepsilon_b = 1)) = (\varepsilon_{sub} + 1)/\{\varepsilon_*^2(\varepsilon_{sub} + \varepsilon_b)\})$. We see the presence of a dielectric medium by no means weakens the dipole-dipole interaction. This is due to the pulling of electrostatic lines into the interfacial layer with the reduced dielectric constant, ε_*.

It should also be mentioned that in a nonlocal medium instead of the general law (at $R \gg z_d, z_a$), Eq.(8), we have a more complicated behavior of the dipole-dipole interaction. This behavior shows a significant change of the form of $T_{zz}(R)$ at a new characteristic length $g\Lambda$. The dependencies of the interaction on the distance between dipoles, R, calculated over the whole range of distances R and for different values of the system parameters using the exact equation for T_{zz} are shown on Fig.2 .

The interaction between dipoles with other orientations of the dipole moments μ_d and μ_a can be found similarly. In the case of dipoles with dipole moments *parallel* to the surface plane and to the vector **R** and placed at small distances from the substrate ($z_d \approx z_a < \Lambda$), the long range asymptotic behavior of the interaction has the form

$$T_{RR} = -\frac{4}{R^3}\frac{1}{(\varepsilon_{sub}+\varepsilon_b)} - 24\frac{z_d^2}{R^5}\frac{\varepsilon_{sub}-\varepsilon_b}{\varepsilon_b(\varepsilon_{sub}+\varepsilon_b)} + 24\frac{\Lambda^2}{R^5}\frac{\varepsilon_{sub}(\varepsilon_{sub}-\varepsilon_b)(\varepsilon_*-\varepsilon_b)^2}{\varepsilon_*^2\,\varepsilon_b(\varepsilon_{sub}+\varepsilon_b)^2}$$

$$\times\left\{1 + 2\frac{z_d}{\Lambda}\frac{\varepsilon_b^2}{(\varepsilon_{sub}-\varepsilon_b)(\varepsilon_b-\varepsilon_*)} - \frac{z_d^2}{\Lambda^2}\frac{2\varepsilon_b^2+\varepsilon_{sub}(\varepsilon_b+\varepsilon_*)}{(\varepsilon_{sub}-\varepsilon_b)(\varepsilon_b-\varepsilon_*)}\right\}, \quad R \gg g\Lambda \quad (11)$$

The first two terms in Eq.(11) constitute the classical result describing an attraction (for $\varepsilon_{sub} > \varepsilon_b$) of two dipoles parallel to the interface between local media. The third term takes into account the influence of the liquid structure as mimicked by the length Λ. In contrast to the previous case, Eq.(6), here the interaction energy depends on the correlation length, Λ, even at large distances between dipoles. A comparison of Eqs.(6) and (11) shows that for large values of dielectric constants of the liquid, ε_b, or of a substrate, ε_{sub}, the interaction between dipoles with moments parallel to the surface is weaker than the interaction between dipoles perpendicular to the surface. In the limiting cases $\varepsilon_{sub} \gg 1$ or $\varepsilon_b \gg 1$ Eq.(11) shows R^{-5} dependence of the interaction on distance between dipoles, different from the case of dipoles perpendicular to the surface where interaction energy decreases as R^{-3}. Again the inclusion of nonlocal dielectric properties leads to an enhancement of dipole-dipole interaction. But in this case the effect of enhancement is not so pronounced. In contrast to the case of perpendicular dipoles when for all values of parameters (ε_b, ε_{sub}, ε_*, Λ, R, z_d and z_a) we had the usual repulsion between dipoles now the type of interaction (repulsive or attractive) depends on these parameters. For given parameters characterizing the liquid and substrate (ε_b, ε_{sub}, ε_*, Λ) the interaction energy may change sign as a function of the distance between dipoles, R. The typical dependencies of the interaction between dipoles parallel to the surface on the distance R are presented at Fig.3. Interaction between dipoles with other directions of dipole moments can be considered in a similar way.

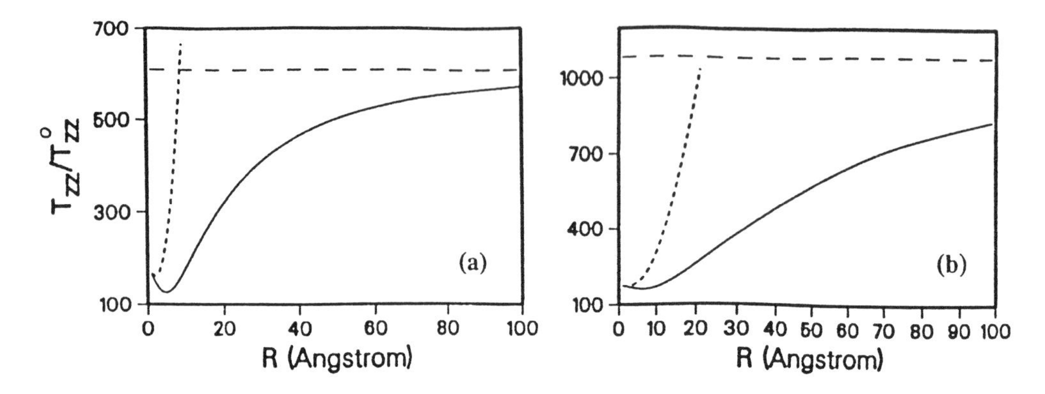

Figure 2. Dependence of the energy of interaction between dipoles perpendicular to the interface, $U = \mu_d T_{zz} \mu_a$, on the distance between them. The energy is normalized by the function $U = \mu_d T^0_{zz} \mu_a$ representing the traditional expression for the energy of dipole-dipole interaction (in perpendicular orientation) inside a local medium near a substrate (T^0_{zz} is given by Eq.(8)). Solid lines are the results of exact calculations ; dashed lines are long range asymptotes (4) and dotted lines are intermediate asymptotes (5) with account of direct interaction between dipoles (3). The calculations were carried out for the following values of parameters: $\varepsilon_{sub} = 10$, $\varepsilon_* = 2$, $\varepsilon_b = 80$, $z_d = z_a = 1$ Å, (a) - $\Lambda = 2$ Å and (b) - $\Lambda = 5$ Å.

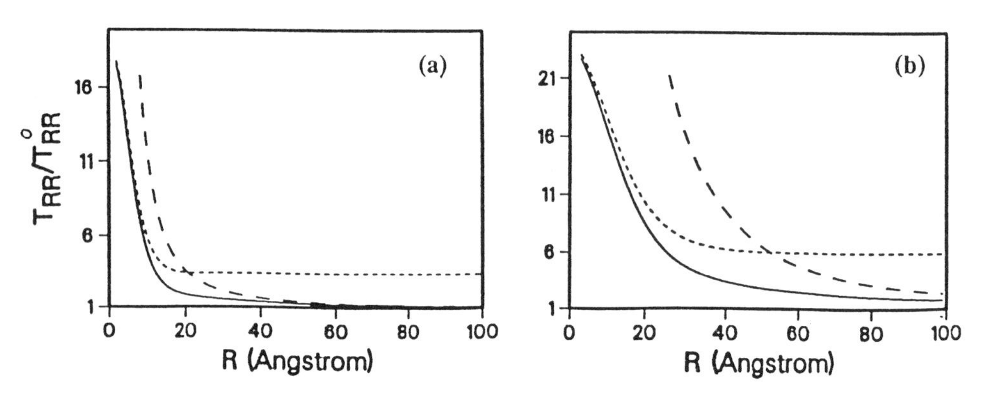

Figure 3. Dependence of the energy of interaction between dipoles parallel to the interface, $U = \mu_d T_{RR} \mu_a$, on the distance between them. The energy is normalized by the function $U^0 = \mu_d T^0_{RR} \mu_a$ representing the the energy of interaction between dipoles in planar orientation inside a local medium near a substrate (T^0_{RR} is given by the first two terms in Eq.(11)). Solid lines are the results of exact calculations.; dashed lines are long range asymptotes (11) and dotted lines are intermediate asymptotes. The calculations were carried out for the following values of parameters: $\varepsilon_{sub} = 10$, $\varepsilon_* = 2$, $\varepsilon_b = 80$, $z_d = z_a = 1$ Å, (a) - $\Lambda = 2$ Å and (b) - $\Lambda = 5$ Å.

3. SUBSTRATE MODIFICATION

There has been some experimental and numerical evidence, that a liquid near a substrate may have modified properties [2,4,20,21]. Close to the interface the liquid may develop local structure due to interaction with the substrate, a structure which is characterized by different dielectric properties and can therefore manifest itself in different equilibrium and dynamic behaviors.

Here we extend our results of the previous section and describe the liquid side by two dielectric functions which correspond to the modified liquid in the close vicinity of the boundary, $\varepsilon_s(\omega)$, and the bulk solution, $\varepsilon_b(\omega)$.

Following the same type of calculations as in previous section, we arrive at the following expression for the dielectric tensor, **T**, for dipoles in the surface layer

$$\mathbf{T}^{(dip)}_{ij} = -\frac{3\, r_i r_j - r^2\delta_{ij}}{\varepsilon_s(\omega)\, r^5} \tag{12}$$

and

$$\mu_d \mathbf{T}^{(ind)} \mu_a = \frac{1}{2\pi\varepsilon_s(\omega)} \int d\mathbf{k}_1 k_1^{-1}[1 + \omega_{12}\omega_{23}\exp(-2k_1 d)]^{-1}\Big\{(\mu_a\mathbf{k}')(\mu_d\mathbf{k}')\omega_{12}\omega_{23}$$

$$\times\exp(-k_1(2d + z_d - z_a)) + (\mu_a\mathbf{k})(\mu_d\mathbf{k})\omega_{12}\omega_{23}\exp(-k_1(2d - z_d + z_a)) +$$

$$+ (\mu_a\mathbf{k}')(\mu_d\mathbf{k})\omega_{12}\exp(-k_1(2d - z_d - z_a)) - (\mu_a\mathbf{k})(\mu_d\mathbf{k}')\omega_{23}\exp(-k_1(z_d + z_a))\Big\}\exp(i\mathbf{k}_1 R) \tag{13}$$

where d is the layer thickness and

$$\omega_{ij} = \frac{\varepsilon_i - \varepsilon_j}{\varepsilon_i + \varepsilon_j}, \quad i,j = 1,2,3 \quad \text{and} \quad \varepsilon_1 = \varepsilon_b, \ \varepsilon_2 = \varepsilon_s, \ \varepsilon_3 = \varepsilon_{sub} \tag{14}$$

The asymptotic behavior of Eq.(13) can be studied along the same procedure as in the previous section. For instance, the interaction tensor for two dipoles perpendicular to the surface in the region of large distances R, between them, has the form

$$T_{zz} \approx \frac{2}{R^3} \frac{\varepsilon_{sub}\,\varepsilon_b}{\varepsilon_s^2(\varepsilon_{sub}+\varepsilon_b)}, \qquad R \gg d\varepsilon_b/\varepsilon_s \tag{15}$$

which behaves as the corresponding Eq.(6) obtained in the framework of nonlocal description of a liquid. The comparison of these two equations shows again that the inclusion of a liquid structure even on a phenomenological level leads to the creation of the surface layer of modified liquid with dielectric constant, ε_s and with the thickness, d. This means that the nonlocal case discussed in the previous section leads essentially to an equivalent behavior but with a surface layer of a thickness of the order of the correlation length Λ with a reduced dielectric constant $\varepsilon_s = \varepsilon_*$. Other limits of Eq.(13) can be derived as well following the previous sections.

4. DIPOLE RELAXATION NEAR BOUNDARIES

The formalism described in this paper has been recently applied [10] to the study of the rotational relaxation of a time-dependent point dipole in a liquid near a substrate. The relaxation properties of a dipole at the interface can be expressed through its effective polarizability and dielectric friction, $\xi_D(\omega)$. In order to calculate these quantities one has to find the field induced by an oscillating dipole at the point of its location, $\mathbf{E}(\mathbf{r}_0,\omega)$.

The dielectric friction relates [14,32] the frictional torque, **T**, to the angular velocity, Ω, of a dipole:

$$\mathbf{T}(\omega) = -\,\xi_D(\omega)\,\Omega(\omega) \tag{16}$$

A frequency dependent rotational time, $\tau_R(\omega)$, can be defined through the dielectric friction:

$$\tau_R(\omega) = \xi_D(\omega)/2kT \tag{17}$$

The net dielectric friction, $\xi_D(\omega)$, is the sum of the bulk, $\xi_D^{(b)}(\omega)$, and the surface, $\xi_D^{(s)}(\omega)$, terms. The influence of boundary on the rotational relaxation can be obtained by studying the change in the friction due to an interface which is also derived from the induced field.

For a dipole perpendicular to the interface and located at the distance z_0 from the substrate we found [10] the following limiting behaviors of the surface component of the dielectric friction $\xi_D^{(s)}(\omega)$:

(1) $z_0 \gg \Lambda$

$$\xi_D^{(s)}(\omega) = \frac{\mu^2}{4\ I\ z_0^3} \operatorname{Im} \left\{ \frac{1}{\varepsilon_b(\omega)} \frac{\varepsilon_b(\omega) - \varepsilon_{sub}}{\varepsilon_b(\omega) + \varepsilon_{sub}} \right\} \tag{18}$$

where I is the moment of inertia of the dipole. Eq.(18) is the classical result obtained [33] for a local representation of the liquid in contact with a substrate and is due to the effect of image charges.

(2) $z_0 \ll \Lambda$

$$\xi_D^{(s)}(\omega) = \frac{\mu^2}{4\ I\ z_0^3} \operatorname{Im} \left\{ \frac{1}{\varepsilon_*(\omega)} \frac{\varepsilon_*(\omega) - \varepsilon_{sub}}{\varepsilon_*(\omega) + \varepsilon_{sub}} \right\} \tag{19}$$

This result originates from the nonlocal description of the liquid and is again insensitive to the details of the boundary. We see that at small distances from the substrate the dielectric friction has the same "image force" form as for large distances, Eq.(18), but with a reduced effective dielectric constant, ε_*.

Comparing Eqs.(18)-(19) with the expression for the dielectric friction in the liquid bulk [10] we can conclude that in the two limiting cases, $z_0 < \Lambda$ and $z_0 > \Lambda$, the boundary corrections to $\xi_D^{(b)}(\omega)$ are small, being of the relative order of R_c^3/z_0^3 ; where R_c is the radius of the cavity around a dipole molecule. The radius of the cavity is estimated to be of the order of few molecule radii which limits the contribution of the boundary to the dielectric friction and consequently to the rotational time. Larger contributions can arise,as discussed in the previous section, when the properties of the liquid itself are changed due to the presence of the interface [10]. One also expects the surface contribution to be significant for macromolecules near an interface where R_c is larger than the radius of the liquid molecules. For such cases a more realistic approach is required in order to account of the detailed charge distribution in molecules.

5. CONCLUSIONS

We have investigated the influence of a nonmetallic interface on the interaction between point dipoles located near the interface in the liquid side and on dipole relaxation. Both the liquid and the boundary are described in terms of the continuum approach by their dielectric properties. We assumed that the substrate is given by a local dielectric function and the liquid by a nonlocal dielectric function which introduces a typical length Λ into the problem. The results show some new limits of the dipole-dipole interaction which originate from the nonlocal nature of the liquid. The results strongly depend on the embedding and neighboring dielectric functions and display a rich range of behaviors which may be amenable to experimental tests.

Our studies demonstrate that in order to provide a correct description of the interaction between dipoles in a liquid near an interface it is necessary to take into account the influence of the bulk liquid and not only the first few layers. There is a difference between dipoles interacting inside one monolayer of liquid molecules on a substrate-vacuum interface, and in the first layer at a substrate-liquid interface. The polarization of the region in the liquid with thickness of the order of the distance between dipoles, R, may contribute significantly to the interaction between two dipoles at a surface. This fact should be taken into account in numerical simulations of the interfacial properties of liquid.

The approach introduced in the paper can be used also in describing liquid-liquid interfaces where one takes into account the nonlocal properties of both liquids in terms of their structure parameter Λ. In such cases the functional form of the interaction tensor, Eqs.(4)-(7) is retained. The parameter Λ should, however, be replaced by an effective length characterizing the thickness of the surface layer of both liquids.

Examples for cases where the modification of the dipole-dipoles interaction due to the presence of a boundary directly related to experimental observables are direct electronic energy transfer of the Förster type [34] adsorption isotherms of molecules carrying dipole moments and may be even the organization of such molecules. The relationship between the energy transfer rate, w_{tr}, from a donor molecule to an acceptor and the nature dipole-dipole interaction can be obtained on the basis of the Golden Rule expression (to be summed over all possible transitions)

$$w_{tr} = (2\pi/\hbar)\ |\mu_d\ \mathbf{T}\ \mu_a|^2\ \delta(E_f - E_i) \tag{20}$$

Here μ_d is the dipole moment for the transition from the donor state $|\Psi_{di}>$ to the acceptor state $|\Psi_{df}>$, μ_a is the transition dipole moment of the acceptor; $E_f - E_i$ is the net change in total energy of the donor-acceptor pair and $\mathbf{T}$ is the dipole-dipole interaction energy tensor. All those limits discussed above which lead to an incoherent energy transfer process (which excludes the slowly decaying, Coulomb-like behavior) should apply to the energy transfer calculations through Eq.(20) and may therefore make it a method to relate the microscopic process of donor to acceptor energy transfer to the macroscopic dielectric behavior at the interface. However for most realistic cases the contribution of the dipole-dipole interaction tensor to Eq.(20) will be in the high frequency range where the effects of liquid structures and molecules are less pronounced. This may explain recent observation on energy transfer at silica interfaces which do not show marked differences when compared to energy transfer in bulk liquids [35]. The effect of the dipole-dipole interaction on adsorption isotherms comes through the contribution to the chemical potential of the surface layer (Λ or d). For low concentrations one expects therefore that the slope of the isotherm (surface coverage vs. concentration) will depend on the nature of the liquid through Λ and the dielectric functions in the interface region. For dipoles perpendicular to the surface the slope should decrease as a result of the nonlocal nature of the liquid.

REFERENCES

1. J.Klafter and J.M.Drake (eds.), Molecular Dynamics in Restricted Geometries, (John Wiley, New York, 1989).
2. J.M.Drake, J.Klafter and R.Kopelman (eds.), Dynamics in Small Confining Systems, (MRS, Pittsburgh, 1990).
3. R.Evans, J. Phys.: Condens. Matt. **2** (1990) 8989 .
4. D.D.Awschalom and J.Warnock, In ref 1.
5. M.Watanabe, A.M.Brodsky, and W.P.Reinhardt, J. Phys. Chem. **95** (1991) 4593.
6. M.A.Vorotyntsev, and A.A.Kornyshev, Sov. Phys. JETP **51** (1980) 509.
7. M.A.Vorotyntsev, in Advances of Science and Engineering, Electrochemistry, **26** (1988) 3 (in Russian).
8. J.M.Drake and J.Klafter, Physics Today **43** (1990) 46.
9. T.W.Zerda and Y.Shao, Chem. Phys. Lett. **209** (1993) 347.
10. M.Urbakh, and J.Klafter, J. Phys. Chem. **96** (1992) 3480.
11. M.Urbakh, and J.Klafter, J. Phys. Chem. **97** (1993) 3344.

12. V.M.Agranovich, and V.L.Ginzburg, Spatial Dispersion in Crystal Optics and the Theory of Excitons, (Interscience, New York, 1976).
13. See, e.g.: The Chemical Physics of Solvation, edited by R.R.Dogonadze, E.Kalman, A.A.Kornyshev and J.Ulstrup., (Elsevier, Amsterdam, 1988), Parts A and C.
14. Tsu-Wei Nee and R.Zwanzig, J. Chem. Phys. **52** (1970) 6353.
15. D.S.Alavi and D.H.Waldeck J. Chem. Phys. **94** (1991) 6196.
16. J.E.Enderby and G.W.Neilson, Rep. Progr. Phys. **44** (1981) 953.
17. E.L.Pollock and B.J.Alder, Phys. Rev. Lett. **46** (1981) 950.
18. M.Revere, R.Miniero, M.Parinello and M.P.Tosi, Phys. Chem. Liq. **9** (1979) 11.
19. M.Warner and M.E.Cates, J. Phys. II (France) **3** (1993) 503.
20. J.N.Israelachvili, Intermolecular and Surface Forces with Applications to Colloidal and Biological Systems, (Academic Press, London, 1985).
21. (a) H.T.Davis, S.A.Somers, M.Tirrell and I.Bitsanis, in Danamics in Small Confining Systems, Extended Abstract of 1990 Fall Meeting of the MRS, edited by J.M.Drake, J.Klafter and R.Kopelman, p 73. (b) M.Lupkowski, and F.van Swol, In Dynamics in Small Confining Systems, Extended Abstract of 1990 Fall Meeting of the MRS, edited by J.M.Drake, J.Klafter and R.Kopelman, p 19.
22. M.A.Vorotyntsev, in The Chemical Physics of Solvation, edited by R.R.Dogonadze, E.Kalman, A.A.Kornyshev and J.Ulstrup, (Elsevier, Amsterdam, 198)8, Part C, p 401.
23. S.Manne ,P.K.Hasma, J.Massie, V.B.Elings and A.A.Gewirth, Science **251** (1991) 133.
24. F.Goodman and N.Garcia, Phys. Rev. B **43** (1991) 4728.
25. N.Garcia and Vu Thien Binh, Phys. Rev. B **46** (1992) 7946.
26. R.P.Rand and V.A.Parsegian, Biochim. Biophys. Acta, **988** (1990) 351.
27. G.Cevc and A.Kornyshev, J. Electroanal. Chem., **330** (1992) 407.
28. W.Harrison, Solid State Theory, (McGraw-Hill, New York, 1970).
29. A.A.Kornyshev, A.I.Rubinstein and M.A.Vorotyntsev, J. Phys. C. Solid State Phys. **11** (1978) 3307.
30. A.A.Kornyshev and S.Leikin, Phys. Rev. B **40** (1989) 6431;
M.K.Granfeldt and Bo Jonsson, Chem. Phys. Lett. **195** (1992) 174.
31. A.A.Kornyshev, J. Electroanal. Chem. **255** (1988) 297.
32. C.J.F.Bottcher and P.Bordewijk, Theory of Electric Polarization, (Elsevier, Amsterdam, 1979)
34. G. van der Zwan and R.M.Mazo, J. Chem. Phys. **82** (1985) 3344
35. V.M.Agranovich and M.D.Galanin, Electronic Excitation Energy Transfer in Condenced Matter, (North-Holland, Amsterdam, 1982).
36. P.Levitz, J.M.Drake and J.Klafter, J. Chem. Phys. **89** (1988) 5224.

MICROCHEMISTRY
Spectroscopy and Chemistry in Small Domains
Edited by H. Masuhara et al.

Picosecond fluorescence dynamics in solid-solution interface layers

S. Hamai[*,#], N. Tamai[‡] and M. Yanagimachi[§]

Microphotoconversion Project[†], ERATO, Research Development Corporation of Japan, 15 Morimoto-cho, Shimogamo, Sakyo-ku, Kyoto 606, Japan

By means of time-resolved total internal reflection (TIR) fluorescence spectroscopic measurements, effects of a sapphire/liquid interface on physicochemical properties of solvents and fluorescent molecules have been investigated. The rates of solvation dynamics of coumarin 460 in 1-butanol, the proton transfer reaction of 1-naphthol in water, the excitation energy relaxation of a merocyanine dye in 1-butanol, and the pyrene excimer formation in toluene containing poly(methyl methacrylate) were retarded in the sapphire/liquid interface layers with submicrometer thickness compared to bulk solutions.

1. INTRODUCTION

Solid/liquid interface layers have attracted many researchers who have studied in the fields including macromolecular chemistry, photochemistry, photophysics, and other pure and applied sciences. Total internal reflection (TIR) spectroscopy has been utilized to investigate physicochemical properties of solutions or solutes at solid/liquid interface layers, because those properties cannot be investigated by means of other conventional spectroscopic methods. The principle of a time-resolved TIR fluorescence spectroscopy is already reviewed in Part IV of this Volume. We describe here on its application to some investigations regarding the dynamic properties of liquids and fluorescent probes at sapphire/liquid interfaces layers.

When TIR occurs under the conditions that the incidence angle is greater than the so-called critical angle (θ_c), there is an evanescent wave whose amplitude drops off exponentially as it penetrates a less dense medium (liquid) from a dense medium (sapphire). A penetration depth (d_p) defined by the following equation represents a

---

[*] To whom correspondence should be addressed.
[#] Present address: Department of Chemistry, College of Education, Akita University, Tegata Gakuen-machi 1-1, Akita 010, Japan
[‡] Present Address: Light and Material Group, PRESTO, JRDC, Department of Chemistry, Faculty of Science, Kwansei Gakuin University, Uegahara, Nishinomiya 662, Japan
[§] Present address: Central Research Institute, Mitui Toatsu Chemicals, Inc., Yokohama 247, Japan
[†] Five-year term project: October 1988 ~ September 1993.

distance from the interface with the intensity of the evanescent wave being reduced to 1/e of that at the interface:

$$d_p = \lambda_i/[4\pi n_1(\sin^2\theta_i - \sin^2\theta_c)^{1/2}], \qquad \sin\theta_c = n_2/n_1, \tag{1}$$

where λ_i is the wavelength of the excitation light, n_1 and n_2 are the refractive indices of a sapphire prism and a liquid at λ_i, respectively, and θ_i is the incidence angle.

TIR fluorescence spectroscopy is a very powerful tool in examining liquid properties, inter- and intra-molecular interactions of solutes in interface layers, and interactions of a probe molecule with the interface itself.

2. SOLVATION DYNAMICS OF A COUMARIN DYE AT A SAPPHIRE/1-BUTANOL INTERFACE LAYER

Solvation dynamics in which polar solvent molecules reorient around a solute molecule immediately after a photo-excitation are closely related to interactions between molecules. Since a solid/liquid interface influences movements of solvent molecules existing in the vicinity of an interface, the solvation dynamics in an interface layer is expected to be different from those in a bulk solution. From an aspect of dielectric properties of liquids, Urbakh and Klafter have theoretically studied the role of a boundary (solid/liquid interface) in modifying the relaxation behavior of a dipole embedded in a liquid [1]. Corrections of the dipole relaxation by the presence of the boundary have been shown to be small. Larger corrections have been introduced by postulating structural changes in the nature of the liquid near the boundary.

There are three solvent relaxation processes for alcohols [2,3]. The fastest solvent relaxation is attributed to a rotation of a hydroxyl group about a C-O bond of alcohol. The intermediate is due to a rotation of a free monomeric molecule itself, and the slowest is due to the breaking of the hydrogen bonding network in alcohol aggregates associated with the reorientation of alcohol clusters. The solvent relaxation times of 1-butanol have been evaluated to be 100 ps (53 %) and 17 ps (47 %) at 298 K [4]. The longer relaxation time (100 ps) is due to the breaking of the hydrogen bonds in alcohol clusters, and the shorter one (17 ps) is due to the rotations of a free alcohol molecule itself and a hydroxyl group about a C-O bond in alcohol. Furthermore, the longest dielectric relaxation time of 1-butanol has been estimated to be 72 ps, which is also intimately related with solvent motion [5].

Fluorescence spectra of coumarin 460 (10^{-4} - 10^{-3} mol dm^{-3}) in 1-butanol were found to be nearly the same under the normal and TIR excitation conditions, indicating that the adsorption of coumarin 460 on a sapphire surface is negligible [6]. In time-resolved fluorescence spectra of coumarin 460 in bulk 1-butanol, the fluorescence peak in the initial time region from 0 to 60 ps was located at 425 nm. The peak position gradually shifted to the red with time, and finally attained 440 nm (240 - 300 ps). This finding shows that a reorientation of solvent molecules takes place within 200 ps. The time-resolved fluorescence Stokes shift was analyzed by the correlation function:

$$C(t) = [\nu(t) - \nu(\infty)]/[\nu(0) - \nu(\infty)], \tag{2}$$

where $\nu(0)$, $\nu(\infty)$, and $\nu(t)$ are the optical frequencies of the maximum fluorescence intensity at null time, at infinite time, and at time t, respectively. In a simple continuum model, C(t) is expressed by the sum of exponentials with a longitudinal relaxation time constant (τ_L):

$$C(t) = \sum_{i=1}^{n} A_i \exp(-t/\tau_{Li}) \tag{3}$$

where n is the number of solvent relaxation processes and A_i is the amplitude of τ_{Li}. From eqs. 2 and 3, the relaxation time of the time-dependent Stokes shift was estimated to be approximately 80 ps, which was in good agreement with the longest longitudinal relaxation time of 1-butanol (72 ps).

Figure 1 shows fluorescence rise and decay curves observed at 400 and 480 nm under both normal and TIR excitation conditions. A slow decay component of ≈ 3.5 ns is attributable to a fluorescence lifetime of coumarin 460 in 1-butanol. In addition to the slow decay component, there are fast decay and rise components in the time profiles observed at 400 and 480 nm, respectively. It should be noted that both fast components obtained under the TIR excitation condition are slower in rate than those under the normal condition. This finding indicates that the dynamic Stokes shift in the sapphire/1-butanol interface layer occurs more slowly than that in bulk 1-butanol. A rotational relaxation time and an initial anisotropy of a coumarin dye

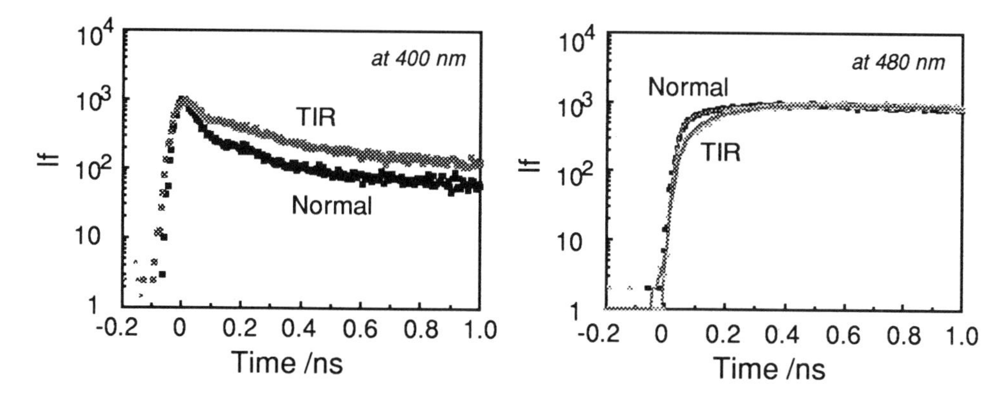

Figure 1. Fluorescence rise and decay curves of coumarin 460 in 1-butanol under the normal and TIR excitation conditions. The best fitting curves at 480 nm with a multi-exponential function similar in form to eq. 3, are also shown: τ_1 = 80 ps (A_1 = -0.569) and τ_2 = 3.57 ns (A_2 = 1.00) for the fitting curve under the normal condition. τ_1 = 90 ps (A_1 = -0.457), τ_2 = 170 ps (A_2 = -0.543), and τ_3 = 3.50 ns (A_3 = 1.00) for the curve under the TIR condition.

excited at 405 nm have been reported to be 250 ps and 0.23, respectively [7]. Such a rotational relaxation may be superimposed on the intrinsic fluorescence decay curves of coumarin 460. For coumarin 460 excited at 295 nm, however, the initial anisotropy was negligibly small. Consequently, we neglected the effect of the rotational relaxation on the observed time profiles of the coumarin fluorescence.

Fluorescence rise and decay curves observed at 480 nm in the normal excitation mode were analyzed by a two-exponential function. Time constants of the rise and decay components thus obtained were 80 ps and 3.4 ns, respectively. The rise time of 80 ps which was evaluated from the fluorescence rise and decay curves was identical to the relaxation time of the dynamic Stokes shift in bulk solution which was previously estimated on the basis of eq. 3. A rise component due to the fast solvation process (17 ps) appears to be difficult to detect because a time resolution in our single photon timing experiments was 5.4 ps per channel.

Fluorescence rise and decay curves obtained in the TIR excitation mode (d_p = 33 nm) could be fitted by a sum of three exponentials with two rise and one decay components but not a two-exponentials. Even in the TIR mode, the emission from a bulk solution as well as from an interface layer contributes to the observed fluorescence. Thus, we analyzed the rise and decay curves in the TIR mode of d_p = 33 nm using a variable time constant for the interface layer and two fixed time constant for the bulk solution (fast rise time of 80 ps and slow decay component of 3.5 ns). From this analysis, the rise time constant for the interface layer was determined to be 170 ps. For other rise and decay curves in the TIR mode where the penetration depth was greater than 33 nm, analyses were made on the basis of a two-layers model of a solvent; one fixed time constant (170 ps) for the interface layer and two fixed time constants (80 ps and 3.5 ns) for the bulk were employed with variable amplitudes.

In Figure 2, the amplitude ratios of the additional slow rise component (170 ps) to the fast one (80 ps) and the χ^2 and Durbin-Watson (DW) parameters for the fitting curve are given as a function of penetration depth. Both the amplitude ratio and χ^2 increase with a decrease in the penetration depth, whereas the DW value decreases. The fittings ($\chi^2 \geq 1.3$ and DW ≤ 1.7) of the fluorescence time profiles for the interface layer less than 50 nm thickness are not so good except for the data of the interface layer with d_p = 33 nm which was utilized to determine the time constant (170 ps) of the additional rise component.

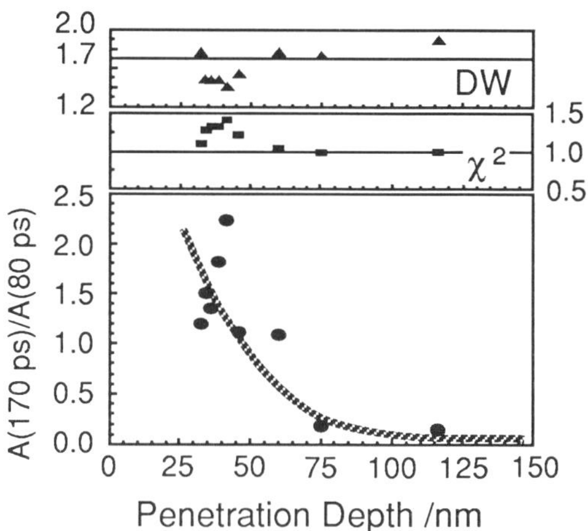

Figure 2. Penetration-depth dependences of the amplitude ratio of the slow [A(170 ps)] to the fast [A(80 ps)] components, χ^2, and Durbin-Watson (DW) parameters for the curve fitting.

The amplitude ratio has a maximum at a penetration depth between 33 and 50 nm, suggesting that the two-layers model with time constants of 170 ps (interface layer) and 80 ps (bulk) may not fully reproduce the solvation process in the interface layer. The solvent relaxation time seems to gradually change from the very proximity of the interface to the bulk solution. Hydrogen bonding between a hydrophilic part of the sapphire surface and a hydroxyl group of 1-butanol may play an important role in the solvation dynamics at the interface. Although hydrogen bonding at the interface is rigid, the hydrogen bonding network becomes flexible and the solvation process becomes faster as the distance of the network from the sapphire surface is increased. This idea is consistent with the nature of the solvent longitudinal relaxation. Because the breaking of the hydrogen bond in the network involves a number of 1-butanol molecules, a wall effect propagates up to a few nm orders from the interface [8]. The result on the solvation dynamics for coumarin 460 suggests that the solvent relaxation in the interface layer is evidently different from that in the bulk.

3. EXCITED-STATE PROTON TRANSFER OF 1-NAPHTHOL AT A SAPPHIRE/WATER INTERFACE LAYER

Photoinduced proton transfer reactions have so far been extensively examined because these reactions play important roles in many processes of chemical and biological systems. The proton transfer reaction of 1-naphthol is expressed by

$$\begin{array}{ccc} \mathrm{NaphOH^{*}} + \mathrm{H_2O} & \underset{k_{-1}[\mathrm{H_3O^+}]}{\overset{k_1}{\rightleftharpoons}} & \mathrm{NaphO^{-*}} + \mathrm{H_3O^+}, \\ \downarrow k_0 \ \downarrow k_q[\mathrm{H_3O^+}] & & \downarrow k_0' \ \downarrow k_q'[\mathrm{H_3O^+}] \end{array} \qquad \text{(scheme 1)}$$

where NaphOH and $NaphO^-$ represent neutral and anionic forms of 1-naphthol, and k_1 and k_{-1} are the proton transfer and recombination rate constants, respectively. k_0 and k_q are the intrinsic and the diabatic bimolecular quenching rate constants of the neutral form of excited 1-naphthol, respectively, and k_0' and k_q' are those of the anionic form of excited 1-naphthol, respectively. These processes shown in scheme 1 are strongly affected by solvent properties. As a pH of aqueous solution is lowered, the diabatic quenching and recombination reactions by a proton are accelerated [9]. Recently, Fillingim et al. have studied fluorescence decays of 1-naphthol in aqueous solution contained in a confining volume between two parallel quartz plates with a path length of 10 - 20 nm [14]. They found that the fluorescence decay of 1-naphthol in the confining space was slower than that in 1 mm cell. In the confining space, water seems to be more structured and orientationally stiffer than bulk water [14,15]. The increase in the fluorescence lifetime of 1-naphthol has been interpreted in terms of the structured interfacial water.

Since, at a pH of 7, a neutral form of 1-naphthol is a predominant ground-state species (99.4 %) [16], a direct photo-excitation of anionic 1-naphthol can be ignored in our experiments using purified water [17]. An emission of 1-naphthol in aqueous solution is assigned to the fluorescence from the excited state of anionic 1-naphthol, which is generated from the protonated form of excited 1-naphthol [9-11]. Nearly the same fluorescence spectra of 1-naphthol (10^{-4} - 10^{-3} mol dm^{-3}) were obtained under the normal and TIR conditions, suggesting no adsorption of 1-naphthol on a sapphire surface [17]. In time-resolved fluorescence spectra of 1-naphthol under the normal condition, a fluorescence peak at the initial time was located at approximately 370 nm. The 370-nm band disappeared very rapidly within 100 ps, and another band due to anionic 1-naphthol appeared at 470 nm and grew up with time. Because the maximum wavelength (370 nm) of the fluorescence in the initial time region is in good agreement with that for 1-naphthol in aqueous solution of a low pH value, neutral 1-naphthol is responsible for the 370-nm band [18,19]. Fluorescence decays observed at a short wavelength of 370 nm, therefore, can be used to determine the time constant of the proton transfer reaction of 1-naphthol in the excited state.

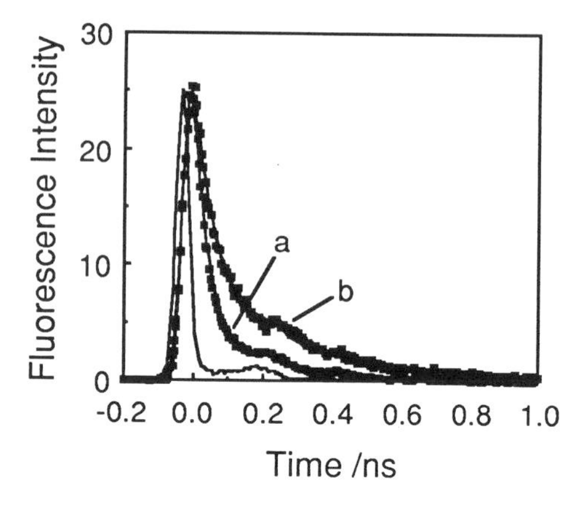

Figure 3. Fluorescence decay curves of 1-naphthol in bulk aqueous solution (a) and an interface layer (b) (d_p = 29 nm) observed at 360 nm. A solid line is a system response function at 300 nm.

Figure 3 illustrates fluorescence decay curves of 1-naphthol observed at 360 nm in the normal and TIR excitation modes. For 1-naphthol-3,6-disulfonate, a fluorescence decay profile does not show a single exponential [20]. This has been attributed to a reversible geminate recombination process. Although this process has been analyzed by a nonexponential function, we deconvoluted the fluorescence decay curves of 1-naphthol by the sum of three exponentials as a first approximation. A lifetime of the slowest decay component that is evaluated to be $\approx$ 7.3 ns with a small contribution (< 2.9 %) is almost identical to a lifetime observed at the peak (470 nm) of the 1-naphthol fluorescence, which is due to 1-naphthol anion [11,16,19]. An intermediate decay component is due to the geminate recombination process, and its lifetime is estimated to be 150 ps (< 5 %) which is comparable with a reciprocal (133 ps/Å) of the geminate recombination rate constant for 1-naphthol-3,6-disulfonate [20]. In the normal excitation mode, a lifetime (τ_1) of the fastest decay component is estimated to be 30 $\pm$ 4 ps, which is similar to lifetimes determined by Webb et al. [16,19] and Shizuka et al [11]. In the TIR excitation mode (d_p = 29 nm), on the other hand, τ_1 is estimated to be 70 $\pm$ 8 ps which is about two times longer than that (30 $\pm$ 4 ps) in the normal excitation mode. This finding indicates that the proton transfer reaction is considerably slower in the interface layer than in the bulk solution. Because proton-dependent bimolecular processes involving a reverse proton transfer reaction are negligible in neutral aqueous solution, the rate constant (k_1) of the proton transfer reaction is given by

$$k_1 = 1/\tau_1 - 1/\tau_0, \tag{4}$$

where τ_0^{-1} is the intrinsic rate constant of neutral 1-naphthol. Using the τ_0^{-1} value of $(6.8 \pm 1.4) \times 10^9$ s^{-1} [19], the rate constants of proton transfer reaction for the bulk solution and the interface layer (d_p = 29 nm) were calculated to be 2.70×10^{10} and 0.76×10^{10} s^{-1}, respectively.

The k_1 values evaluated at various penetration depths are shown in Figure 4. As the penetration depth is decreased, k_1 is decreased. For 1-naphthol in aqueous solution that was contained in an ultra-thin cell with a path length of 14 ± 2 nm, fluorescence decays have been well reproduced by a two-exponential with lifetimes of ≈ 600 ps (60 %) and ≈ 5.8 ns (40 %) [14]. The fast decay (600 ps) which is attributed to a time constant of the proton transfer reaction seems to be due to 1-naphthol adsorbed on a quartz cell [19]. The time constant of 600 ps is extraordinarily longer than τ_1 (70 ± 8 ps) for the interface layer, suggesting that another factor rather than the adsorption plays an important role in decelerating k_1. Hydrogen bonding between water molecules in the vicinity of the interface is influenced by sapphire/water interactions, so that the hydrogen bonding network may be more rigid in the interface layer than in a bulk solution.

Not a single water molecule but a cluster of four water molecules act as a proton trap in a bulk solution [12,13,21]. Addition of NaCl or alcohol results in a decrease in the rate of the proton transfer reaction [10-13]. These results have been attributed to a destruction of water clusters by ions or alcohols. Thus, we examined the salt effect on k_1 of 1-naphthol in water. Figure 5 illustrates the effect of NaCl on k_1 for bulk solutions and the interface layer (d_p = 29 nm). At the same NaCl concentrations, k_1 for the bulk solution is always greater than that for the interface layer. The k_1 values for both the bulk solution and the interface layer are decreased as the NaCl concentration is increased. A slope of the plot for the bulk is negatively greater than that for the interface layer. Because of a more rigid water structure at the interface,

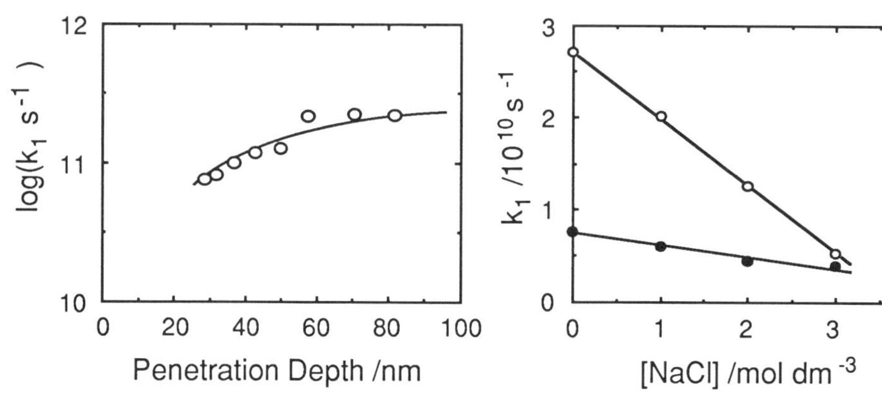

Figure 4. Penetration-depth dependence on k_1 for 1-naphthol in aqueous solution.

Figure 5. Effect of NaCl on k_1 for 1-naphthol in water: bulk (o) and interface layer (o) with d_p = 29 nm.

NaCl may exert a weak effect on the water structure in the interface layer compared to the bulk, resulting in the poor NaCl dependence of k_1 for the interface layer.

4. EXCITATION ENERGY RELAXATION OF A MEROCYANINE DYE AT A SAPPHIRE/1-BUTANOL INTERFACE LAYER

Dyes are widely distributed as many kinds of key compounds in biological systems. In this Section, we have investigated the excitation energy relaxation and the inhomogeneous aggregation of a merocyanine dye (MCD-18) in 1-butanol [22].

Figure 6 depicts fluorescence spectra of MCD-18 (4.6 x 10^{-5} mol dm^{-3}) in 1-butanol under the normal and TIR conditions. The maximum of the MCD-18 fluorescence observed in the TIR mode is slightly shifted to longer wavelengths (≈ 558 nm) compared to that (556 nm) in the normal mode. At the same time, the fluorescence intensity of a shoulder at ≈ 580 nm is reduced as the penetration depth is decreased.

Figure 7 shows the concentration effect of MCD-18 in 1-butanol on its fluorescence spectrum. When the concentration is raised, the fluorescence maximum is shifted to the red accompanied by an intensity reduction of the shoulder at ≈ 580 nm. The spectral changes in Figure 7 are attributable to the aggregation of MCD-18. Since the fluorescence spectral changes in Figure 6 resemble those in Figure 7, MCD-18 aggregates more efficiently in the interface layer compared to the bulk solution.

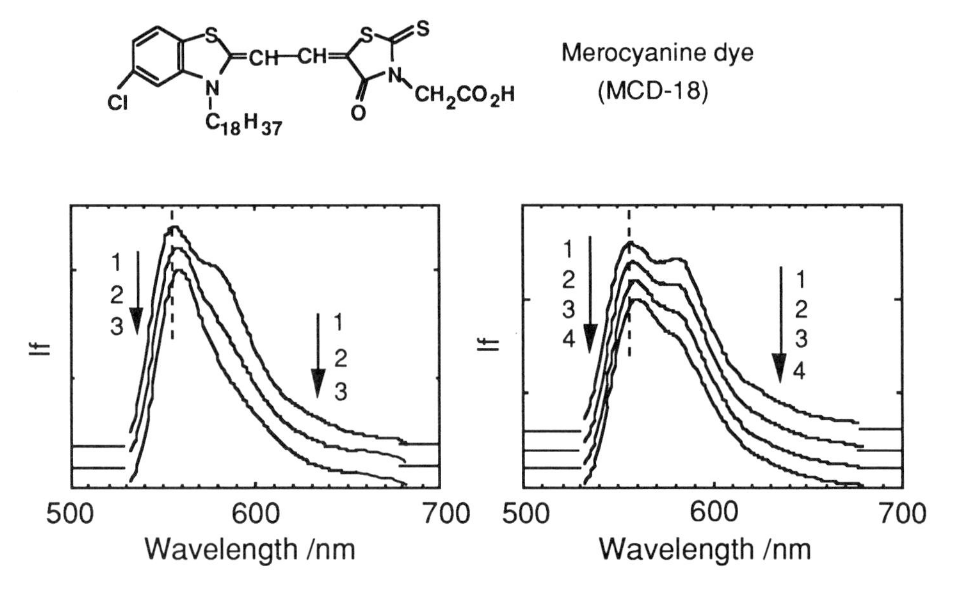

Figure 6. Normalized fluorescence spectra of MCD-18 (4 x 10^{-5} mol dm^{-3}) in bulk 1-butanol (1) and in the sapphire/1-butanol interface layers. Penetration depth: (2) 112 ± 13 and (3) 53 ± 2 nm.

Figure 7. Concentration dependence of the normalized MCD-18 fluorescence spectra in bulk 1-butanol. Concentration of MCD-18: (1) 2.4 x 10^{-6}, (2) 1.1 x 10^{-5}, (3) 4.6 x 10^{-5}, and (4) 1.0 x 10^{-4} mol dm^{-3}.

A fluorescence decay curve of MCD-18 (4.6 x 10^{-5} mol dm^{-3}) in bulk 1-butanol could be analyzed by a two-exponential function with lifetimes of 16 and 150 ps. The amplitude A_1 (97.9 %) of the fast decay component is significantly greater than the other A_2 (2.1 %), indicating that the fast and slow decay components are due to a monomer and an aggregate of MCD-18, respectively. On the other hand, decay curves obtained in the TIR mode were analyzed by three exponentials; an additional component with a lifetime of a few hundreds ps was obtained. The shortest lifetime was attributed to the monomer because, at a penetration depth of about 100 nm, it was nearly the same as the shorter lifetime for a bulk solution.

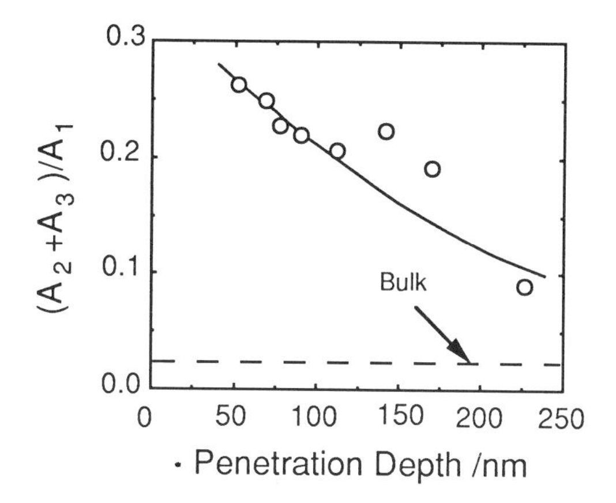

Figure 8. Penetration- depth dependence of $(A_2 + A_3)/A_1$ for MCD-18 (4.6 x 10^{-5} mol dm^{-3}) in 1-butanol. A dashed line represents A_2/A_1 for the bulk solution.

All the lifetimes became larger with decreasing the penetration depth (≤ 100 nm). As the penetration depth was decreased, the amplitude of the monomer, A_1, was decreased in contrast to those (A_2 and A_3) of the aggregates.

Figure 8 shows the amplitude ratio of aggregates to the monomer, $(A_2 + A_3)/A_1$, as a function of penetration depth. In Figure 8, also shown is the A_2/A_1 value for a bulk solution as a dashed line. With a decrease in the penetration depth, $(A_2 + A_3)/A_1$ is increased, indicating that the aggregation is remarkably promoted in the interface layer. Such an inhomogeneous aggregation is probably due to an enhancement of equilibrium constants for the aggregation at the interface relative to a bulk solution; water adsorbed on the surface may be responsible for the enhanced equilibrium constants. Another possible explanation is that MCD-18 adsorbed on the sapphire surface exists as an aggregate whose contribution to the fluorescence is extremely higher than that of aggregates in a bulk solution. Thus, we examined the fluorescence of MCD-18 adsorbed on sapphire under the atmosphere. The fluorescence of the adsorbed MCD-18 was found to have a peak at ≈ 562 nm, and is most likely to be attributed to an aggregate because of a disappearance of the shoulder at ≈ 580 nm. It is worth noting that the fluorescence of the adsorbed MCD-18 is slightly red-shifted (≈ 4 nm) relative to that of MCD-18 in the interface layer. When the adsorbed MCD-18 is hemispherically surrounded by solvent 1-butanol molecules possessing a large dielectric constant, the fluorescence peak of the adsorbed MCD-18 is expected to be shifted to much longer wavelengths compared to that of adsorbed one surrounded by atmosphere. This is not the case. Consequently, it is concluded that the fluorescence changes in the TIR mode shown in Figure 6 are not due to MCD-18 aggregates adsorbed on the sapphire surface but due to those in the interface layer.

Our results on the excitation energy relaxation of the merocyanine dye strongly suggest that the sapphire surface affects some physical and/or chemical properties of liquids to a great extent.

5. PYRENE EXCIMER FORMATION AT A SAPPHIRE/POLYMER SOLUTION INTERFACE LAYER

Fluorescence properties of pyrene doped in polymer films were investigated employing time-resolved TIR fluorescence spectroscopy [23-25]. For segmented poly(urethane urea) (SPUU, copolymer of toluene-2,4-diisocyanate (TDI) and poly(propylene oxide) (PPO)) and poly(*o*-hydroxystyrene) (PHST) films, both the rise and decay of the excimer fluorescence were faster in the TIR excitation mode than in the normal excitation mode. Although the local concentration of pyrene in the interface layer was higher than that in the bulk layer, the excimer formation was less feasible in the interface layer than in the bulk layer. This result was attributed to the formation of a non-emissive ground-state pyrene dimer in the interface layer. For polymer solutions, on the other hand, Rondelez et al. have shown the existence of depletion layers in which the local concentrations of a copolymer of styrene-methyl methacrylate and a xanthan polymer are decreased relative to bulk solutions [26-28]. It has been pointed out that poly(4-butoxycarbonylmethyl urethane) is adsorbed onto a fused silica surface in solutions of 2-methyl-tetrahydrofuran and a chloroform-hexane mixture [29]. Poly(methyl methacrylate) (PMMA) in toluene has been found to be adsorbed on a sapphire surface [30].

In spite of these investigations, photochemical and photophysical behaviors of a polymer at an interface in solutions are little known. Thus, employing pyrene as a fluorescent probe, we investigated effects of a sapphire/toluene interface on physicochemical properties of a solvent containing PMMA through time-resolved TIR fluorescence spectroscopy.

Figure 9 shows normalized fluorescence spectra of pyrene (4.1 x 10^{-3} mol dm^{-3}) in aerated toluene containing 3.3 x 10^{-4} mol dm^{-3} of PMMA (M.W. = 120000) under the normal and TIR excitation conditions. The pyrene excimer fluorescence (λ_{max} = 480 nm) as well as the monomer fluorescence (λ_{max} = 397 nm) in the TIR mode exhibits no peak shift relative to that in the normal mode. The fluorescence intensities of the pyrene monomer bands at 372 and 385 nm are lower in the normal mode than in the TIR mode. This finding can be interpreted in terms of a reabsorption effect of high-concentration pyrene because nearly the same band intensities were observed for a dilute solution of

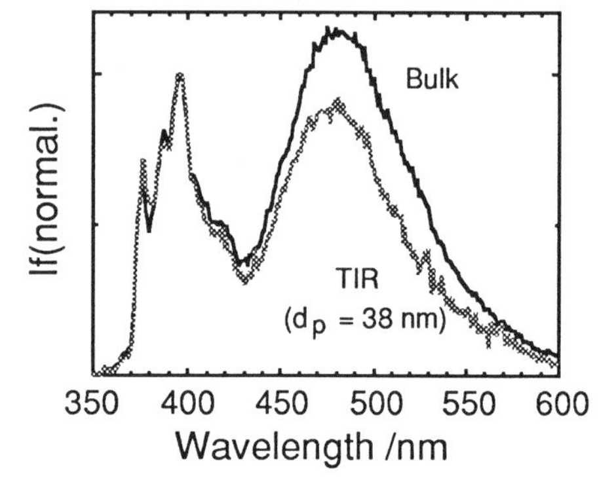

Figure 9. Fluorescence spectra of pyrene in toluene containing 3.3 x 10^{-4} mol dm^{-3} PMMA (M.W. = 120000) under the normal and TIR excitation conditions. Fluorescence spectra are normalized at the monomer fluorescence peak (397 nm).

pyrene in the both modes. On the other hand, the fluorescence intensity of the excimer is decreased in the interface layer compared to that in the bulk solution, indicating that the excimer formation is less feasible in the interface layer. The reduction of the excimer fluorescence cannot be attributed to the effective quenching of the excimer by oxygen dissolved in the interface layer because a decrease in the excimer fluorescence intensity was similarly observed for an Ar bubbled pyrene solution with PMMA.

Fluorescence decay and rise curves of pyrene in toluene with PMMA (3.3×10^{-4} mol dm^{-3}), observed at 385 and 500 nm, were analyzed according to the Birks kinetics [31]:

$$I_M(t) = k_{FM}(\lambda_2 - X)[\exp(-\lambda_1 t) + A \exp(-\lambda_2 t)]/(\lambda_2 - \lambda_1), \quad (5)$$

$$\text{with } \lambda_{2,1} = [X + Y \pm \{(Y - X)^2 + 4k_{MD}k_{DM}[^1M]\}^{1/2}]^2,$$

$$X = k_M + k_{DM}[^1M] + k_{QM}[O_2], \qquad Y = k_D + k_{MD} + k_{QD}[O_2],$$

$$\text{and } A = (X - \lambda_1)/(\lambda_2 - X),$$

where $I_M(t)$ is the monomer fluorescence intensity at time t, k_{FM}, k_{MD}, k_{DM}, k_{QM}, and k_{QD} are the rate constants for the monomer fluorescence, dissociation of the excimer, association between the excited and ground-state monomers, oxygen-quenching of the monomer, and that of the excimer, respectively, k_M and k_D are the reciprocals of the fluorescence lifetimes of the monomer and the excimer in a deoxygenated solution, respectively, and $[^1M]$ and $[O_2]$ are the concentrations of the ground-state monomer and oxygen, respectively.

Observed decay curves of the monomer fluorescence at 385 nm were deconvoluted based on eq. 5. For a toluene solution containing pyrene and PMMA (3.3×10^{-4} mol dm^{-3}), reciprocal values of λ_1 and λ_2 were evaluated as a function of penetration depth, and are given in Figure 10. The λ_1^{-1} and λ_2^{-1} values are increased as the penetration depth is decreased in the range less than $\approx$ 60 nm, indicating that both the formation and decay rates of the pyrene excimer are decelerated in the close proximity of a sapphire surface. The reduction of the excimer fluorescence and the enhancement of λ_1^{-1} and λ_2^{-1} in the interface layer suggest (1) an increase in the viscosity of the interface layer and/or (2) a decrease in the pyrene concentration in the interface layer. To clarify the cause(s) of the above results on the excimer fluorescence and λ_1^{-1} and λ_2^{-1}, we estimated k_{DM} and k_{MD} from the pyrene concentration dependence of λ_1 and λ_2 under the assumption that the pyrene concentration in the interface layer is the same as that in the bulk solution. The values of k_{DM} and k_{MD} for the interface layer of a penetration depth of 38 nm were 8.1×10^9 mol^{-1} dm^3 s^{-1} and 1.0×10^7 s^{-1}, respectively, while those for the bulk solution were 8.9×10^9 mol^{-1} dm^3 s^{-1} and 3.3×10^7 s^{-1}, respectively. It is known that k_{DM} and k_{MD} are dependent on the solvent viscosity. In addition, k_{MD} is independent of the monomer concentration and directly correlates to the microscopic viscosity. The finding that k_{MD} for the interface layer is about one third of that for

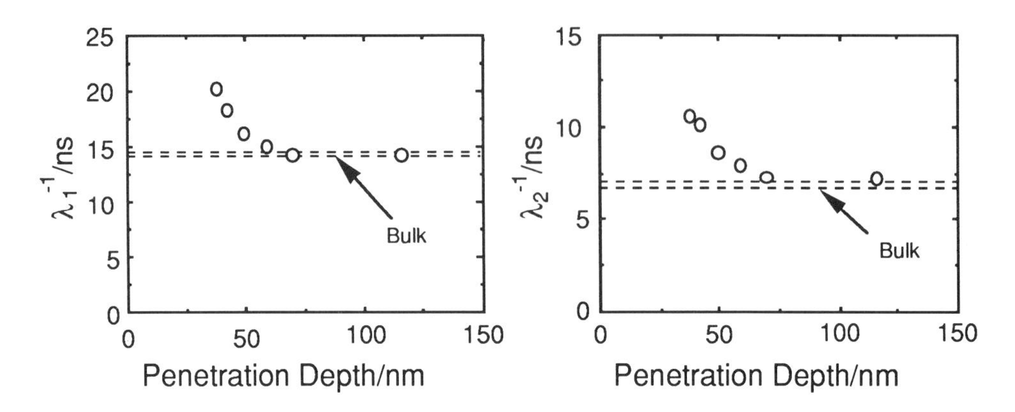

Figure 10. Penetration-depth dependence of λ_1^{-1} and λ_2^{-1} for pyrene in toluene containing 3.3×10^{-4} mol dm^{-3} of PMMA (M.W. = 120000).

the bulk solution evidently indicates that the viscosity of the interface layer is greater than that of the bulk solution. On the other hand, k_{DM} for the interface layer was only about 10 % less than that for the bulk solution in spite of the enhancement of the viscosity of the interface layer. A much less k_{DM} value than the estimated k_{DM} value is deduced from the increased viscosity of the interface layer. As stated previously, k_{DM} was evaluated under the assumption of the pyrene concentration being held constant over the range from the bulk to the interface. Because k_{DM} is calculated from a value of k_{DM} multiplied by the pyrene concentration, an estimation of k_{DM} inversely depends on the pyrene concentration. When the pyrene concentration in the interface layer is high compared to that of the bulk solution, k_{DM} takes a value less than that previously estimated. Therefore, the concentration of pyrene in the interface layer seems to be high compared to that in the bulk solution. Our analyses concerning the incidence- and observation-angle dependence of the fluorescence intensity for a pyrene solution with PMMA also suggest that the pyrene concentration is higher in the interface layer than in the bulk solution.

The incidence-angle dependence is shown in Figure 11 where a best fit curve is simulated using a step function of the relative pyrene concentration in a bulk solution ([pyrene] = 1) and that in the interface layer ([pyrene] = 14) with a thickness of 25 nm. These parameters seem to be consistent with our model that the viscosity and pyrene concentration for the interface layer are high compared to those for the bulk solution in spite of the slightly high pyrene concentration in the interface layer than expected. Although the pyrene concentration is high in the interface layer compared to that in the bulk solution, the excimer fluorescence for the interface layer is reduced relative to the bulk solution as shown in Figure 9. The effect of the enhanced viscosity of the interface layer predominates over the effect of the increased pyrene concentration of the interface layer, resulting in the reduction of the excimer emission.

Caucheteux et al. have pointed out that PMMA chains are adsorbed from toluene on a sapphire surface and that the concentration of PMMA in an adsorption layer is greater than that in a bulk solution [30]. For PMMA of molecular weights of 12000 and 1 x 10^7, thicknesses of an adsorption layer of PMMA have been estimated to be 8.2 ± 0.7 and 16.3 ± 0.7 nm, respectively. Since the translational and rotational motions of adsorbed PMMA are restricted owing to the high concentration of PMMA in the adsorption layer, it is most likely that the viscosity of the adsorption layer is enhanced relative to a bulk solution.

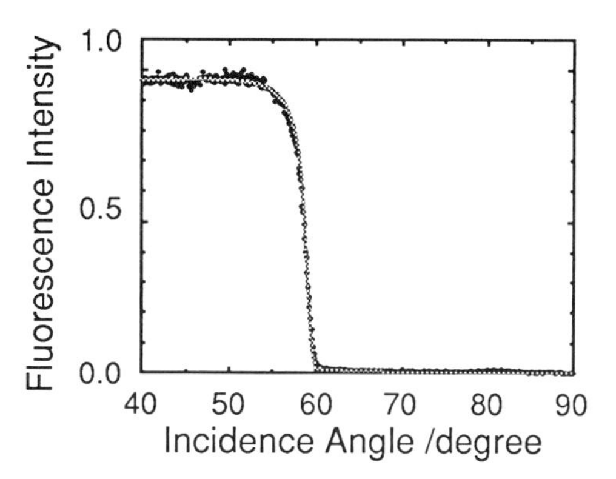

Figure 11. Incidence-angle dependence of the pyrene fluorescence intensity in toluene containing 3.3 x 10^{-4} mol dm^{-3} PMMA (M.W. = 120000). A best fit curve was calculated according to a step function for the relative pyrene concentration of a bulk solution ([pyrene] = 1) and that of the interface layer ([pyrene] = 14) with a thickness of 25 nm.

For toluene without PMMA, there were no differences in the fluorescence spectrum and decay kinetics of pyrene between the interface layer and the bulk solution, indicating that the sapphire/toluene interface does not affect the microscopic viscosity of the interface layer. Furthermore, this fact provides an evidence for no adsorption of pyrene on a sapphire surface in toluene.

The enhancement of λ_1^{-1} and λ_2^{-1} for the PMMA solutions is contrary to the observation of the faster rise and decay of the pyrene excimer fluorescence for the interface layers of SPUU and PHST films. With respect to the translational diffusion between polymer chains, the interface effects for liquids may be different from those for the solid phase.

As evidenced by our results, the adsorption of PMMA increases the viscosity of the sapphire/toluene interface layer, and photophysical processes such as a translational diffusion of pyrene and a deactivation of the pyrene excimer are modified in the interface layer. In addition, the pyrene concentration gradient occurs in the vicinity of the interface.

ACKNOWLEDGMENTS

The authors thank Prof. H. Masuhara (Osaka Univ.) for the collaborations and discusions during the research period of the Microphotoconversion Project.

REFERENCES

1. M. Urbakh and J. Klafter, J. Phys. Chem., 96, (1992) 3480.
2. E. G. Castner Jr., B. Bagchi, M. Maroncelli, S. P. Webb, A. J. Ruggiero and G. R. Fleming, Ber. Bunsenges. Physik. Chem., 92 (1988) 363.

3. S. K. Garg and C. P. Smyth, J. Phys. Chem., 69 (1965) 1294.
4. E. W. Castner Jr., M. Maroncelli and G. R. Fleming, J. Chem. Phys., 86 (1987) 1090.
5. E. M. Kosower and D. Huppert, Ann. Rev. Phys. Chem., 37 (1986) 127.
6. M. Yanagimachi, N. Tamai and H. Masuhara, Chem. Phys. Lett., 200 (1992) 469.
7. M. Maroncelli and G. R. Fleming, J. Chem. Phys., 86 (1987) 6221.
8. C. Y. Lee, J. A. McCammon and P. J. Rossky, J. Chem. Phys., 80 (1984) 4448.
9. C. M. Harris and B. K. Selinger, J. Phys. Chem., 84 (1980) 1366.
10. C. M. Harris and B. K. Selinger, J. Phys. Chem., 84 (1980) 891.
11. H. Shizuka, T. Ogiwara, A. Narita, M. Sumitani and K. Yoshihara, J. Phys. Chem., 90 (1986) 6708.
12. J. Lee, R. D. Griffin and G. W. Robinson, J. Chem. Phys., 82 (1985) 4920.
13. G. W. Robinson, P. J. Thistlethwaite and J. Lee, J. Phys. Chem., 90 (1986) 4224.
14. T. G. Fillingum, S. -B. Zhu, S. Yao, J. Lee and G. W. Robinson, Chem. Phys. Lett., 161 (1989) 444.
15. W. Drost-Hansen and J. S. Clegg, eds., Cell-associated water (Academic Press, New York, 1967).
16. S. P. Webb, S. W. Teh, L. A. Philips, M. A. Tolbert and J. H. Clark, J. Am. Chem. Soc., 106 (1984) 7286.
17. M. Yanagimachi, N. Tamai and H. Masuhara, Chem. Phys. Lett., 201 (1993) 115.
18. K. Tsutsumi and H. Shizuka, Z. Physik. Chem. N. F., 122 (1980) 129.
19. S. P. Webb, L. A. Philips, S. W. Yeh, L. M. Tolbert and J. H. Clark, J. Phys. Chem., 90 (1986) 5154.
20. A. Masad and D. Huppert, Chem. Phys. Lett., 180 (1991) 409.
21. H. Shizuka, Accounts Chem. Res., 18 (1985) 141.
22. S. Hamai, N. Tamai and H. Masuhara, Chem. Lett., (1983) 1105.
23. M. Yanagimachi, M. Toriumi and H. Masuhara, Chem. Materials, 3 (1991) 413.
24. M. Toriumi and H. Masuhara, Spectrochim. Acta Rev., 14 (1991) 353.
25. M. Toriumi, M. Yanagimachi and H. Masuhara, Advances in Resist Technology and Processing VIII (SPIE), 1466 (1991) 458.
26. C. Allain, D. Ausserre and F. Rondelez, Phys. Rev. Lett., 49 (1982) 1694.
27. D. Ausserre, H. Hervet and F. Rondelez, Macromolecules, 19 (1986) 85.
28. D. Ausserre, H. Hervet and F. Rondelez, Phys. Rev. Lett., 54 (1985) 1948.
29. G. Rumbles, A. J. Brown, D. Phillips and D. Bloor, J. Chem. Soc., Faraday Trans., 88 (1992) 3313.
30. I. Caucheteux, H. Hervet, R. Jerome and F. Rondelez, J. Chem. Soc., Faraday Trans., 86 (1990) 1369.
31. J. B. Birks, Photophysics of aromatic molecules (Wiley-Interscience, London, 1970).

MICROCHEMISTRY
Spectroscopy and Chemistry in Small Domains
Edited by H. Masuhara et al.

Photophysics and photochemistry of individual microparticles in solution

M. Koshioka,# U. Pfeifer-Fukumura,‡ S. Funakura,‡‡ K. Nakatani,$ and H. Masuhara*,§

Microphotoconversion Project,† ERATO, Research Development Corporation of Japan, 15 Morimoto-cho, Shimogamo, Sakyo-ku, Kyoto 606, Japan

Fluorescence and transient absorption spectra and their rise and decay curves were measured for individual microparticles dispersed in solution. Functionalized polymeric microspheres were characterized, and the surface micropolarity and local condition were elucidated by fluorescence spectroscopy. Analysis of excimer kinetics in microcapsules and oil droplets gave information on solute concentration distribution and viscosity of the inner solution. The viscosity in the oil droplet was also considered on the basis of T-T annihilation rate constants. It was emphasized that measurement of a single microparticle is indispensable and important for understanding its structure, dynamics, and nature.

1. INTRODUCTION

Photophysical and photochemical dynamics are expected to reflect characteristics of μm dimension, which are due to structural, diffusional, and optical origins [1]. In the case of hydrogen-bonding solutions such as water and alcohols, and polymer solutions, mutual orientation and association of molecules and polymer conformations are much affected by interface and surface, and the effects may extend to sub μm

* To whom correspondence should be addressed.
# Present Address: Kaneka Corporation, Elmech Business Development Section R&D Group, 5-1-1, Torigai-Nishi, Settsu, Osaka 566, Japan.
‡ Present Address: Institute of Inorganic Chemistry and Analytical Chemistry, University of Mainz, 55122 Mainz, F. R. Germany.
‡‡ Present Address: Chemicals Division, Dainippon Ink and Chemicals, Inc., Higashi-fukashiba-18, Kamisu-machi, Ibaraki 314-02, Japan.
$ Present Address: Department of Chemistry, Faculty of Science, Hokkaido University, Sapporo 060, Japan.
§ Present Address: Department of Applied Physics, Osaka University, Suita, Osaka 565, Japan.
† Five-year term project: October 1988 ~ September 1993.

dimension [2]. When molecules have no appreciable mutual interactions and undergo rotation and diffusion independently, such a structural effect is not expected. However , it is worth noting that molecular diffusion in μm dimension is completed rapidly in sub ms time region [3]. This may also result in chemical phenomena characteristic of the μm dimension. Furthermore, optical cavity formed in the μm domains confines the light in itself and enhances photophysical and photochemical processes [4].

Microparticles are one of such interesting μm systems, however, the nature cannot be clarified by applying conventional time-resolved fluorescence and absorption spectroscopy to neat powders and dispersed solutions. Fluorescence and absorption spectra as well as their rise and decay curves are always due to an ensemble of microparticles, hence properties analyzed from the data are an average of microparticles and/or a sum of them. On the other hand, shape, size, and chemical composition of microparticles take various kinds of distribution, indicating that a simple analysis by the conventional method will never give a real picture of microparticles.

Recently, we have developed some dynamic microspectroscopy systems giving various information on dynamics in μm small domains. Combining optical microscope, pulsed laser, and fast-response detector, we have succeeded in constructing space- and time-resolved fluorescence and absorption spectroscopy, where sub μm space-, ps time-, and nm energy-resolutions are satisfied simultaneously [5-7]. This is closely related to total internal reflection fluorescence spectroscopy [8,9] and transient grating spectroscopy [10], by which ps dynamics in surface and interface layers with thickness of sub μm is available. Among them 3 dimensional sub μm space- and ps time-resolved fluorescence and absorption spectroscopy is quite useful for analyzing ps dynamics of a single microparticle. This is quite unique, since photophysical and photochemical dynamics in an individual particle can be investigated and correlated to its size, shape, and chemical composition. We believe this is a new stage of colloid and interface science and summarize here our recent results on polymeric microspheres, microcapsules, and oil droplets.

2. FLUORESCENCE CHARACTERIZATION OF POLYSTYRENE MICROSPHERE SURFACE

Polymeric microspheres modified with functional molecules are useful for analytical work in biomaterial and biomedical fields. Molecules in solution interact with functional moieties on the microsphere surface and form complexes in solution, giving chemical and spectroscopic signals. Since functional microspheres are prepared, separated from the residuals, treated with other solutions, dried, titrated, and so on, precise characterization of microspheres is indispensable. When dispersed in solution, however, microspheres undergo Brownian motion, preventing precise measurement of a single microsphere under a microscope. To overcome the problem laser trapping method is indispensable. Additional introduction of 1064 nm laser beam into the microscope makes it possible to fix a Brownian microsphere at a certain position, then time-resolved fluorescence microspectroscopy can be done without any difficulties [11].

Fluorescence analysis of microspheres needs fluorescent probe molecules giving information on microenvironmental conditions. Polystyrene particles containing carboxyl groups on the surface were chosen in our experiment, because it has a long history in the relevant research field and can be easily modified with various amines via amide bondings. Amines are fruitful as a substituent, since further chemical modification of microspheres can be done for specified purposes. Molecular structures and abbreviations of modified spheres, and fluorescence data are listed in Table 1 [12]. 8-Anilino-1-naphthalene sulfonic acid (ANS) was used as a probe, since its fluorescence spectral shape and intensity are sensitive to micropolarity and flexibility of surrounding environment.

Fluorescence spectra of ANS interacting with a single microsphere dispersed in phosphate buffered saline (PBS) pH 7.4 were measured and analyzed. The spectrum was broad and structureless as in homogeneous organic solvents. Fluorescence of ANS dissolved in PBS buffer solution without microspheres was weak, while its intensity was increased on the microsphere, indicating that ANS is interacting with the modified surface is fluorescent. Referring to polarity dependence of ANS fluorescence, it is considered that the ANS-probed surface is less polar compared to water. The dye should interact with both the surface and water even when adsorbed strongly on the surface, so that micropolarity discussed here is of course an effective parameter. In the case of *hexyl* fluorescence peak was in the longest wavelength compared to other modifications, indicating that micropolarity is high. This seems inconsistent with the chemical structure of the substituent on the surface. We consider that the hydrophobic alkyl chain is not extended into the water but folded on the surface. ANS could not be incorporated into the aggregated substituents, but loosely bound on the surface and partly contacted with water. When the end of the hydrocarbon chain is bonded to polar groups like -COOH and -OH, the chains are extended and can incorporate the dye molecule. Indeed, ANS fluorescence of *aOH* and *amCOO* showed blue shift compared to *hexyl*. Shortening the chain length (*etham* and *gly*), fluorescence spectrum was shifted to blue and its intensity was enhanced. The amido group may cooperate with the end groups of -OH and -COOH, increasing the surface affinity with ANS and preventing the latter to contact with water. Thus, the microenvironment condition of the surfaces in *etham* and *gly* was less polar than that of other particles.

Referring to the fluorescence maximum of ANS in some organic solvents, we can estimate an effective micropolarity of the functionalized polystyrene surface. Since ANS fluorescence gives 464 nm in n-octanol and n-butal and 472 nm in dioxane as its peak, respectively, the surface polarity of *aOH* and *amCOO* corresponds to that of the former alcohols, and *hexyl* to dioxane. Such microenviromental information has never been obtained before, because an scattering effect in dispersed solution disturbs precise fluorescence spectral measurement and fluorescence peak was not determined.

Fluorescence characterization of individual polymeric microspheres by 3 dimensional space- and time-resolved spectroscopy can confirm the reproducibility and reliability of the functionalization procedures. This is made possible to measure fluorescence from microsphere to microsphere and is one important advantage in the

Table 1
Abbreviation, chemical structure, and ANS fluorescence data of modified polystyrene microspheres

abbreviation	chemical structure	fluorescence maxima (nm)	relative fluorescence intensity
hexyl	⊘–CONH$(CH_2)_5CH_3$	476 - 484	22
aOH	⊘–CONH$(CH_2)_6$OH	463 - 465	30
amCOO	⊘–CONH$(CH_2)_5$COOH	463 - 466	20
gly	⊘–CONHCH_2COOH	451 - 454	35
etham	⊘–CONH$(CH_2)_2$OH	444 - 450	67
(Reference)	ANS - PBS	525	< 2

Polystyrene microsheres (diameter, 6.5 ± 0.19 μm; COOH groups per g polymer, 0.12 meq) were from Polyscience.
ANS: 8-Anilino-1-naphthalene sulfonic acid

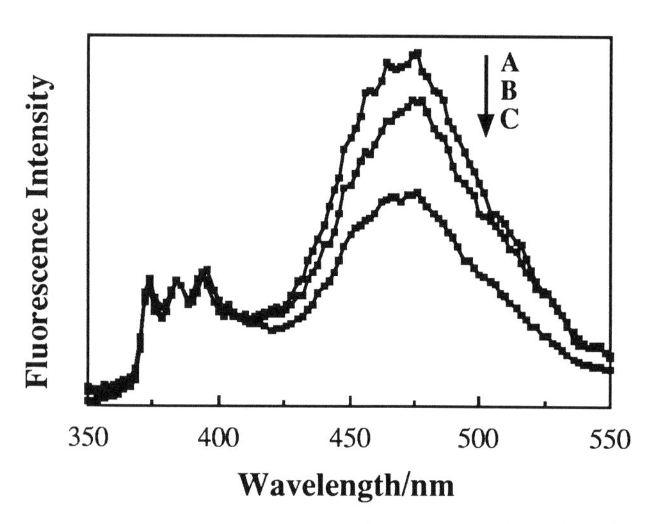

Figure 1. Normalized fluorescence spectra of pyrene in individual microcapsules dispersed in water. The diameter of the capsule A, B, and C is 19.2, 6.5, and 6.8 μm, respectively.

relevant study. Thus, the surface of each particle is well characterized and modification process can be confirmed to be perfect and reliable.

3. EXCIMER FORMATION DYNAMICS IN MICROCAPSULES

Analysis of fluorescence dynamics of a single microparticle was first measured for microcapsules dispersed in aqueous solution in ns and ps time regions as has been done in bulk solution. As excimer dynamics of aromatic molecules in solution is dependent on the concentration and solvent viscosity, its precise measurement and analysis of a single microparticle is expected to give an information on properties of individual microparticles. We describe here excimer formation kinetics in a microcapsule containing a toluene solution of pyrene [13]. Microcapsules posses unique geometrical structures with a solvent being encapsulated by the thin polymer resin wall, and are widely used for industrial applications. Its physical and chemical nature is determined by chemical composition of polymer, contained molecules, and their concentration. The microcapsules were dispersed in pure water, and pyrene fluorescence was measured by a space- and time-resolved fluorescence spectroscopy.

First, we summarize pyrene fluorescence dynamics of the solution obtained by spatially unresolved (conventional) spectroscopy. The concentration of the mother toluene solution was adjusted to be 8.1 x 10^{-3} M. Both monomer and excimer fluorescence were observed with the excimer (475 nm) and monomer (384 nm) fluorescence intensity ratio (I_E/I_M) of 2.4. The I_E/I_M value corresponds to the concentration of pyrene, and it was estimated to be 1.14 x 10^{-2} by the concentration dependence of I_E/I_M in air-saturated toluene. This was slightly higher than that of the mother solution. Since the sample solution was not deaerated, the excited pyrene is quenched by oxygen, showing relatively fast decay. Actually the excited pyrene monomer showed a decay of 17.5 ns in diluted solution.

In homogeneous solution, the pyrene excimer formation is known to proceed via the following Birks kinetics model [14],

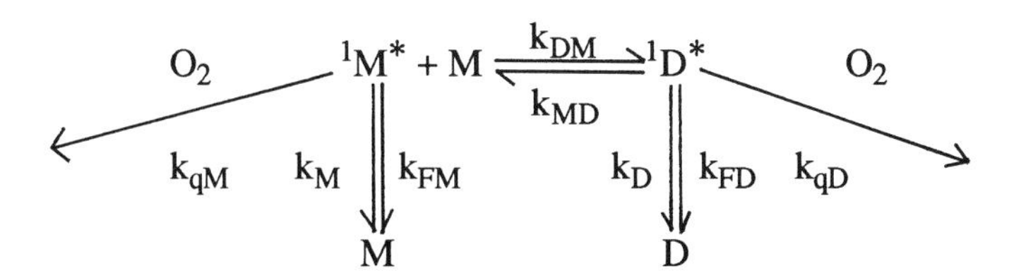

where k_{DM} and k_{MD} are the rate constants of the excimer ($^1D^*$) formation and the dissociation of $^1D^*$ to the excited ($^1M^*$) and ground state pyrene (M), respectively. k_{FM} and k_{FD} are the radiative rate constants of $^1M^*$ and $^1D^*$, respectively. k_D and k_M are the nonradiative decay rate constant of $^1D^*$ and $^1M^*$, respectively. k_{qM} and k_{qD} are the quenching rate constants by O_2. According to this scheme, the time response of the monomer ($i_M(t)$) and excimer fluorescence ($i_D(t)$) can be expressed as in eqs. (1) and (2), respectively.

$$i_M(t) = \frac{k_{FM}(\lambda_2 - X)}{(\lambda_2 - \lambda_1)}(e^{-\lambda_1 t} + C\, e^{-\lambda_2 t}) \tag{1}$$

$$i_D(t) = \frac{k_{FD}k_{DM}[^1M^*]}{(\lambda_2 - \lambda_1)}(e^{-\lambda_1 t} - e^{-\lambda_2 t}) \tag{2}$$

where

$$\lambda_{1,2} = \frac{1}{2}[X + Y \mp \{(Y-X)^2 + 4k_{MD}k_{DM}[^1M]\}^{1/2}]$$
$$C = (X-\lambda_1)/(\lambda_2-X)$$
$$X = k_M + k_{FM} + k_{qM}[O_2] + k_{DM}[^1M]$$
$$Y = k_{FD} + k_D + k_{MD} + k_{qD}[O_2]$$

Eqs. (1) and (2) indicate that both curves of the monomer decay and excimer decay/rise are characterized by the same time constants of λ_1 and λ_2 if the Birks model prevails for the pyrene excimer formation in the capsules. For the data obtained by spatially unresolved spectroscopy, however, $i_M(t)$ and $i_D(t)$ did not obey double-exponential functions and the excimer fluorescence showed a relatively slow rise. Phenomenologically, it is clear that the pyrene excimer formation observed for the microcapsules by spatially-unresolved spectroscopy cannot be explained based on the Birks kinetics model.

Now the fluorescence spectrum and its dynamics for pyrene in a single microcapsule is measured by three-dimensional space- and time-resolved fluorescence spectroscopy. Fluorescence spectra of three different microcapsules are shown in Figure 1. It was clearly demonstrated that the efficiency of the excimer formation was quite different between the capsules. $i_M(t)$ and $i_D(t)$ relevant to the spectra in Figure 1 are shown in Figure 2. The microcapsule with relatively large I_E/I_M showed the faster monomer decay as well as the faster excimer rise as compared with that showing smaller I_E/I_M. In contrast to the results on an ensemble of microcapsules, furthermore, $i_M(t)$ and $i_D(t)$ for individual microcapsules were fitted by double-exponential functions. Analogous results were obtained for a number of the microcapsules. These results clearly indicate that the pyrene excimer formation in an individual microcapsule can be explained by the Birks kinetics model as found in homogeneous solutions. On the other hand the results for a solution of the dispersed capsules cannot be expressed by the simple excimer kinetic model. The origin of the multi-exponential behavior is due to the observation of the sum of various microcapsules with different fluorescence dynamics.

Before discussing the excimer formation dynamics in individual microcapsules, possible origins for the variation of the excimer formation efficiency (I_E/I_M) with the capsules are to be considered. The factors influencing the excimer formation are viscosity of the inner solution and concentration of pyrene in each capsule. The viscosity of the inner toluene solution might increase with decreasing the size of the capsule owing to an increase in the contribution of the surface forces between the melamine-resin wall and toluene to the viscosity. To test such possibilities, we selected various sizes of the microcapsules and determined I_E/I_M of each capsule as shown in Figure 3. I_E/I_M was not correlated to the size of the capsule, indicating that the viscosity of the inner toluene solution is not a factor governing I_E/I_M of each

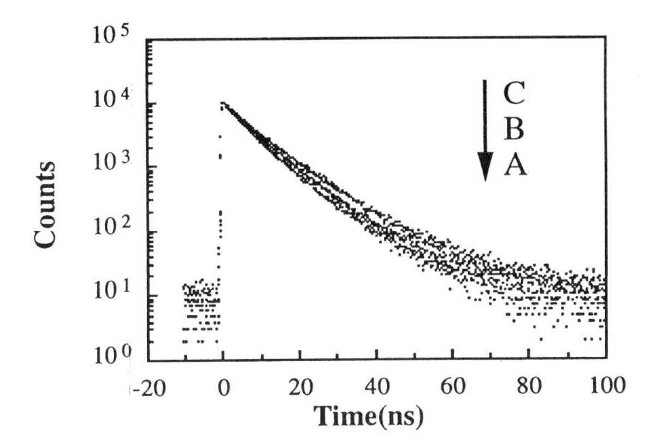

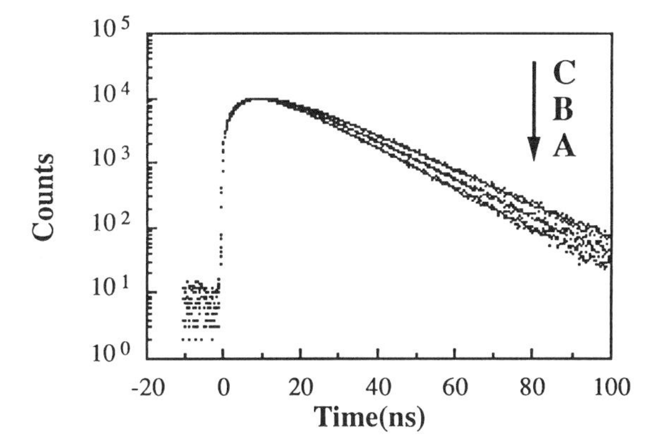

Figure 2. Fluorescence rise and decay curves of the monomer (upper) and the excimer (lower) in individual microcapsules dispersed in water. A, B, and C correspond to those in Figure 1.

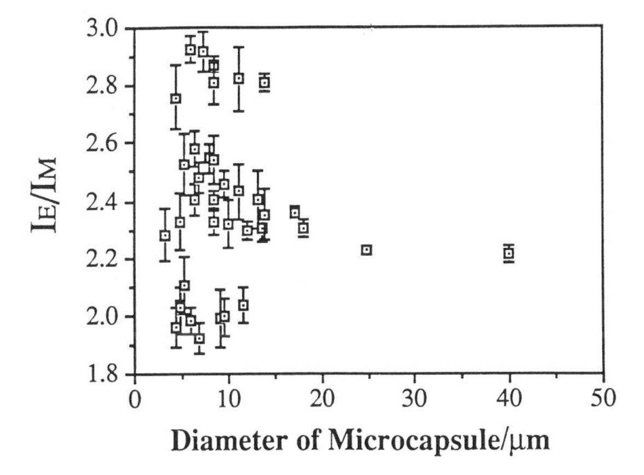

Figure 3. Excimer to monomer fluorescence intensity ratio (I_E/I_M) of microcapsules as a function of the diameter.

capsule. The most probable explanation for the variation of I_E/I_M and the fluorescence dynamics with the capsules will be a scatter of the pyrene concentration. I_E/I_M of 1.9 ~ 2.9 in Figure 3 corresponds to the distributions of the pyrene concentration to be $8.5 \times 10^{-3} \sim 1.3 \times 10^{-2}$ M.

Knowing that the variation of I_E/I_M with the capsules is primarily ascribable to that of the pyrene concentration, we analyzed the monomer decay ($i_M(t)$) and excimer rise/decay ($i_D(t)$) profile in Figure 2 based on eqs. (1) and (2), respectively. Double exponential fittings of $i_M(t)$ and $i_D(t)$ were not fortuitous and the excimer formation in every single capsule was analyzed by eqs. (1) and (2). λ_1 and λ_2, and their pre-exponential factors (A_1 and A_2, respectively) were thus obtained for three microcapsules. λ_1 and λ_2 determined at 384 nm (monomer) coincide fairly well with the corresponding values observed for the excimer fluorescence at 475 nm for a given capsule. Furthermore, A_1 and A_2 are also in good agreement with the expectations from eqs. (1) and (2); $(A_1 + A_2) \sim 1.0$ and $(A_1 - A_2) \sim 0$ for $i_M(t)$ and $i_D(t)$, respectively.

The rate parameters for the excimer dynamics, k_{DM} and k_{MD}, in the capsule could be obtained based on eqs. (1) and (2) as in Table 2, since the concentration of pyrene was estimated and $(k_M + k_{FM} + k_{qM}[O_2])$ in each capsule were measured. $(k_M + k_{FM} + k_{qM}[O_2])$ was determined to be 5.7×10^6 s^{-1} in a diluted toluene solution of pyrene (1.0×10^{-5} M) in air. k_{DM}, k_{MD}, and $(k_D + k_{FD} + k_{qD}[O_2])$ were thus determined to be $(5.8 \sim 6.7) \times 10^9$ M^{-1} s^{-1}, $(5.4 \sim 7.1) \times 10^6$ s^{-1}, and $(6.7 \sim 7.7) \times 10^7$ s^{-1}, respectively, as listed in Table 2. The values estimated for each capsule are comparable to those determined for a homogeneous toluene solution of pyrene (6.4×10^{-3} M, see Table 2). The variation of I_E/I_M with the capsules are therefore reasonably explained by that of the pyrene concentration in the capsule, and the excimer formation dynamics in each microcapsule is concluded to be similar to that in a homogeneous bulk solution.

One possible origin of the concentration distributions between the capsules may be inhomogeneous location of pyrene. Its precipitation in inner toluene solution and adsorption/incorporation in the polymeric capsule wall are denied on the basis of data for $i_M(t)$ and $i_D(t)$. The concentration distribution could be determined during the synthetic procedures of the microcapsules. Evaporation of toluene during vigorous stirring of a pyrene toluene solution in water will account for the higher I_E/I_M values than the average (> 2.3). For the capsules with $I_E/I_M < 2.3$, on the other hand, we suspect that partition of water into the toluene droplets, leading to microcrystallization of pyrene followed by its exclusion from the droplets, is responsible. An evaporation rate of a toluene droplet and partition of water into the toluene layer will be also influenced by polymerization rate and wall thickness of the capsule. It is worth noting that the polymerization rate and the thickness of the capsule wall will be scattered among the capsules, since emulsion polymerization of the reactions proceeds in inhomogeneous toluene-water solution. All these factors should be related to each other and result in the present fluorescence characteristics of microcapsules.

4. EXCIMER FORMATION DYNAMICS IN OIL DROPLETS

Table 2
Rate parameters for pyrene excimer formation in a single microcapsule

Sample[a)]	I_E/I_M	Concentration (10^{-2}M)	k_{DM} ($\times10^9 M^{-1}s^{-1}$)	k_{MD} ($\times10^6 s^{-1}$)	$k_D + k_{FD} + k_{qD}[O_2]$ ($\times10^7 s^{-1}$)
A	2.8	1.34[b)]	6.3	7.1	7.7
B	2.5	1.18[b)]	5.8	5.4	7.3
C	1.7	0.82[b)]	6.7	5.7	6.7
Solution	1.3	0.64	6.4	6.6	6.9

a) Samples, A, B, and C are the same as those in Figures 1 and 2.

b) The concentration was estimated from its dependence of I_E/I_M in toluene (separate experiments).

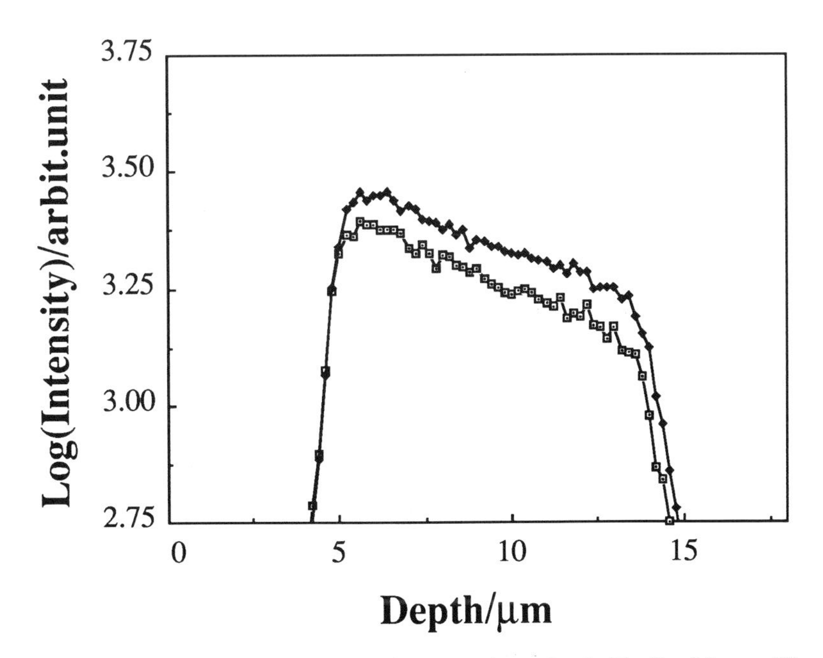

Figure 4. Fluorescence depth profiles of pyrene in a single E_1 liquid paraffin droplet monitored by the monomer (▲) and the excimer (❒). See text

Since polymerized microcapsule wall is considered to affect rapid re-equilibrium between microparticles and water, studies on pyrene excimer formation in oil-in-water emulsions are worth studying [15]. Because of direct contact of liquid-liquid at the interface, pyrene distribution between droplet and water is considered to be balanced through fast re-equilibrium. The emulsions were prepared by stirring a mixture of liquid paraffin, pyrene, water, gelatin, and sodium dodecyl sulfate (SDS). The emulsion (0.10 g) with the initial pyrene concentration in the liquid paraffin ($[Py]_0$) of 8.80×10^{-3} or 2.64×10^{-2} M was redispersed in aqueous gelatin (the prepared systems are abbreviated as E_0 and E_1, respectively). As the diffusion motion of the droplets prepared here was completely frozen at room temperature, no Brownian motion was observed.

Fluorescence spectrum of pyrene in the droplet showed an efficient excimer formation in addition to the monomer fluorescence. For characterizing the individual droplets, fluorescence depth profile of the droplet was measured for the excimer (I_E at 475 nm) and the monomer fluorescence (I_M at 384 nm). The results are shown in the Figure 4, where both I_E and I_M in the logarithmic scale decreased linearly with the depth of the droplet. This means that the Lambert-Beer's law held well and pyrene was distributed homogeneously inside the droplet. Since the slopes for I_E and I_M agreed with each other, inhomogeneous aggregation formation giving the excimer emission was excluded. According to the Lambert-Beer's law, the slope of the profile should correspond to $[P_y]\varepsilon_{333}$, where $[P_y]$ and ε_{333} are the concentration of pyrene in the droplet and its molar extinction coefficient at 333 nm, respectively. ε_{333} was assumed to be equal to that determined for a thin liquid paraffin film of pyrene (2.64×10^{-2} M) under the same experimental conditions of microscope optics; $\varepsilon_{333} = 2.4 \times 10^4$ M^{-1} cm^{-1}. ε_{333} also agreed with that observed for the same solution by a conventional absorption spectrometer without a microscope. From the slope of the profile and ε_{333}, $[P_y]$ in the droplet was obtained to be 1.3×10^{-3} M, which was much lower than $[P_y]_0 = 2.64 \times 10^{-2}$ M.

As similarly in the microcapsules described above, the fluorescence intensity ratio of the excimer to the monomer (I_E/I_M) was plotted against the diameter for individual paraffin droplets in Figure 5. It was clearly demonstrated that I_E/I_M was almost independent of the droplet size and constant at ~1 or ~0.4 for E_1 or E_0, respectively. This is quite contrasting to the results for microcapsules. In the present system, since the droplets were surrounded by aqueous solution without a polymeric resin wall, distribution of pyrene between water and oil phases should have been sufficiently equilibrated during the sample preparation procedures. Although the solubility of pyrene in pure water is poor, pyrene is soluble in aqueous solution of SDS and gelatin. Furthermore, the weight percentage of liquid paraffin in emulsion is very low (7.8×10^{-3} %), so that pyrene is likely to partition to the aqueous phase during mixing processes. The lower pyrene concentration in the droplets ($[P_y] = 1.3 \times 10^{-2}$ M for E_1) relative to $[P_y]_0 = 2.64 \times 10^{-2}$ M proves this. The distribution coefficient, $P = C_{oil}/C_{gel}$, where C_{oil} and C_{gel} are the weight molar concentrations of pyrene in the droplet and the gelatin, respectively, is calculated to be 1.3×10^3. The equilibrium concentration of pyrene in water partitioned with a liquid paraffin solution of pyrene ($[P_y]_0 = 2.64 \times 10^{-2}$ M) is 4×10^{-8} M ; $P = C_{oil}/C_{H_2O} = 7 \times 10^5$. In the presence of SDS (3.9×10^{-3} %), the corresponding P values decreases to 2×10^5. Gelatin consists of

Table 3
Rate parameters for pyrene excimer formation in a paraffin droplet

Sample	k_{DM} ($\times 10^8$ M^{-1}s^{-1})	k_{MD} ($\times 10^6$ s^{-1})	$k_D + k_{FD} + k_{qD}[O_2]$ ($\times 10^7$ s^{-1})
droplet	3.5	3.8	1.8
solution	1.7	3.8	2.4

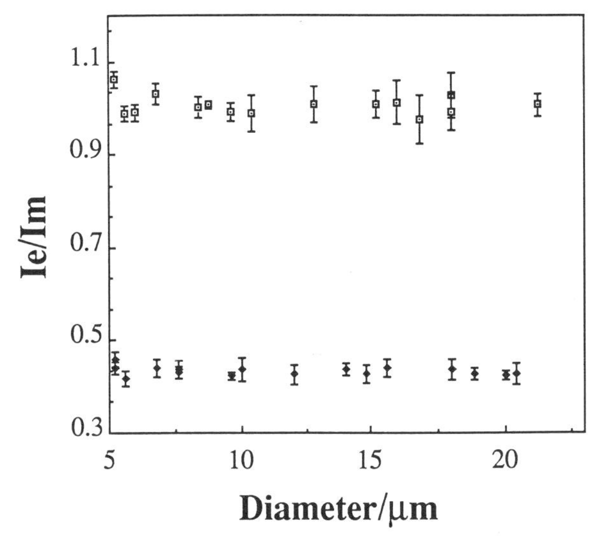

Figure 5. Excimer to monomer fluorescence intensity ratio (I_E/I_M) of droplets E_1 (◆) and E_2 (❐) as a function of the diameter.

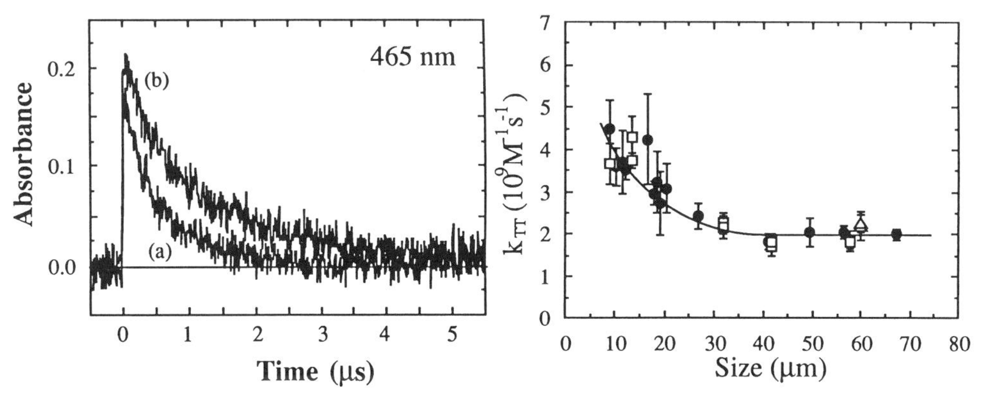

Figure 6. Decay curves of ZnTPP $T_n \leftarrow T_1$ absorption of (a) in a TBP droplet (d = 20 μm) and (b) in TBP film (d = 20 μm). $[ZnTPP]_0$ = 4 mM.

Figure 7. T-T annihilation rate constant of ZnTPP in a TBP droplet as a function of the diameter.

various proteins and amino acids, hence the gelatin in water will further induce a partition of pyrene with the aqueous phase through hydrophobic interaction. If one takes such circumstances in the emulsion into account, $P = C_{oil}/ C_{gel} = 1.3 \times 10^3$ is a reasonable consequence and no droplet size-dependence of I_E/I_M is expected.

Pyrene excimer formation dynamics in individual droplets and in the bulk paraffin solution was measured and compared with each other. Even in the bulk solution, the dynamics was unusual and simple bi-exponential behavior as in toluene and cyclohexane solution was not observed. This is partly ascribed to small contribution of diffusion limited-excimer formation due to high viscosity ($\eta \sim 100$ cp at 20 °C) and to dissolved O_2 which was not removed. Furthermore, O_2 concentration was changed in the droplet due to its consumption by impurities in gelatin, so that the analysis of dynamics was done in the same way as in microcapsules under some assumptions. The obtained rate constants of individual droplets and bulk solution are listed in Table 3. k_{DM} and k_{MD} in droplets and paraffin solution are smaller than those in cyclohexane, which is quite reasonable in viewpoint of viscosity. $k_D + k_{FD} + k_{qD}[O_2]$ is considered to be almost independent of solvent viscosity, so that the present results are in good agreement with $k_D + k_{FD} + k_{qD}[O_2]$ in cyclohexane if O_2 effect is considered. It is worth noting that k_{DM} of the droplet is larger than that of the bulk and its difference is larger than that of $k_D+k_{FD} + k_{qD}[O_2]$. One possible reason for this is a decrease in η, which is caused by penetration of water and/or SDS to the liquid paraffin. Indeed, the droplets E_1 showed a larger I_E/I_M (~ 1.0) than the corresponding bulk solution (~ 0.6) of a similar [Py] of ~ 1.3×10^{-2} M. Thus, analyses of excimer dynamics in individual droplets give characteristic solution properties of the droplets and equilibrium distribution of solute between droplets and surrounding solvents.

5. TRIPLET-TRIPLET ANNIHILATION IN INDIVIDUAL OIL DROPLETS

An investigation on triplet-triplet (T-T) annihilation process also gives an information of viscosity of microdroplets, since it is limited by molecular diffusion. Space- and time-resolved absorption spectroscopy using a microscope was applied to individual droplets [7]. In view of spectroscopic characteristics and solubility Zinc tetraphenylporphyrin (ZnTPP) was dissolved in tri-n-butyl phosphate (TBP) microdroplets, which gave a broad T-T absorption band at 465 nm and a depletion at 556 and 597 nm upon excitation [7]. The decay profiles of ZnTPP $T_n \leftarrow T_1$ absorbance in a microdroplet and the reference TBP solution are shown in Figure 6. Although the lifetime of the ZnTPP triplet state was reported to be 3 ms, the present decay was completed in a few μs. The inverse plot of $T_n \leftarrow T_1$ absorbance with time showed a linear relation, meaning that the bimolecular decay of T-T annihilation was confirmed. Using molar extinction coefficient of ZnTPP $T_n \leftarrow T_1$ transition ($\varepsilon = 9000$ $M^{-1}cm^{-1}$ at 465 nm) and assuming the diameter as the path length, the T-T annihilation rate constant (k_{TT}) was estimated to be $(2.9 \pm 0.7) \times 10^9$ $M^{-1}cm^{-1}$ for a droplet of 20 μm diameter. Systematically, k_{TT} was measured as a function of the droplet diameter and plotted in Figure 7. Similarly k_{TT} of a solution film of the 20 μm thickness was examined as a reference. It is worth interesting that k_{TT} for the droplet is always larger than that of the solution and increases with decreasing the diameter.

The T-T annihilation process in solution is the diffusion controlled process, and its rate constant is expressed as follows.

$$k_{TT} = k_{diff} = \frac{8RT}{3000\,\eta} \qquad (3)$$

Here k_{diff} is the rate constant of the diffusion-controlled reactions, R is gas constant, T is temperature, and η is viscosity. In the case of the TBP solution film, analysis of k_{TT} with eq. (3) gives 3.8 ~ 5.9 cp as η at 20 °C. Bulk viscosity of TBP is 3.8 cp, hence the present data are concluded to be reliable. On the other hand, the viscosity of a single microdroplet with the diameter of 20 μm was obtained to be 1.8 ~ 3.2 cp at 20 °C. Thus, the interesting difference of k_{TT} between microdroplet and solution is now ascribed to the difference in viscosity. One possible explanation of the difference is to assume water penetration into the droplet. This was confirmed directly by measuring k_{TT} of the thick (60 μm) TBP solution film saturated with aqueous SDS solution. The obtained value of $(2.1 \pm 0.3) \times 10^9\ M^{-1}s^{-1}$ is larger than that of the TBP film. It was already reported that the water solubility in TBP droplets was as large as 23 %, and the TBP droplet was swollen by water.

The micrometer size effect on the viscosity in TBP droplets is possibly examined by the following idea. One is that partition degree of water in the droplet increases as the diameter decreases. This seems probable because of the larger surface-to-volume ratio for the smaller droplet. The ground state absorption spectroscopy of ZnTPP, however, indicated that the water content is irrespective to the droplet size [17], denying the interpretation. Alternatively, specially ordered structure is formed in the droplet and the viscosity cannot be simply explained in terms of water content. For example, the contained water molecules accompanying SDS form reverse micelles in TBP droplets [18], and the micellar volume becomes larger as the diameter decreases. In the case solvent viscosity can be given in the following equation based on solvent hole theory [19].

$$\eta = A \cdot \exp(B(\upsilon_f/\upsilon_0)) \qquad (4)$$

where A and B are characteristics constants of the solvent, υ_f is free volume in the liquid, and υ_0 is the volume of the liquid which is obtained by virtually extrapolating it to 0 °k without phase change. As the free space, υ_f/υ_0, is larger, the viscosity is lower. The reverse micelles with the diameter of around 2 nm are assumed to be formed in TBP solution [17] and they may produce such free space in the microdroplet, leading to the lowering of the viscosity. The reduction of the droplet diameter, which may be accompanied with the volume increase of the reverse micelle, leads to the lower viscosity. This becomes distinct for the droplet with the diameter less than 20 μm.

6. CONCLUSION

Photochemical and photophysical processes occurring in a single microparticle can be analyzed by spatially resolved spectroscopic techniques. Physical and chemical properties in individual particles are measured directly, and now statistical treatment of the data for an ensemble of particles is not needed. Fluorescence characterization of functionalized polymeric microspheres, solute concentration distribution in microcapsules and oil droplets, and size-dependent viscosity of oil droplet were elucidated for the first time. Studies on photophysical and photochemical process of individual microparticles are important approach for understanding structure, dynamics, and chemical composition of the microparticle.

ACKNOWLEDGMENTS

Drs. N. Kitamura and H. Misawa are gratefully acknowledged for their collaboration.

REFERENCES

1. H. Masuhara, Microchemistry by laser and microfabrication techniques, thisvolume.
2. S. Hamai, N. Tamai, and M. Yanagimachi, Time-resolved total internal reflection fluorescence spectroscopy at solid-solution interface layers, this volume.
3. H. Masuhara, Microphotoconversion: exploratory chemistry by laser and microfabrication, *in* Photochemical processes in organized molecular systems (ed. by K. Honda et al.), Elsevier, Amsterdam, 1991 p.491.
4. K. Kamada, K. Sasaki, R. Fujisawa, and H. Misawa, Picosecond lasing dynamics of an optically trapped microparticle, this volume.
5. K. Sasaki, M. Koshioka, and H. Masuhara, Appl. Spectrosc., 45 (1991) 1041.
6. N. Tamai, T. Asahi, and H. Masuhara, Rev. Sci. Instr., 64 (1993) 2496.
7. N. Tamai, S. Funakura, and C. Porter, Transient absorption microspectroscopy of a single optically trapped particle, this volume.
8. H. Masuhara, N. Mataga, S. Tazuke, T. Murao, and I. Yamazaki, Chem. Phys.Lett., 100 (1983) 415.
9. M. Toriumi and H. Masuhara, Spectrochimica Acta Rev., 14 (1991) 353.
10. N. Tamai, T. Asahi, and H. Masuhara, Chem. Phys. Lett., 198 (1992) 413.
11. N. Kitamura, K. Sasaki, H. Misawa, and H. Masuhara, Optical harmony of microparticles in solution, this volume.
12. U. Pfeifer-Fukumura, H. Misawa, H. Fukumura, and H. Masuhara, Chem. Lett., (1994) submitted.
13. M. Koshioka, H. Misawa, K. Sasaki, N. Kitamura, and H. Masuhara, J. Phys. Chem., 96 (1992) 2909.
14. J. B. Birks, Photophysics of Aromatic Molecules, Wiley Interscience, New York, 1970.
15. K. Nakatani, H. Misawa, K. Sasaki, N. Kitamura, and H. Masuhara J. Phys. Chem., 97 (1993) 1701.
16. S. Funakura, K. Nakatani, H. Misawa, N. Kitamura, and H. Masuhara, in preparation.
17. S. Funakura, K. Nakatani, H. Misawa, N. Kitamura, and H. Masuhara, J. Phys. Chem. (1993) submitted.

MICROCHEMISTRY
Spectroscopy and Chemistry in Small Domains
Edited by H. Masuhara et al.

Stimuli-responsive polymer gels: an approach to micro actuators

M. Irie

Institute of Advanced Material Study, Kyushu University, Kasuga-Koen 6-1
Kasuga, Fukuoka 816, Japan

Stimuli-responsive polymer solution and gel systems, which undergo isothermal phase transitions by external stimulation, such as photons or chemicals, have been constructed.

1. INTRODUCTION

Conformation of polymers governs their various physico-chemical properties. When the conformation is reversibly controlled by external stimulation, such as photons or chemicals, the conformation changes alter the size of polymer gels. The gels may be utilized as micro actuators (or artificial muscles), because the driving force for the size change depends on structural or chemical property changes of molecules and the size can be minimized to a micrometer scale. So far, various kinds of polymers which change their conformation reversibly by photoirradiation have been reported[1,2]. When we intend to use the polymers or gels for micro actuators, they should have high efficiency of energy conversion and rapid response time. The efficiency in the polymers, however, is rather low and the response is slow. They need many photons and a high content of photo-isomerizable chromophores in the polymers to induce a large conformation change, and it takes more than several tens minutes for the gels to expand. We tried to make sensitive stimuli-responsive polymer gels by using phase transitions of polymer systems.

2. STIMULI-RESPONSIVE PHASE TRANSITIONS

One possible way to make a sensitive photoresponsive polymer, i.e. one which responds to a fewer photons, is to utilize the phase transition of polymer systems. At temperatures close to the phase transition the system is in unstable state and a small perturbation induces large property changes. The stimuli-responsive phase transition is illustrated in Fig. 1. In the absence of external stimulation the polymer system changes the state from X to Y at a temperature Ta. We assume that the phase transition temperature will rise to Tb in the presence of external stimulation. Then, if the external stimulation is applied to the system at T ($Ta<T<Tb$), the state will change from Y to X at a certain value of the external stimulation, Cc, as shown in Fig. 1b. This principle is useful for constructing efficient stimuli-responsive polymers.

For constructing such polymer systems it is required to incorporate

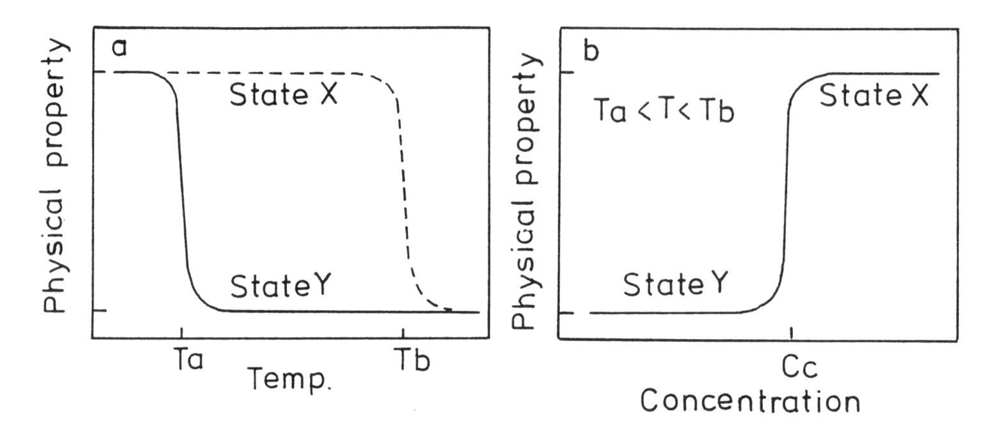

Figure 1. Schematic illustration of stimuli-responsive phase transition of a polymer system. Thermal phase transition from state X to Y in the presence (----) and absence (——) of external stimulation (a) and isothermal phase transition by the external stimulation.

receptor molecules which reversibly change the molecular properties by external stimulation to the polymers. Many molecules are known to be reversibly transformed to other forms by external stimulation, such as photons, electrons or chemicals. Table 1 lists a few examples. Azobenzene isomerizes from the trans to cis form by ultraviolet irradiation, and the dipole moment increases from 0.5 to 3.1 debye. The polar cis form returns to the less polar trans form by visible irradiation. Electrochemical oxidation of ferrocene changes the hydrophilicity. When it is oxidized from Fe(II) to Fe(III), the hydrophilicity increases. Host molecules also change the properties in the presence of suitable guest ions. Benzo[18]-crown-6, for example, captures potassium ion in the cavity, and increases the hydrophilicity.

When these chromophores are incorporated into the pendant groups of a polymer-A, the polymer changes the properties and converts to a polymer-B by photons, electrons or chemicals.

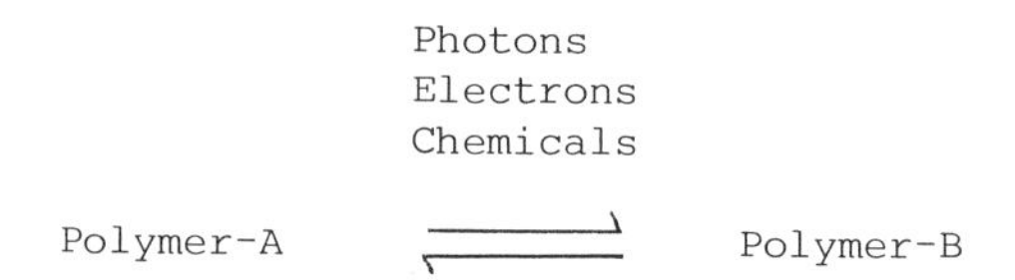

Consequently, the stimulation alters the phase transition temperature from Ta of the polymer-A to Tb of the polymer-B.

We used poly(N-isopropylacrylamide), PNIPAM, as the model polymer and incorporated the receptor molecules into the pendant groups. An aqueous solution of PNIPAM is known to undergo the phase separation upon heating above 31°C. The aqueous solution is homogeneous below 31°C, while it turns

Table 1 Receptor molecules

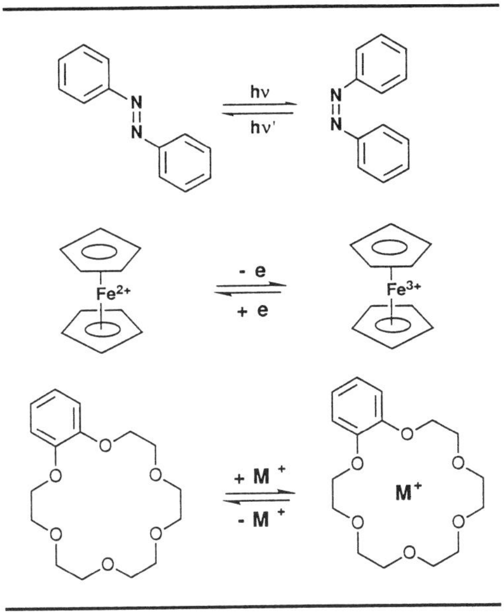

turbid above 31°C[3]. The phase separation temperature Tc depends on the subtle balance between the ability of the polymer to form hydrogen bonds with water and the intermolecular hydrophobic forces. The hydrophobic interaction is expected to be enforced or reduced by the property changes of the pendant receptor chromophores by the external stimulation. Thus,the phase separation temperature is changed.

2.1. Photostimulated phase separation

Azobenzene chromophores were incorporated into the pendant groups of PNIPAM as follows[4].

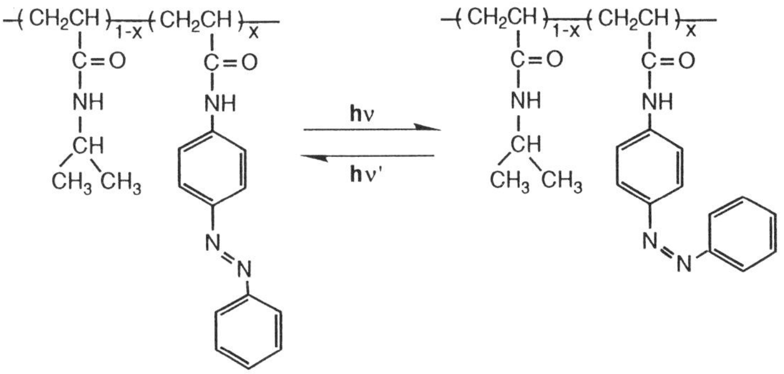

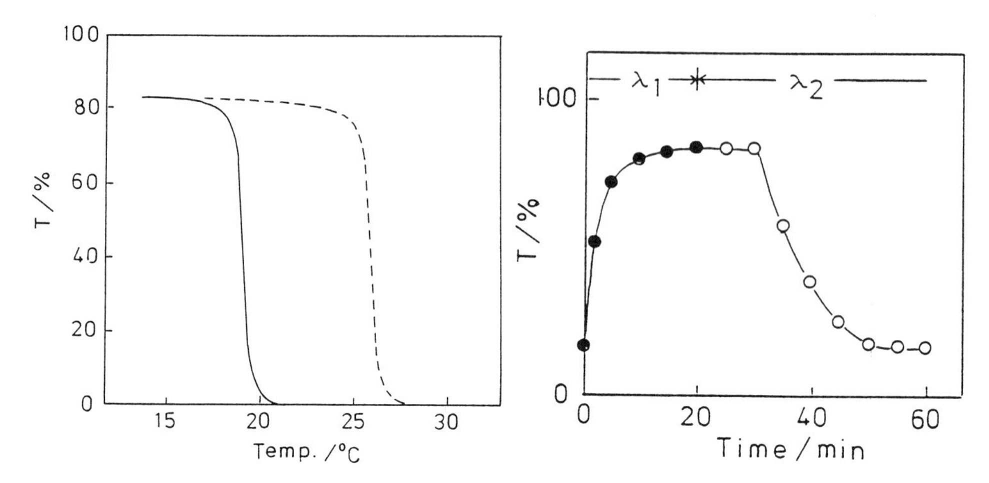

Figure 2. Transmittance changes at 750 nm of a 1 wt% aqueous solution of PNIPAM with pendant azobenzene groups (2.7 mol%) when heated at a rate of 2°C/min (—) before photoirradiation (---) in the photostationary state by irradiation with UV (350< λ <410 nm) light.

Figure 3. Photostimulated isothermal phase separation of the 1 wt% aqueous solution of PNIPAM with pendant azobenzene groups (2.7 mol%) by alternate irradiation with UV (350< λ <410 nm) and visible (λ >470 nm) light at 19.5°C.

Figure 2 shows the transmittance changes at 750 nm, which were observed when a 1 wt % aqueous solution of the above copolymer(x=0.027) was heated. Both isomers of the azobenzene chromophore, trans and cis forms, have no absorption at 750 nm. In the dark before photoirradiation, the solution turned opalescent at 18.5°C, and the transmittance decreased to a half the initial value at 19.4°C. The solution became completely opaque above 22°C. When cooled below 18°C the solution again became transparent. The solution was irradiated with UV light (350< λ <410 nm) for 10 min at 10°C. Then, the pre-irradiated solution showed the phase separation at 26.0°C. On standing overnight in the dark or visible (λ > 470 nm) irradiation, Tc again returned to 19.4°C.

The phase separation temperature shift by photoirradaition implies that at a temperature between 19.4 and 26.0°C, ultraviolet irradiation solubilizes the polymer,while visible irradiation decreases the solubility and leads to the phase separation. Figure 3 shows the reversible phase separation at 19.5 C upon alternate irradiation with UV and visible light.

Here, it is worthwhile to note the difference in the response times of the dissolution and phase separation processes. In the dissolution process a small number of photons were enough to induce the phase transition,while in the phase separation process it needed some induction period for the polymer to start the phase separation. The phase separation process exhibited a non-linear response to the irradaition time, or the number of photons. The photo-stimulated phase separation/dissolution cycle was not observed below 19.4 and above 26.0°C. The maximum difference in Tc was observed at a very small azobenzene content of 2.7 mol%. Below and above this content the phase separation temperature was not affected by photo-irradiation.

2.2. Chemical-induced reversible phase separation

Chemical-induced phase transition systems, which change the conformation reversibly in response to special chemicals, was constructed by incorporating host molecules into the pendant groups. Host molecules, such as crown ethers or cyclodextrins, are known to change the property by capturing guest chemicals in their cavity. We synthesized the following polymers[5].

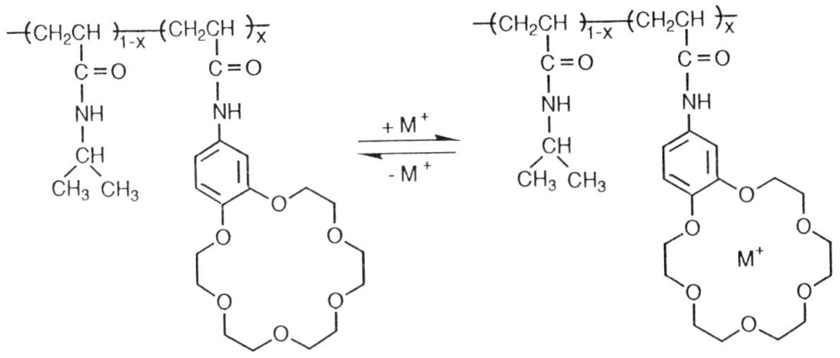

When the pendant crown ether groups bind metal ions, the phase separation temperature is expected to rise because the hydrophilicity of the polymer increases. The transmittance change of a 1 wt% aqueous solution of PNIPAM containing pendant crown ether groups (11.6 mol%) was measured in the presence and absence of potassium chloride. Upon being heated with the rate of 0.3°C/min, the solution turned lactescent at 30°C in the absence of the metal ion and the transmittance at 500 nm decreased to a half the initial value at 31.5°C. In the presence of potassium chloride (1.05×10^{-1} M), on the other hand, the aqueous solution remained transparant even when the solution temperature was raised to 31.5°C. The solution began opaque above 37°C. Tc increased as much as 7.4°C by the addition of potassium chloride.

The increase in Tc depended on metal ions. A large temperature increase was observed for potassium chloride, while the increase was only 1.5°C when sodium chloride was added instead of potassium chloride. The temperature increase was not observed for the solution containing lithium chloride and cesium chloride.

Figure 4 shows the relation between the metal ion diameter and the increase in Tc. The binding affinity of a crown ether with metal ions depends on the cavity size [6]. When the cavity size fits ion diameter, the ion is captured by the crown ether. The cavity size of benz[18]crown-6 is known to fit the diameter of K^+. The relative Tc increase shown in Fig. 4 correlates well with the binding affinity of benzo[18]crown-6 to metal ions. K^+ which efficiently binds to pendant benzo[18]crown-6 most

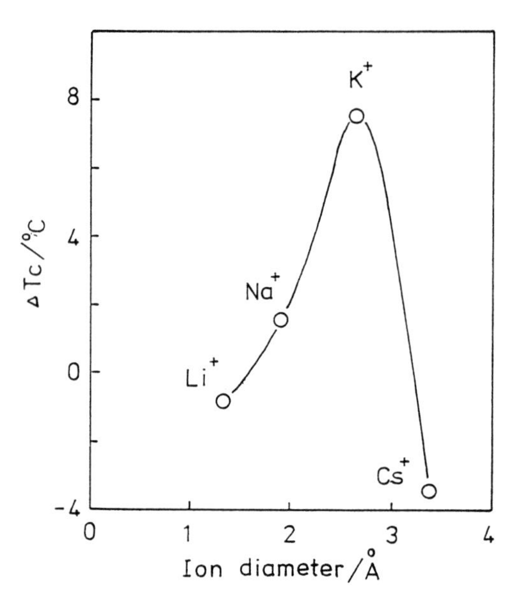

Figure 4. Phase separation temperature changes of the aqueous solution of PNIPAM containing 11.6 mol% crown ether groups by the addition of metal chloride.

pronouncedly increased Tc, while Li^+ and Cs^+ which are scarcely captured by the crown ether groups could not affect Tc.

If Tc increases from Ta to Tb by ion binding of the pendant crown ether groups, the phase transition is induced isothermally at T (Ta<T<Tb) by the addition of metal ions, as shown in Fig. 1. The polymer chain conformation is also expected to expand in the process. Figure 5 shows the phase transition at 32°C and 37°C by the addition of potassium and sodium chlorides. At 32.0°C both K^+ and Na^+ induced the phase transition from the phase separated to homogeneous state. The ion concentration necessary to induce the phase transition, however, depended on the ions. A very small amount was enough for K^+ to induce the transition, while 5.0×10^{-2} M was necessary for Na^+. A critical concentration depended on the ions. When the measuring temperature was raised to 37°C, the phase transition from the phase separated to homogeneous state was not observed by the addition of sodium chloride. Only K^+ could induce the transition. 4.0×10^{-2} M K^+ was required to induce the transition. The ion with concentration lower than 4.0×10^{-2} M did not cause any effect. The isothermal phase transition was induced by specific ions depending on measuring temperature.

Alternate addition of potassium chloride and low molecular weight [18]crown-6 induced reversible phase transition at 37°C. The addition of K^+ caused the increase of the transmittance at 500 nm, while [18]crown-6 decreased the transmittance. The hydrophilicity increase of the pendant groups by the ion binding expands the polymer chains. Subsequent addition of low molecular weight [18]crown-6 extracted the metal ions from the pendant groups, and this resulted in the contraction of the polymer chain.

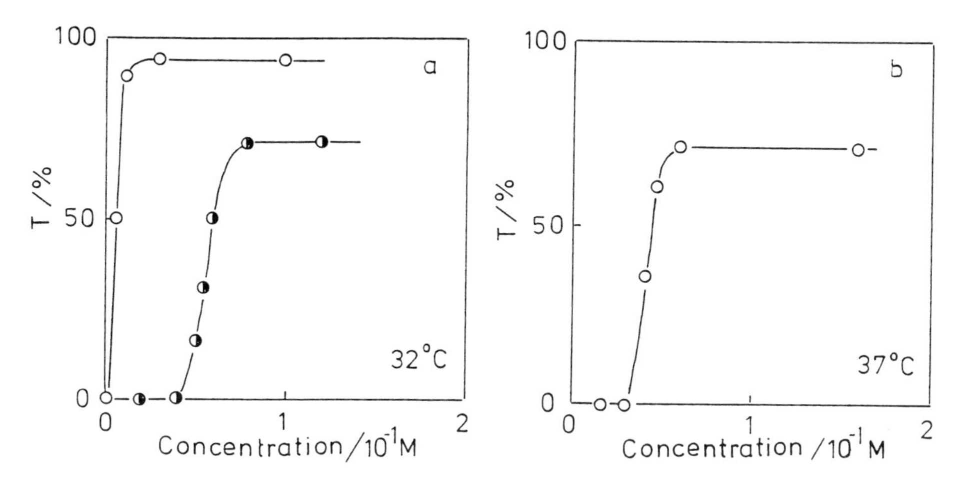

Figure 5. Isothermal phase transition from the phase separated to homogeneous states of the aqueous solution of the copolymer containing 11.6 mol% pendant crown ether groups by the addition of potassium (○) and sodium (◑) at 32°C (a), and potassium (○) at 37°C (b).

3. PHOTO- and CHEMICAL-INDUCED VOLUME PHASE TRANSITIONS OF PNIPAM GELS

By applying the concept of stimuli-responsive phase separation of aqueous polymer solutions to gel systems, it becomes possible to construct photo- and chemical-responsive gels. PNIPAM gels containing triphenylmethane leuconitriles, azobenzene or crown ether groups were synthesized.

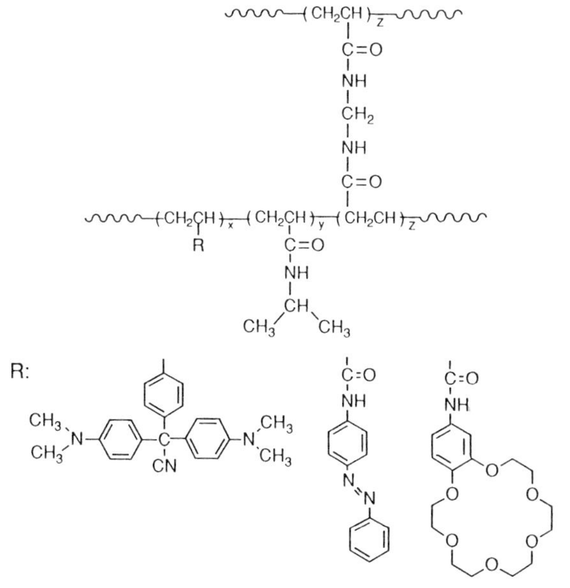

Figure 6 shows equilibrium volumes of the gel containing pendant triphenylmethane leuconitrile groups measured as a function of temperature in the dark and under UV irradiation[7]. When the gel was not irradiated with ultraviolet light, the gel underwent a sharp but continuous volume change at around 30.0°C. Upon irradiation the gel showed a discontinuous volume transition. At 32.6°C, the volume of the gel suddenly decreased by approximately ten time. Above the transition temperature the gel did not change markedly. When the temperature was fixed at 32°C, the gel underwent a discontinuous swelling-shrinking switching upon irradiating and removing ultraviolet light.

The aqueous solution of PNIPAM having pendant azobenzene groups showed sharp phase separation upon heating, as shown in Fig. 2. In the gel phase, however, the volume change of the gel upon heating was continuous. The discontinuous deswelling of the gel was not observed in the dark as well as under UV irradiation. The UV irradiated gel showed a higher swelling degree in the dark below 30°C. This indicates the possibility to control both swelling and deswelling processes by UV and visible irradiation. In fact, the volume of the gel was reversibly expanded as much as 25 % at 15 °C by alternate irradiation of UV ($350<\lambda<410$ nm) and visible ($\lambda>450$ nm) light.

The PNIPAM containing benzo[18]crown-6 described in section 3 which underwent metal ions induced phase separation in aqueous solution can be applied for constructing metal ion responsive polymer gels. When the PNIPAM is crosslinked to form a gel network, the conformational change is expected to result in the volume phase transition.

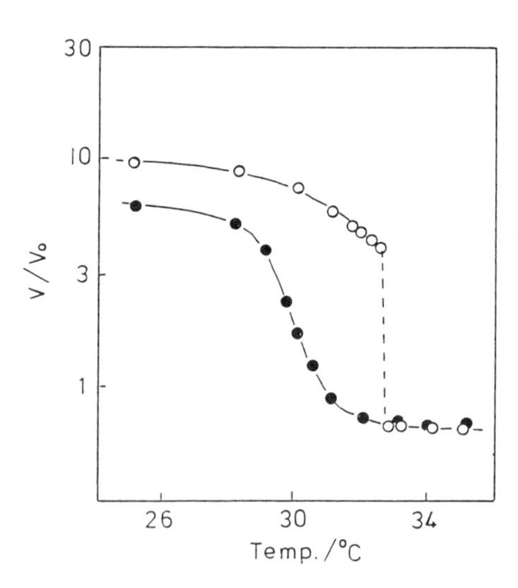

Figure 6. Volume changes of a PNIPAM gel containing pendant triphenylmethane leuconitrile groups as a function of temperature before (●) and after (○) UV irradiation.

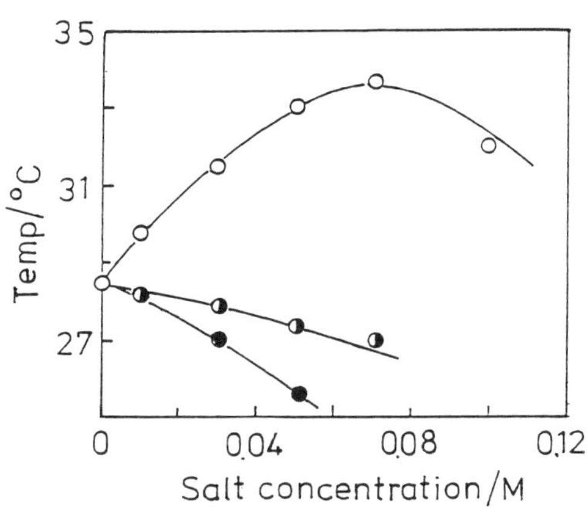

Figure 7. Concentration dependence of the transition temperature at which a PNIPAM gel containing benzo[18]crown-6 underwent a sharp volume changes in the presence of potassium (○), sodium (◑) and lithium (●) sulfates.

PNIPAM microsphere gels with diameter of 100 - 200 μm were prepared by emulsion polymerization. The gel containing 12 mol % benzo[18]crown-6 was immersed in water and the diameter change of the gel was measured during heating with a rate of 0.3°C/min. The gel was swollen below 25°C. In the absence of metal ions, the gel started to shrink at 26 C and showed a sharp volume change at 28.4°C. Finally the volume decreased by as much as ten times the original volume.

The deformation process was remarkably affected by the addition of metal ions. Figure 7 shows the temperature at which the gel underwent a sharp volume change in the presence of various metal ions. The addition of potassium sulfate increased the temperature, while other metal sulfates, such as sodium and lithium sulfates, decreased the temperature. When potassium ions are bound to the pendant crown ether groups, the hydrophobic interaction among pendant groups is diminished because of the hydrophobic nature of the ions. Consequently, the ion caused a raise in the transition temperature. The PNIPAM gel expanded as much as ten times by the addition of 3.0×10^{-2} M potassium sulfate at 32°C.

4. CONCLUSION

The concept of photostimulated and chemicals-induced phase transition described above can be applied to other systems, such as proton-induced or electrochemical phase transitions. All of these systems have high energy conversion efficiency in principle and the stimuli-responsive polymer systems are sensitive. When gels are made of these polymers, the gels act as micro actuators. The shapes or sizes of the gels can be controlled at will by external stimulation, such as photons, electrons or chemicals. The driving forces inherently originate from property changes of molecules. Therefore, the size of the actuators can be minimized to a micrometer scale. This is one of possible approaches for constructing micro actuators.

REFERENCES

1. M. Irie, Adv. Polym. Sci., 94 (1990) 27
2. M. Irie, Pure Appl. Chem., 62 (1990) 1495
3. S. Fujishige, K. Kubota and I. Ando, J. Phys. Chem., 93 (1989) 3311
4. M. Irie and D. Kungwatchakun, Proc. Jpn. Academy, 68-B (1992) 127
5. M. Irie, Y. Misumi and T. Tanaka, Polymer, 34 (1993) 4531
6. F. Vogtle (ed), Host Guest Chemistry, Springer-Verlag, Berlin, 1981
7. A. Mamada, T. Tanaka, D. Kungwatchakun and M. Irie, Macromolecules, 23 (1990) 1517

MICROCHEMISTRY
Spectroscopy and Chemistry in Small Domains
Edited by H. Masuhara et al.

Photoinduced volume change of polyacrylamide microgels ; micrometer size effects and kinetic analyses

M. Ishikawa[#] and N. Kitamura[*,‡]

Microphotoconversion Project,[§] ERATO Program, Research Development Corporation of Japan, 15 Morimoto-cho, Shimogamo, Sakyo-ku, Kyoto 606, Japan

Photoinduced volume change of polyacrylamide microgels having triphenylmethane leuco cyanide (TPMCN) in water was studied in detail and, the response time of the volume change was shown to be much improved by reducing the gel radius from millimeter to micrometer. Kinetics of the volume change processes and the origin of the micrometer size effects on the photoresponse time were discussed based on simultaneous measurements of the volume change and the degree of photoionization of TPMCN to triphenylmethane cation (TPM^+). Microviscosity change in the gel during the photoinduced volume change was also studied through picosecond fluorescence lifetime measurements of TPM^+.

1. INTRODUCTION

Photoresponsive polyacrylamide (AA) gels containing triphenylmethane leuco derivatives (TPMX where X is CN^- or OH^-) will play important roles in various research fields [1, 2], since the gel exhibits a large volume change upon UV photoirradiation in water owing to ionization of the excited-state TPMX chromophore to triphenylmethane cation (TPM^+) and X^- (CN^- or OH^-) [1]. Actually, a large volume change of the gel is expected to be applied to actuators, artificial muscle, and so forth [2], and ejection of X^- from the gel will be utilized as a catalyst through arbitrary release of a nucleophile (i.e., CN^- or OH^-) upon photoirradiation. For millimeter-sized, disk-shaped AA-TPMCN gels, it has been so far reported that the gel expands as large as ~ 18 times of the initial volume upon UV irradiation in

---

* To whom all correspondence should be addressed.
# Present address ; Research Center, Nissan Motor Co. Ltd., 1 Natsujima-cho, Yokosuka 237, Japan.
‡ Present address ; Department of Chemistry, Faculty of Science, Hokkaido University, Sapporo 060, Japan.
§ Five-year term project ; October 1988 - September 1993.

water [1, 3, 4] and, therefore, the gels certainly act as actuators as mentioned above. However, the time necessary to reach an equilibrium gel volume is very slow with the response time of ~ 1 h. For further applications of the AA-TPMCN gels to photoresponsive materials, improvement of the photoresponse time is highly expected. For basic understandings of the phenomena, on the other hand, theoretical considerations on the volume change of polymer gels upon external stimulus such as a pH change or solvent change have been reported by several research groups [5, 6]. Nevertheless, since the degree of ionization (α) of the chromophores in gels cannot be determined experimentally for ordinary polymer gels such as polyacrylamide and poly(acrylic acid), the volume change processes of these gels have never been discussed based on the α value, which is a very important factor governing the gel volume. If one can measure the α values in gels, kinetics and mechanistic studies on the volume change will be advanced and, the available theories will be more explicitly compared with the experimental observations.

As a very unique property of the AA-TPMCN gel, TPM^+ produced upon UV photoirradiation shows strong absorption in the visible region, so that the degree of photoionization of the TPMCN chromophore in the gel can be directly determined by absorption spectroscopy. The AA-TPMCN gels are therefore very suitable for the studies on the volume change processes in detail. An another important property of the gel is fluorescence characteristics of the TPM^+ chromophore. It is well known that triphenylmethane cation dyes represented by malachite green (MG^+) as a reference compound of TPM^+ shows a viscosity-dependent fluorescence lifetime in picosecond time region and, the lifetime decreases considerably with increasing the viscosity of a medium [7]. Microviscosity change of the gel during the photoinduced volume change could be also studied by the fluorescence lifetime measurements TPM^+ of in the AA-TPMCN system.

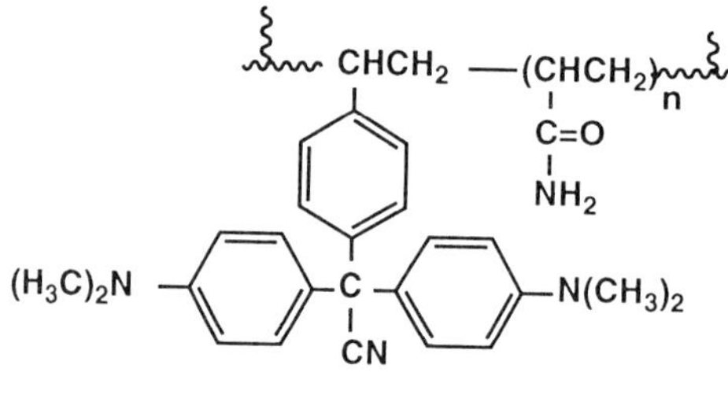

AA-TPMCN 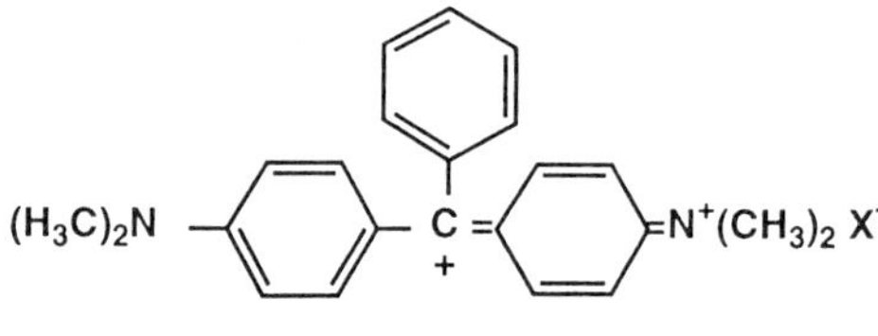 MG^+

In this article, we describe micrometer-size effects on the photoinduced volume change of the AA-TPMCN gel in water and, discussed on the kinetics and mechanisms of the volume change processes in detail on the basis of the results of simultaneous measurements of the gel volume and the degree of photoionization of the TPMCN chromophore [8 - 10]. A picosecond fluorescence spectroscopic study on TPM^+ is also demonstrated to show the effects of the three-dimensional polymer network on microviscosity in the gel [11].

2. MICROMETER SIZE EFFECTS ON PHOTOINDUCED VOLUME CHANGE

In order to study effects of the size of AA-TPMCN gels on photoinduced volume change in water, we prepared a series of rod-shaped microgels with different radii by free radical polymerization of a dimethyl sulfoxide solution (3 mL) containing acrylamide (AA, 10 mmol), a vinyl derivative of TPMCN (0.1 mmol), N,N'-methylenebis(acrylamide) (0.12 mmol), and 2,2'-azobis(iso-butyronitrile) (10 mg) in microcapillary glass tubes at 60 °C for 2.5 h. After the polymerization, the microcapillary glass tube was washed with acetone and one side of the glass tube was broken to expose the rod-shaped gel. The gel was then washed with enough pure water prior to measurements. The length of the gel used in this study was much larger than the diameter (length : diameter = 10 :1), so that the gels were regarded as cylindrical microgels. Photoirradiation to the microgel in water was performed with a 500 W high-pressure Hg lamp under an optical microscope and the volume change of the gel was monitored by a CCD camera equipped to the microscope [8].

The microgels prepared in this study with the initial radii (R_0) of 11 ~ 235 μm showed photoinduced volume change in water. The volume change of the gel was isotropic, so that a volume expansion ratio, V/V_0, where V_0 and V are the volume of the gel at irradiation time (t) of t = 0 and t = t', respectively, was calculated based on the change in the gel diameter. Time-response profiles of V/V_0 determined for several microgels are summarized in Figure 1. The results demonstrate that the photoinduced volume change of the AA-TPMCN gels is strongly dependent on R_0. For the microgel with R_0 = 11 μm, as an example, the volume change was almost finished

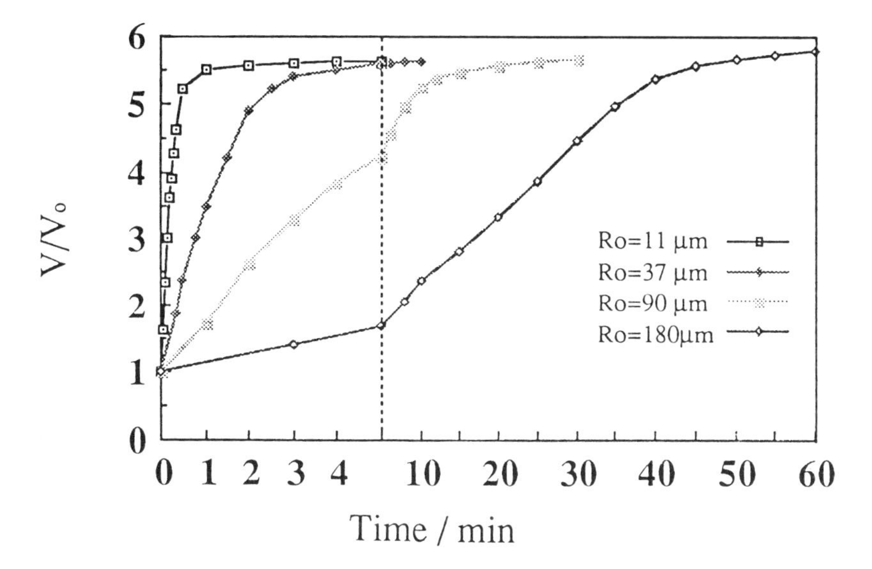

Figure 1. Micrometer size effects on the photoinduced volume change of the AA-TPMCN gels in water.

within 1 min, while that continued over 30 min for the gel with R_0 = 180 μm. The smaller the gel size, the faster the photoinduced volume change. It is noteworthy, furthermore, that the equilibrium V/V_0 at $t = \infty$ is almost independent of R_0 with the value of 5.6 ~ 6.0. This is in marked contrast to the previous results on millimeter-sized AA-TPMCN gels, which show the photoresponse time of ~ 1 h [3, 4]. The smaller V/V_0 ($t = \infty$) value of 5.6 ~ 6.0 compared with that for the millimeter-sized gels (~18) is simply explained by the lower molar ratio of TPMCN to AA (1 : 100) and, will be improved in the microgel system by optimizing the molar ratio. This is the first demonstration of the micrometer size effects on the photoinduced volume expansion of the AA-TPMCN gels in water [8].

3. KINETICS OF PHOTOINDUCED VOLUME CHANGE

3.1. Simultaneous Measurements of Volume Change and Degree of Photoionization of TPMCN Chromophore

The time response profiles of V/V_0 in Figure 1 cannot be fitted by a simple exponential function, so that the volume expansion of the gel is supposed to involve several processes. The origin of the micrometer size effects on the photoinduced volume expansion should be therefore studied based on more elaborated experiments. The photoinduced volume change of the AA-TPMCN gels in water is induced by photoionization of the TPMCN chromophore to TPM^+ and, this leads to a change in the osmotic pressure between the gel and the surrounding water phase [3, 4]. For equilibration of the osmotic pressure change, the gel absorbs a considerable amount of water, resulting in the volume expansion of the gel (i.e., swelling). In order to analyze the volume change processes, simultaneous measurements of the degree of photoionization of TPMCN and the volume change of the gel are absolutely necessary.

For simultaneous measurements of the visible absorption of TPM^+ and volume changes of the AA-TPMCN microgel during photoirradiation, we developed a microspectroscopic system as schematically shown in Figure 2 [9]. A quartz sample cell containing a AA-TPMCN gel and water was set on the stage of an inverted optical microscope and, the gel was irradiated with a 500 W Hg-Xe lamp through a glass filter (λ = 255 - 380 nm). For absorption measurements, monitoring light from a 100 W Hg-lamp was reflected by a dichroic mirror and focused to the gel through an objective lens. Light passed through the gel (I) was collected by a condenser lens and introduced to a monochromator-photomultiplier-computer system via an optical fiber to monitor absorbance of TPM^+ in the gel. The incident monitoring light intensity (I_0) was determined under the same conditions without the gel. An experimental error in determining I or I_0 was ~ ± 5 %, which was mainly due to fluctuation of the Hg-Xe lamp intensity with time. The volume change was observed by monitoring the gel diameter.

A typical example of the time profiles of the radius (R(t)) and absorbance (Abs(t)) of TPM^+ for the gel with R_0 = 98 μm under continuous UV irradiation is shown in Figure 3. Since the molar absorption coefficient (ε) at the absorption maximum wavelength of TPM^+ (λ_{max} = 623 nm, $\varepsilon = 1.8 \times 10^5$ M^{-1} cm^{-1}) was too large to follow

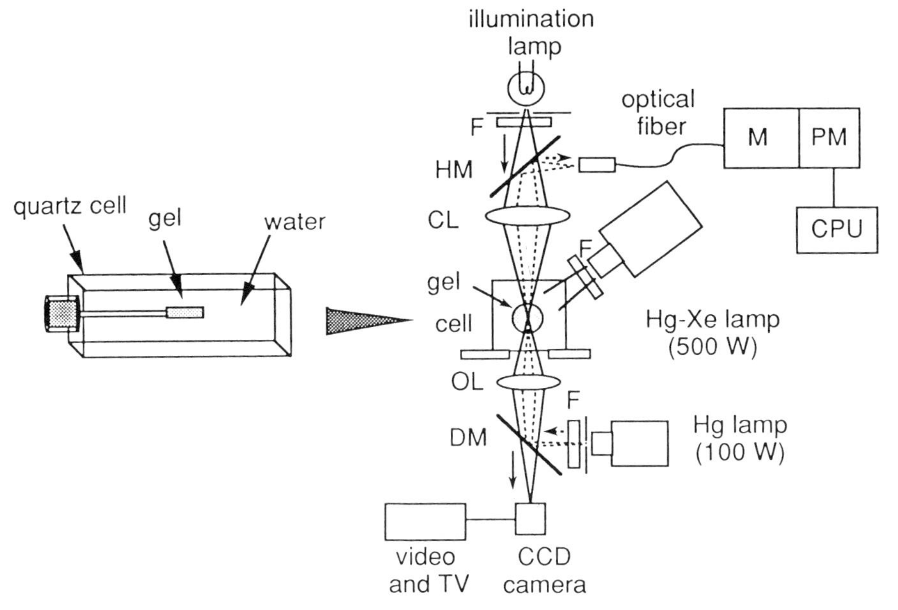

Figure 2. Experimental setup for simultaneous measurements of absorbance and the radius of AA-TPMCN microgel. HM ; half mirror, CL ; condenser lens, OL ; objective lens, DM ; dichroic mirror, F ; glass filter, M ; monochromator, PM ; photomultiplier.

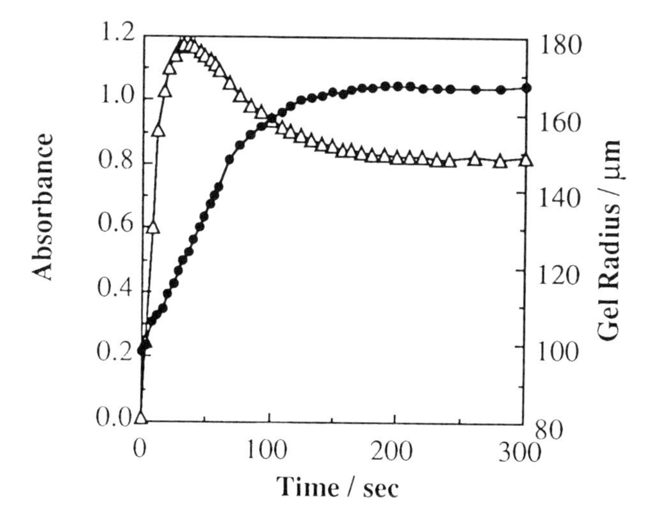

Figure 3. Time-response profiles of absorbance of TPM^+ at 535 ± 2 nm (triangles) and the radius of the gel (circles) under UV photoirradiation in water.

Abs(t), absorbance at (535 ± 2) nm was monitored ($\varepsilon = 1.04 \times 10^4\ M^{-1}\,cm^{-1}$). ε for the gel was assumed to be the same as that for a water soluble AA-TPM^+ copolymer. As seen in Figure 3, the absorbance of TPM^+ in the gel sharply increases with photoirradiation time and reaches a maximum value around t = 30 s. Further prolonged irradiation rendered a slow decrease in the absorbance and showed a plateau at t > 150 s. On the other hand, the radius of the gel increases gradually with time and reaches an equilibrium value at t > 150 s. The time necessary to reach the equilibrium absorbance coincides very well with that necessary to attain the equilibrium gel radius ; t ~ 150 s. Analogous results were obtained for the gels with various R_0 (98 ~ 235 μm), while the time responses of the absorbance and the radius were slower for the gel with larger R_0. Figure 3 explicitly proves that R(t) and Abs(t) do not agree with each other and, the time course of the gel radius is much slower than that of the degree of photoionization of TPMCN.

The volume change processes of the AA-TPMCN microgels cannot be directly explained based on the observed Abs(t) and R(t), since Abs(t) is a function of the concentration of TPM^+ in the gel ($C^+(t)$) as well as of the optical path length of the monitoring beam (d(t)). The optical path length of the monitoring beam is assumed to be equal to the diameter of the gel (i.e., d(t) = 2R(t)). Therefore, the concentration of the TPM^+ in the gel at given photoirradiation time can be calculated by eq. (1),

$$C^+(t) = Abs(t) / 2R(t)\,\varepsilon \qquad (1)$$

On the other hand, the degree of photoionization of TPMCN in the gel ($\alpha(t)$) is given by eq. (2),

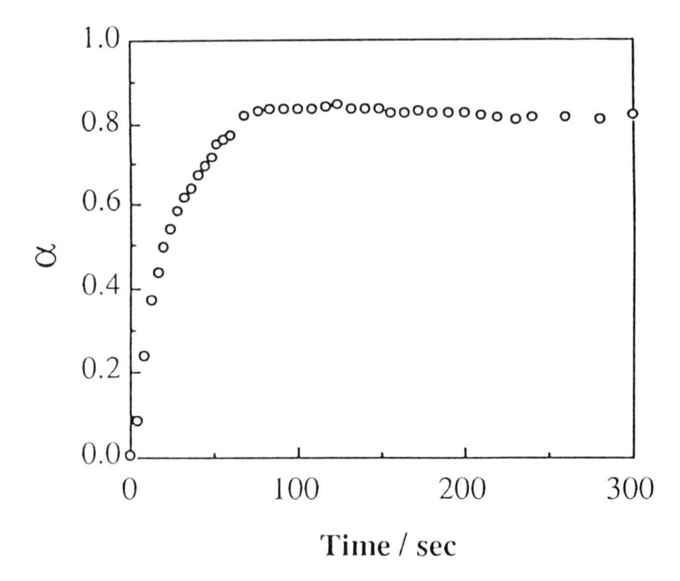

Figure 4. Time-response profile of the degree of photoionization of TPMCN in the microgel.

$$\alpha(t) = (C^+(t) / C_0)\ (R(t)^3 / R_0^3) \qquad (2)$$

where C_0 represents the concentration of TPMCN in the gel at $t = 0$ and can be estimated on the basis of the dry weight of the gel and the molar ratio of TPMCN to AA (1 : 100). From eqs. (1) and (2), we obtained eq. (3),

$$\alpha(t) = Abs(t)\ R(t)^2 / 2C_0R_0^3\ \varepsilon \qquad (3)$$

$\alpha(t)$ thus obtained for the gel with $R_0 = 98$ μm and $C_0 = 1.36 \times 10^{-2}$ M is shown in Figure 4. The α value reaches the equilibrium value (~ 0.8) at t ~ 60 s and does not decrease even upon prolonged UV photoirradiation in contrast to Abs(t) in Figure 3. The curious behavior of the observed Abs(t) can be explained by the changes in both $R(t) = d(t)/2$ and $\alpha(t)$. Ionic photodissociation of TPMCN in the gel almost finishes at $t < 60$ s, while the radius of the gel gradually increases up to t ~ 150 s. Therefore, Abs(t) shows a maximum and subsequent decrease to a certain value during the volume expansion as easily expected from eq.(3). The results demonstrate that the volume expansion of the gel should be analyzed by taking time dependencies of Abs(t) and R as revealed from Figures 3 and 4. To the best of our knowledge, this is the first example of direct measurements of $\alpha(t)$ for the volume change of polymer gels [9]. The AA-TPMCN gels are certainly useful to study the kinetics and mechanisms of the photoinduced volume change.

3.2. Analysis of Photoinduced Volume Change Processes

Close inspection of the data in Figures 3 and 4 indicates that there appears a time-lag between $\alpha(t)$ and R(t) with the latter being delayed compared to $\alpha(t)$. This is in particular importance for the analysis of the photoinduced volume change of the gel. For poly(acrylic acid) gels, for example, the volume change by an external stimulus such as a pH or solvent change proceeds from the skin part of the gel and then, the change in the inner gel network follows. Therefore, the volume change is solely governed by the diffusion of protons or solvent molecules to the gel from the surrounding solution. For the present AA-TPMCN gels, on the other hand, the gel is uniformly irradiated by UV light so that the gel network expands homogeneously in both the skin and inner parts of the gel. The volume expansion rate of the AA-TPMCN gel is thus dependent on both the diffusion coefficient of the gel network and $\alpha(t)$. This is the one of the important characteristics of the photoinduced phenomenon and is different from the responsiveness of other polymer gels by a pH or solvent change. Photoionization of TPMCN in the gel does not proceed instantaneously, but takes several tens of second as seen in Figure 4. A convolution analysis of the data in Figures 3 and 4 is thus necessary to elucidate the time-lag between R(t) and $\alpha(t)$.

For given photoirradiation time, an equilibrium gel radius (r_e) is determined by the relevant α value. The time response of the volume expansion ratio (r(t)), defined as $r(t) = [\{R(t) - R_0\} / R_0]$, should be related to $dr_e(t)/dt$ and the response function for the diffusion of the gel network (G(t)) as discussed above, where $r_e(t)$ is a time response of the equilibrium gel radius. If this is the case, r(t) can be given by a following convolution equation,

$$r(t) = \int (dr_e(t) / dt)\, G(t - t')\, dt' \tag{4}$$

According to Tanaka and his co-workers [5], G(t) is expressed as in eqs. (5) and (6),

$$G(t) = 1 - [\Sigma \{n^{-2} \exp(-n^2 t / \xi)\} / \Sigma n^{-2}] \tag{5}$$

$$\xi = R_0^2 / \pi^2 D \tag{6}$$

where D is a diffusion coefficient of the gel network. A relationship between $r_e(t)$ and $\alpha(t)$ has been obtained by separate experiments and is given by eq. (7),

$$r_e(t) = A\, \alpha(t) + B \tag{7}$$

where A ~1.04 and B ~ 0 [10]. Equation (4) is now rewritten as in eq. (8).

$$r(t) = A \int (d\alpha(t) / dt)\, G(t - t')\, dt' \tag{8}$$

Equations (5) and (8) indicate that r(t) is solely determined by the known parameters of $\alpha(t)$ and t. The experimental results can be therefore compared with the theoretical prediction.

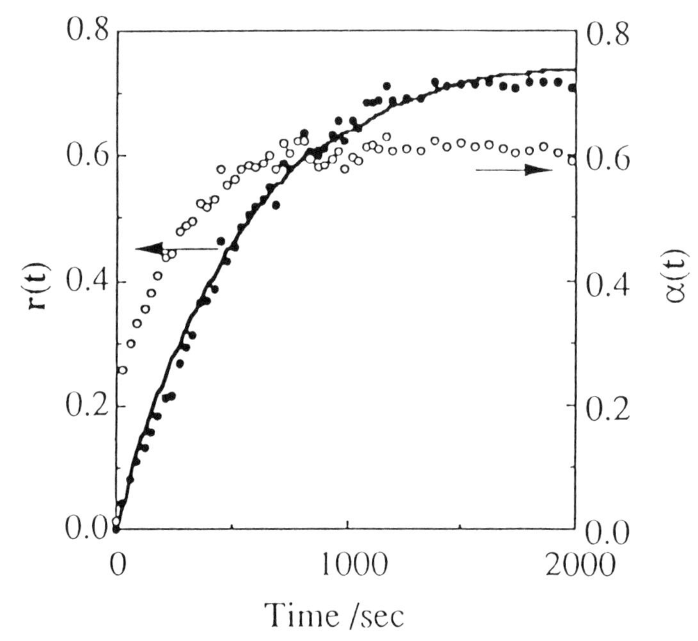

Figure 5. Simulation of the time-response profile of the gel radius. Closed and open circles represent the observed gel radius and the degree of photoionization of TPMCN, respectively. Solid curve is the simulation curve (see text in detail).

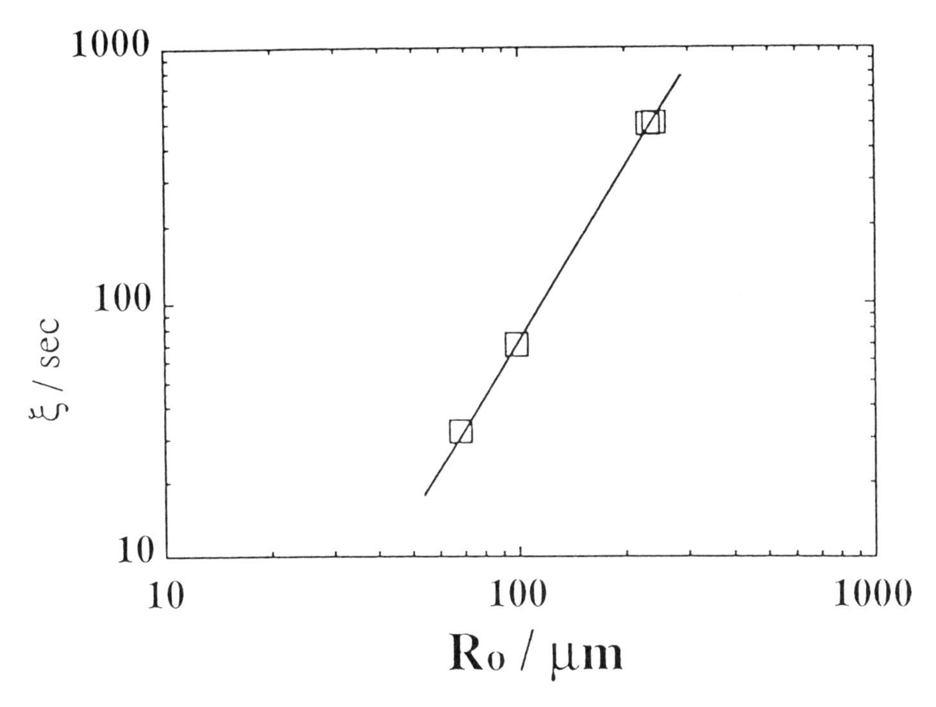

Figure 6. Micrometer size effects on the response time of the photoinduced volume change of the AA-TPMCN gels in water.

For the gel with R_0 = 235 μm, the observed r(t) curve (closed circles) was fitted by eqs. (5) and (8) with the experimentally observed $\alpha(t)$ data (open circles) as shown by the solid curve in Figure 5. As clearly seen in the figure, the simulated curve satisfactorily agreed with the observed one when ξ in eq. (5) was assumed to be 550 s. This demonstrates that the present convolution analysis correctly explains the photoinduced volume change of the AA-TPMCN microgels. The analogous analysis of the data for other AA-TPMCN microgels with different R_0 was successful and, we obtained the ξ value characteristic to each AA-TPMCN microgel. ξ was then plotted against R_0 of the microgel as shown in Figure 6. Although the number of the experimental points is still limited, the ξ value is shown to be proportional to the square of the initial radius of the gel as expected from eq. (6). The linear relationship between ξ and R_0 in Figure 6 affords D to be (1.1~1.5) x10^{-7} cm^2 s^{-1}, which is slightly smaller than the value for polyacrylamide gels ((2.4 ~ 4.0) x 10^{-7} cm^2 s^{-1}) [12, 13]. However, Li and Tanaka have reported that the diffusion coefficient of the gel network for a rod-shaped gel is 2/3 of that for a relevant spherical gel [13]. We conclude, therefore, that the photoinduced volume expansion of the AA-TPMCN microgels is reasonably explained in terms of the diffusion theory of the gel network [10].

4. PICOSECOND FLUORESCENCE SPECTROSCOPIC STUDY ON PHOTOINDUCED VOLUME CHANGE PROCESSES

In addition to the UV absorption change, TPM^+ bound to the gel shows a viscosity-dependent fluorescence lifetime (fluorescence maximum wavelength ~ 650 nm), whereas TPMCN fluoresces in the UV-VIS region shorter than 500 nm. During the volume expansion of the gel, the diffusion of the gel network will lead to changes in microviscosity in the vicinity of the TPMCN/TPM^+ chromophore, so that the fluorescence lifetime of TPM^+ will be strongly dependent on the volume change ratio (V/V_0). The photoinduced volume expansion processes of the gel can be thus discussed on the basis of fluorescence spectroscopy. In order to perform the experiments, thin-film AA-TPMCN gels with the thickness of 20 μm were prepared on quartz plates and, the fluorescence lifetime of TPM^+ in the gel was measured by a picosecond single photon timing technique [11].

In order to evaluate the microviscosity in the AA-TPMCN gels before and after the volume expansion, we measured fluorescence decays of malachite green (MG^+) as a reference compound of TPM^+ in various alcohols with different viscosity. A fluorescence decay of MG^+ in a low-viscosity solvent such as 1-butanol or water was best analyzed by a single exponential function, while that in more viscous solvents (1-octanol, 1-decanol, and so on) was fitted by a sum of two- or three-exponentials. Two-exponential fluorescence decays in viscous solvents have been reported for MG^+ in a mixture of glycerol and water [7], and interpreted in terms of rotational relaxation of the phenyl ring(s) of the compound in the excited state. Since the rotational relaxation of the phenyl ring(s) is unfavorable in high-viscosity solvent

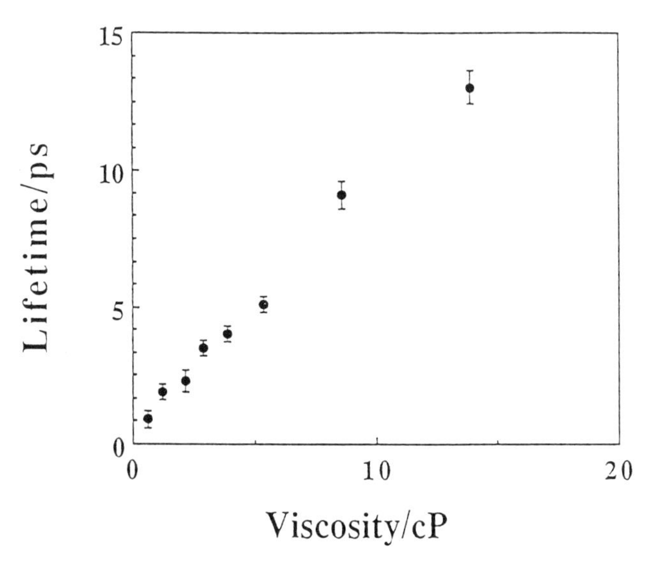

Figure 7. The shortest fluorescence lifetimes of MG^+ in various alcohols analyzed by one exponential neglecting the longest component as a function of solvent viscosity.

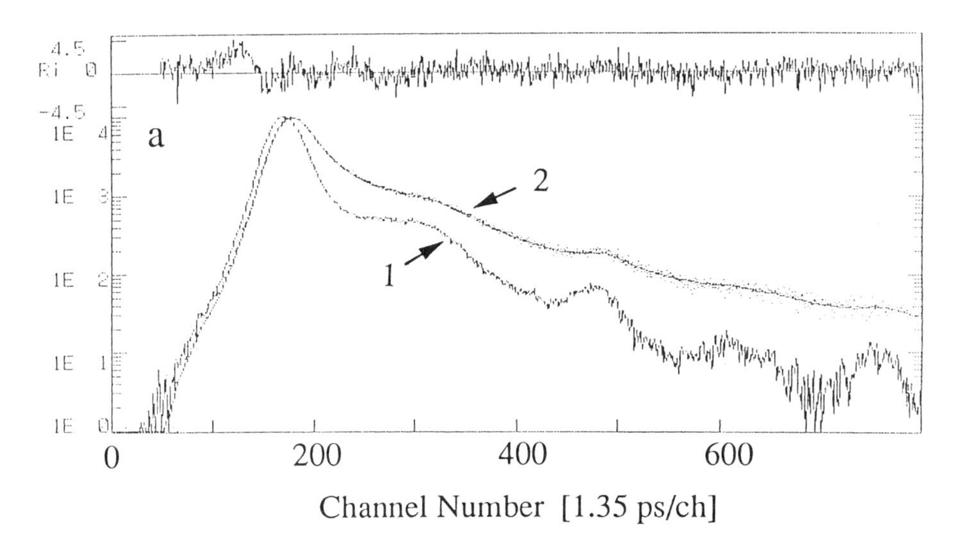

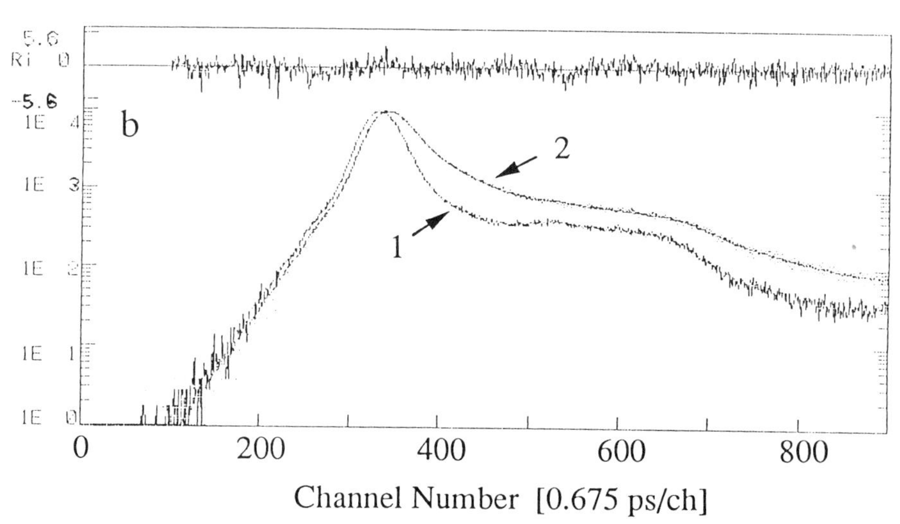

Figure 8. Fluorescence decay curves of TPM^+ in the thin-film AA-TPMCN gel before (a ; upper trace) and after UV photo-irradiation (b ; lower trace). System response function (solid curve 1) at the excitation wavelength (580 nm) was used for the decay curve analyses. Best fit to the decay data is shown as a solid curve 2. The residual (R_i) are plotted against the channel number.

solvents, the fluorescence lifetime of MG^+ becomes longer with increasing solvent viscosity (η). Actually, we found that the short fluorescence lifetimes (τ) analyzed by a single or a sum of two exponentials in various alcohols were correlated with η as shown in Figure 7 and, obtained an empirical relation of $\tau \propto \eta^{-0.8}$ in the η range of 0.6 ~ 14 cP [11].

Figure 8 shows typical fluorescence decays of TPM^+ bound to the thin-film AA-TPMCN gels before and after UV irradiation (λ = 250 - 400 nm) in water. Fluorescence of TPM^+ in the gel decayed much slower than that of MG^+ in an aqueous solution owing to restricted rotation of the phenyl ring(s) of TPM^+ in the gel and, therefore, to the microviscosity in the vicinity of the chromophore. Before UV irradiation to the gel (a in Figure 7), actually, fluorescence decay of TPM^+ can be analyzed by a sum of three exponentials with a small contribution of the third component ; $\tau_1 = 9.8 \pm 1.8$ ps ($A_1 = 0.920$), $\tau_2 = 50 \pm 8$ ps ($A_2 = 0.075$), $\tau_3 = 290 \pm 90$ ps ($A_3 = 0.005$). In equilibrium conditions after prolonged UV irradiation, where the volume is expected to expand as large as six times the initial volume, the fluorescence decay of TPM^+ becomes faster and can be analyzed by a two-exponential function ; $\tau_1 = 7.4 \pm 1.0$ ps ($A_1 = 0.96$), $\tau_2 = 46 \pm 10$ ps ($A_2 = 0.04$). Two-exponential decays of TPM^+ in the gel after UV irradiation are quite similar to the results of MG^+ in 1-octanol or 1-decanol, indicating that the microenvironment in the swollen gel is rather uniform and the viscosity is relatively low as compared with that before UV irradiation.

Analogous experiments were performed for the AA-TPMCN gels with various V/V_0 values, where V/V_0 was estimated from the absorbance of TPM^+ in the gel, $\alpha(t)$, and the relation between $\alpha(t)$ and V/V_0 as described in the preceding section. The

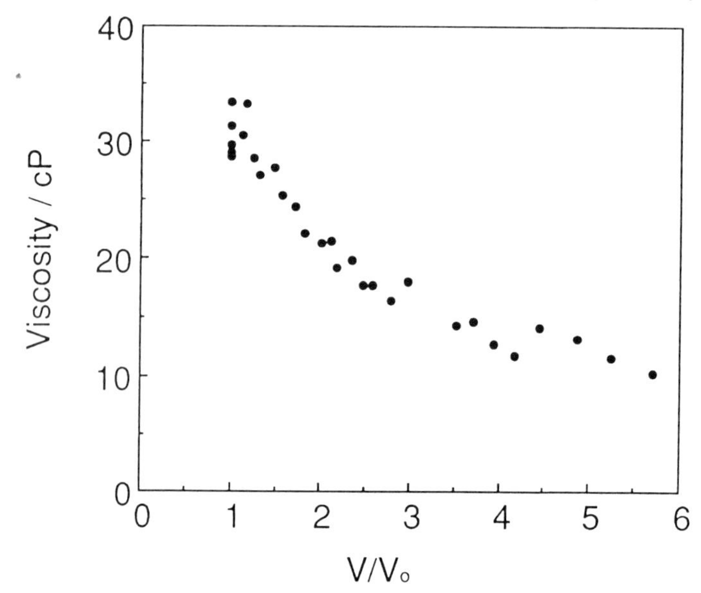

Figure 9. Microviscosity change in the gel during photoinduced volume change of AA-TPMCN gel in water.

microviscosity change in the gel during the photoinduced volume change was roughly estimated on the basis of the observed fluorescence lifetime of TPM^+ in the gel and the η dependence of the fluorescence lifetime of MG^+ (Figure 7) as shown in Figure 9. Before photoirradiation, the microviscosity in the gel is as high as ~ 30 cP. With proceeding the volume change, the microviscosity in the AA-TPMCN gel gradually decreases and that of the fully swollen gel (V/V_0 ~ 6.0) is ~ 7 cP. The microviscosity of ~ 7 cP is still seven times higher than that in the bulk water (1 cP). The polymer network in the gel prevents free rotation of the chromophore even upon swelling in water. Fluorescence spectroscopy is highly potential to study microenvironmental changes during the macroscopic shape change of the gel.

5. CONCLUSION

A study of the photoinduced volume change of the AA-TPMCN microgels based on absorption and fluorescence spectroscopy has provided an inside look at the kinetics and mechanisms of the micrometer size effects on the photoresponse time and the diffusion of the polymer network in the gel. This was only possible by direct measurements of the degree of photoionization of the TPMCN chromophore in the gel through UV-VIS absorption spectroscopy and fluorescence lifetime measurements during the volume change. Such the study has never been explored owing to a lack of a determination method of the α value for ordinary polymer gels. The TPMCN chromophore will therefore play an important role for studies on swelling and shrinking processes of various polymer materials. Furthermore, microgels are certainly superior to millimeter-order gels in respect to the response time against an external stimulus. Functionalized materials such as actuators, artificial muscles, or catalysts with higher response abilities will be realized through miniaturization and, such materials are highly promising to conduct chemistry in micrometer dimension as well.

ACKNOWLEDGMENT

The authors wish to express sincere thanks to Dr. N. Tamai (Research Development Corporation of Japan) and Prof. H. Masuhara (Osaka University) for collaboration of picosecond fluorescence measurements and various discussions, respectively, during the research period of the Microphotoconversion Project. Special thanks are also due to Prof. M. Irie (Kyushu University) for valuable comments.

REFERENCES

1. M. Irie, Advanced in Polymer Science, H. -J. Cantow et al. (eds), Springer-Verlag, Berlin, 94 (1990) 27.
2. Proceedings of the 1st International Conference on Intelligent Materials, T.

Takagi et al. (eds), Technomic Publishing, Lancaster, (1992).
3. M. Irie and D. Kunwatchakun, Macromolecules, 19 (1986) 2476.
4. A. Mamada, T. Tanaka, D. Kunwatchakun and M. Irie, Macromolecules, 23 (1990) 1517.
5. T. Tanaka and D.J. Fillmore, J. Chem. Phys., 70 (1979) 1214.
6. P. J. Flory, Principles of Polymer Chemistry, Cornel Univ. Press, Ithaca, (1953).
7. M. D. Hirsch and H. Mahr, Chem. Phys. Lett., 60 (1979) 299.
8. M. Ishikawa, N. Kitamura, H. Masuhara and M. Irie, Makromol. Chem., Rapid Commun., 12 (1991) 687.
9. M. Ishikawa, H. Misawa, N. Kitamura, M. Irie and H. Masuhara, Chem. Lett., (1992) 311.
10. M. Ishikawa, N. Kitamura, M. Irie and H. Masuhara, in preparation.
11. N. Tamai, M. Ishikawa, N. Kitamura and H. Masuhara, Chem. Phys. Lett., 184 (1991) 398.
12. A. Peters and S. J. Candau, Macromolecules, 21 (1988) 2778.
13. Y. Li and T. Tanaka, J. Chem. Phys., 92 (1990) 1364.

MICROCHEMISTRY
Spectroscopy and Chemistry in Small Domains
Edited by H. Masuhara et al.

Fluorescence decay studies of polymer diffusion across interfaces in latex films

Mitchell A. Winnik, Lin Li and Yuan Sheng Liu

Department of Chemistry and Erindale College, University of Toronto, 80 St. George Street, Toronto, Ontario, Canada M5S 1A1

Films are prepared by compression molding of a mixture of donor- and acceptor labeled poly(methyl methacrylate) microspheres. When these films are heated, polymer molecules diffuse across the particle boundaries, bringing the donor and acceptor groups into proximity. Fluorescence decay measurements of the films allow the extent of diffusion to be monitored.

1. INTRODUCTION

In this paper, we examine the use of single photon timing [1] measurements to follow the diffusion of polymer molecule across a polymer-polymer interface in melt-pressed poly(methyl methacrylate) (PMMA) films. These films are prepared by compression molding of PMMA powder, in which the individual particles are present in the form of polymeric nanospheres, monodisperse in size. The films are prepared from a 1:1 mixture of two polymer nanosphere samples, one containing PMMA labeled with a dye (phenanthrene, Phe) which can act as a donor in direct non-radiative energy transfer (DET) [2], and the other labeled with a second dye (anthracene, An) which can act as the corresponding acceptor. A drawing depicting the compression molding process is shown in Figure 1.

If the compression process is carried out at a temperature not too much greater than the glass-to-rubber transition temperature (T_g) of the polymer, little if any polymer diffusion will occur. This means that one can prepare a system with the polymer components in intimate contact, but with the donor- and acceptor-labeled polymers confined to opposite sides of the interface. If this film sample is now placed in an

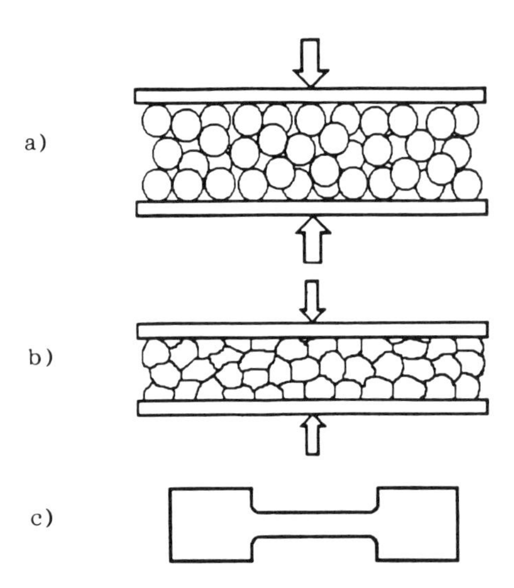

Figure 1. A cartoon depicting the compression molding of latex particles: a) hot-pressing of latex particles; b) particles deform to form a void-free film; and c) dogbone-like films were annealed for various periods of time.

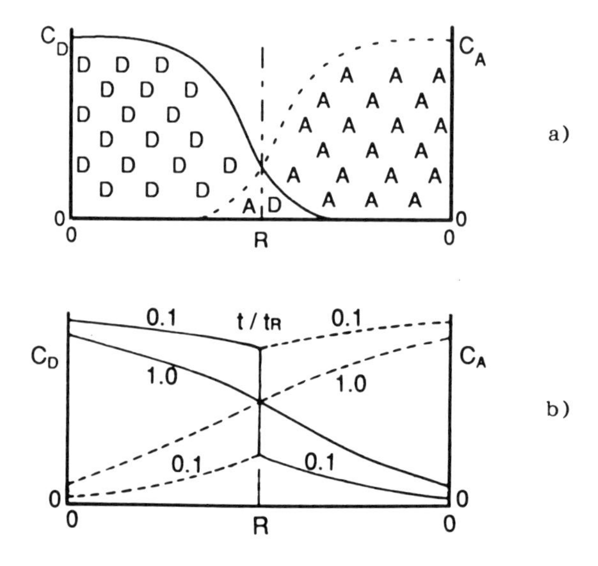

Figure 2. Concentration profiles at particle interface, dominated by a) the Fickian diffusion and by b) the reptation motion, respectively. R is the particle radius and t_R is the reptation time.

oven and heated, polymer diffusion will begin. Diffusion across the particle-particle boundary leads to mixing of donor [D] and acceptor [A] substituents. This in turn should lead to a large increase in energy transfer in the sample.

This process is conceptualized in Figure 2a. Here we depict a sharp interface separating the D- and A-labeled polymer in the nascent film. Upon annealing, polymer interdiffusion mixes the two components, generating concentration profiles $C_D(r,t)$ and $C_A(r,t)$ for the D- and A-labeled polymers, respectively, where r is the diffusion distance. Several years ago, Prager and Tirrell [3] pointed out that in such an experiment, the extent of interdiffusion (the integral over the product of the concentration profiles) is related to the quantum efficiency of energy transfer in the system. This insight is the starting point for the use of energy transfer measurements to study polymer diffusion across interfaces.

From a polymer physics point of view, the issue of greatest significance is the detailed shape of the concentration profile. If the polymer molecular weights are low, then one expects the diffusion to be described by Fick's laws of diffusion [4]. If, however, the polymer molecular weights are sufficiently large, new effects come into play which are expected to dominate the interdiffusion process. These factors were first recognized by de Gennes [5] and were subsequently examined in detail by Tirrell [6] and by Wool [7]. The ultimate challenge for someone examining interdiffusion by DET is to see if these ideas can be tested.

There are three parts to the de Gennes-Tirrell-Wool (GTW) analysis. The first is that entanglements suppress large amplitude lateral motion of the polymer molecules. Diffusion should occur by reptation [8], and not follow Fick's Laws. The second is that the kinetics of reptative interdiffusion should be extremely sensitive to the location of the chain ends. These may be randomly distributed in the matrix or may actually be concentrated in the interface. The third is related to the fact that long chains adjacent to the interface have their shapes distorted from the most probable (Gaussian) distribution of conformations. These chains have an added entropic driving force for interdiffusion. As a consequence, a discontinuity persists in the concentration profile at the interface for diffusion times less than the reptation time, t_R [6-9]. The shape of $C_D(r,t)$ and $C_A(r,t)$ appropriate to this type of interdiffusion is shown in the lowermost part of Figure 2b [11]. At times greater than t_R, the boundary chains have randomized their shapes, and the number of chains crossing the interface has reached its asymptotic value.

To approach this problem from the perspective of energy transfer experiments, one begins with the idea of determining quantum efficiencies ϕ_{ET} for DET. These efficiencies are in principle available through steady-state fluorescence measurements. One should be able to measure relative intensities of Phe and An fluorescence, exciting the sample at a wavelength (eg 300 nm) where Phe is selectively excited, in film samples annealed for various periods of time. In our hands with film samples, these types of measurements have never worked well, largely due to the contribution of radiative energy transfer. To avoid these problems, we use single photon timing (SPT) to obtain donor fluorescence decay profiles ($I_D(t')$). This has two important advantages.

First, we can compute the areas under the $I_D(t')$ profiles by fitting these profiles to arbitrary functions which are easy to integrate. These integrated areas can be used to calculate ϕ_{ET} . This approach avoids the need for rigorous physical models for fitting the $I_D(t')$ profiles and still provides quantitative information about the extent of polymer interdiffusion. It does, however, discard information about the donor-acceptor distribution in the interphase which are contained in the detailed shape of each $I_D(t')$ profile. The second advantage of time-resolved measurements is that it should be possible to develop models for fitting these decay curves that would allow information about the D/A concentration profiles to be obtained. Since the D/A pair distribution is related to the polymer segment distribution across the interphase, one could imagine that it is in terms of these profiles that the GTW ideas could be tested.

In the sections that follow, we describe the preparation of polymer samples and examine the fluorescence decay signals that they generate. We emphasize that while a significant amount of interesting information has already been obtained, we are just beginning to appreciate how much more information about the D/A pair distribution can be extracted from the $I_D(t')$ profiles. Thus we are still far from our hope of being able to use DET to provide quantitative tests for theoretical predictions about polymer diffusion across interfaces.

2. EXPERIMENTAL

2.1 Emulsion polymerization

All particles used in this study were prepared by a two-stage emulsion polymerization of MMA with either 9-vinylphenanthrene (VPhe) or 9-anthrylmethacrylate (AnMA) as fluorescent comonomers. Details are given in our earlier publications[12,13]. A typical recipe is given in Table I. In the first stage, latex seeds (diameter = 50 nm) were formed from a small amount of MMA. Then, initiator and surfactant in water, MMA plus fluorescent comonomers, were added simultaneously into the reactor under monomer-starved conditions.

Table I
Recipe and conditions for emulsion polymerization.

	First Stage (seed)	Second Stage
Water (ml)	135	75
MMA (ml)	10.5	85
VPhe or AnMA (g)		1.62(Phe) or 2.04(An)
KPS (g)[a]	0.14	0.13
SDS (g)[b]	0.31	1.80
$NaHCO_3$ (g)	0.24	
Temp. (°C)	80	80
Feed time (hr)		12
Total time (hr)	1	21

a. potassium persulfate. b. sodium dodecyl sulfate

2.2 Characterization

The latex particle size was determined by dynamic light scattering. Molecular weights and molecular weight distributions were determined by gel permeation chromatography coupled with tandem refractive index and fluorescence detectors, the latter also used to monitor for the presence of fluorescent monomers, either unreacted during polymer synthesis or formed by fragmentation from the polymer during annealing of film samples. UV measurements were carried out to determine the fluorescent chromophore content of the polymer. Table II lists the chromophore content, molecular weight, molecular weight distribution, and particle size of the samples used in this study.

Table II
Characterization of labeled PMMA latexes.

Sample	Chromophore content (mol%)	10^{-3} M_w	M_w/M_n	Particle diameter (nm)
PhePMMA	0.63	289	3.94	125
AnPMMA	0.88	274	3.73	119

2.3 Latex cleaning

In general, various electrolytes such as initiator, surfactant and buffer are retained in latex dispersions. These impurities were removed by serum replacement using a membrane filtration (Millipore Pellicon^{TM}) system until the conductivity of the dispersion (10 wt% solids) was less than 10 ppm. To ensure proper mixing of latex particles, samples of PhePMMA and AnPMMA dispersion containing equal numbers of both particles were mixed prior to purification. Finally, the cleaned latex dispersions were freeze-dried over 48 hrs. We refer to this solid as "latex powder".

2.4 Film preparation

In our early experiments, films (ca 100 μm thick) were prepared simply by melt-pressing ca 100 mg samples of the latex powder between Teflon sheets in a Carver Press at 115°C and 6.6 MPa The experiments reported here were designed to allow for tensile strength testing as a function of annealing. "Dogbone"-shaped films samples were prepared by compression molding using an ASTM standard mold [13] inserted in the Carver Press and heated at 140 °C for various times under a pressure of 6.9 MPa. These films were 0.48 mm in neck width, 22 mm in neck length, and about 0.6 mm in thickness. About 15 min of preheating time for temperature equilibrium was required before subjecting the sample to pressure. The onset of pressure defines our time zero from which we measure the molding time, which is also taken as the diffusion time. As soon as the desired times were satisfied, the films were quenched by a 8-10°C water cooling system down to $T < T_g$ to stop diffusion. No significant interdiffusion occurs during these sample preparation conditions.

The differences in sample preparation conditions for the two types of films are related to the ability of the thin melt-pressed films to flow. This relieves internal stresses. Corresponding physical relaxation of the thicker dogbone samples is more difficult. While one can prepare transparent films at lower temperatures, these become turbid upon heating. Turbidity indicates the formation of voids within the film, indicating that residual stresses associated with particle deformation are stronger than adhesion forces at the particle-particle interface.

Fluorescence measurements

Fluorescence decay profiles were measured by SPT [1]. Film pieces were placed inside quartz test tubes, sealed with a rubber septum and flushed with O_2-free Argon in order to prevent possible O_2 quenching. Samples were excited at 298 nm and emission was detected at 366 nm. All measurements were made at 22 °C.

3. RESULTS AND DISCUSSION

Under the influence of heat and pressure, the latex powder forms a void-free transparent film. The individual polymeric nanospheres are deformed into space-filling polyhedra. Since we expect random closepacking in the powder before compression, the shapes of the deformed spheres will be those of random Wigner-Seitz cells [14], typically 14- to 17-sided polyhedra with the faces of adjacent particles in intimate contact.

We wish to prepare film samples under sufficiently mild conditions that there is no significant polymer diffusion across the interfaces during sample preparation. This task is complicated by elastic forces associated with particle deformation. These must be allowed to relax by maintaining the samples for a sufficient time under pressure at $T>T_g$. If the stresses are not sufficiently relaxed, voids will form at the interfaces when the films are heated. Since polymers cannot diffuse across voids, the DET measurements will underestimate the kinetics of interdiffusion.

In properly prepared samples, the films remain transparent throughout the annealing process. In such samples we find that the initial fluorescence decay profiles are nearly exponential, exhibiting small deviations at early times. (Decays are strictly exponential in films prepared from the Phe-labeled polymer alone.) After heating the film, the rapidly decaying component increases in magnitude, and becomes more prominent the longer the sample is annealed. An example is shown in Figure 3. These traces give clear indication of the growing importance of fluorescence quenching through energy transfer with increased annealing time. Each decay curve represents a snapshot of the system at the time of its removal from the annealing oven.

3.1 Modeling the decay profiles

In the simplest model of the interdiffusion process, the sample can be divided into three classes of domains. There is a mixed domain containing both D and A groups, and there are two sets of domains which have yet to undergo mixing. One contains only A groups; and the other, only D groups. If D is chosen so that its decay profile in the matrix is strictly exponential (hence the choice of D = Phe), then the D groups in the unmixed domain will contribute an exponential component to the $I_D(t)$ profile. If the size of the mixed domain is sufficiently large that D and A can be considered to be uniformly distributed, then the signal from this region will follow Förster's rate law [2]. The total signal will be a superposition of the two contributions, and the relative magnitudes of the two contributions are given by the prefactors of the two terms in eq. 1 [15,16].

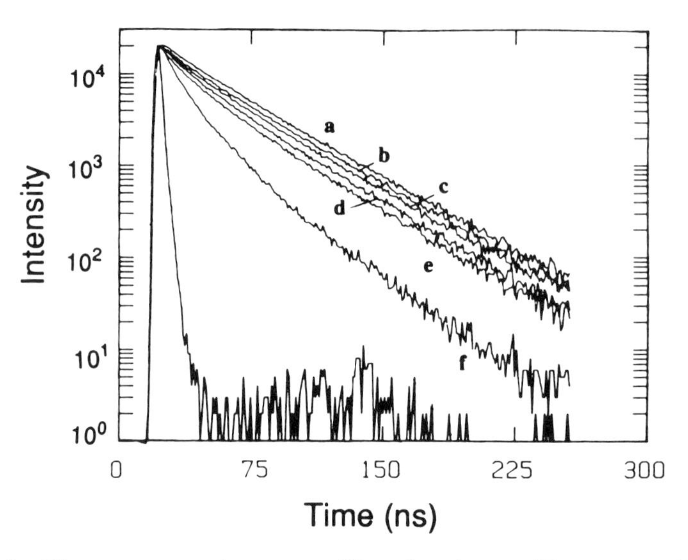

Figure 3. Fluorescence decay profiles for latex films annealed at 140 °C for various periods of time (minutes): a) 10; b) 60; c) 180; d) 480; e) 2530; and f) infinite (a solvent-cast film).

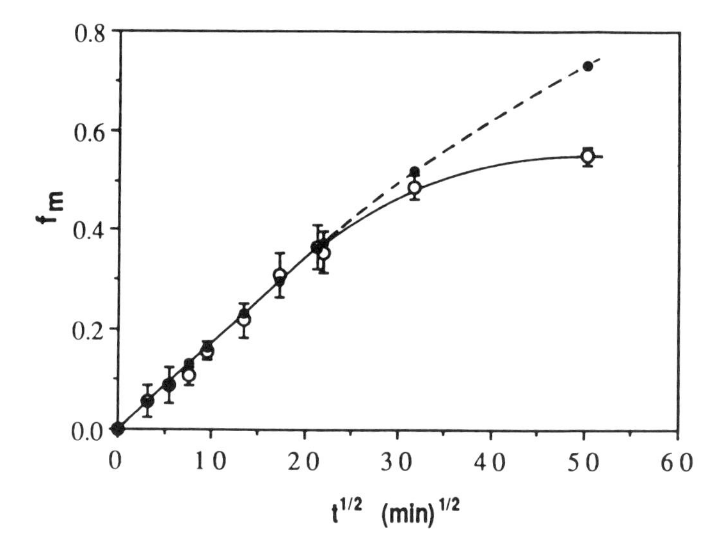

Figure 4. Experimentally determined f_m (solid line) as a function of $t^{1/2}$ at 140 °C. The dashed line indicates theoretical values, $f_s(t)$, calculated by solving the Fickian spherical diffusion equation in terms of a single diffusion coefficient, $D = 6.48 \times 10^{-17}$ cm^2/s.

$$I_D(t') = A_1 \exp\left[-\left(\frac{t'}{\tau_D}\right) - P\left(\frac{t'}{\tau_D}\right)^{1/2}\right] + A_2 \exp\left(-\frac{t'}{\tau_D}\right) \quad (1)$$

where τ_D is the unquenched donor decay time, and P is a parameter related to the concentration of acceptor in the microdomains where diffusive mixing has occurred. Eq. 1 is a powerful fitting function. It might be expected to give good fits (in terms of χ^2) to individual decay curves, even when assumptions in the construction of eq. 1 are not valid.

In the traditional energy transfer analysis [2,15]

$$P = \beta [A]/[A]_o \quad (2)$$

where β is an orientational parameter and $[A]_o$ is a critical acceptor concentration related to the characteristic energy transfer distance R_o by the expression

$$[A]_o = \frac{3000}{\pi^{3/2} N_A R_o^3}. \quad (3)$$

where N_A is Avogadro's number. For Phe/An, R_o is approximately equal to 23 Å [15].

In terms of our model, the apparent extent of mixing of Phe- and An-labeled polymer, f_m', can be calculated directly from A_1 and A_2:

$$f_m'(t) = \frac{A_1(t)}{A_1(t) + A_2(t)} \quad (4)$$

Note that in eq. 4, t refers to the annealing time (min or hr), whereas in eq. 1, t' refers to the timescale for fluorescence decay (ns). The term $f_m'(t)$ has to be corrected for energy transfer across the particle interface, $f_m'(0)$, which is not due to interdiffusion, and for the possibility of incomplete mixing in the limit of infinite diffusion time ($f_m'(\infty)$). The true fraction of mixing is given by

$$f_m = \frac{f_m'(t) - f_m'(0)}{f_m'(\infty) - f_m'(0)} \quad (5)$$

Here $f_m'(\infty)$ can be obtained directly from a film cast from a solution (THF or toluene) of the mixture. In these systems, it always takes a value close to unity, indicating that the Förster model is valid once interdiffusion is complete.

Implicit in this data analysis is that the magnitude of the parameter P in eq. 1 is a constant throughout the interdiffusion process. As we see in Figure 2, this assumption is not valid. This result points to compositional heterogeneities in the mixed domain that reflect details of the $C_A(r,t)$ and $C_D(r,t)$ concentration profiles. For prolonged annealing, P values level off and attain the magnitude predicted from eq. 2 taking into account dilution of the An labels through mixing. This result, plus the finding that at long times $f_m'(\infty)$ reaches unity, indicates that eq. 1 provide a reasonable description of mixing once substantial interdiffusion has taken place.

In this analysis, all details of the polymer concentration profile at the interface, contained in the $I_D(t)$ profile, are lost. This is the same type of result that would be obtained from quantum efficiency measurements. If we wish to use the fluorescence decay profiles to obtain the quantum yields of energy transfer, we must integrate over the decay profiles [17]. In this case, f_m is defined as

$$f_m = \frac{\text{Area}(t) - \text{Area}(0)}{\text{Area}(\infty) - \text{Area}(0)} \qquad (6)$$

where $\text{Area}(t) = \int_0^\infty I(t, t')dt'$. When P is a constant, eq. 6 is identical to eq. 5. Since the P values vary seriously with annealing time, all data were analyzed with eq. 6.

3.2 Monitoring interdiffusion

Values of the extent of interdiffusion f_m are plotted as a function of $t^{1/2}$ in Figure 4 for samples compression molded at 140°C. Each point represents the average value of 3 to 5 separate samples. The extent of interdiffusion is a monotonically increasing function of time. The most meaningful parameter to describe diffusion processed is the diffusion coefficient D. These can be calculated by recourse to a diffusion model. The line labeled f_s in Figure 4 is the expected extent of interdiffusion calculated from a Fickian spherical diffusion model assuming a value of $D = 6.48 \times 10^{-17}$ cm^2/s. The differences between f_m and f_s may be related to the molecular weight polydispersity of the polymer, which should lead to a distribution of D values.

The focus of this paper is not so much on the interpretation of D values, but on attempts to obtain further information about the D/A concentration profile at the interface. In energy transfer processes in which no mass transfer occurs during the excited state lifetime, the donor survival probability is the integral over the donor acceptor pair distribution $\rho(r)$. One approach to obtaining more detailed information about the concentration profile across the interface would be to attempt to extract $\rho(r)$ from the $I_D(t)$ decay curves. We take a somewhat different approach here. We begin by asserting that the intermixed domain can be arbitrarily divided into regions of different local acceptor concentrations [18]. This is akin to looking at the pair distributions within this domain on a more coarse-grained scale. The observed decay profile represents a superposition of donor decays from each of these microdomains.

To proceed, we rewrite eq. 1 to emphasize that P in the Förster analysis is comprised of a constant (B) times the concentration of A (C_A).

$$I_D(t') = A_1\left[\exp\left(-\frac{t'}{\tau_D} - C_A B\left(\frac{t'}{\tau_D}\right)^{1/2}\right)\right] + A_2\exp\left(-\frac{t'}{\tau_D}\right) \tag{7}$$

When we have a distribution of acceptor concentrations, we can rewrite eq. 7 as

$$I_D(t') = \sum_{i=1}^{\infty} A_i\left[\exp\left(-\frac{t'}{\tau_D} - C_{A,i}\,B\left(\frac{t'}{\tau_D}\right)^{1/2}\right)\right] \tag{8}$$

where the A_2 term in eq. 7 is included as the first term in eq. 8, in which $C_{A,1} = 0$. For a continuous distribution of donor population vs local acceptor concentration, eq. 8 can be rewritten as

$$I_D(t') = \int_c f(c)\,\exp\left[-\frac{t'}{\tau_D} - c\,B\left(\frac{t'}{\tau_D}\right)^{1/2}\right] dc \tag{9}$$

The decay curves shown in Figure 3 can then be analyzed in terms of eq. 8 or eq. 9.

Results of this analysis are shown in Figure 5. Here we plot on the y-axis the relative population of donors that have sufficient local concentration of A to undergo energy transfer. The x-axis represents the span of local acceptor concentrations sampled in the experiment. What we observe is that in newly prepared films, only a small fraction

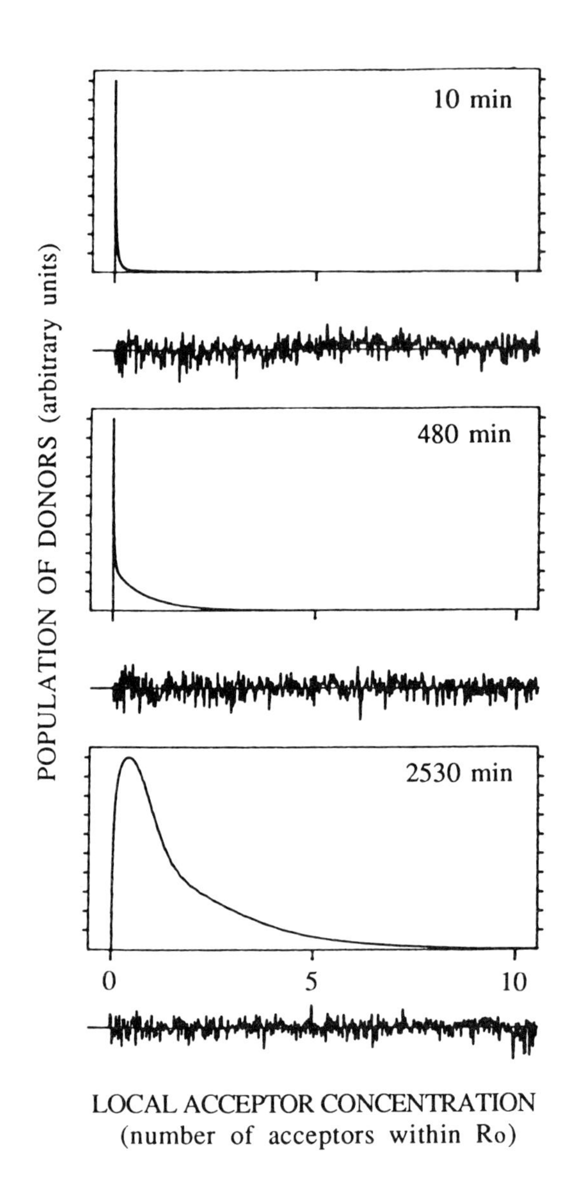

Figure 5. The recovered acceptor concentration distribution in the experiments involving films prepared by compression molding PMMA powders at 140 °C. The powdered sample is a mixture of PMMA microspheres, half labeled with Phe, half labeled with An. The residual profiles for the fit are shown underneath each figure.

of donors exhibit energy transfer, and the distribution of $C_{A,i}$ values is over low local concentrations. With increased annealing time, the number of D which transfer energy increases, as does the distribution of $C_{A,i}$ values. At very long annealing times, the distribution evolves to give an average value corresponding to bulk concentration of acceptor.

The next step in this research will be to try to use these $C_{A,i}$ distributions to calculate the polymer segment density profiles at the interface. Work to this end is in progress.

ACKNOWLEDGEMENTS

The authors thank the Ontario Centre for Materials Research and NSERC Canada for their support of this research.

REFERENCES

1. D. V. O'Connor and D. Phillips, Time-Correlated Single Photon Counting, Academic, New York (1984).
2. (a) Th. Förster, Ann. Phys. (Leipzig), 2 (1948) 55 ; (b) J.B. Birks, Photophysics of Aromatic Molecules, Wiley-Interscience, London, 1970.
3. M. Tirrell, D. Adolf, S. Prager, Springer Lecture Notes Appl. Math., 37 (1984) 1063.
4. J. Crank, The Mathematics of Diffusion, Clarendon, Oxford, 1975.
5. P. G. de Gennes, C. R. Acad. Sci. Ser. B, 291 (1980) 219.
6. (a) S. Prager, M. Tirrell, J. Chem. Phys., 75 (1981) 5194;
   (b) S. Prager, D. Adolf, M. Tirrell, J. Chem. Phys., 78 (1983) 7015;
   (c) S. Prager, D. Adolf, M. Tirrell, J. Chem. Phys., 84 (1986) 5152.
7. (a) R. P. Wool, K. M. O'Conner, J. Appl. Phys., 52 (1981) 5194;
   (b) Y. H. Kim, R. P. Wool, Macromolecules, 16 (1983) 1115.
8. P. G. de Gennes, *Scaling Concepts in Polymer Physics,* Cornell Univ. Press, Ithaca, N.Y., 1979.
9. In addition to the description of the kinetics of interdiffusion, these authors in [6,7] also make specific predictions about the growth in mechanical strength of the joint that is formed [10].
10. (a) R. P. Wool, B. L. Yuan, O. J. McGarel, Polym. Eng. Sci., 29 (1989) 1341; (b) H. H. Kausch, M. Tirrell, Annu. Rev. Mater. Sci., 19 (1989) 341.
11. H. Zhang, R. P. Wool, Macromolecules, 22 (1989) 3018.
12. Y. Wang, M. A. Winnik, Macromolecules, 26 (1993) 3147.

13. L. Li, J. Feng, M. A. Winnik, H. Yan, T. North, to be published.
14. R. Zallen, The Physics of Amorphous Solids, Wiley-Interscience, New York, 1983.
15. Y. Wang, C.-L. Zhao, and M. A. Winnik, J. Chem. Phys., 95 (1991) 2143.
16. Y. Wang, M. A. Winnik, F. Haley, J. Coatings Technol., 64(811) (1992) 51.
17. H-B. Kim, Y. Wang, M. A. Winnik, Polymer, in press, 1993.
18. Y.S. Liu, L. Li, S. Ni, M.A. Winnik, Chem. Phys., in press, 1994.

MICROCHEMISTRY
Spectroscopy and Chemistry in Small Domains
Edited by H. Masuhara et al.

Direct measurement of surface forces of supramolecular systems: Structures and interactions

Kazue Kurihara

Department of Applied Physics, School of Engineering, Nagoya University, and PRESTO, JRDC, Furo-cho, Chikusa-ku, Nagoya 464-01, Japan

Nanometer range of interactions and structures of supramolecular assemblies have been studied from force-distance profiles measured in a surface forces apparatus. Langmuir-Blodgett technique was employed for the deposition of functionalized amphiphilic layers onto mica surface. This makes us possible to investigate surface interactions at the nanometer resolution, and systematically in terms of chemical structures of surfaces. Typical systems studied were: 1) structures of anchored poly(methacrylic acid) layers, 2) structures of adsorbed layers of poly(styrene sulfonic acid) onto a cationic surface of an ammonium monolayer, 3) submicron-range attraction found in water between hydrophobic surfaces of a polymerized ammonium amphiphile, and 4) long-ranged attraction between surfaces of nucleobase functionalized surfaces. Advantages of the surface forces measurement in characterization of supramolecular systems are discussed.

1. INTRODUCTION

Nanometer range of characterization of molecular architectures is the subject of increasing number of scientific publication because of their importance in such areas as nanotechnology and biological science. Molecular aggregates, particularly *dimensionally and structurally controlled supramolecular assemblies* behave differently than their constituent *individual* molecules.

Surface forces measurement [1] is unique and essential among techniques employed to characterize supramolecular assemblies. Because many physical events occurring at surfaces and near surfaces are determined by forces acting between surface-surface, surface-solvent and surface-solute, importance of measuring these interactions is obvious. This technique measures forces between two surfaces directly as a function of the surface separation, which allows us to analyse origins of observed forces (van der Waals, electrostatic, and others). When two surfaces approach to nanometer-ranged separations, the size effect of surface molecules becomes crucial. Therefore, one is able to study structural forces of solvents [2] and of polymer chains attached to the surface [3], which are otherwise difficult to examine. Surface forces measured are often different from conventional pictures of surface interactions. This should be (at least partly) because *supramolecular assemblies multiply the effect of interaction among their constituent molecules*, although we do not understand the exact mechanism yet.

Systematic investigations have been launched in our laboratories for obtaining an insight into nanometer range of interactions and structures of supramolecular assemblies. These knowledges are essential to design materials with microscopically controlled structures as well as to elucidate specific properties of these supramolecular systems. In order to prepare well-defined surfaces, Langmuir-Blodgett (LB) technique was employed for the deposition of amphiphilic layers onto mica surfaces [4]. LB films can be characterized by various physico-chemical techniques, enabling us to investigate surface interactions in terms of actual chemical structures of surfaces.

This article outlines how force measurement can be applied to study structures of brush layers of polyelectrolytes and of adsorbed layer of polyions in combination with other characterization methods. Unconventional forces, long-ranged attraction found in water between hydrophobic surfaces and between nucleobase monolayers, are also described.

2. SURFACE FORCES MEASUREMENT

2.1. Measurement

The surface forces between LB modified surfaces have been measured directly by using the surface forces apparatus (Mark 4, ANUTECH). Molecularly smooth mica is used as a substrate. Modified mica sheets on cylindrical lenses (radius, $R \approx 20$ mm) are mounted as crossed cylinders in the apparatus. The surface separation D is measured by use of multiple-beam interferometry. The force (F) is determined from deflection of a double cantilever spring (spring constant K: 100 - 400 N/m) on which one surface is mounted. The jump-in and jump-out distances are also used to evaluate attraction and adhesion, respectively (Figure 1). The measured force is

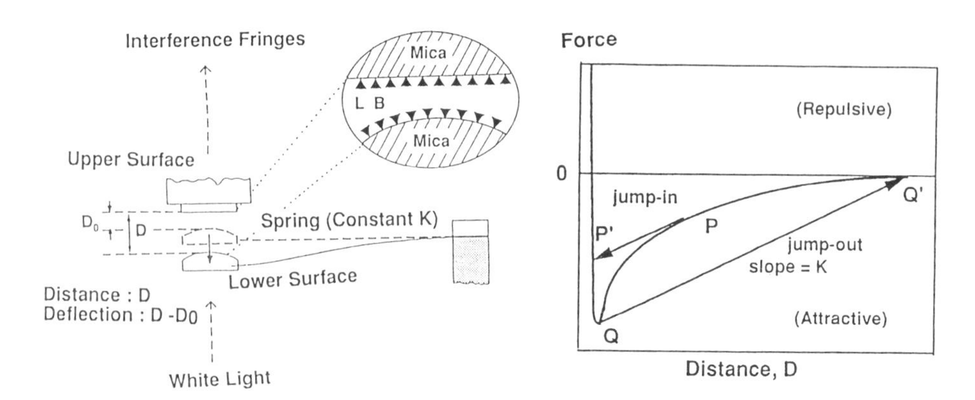

Figure 1. Schematic drawing of forces measurement. Experimental set-up (left). Forces are measured from deflection of the spring and from jump-in and -out distances (right).

normalized by mean radius R of the surface curvature. This quantity is proportional to the free energy of interaction of flat surface Gf according to the formula, $F/R = 2\pi Gf$.

2.2. What can be monitored?

This technique measures attraction or repulsion between two surfaces directly as a function of the surface separation, which allows us to analyse origins of observed forces (van der Waals, electrostatic, and others). When two surfaces approach to nanometer-ranged separations, the size effect of surface molecules becomes crucial. Various chemical processes occurring at surfaces of supramolecular assemblies can be monitored in water or organic solvents by the forces measurement. Figure 2 schematically draws these processes: (a) interactions between hydrophilic (functional) groups, (b) interactions between hydrophobic groups, (c) structural changes of polymer assemblies, (d) adsorption of polymers, (e) change of surface properties upon adsorption of second species, and (f) ordering of solvent molecules between surfaces. The two-dimensional organization of molecules should enhance surface interactions of supramolecular assemblies compared to those between constituent molecules. Specific orientation of molecules at surfaces should induce ordering of solvents and/or molecules on opposing surfaces, and bring out unconventional interactions of surfaces.

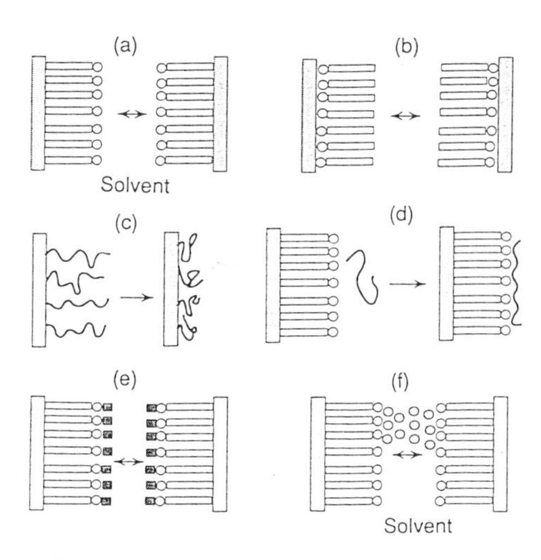

Figure 2. Modes of interactions.

3. BRUSH LAYERS OF POLYELECTROLYTE

Structures and interactions of polyelectrolyte layers play important roles in polyelectrolyte gels, colloidal stability and biological organizations, but they are still poorly understood. Direct measurements of surface forces should be useful to provide information about concrete structures of polyelectrolyte layers. Previously, measurements of surface forces limited to only adsorbed layers of polyelectrolytes such as poly(L-lysine) and gelatin [5, 6]. Although structural information on these adsorbed layers are important, their force profiles are often complicated because their structures are modified dominantly by interactions of polyelectrolyte adsorbates with the mica substrate. Studies on simple brush layers are necessary [7]. We used LB (Langmuir-Blodgett) films to prepare layers of chain-end-anchored poly-(methacrylic acid). Structures of polyelectrolyte layers were investigated based on the force profiles together with FT-IR spectra of LB films and the surface pressure - area isotherms of monolayers at the air-water interface.

3.1. Preparation of polyelectrolyte brush layers

Monolayers of poly(methacrylic acid) amphiphile **1** (n = 55, M_w/M_n = 1.89) formed on pure water, were transferred at 35 mN/m and 10 mm/min in the down-stroke mode onto hydrophobized mica sheets [3]. Mica surfaces were hydrophobized by transferring a monolayer of dioctadecyldimethylammonium bromide at 35 mN/m and 10 mm/min in the up-stroke mode. The density of **1** on mica was 2.7 polyelectrolyte chains/nm^2.

The pK_a values of monolayer **1** were estimated to be 8.5 in the absence of added salt and 6.8 in 10 mM aqueous NaBr from relative FT-IR peak intensities at 1726 cm^{-1} (C=O) and at 1556 cm^{-1} (COO^-) of LB films of **1** (reflection method).

$C_{18}H_{37}OC(=O)CHNHC(=O)-CBr_2-(-CH_2C(CH_3)(COOH)-)_{55}$
$C_{18}H_{37}OC(=O)(CH_2)_2$

1

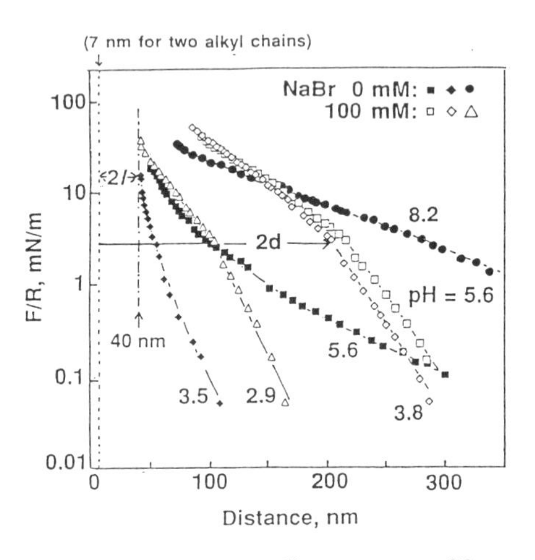

Figure 3. Force-distance profiles for monolayers **1** at various pH's without and with NaBr. The pH of the aqueous phase was adjusted by H_2SO_4 and NaOH.

3. 2. Force profiles at various pH's and NaBr concentrations

Typical profiles of force between monolayer **1** are shown in Figures 3 and 4. Surface separations, D, in Figures 3 and 4 are distances from the hydrophobized mica surfaces. The forces are reversible and reproducible over the several compression and decompression cycles. Repulsion extends to distances of longer than 350 nm in pure water (pH ~ 6), and increases significantly at pH 8.1, while decreases at lower pH's (Figure 4). Poly(methacrylic acid) is known to undergo a conformational transition from a rod-like form to a hyper-coiled form, when pH decreases below the critical value (pH ~ 6 in bulk aqueous solutions where the degree of ionization α is $0.1 < \alpha < 0.3$) [7]. Therefore, the pH dependence of repulsive forces must be associated with changes in the ionization state and the

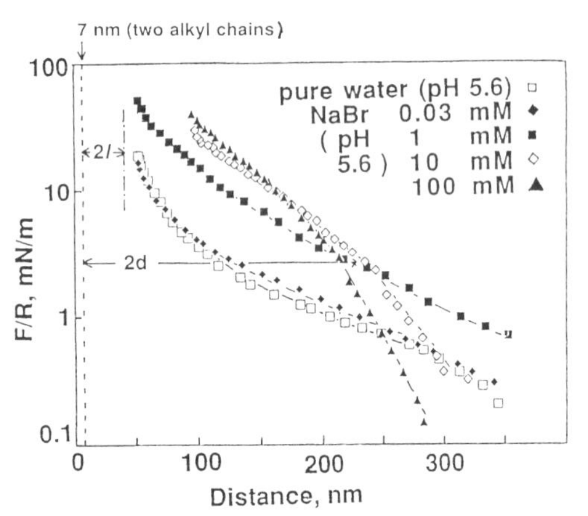

Figure 4. Force profiles for monolayer **1** at various NaBr concentrations at pH 5.6.

consequent conformation change of the polyelectrolyte head group. Addition of a monovalent salt, NaBr, enhances the ionization as indicated in the measured pKa values, thus increasing the repulsion significantly at distances shorter than a critical distance (ca. 230 nm at pH 5.6), where the force curve exhibits deflections at high NaBr concentrations (higher than 10 mM at pH 5.6, Figure 4). The repulsion saturates at concentrations above 10 mM at pH 5.6. On the other hand, at distances longer than this critical distance, the repulsion is suppressed markedly with increasing NaBr concentrations from 1 mM to 100 mM. Its decay lengths (D_0), defined by $F/R = A \exp(-D/D_0)$ (A: intensity parameter), are 85 nm, 45 nm, and 25 nm at NaBr concentrations of 1 mM, 10 mM, and 100 mM, respectively. These lengths are one to two orders of magnitude larger than the calculated Debye lengths.

Force curves obtained here cannot be described by the diffuse electrical double-layer model for normal solid surfaces [8], except for the curve at pH 3.5 in the absence of NaBr. The latter curve is close to a profile computed by using a surface potential of 180 mV (i.e. 1 charge / 20 nm^2) and a Debye length of 17 nm. Apparently, at pH 3.5 where the polyelectrolyte chain assumes a hyper-coiled form with very limited charges, the monolayer surface must behave as a rigid charged surface like that of mica. This locates at 40 ± 6 nm from the hydrophobic surface. The thickness (l) of the hyper-coiled polyelectrolyte layer can be estimated as 17 ± 3 nm by subtracting 3.5 nm (length of the hydrocarbon chain) from 40/2 nm.

3.3. Structure of polyelectrolyte brush layer

An increase in pH and/or NaBr concentration enhances ionization of the carboxylic acid group, causing extension of the polyelectrolyte chain due to Coulombic repulsion along the chain. The repulsion between two surfaces would then increase and extend to longer distances. At critical pH's and NaBr concentrations where the degree of ionization exceeds ca. 0.1, the repulsion reaches saturation. Counterions should penetrate and condense among the extended polyelectrolyte chains to maintain the intralayer effective charge low and constant, thus leading to saturation of the repulsion. This effect is visible on force profiles at NaBr concentrations higher than 1 mM. The maximum charge density estimated from the surface pressure - area isotherms of **1** by using an equation derived for the flat electrostatic double layer is 1.4 charges/molecule at a molecular area of 1 nm^2/molecule at pH 5.6 at 10 mM NaBr, while the degree of dissociation estimated from an FT-IR spectrum of its LB film is 0.17 (9.4 charges/molecule in average). Therefore, less than 15 % of the total charge is effective [6]. This

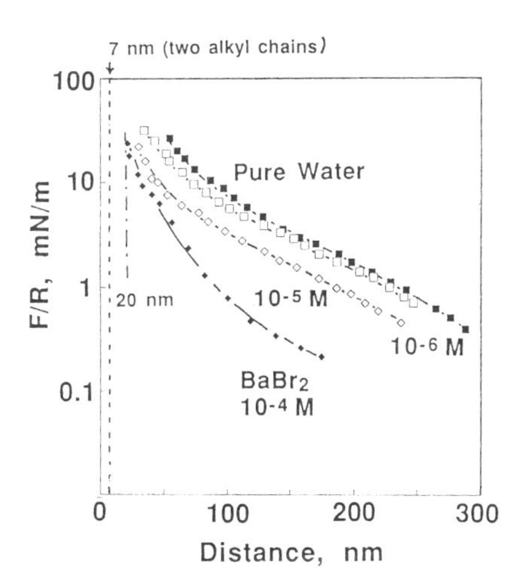

Figure 5. Force profiles of monolayer **1** at various $BaBr_2$ concentrations at pH 5.6.

again supports the concept of counterion condensation which lessens the amount of effective charges. Gradual increase in repulsion upon compression at separations below deflection points indicates that the polyelectrolyte layer with condensed counterions is elastic. Outside of this layer, the repulsion depends markedly on the NaBr concentration. Isolated charges of polyelectrolyte chain terminals and their accompanying counterions can be counted as plausible causes of this part of repulsion. The decay length larger than the Debye length has been predicted for an inward shift of the plane of charge for *soft* surfaces [9].

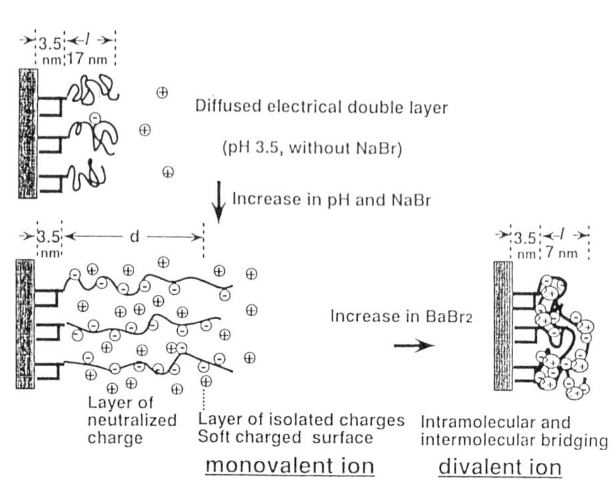

Figure 6. Probable structures of the poly(methacrylic acid) layer

The effective thickness of the condensed counterion layer (d) is estimated from separations of the deflection points to be 50 nm, 95 nm and 100 nm at pH 2.9, 3.8 and 5.6, respectively, at 100 mM NaBr. These values indicate that the polyelectrolyte layer is lengthened three to six times from the hyper-coiled structure.

3.4. Effect of divalent ion

Addition of a divalent salt, $BaBr_2$, completely changes the force profiles as shown in Figure 5. Now, the repulsion decreases at all distance ranges and at concentrations from 0.01 to 0.1 mM $BaBr_2$. The polyelectrolyte layers can be compressed much further: the separation exhibiting a repulsion of 20 mN/m at 0.1 mM $BaBr_2$ is 21 nm, which corresponds to the layer thickness of 7 nm. This value of the thickness is close to the layer thickness expected for the average polymerization degree of **1** (n = 55). Bridging of the carboxylic acid groups of adjacent chains must be responsible for this condensation of the polyelectrolyte layers.

Probable structures of polyelectrolyte layers are drawn schematically in Figure 6. The swelling of poly(methacrylic acid) gel is known to be affected by the presence of salts, while its mechanism has not been well understood. The swollen volume of a ionized poly(methacrylic acid) gel reaches values of several hundreds times of the dry volume when sodium ions are present. On the other hand, with Ba^{2+} ions, it shrinkes to a volume close to the dry volume [10]. We observed a similar effect of salts on brush layers **1**, indicating that molecular mechanisms involved in two processes should be closely related and that our system of polyelectrolyte brushes in contact is a good model of polyelectrolyte gel.

4. ADSORPTION OF POLYANION TO CATIONIC MONOLAYER

The adsorption of polyions to monolayer surfaces has been studied increasingly in recent years. Polyions efficiently stabilize monolayers with opposite electronic charges, decreasing solubility of monolayers in water and enhancing their mechanical stability [11, 12]. Designing composite materials by complexation of polyions and monolayers is becoming a popular strategy [13]. However, little is known about the structure of the polyion-monolayer complexes. Direct measurement of surface forces is one of powerful means to study such structures at a nanometer resolution. We employed this technique to investigate the mechanism of poly(styrenesulfonate) **2** adsorption onto a fluorocarbon ammonium monolayer **3** [14].

2 $[-CH_2-CH(C_6H_4SO_3^- Na^+)-]_n$

3 $CF_3C_9F_{18}CH=CHC_9H_{18}O-C(=O)-CH(CH_2CH_2C(=O)-OC_9H_{18}CH=CHC_9F_{18}CF_3)-NH-C(=O)-C_{10}H_{20}N^+(CH_3)_3\ Br^-$

4.1. Stoichiometric adsorption of poly(styrenesulfonate) at low concentrations

A force-distance profile between monolayers of ammonium amphiphile **3** (Figure 7 - 1) in pure water shows long range electrostatic repulsive force as expected for a charged monolayer. A layer of fluorocarbon ammonium amphiphile **3** was transferred onto hydrophobized mica surfaces in the down-stroke mode at a surface pressure of 30 mN/m and a deposition rate of 10 mm/min. The transfer ratio of the amphiphile was found to be 0.8 - 1.0. The curve of Figure 7 - 1 can be fitted [8] by using an apparent surface potential of 70 mV and decay length of 48 nm.

Adsorption of a charged polymer to an oppositely charged surface should alter the charge density on the surface. These alternations can be detected by surface forces measurement. Polymer concentrations were chosen with reference to charges of the surface pressure-area isotherms of **3** over aqueous solution of **2** the monolayer studies. The surface forces data in the presence of the polymer with molecular weight of 5 × 10^5 are presented in Figure 7.

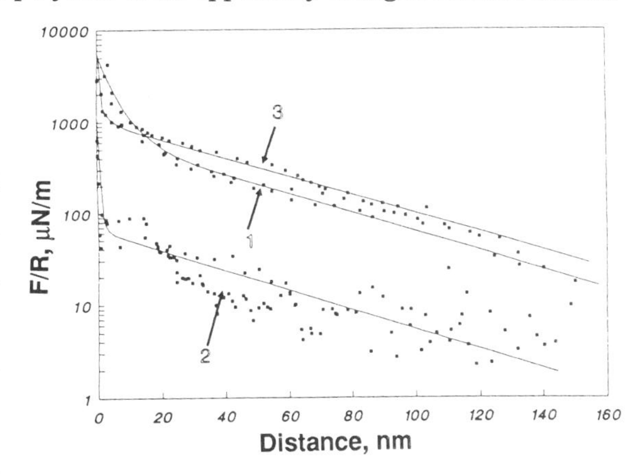

Figure 7. Force-distance profiles for monolayers **3** in pure water (1) and in aqueous solutions of 0.7 mg/L of **2** (2) and 7.0 mg/L of **2** (3).

At concentrations below 0.088 mg/L (0.17 nM) only small changes were observed. A more drastic effect was observed as the amount of polyelectrolyte became close to that of the amphiphile in the deposited layer. At a polymer concentration of 0.7 mg/L (1.4 nM, equivalent to the addition of 0.7 nmol of polymer into the aqueous phase), the force decreased dramatically to a barely detectable extent. Over the whole range of separations from 5 Å to 100 nm, the force decreased more than 1 order of magnitude and did not exceed 100 μN/m. Similar results were obtained for the polymer with a smaller molecular weight of 1×10^4.

The decrease in the force between the surfaces by polymer adsorption can be interpreted in terms of the Poisson-Boltzmann formalism. It is obvious that our observation reflects a decrease in the net surface charge density. Using the data for 1.4 nM poly (styrenesulfonate) (Figure 7 - 2), we estimate that the residual apparent surface potential is less than 20 mV. This potential corresponds to the charge density 1 charge per 300 nm^2. Since the charge density of the amphiphile monolayer on mica is 1 charge per 1.5 nm^2, only one of 200 surface charge is left after polymer adsorption; therefore, more than 99% of the initial surface charge are masked.

4.2. Mechanism of surface charge neutralization

In order to explain nearly complete disappearance of the surface net charge, the number of negative charges brought to the surface by the polymer must be exactly equal to the number of positive charges on the surface, e. g. one molecule of the 10^4 M_w polymer (50 charges/molecule) bound to 50 amphiphile molecules.

The complete disappearance of charge also requires that the polymer must be bound in a configuration that prevents adsorption of excess polymer (the discharged state of the surface is stable longer than 24 h). If the polymers were in a random coil conformation, the adsorbed polymer layer should carry a large number of excess charges. This is contrary to our observation. Our force-profiles indicate that the thickness of the adsorbed layer of the polymer with M_w of 5×10^5 is in the range of 1.5-2.5 nm and that of the adsorbed polymer of 1×10^4 M_w is less than 1 nm. These data strongly support *flat and stoichiometric* adsorption of poly (styrenesulfonate) on the ammonium monolayer surface. Figure 8A illustrates flat adsorption of polyanions onto the cationic monolayer surface.

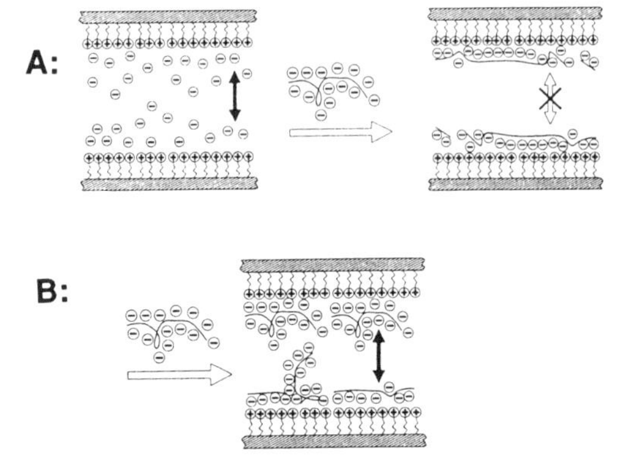

Figure 8. Schematic illustration of adsorption of poly(styrenesulfonate) to a cationic monolayer surface.

4.3. Additional adsorption of poly(styrenesulfonate) onto the neutralized surface

Increased concentrations of poly(styrenesulfanate) at about 1-2 mg/L ((5-10) $\times 10^{-6}$ M) lead to increases in the force to values comparable to or even larger than those found for a surface of fluorocarbon amphiphile alone (Figures 7, curve 3). The origin of this force was found to be electrostatic. Formation of an electrostatic double layer was proven by suppression of the forces upon addition of simple electrolyte in the aqueous phase. Recharging of the surface is produced by additional adsorption of the polymer (Figure 8B), leading to net charge formation at the surface.

The theoretical curve for the electrostatic interaction fits the experimental data satisfactorily except in the range below 5 nm. At closer distances an additional repulsive force appears to be present. We may identify this force as the structural force caused by the adsorption of a thicker polymer layer.

5. SUBMICRON-RANGE ATTRACTION BETWEEN HYDROPHOBIC SURFACES IN WATER

The "*hydrophobic effect*" is commonly used to express specific properties of nonpolar molecules (and surfaces) in water. Recent reports suggest that interactions between *nonpolar (hydrophobic)* surfaces in water are more complicated than have been thought. The attractive force, which is much stronger and in longer distance range than the conventional van der Waals force was first reported between hydrophobic adsorbed layers of trimethylcetylammonium bromide on mica [15]. Separation where the attraction was detected increased ever since, to 70 nm for the attraction observed between LB films of hydrocarbon and fluorocarbon surfaces [16]. The origin of the long-range attraction is not yet fully understood, although several mechanisms have been proposed. All of these previous force studies suffer from the presence of surface changes probably due to formation of the second layer in the case of the adsorbed monolayers [15], or due to instability of LB layers especially in the presence of salt [16].

In the present study, we employed stable monolayers of polymerized ammonium amphiphile **4**, and performed direct measurement of forces between layers deposited on molecularly smooth mica sheets[17,18]. Owing to the improved stability of the

$C_{16}H_{33}OC_2H_4OC(=O)CHNHC(=O)C_{10}H_{20}N^+(CH_3)_2C_2H_4OC(=O)$–$[CH(CH_2)]_n$; $C_{16}H_{33}OC_2H_4OC(=O)(CH_2)_2$; Br^-

4

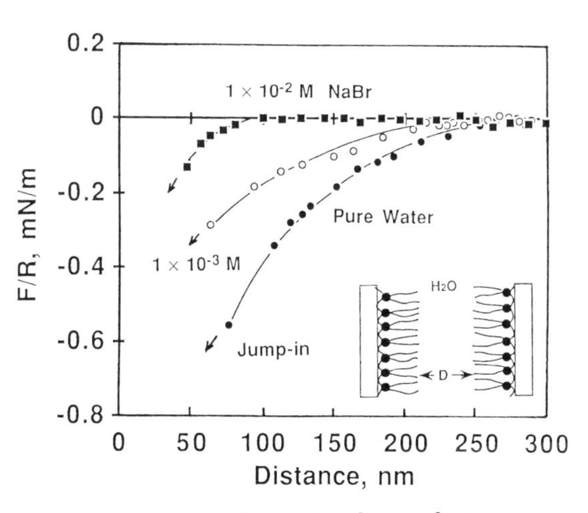

Figure 9. Attractive forces between monolayers **4** deposited on mica in down-stroke mode.

monolayer, we could analyze force curves under varied conditions.

5.1. Long-range attraction between hydrophobic layers of polymerized amphiphile

Polymerized amphiphile **4** forms a stable monolayer on pure water, providing a similar limiting area in the condensed phase as to that of the surface pressure-area isotherm of the precursor monomer. This indicates that an ordinary molecular orientation is maintained in the condensed phase of polymerized monolayer **4**. The monolayer of **1** was transferred successfully in the *down-stroke* mode onto bare mica surfaces (transfer ratio: 0.8). Very long-range attraction extending to a separation of about 300 nm was found between these surfaces in pure water (Figure 9), and the surfaces jumped-in to contact from a separation of 76 nm (spring constant, $K \sim 100$ N/m). The pull-off force was 280 ± 80 mN/m (interfacial energy γ: 30 ± 8 mJ/m^2). Double-layer repulsion was not detected at any salt concentration, indicating that the surfaces were maintained essentially uncharged.

Phenomenologically, these force curves can be described in the from of,

$$F(D)/R = -A \exp(-D/D_0) \tag{1}$$

where A denotes the intensity factor corresponding to the force at $D = 0$ and D_0 is the decay length. When plotted according to eq 1, all data points of Figure 9 give straight lines except in the weak force region (< ca. 0.05 mN/m), and values obtained for A and D_0 are 1.7 ± 0.5 mN/m and 62 ± 4 nm, respectively, in pure water. The addition of salt reduces the intensity parameter A to 0.40 ± 0.14 mN/m (1 mM NaBr) and 0.25 mN/m (10 mM NaBr), while the decay length D_0 is virtually independent of salt concentrations.

Table 1
Attractive forces between hydrophobic surfaces

medium	$F(D)/R = -A\exp(-D/D_0)$: A/mN m^{-1}	D_0/nm	pull-off force, mN m^{-1} (interfacial energy, mJ m^{-2})
Monolayers of **4** Transferred in the Down-Stroke Mode (Transfer Ratio: 0.8)			
From Deflection			
pure water	1.7 ± 0.5	62 ± 4	
1 mM NaBr	0.40 ± 0.14	63 ± 17	
10 mM NaBr	0.25	42 ± 22	
From Jump-In			
pure water	0.59	72	280 ± 80 (30 ± 8)
1 mM NaBr	0.34	57	220 ± 50 (23 ± 5)
10 mM NaBr	0.18	61	190 ± 20 (20 ± 2)

The attractive mica surface held by the spring jumps to contact with the fixed surface, once the slope of the force-distance curve equals or slightly exceeds the spring constant (K).

$$dF(D)/dD \geq K \tag{2}$$

Thus, the attractive force profiles can be obtained by plotting spring stiffness as a function of jump-in distance. Intensity parameter A and decay length D_0 obtained by the jump-in method are identical to those by the deflection method within experimental errors.

The attractive interaction was similarly observed when the surface were modified in the *up-stroke* mode (ordinary hydrophobic modification mode, transfer ratio: 1.0) and returned into pure water. The distance range where the attraction appeared was considerably shorter. The attraction caused jump-in of the surface to contact at 24 ± 6 nm. The net interaction was virtually zero at the distance range longer than this jump-in distance. In 1 mM aqueous NaBr, the net interaction turned to weak repulsion in the longer distance range, and the attractive jump-in distance decreased to 15 ± 3 nm. Further addition of salt (10 mM) strengthened the repulsion, but the jump-in distance changed only a little to 12 ± 2 mN/m.

On the other hand, pull-off forces were close to each other among surfaces prepared in both of the up-stroke and down-stroke modes: ~ 200-400 mN/m (γ : 21-42 mJ/m^2) in pure water as well as in aqueous NaBr. This indicates that surfaces prepared in the two different modes are similar and basically hydrophobic. Therefore, remarkable differences in the extent of long-ranged attraction between the two different preparations must be attributed to *relatively small (though important) variations in surface structures.*

5.2. Interactions between hydrophilic layers of polymerized amphiphile

Bilayers of polymerized amphiphile **4** were prepared by transferring monolayer **1** onto mica first in the up-stroke mode (35 mN/m and 10 mm/min, transfer ratio: 1.0), then in the sown-stroke mode (transfer ratio: 0.6). This procedure would produce hydrophilic surfaces. In fact, forces between these bilayer surfaces were repulsive, displaying electrostatic double-layer repulsion similar to that found between hydrophilic surfaces of monomeric ammonium amphiphiles. This force is quite different from the long-range attraction observed between monolayers of **4** deposited in the down-stroke mode.

5.3. Origin of long-range attraction

This work can provide the first salt dependence data of the long-range attraction, since we do not observe any appearance of the surface charge which may arise from instability of LB layers [15, 16]. Direct analysis of the attraction was difficult in the past studies, because the electrostatic repulsion shielded long-range components of the attraction. In our case, the polymerized layer was much stabler and the attraction between these layers was much longer ranged. Thus, it was possible to examine the salt effect without interferences of the electrostatic effect and undesirable overlapping with the short-range component of the hydrophobic attraction. Intensity parameter A was found to decrease with increasing NaBr concentrations from ca. 1 mN/m (pure water) to 0.2 mN/m (10 mM NaBr), whereas decay length D_0 remained the same at around 60 nm (see Table I for detailed values).

Observation of long-ranged attraction by several groups including us has occasioned active search over a possible origin of the attraction. This unprecedented attraction cannot be easily accommondated previous explanations of attraction. Several mechanisms have been proposed, although none of them finds consensus. Mechanisms mainly discussed are (1) cavitation (solvent density) effect, (2) water structure effect, and (3) electrostatic correlation effect [18, 19]. Although further investigation is necessary to define a molecular mechanism of the attraction, all of these mechanisms consider specific effect of *two-dimentinally assembled, oriented molecular arrangement at the surface* : (1) density gradient of water on smooth hydrophobic surfaces, (2) structure formation of water in the visinity of smooth hydrphobic surfaces, (3) correlation of oriented dipole moments.

6. INTERACTIONS BETWEEN TWO-DIMENTIONALLY ORGANIZED COMPLEMENTARY AND NONCOMPLEMENTARY NUCLEOBASES

Nucleobase interactions occupy a special position among the fundamental biological molecular interactions. The high but not yet ultimate specificity of base-base pairing in nucleic acids proved to be one of the most efficient mechanisms of accumulating, storing, reproducing, and evolving genetic information. In order to shed light on these interactions, we have measured interaction between complementary and noncomplementary pairs of nucleobase monolayers in water [20].

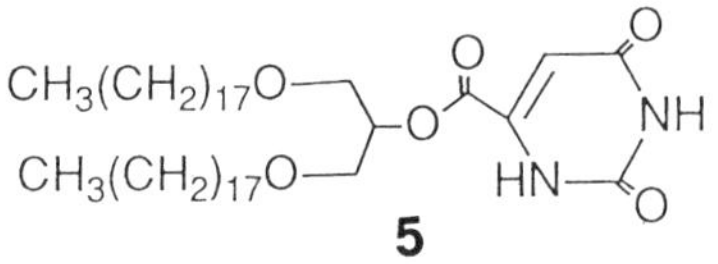

Langmuir-Blodgett films of amphiphiles with an uracyl head group **5** or a complementary adenine headgroup **6** have been prepared on mica surfaces and investigated by surface forces measurement. Electrolyte-dependent electrostatic repulsion was found at pH close to the pK_a of the head group. In the absence of strong electrostatic forces, weak long range attraction was detected. At close distances, strong attractive forces made the surfaces jump into a contact position. The adhesive force between the complementary layers was found to be 100 mN/m.

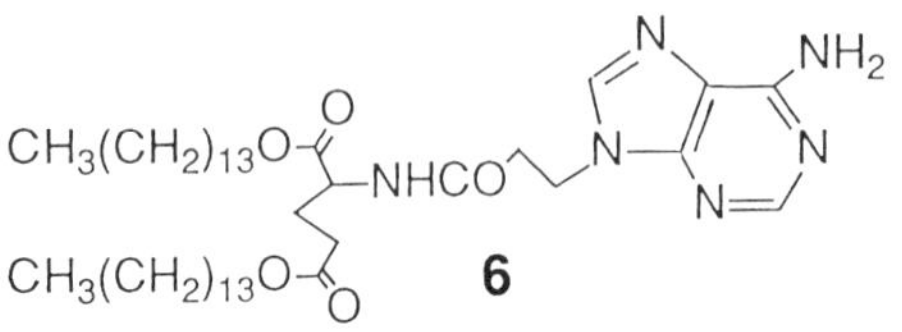

$CH_3(CH_2)_{17}O$
$CH_3(CH_2)_{17}O$
O O NH HN O
5

$CH_3(CH_2)_{13}O$
$CH_3(CH_2)_{13}O$
O NHCO N N NH$_2$ N N O
6

The attraction found between nucleobase surfaces are made of two components. One is observed in the region from 20 to 100 nm and changes its intensity drastically depending on conditions. This type of force is found both for complementary and noncomplementary surfaces, and is similar to the very long range attraction reported for hydrophobic surfaces. The other appearing at around 20 nm is characteristic of complementary pairs, and remains unchanged for pH changes and in the presence of a salt (KBr).

Origins of these long-ranged attractions, again, have not been elucidated. One may state that two-dimensional organization of functional groups should play a crucial role for bringing out these unconventional attraction.

7. CONCLUSION

This article has described strength of the direct measurement of surface forces for exploring *surface structures and interactions* of molecular assemblies at *nanometer (molecular)* resolution. *In situ* determination of molecular structures at surfaces of these assemblies, polyelectrolyte brushes and complexes of polyions with charged monolayers, has been demonstrated. The surface forces measurement is sensitive to differences in steric structures of surface molecules, whereas other techniques such as ellipsometry monitor average macroscopic properties (thickness, refractive index etc.) of surface films. Observed forces are often surprisingly different from the conventional pictures of surface interactions, when we measure interactions between *well-defined surfaces as a function of the surface separation*. Information derived from bulk measurements of surface interactions, colloidal stability and others, is imperfect. One may stress that assembled, oriented molecular arrangement at the surface of supramolecular systems should multiply the effect of interaction among their constituent molecules. This knowledge must be important in elucidation of self-assembling phenomena and biological functions occurring on cell surfaces.

Advantages of direct measurement of surface forces has become widely acknowledged, however, only limited variety of surfaces have been studied previously [3]. This can be attributed to that the availability of differently modified, well-defined surfaces is restricted because surfaces are modified commonly by spontaneous adsorption. In order to open the door for wide application of this method in areas such as physics, chemistry, material science and biology, it is necessary to promote systematic investigations on well-defined surfaces, which are characterized by complementary methods. By using LB modification, we are able to present a new approach to get insight into chemically and/or biologically important events occurring at the interface. We continue to develop techniques for chemical modification and characterization of mica surfaces in our laboratory. This type of approach promises a novel rich field of surface forces research, and will become an essential technique for exploitation of nanotechnology where surfaces are key elements.

ACKNOWLEDGMENT

Measurements described here were conducted at Molecular Architecture Project (ERATO program, JRDC) directed by Prof. Kunitake of Kyushu University. The research has been continued at Nagoya University with support from PRESTO program of JRDC.

REFERENCES

1. J. N. Israelachvili, Intermolecular and Surface Forces (Second Edition), Academic Press, London, 1992.
2. R. J. Horn and J. N. Israelachvili, J. Chem. Phys., 75 (1988) 1400.
3. S. J. Patel and M. Tirrell, Ann. Rev. Phys. Chem., 40 (1989) 597.
4. K. Kurihara, T. Kunitake, N. Higashi and M. Niwa, Thin Solid Films, 210/211 (1992) 681.
5. P. F. Luckham and J. Klein, J. Chem. Soc. Faraday Trans. 1, 80 (1984) 865.
6. N. Kawanishi, H. K. Christenson and B. Ninham, J. Phys. Chem., 94 (1990) 4611.
7. K. Kurihara, T. Kunitake, N. Higashi and M. Niwa, Langmuir, 8 (1992) 2087.
8. D. Y. C. Chan, R. D. Pashley and L. R. White, J. Colloid Interface Sci., 77 (1980) 283.
9. S. J. Miklavic and S. Marcelja, J. Phys. Chem., 92 (1988) 6718.
10. A. Katchasky and M. Zwick, J. Polymer Sci., 25 (1955).
11. M. Shimomura and T. Kunitake, Thin Solid Films, 132 (1985) 243.
12. M. E. Gomez, J. Li and A. E. Keifer, Langmuir, 7 (1991) 1571.
13. G. Decher, J. D. Hong and J. Schmitt, Thin Solid Films, 210/211 (1992) 831.
14. P. Berndt, K. Kurihara and T. Kunitake, Langmuir, 8 (1992) 2486.
15. J. N. Israelachvili and R. M. Pashley, J. Colloid Interface Sci., 98 (1984) 500.
16. P. M. Claesson and H. K. Christenson, J. Phys. Chem., 92 (1988) 1650.
17. K. Kurihara, S. Kato and T. Kunitake, Chem. Lett., (1990) 1555.
18. K. Kurihara and T. Kunitake, J. Am. Chem. Soc., 114 (1992) 10927.
19. Y-H. Tsao, D. F. Evans and H. Wennerström, Langmuir, 9 (1993) 779.
20. P. Berndt, K. Kurihara and T. Kunitake, manuscript in preparation.

MICROCHEMISTRY
Spectroscopy and Chemistry in Small Domains
Edited by H. Masuhara et al.

Luminescence spectroscopy and microdomains

A contribution from the "Laboratorium Moleculaire Dynamica en Spectroscopie", K.U. Leuven, Belgium

Frans C. De Schryver[a], Jan van Stam[b], Marcelo H. Gehlen[c], Mark Van der Auweraer[a], Noël Boens[a], Steven Reekmans[a], R. Martin Negri[a], Nadine Wittouck[a], Delia Bernik[d], Marcel Ameloot[e], Herman Faes[a], and Dimitri Noukakis[a]

[a]Chemistry Department, Katholieke Universiteit Leuven, Celestijnenlaan 200F, 3001 Heverlee, Belgium
[b]Department of Physical Chemistry, University of Uppsala, POB 532, 751 21 Uppsala, Sweden
[c]Instituto de Física e Quimíca de São Carlos, Universidade de São Paulo, 13560 São Carlos, Brazil
[d]Instituto de Investigaciones Fisicoquímicas Teóricas y Aplicadas, La Plata 1900, Argentina
[e]Limburgs Universitair Centrum, Universitaire Campus, 3590 Diepenbeek, Belgium

Recent developments in fluorescence spectroscopy in relation to the analysis of microdomains is discussed. It is shown that dynamic fluorescence methods in particular can be successfully used to determine the distribution of probe and quencher, complex formation, the polydispersity and the eventual migration of probe and/or quencher.

1. INTRODUCTION

Dynamic fluorescence methods have been widely employed to describe static and dynamic properties of microstructures, such as surfactant systems forming micelles, lamellae, bilayers, surfactant-polymer systems where the surfactant and the polymer interact with each other, and polymer systems forming polymeric aggregates and gels [1-3]. The methods take advantage of the fact that a probe molecule, which is excited by a light-pulse of suitable energy, can relax to its ground state by different radiative or non-radiative processes. According to which processes are at hand in a specific system, information about aggregate sizes, polydispersity, order parameters, equilibrium processes in the excited state, migration processes between and within the aggregates, and complexation processes can be extracted.

In this paper, we want to focus on some of the possibilities offered by the recent developments in the research group in the field of the dynamic fluorescence analysis and exemplify these with some recent experimental results. The examples will also clearly show the capability to discriminate between different models.

2. ANALYSIS OF EXPERIMENTAL DATA

2.1 The global analysis approach

The most common way to compute estimates of model parameters is based on the non-linear weighted least-squares Marquardt [5] algorithm. The fits can be judged by a statistical residual analysis, e.g., the reduced χ_ν^2 and its normal deviate $Z(\chi_\nu^2)$ [4,6].

When analysing data in general, it has been shown that a global approach [7-9] provides better estimates of model parameters as compared with individual curve analysis. This is so not only for dynamic fluorescence data, but a general feature. Whenever it is possible to link at least one model parameter over different experiments one should simultaneously analyse the data. In the case of dynamic fluorescence measurements this is an obvious choice, as in most of the cases one or more of the parameters can be held common over the whole or a part of the experimental data surface. For example, in a dynamic fluorescence quenching experiment (Section 3), the quenching rate constant can be supposed to be common in all measurements performed (differing in quenching concentration), and should therefore be linked over the whole decay surface. In the case of dynamic fluorescence quenching measurements on micellar systems, and if both the probe and the quencher are stationary in their host micelles during the lifetime of the excited state of the probe, the natural decay time of the probe could be treated as a global parameter. To take full advantage of the internal relationships between the different decay curves, it is highly recommended that the data analysis will be performed using the global analysis approach. This is certainly the case when the system studied exhibits more than one decay time, i.e., when more than one photophysically active species is present, and/or when these decay times are close to each other. A single curve analysis is also likely to fail in the case when one of the observed decay times has a very low contribution to the total decay. In the case of dynamic fluorescence quenching measurements, systems with a very low quencher concentration (which is a necessity for the determination of the polydispersity effects in the system, see Section 4.3) will be difficult to resolve without a global analysis approach.

2.2 Global compartmental analysis

The recently developed compartmental analysis method of fluorescence decay surfaces [10] requires a global analysis approach. In a compartmental analysis, the system considered is treated as a system of different compartments for both the ground and the excited states. An aqueous micellar system, for example, can be considered as a compartmental system where the probe can be solubilized either in the micellar compartment or in the aqueous bulk compartment. Compartmental analysis allows the determination of the rate constants for exchange between the different compartments, including the rate constants for fluorescence deactivation and quenching as well as spectral parameters.

3. DYNAMIC FLUORESCENCE QUENCHING MEASUREMENTS

In a dynamic fluorescence quenching (DFQ) measurement a fluorescent probe molecule and a quencher molecule are added to the system of interest. The fluorescence decays of the probe in the presence of different concentrations of the quencher are recorded and analysed by appropriate kinetic models to recover the relevant parameters of the system. Two main situations can be distinguished, one with the probe molecule completely solubilized in the aggregate sub-phase and stationary in its host aggregate during the excited state lifetime of the probe (Section 3.1), and in the other case one with the probe migrating between the aggregates during its excited state lifetime (Section 3.2).

3.1 In the absence of probe migration

For micelles and similar aggregates, the basic relation is developed under the assumptions that the micellar aggregates are of equal size, that the fluorescence probe is stationary in its host micelle during the lifetime of the excited state, that the probe and the quencher molecules are Poissonian distributed among the micelles, and that the quencher molecules do not interact with each other [11,12].

In the case of an immobile quencher [11-13], i.e., the quenching rate is much faster than the quencher molecule exit rate, and the fluorescence decay rate is faster than the product of the quencher concentration in the bulk phase and the rate for quencher entry from the bulk into an aggregate, the fluorescence δ-response function takes a simple form,

$$F_t = A_1\exp[-A_2t - A_3\{1 - \exp(-A_4t)\}] \tag{1}$$

with the following expressions for A_1 - A_4:

$A_1 = F_0$ (2a)
$A_2 = k_0$ (2b)
$A_3 = \langle n \rangle$ (2c)
$A_4 = k_q$ (2d)

where F_0 is the intensity at time t=0, k_0 is the first-order rate constant for fluorescence deactivation in the absence of a quencher (k_0 is often given as the decay time τ_0; $\tau_0 = 1/k_0$), $\langle n \rangle$ is the average occupation number of quenchers in a micelle; $\langle n \rangle = Q_m / M$, with the subscript m denoting the micellar phase and M the micelle concentration, and k_q is the first-order rate constant for quenching in a micelle containing one quencher.

3.2 In the presence of probe migration

The immobility of the probe is one of the conditions in the original model. In many cases this is true, e.g., when a highly hydrophobic probe is introduced to an aqueous micellar solution of moderate micelle concentration. In other cases, however, this condition may not hold, as in the case of highly concentrated micellar solutions, reversed micellar solutions or when the probe has an amphiphilic character [14-17].

For the case of migrating probes and/or quenchers, generalised versions of the model have been developed [18,19]. In a first approach [18] it was shown that the fluorescence δ-response function will still be described by an equation similar to Eq. 1, but with a generalised interpretation of the A_i parameters. According to this approximate solution, the different A_i's can be expressed as Eqs. 3 (A_1 is the same in all models):

$A_2 = k_0 + \langle x \rangle_s k_q$ (3b)
$A_3 = \langle n \rangle (1 - \langle x \rangle_s / \langle n \rangle)^2$ (3c)
$A_4 = k_q / (1 - \langle x \rangle_s / \langle n \rangle)$ (3d)

where $\langle x \rangle_s$ is the long time value of $\langle x \rangle$, the average number of quenchers in a micelle with an excited probe, and all other parameters have the same meaning as above. It follows that the observed fluorescence decay becomes single-exponential if, and only if, $\langle x \rangle$ reaches a constant value $\langle x \rangle_s$ within the time-window studied.

It has been shown that this approximation holds only at low average numbers of quencher per micelle, $\langle n \rangle$ [20]. Alternatively, on the basis of a stochastic approach, an exact solution for this problem was presented [19]. It could be shown that in this approach $\langle x \rangle_s$ can be evaluated by the following expression [19]

$$\langle x \rangle_s = (k_p M / k_q)\{1 - \exp[-A_3 A_4 / (A_4 + k_p M)]\} + \langle n \rangle k_{exq} / A_4 \quad (4)$$

where k_p is the second-order rate constant for the probe migration process and k_{exq} is the generalised quencher exchange rate constant. The product $k_p M$ can be regarded as the reciprocal of the first passage time between two micelles when migration through the bulk phase is considered [19].

From Eq. 4, Eqs. 3 can be rewritten [19,21,22] to give

$A_2 = k_0 + k_p M\{1 - \exp[-\langle n \rangle k_q / (k_q + k_p M)]\}$ (5b)
$A_3 = \langle n \rangle \{1 - k_p M(1 - \exp[-\langle n \rangle k_q / (k_q + k_p M)]) / k_q \langle n \rangle\}^2$ (5c)
$A_4 = k_q \{1 - k_p M(1 - \exp[-\langle n \rangle k_q / (k_q + k_p M)]) / k_q \langle n \rangle\}^{-1}$ (5d)

or, when the probe is considered to be immobile;

$A_2 = k_0 + \langle n \rangle k_{exq} / A_4$ (6b)
$A_3 = \langle n \rangle k_q^2 / A_4^2$ (6c)
$A_4 = k_q + k_{exq}$ (6d)

which will reduce to Eqs. 3 at low <n>-values.

The use of Eqs. 5 or 6 in the analysis of the fluorescence decay permits the determination of the different model parameters under the condition that both probe and quencher migrations are allowed.

4. Experimental results

Compartmental analysis can be used as a tool for the determination of the distribution of a probe molecule between the aggregate phase and the bulk phase if it has a reasonable solubility in both phases (Section 4.1). In some situations, the probe and quencher molecules may form ground-state charge transfer complexes. In such a case, a combination of static and dynamic fluorescence methods can be used to determine the complexation constant (Section 4.2). The effect of aggregate polydispersity on the determination of aggregation numbers (Section 4.3), the possibility of probe and quencher migration (Section 4.4), and the interaction between surfactants and polymers (Section 4.5) could be evaluated. Finally, the use of fluorescence depolarisation or anisotropy when studying reverse micellar systems and determining, e.g., order parameters to obtain detailed information on a molecular level of the system will be discussed (Section 4.6).

4.1 Distribution of the probe between the aqueous bulk and the aggregates

The compartmental analysis method [10] was used to determine the distribution of naphthalene between cetyltrimethylammonium chloride (CTAC) micelles and the aqueous bulk phase [21].

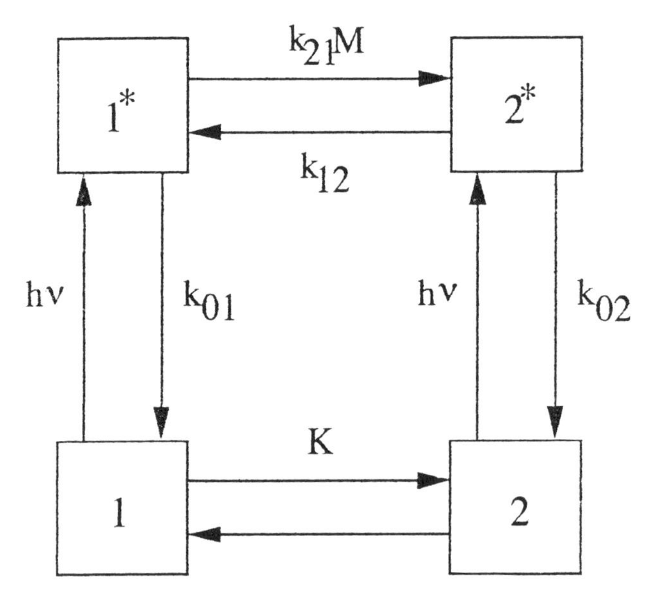

Figure 1. Schematic description of a bicompartmental system. See text for more details.

To apply compartmental analysis, one has to consider the system as composed of two different compartments, as is shown in Fig. 1. Compartment 1 is the aqueous bulk phase and compartment 2 the micellar phase. The rate constants k_{21} and k_{12} describe the entry and the exit of the probe to and from the micellar phase, respectively. The deactivation rate constants include, beside fluorescence deactivation, all non-radiative deactivation paths. Applying compartmental analysis to the CTAC/water/naphthalene system gave information about these rate constants and, thus, also information about the equilibrium process in the excited state. Furthermore, information about the spectral properties, i.e., the relative amount of light absorbed and emitted from each compartment, is also obtained. From this information, together with the known total absorption and the overall naphthalene concentration, it is possible to calculate the equilibrium constant for the ground-state distribution of the probe between the two phases, e.g., the concentration of the probe in the bulk phase, as is visualised in Fig. 2.

4.2 Fluorescence quenching when ground-state complexes are formed

If the probe and quencher used have the ability to form non-fluorescent charge-transfer ground-state complexes, it will affect the model for evaluation of data. The presence of ground-state complexes will, as an example, lower <n> below Q_m/M.

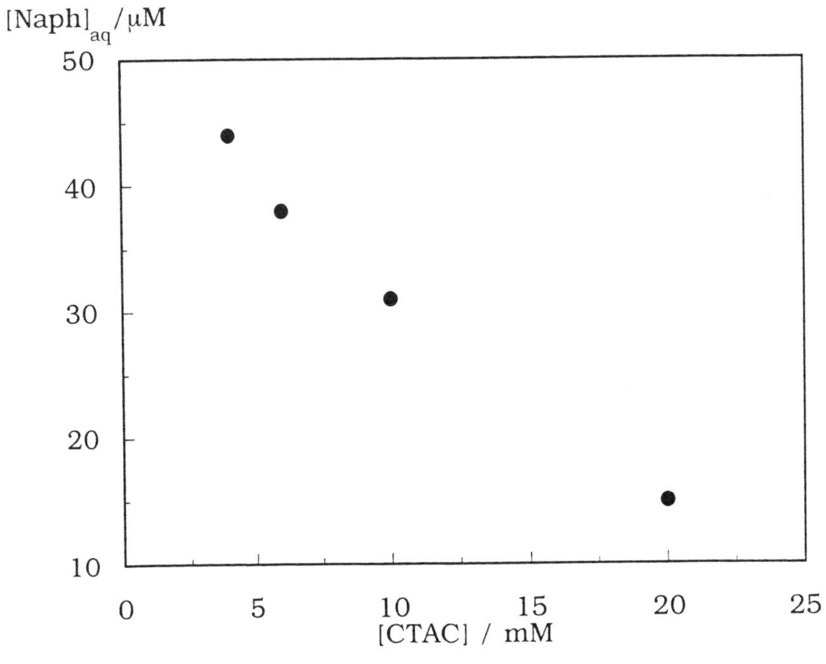

Figure 2. Concentration of naphthalene in the aqueous bulk phase as a function of [CTAC]

The DFQ method was combined with static fluorescence measurements to determine the apparent complexation constant between the probe pyrene and the quencher methylviologen (MV^{2+}) in sodium dodecylsulfate (SDS) micelles [23]. Upon formation of a non-fluorescent ground-state complex, the system will exhibit both dynamic and static quenching. The stationary relative fluorescence intensity is given by [24]

$$I_0 / I = (1 + K_{SV}Q_t)(1 + K_C Q_t) \quad (7)$$

where I_0 and I are the fluorescence intensities in the absence and presence of quencher, respectively, Q_t is the total quencher concentration, and K_{SV} and K_C denote the Stern-Volmer and complexation constants, respectively.

For the case of both static and dynamic quenching, an expression was derived for the relative fluorescence intensity assuming that the probe concentration is low in comparison with the quencher concentration, that <n> ≤ 2, and that the number of probe-quencher complexes per micelle is not more than one [23]

$$I_0 / I = (1 + K_{app}\langle n\rangle)\left(\sum_{x=0}\frac{\langle n\rangle^x e^{-\langle n\rangle}}{x!(1+xk_q\tau_0)}\right)^{-1} \qquad (8)$$

where K_{app} is an apparent equilibrium constant of complexation [25]. Eq. 8 consists of a dynamic part (the Poisson weighted series of $(1 + xk_q\tau_0)$ and a static part, $(1 + K_{app}\langle n\rangle)$. Among the parameters in Eq. 8, τ_0 ($k_0 = 1/\tau_0$), k_q, and <n> can be determined from DFQ measurements to calculate the dynamic factor. A plot of the static quenching factor (SQF), i.e., the product of I_0/I obtained from static fluorescence measurements and the calculated dynamic factor, versus <n> should yield a straight line with slope equal to K_{app}, as illustrated in Fig. 3.

The evaluation of the data in the pyrene/MV^{2+}/SDS-system gave an apparent complexation constant of 42±3, within the same order as reported [26,27].

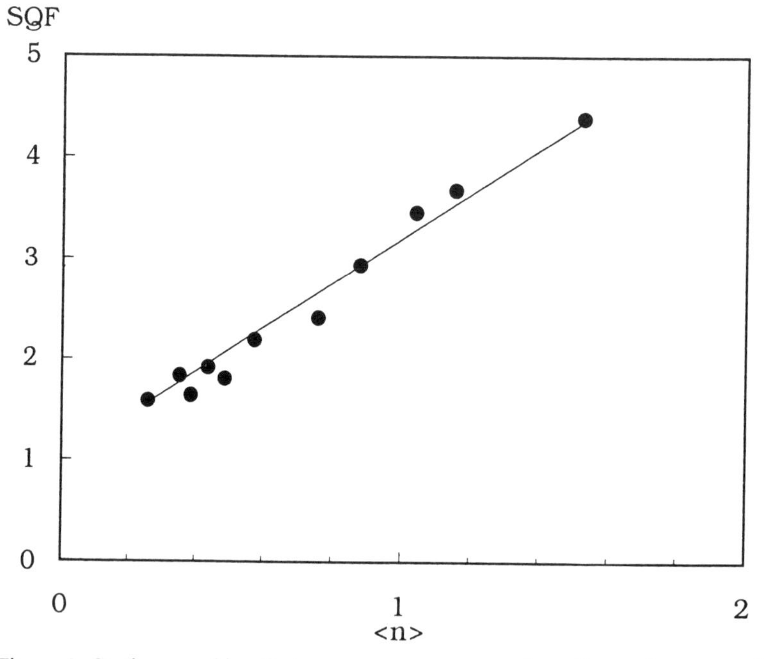

Figure 3. Static quenching factor versus average number of quenchers per micelle, <n>.

Evaluation of the DFQ data using of a modified form of Eqs. 6 [23], and assuming a mobile quencher and immobile probe, the SDS aggregation numbers were determined to increase from 60 to 81 in the SDS concentration range of 35 mM to 156 mM. This growth of SDS micelles at higher SDS concentrations arises from an increased electrostatic shielding due to the increased ionic strength as reported previously [28].

4.3 Polydispersity effects

In a monodisperse system, one can make use of the relation between <n> and Q_m, as a plot of the former versus the latter should yield a straight line through the origin with a slope equal to 1/M. From this slope, k_m, the aggregation number for the monodisperse micelles, $<a>_m$, can be calculated as $<a>_m = k_m S_m$.

If the system is not monodisperse, but shows a broader size distribution, it is shown [28] that the fluorescence _-response function still is well described by an expression like Eq. 1. Within each subset of the system, differing in aggregation number, the probe and quencher molecules will be distributed in a Poissonian way. The distribution of molecules between the different subsets, however, will be weighted by the relative volume of each subset, i.e., at low quencher to micelle ratios and at low surfactant concentrations, the quencher molecules will be preferably solubilized in the larger aggregates. The obtained aggregation numbers will in such a case be dependent on the quencher concentration, and should be treated as a quencher-averaged aggregation number, $<a>_q$ [29,30]. From $<a>_q$ it is possible to calculate a weight-averaged aggregation number, $<a>_w$, which is independent of the quencher concentration, by

$$<a>_q = <a>_w - 1/2\sigma^2\eta + 1/6\kappa\eta^2 - \ldots \qquad (9)$$

where σ^2 is the variance and κ the third cumulant, giving the skewness, of the size distribution. h is the ratio of micellized quencher molecules to micellized surfactant molecules, i.e., $\eta = Q_m/S_m$.

The ratio $\sigma/<a>_w$ can be used as a "polydispersity index", showing trends in the size distribution.

If a microheterogeneous solution shows a pronounced polydispersity, this feature will highly influence the results from DFQ measurements. The effects of the choice of model to fit to the

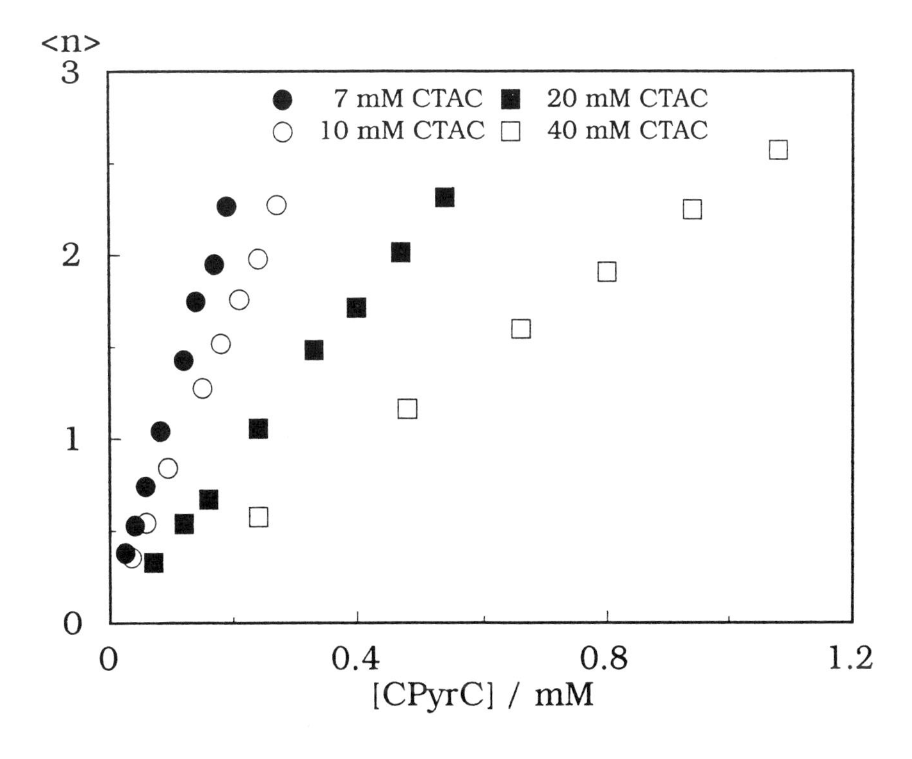

Figure 4. Estimated quencher occupancy number, <n>, versus quencher concentration.

obtained data was clearly demonstrated in the results obtained from the micellar CTAC system with 1-methylpyrene as probe and cetylpyridinium chloride (CPyrC) as quencher [31]. In this system it turned out that both the probe and the quencher did not migrate, allowing the use of Eqs. 2. Treating the system as being monodisperse gave the impression that the micelles were growing in aggregation number upon increasing surfactant concentration, Figs. 4 and 5. If, on the other hand, polydispersity was assumed, making use of Eq. 9, it turned out that the aggregation number remained constant within the CTAC concentration range studied, see Fig. 5. Furthermore, the initially pronounced polydispersity decreased with increasing surfactant concentration, as can be seen in Fig. 6.

It was also found that the quenching rate constant k_q decreased with increasing surfactant concentration. This indicates that the quenching occurs in larger aggregates (k_q is assumed to be, approximately, inversely proportional to the aggregation number for small, spherical, micelles), and the conclusion drawn was that there is an aggregation number increase when more surfactant is added to the system. This increase, however, is selective: preferably the smaller aggregates in the initially polydisperse system grow. This would simultaneously lead to a slower quenching rate as well as a decreased polydispersity.

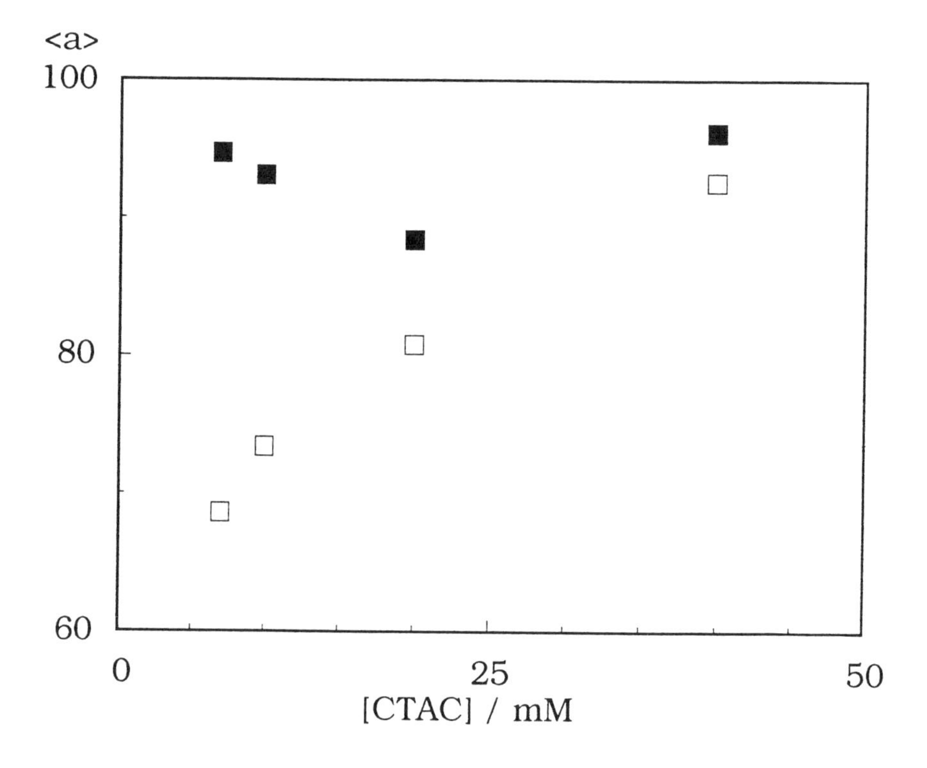

Figure 5. Estimated aggregation numbers: $\langle a \rangle_w$ (filled) and $\langle a \rangle_m$ (open).

As expected from thermodynamics for a system where the mean aggregation number does not increase with temperature, increased temperature leads to an increased polydispersity [32].

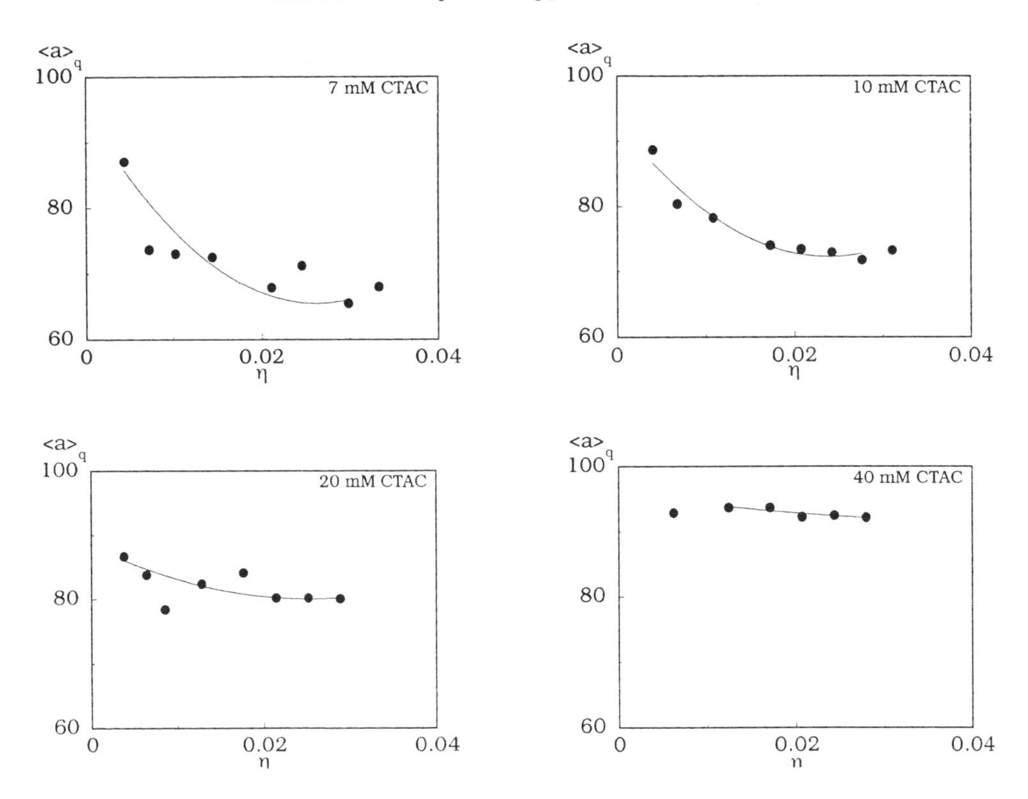

Figure 6. $<a>_q$ versus η at 7, 10, 20, and 40 mM CTAC with the fits by use of Eq. 9

4.4 Migration and exchange processes

In a reverse micellar system of N-benzyl-N,N-dimethyltetradecylammonium chloride (TBDAC) in toluene with pentanol as cosurfactant, the intermicellar migration of the probe 1-pyrene sulfonate (PSA) and the quencher N-tetradecylpyridinium chloride (TPyrC) was investigated by DFQ measurements at different water-to-surfactant ratios [22]. From a comparison between model fittings assuming mobile species or immobile species, Eqs. 5 and 6 and Eqs. 2, respectively, it turned out that the former was more appropriate to describe the system. The decay time for PSA in the reverse micelles, 120 ns, increased as compared with the decay time in water, 63.5 ns. This indicates that the probe is located preferably in the surfactant layer of the reverse micelle, as could be expected from electrostatic reasons.

Quite surprisingly, it was found that only the probe was migrating. To rationalise this, it was assumed that the electrostatic potential should be perturbed by the presence of an approaching micelle. The change in potential should have a greater impact on the counter-ion (probe) than on the co-ion (quencher) exchange. The long aliphatic tail of the quencher may also impose a slower exchange rate.

Increasing the water content, w, initially increases the size of the reversed micelles, given as aggregation numbers, but the size rapidly reaches a limiting value, see Fig. 7.

To compare the obtained values on k_p with the diffusion controlled limit, the bimolecular rate constant for diffusion k_d - calculated to be approximately 1.1×10^{10} $M^{-1}s^{-1}$

in this system [22] - was multiplied with the experimentally obtained micelle concentrations. The results are summarised in Fig. 8.

It could be concluded that at higher water content, k_p exceeds the diffusion controlled limit. This strongly supports clustering of the micelles, leading to a locally higher micelle concentration.

Under the condition that clusters are formed, $1 / k_p$ should in the saturation limit give the mean first passage time. Experimentally it was found to be approximately 65 ns. This should be compared with the calculated value 64 ns, which, assuming that the probe and the quencher have diffusion coefficients of about the same order of magnitude [22] was obtained by the relations:

$$1/ k_p = \langle l\rangle^2 / D_p \qquad (10a)$$
$$R_w = [3\langle a\rangle w V_{aq} / 4\pi]^{1/3} \qquad (10b)$$
$$1/k_q = \{R_w / D\}[\Omega \ln\Omega / \Omega\text{-}1) - 1] \qquad (10c)$$
$$\Omega = [2R_w / (R_q + R_p)]^2 \qquad (10d)$$

where $\langle l\rangle$ is the aliphatic carbon tail length, D_p is the diffusion coefficient of the excited probe, R_w is the radius of the water nanodroplet, V_{aq} is the volume of one water molecule, D is the mutual diffusion coefficient of the probe and the quencher for diffusion on the surface of a reversed micelle, and R_q and R_p are the molecular radii of the excited quencher and probe, respectively [22]. The validity of Eq. 10c is restricted to $\Omega \gg 1$ [12].

Assuming cluster formation, it is relevant to determine the average number of micelles per cluster. Taking the volume of a cluster as a multiple of the volume of one micelle, assuming spherical aggregates, and calculating the minimum cluster-to-micelle radii ratio with Eq. 10c, the following expression for $\langle x\rangle$, the average number of micelles per cluster was obtained [22]:

$$\langle x\rangle = [\langle k_q\rangle_m / \langle k_q\rangle_c]^{3/2} \qquad (11)$$

where indices m and c denote micelle and cluster, respectively. The resulting numbers of micelles per cluster was found to be approximately 2.2, 8.4, and 23 at w equal to 16, 24, and 34, respectively.

It has been suggested by Jada et al. [33] and Lang et al. [33] that clustering should occur only above the percolation threshold. This statement have been criticised by Jóhannsson et al. [38], as the kinetic model used by Jada and Lang [33] does not include micelle clusters below the percolation threshold. The results obtained in our analysis [22] clearly indicate the presence of clusters well below the percolation threshold. It is appropriate to mention that the results obtained by Jada, Lang and co-workers all emanate from assuming an immobile probe model. This model always gives much larger aggregation numbers than if mobility of the probe is assumed, see Fig 7 (note the different scales)

In order to investigate what the results from a fit of an assumed immobile probe's fluorescence d-response function actually means, the number of micelles per cluster was multiplied with the aggregation number of each micelle as obtained with the mobile probe model. Assuming that the droplet size remains unchanged when the volume fraction is increased at constant w-value, the obtained number of surfactant molecules per cluster are about the same as reported by Jada et. al [33a].This indicates that applying the model of immobile species to a microemulsion containing clusters will yield not the aggregation number of the individual micelles within the cluster, but the cluster aggregation number. A requirement is, of course, that the probe is mobile and has a sufficiently long-lived excited state to allow diffusion over the cluster volume, otherwise the size monitored will be regarded as infinite.

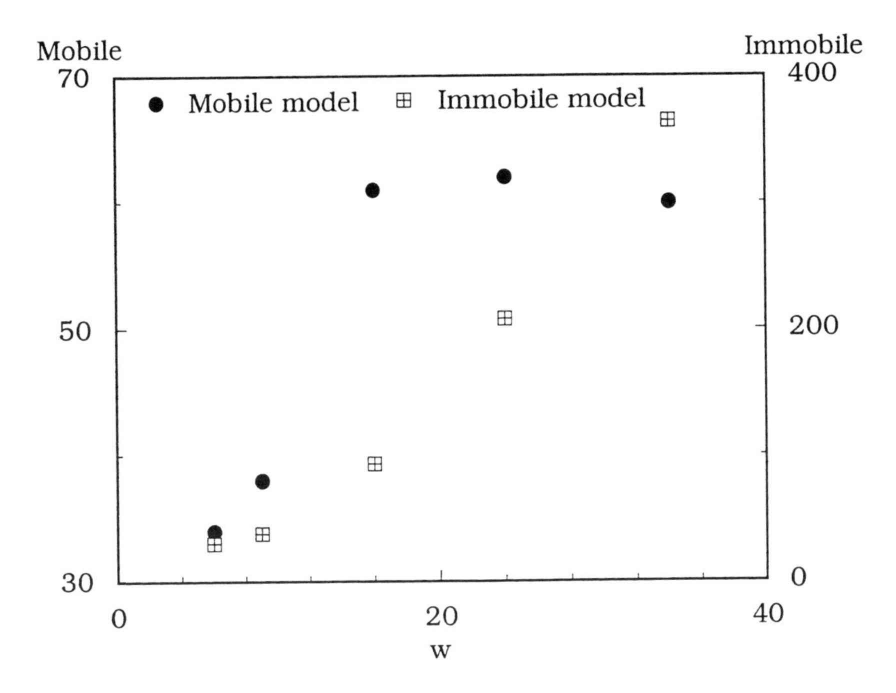

Figure 7. Aggregation numbers using the immobile and the mobile models.

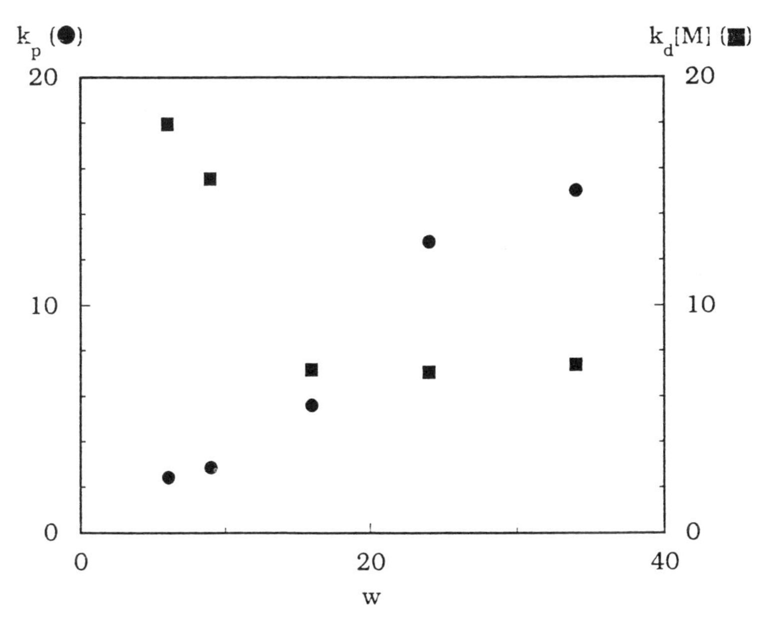

Figure 8. Dual plot of probe migration constant k_p and the micellar collision frequency in the diffusion controlled limit, k_dM versus water content, w.

4.5 Surfactant-polymer interactions

The interactions between surfactants and polymers have been intensively studied by classical methods [35]. The use of dynamic fluorescence quenching, however, has appeared as method for the study of these kind of systems quite recently [36,37], and has become an important tool in this field. The interactions between polyelectrolytes and surfactants are in principle determined by the balance between electrostatic and hydrophobic forces, i.e., the interaction between the polyelectrolyte and the surfactant starts as a counterion binding of the surfactant to the polyelectrolyte, while the driving force for aggregate formation is hydrophobic. For systems containing a nonionic polymer, the hydrophobic force is the most important. It is well-known that the interactions between anionic surfactants and nonionic polymers are much stronger as compared to systems with a cationic or nonionic surfactant [35].

It was, however, shown in a DFQ study of the cationic surfactant CTAC and the nonionic poly(vinyl alcohol)-poly(vinyl acetate), PVA-Ac, that also cationic surfactants interact with nonionic polymers [21]. The interaction caused the formed aggregates to have a smaller aggregation number as compared to ordinary CTAC micelles. It was also concluded that the polymer strand coils around the aggregates in very much the same way as has been suggested for anionic surfactants interacting with nonionic polymers [37b]. The association of ionic surfactants to the polymer causes the polymer-surfactant complex to behave more and more as a polyelectrolyte, as confirmed by viscosity measurements.

With the use of compartmental analysis [10], it was possible to evaluate the solubility capacity of the surfactant aggregates. Introducing PVA-Ac to an aqueous CTAC solution increases the solubility capacity of small, hydrophobic, molecules. The equilibrium constant for partitioning between aggregates and the aqueous bulk increased by approximately 50% when the polymer was introduced, see also Section 4.1. This was rationalised by the reduced polarity of the aggregate Stern layer, a reduction also monitored by the ratio of the first and third vibronic peaks in the pyrene steady-state emission spectrum [38].

The presence of a polymer in a micellar solution may also alter the migration rates. It was found that, based on the fluorescence decay of PSA in the presence of the quencher TPyrC, the probe migration increased by the presence of the polymer compared with a polymer-free solution at the same surfactant concentration. Furthermore, increasing the polymer concentration, at a fixed surfactant concentration, leads to a decrease in migration rate and, equivalently, increasing the surfactant concentration at constant polymer concentration increases the migration rate. This was accounted for by the higher local concentration of micelles interacting with a polymer chain.

Supporting results were reported for the system CTAB (B denotes bromide as counter-ion) in the presence of the highly charged polyelectrolyte poly(styrene sulfonate) (PSS) [37d] when pyrene was employed as probe and dimethylbenzophenone (DMBP) as quencher. In this case the probe is stationary within the time-scale studied, but at elevated CTAB concentrations the quencher, DMBP, started to migrate significantly. The conclusion drawn that the migration emanates from interaggregate exchange between micelles attached to the same PSS-strand was furthermore supported when Cu^{2+} ions were used as quencher [37d].

4.6 Fluorescence depolarisation

The reversed micellar system AOT/water/oil, with heptane or dodecane as organic phases and with cresyl violet as probe was studied using fluorescence depolarisation measurements [39].

The δ-response of the total fluorescence was found to be monoexponential, when globally analysed (see also Section 2.2). The decay time decreased with increasing water-to-AOT ratio, w, due to the increasing polar head hydration.

With increasing w, the steady-state anisotropy value decreased, but was always larger then the value found in pure water solutions, even at high water content (w=42).

Global analysis of the anisotropy decay [40] showed that the dynamic fluorescence anisotropy was described by a bi-exponential decay. The largest correlation time was found to be different from the overall correlation time of the micelle. AOT reversed micelles can be considered as both monodisperse and spherical [41], but they are not rigid structures.

The two dynamic processes, beside the rotation of the micelle as whole, that can have a significant impact on the depolarisation of the probe are:
1) lateral diffusion of the probe over the curved surface of the water pool
2) transversal diffusion perpendicular to the micellar surface

The sum of the inverse of the lateral and transversal correlation times differs in the two hydrocarbon solvents. This sum was, however, not affected by the water content, leading to the conclusion that the lateral diffusion has no, or little, influence on the depolarisation process in this system.

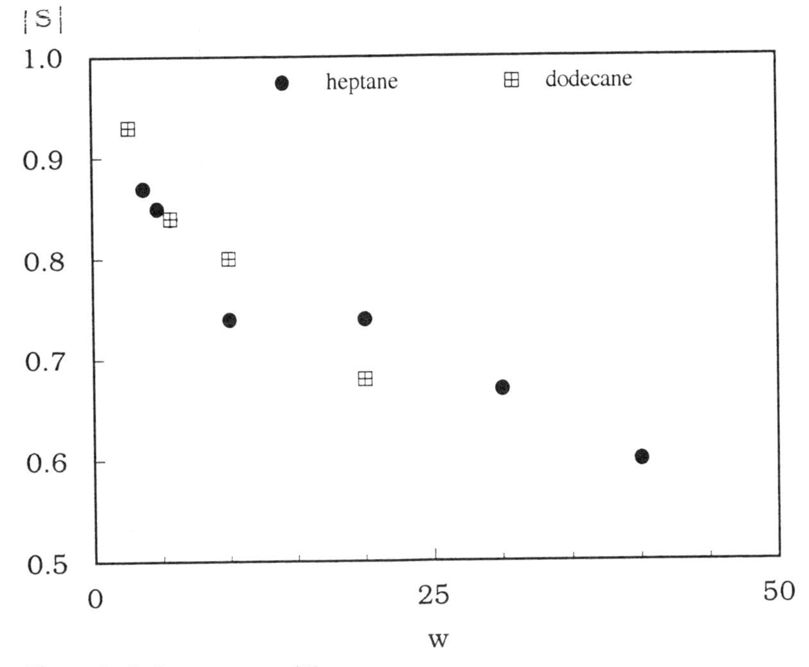

Figure 9. Order parameter |S| versus water content w

The two dynamic processes, beside the rotation of the micelle as whole, that can have a significant impact on the depolarisation of the probe are:
1) lateral diffusion of the probe over the curved surface of the water pool
2) transversal diffusion perpendicular to the micellar surface

The sum of the inverse of the lateral and transversal correlation times differs in the two hydrocarbon solvents. This sum was, however, not affected by the water content, leading to the conclusion that the lateral diffusion has no, or little, influence on the depolarisation process in this system.

It was also possible to determine an order parameter, |S|, which can take values between 0 and 1. |S| = 1 indicates that the probe is totally restricted in its motions, while when |S| = 0, the probe moves totally free. The obtained |S|-values, graphically shown in Fig 9, decreases with increasing water content, in agreement with the assumption that the probe is located near the micelle inner surface attached to one or more surfactant molecules.

In conclusion, it was found that two correlation times were needed to describe the dynamic anisotropy, which was explained by a hindered, internal, rotation.

ACKNOWLEDGEMENTS

JvS is a visiting post-doctoral fellow at K.U. Leuven who thanks K.U. Leuven and The Swedish Institute for financial support. MHG thanks CNPq (Brazil) and K.U. Leuven for financial support. MVdA is an "Onderzoeksleider" of the Belgian "Fonds voor Wetenschappelijk Onderzoek" (NFWO). NB is an "Onderzoeksleider" of the Belgian "Fonds voor Geneeskundig Wetenschappelijk Onderzoek" (FGWO). RMN is a post-doctral fellow at K.U. Leuven who thanks K.U. Leuven for financial support. NW and HF thank the IWONL for a pre-doctoral fellowship. DB thanks Comisión de Investigaciones Científicas (CIC) de la Provincia de Buenos Aires (Argentina) and K.U. Leuven for financial support. DN is a fellow in the European Human Capital and Mobility Scheme. The support of the FKFO (Belgium) and the Ministry of Scientific Programming in Belgium (through Grants IUAP-II-16 and IUAP-III-040) is gratefully acknowledged. The authors want to thank Mr. Johan Hofkens and Mrs. Sigrid Depamelaere for their maintenance of the single-photon timing systems.

REFERENCES

(1) a. Almgren, M. In Kinetic and Catalysis in Microheterogeneous Systems; Grätzel, M., Kalyanasundaram, K., Eds.; Marcel Dekker: New York; p. 63, 1991.
b. Almgren, M. Adv. Coll. Int. Sci. , 41(1992) 9.
(2) a. Van der Auweraer, M.; De Schryver, F. C. In Inverse Micelles, Studies in Physical and Theoretical Chemistry; Pileni, M. P., Ed.; Elsevier: Amsterdam; p. 70, 1990.
b. Gehlen, M.; De Schryver, F. C. Chem. Rev. 93 (1993) 199.
(3) a. Zana, R. In Surfactant Solutions. New Methods of Investigation; Zana, R., Ed.; Marcel Dekker: New York & Basel; p. 241, 1987.
b. Zana, R.; Lang, J. Coll. Surf. , 48 (1990) 153.
(4) Boens, N. In Luminiscence techniques in chemical and biochemical analysis; Baeyens, W.R.G., Keukeleire, D., Korkidis, K., Eds.; Marcel Dekker: New York, p. 21,1991 and references therein.
(5) Marquardt, D.W. J. Soc. Ind. Appl. Math., 11 (1963) 431.
(6) O'Connor, D:V.; Phillips, D. Time-correlated single photon counting; Academic Press; New York, 1984.
(7) a. Knutson, J.R.; Beechem, J.M.; Brand, L. Chem. Phys. Lett., 102 (1983) 501.
b. Beechem, J.M.; Knutson, J.R.; Brand, L. Photochem. Photobiol., 37 (1983) 520.
c. Beechem, J.M.; Ameloot, M.; Brand, L. Anal. Instrum., 14 (1985) 379.
d. Beechem, M; Ameloot, M.; Brand, L. Chem. Phys. Lett., 120 (1985) 466.
e. Ameloot, M.; Beechem, J.M.; Brand, L. Chem. Phys. Lett., 129 (1986) 211.
(8) a. Löfroth, J.-E. Anal. Instrum., 14 (1985) 403.
b. Löfroth, J.-E. J. Phys. Chem., 90 (1986) 1160.
(9) a. Boens, N.; Janssens, L.D.; De Schryver, F.C. Biophys. Chem., 33 (1989) 77.
b. Janssens, L.D.; Boens, N.; Ameloot, M.; De Schryver, F.C. J. Phys. Chem., 94 (1990) 3564.
c. Ameloot, M.; Boens, N.; Andriessen, R.; Van den Bergh, V.; De Schryver, F.C. J. Phys. Chem., 95 (1991) 2047.
(10) a. Ameloot, M.; Boens, N.; Andriessen, R.; Van den Bergh, V.; De Schryver, F.C. J. Phys. Chem., 95 (1991) 2041.
b. Andriessen, R.; Boens, N.; Ameloot, M.; De Schryver, F.C. J. Phys. Chem., 95 (1991) 2047.
c. Andriessen, R.; Ameloot, M.; Boens, N.; De Schryver, F.C. J. Phys. Chem., 96 (1992) 314.
d. Boens, N.; Andriessen, R.; Ameloot, M.; Van Dommelen, L.; De Schryver, F.C. J. Phys. Chem., 96 (1992) 6331.
e. Boens, N.; Ameloot, M.; Hermans, B.; De Schryver, F.C.; Andriessen, R. J. Phys. Chem., 97 (1993) 799.
(11) a. Infelta, P. P.; Grätzel, M.; Thomas, J. K. J. Phys. Chem., 78 (1974) 190.

b. Infelta, P. P.; Grätzel, M. J. Chem. Phys., 78 (1983) 5280.
(12) a. Tachiya, M. Chem. Phys. Lett., 33 (1975) 289.
b. Tachiya, M. J. Chem. Phys., 76 (1982) 340.
c. Tachiya, M. J. Chem. Phys., 78 1983) 5282.
d. Sano, H.; Tachiya, M. J. Chem. Phys., 75 (1981) 2870.
(13) Dederen, J.C.; Van der Auweraer, M.; De Schryver, F.C. Chem. Phys. Lett., 68 (1979) 451.
(14) Malliaris, A.; Lang, J.; Sturm, J.; Zana, R. J. Phys. Chem., 91 (1987) 1475.
(15) Luo, H.; Boens, N.; Van der Auweraer, M.; De Schryver, F.C.; Malliaris, A. J. Phys. Chem., 93 (1989) 3244.
(16) Fletcher, P.D.I.; Robinson, B.H. Ber. Bunsen-Ges. Phys. Chem., 85 (1981) 863.
(17) a. Fletcher, P.D.I.; Howe, A.M.; Robinson, B.H. J. Cem. Soc., Faraday Trans. 1 83 (1987) 985.
b. Howe, A.M.; Mc Donald, J.A.; Robinson, B.H. J. Chem. Soc., Faraday Trans. 1 83 (1987) 1007.
(18) Almgren, M.; Löfroth, J.-E.; van Stam, J. J. Phys. Chem., 90 (1986) 4431.
(19) a. Gehlen, M.H.; Van der Auweraer, M.; Reekmans, S.; Neumann, M.; De Schryver, F.C., J. Phys.Chem., 95 (1991) 5684.
b. Gehlen, M.H.; Van der Auweraer, M.; De Schryver, F.C.; Photochem. Photobiol., 54 (1991) 613.
c. Gehlen, M.H.; Van der Auweraer, M.; De Schryver, F.C. Langmuir, 8 (1992) 64.
d. Gehlen, M.H.; Boens, N.; De Schryver, F.C.; Van der Auweraer, M.; Reekmans, S. J. Phys. Chem., 96 (1992) 5592.
(20) Tachiya, M. Can. J. Phys., 68 (1990) 979.
(21) Reekmans, S.; Gehlen, M.; De Schryver, F.C.; Boens, N.; Van der Auweraer, M. Macromolecules, 26 (1993) 687.
(22) Gehlen, M.H.; De Schryver, F.C.; Boens, N.; Van der Auweraer, M.; van Stam, J.; Noukakis, D. submitted.
(23) Gehlen, M.H.; De Schryver, F.C: J. Phys. Chem., in press.
(24) Vaughan, W.M.; Weber, G. Biochemistry, 9 (1970) 464.
(25) Hatlee, M.D.; Kozak, J.J. J. Chem. Phys., 74 (1981) 1098.
(26) Martens, F.M.; Verhoeven, J.W. J. Phys. Chem., 85 (1981) 1773.
(27) Fornasiero, D.; Grieser, F.J. J. Chem. Soc., Faraday Trans., 86 (1990) 2955.
(28) Almgren, M.; Löfroth, J.-E. J. Coll. Int. Sci., 81 (1981) 486.
(29) Almgren, M.; Löfroth, J.-E. J. Chem. Phys., 76 (1982) 2734.
(30) a. Warr, G.; Grieser, F. J. Chem. Soc., Faraday Trans. I, 82 (1986) 1813.
b. Warr, G.; Grieser, F.; Evans, D. F. J. Chem. Soc., Faraday Trans. I, 82 (1986) 1829.
(31) Reekmans, S.; Bernik, D.; Gehlen, M.; van Stam, J.; Van der Auweraer, M.; De Schryver, F.C. Langmuir, 9 (1993) 2289.
(32) Israelachvili, J. Intermolecular & Surface Forces, 2nd ed., Academic Press, Chapter 16, 1991.
(33) a. Jada, A.; Lang. J.; Zana, R. J. Phys. Chem., 93 (1989) 10.
b. Jada, A.; Lang, J.; Zana, R.; Makhloufi, R.; Hirsch, E.; Candau, S.J. J. Phys. Chem., 94 (1990) 387.
c. Lang, J.; Lalem, N.; Zana, R. J. Phys. Chem., 96 (1992) 4667.
(34) Jóhannsson, R.; Almgren, M.; Alsins, J. J. Phys. Chem., 95 (1991) 3819.
(35) a. Goddard, E.D. Coll. Surf., 19 (1986) 255.
b. Goddard, E.D. Coll. Surf., 19 (1986) 301.
(36) a. Zana, R.; Lang, J.; Lianos, P. In Microdomains in polymer solutions; Dubin, P.L., Ed., Plenum Press; New York, pp. 357, 1985.
b. Zana, R.; Lianos, P.; Lang, J. J. Phys. Chem., 89 (1985) 41.
(37) a. Almgren, M.; van Stam, J.; Lindblad, C.; Li, P.; Stilbs, P.; Bahadur, P. J. Phys. Chem., 95 (1991) 5677.
b. van Stam, J.; Almgren, M.; Lindblad, C. Progr. Colloid Polym Sci., 84 (1991) 13.

c. Thalberg, K.; van Stam, J.; Lindblad, C.; Almgren, M.; Lindman, B.; J. Phys. Chem., 95(1991) 8975.
d. Almgren, M.; Hansson, P.; Mukhtar, E.; van Stam, J. Langmuir, 8 (1992) 2405.
e. van Stam, J.; Brown, W.; Fundin, J.; Almgren, M.; Lindblad, C. In Colloid-Polymer Interactions; Dubin, P.L., Tong, P., Eds.; ACS Symposium Series; American Chemical Society: Washington, DC, in press.
(38) Kalyanasundaram, K.; Thomas, J.K. J. Am. Chem. Soc., 99 (1977) 2039.
(39) Negri, R.M.; Ameloot, M.; Wittouck, N.; De Schryver, F.C. in preparation
(40) Crutzen, M.; Ameloot, M.; Boens, N.; Negri, R.M.; De Schryver, F.C. J. Phys. Chem., 97 (1993) 8133.
(41) Almgren, M.; Jóhannsson, R. J. Phys. Chem., 96 (1992) 9512.

MICROCHEMISTRY
Spectroscopy and Chemistry in Small Domains
Edited by H. Masuhara et al.

Site-selected excitation energy transport in Langmuir-Blodgett multilayer films

I. Yamazaki, N. Ohta, S. Yoshinari and T. Yamazaki

Department of Chemical Engineering, Faculty of Engineering, Hokkaido University, Sapporo 060, Japan

Interlayer excitation transfer has been studied with LB multilayer films. Site-selection was observed in the fluorescence spectra of acceptors, and was interpreted as due to a contribution of a non-equilibrium excitation transfer from initially-excited vibrational levels of donor to the isoenergetic vibrational levels of acceptor.

1. INTRODUCTION

Special attention has recently been paid to the excitation energy transfer in restricted molecular geometries of biological and artificial molecular assemblies. The photosynthetic light-harvesting antenna in plants, as an example, is characterized by highly efficient absorption and transport of photonic excitation energy through a stack of several kinds of chromoproteins to the reaction center [1]. Functional dyes are incorporated within polypeptide networks with intermolecular distances being 20~40 Å. An artificial analogue of such biological system can be obtained with the Langmuir-Blodgett (LB) technique. The LB film is a mono- or multi-layered molecular assembly which is prepared by transferring a compressed monolayer spread on a water surface onto a solid substrate [2]. With the LB multilayer film, one can obtain a stacking molecular architecture in which different kinds of dyes as donor and/or acceptor are stacked sequentially 20~30 Å apart such that the excitation energy is transported layer-to-layer through the Förster dipole-dipole interaction mechanism. A schematic illustration for this transport is shown in Figure 1. Such a vectorial excitation transfer can be monitored by detecting the fluorescences spilled out from respective layers in a femto- or pico-second time scale.

The excitation energy transfer studied here is basically due to the Förster dipole-dipole interaction mechanism, where the spectral overlap between the donor fluorescence and the acceptor absorption, the molecular orientation, and the molecular distance between donor and acceptor are essential factors for determining the transfer rate [3]. Most of previous works have been made for 3-D systems like solutions and glasses in which acceptor molecules are distributed uniformly and randomly around a donor molecule. In the stacking multilayer systems, acceptors are located closely to donors at approximately constant distance and molecular orientation. Then one can expect to meet kinetic behaviors different from 3-D systems of random dispersion.

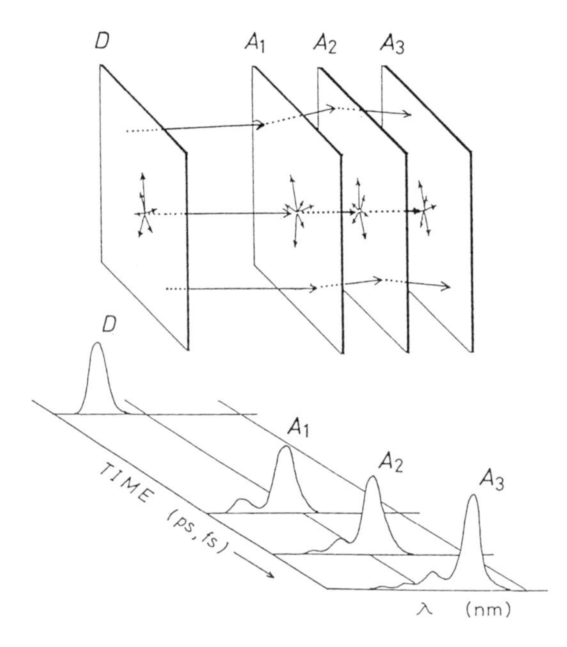

Figure 1. Schematic illustration of the sequential energy transport in the LB multilayers and its observation by means of the time-resolved fluorescence spectra. Following a ps laser excitation at D layer, the fluorescence spectra appear sequentially from D, A_1, A_2 and A_3 layers.

In our previous papers, the general features of the sequential excitation transfer in LB multilayers were reported in comparison with the biological antenna [4,5]. In this study, we will focus attention on changes of emission characteristics by changing excitation wavelength. We will discuss on (1) a site selection in the excitation energy relaxation in a single monolayer, (2) a site-selection in the interlayer transfer, and (3) enhancement in the transfer efficiency in LB mutilayers. We propose a hypothetical mechanism involving a non-equilibrium excitation transfer, for elucidating some unusual behaviors of the sequential interlayer transfer.

2. EXPERIMENTAL

Three groups of LB films have been prepared; (1) single monolayer films containing oxacyanine, (2) two-layer films of donor and acceptor, D-A, and (3) multilayer films of sequentially stacking donor and/or acceptor, D-A_1, D-A_1-A_2, and D-A_1-A_2-A_3.

As pigment molecules for D, A_1, A_2 and A_3 layers, we used, respectively, 10-(3,6-hexadecanylcarbazol-9-yl)decanyl trimethyl ammonium bromide (synthsized by Prof. Y. Kunitake and Prof. N. Kimizuka of Kyushu University [6], and received as a gift from them), 1,1'-dioctadecyloxacyanine (Nippon Kanko Shikiso Kenkyusho Co., Okayama, Japan), 1,1'-dioctadecylthiacyanine (Nippon Kanko Shikiso), 1,1'-dioctadecyl-3,3,3',3'-tetramethylindocarbocyanine (Molecular Probes, U.S.A) (see the next page for structures). Palmitic acid (Wako Chemical Co., Osaka) was purified five times by recrystallization from ethanol. Mixtures of palmitic acid and small amounts of pigment dissolved in chloroform were spread onto the surface of water subphase containing 3×10^{-4} M $CdCl_2$. The subphase conditions were adjusted to be 17°C and pH 6.3 by adding $NaHCO_3$ buffer solution. Nonfluorescent quartz plates were precoated with five layers of palmitic acid cadmium salt to make a surface uniform and hydrophobic. Mixed monolayers were deposited on a precoated quartz plate under a constant surface pressure of 2.5×10^{-2} N m^{-1}. The concentration of pigment molecules in each layer was 5 mol% as standard. The interlayer distance between donor and acceptor chromophores is 25 Å which corresponds to the hydrocarbon-chain length of arachidic acid molecules. The surface was further coated with two layers of cadmium arachidate to prevent the multilayer structure of the LB film from being destroyed.

$CH_3(CH_2)_{15}$ / $CH_3(CH_2)_{15}$ — $N(CH_2)_{10}\overset{+}{N}(CH_3)_3\ Br^-$ *D*

$C_{18}H_{37}$ $C_{18}H_{37}$ *A*$_1$

$C_{18}H_{37}$ $C_{18}H_{37}$ *A*$_2$

CH=CH–CH $C_{18}H_{37}$ $C_{18}H_{37}$ *A*$_3$

3. EXCITATION ENERGY RELAXATION IN A SINGLE MONOLAYER

The excitation energy relaxation in single monolayer LB films has been studied for various types of dyes and aromatic hydrocarbons [7-9]. These studies demonstrated dynamical aspects on hopping migration and quenching of excitation energy, and excimer formation in a monolayer. Most of the previous studies were carried out by using light sources at a particular wavelength. In the present study, we are concerned with the excitation-wavelength dependence on the fluorescence spectrum.

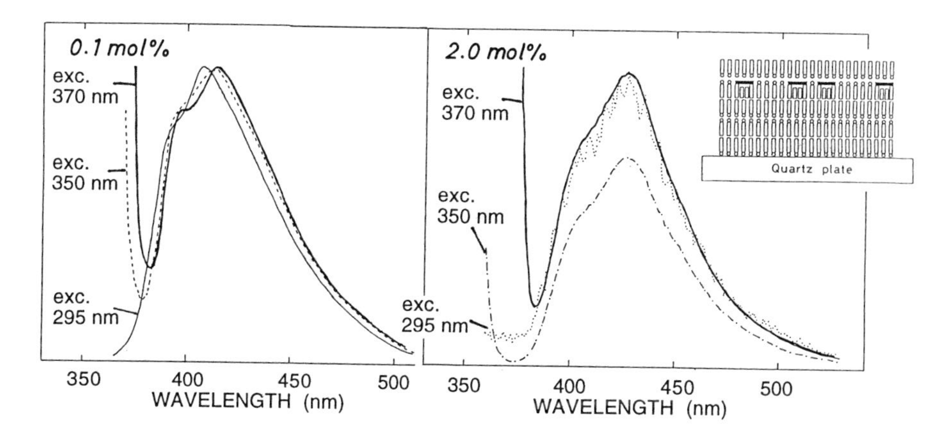

Figure 2. Fluorescence spectra of oxacyanine LB monolayer films. The excitation wavelengths are 295, 350 and 370 nm.

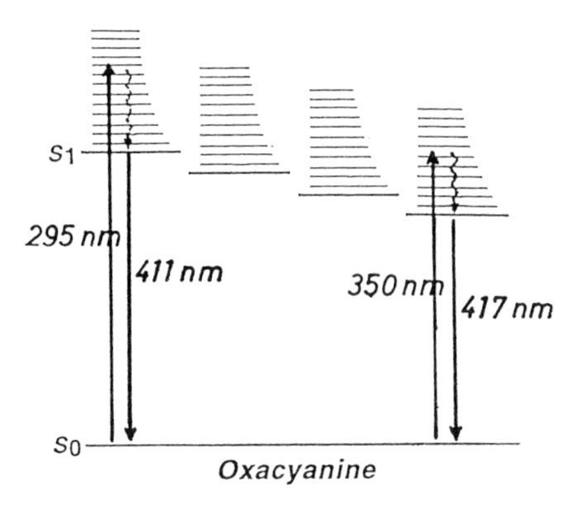

Figure 3. Energy level diagram of the ground (S_0) and excited singlet (S_1) states of oxacyanine in LB monolayer films. The energy levels of various sites are shown. In lower concentration, excited state at respective sites undergoes relaxation independently.

Figure 2 shows fluorescence spectra of single monolayer films of oxacyanine (A_1) taken at different excitation wavelengths. In higher concentration, 2.0 mol%, the fluorescence spectrum is centered at 428 nm and is constant irrespective of excitation wavelength. This spectrum is recognized as due to a dimer formed in the ground state [7,8]. On the other hand, in low concentration, 0.1 mol%, the spectrum and its maximum depend on the excitation wavelength; the fluorescence maximum is at 411 nm with excitation at 295 nm, while it is at 417 nm with excitation at 370 nm.

We assume that molecules in LB monolayer form various sites which are different in their electronic (S_1) energy, and that molecules in respective sites undergo the energy relaxation independently in low concentration. This is shown in an energy level diagram in Figure 3. The experimental result can be interpreted as follows: photoexcitation at short (or long) wavelength results in higher- (or lower-) energy site excitation and gives a blue- (or red-) side fluorescence.

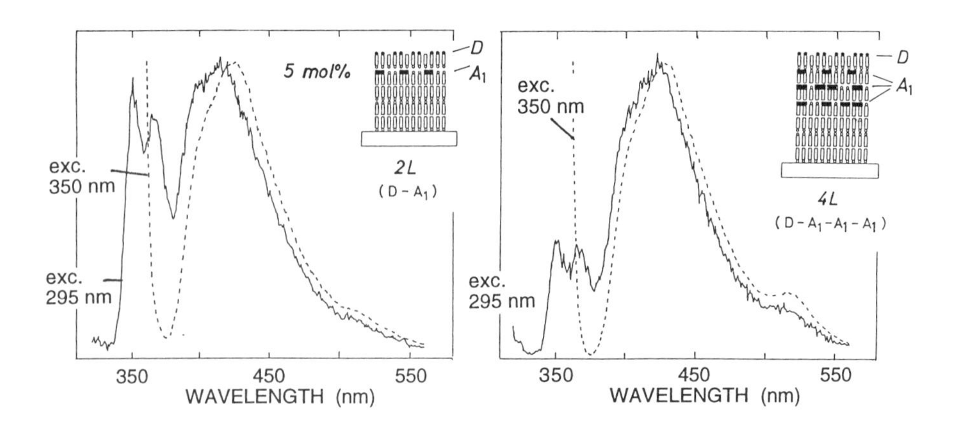

Figure 4. Fluorescence spectra of D-A_1 and D-nA_1 (n=3). The excitation wavelengths are 295 and 370 nm.

4. SITE-SELECTION IN THE INTERLAYER EXCITATION TRANSFER

In the two layer system D-A_1, the interlayer excitation transfer becomes dominant over the relaxation processes inside a monolayer. A *site selection* was observed in the A_1 fluorescence spectrum; the spectra obtained by the energy transfer from D are different from those obtained by the direct photoexcitation of A_1.

Figure 4 shows the fluorescence spectra of D-A_1 and D-nA_1 taken with excitation of D layer at 295 nm and with excitation of A_1 layer at 350 nm. In the A_1 spectrum, the direct excitation of A_1 gives a spectrum centered at 424 nm, while the excitation following the energy transfer from D layer gives a blue-shifted spectrum centered at 414 nm. Similar spectral difference can be seen also by changing the stacking number of A_1 layers up to four, i.e., D-nA_1 (n = 2, 3 and 4). The corresponding shift (ca. 6 nm) between the two spectra is somewhat smaller than that in D-A_1. In this case, the excitation at D layer transfers to nA_1 acceptor layers much more effectively than in D-A_1, and thus the D fluorescence intensity decreases furthermore.

In the three layer system, D-A_1-A_2, essentialy the same spectral difference appears in the A_2 spectrum, as is shown in Figure 5. The two spectra differ in peak position by 12 nm, depending on the direct excitation or the energy transfer excitation.

Note that, in D-A_1, the blue-shifted fluorescence of A_1 obtained with energy transfer from D is corresponding to the spectrum of A_1 monolayer obtained with 295-nm excitation, as has been shown in Section 3. In this regard, we have studied with the LB multilayers having a spacer between D and A_1 by stacking 3~5 layers of arachidic acid between D and A_1. Even an excitation of D at 295 nm gives only a red-shifted spectrum, not a blue-shifted spectrum. In this case, the transfer rate and efficiency must be lowered significantly due to long distance between D and A_1.

For elucidating above experimental results, we propose a hypothetical mechanism of the interlayer excitation transfer, as is shown in Figure 6. Following excitation of D, the excitation energy relaxes in two competitive pathways; (*a*) a vibrational

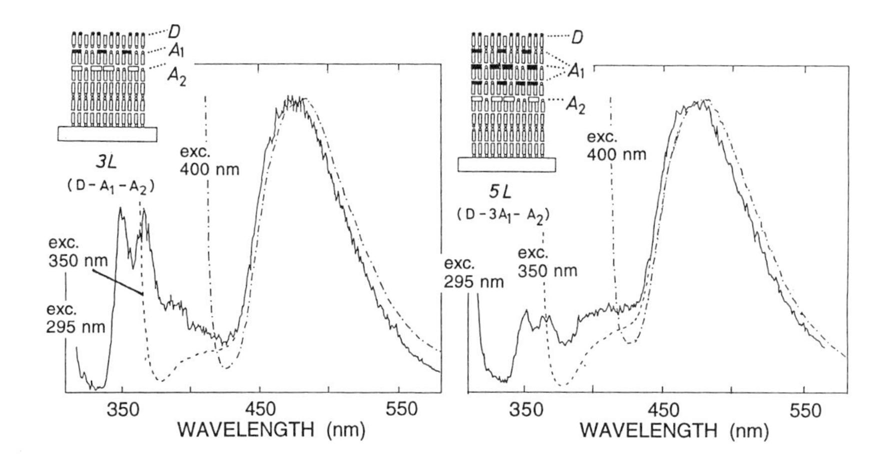

Figure 5. Fluorescence spectra of D-A_1-A_2 and D-nA_1-A_2 (n=3). The excitation wavelengths are 295 and 370 nm.

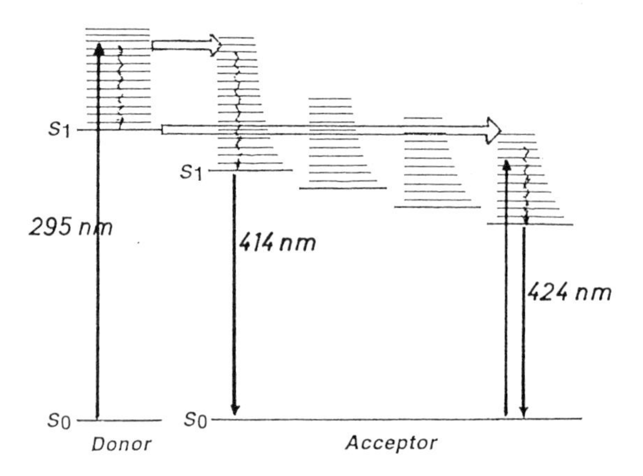

Figure 6. Energy level diagram of D and A_1 in the D-A_1 system. In lower concentration, the excitation energy transfers from highly vibrational levels of D to the isoenergetic vibrational levels of A_1.

relaxation in the singlet manifold of D, and (*b*) an energy transfer to vibrationally excited levels of the acceptor layer A_1 isoenergetic to the vibrational level of D initially photoexcited. It is well known that, in homogeneous systems such as solutions and glasses, the process (*a*) is much faster than the process (*b*). It should be probable, however, that the process b is not negligible in the LB multilayers where donor and acceptor molecules are stacked sequentially in short distances. This will be discussed in Section 6.

According to this mechanism, the energy transfer $D \rightarrow A_1$ involves a direct transfer from highly vibrational levels of D before thermal equilibriation, and produces an excited state of higher energy sites of A_1. In the LB films having a spacer between D and A_1, the direct excitation transfer should become much slower than the vibrational relaxation. Then the energy transfer occurs from thermally equilibriated levels of $D(S_1)$ to lower energy sites of $A_1(S_1)$ from which a red-shifted fluorescence emits.

Probably such a site selection or a non-equilibrium energy transfer will be related to the enhancement in the transfer efficiency in multilayers which is shown in the next section.

5. ENHANCEMENT OF THE EXCITATION TRANSFER EFFICIENCY IN MULTI-LAYERS D-A_1-A_2-A_3

A representative example of experimental data on the sequential energy transfer is shown in Figure 7. The time-resolved fluorescence spectrum changes with time ; following excitation of the D layer at 295 nm with a 2-ps laser pulse, a fluorescence band of D appears at 350 nm and the A_1 band weakly at 420 nm in -15 ~ 0 ps, and then the A_2 band appears at 470 nm after 40 ps, and finally the A_3 band at 580 nm after 600 ps. It is seen that the fluorescence from the inner layer rises more slowly than those of the outer layers. From the kinetic analysis for these results, we have found that the energy transfer takes place sequentially from the outer to the inner layers, in a similar way as in the photosynthetic light-harvesting antenna pigment system [4].

The energy transfer efficiency of respective pairs of donor and acceptor can be estimated from inspection of fluorescence intensity changes. Figure 8 shows the

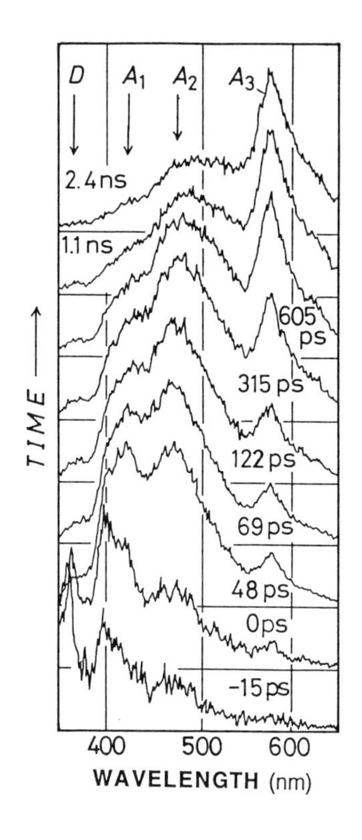

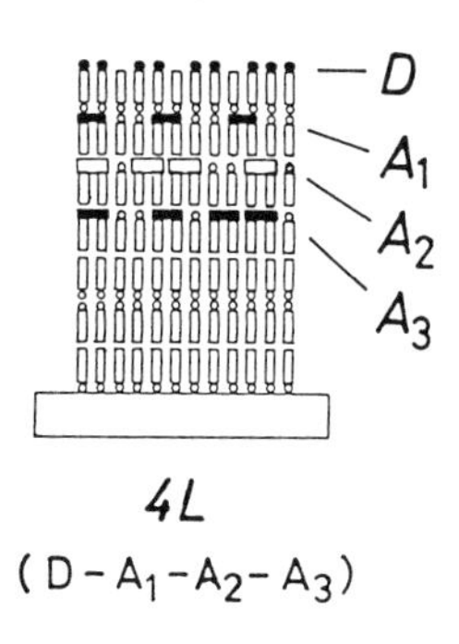

Figure 7. Time-resolved fluorescence spectra of the LB multilayer film of D-A_1-A_2-A_3. The excitation wavelength is 295 nm. The time zero corresponds to the time in which the excitation laser reaches maximum intensity.

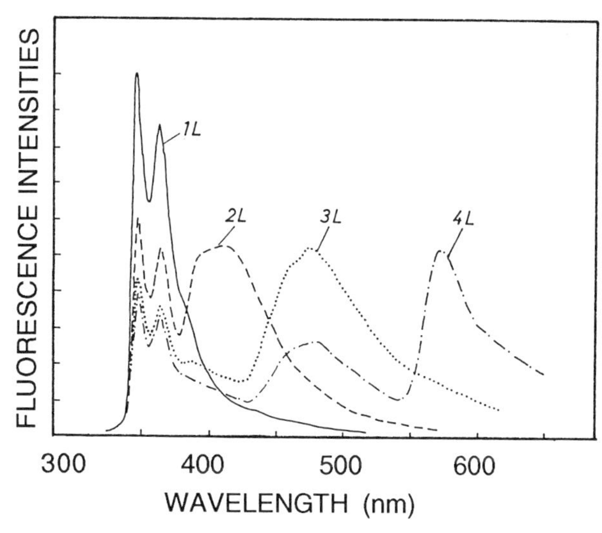

Figure 8. Fluorescence spectra of D (1L), D-A_1 (2L), D-A_1-A_2 (3L) and D-A_1-A_2-A_3 (4L). The excitation wavelength is 295 nm.

fluorescence spectra obtained with excitation at 295 nm for four types of LB multilayers, i.e., D (1L), D-A_1 (2L), D-A_1-A_2 (3L) and D-A_1-A_2-A_3 (4L). In 2L, the intensity of D fluorescence is reduced to about a half of that in 1L, and the A_1 fluorescence appears owing to the excitation transfer D $\rightarrow$ A_1. In 3L, the A_1 fluorescence disappears and the A_2 band appears. Similarly in 4L, the A_2 fluorescence is reduced to about a half of that in 3L, and the A_3 fluorescence appears. From the successive relationship with respect to reduction of the fluorescence intensity, the energy transfer efficiency Φ_{ET} can be derived from the following equations:

$$\Phi_{ET}(D\text{-}A_1) = \{\Phi_F(D,1L) - \Phi_F(D,2L)\} / \Phi_F(D,1L) \tag{1}$$

$$\Phi_{ET}(A_1\text{-}A_2) = \{\Phi_F(A_1,2L) - \Phi_F(A_1,3L)\} / \Phi_F(A_1,2L) \tag{2}$$

$$\Phi_{ET}(A_2\text{-}A_3) = \{\Phi_F(A_2,3L) - \Phi_F(A_2,4L)\} / \Phi_F(A_2,3L) \tag{3}$$

where $\Phi_F(A_1,2L)$, for example, denotes the fluorescence quantum yield of A_1 in 2L. The values of Φ_{ET} for respective steps are summarized in Table 1. One should note in this table that the Φ_{ET} value depends on stacking structure among 2L, 3L and 4L; $\Phi_{ET}(D\text{-}A_1)$ of 0.40 in D-A_1 is increased to 0.57 in D-A_1-A_2; and similarly $\Phi_{ET}(A_1\text{-}A_2)$ of 0.76 in D-A_1-A_2 is increased to 0.84 in D-A_1-A_2-A_3. It is very unusual that the transfer efficiency in D$\rightarrow$$A_1$ or $A_1$$\rightarrow$$A_2$ is enhanced by further stacking a third acceptor layer. A fourth-layer stacking, however, appears to give only a small or negligible effect. This is not a surprising result, given the fact that the excitation transfer occurs between donor and acceptor in non-equilibrium states, as described in the previous section (Figure 6).

Great care was taken to avoid trivial effects like a direct transfer from D to A_2 layers. We checked it by using a reference sample consisting of D-(spacer)-A_2 with an arachidic acid monolayer as a spacer, in which A_2 is 50 Å apart from D, and found that the D-layer fluorescence no longer decreases in the presence of A_2. This means that the direct transfer D$\rightarrow$$A_2$ is negligible. Also we studied for a series of samples of D-nA_1 and D-nA_1-A_2 (n = 2,3 and 4), and found that there is no further decrease of the D fluorescence in the presence of A_2. The enhancement effect in the transfer efficiency of D-A_1 is seen only in D-A_1-A_2, not in D-nA_1-A_2.

The enhancement effect is seen also in the fluorescence decay curves. Figure 9

Table 1.
Efficiencies of the interlayer energy transfer in respective donor-acceptor layers

Samples	$\Phi_{ET}(D\text{-}A_1)$	$\Phi_{ET}(A_1\text{-}A_2)$	$\Phi_{ET}(A_2\text{-}A_3)$
D-A_1-A_2-A_3	0.62 ± 0.06	0.84 ± 0.01	0.38 ± 0.02
D-A_1-A_2	0.57 ± 0.08	0.76 ± 0.07	
D-A_1	0.40 ± 0.08		

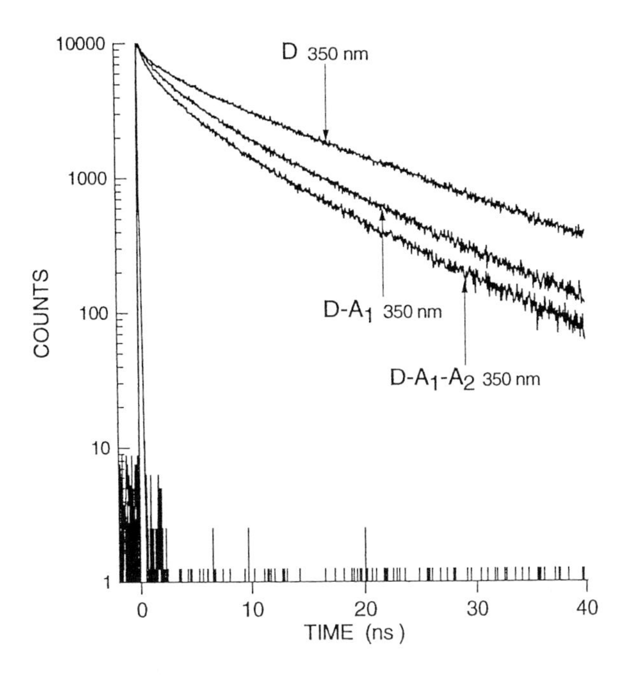

Figure 9. Fluorescence decay curves of D in D, D-A_1 and D-A_1-A_2, obtained with 2-ps laser excitation. The excitation wavelength is 295 nm, and the monitoring wavelength is 340 nm.

shows the fluorescence decay curves of D for D, D-A_1 and D-A_1-A_2. The decay of D becomes faster by stacking A_1 due to the usual transfer to A_1. It should be noted that this decay is furthermore affected by the presence of A_2, but further stacking of A_3 affects no longer the decay curve. Among a series of D-nA_1-A_2, the fluorescence decay curve is affected most in n=1, but negligible in n≥2. These changes in fluorescence decay are consistent with those in fluorescence yield mentioned above, and can be interpreted in terms of the non-equilibrium excitation transfer mechanism.

6. CONCLUDING REMARKS

The experimental results presented in Sections 3 and 4 are summarized as follows: (1) The fluorescence spectrum of single monolayer film depends slightly on excitation wavelength; and (2) the fluorescence spectrum of acceptor layer, A_1 in D-A_1 or A_2 in D-A_1-A_2, changes depending on excitation paths, i.e., a direct optical excitation of A_1 or an excitation by the energy transfer D→A_1. These behaviors can be accounted for only if the S_1 energy level of dyes is distributed among various sites in a monolayer film, and the energy transfer D→A_1 occurs from initially excited vibrational levels of D to higher energy sites of A_1 (Figure 6). This model is appropriate only for the molecular system in which molecules undergo the excitation transfer on a time scale faster compared to the time scale of vibrational relaxation in

S_1 of D. The rate of energy transfer between D and A_1 25 Å apart is estimated to be 5 ps. Note that the rate of vibrational relaxation in S_1 is recognized to occur within 10 ps in condensed media [10]. Then it is possible that the energy transfer before the vibrational relaxation (the *non-equilibrium* energy transfer) can contribute as a process competing with the vibrational relaxation.

Enhancement of the transfer efficiency in multilayers, which is presented in Section 5, can be interpreted also as due to a contribution of the non-equilibrium energy transfer process. The vibrational relaxation, i.e., the thermal equilibration process, should associate with some energy loss due to excitation migration and quenching at trap sites. This makes the efficiency of D-A_1 transfer lower significantly.

An important implication of this transfer scheme is directed toward the excitation transfer in a photosynthetic light-harvesting antenna. The ultrafast and highly effective transfer should involve the non-equilibrium energy transfer. Antenna pigment systems of red and blue-green algae contain an accesory pigment, *phycobilisomes*, which are attached on the surface of thylakoid membranes containing chlorophyll systems and reaction centers [4]. Phycobilisome consists of three types of chromoproteins, *phycobiliproteins*: phycoerythrin (PE), phycocyanin (PC) and allophycocyanin (APC) [11]. A long-distance (500-700 Å) energy transfer takes place through these phycobiliproteins and chlorophyll *a* (Chl *a*) toward the reaction center (RC). Recently we studied dynamical aspects of this sequential energy transfer (PE→PC→APC→Chl *a*) with a picosecond time-resolved fluorescence spectrometer [4]. It was found that the transfer rate of the process PC→APC depends on whether PC is excited by direct optical excitation or by the energy transfer PE→PC: the rate constant is larger in the energy transfer excitation by a factor of three than those of the direct photoexcitation of PC. This result suggests partial contribution of the non-equilibrium energy transfer to the sequential transfer PE→PC→APC.

REFERENCES

1. J. Klafter and J. M. Drake (eds.), Molecular Dynamics in Restricted Geometries, Wiley Interscience, New York, 1989.
2. H. Kuhn, D. Möbius and H. Bücher, in Techniques of Chemistry, A. Weissberger and B. W. Rossiter (eds.), Vol. 1, Part 3B, Wiley, New York, 1972, pp. 577-702.
3. Th. Förster, Z. Naturforsch., A, Astrophys., Phys. Phys. Chem., 4 (1949) 321.
4. I. Yamazaki, N. Tamai, T. Yamazaki, A. Murakami, M. Mimuro and Y. Fujita, J. Phys. Chem., 92, (1988) 5035.
5. I. Yamazaki, N. Tamai and T. Yamazaki, J. Phys. Chem., 94, (1990) 516.
6. N. Kimizuka, M. Tsukamoto and T. Kunitake, Chem. Lett., (1989) 909.
7. N. Tamai, T. Yamazaki and I. Yamazaki, Can. J. Phys., 68, (1990) 1013.
8. N. Tamai, H. Matsuo, T. Yamazaki and I. Yamazaki, J. Phys. Chem., 96 (1992) 6550.
9. N. Ichinose, Y. Nishimura and I. Yamazaki, Chem. Phys. Lett., 197 (1992) 364.
10. R. M. Hochstrasser and R. B. Weisman, in Radiationless Transitions, S. H. Lin (ed.), Academic Press, S. H. Lin (ed.), New York, 1980, pp. 317-362.

MICROCHEMISTRY
Spectroscopy and Chemistry in Small Domains
Edited by H. Masuhara et al.

Binary self-assembled monolayers: spectroscopy and application to liquid crystal alignment

Jean Y. Yang, Klemens Mathauer and Curtis W. Frank*

Department of Chemical Engineering, Stanford University
Stanford, California 94305-5025, U. S. A.

In this paper, the self-assembled monolayers (SAMs) of alkyltrichlorosilanes are employed as substrates for the study of liquid crystal (LC) alignment. By co-adsorption of 11-(2-naphthyl)-undecyltrichlorosilane with octadecyltrichlorosilane or decyltrichlorosilane, we were able to produce binary monolayers with surface chemistry spanning the range set by the homogeneous SAMs of each of the two functional groups used. By so doing, we successfully created a model system which can be used as a chemically tuned surface to examine the chemical interactions of a LC overlayer with the SAM surface. This system represents an improvement over the polymer films commonly used in the study of LC alignment in that it allows gradual changes in surface chemistry while maintaining a well-defined surface geometry. To demonstrate the feasibility of the alkyltrichlorosilane SAM as a model system for LC alignment studies, we examined the alignment of 4-(cyano)-4'-pentylbiphenyl (5CB) on the surface of two types of alkyl terminated SAMs and one naphthalene terminated SAM. We observed homeotropic alignment of 5CB on the two alkyl SAM surfaces and a tilted, near-planar alignment of 5CB on the naphthyl SAM surface. This result confirms the feasibility of using alkyltrichlorosilane SAMs for measuring the chemical interactions of a LC overlayer with a surface.

1. INTRODUCTION

There is significant experimental evidence that the nature of the adjacent surface can dramatically affect the orientation of liquid crystal (LC) molecules at the surface, which propagates into bulk LC alignment, especially in the case of nematic liquid crystals [1]. For example, a recent study describes the use of scanning tunneling microscopy (STM) to investigate LC orientation on single-crystal surfaces and shows surface-induced positional ordering of liquid crystals in not just one but two dimensions[2]. Second harmonic generation also shows that orientation of LC molecules at the surface to be dependent on surface conditions [3]. In spite of these results and many others, there is no universally accepted molecular-level explanation for the phenomenon.

1.1. Liquid crystal alignment

Liquid crystals were first discovered in 1888 [4], and the physics of the ordering process was intensely studied during the subsequent decades. Interest in low-molecular weight liquid crystals diminished somewhat with time until the late 1960's when their potential use in display devices was recognized.

When a drop of LC is placed on a solid substrate, the molecules at the interface will orient at a specific angle with respect to the surface, a phenomenon known as surface alignment or surface anchoring. This orientation propagates through the bulk LC, especially for nematic liquid crystals. It is generally agreed that the phenomenon of surface anchoring is presently not well-understood in spite of many studies [5,6], but its control is central to LC display technology.

Microscopic studies of the alignment phenomenon have taken two approaches to understand the process. The first concentrates on the chemical nature of the surface, since it determines the types of interfacial interactions [7]. These include van der Waals, hydrogen bonding and polar forces that are determined by the specific chemical moieties in the LC and that constitute the surface of the substrate. Such interactions will depend on the distance of separation as well as the relative orientation of the substrate on the molecular level. As a complement to the experimental work on chemically different surfaces, molecular modeling studies involving the addition of various intermolecular forces to the hard core repulsion have been used to predict the anchoring direction [8,9]. Unfortunately, the results have been conflicting and difficult to verify [6].

The second approach addresses the geometric or topological nature of the surface on a more macroscopic scale. For example, a continuum description of the nematic LC elastic free energy developed by Berreman attributes the anchoring direction to minimization of elastic free energy [10]. In a macroscopic approach to this problem, the form of the surface energy is assumed to be a function of the anchoring parameter and the order parameter of the LC. The anchoring energy can thus be parameterized, but attempts to understand its theoretical basis have been inconclusive [5].

Part of the mystery behind the mechanism of surface alignment stems from the fact that most of the experimental studies have concentrated on the rubbed polyimide surface, which is the predominant substrate used industrially [11,12]. Polyimides, as a class of polymers, have been extensively studied in recent years. Nevertheless, there still remains a lack of understanding of the details of the bulk molecular organization in these polymers and of the important interchain interactions that impact the molecular organization. Given the lack of understanding of the bulk, it is not surprising that the polyimide surfaces are very difficult to quantify in terms of their chemical and geometrical contributions to LC anchoring.

1.2. Self-assembled monolayers

Self-assembled monolayers (SAMs) represent a new class of well-defined surfaces that are especially suited to examine complex surface phenomena such as LC alignment. Generally, SAMs are formed through the thermodynamically-driven adsorption of a particular class of amphiphilic molecules on several types of solid substrates. The formation of covalently bound silane monolayers on surfaces with free OH groups was postulated as early as 1968 [13]. However, Sagiv demonstrated much later that long chain alkyltrichlorosilanes will form true monolayers of high degree of perfection in orientation and packing on hydroxylated surfaces [14]. The order in the monolayer stems partially from the van der Waals interactions between adjacent chains, which contribute a few kcal/mol of energy to the monolayer formation process [15]. The resulting monolayers are therefore highly organized and ideal model surfaces for the study of complex surface interactions. Thus, by molecular-level characterization of the SAM model surface, we may be able to predict LC alignment on more complex surfaces, such as the aromatic polyimides.

It is important to note the distinction between these alkylsilane SAMs and other silane modified surfaces that may be chemically defined but are topologically extremely ill-defined. Some silane modified surface have been used for the study of LC alignment [16,17] including several second harmonic generation studies [18,19]. More recently, Ichimura and co-workers employed azobenzene terminated alkoxysilane films to produce surfaces capable of radiation driven topological variations and studied their effects on liquid crystal alignment [20,21]. It is interesting to note that they found the alignment regulation to be markedly influenced by the hydrophobicity of the substituent groups on the azobenzene. However, these silane modified surfaces suffer from similar problems as polymer surfaces in that they lack the well-defined topology of SAMs.

For our study, we have chosen to use single-component and binary systems of alkyltrichlorosilane SAMs for our model surfaces for several reasons. First, of course, is the well-defined nature of the SAM. The densely packed structure of these films allows close control of surface topology, which is known to drastically affect LC alignment. The second reason is the chemical stability of such SAMs. Alkyltrichlorosilane SAMs are believed to be covalently bound to the substrate, resulting in more thermally and chemically stable films than those obtained by other monolayer methods such as Langmuir-Blodgett. Thus, monolayer molecules cannot desorb from the surface and migrate through the LC overlayer. Lastly, we have also chosen alkyltrichlorosilane SAMs over the other well-known SAM system, alkane thiol on gold, for availability of transparent substrates, which is necessary for examining LC alignment.

In this paper, we present data which show that the alkyltrichlorosilane SAM is an excellent new tool for the study of LC alignment by surfaces. We examined the structures of single component and binary SAMs as a chemically tuned surface with which to study chemical interaction contributions to LC alignment. Binary SAMs with naphthyl and alkyl termination were examined using UV, FTIR, and fluorescence spectroscopies as well as contact angle goniometry. Based on these measurements, we conclude that we can produce homogeneous binary SAMs containing these two components in all concentrations.

To determine the feasibility of the SAM model system for LC alignment studies, single component SAMs were used to study LC alignment with the liquid crystal, 5CB. We expected to find different LC anchoring directions for

SAMs of the two types of terminal groups, alkyl and naphthyl because the end groups dominate chemical interactions with the LC overlayer. We will discuss our results in terms of a critical surface tension model and assess the strengths and weaknesses of this model.

2. EXPERIMENTAL

2.1. Self-assembled monolayers

Substrates. UV grade fused silica slides were first rinsed with chloroform then treated with a mixture of concentrated H_2SO_4 and H_2O_2 (7/3 v/v) for 30 minutes at 90°C. After cooling to room temperature, the slides were rinsed with deionized water and dried with a stream of nitrogen.

Monolayer preparation. Freshly cleaned fused silica slides were immediately transferred to nitrogen atmosphere where they are immersed in a 2mM solution of alkyltrichlorosilanes in hexadecane for 20 hours, then cleaned according to the procedure outlined in a previous paper [22]. Octadecyltrichlorosilane (OTS) and decyltrichlorosilane (DTS) were purchased from Hüls America. 11-(2-naphthyl)-undecyltrichlorosilane (2-Np) was synthesized [22]. Single component monolayers were formed from OTS, DTS and 2-Np solutions. Co-adsorbed monolayers were formed from solutions of OTS and 2-Np as well as DTS and 2-Np.

UV-spectroscopy. UV-spectra of monolayers containing 2-Np were obtained with a Varian CARY3 UV-VIS spectrometer with a freshly cleaned fused silica slide as reference.

Transmission FTIR-spectroscopy. FTIR transmission spectra of monolayers on fused silica slides were recorded using a BIORAD Digilab FTS-60A single beam spectrometer. Spectra were recorded at $4cm^{-1}$ resolution and with 1024 scans. Again, a freshly cleaned fused silica slide was used to get the background spectrum.

Fluorescence spectroscopy. Steady-state fluorescence spectra were recorded on a SPEX Fluorolog 212 spectrometer equipped with a DM3000F data system. Corrected emission spectra were recorded in the front-face mode in atmospheric conditions with an excitation wavelength of 285 nm and slit width of 4mm.

Contact angle. Advancing contact angles were determined by a Rame-Hart Model 100 contact angle goniometer at room temperature. High purity water was used for these measurements.

2.2. Liquid crystal alignment

Liquid crystal cells. Monolayer covered slides were examined for LC anchoring by converting the substrates into LC sandwich cells. To produce monodomain LC cells, the SAM substrates were heated to 60°C, then rubbed with velvet in one direction using a translational stage and velvet covered rollers to ensure even and consistent rubbing. The substrates were measured for UV and IR absorbance before and after rubbing to ensure that the monolayers were not damaged or removed by the rubbing process. A single component LC consisting of 4-(cyano)-4'-pentylbiphenyl (5CB) purchased from Merck was used. The LC sandwich cell has SAM substrates on both sides and

a Teflon spacer of approximately 75 microns. The LC cell was heated to above the clearing point for 5CB (35.4°C), then cooled to room temperature to remove any alignment effects arising from the cell assembly process.
Polarized light microscopy. The LC cells were examined using a Nikon Optiphot polarized light microscope equipped with a Hamamatsu photo diode system. Domains in the LC were clearly visible with cross-polarized optics. The intensity of light was measured as a function of LC cell orientation with respect to the polarizers, and a sinusoidal change in intensity indicated birefringence in the plane of view.
Conoscopy. Conoscopy was conducted using the same microscope, but with a Bertrand lens in place.

3. RESULT

3.1. Characterization of binary self-assembled monolayers

The detailed structure of a binary SAM was examined by the co-adsorption of two pure components, alkyl and naphthyl, by using various concentrations of DTS or OTS and 2Np. As a new class of surfaces, the binary SAM structure is interesting in itself, and also important in its use for LC alignment studies, because homogeneous monolayers are necessary to accurately observe the relevant chemical contributions to LC alignment. Also, the ability to make binary SAMs in all concentrations is important to form a chemically tuned surface having specific characteristics. The following are data pertaining to the structure and composition of the binary SAM.
UV spectroscopy. UV-spectra were taken of coadsorbed monolayers over the whole range of composition for OTS/2-Np and DTS/2-Np, including single component monolayers. The absorption of the ^1A-^1B_b transition of naphthalene at 280 nm and the ^1A-^1B_b transition at 224 nm show that the composition of the monolayers is approximately equal to the solutions from which they coadsorb. UV absorption for monolayers of OTS and 2-Np increases as a linear function of naphthalene composition. DTS and 2-Np monolayers show similar behavior with only a slight positive deviation from linearity (see Figure 1). Previous work has shown that OTS and 2-Np adsorb at the same rate [22]; our study confirms that finding for DTS and 2-Np as well. The slight positive deviation for the DTS/2-Np monolayers reveals interesting information about the monolayer structure. We know that the transition dipole moment for the ^1A-^1B_b transition (280 nm) is parallel to the short axis of naphthalene [23,24], whereas the ^1A-^1B_b transition (224 nm) is parallel to the long axis. For transmitting light, the electric field lies parallel to the substrate surface, so the positive deviation suggests that the naphthalene chromophore has some degree of orientational freedom when mixed with the shorter alkyl chain (DTS). On the other hand, the orientation would be restricted when mixed with the longer OTS. The orientational freedom of naphthalene in the DTS/2-Np monolayers appears to be a function of naphthalene concentration. We conclude from this result that the alkyl and naphthyl components in binary SAMs are very homogeneously distributed.

Also interesting is the $^1B_b/^1L_a$ absorption ratio of the coadsorbed monolayers, which is another indicator of local orientation and orientational

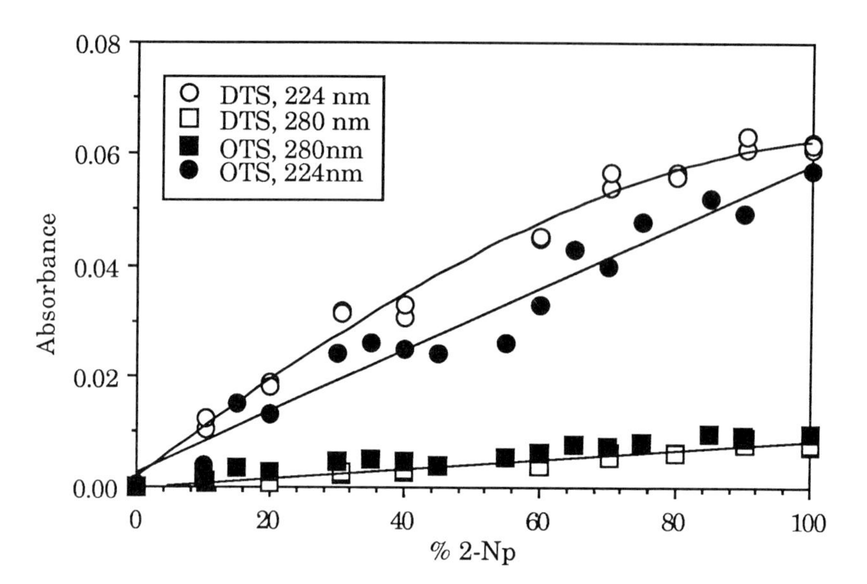

Figure 1. UV absorbance for DTS/2-Np and OTS/2-Np co-adsorbed monolayers showing differences in local orientational order for the naphthalene chromophore.

freedom for the naphthalene chromophore. We note that the $^1B_b/^1L_a$ absorption ratio is much higher in solution where there is random local orientation (see Figure 2). Therefore, the ^1B_b absorption, which has its transition dipole moment parallel to the long axis of the naphthalene, is severely restricted in the monolayer, suggesting that the naphthalene is oriented with its long axis nearly perpendicular to the surface, as we have previously found [22]. Also, the $^1B_b/^1L_a$ absorption ratio for the OTS/2-Np stays constant as a function of composition but changes for the DTS/2-Np monolayers. The $^1B_b/^1L_a$ absorption ratio increases with decrease in naphthalene concentration suggesting an increase in liquid-like behavior. This result is similar to the absorption behavior of successively adsorbed monolayers [25] and is more evidence confirming the homogeneity of the binary SAMs. Thus, in the individual UV absorption behavior and the $^1B_b/^1L_a$ absorption ratios for the naphthalene chromophore in binary SAMs, we find that the longer OTS will restrict the local orientation of the naphthalene chromophore, while the shorter DTS allows some freedom. The change in local orientational freedom in the DTS/2-Np monolayers as a function of naphthalene composition also indicates a homogeneous distribution of the two components in the co-adsorbed monolayers.

Transmission FTIR-spectroscopy. The IR absorbance measurements of the monolayers also show the same trend as the UV measurements. For monolayers of DTS and 2-Np, we find that the CH_2 asymmetric stretching

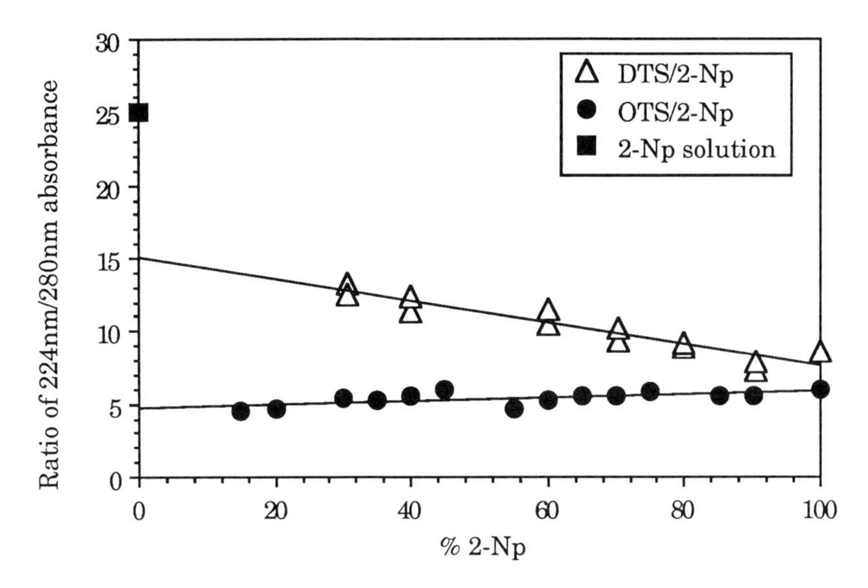

Figure 2. UV absorbance ratios for DTS/2-Np and OTS/2-Np co-adsorbed monolayers, also showing a difference in local orientational order for the naphthalene chromophore.

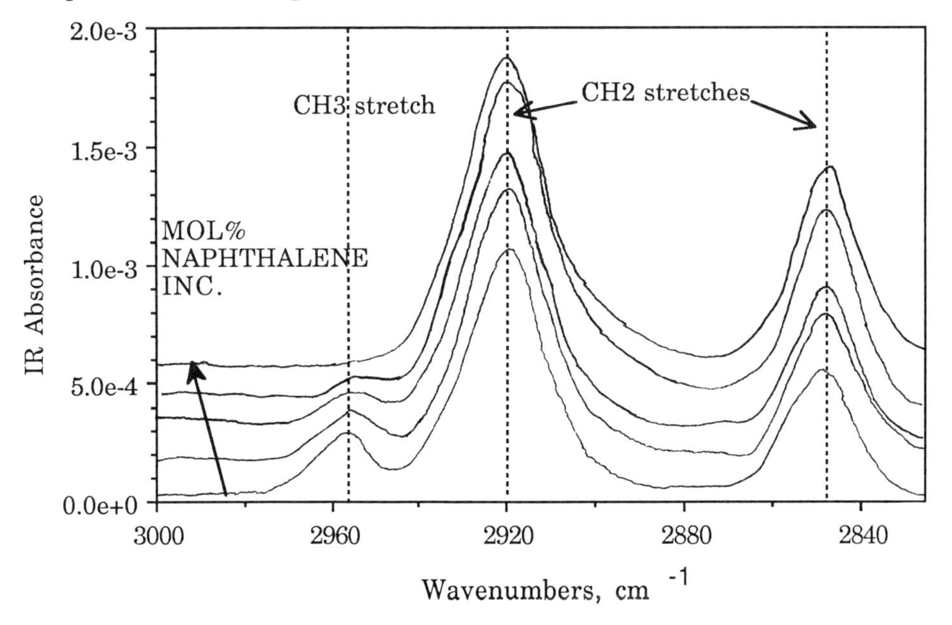

Figure 3. FTIR absorbance spectra for DTS/2-NP co-adsorbed monolayers showing a decrease in CH3 stretch peak as mol% of naphthalene increases.

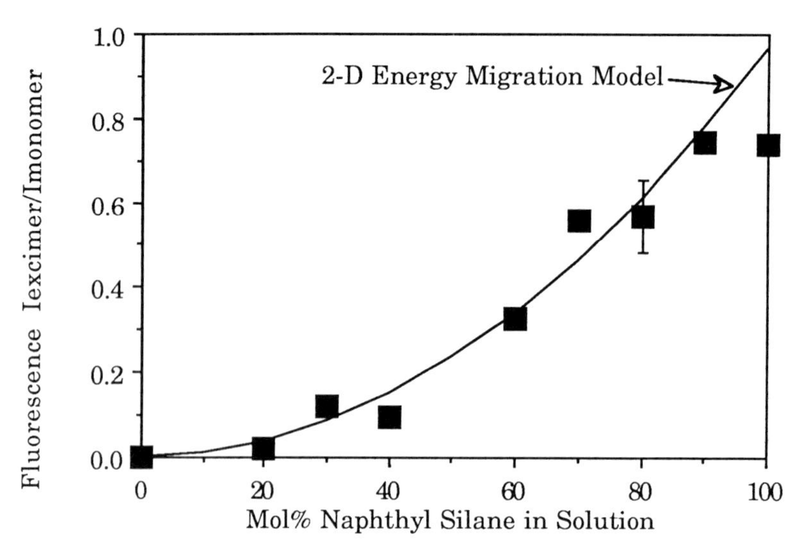

Figure 4. Fluorescence excimer to monomer peak height ratios for DTS/2-Np co-adsorbed monolayers

vibration at 2920 cm^{-1} increases in intensity with increase in 2-Np composition, as expected (see Figure 3). We also find the asymmetric CH_3 stretching vibration at 2958 cm^{-1} decreases in intensity as the 2-Np composition increases, indicating that the corresponding percentage of DTS is also decreasing. FTIR measurements thus show the alkyl portions of the monolayers to follow the compositions of the adsorption solutions and confirm the UV absorption results of monolayer composition.

Fluorescence spectroscopy. Whereas UV and IR spectroscopies provide information on the composition and orientation of the alkyl and aromatic components, fluorescence spectroscopy provides details on the distribution of the aromatic groups within the monolayer. Since the formation of fluorescent excimers requires adjacent chromophores in specific geometries, information about the morphology of these binary SAMs can be derived from the excimer-to-monomer ratio (I_d/I_m) of the fluorescence spectra. Each spectrum is deconvoluted into monomer and excimer peaks at 320 and 370 nm, respectively. It is worth noting that the excimer found here is the partial overlap high-energy excimer for naphthalene, which arises from the restricted geometry imposed by the monolayer [22]. The ratio of the excimer to the monomer peak height increases very slowly with composition until the naphthalene composition exceeds 50% (see Figure 4). Excimer formation is proportional to the local chromophore concentration, so this behavior is contrary to what is expected for a phase segregated system in which 2-Np is clustered into islands at low concentrations. Therefore, the fluorescence data indicate that the distribution of the naphthyl chromophores is homogeneous over the whole range of composition.

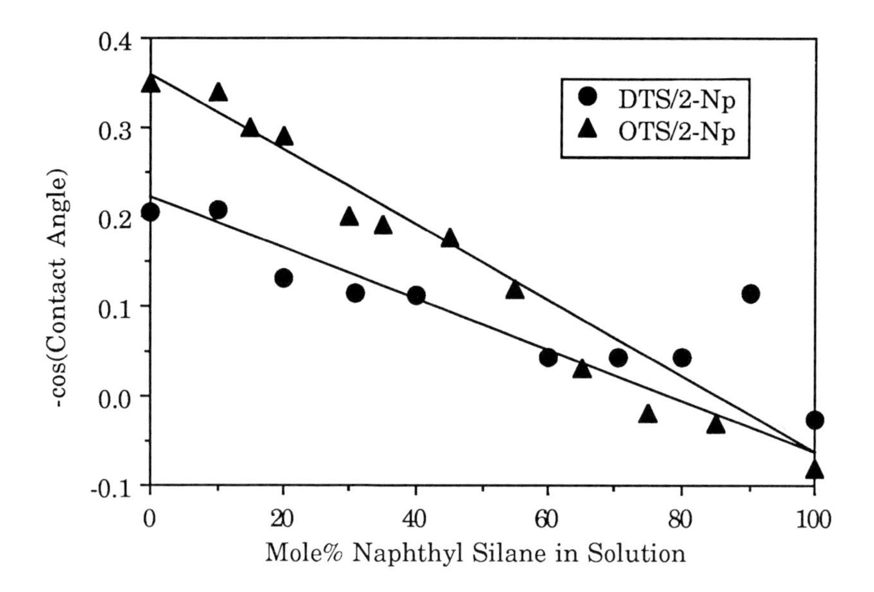

Figure 5. Contact angles of water on co-adsorbed monolayers showing surface energy change as a direct function of monolayer composition.

In our previous study, a two-dimensional lattice model taking energy migration into account was developed to compare with the fluorescence spectra in monolayer systems [22]. The model assumes molecularly random distribution of the chromophores in the monolayer. Since not all of the experimental constants are available, this model cannot be applied strictly to the data, but the dependence of fluorescence I_d/I_m on naphthalene concentration can be described by the model with theoretically acceptable constants. The explanation of the I_d/I_m ratio dependence to chromophore concentration offers quantitative support for the homogeneity of the binary SAM.

Contact angles. Lastly, contact angle measurement, which is an indication of surface energies, shows a gradual variation between the two pure components as the composition of the monolayer is varied (see Figure 5).

In summary, spectroscopic examination of the alkyl/naphthyl coadsorbed SAMs demonstrates that we can make homogeneous SAMs of intermediate composition and intermediate surface chemistry, thus allowing binary SAMs to be used as model surfaces in the examination of LC surface anchoring.

3.2. Liquid crystal anchoring on a single-component SAM surface

The liquid crystal 5CB, which has been widely studied, was used to verify the applicability of a SAM as a model surface for the study of LC alignment. Three types of single-component SAMs were made containing OTS, DTS and 2-Np.

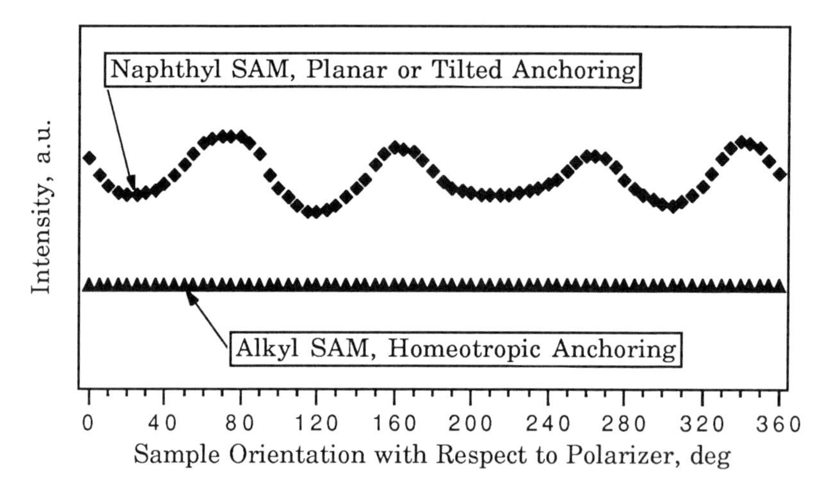

Figure 6. Light transmission intensity through LC cells through cross polarizers. Naphthyl SAM LC cells show birefringence in the plane as indicated by sinusoidal change in intensity, suggesting planar or tilted LC alignment. Alkyl SAM LC cells show no in-plane birefringence, indicating homeotropic LC alignment.

Polarized light microscopy. The LC anchoring on these three types of surfaces was examined using polarized light microscopy. Light transmission of the 5CB liquid crystal sandwiched between each of the three SAM surfaces wasmeasured as a function of sample rotation between fixed, crossed polarizers (see Figure 6). Since nematic liquid crystals are highly birefringent, the fluctuations in light transmittance for an oriented sample can be detected visually. Liquid crystal cells of the alkyl SAMs (OTS and DTS) behaved the same and are not birefringent in the plane of the surface, which indicates that the LC molecules are oriented perpendicular to the surface in homeotropic alignment(see Figure 6). This is called homeotropic alignment. On the other hand, the naphthyl SAM made from 2-Np caused the 5CB LC cell to exhibit sinusoidal light transmission, indicating birefringence in the plane of the surface. Thus, the 5CB must be anchored at an angle less than 90 degrees to the surface.

Conoscopy. Oriented nematic liquid crystals obey general uniaxial crystal optics, as described by Born and Wolf [26], thus allowing for the identification of molecular orientation based on the resulting interference pattern caused by the birefringence of the LC. Conoscopy, which is the examination of this interference pattern, was also performed to detect LC anchoring. The resulting conoscopic patterns confirmed the homeotropic anchoring on the alkyl SAM and the tilted, near planar anchoring on the naphthyl SAM (see Figure 7). This preliminary result demonstrates that we are able to produce different LC anchoring using SAMs of different functionalities.

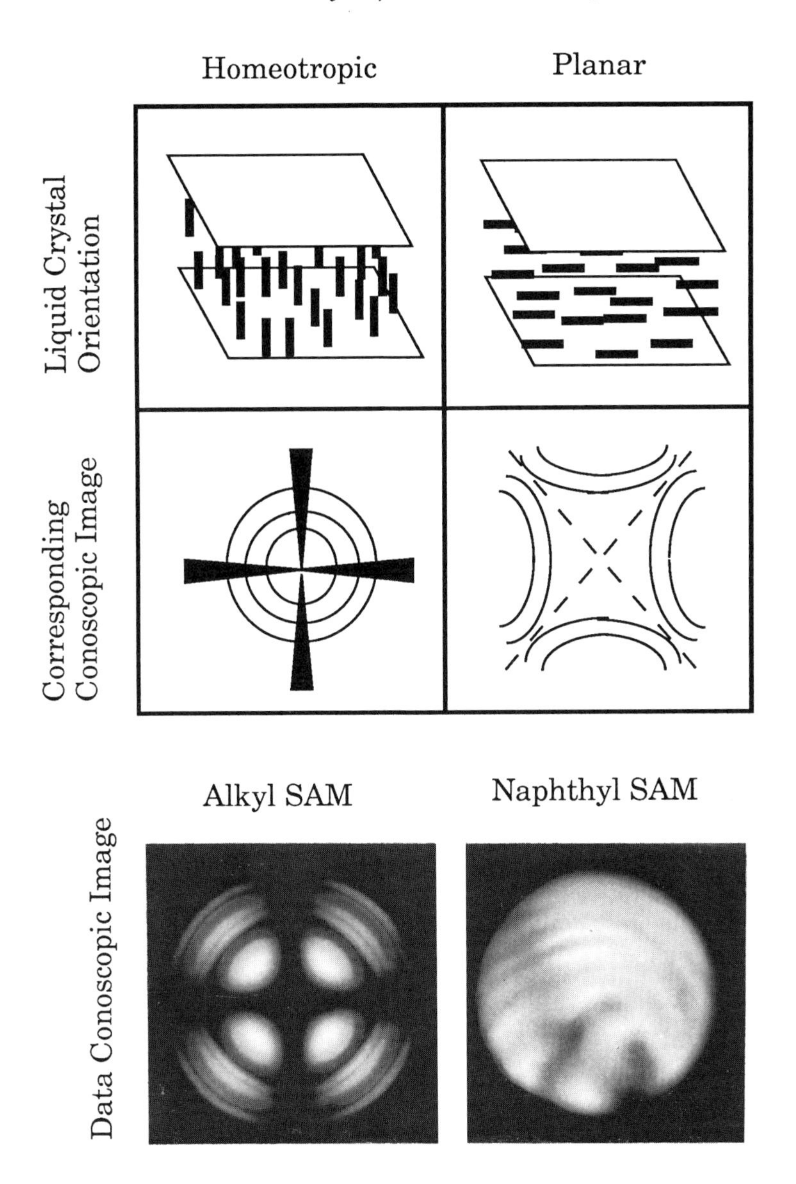

Figure 7. Comparison of conoscopic images from experimental liquid crystal cells with conoscopic images for liquid crystal cells with known alignment. Result shows that alkyl SAMs cause homeotropic alignment of 5CB while a naphthyl SAM causes a tilted, but near planar alignment.

4. DISCUSSION

The successful formation of coadsorbed SAMs in all compositions for naphthyl and alkyl silanes was shown by UV, FTIR, fluorescence and contact angle measurements. These coadsorbed monolayers provide well-controlled chemically tuned surfaces with which we can measure many surface-induced phenomena. We have shown one example of its use, where the surface alignment of 5CB is examined using SAMs of alkyl and naphthyl terminal groups. The resulting anchoring directions were different for the two types of terminal groups, while the two alkyl SAMs both show the same 5CB anchoring. This indicates that the terminal group strongly influences the orientation of 5CB on SAM surfaces

We interpret this preliminary result in terms of a simple critical surface tension model [27]. The model assumes that surface wetting and anchoring behaviors are correlated. The anchoring direction is thus attributed to the difference between the surface free energy, as measured by the critical surface tension, and the surface tension of the LC. This model accounts for only the chemical variations in the surface, but it lends itself to a system where the chemical characteristics of the aligning layer are changed in a systematic way.

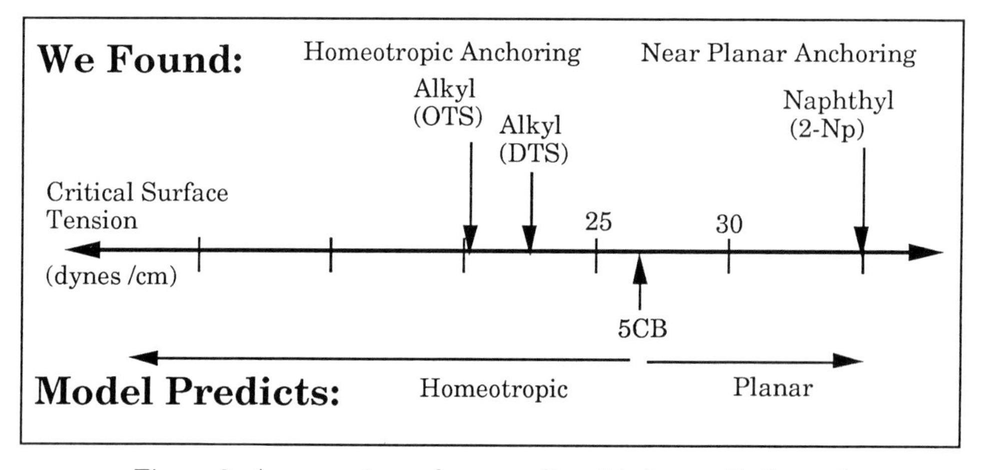

Figure 8. A comparison of our results with the predictions of the Cregh and Kmetz model for liquid crystal alignment.

As a simple, one-parameter treatment, the critical surface tension model is not able to account equally for the different types of interfacial forces, such as hydrogen bonding and van der Waals forces. Therefore, as Haller has shown [28], it will not be useful for comparing the anchoring behavior of liquid crystals with different chemical structure because of the different interfacial forces that can be involved. As a purely chemical model, it also cannot account for geometric variations on the molecular scale, such as those formed by monolayers of various packing densities or surface roughness [29]. However, these two variables can be controlled in the SAM system that we have described. By synthesis, we have created SAMs with individual functional groups, thus bypassing the question of multiple interfacial forces. Also, by the nature of SAM formation, we can create close-packed monolayer surfaces, which eliminates the problem of varying surface geometry.

We found that the alkyl SAM produces homeotropic 5CB anchoring, while the naphthyl SAM produces tilted, near planar anchoring. Compared to the expected critical surface tensions of these SAMs, our result agrees qualitatively with the predictions of the critical surface tension model (see Figure 8). A more in-depth study of the transition in anchoring between the two types of surfaces is required. The important point is that the binary SAM approach may allow the creation of tailored model surfaces, such that at least the chemical contributions to the LC alignment process may be better understood.

5. ACKNOWLEDGMENTS

This work was supported in part by the Office of Naval Research and in part by Raychem Corporation. J.Y.Y. acknowledges a graduate fellowship from the National Science Foundation.

REFERENCES

1. J. Cognard, Mole. Cryst. Liq. Cryst., 78, (1982) Supple.1.
2. M. Hara, et al., Nature, 344, (1990) 228.
3. Y.R. Shen, Nature, 337, (1989) 519.
4. A. Rapini, M. Papoular, J. Physiques Coll., 30, (1969) 54.
5. W.J.A. Goossens, Mole. Cryst. Liq. Cryst., 124, (1985) 305.
6. B. Jerome, Rep. Prog. Phys., 54, (1991) 391.
7. J.A. Castellano, Molecular Cystals and Liquid Crystals, 94, (1983) 33.
8. J.D. Parsons, J. de Phys., 37, (1976) 1187.
9. B. Tjipto-Margo, D.E. Sullivan, J. Chem. Phys., 88, (1988) 6620.
10. D. Berreman, Mole. Cryst. Liq. Cryst., 23, (1973) 215.
11. M. Murata, H. Awaji, M. Isurugi, M. Uekita, Y. Tawada, Jpn. J. Appl. Phys., Part 2, 31, (1992) L189.
12. S. Ishihara, H. Wakemoto, K. Nakazima, Y. Matsuo, Liq. Cryst., 4, (1989) 669.
13. L.-H. Lee, J. Col. Interf. Sci., 27, (1968) 751.
14. J. Sagiv, J. Am. Chem. Soc., 102, (1980) 92.
15. A. Ulman, An Introduction to Ultrathin Organic Films: from Langmuir-Blodgett to Self-Assembly, (Academic Press, Inc., San Diego, 1991).
16. K. Miyano, J. Chem. Phys., 71, (1979) 4108.
17. J.C. Tarczon, K. Miyano, J. Chem. Phys., 73, (1980) 1994.
18. C.S. Mulline, P. Guyot-Soinnest, Y.R. Shen, Phys. Rev. A. RC., 39, (1989) 3745.

19. M. Barmentlo, F.R. Hoekstra, H.P. Willard, R.W. Hollering, Phys. Rev. A. RC., 43, (1991) 5740.
20. K. Aoki, A. Hosoko, K. Ichimura, Langm., 8, (1992) 1007.
21. K. Ichimura, Y. Hayashi, Langm., 9, (1993) 857.
22. K.A. Mathauer, C.W. Frank, Langm., (in press), (1993) a.
23. J. Platt, J. Chem. Phys., 17, (1949) 484.
24. H.B. Klevens, J. Platt, J. Chem. Phys., 17, (1949) 470.
25. K.A. Mathauer, C.W. Frank, Langm., (in press), (1993) b.
26. M. Born, E. Wolf, Principles of Optics: Electromagnetic Theory of Propagation, Interference and Diffrection of Light, (Pergamon Press, Oxford, 1986).
27. L.T. Creagh, A.R. Kmetz, Mole. Cryst. Liq. Cryst., 24, (1973) 59.
28. I. Haller, Appl. Phys. Let., 24, (1974) 346.
29. K. Hiltrop, H. Stegemeyer, Mole. Cryst. Liq. Cryst., 49 (letters), (1978) 61.

MICROCHEMISTRY
Spectroscopy and Chemistry in Small Domains
Edited by H. Masuhara et al.

Conductivity and photoconductivity in nanosize conductors

M. Van der Auweraer[a], C. Catry[a], F.C. De Schryver[a], H. Bengs[b], O. Karthaus[b], L. Häussling[b], H.Ringsdorf[b]

[a]Chemistry Department K.U. Leuven
Celestijnenlaan 200F 3001 Leuven Belgium

[b]Institut Für Organische Chemie, Johannes Gutenberg Universität Mainz
J.J. Becherweg 21, 6500 Mainz B.R.D.

Spreading of a solution of monomers and oligomers of hexa-alkoxytriphenylenes, which form bulk discotic mesophases, on a Langmuir trough allows to obtain stable monolayers. By successive deposition of monolayers it is possible to obtain multilayers, with a thickness between 2 and 80 nm which are characterized by a columnar packing of the aromatic moieties. When the Langmuir-Blodgett films are deposited on interdigiting electrodes with a spacing of 100 μm, it is possible to obtain a dark current which is one order of magnitude larger for the oligomer compared to the monomer. The dark currents are proportional to the number of layers and depend in a superlinear way on the applied electric field. Upon illumination with ultraviolet light a photocurrent proportional to the applied field and the square of the incident light intensity can be observed. The action spectrum of the photocurrent corresponds to the absorption spectrum of the Langmuir Blodgett films. The photocurrent is enhanced in the presence of oxygen. As observed for the dark current the photocurrent is an order of magnitude larger for the oligomer compared to the monomer. A similar behaviour is observed for the emission spectra of the Langmuir Blodgett films which consist for the monomer of a structured band and resemble those of a dilute solution. For the oligomer they consist mainly of a bathochromic structureless band suggesting efficient excimer or dimer formation.

1. INTRODUCTION

In spite of the long standing interest the investigation of the conductive [1, 2] and photoconductive [3, 4] properties of Langmuir Blodgett films is still an expanding and sometimes even controversial field. While the conductivity [5] and photoconductivity [6] perpendicular to Langmuir Blodgett films is sometimes attributed to pinholes and photo ejection respectively the interpretation of the in plane conductivity [7-20] is less controversial. As the large conductivity and photoconductivity observed in those systems were attributed to a stacked arrangement and a small intermolecular distance leading to large carrier mobilities [21-27] it was attempted [14, 19, 20] to investigate the conductive and photoconductive properties of molecules which formed discotic bulk phases [28, 29]. Beside their tendency to adapt a stacked organization in crystalline phases or discotic mesophases [30, 31] or monolayers [32-34] hexaalkoxytriphenylenes are characterized by a low oxidation potential [35, 36] which would favour injection of charge carriers from an electrode by a thermal process [37, 38] or by the dissociation of an exciton [39, 40]. As the symmetric hexa-alkoxytriphenylenes do not form stable condensed Langmuir-Blodgett films [32, 41] a oligomeric hexa-alkoxytriphenylene T(EO)$_4$ forming stable condensed Langmuir

Blodgett films was synthesized. The conductive and photoconductive properties of these films were determined and compared to those of other Langmuir-Blodgett films containing polycyclic or heterocyclic compounds. The different photophysical behaviour of the oligomer compared to the monomer [20] was related to the different conductivity and photoconductivity.

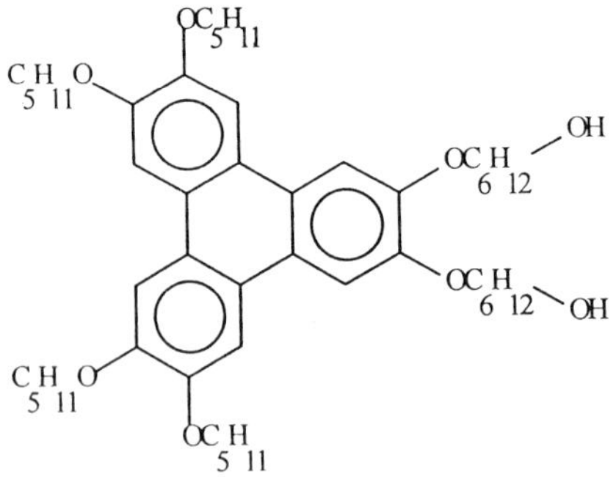

T(OH)$_2$ 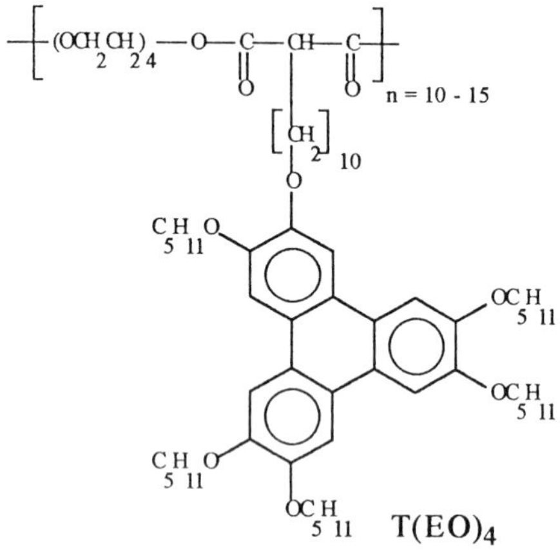 T(EO)$_4$

2. EXPERIMENTAL

The polymalonate T(EO)$_4$ was synthesized by condensation of the diethyl 10-(3,6,7,10,11-pentapentyloxy-2-triphenylenyloxy)-decyl malonate with 1,12-Bis-[2-(2-hydroxyethoxy)-ethyl]-ether using the procedure described by Karthaus [31. Its number average molecular weight determined by GPC amounted to 15.000 (standard: polystyrene, solvent $CHCl_3$). The DSC curve, recorded on a Perkin Elmer DSC-2 at a scan rate of 20 K/min, showed at -20 °C a transition from a glassy phase to a liquid crystalline phase followed at 20 °C and 35 °C by a phase transition to a second liquid crystalline phase and an isotropic phase respectively. In pressure area diagrams of a solution of oligomer T(EO)$_4$ in $CHCl_3$, determined using a KSV 5000 ALT trough on a subphase of pH 5.5, the onset of the compression can be observed at a molecular area of 100 $Å^2$/molecule. The pressure area diagram [20] is characteristic for a condensed monolayer. The collapse pressure amounted to 35 mN/m at a molecular area of 70$\pm$2 $Å^2$/molecule. The multilayers of oligomer T(EO)$_4$ were prepared at a surface pressure of 25 mNm^{-1} on a subphase of pH 5.5. The slides covered by the multilayers were stored at least 48 hours in the dark and in a dry atmosphere before the investigation of the conductive and photoconductive properties.

Absorption spectra of solutions and multilayer assemblies were obtained by a Perkin Elmer Lamda-5 spectrophotometer. Corrected emission and excitation spectra were obtained on a Spex Fluorolog. While for liquid solutions a rectangular geometry was used the spectra of multilayer assemblies were determined using front face excitation. To avoid quenching by O_2 and photooxidation the spectra of the multilayer assemblies were obtained at reduced pressure (1 torr) [42, 43]. The phosphorescence spectrum was determined by immersing a cylindrical quartz cell (ϕ 2 mm) in a quartz Dewar filled with liquid nitrogen.

On a quartz slide cleaned using the procedures described elsewhere [43] an array of interdigited gold or aluminium electrodes [20] was evaporated to determine the conductivity and photoconductivity in a gap arrangement. Stationary dark and photocurrents were determined using a set-up described elsewhere [20]. The sample compartment was equipped with a temperature control system which allowed to change the temperature from -15 to + 70 °C. Except when indicated otherwise the results of the photocurrents refer to data obtained at reduced pressure (0.1 torr) and using aluminium electrodes. The knowledge of the optical density, the photocurrent and the incident intensity allows to calculate the quantum yield of the photocurrent (Φ_{ph}) using following expression

$$\Phi_{ph} = \frac{1.238\text{x}10^3 i_{ph}}{I_o \lambda_{exc} (1 - 10^{-Abs}) \cdot A} \quad (1)$$

In equation 1 I_o, Abs, λ_{exc}, i_{ph} and A correspond to the incident light intensity (in Wm^{-2}), the absorbance of the sample at the excitation wavelength, the excitation wavelength (in nm), the observed photocurrent (in A) and the illuminated area of the Langmuir Blodgett film between the metal electrodes (m^2). In the photocurrent action spectra the ratio of the photocurrent to the incident light intensity was plotted versus the wavelength. The incident light intensity was determined using a power meter (IL700) with a PT171C detector from International Light.

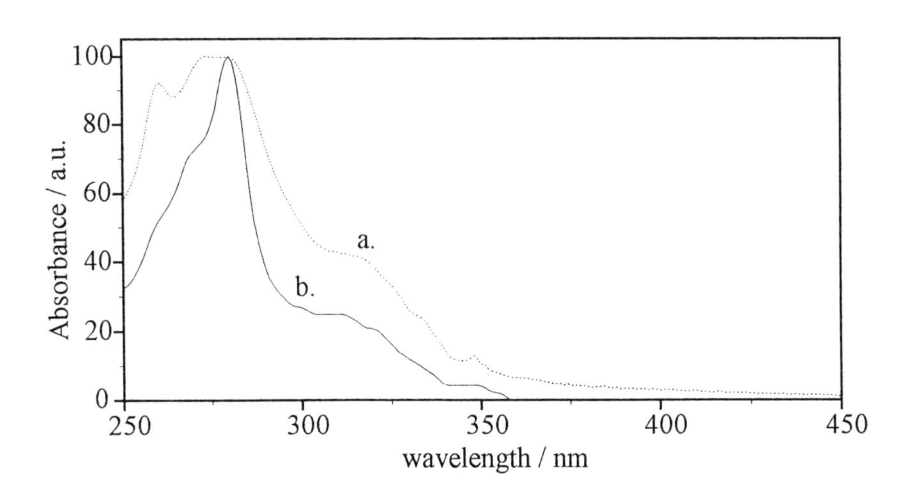

Figure 1: Absorption spectra of a Langmuir Blodgett film (15 layers) of $T(OH)_2$ (a) and of $T(OH)_2$ in chloroform (b). The spectra are normalized at the maximum.

3. RESULTS

3.1. Spectroscopic properties of $T(OH)_2$ and $T(EO)_4$

The absorption spectrum of a hydrophobic quartz slide covered on both sides by fifteen layers of $T(OH)_2$ is characterized by a maximum at 278 nm and a shoulder at 310 and 350 nm. For an area per molecule of 80±2 Å² the molar extinction coefficient of the hexa-alkoxytriphenylene moiety in the Langmuir Blodgett film amounts to

16700±2000, 49200±2000 and 119000±2000 $lmol^{-1}cm^{-1}$ at 350, 310 and 278 nm respectively (Figure 1). This resembles the absorption spectrum of a dilute solution of a hexa-alkoxytriphenylene in chloroform which is characterized by a main maximum at 278 nm and less intense maxima at 306 and 347 nm. The molar extinction coefficient at 280 nm amounts to 97000 $lmol^{-1}cm^{-1}$. The broadening of the maximum of the band at 278 nm is reproducible and is probably due to intermolecular interaction. The absorption spectrum of a multilayer of $T(EO)_4$ is within the experimental error identical to that of a multilayer of $T(OH)_2$.

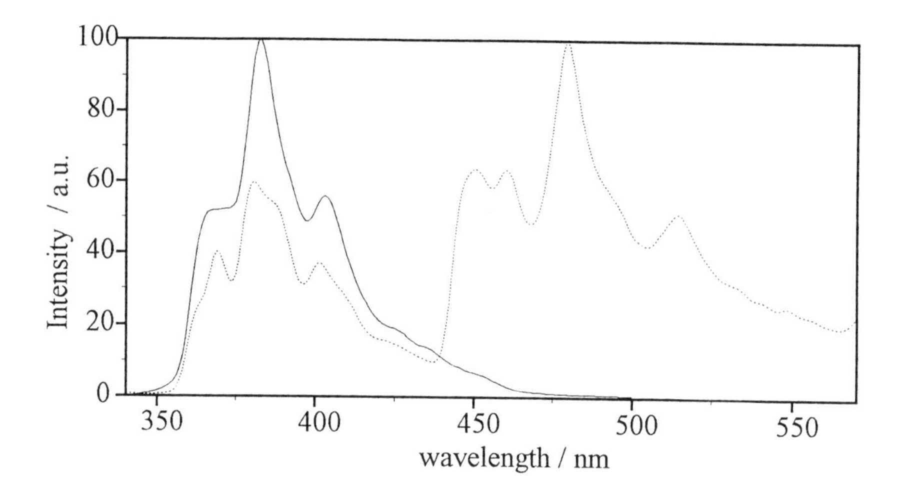

Figure 2: Emission spectrum of $T(OH)_2$, excitation occurred at 280 nm; ———: in chloroform at 298 K, ·······: in methylcyclohexane at 77 K.

In spite of the large oscillator strength the splitting (800 cm^{-1}) is much smaller than the hypsochromic shift observed for the ^1B_b-transition [42, 45] of anthracene which has a similar oscillator strength. The change of features the band at 278 nm is even less pronounced than observed for the ^1L_a-transition of anthracene [42, 45] which is characterized by an oscillator strength that is one order of magnitude smaller. The absorption spectrum, which is determined by the product of the concentration and the molar extinction coefficient of the different species present, suggests that most of the triphenylene chromophores are not involved in strongly interacting dimers or larger aggregates. In this case the absorption spectrum of the Langmuir Blodgett films would differ from that of a dilute solution as is e.g. observed for the ^1B_b-transition of anthracene or for cyanine dyes [42, 43, 45, 46].

The emission spectrum obtained upon excitation at 280 nm of a solution of $T(OH)_2$ in chloroform consists of a maximum at 384 nm and shoulders at 370 nm, 405 nm and 425 nm (Figure 2). Upon cooling a solution of $T(OH)_2$ in methylcyclohexane to 77 K a second emission band with a 0-0 transition at 450 nm and a maximum at 480 nm is observed. This band is attributed to phosphorescence of $T(OH)_2$. Upon excitation at 275 nm the fluorescence spectrum of fifteen layers of $T(OH)_2$ deposited on hydrophobic quartz is characterized by a structured spectrum with a maximum at 386 nm and shoulders at 370, 407 and 430 nm (Figure 3) and a weak and broad emission band with a maximum at 540 nm. The structured part of the emission spectrum resembles that of $T(OH)_2$ in chloroform or hexane. In the Langmuir Blodgett film and in chloroform or hexane the fluorescence excitation spectrum matches the absorption spectrum.

The broad band at 540 nm could be due to aggregate or excimer emission. In spite of the extensive energy migration occurring at this high local concentration of chromophores [47] most of the emission is still due to species of which the emission spectrum resembles that of isolated molecules. This is only possible if after excitation no rearrangement to an excimer geometry is possible and if the concentration of aggregates, characterized by a low lying relaxed singlet excited state, is very small. In this aspect the results obtained for the Langmuir Blodgett films differ slightly from those obtained for bulk liquid crystalline phase where the fine structure of the emission spectrum becomes blurred and the emission maximum is shifted to 393 nm.

The emission maximum of a multilayer of $T(EO)_4$ is characterized by a broad structureless band with maximum at 530 nm and a shoulder between 380 and 390 nm. If the broad band is due to an excimer or to a dimer then either the excimer formation is much more efficient for $T(EO)_4$ compared to $T(OH)_2$ or energy transfer to dimer sites is much faster in multilayers of $T(EO)_4$ compared to those of $T(OH)_2$. The latter could be due to a higher concentration of dimer sites or to faster hopping of the energy between two neighbouring hexaalkoxytriphenylene units. The smaller tendency of $T(OH)_2$ to give excimer or dimer emission could be correlated with the intracolumnar separation which amounts according to X-ray diffraction of multilayers to 4.6 Å [48].

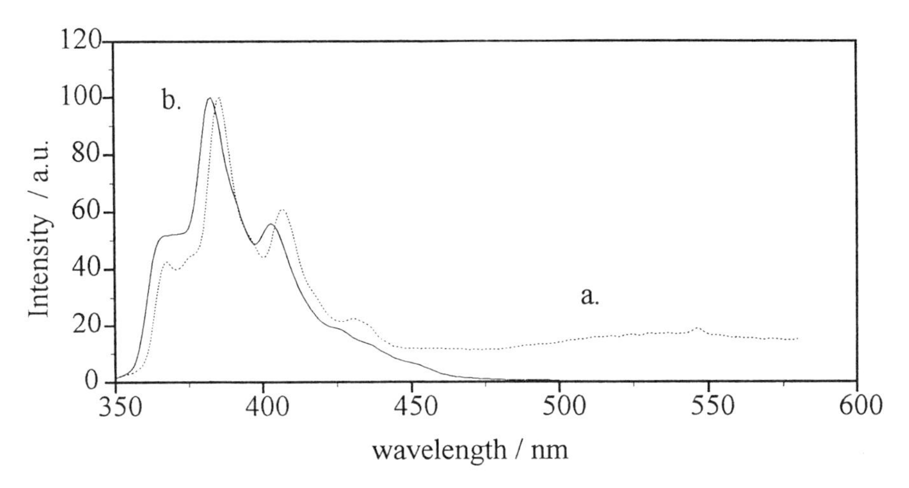

Figure 3: Emission spectra of an Langmuir Blodgett film (15 layers) of $T(OH)_2$ (a) and of $T(OH)_2$ in chloroform (b) at 298 K. The spectra are normalized at the maximum. Excitation occurred at 275 nm.

3.2. Dark conductivity of multilayers of $T(OH)_2$

At reduced pressure a dark current of $7.0 \pm 1.0 \times 10^{-13}$ A could be observed when a potential difference of 100 V (1.0×10^{4} V/cm) was applied on an interdigiting aluminium electrode covered by 20 layers of $T(OH)_2$ at room temperature. This corresponds to a conductivity of $1.2 \pm 0.2 \times 10^{-12}$ Scm^{-1}. For an applied field ranging from zero to 3.0×10^{4} V/cm the dark current increases apparently proportional to the applied field (Figure 4). In the presence of air (1 atm) the dark current increased by 20 %. When the applied electric field is increased to $5.0x10^4$ V/cm, which is only possible in the presence of air a superlinear increase of the dark current is observed. As shown in Table 1 the dark current is proportional to the number of layers. Separating the $T(OH)_2$ multilayer from the aluminium electrodes by ten layers of arachidic acid reduced the dark current at an applied field of $1.0x10^4$ V/cm to $1.2 \pm 1.0x10^{-13}$ A, which is close to the detection limit of the experimental set-up. For an aluminium electrode covered by 12 layers of $T(OH)_2$ the dark current is increased by factor of 80 (Figure 5) upon increasing the temperature from 23 °C to 80 °C. Using an Arrhenius relationship the temperature dependence of the dark current suggests an activation energy of 0.70 eV.

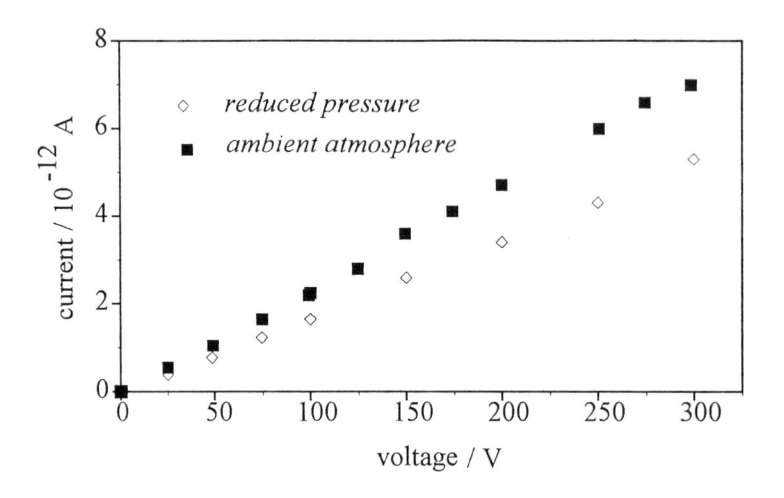

Figure 4: Dark current of an assembly of interdigiting aluminium electrodes covered by 40 layers of $T(OH)_2$; ◊: at reduced pressure, ■: at ambient atmosphere.

3.3. Photocurrents observed for $T(OH)_2$

Illumination of multilayers of $T(OH)_2$ deposited on an aluminium electrode with light with a wavelength below 350 nm results in a photocurrent. A decrease of the wavelength of the excited light leads to a strong increase of the photocurrent. In contrast to the absorption spectrum the action spectrum of the photocurrent is not characterized by a maximum at 270 nm but the photocurrent increases gradually when the wavelength of the exciting light becomes shorter. It is however characterized by a shoulder at 290-300 nm and 320 nm.

After the beginning of the excitation the photocurrent increases slowly until after five to ten minutes it levels off. This is much larger than the RC-time of our set-up amounting between one and ten seconds. The opposite effect is observed when the excitation is stopped. It was also observed that prolonged illumination induced, even in vacuum (10^{-1} torr), an irreversible increase of the dark and photoconductivity. Both at reduced pressure and at ambient atmosphere the photocurrent is proportional to the applied field. In contrast to the observations made for the dark current the photocurrent is increased by a factor of three at ambient atmosphere compared to the values obtained at reduced pressure. The photocurrent is also increased upon increasing the number of layers (Table 1). When the quantum yield of the photocurrent is calculated using equation 1 it is observed that the quantum yield is increased upon increasing the number of layers. In contrast to the observations made for the dark current insulating the $T(OH)_2$ multilayer from the electrode by ten layers of arachidic acid reduces the photocurrent only by 50 %.

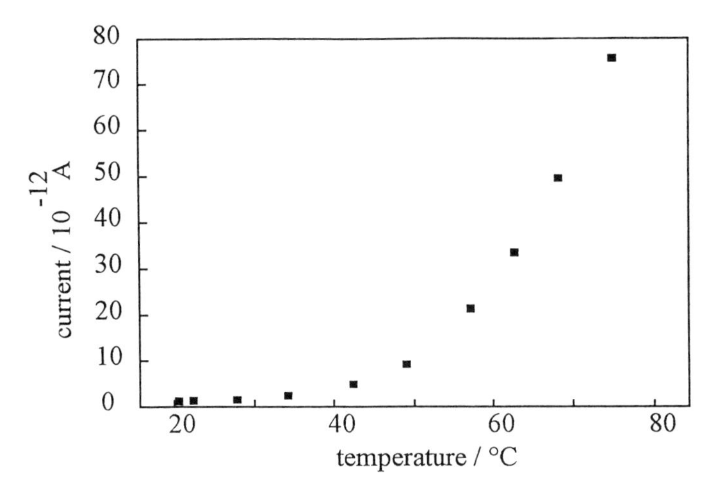

Figure 5: Influence of the temperature on the dark current observed for an aluminium electrode covered by 12 layers of $T(OH)_2$ at reduced pressure. The applied field equals $3.0x10^4$ V/cm.

Table 1
Influence of the number of layers on the dark and photocurrent of an aluminium electrode covered by a multilayer of $T(OH)_2$. The applied field equals $1.0x10^4$V/cm and the samples are excited at 350 nm.

Number of layers	Dark Current (A)	Photocurrent (A)	Quantum yield (Φ_{ph})
20	$7.0 \pm 1.0x10^{-13}$	$1.8 \pm 0.2x10^{-12a}$	$5.2 \pm 1.x10^{-7}$
20[b]	$1.2 \pm 1.0x10^{-13}$	$9.6 \pm 1.0x10^{-13a}$	$2.8 \pm 0.7x10^{-7}$
40	$1.8 \pm 0.2x10^{-12}$	$4.2 \pm 0.4x10^{-12c}$	$1.3 \pm 0.4x10^{-6}$

a) The incident light intensity amounts to $1.3 \pm 0.1x10^{-3}$ W/cm^2, b) The $T(OH)_2$ multilayer is separated from the electrode by 10 layers of arachidic acid, c) The incident light intensity amounts to $0.91 \pm 0.1x10^{-3}$ W/cm^2

Upon excitation a sample consisting of an aluminium electrodes covered by forty layers of $T(OH)_2$ at 350 nm and an applied potential of 300 V it is observed that, at incident light intensity exceeding $0.8x10^{-4}$ W/cm^2, the photocurrent is proportional to the square root of the light intensity (Figure 6). For lower light intensities the photocurrent is apparently proportional to the incident light intensity. A similar behaviour has been observed for the photocurrent perpendicular to mixed multilayers of cadmium arachidate and a merocyanine dye [12, 49].

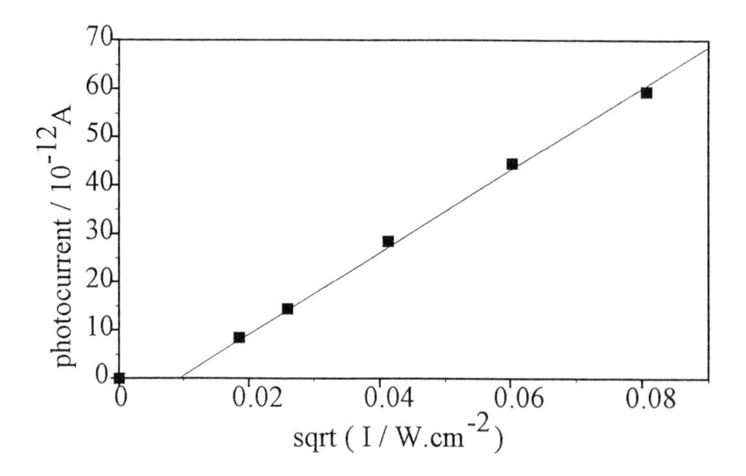

Figure 6: Influence of the incident light intensity on the photocurrent for an aluminium electrode covered by 40 layers of $T(OH)_2$ at reduced pressure. The applied field equals 3.0×10^4 V/cm. Excitation occurred at 350 nm.

Upon increasing the temperature the logarithm of the photocurrent increases proportional to 1/T between 240 K and 270 K, corresponding to an apparent activation energy of 0.41 ± 0.05 eV (Figure 7). At higher temperatures the photocurrent depends only to a minor extent on the temperature and between 270 K and 303 K the activation energy is apparently reduced to 0.04 ± 0.01 eV. While the low temperature activation energy corresponds to that reported for photocurrents perpendicular to mixed Langmuir Blodgett films of cadmium arachidate and a merocyanine dye [49] it is larger than that reported for mixed Langmuir Blodgett films of cadmium arachidate and cyanine and azo dyes [4]. On the other hand the high temperature activation energy is significantly lower than the values (0.5 eV) reported for the latter systems.

4. DISCUSSION

4.1 The dark current

The relationship between the dark current and the number of layers suggests that the dark current is due to the deposited layers and not to the substratum. This excludes that it is due to ions or water adsorbed at the quartz slide. The dark current is not decreased at ambient atmosphere compared to reduced pressure. This makes it improbable that electrons are the charge carriers. When the multilayer is isolated form the electrodes by 10 layers of arachidic acid the dark current is reduced by 80 %. Those arachidic acid layers contain acid protons and have a larger tendency than $T(OH)_2$ to bind water or cations (salt formation). Hence the reduction of the dark current suggests that the dark current is not due to water associated with the multilayer or to adsorbed ions. The reduction of the dark current suggests furthermore that the dark current is not due to holes generated in the multilayer by bandgap excitation or by the presence of impurities acting as electron acceptor. While the bandgap excitation is very improbable due to the large bandgap of aromatic molecules [37] the low oxidation potential [35] could make $T(OH)_2$ very sensitive to oxidizing impurities. The different data suggest that the dark current in $T(OH)_2$ is due to electrons injected from the electrodes, as already observed for $T(EO)_4$ [20]. This dark current could be limited by space charge effects or by the injection process itself. In the first case one would expect that the dark current, $i_{e,h}$, is given by equation 2 [37, 44].

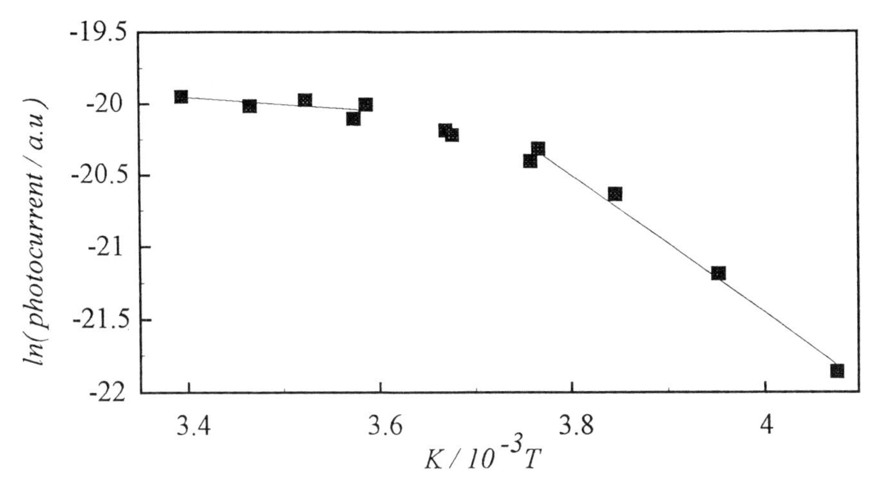

Figure 7: Influence of the temperature on photocurrent of an aluminium electrode covered by 12 layers of $T(OH)_2$ at reduced pressure. The applied field equals 3.0×10^4 V/cm.

$$i_{e,h} = 9A\mu_{e,h}\varepsilon_0\varepsilon_r E^2/8\ell \qquad (2)$$

where ε_0, ε_r and ℓ correspond to the permittivity of vacuum, the static dielectric constant of the sample and the width of the gap between two neighbouring electrodes respectively. In the presence of a large number of traps the mobility, μ, has to be multiplied by Θ, the fraction of electrons or holes present as free carriers in the conduction or valence band. On the basis of equation 2 a value of 3.9×10^{-6} cm^2/Vs (20 layers of $T(OH)_2$ at 100 V) is obtained for $\mu\Theta$ which is corresponds [50] to values observed for disordered molecular solids. It is however three orders magnitude smaller than the values observed by Adam [51] for bulk liquid crystal of hexaalkoxytriphenylenes. When on the other hand the dark current is limited by the rate of injection of charge carriers from the electrode the activation energy would be due [37] to the mismatch between the Fermi level of the metal and the valence band of $T(OH)_2$. While the Fermi level of aluminium is situated at 4.2 eV the valence band is situated [35, 38] at 5.0 eV which matches the observed activation energy of 0.7 eV.

The dark conductivity of $T(OH)_2$ is at room temperature seven times smaller than observed for a multilayer of $T(EO)_4$. Hence it is about one order of magnitude smaller than the dark current observed for undoped liquid crystalline bulk phases of hexa-alkoxy-triphenylenes [28, 29] although the latter results were obtained at an applied field less than 1500 Vcm^{-1}. Comparing the different values one should take into account that the results of Boden and Van Keulen [28, 29] were obtained using tindioxide and platinum black electrodes. The activation energy obtained corresponds to that obtained for hexahexoxytriphenylene using indium oxide electrodes [29]. It is considerably larger than that obtained for hexabutoxy- or hexapentoxytriphenylene.

Assuming a space charge limited current as well as an injection limited current the larger dark currents observed for $T(EO)_4$ could be related to a larger hole mobility. While this is evident for a space charge limited current (2) the hole mobility can also influence the injection limited current when one takes into account the trapping of the injected charge near the electrode due to the effect of an image charge [38, 53-55]. In this situation the photocurrent will be reduced by the competition between the recombination of the injected charge with the conduction band of the metal, which does not depend on the mobility and the escape from the potential well of the image charge which depends on the mobility and the applied field. The larger mobility would correspond to a larger intracolumnar interaction or a more regular packing in $T(EO)_4$. This is also suggested by the more important excimer emission in $T(EO)_4$ suggesting faster energy transfer (larger exciton mobility) to excimer forming sites or a larger intermolecular interaction favouring excimer formation. Also the smaller molecular area observed for $T(EO)_4$ in condensed monolayers would suggest a smaller intracolumnar separation.

Using aluminium electrodes the dark current density amounted to $1.2 \pm 0.3 \times 10^{-8}$ Acm^{-2} (20 layers of $T(OH)_2$) at an applied field of 1.0×10^4 Vcm^{-1}. Those current densities are one order of magnitude smaller than those obtained for multilayers of 9-butyl-10-anthrylpropanoic acid [7] in the plane of the Langmuir Blodgett film. The observed current densities are one to two orders of magnitude larger than observed normal to Langmuir Blodgett films of 9-butyl-10-anthrylpropanoic acid at a field strength of 1.2×10^6 Vcm^{-1}. The dark conductivity observed for $T(EO)_4$ is several orders of magnitude smaller than that observed for a mesoporphyrin diol [9-11] which is furthermore characterized by an ohmic dark current. In addition at 5.0×10^4 Vcm^{-1} the dark conductivity of the Langmuir Blodgett films of $T(OH)_2$ on gold electrodes is about three orders of magnitude smaller than that of Langmuir Blodgett films of t-butyl-phthalocyanine [52]. This could probably suggest that in the latter monolayer assemblies the dark current is due to charge carriers that are not injected from the electrodes but are present in the multilayers either due to an intrinsic process or to reducing or oxidizing molecules acting as donor or acceptor. As observed for $T(EO)_4$ doping the multilayers with electron acceptors as hexa-aza-triphenylene derivatives or 2,4,7-trinitrofluorene-9-one did not increase the dark conductivity.

4.2 Photocurrents

The photocurrent is increased by a factor of three in ambient atmosphere compared to reduced pressure. This suggests that, in analogy to the dark current, the photocurrent is not due to the photogeneration of mobile electrons. The dependence of the photocurrent on the number of layers or on the presence of air suggests furthermore that the photocurrent is not due to photoemission of electrons from the negatively charged electrode (photo-electric effect). At high light intensities the photocurrent is proportional to the square root of the light intensity. This suggests a combination of a monophotonic generation of holes with a second order recombination process. At low light intensities the photocurrent becomes proportional to the incident light intensity, this suggests [56-58] that either:

1) the generation of the photocurrent is a monophotonic process and the generated charge carriers, probably holes, disappear by a first order kinetics (alien or geminate recombination) [55]. In this case charge carriers are probably generated by dissociation of excitons at the electrodes or at electron or hole traps present in the Langmuir Blodgett film [39, 40, 54, 59-62].
2) the generation of the photocurrent is a biphotonic process and the charge carriers

disappear by a second order kinetics. In this case the photogeneration of the charge carriers is a bulk process [58-60, 63-65].

Combination of the results obtained at low and high light intensity suggest that

1) charge generation is a monophotonic process
2) at low light intensities the charge carriers disappear by a first order process (geminate or alien recombination)
3) a high light intensities charge carriers disappear by a second order process.

The observed light intensity dependence of the photocurrent could also be due to the occurrence of space charges. The efficiency for the generation of the photocurrent is one order of magnitude lower than that observed by Sugi normal to merocyanine [12, 13] multilayers (assuming an ohmic photocurrent at $5.0x10^6$ Vm^{-1}). It is however three orders of magnitude below that observed for photocurrents parallel to the Langmuir Blodgett film of merocyanines. In analogy to the results obtained for $T(OH)_2$ Sugi observed that both the parallel and normal photocurrent were proportional to the square root of the light intensity.

In the case of geminate or alien recombination with a first order rate constant k_1 the photocurrent quantum yield is given by

$$\Phi = \frac{g\mu E}{k_1 \ell} = \frac{g}{k_1 t_{tr}} \qquad (\mu Ee << \ell k_1) \qquad (3)$$

or

$$\Phi = g \qquad (\mu Ee >> \ell k_1) \qquad (4)$$

where g, μ, E and t_{tr} correspond to the quantum yield of charge carrier generation, the charge carrier mobility, the applied electric field and the width of the gap between two electrodes. In the case of non-geminate recombination with a second order rate constant k_2 the photocurrent quantum yield is given by

$$\Phi = \frac{g^{1/2}\mu E}{(I_{abs}k_2)^{1/2}\ell} \qquad (5)$$

Where I_{abs} correspond to the number of photons absorbed per second. A geminate or alien recombination can only lead to a stationary photocurrent when the generation of charge carriers occurs by photo-sensitized injection at the electrodes. The intensity dependence at low intensity as well as the reduction of the photocurrent when the $T(OH)_2$ film is isolated from the electrode by twenty layers of arachidic acid suggest this mechanism. If the generation would occur by exciton dissociation of the electrons at the electrodes only photons absorbed within an exciton diffusion length from the electrode can contribute to the photocurrent. As far as the exciton decay time and exciton diffusion coefficient in the Langmuir Blodgett films are similar to the values determined by Markovitsi [47] for the bulk liquid crystalline phase of the symmetric hexaalkoxytriphenylene this exciton diffusion length would amount to 0.2 μm. As the distance between the electrodes amounts to 100 μm only 0.45 % of the absorbed photons could generate a hole in this case. If the generation would be due to the dissociation of triplet excitons at the electrode the observed quantum yields should be multiplied by the ratio of the electrode spacing to the triplet exciton [67] diffusion length to know the efficiency with which triplet excitons inject holes into the Langmuir

Blodgett films. In the framework of this mechanism the difference between $T(OH)_2$ and $T(EO)_4$ can be related to a difference in exciton mobility, decreasing the exciton diffusion length [39, 40] or to a reduced hole mobility leading to a less efficient escape of injected holes from the image potential well near the interface [54-56]. In the framework of this mechanism the photocurrent quantum yield would be given by equation 4 rather than by equation 3 and the observed activation energy would correspond to the sum of activation energy necessary for the escape from the well of the image potential and the activation energy of the exciton mobility.

If the non geminate recombination is important both charge carriers of both polarities should be generated in the bulk of the monolayer assembly. This does not exclude that one type of charge carriers gets trapped very rapidly leading to the formation of recombination centres for carriers of the opposite sign. This mechanism is suggested by the intensity dependence of the photocurrent at high fields and by partial reduction of the photocurrent in the presence of ten layers of arachidic acid between the electrodes and the multilayer assembly of $T(OH)_2$. The quantum yield of the photocurrent is several orders of magnitude lower than the values observed for dispersion of organic molecules in polymers [59-63, 66], or for organic single crystals [64, 65, 67]. It is however of the same order of magnitude as the values observed for molecular crystals of dibenzothiophene. If the generation would occur in the bulk of the Langmuir Blodgett films it is possible that the reduced dimensionality of the system induces a more efficient geminate recombination [68, 69]. Although in this case the difference between $T(OH)_2$ and $T(EO)_4$ [30] could still be related to a difference in mobility it could also be related to a difference of k_2.

The increase of dark and photoconductivity after prolonged illumination has also been observed for polyvinylpyrene and polyvinylcarbazole were it was attributed to acceptors created by photo-oxidation [94]. Even at 0.1 torr oxygen molecules adsorbed [72] to the sample will probably still be able to oxidize excited triphenylene moieties. From the ground state oxidation potential, amounting to 0.5 V [35], and the excitation energy (3.38 eV) an excited state oxidation potential [73] of -2.88 V can be obtained for a hexa-alkoxytriphenylene. To obtain more information on the mechanism of the photogeneration and transport of the charge carriers and to explain field dependence of the photocurrent it would be necessary to determine the transient photocurrents using pulsed excitation.

ACKNOWLEDGMENTS

M.V.d.A. is a "Onderzoeksleider" of the "Fonds voor Kollectief Fundamenteel Onderzoek" (F.K.F.O.). C.C. is a "Aspirant" of the "Nationaal Fonds voor Wetenschappelijk Onderzoek". The authors thank the Belgian Ministry of Science Programmation through IUAP 3-40 and IUAP 2-16, the F.K.F.O. and the E.E.C (contract SCC CT90-0022 TSTS) for financial support. The authors also want to thank M. Yokoyama, Osaka University for providing a model for the mask used to evaporate the electrodes.

REFERENCES

1. B.Mann and H.Kuhn, J. Appl. Phys., 42 (1971) 4398.
2. M.Sugi, K.Nembach, D.Möbius, H.Kuhn, Solid State. Communications, 13 (1973) 603.
3. H.Kuhn, Pure & Appl.Chem., 51 (1979) 341.
4. E.E.Polymeropuolos, D.Möbius, H.Kuhn, J. Chem. Phys., 68 (1978) 3918.

5. N.Gemma, K.Mizushima, A.Miura, M.Azuma, Synthetic Metals, 18 (1987) 809.
6. T.Nakayama, K.Mizushima, S.Egusa, M.Azuma, Synth. Metals, 18 (1987) 803.
7. G.G.Roberts, T.M.McGinnity, W.A.Barlow, P.S.Vincett, Thin Solid Films, 68 (1980) 223.
8. R.Jones, R.H.Tredgold, A.Hoorfar, P.Hodge, Thin Solid Films, 113 (1984) 115.
9. D.Jones, R.H.Tredgold, J.E.O'Mullane, Photochemistry and Photobiology, 32 (1980) 223.
10. R.Jones, R.H.Tredgold, A.Hoorfar, Thin Solid Films, 123 (1985) 307.
11. R.H.Tredgold, S.D.Evans, P.Hodge and A.Hoorfar, Thin Solid Films, 160 (1988) 99.
12. M.Sugi, H.Iizima, Thin Solid films, 68 (1980) 199.
13. M.Sugi, M.Saito, T.Fukui, S.Iizima, Thin Solid Films, 99 (1983) 17.
14. T.Sauer, T.Arndt, D.N.Batchelder, A.A.Kalachev, G.Wegner, Thin Solid Films, 197 (1990) 357.
15. M.Vandevyver, J.Richard, A.Barraud, A.Ruaudel-Teixier, M.Lequan, R.M.Lequan, J. Chem. Phys., 87 (1987) 6754.
16. Y.Nishikata, M.Kakimoto, Y.Imai, J. Chem. Soc. Chem. Com., (1988) 1040.
17. M.Vandevyver, M.Roulliay, J.P.Bourgoin, A.Barraud, V.Gionis, V.C.Kakoussis, G.A.Mousedis, J.P.Morand, O.Noel, J. Phys. Chem., 95 (1991) 246.
18. V.I.Troitsky, T.S.Berzina, P.S.Sotnikov, T.V.Ujinova, O.Y.Neiland, Thin Solid Films, 187 (1990) 337.
19. M.Van der Auweraer, C.Catry, L.Feng Chi, O.Karthaus, W.Knoll, H.Ringsdorf, M.Sawodny, C.Urban, Thin Solid Films 210/211 (1992) 39.
20. C.Catry, M.Van der Auweraer, F.C.De Schryver, H.Bengs, L.Haüssling, O.Karthaus, H.Ringsdorf, Makromolekulare Chemie, in press.
21. A.Ruaudel-Teixier, M.Vandevyver, A.Barraud, Mol. Cryst. Liq. Cryst., 120 (1985) 319.
22. A.Bran, J.P.Frages, Physics Letters, 41A (1972) 179.
23. H.Inokuchi, K.Imaeda, T.Enoki, T.Mori, Y.Maruyama, G.Saito, N.Okada, H.Yamochi, K.Seki, Y.Higuchi, N.Yasuoka, Nature, 329 (1987) 39.
24. F.Wudl, G.M.Schmidt, E.J.Hufnagel, J. Chem. Soc. Chem. Commun., (1970) 1453.
25. P.Calas, J.M.Fabre, M.K.El-Saleh, A.Mas, E.Torreilles, L.Giral, C.R. Acad. Sci., C 281 (1975) 1037.
26. G.S.Bajwa, K.D.Berlin, H.A.Pohl, J. Org. Chem., 41 (1976) 145.
27. V.Enkelmann, Angew. Chemie, 103, 1142 (1991); Angew. Chem. Int. Ed. Engl., 30 (1990) 1121.
28. N.Boden, R.J.Bushby, J.Clements, M.V.Jesudason, P.F.Knowles, G.Williams, Chem. Phys. Lett., 152 (1988) 94.
29. J.Van Keulen, T.W.Warmerdam, R.J.M.Nolte, W.Drenth, Rec. Trav. Chim. Pays Bas, 106 (1987) 534.
30. I.G.Voigt-Martin, H.Durst, V.Berezinski, H.Krug, W.Kreuder, H.Ringsdorf, Angew. Chem., 101 (1989) 332; Angew. Chem. Int. Ed. Engl., 28 (1989) 323.
31. O.Karthaus, H.Ringsdorf, M.Ebert, J.H.Wendorff, Makromol. Chemie., 193 (1992) 507.
32. O.Karthaus, H.Ringsdorf, C.Urban, Macromol. Chem., Macromol. Symp., 46 (1991) 347.
33. A.Laschewsky, Adv. Mater., 1 (1989) 392.
34. E.Orthmann, G.Wegner, Angew. Chem., 98 (1986) 1114; Angew. Chem. Int. Ed. Engl., 25 (1986) 1105.
35. K.Bechgaard, V.D.Parker, J. Am. Chem. Soc., 94 (1972) 4749.
36. D.Markovitisi, H.Bengs, H.Ringsdorf, J. Chem. Soc. Faraday Trans., 88 (1992) 1275.
37. E.A.Silinsh, in "Organic Molecular Crystals", Springer Series in Solid-State

Science, Springer Verlag, Berlin Heidelberg New York 1980.
38. F.Willig in "Advances in Electrochemistry and Electrochemical Engineering", Eds. H.Gerischer, C.W.Tobias, Wiley-Interscience, New York, 12 (1982) 1-106.
39. G.Vaubel, H.Bässler, D.Möbius, Chem. Phys. Lett., 10 (1971) 334
40. V.M.Agranovich, M.D.Galanin in "Electronic Excitation Energy Transfer in Condensed Matter", North Holland Publishing Company, Amsterdam 1982.
41. O.Albrecht, W.Cumming, W.Kreuder, A.Laschewsky, H.Ringsdorf, Colloid & Polymer Science, 264 (1986) 659.
42. G.Biesmans, G.Verbeek, B.Verschuere, M.Van der Auweraer, F.C.De Schryver, Thin Solid Films, 168 (1989) 127-142.
43. G.Biesmans, M.Van der Auweraer, F.C.De Schryver, Langmuir, 6 (1990) 277
44. F.J.Dolezak, Photoconductivity and Related Phenomena, Mort J. & Pai D.M., Editors 1976 (Elsevier Sci. Publ. Co., New York), p27
45. W.S.Durfee, W.Storck, F.Willig and M.von Frieling, J. Am. Chem. Soc., 109 (1987) 1297
46. H.Kuhn, D.Möbius and N.Bücher, in "Physical Methods in Chemistry" Vol. I, Part 3B, A.Weissberger en B.Rossiter eds., Wiley, New York, (1972) 577
47. D.Markovitsi, I.Lécuyer, P.Lianos, J.Malthête, J. Chem. Soc. Far. Trans., 87 (1991) 1785
48. M.Vandevyver, Mol. Cryst. Liq. Cryst., in press (1993).
49. M.Sugi, K.Nembach, D.Möbius, Thin Solid Films, 27 (1975) 205.
50. M.Van der Auweraer, F.C.De Schryver, P.M.Borsenberger, H.Bässler, Adv. Mat., in press
51. D.Adam, F.Closs, D.Funhoff, D.Haarer, H.Ringsdorf, P.Schuhmacher, K.Siemensmeyer, Phys. Rev. Lett., 70 (1993) 457.
52. S.Baker, M.C.Petty, G.G.Roberts, M.V.Twigg, Thin Solid Films, 99 (1983) 53.
53. K.-P.Charlé en F.Willig, Chem. Phys. Lett., 57 (1978) 253
54. B.Korsch, F.Willig, H.J.Gachr, B.Teschke, Phys. Stat. Solidi (a), 33 (1976) 461.
55. F.Willig, Chem. Phys. Lett., 40 (1976) 331
56. H.Meier in "Organic Semiconductors", Monographs in Modern Chemistry, Vol 2, Ed. H.E.Ebel, Verlag Chemie, Weinheim 1974.
57. J.N.Murrell, Quart Rev., (1959) 37
58. T.E.Orlowski, H.Scher, Phys. Rev. B, 27 (1983) 7691.
59. P.M.Borsenberger, L.E.Contois, D.C.Hoesterey, J. Chem. Phys., 68 (1978) 637
60. P.M.Borsenberger, A.I.Ateya, J. Appl. Phys., 49 (1978) 4035
61. P.J.Regensburger, Photochem. Photobiol., 8 (1968) 429.
62. J.Mort, G.Pfister, Polym. Plast. Technol. Eng., 12 (1979) 89.
63. G.Pfister, D.Williams, J. Chem. Phys., 61 (1974) 2416
64. N.Karl, "Organic Semiconductors": in "Festkörperprobleme XIV", Vieweg, Braunschweig, (1974) 261.]
65. L.E.Lyons, K.A.Milne, J. Chem. Phys., 65 (1976) 1474.
66. J.Mort, S.Grammatica, D.J.Sandman, A.Troup, J. Electronic Materials, 9 (1980) 411.
67. G.Klein, R.Voltz, Int. J. Radiat. Chem., 7 (1975) 155.
68. K.M.Hong, J.Noolandi, J. Chem. Phys., 69 (1978) 5026.
69. J.Noolandi, K.M.Hong, J. Chem. Phys., 70 (1979) 3230.
70. D.Markovitsi, F.Rigaut, M.Mouallem, J.Malthête, Chem. Phys. Lett., 135 (1987) 236.
71. A.Itaya, K.-I.Okamoto, S.Kusabayashi, Bull. Chem. Soc. Japan, 52 (1979) 2218.
72. J.J.André, J.Simon, G.Guillaud, B.Boudjema, M.Maitrot, Mol. Cryst. Liq. Cryst., 121 (1985) 277.
73. H.Gerischer, F.Willig, Topics in Current Chemistry, 61 (1973) 31.

MICROCHEMISTRY
Spectroscopy and Chemistry in Small Domains
Edited by H. Masuhara et al.

Organization and spectroscopy of dyes on submicron-sized crystalline solids

R.A. Schoonheydt

Centrum voor Oppervlaktechemie en Katalyse, Materials Research Center, K.U.Leuven, K. Mercierlaan, 92, 3001 Heverlee, Belgium

Cationic dyes, organic as well as inorganic, are adsorbed on layered crystalline solids, such as clay minerals and $Zr(HPO_4)_2.H_2O$, via an ion exchange process. The organization of the dye molecules on the bidimensional surface of clay minerals is characterized by a strong tendency to agglomerate. The spectroscopic properties of the adsorbed dyes reflect the heterogeneous nature of the surface. Clay-dye materials can be prepared with well-defined organization of the clay particles by adsorption of alkylammonium cations or polymers. In these materials, the photochromism of spyropyrans and of viologens, and the photochemical hole burning of dihydroxyanthraquinone have been reported. Clay-dye and $Zr(HPO_4)_2$-dye complexes can also be prepared by direct synthesis. In the case of $Zr(HPO_4)_2$ the dye-modified phosphonates give extremely well ordered assemblies, the properties of which are under investigation.

1. INTRODUCTION

The search for advanced materials is still open. Chemists are particularly involved in this field with submicron- or nano-scale chemistry. Molecules or chemical groups with particular properties are organized into supramolecular assemblies, which show the desired properties of optical, magnetic, electronic, photochemical or electrochemical nature. The ultimate goal is to fabricate devices for sensing, for information storage, for electron transport and for photonics.

One particular way of organization of molecules is to make use of microporous materials as host matrices. The organization of the molecules in supramolecular assemblies is then regulated by the size and geometry of the porous systems. Two types of microporous materials are especially suited for that purpose: zeolites and layered inorganic solids, including clay minerals. The former have a three-dimensional structural network with pores of molecular size in one, two or three directions. Layered inorganic materials have two-dimensional interlayer spaces to accomodate molecules and to arrange them in a two-dimensional supramolecular assembly by a process, generally called intercalation (Figure 1).

Nanochemistry in the zeolitic pore system has been the subject of several recent reviews [1-7]. The field is less developed in the case of layered, crystalline solids (LCS), despite the existence of several types of LCS capable of intercalation (Table 1) [7-9]. In this review attention will be given to two types of LCS, smectites or swelling clays and $Zr(HPO4)_2$ and its organic derivatives.

Table 1:
Examples of Layered Crystalline Solids.

Clay Minerals	Structures
- β-alumina:	$(1+x)Na_2O$ $0.11Al_2O_3$
- Dichalcogenides:	TaS_2, $TaSe_2$...
- Phosphortrichalcogenides:	$FePS_3$, $FePSe_3$...
- Oxide bronzes:	V_2O_5, MoO_3 ...
- Halides and oxyhalides:	$CdCl_2$, $CrCl_3$, VOCl, LiOCl...
- Phosphates and arsenates o tetravalent metals:	$Zr(HPO_4)_2.H_2O$
	$Ti(HPO_4)_2.H_2O$
	$Zn(HAsO_4)_2.H_2O$

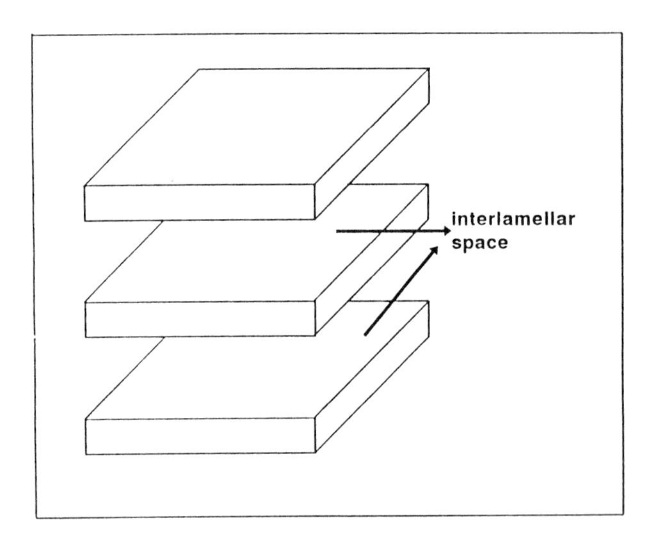

Figure 1. Schematic drawing of a layered crystalline solid, showing the interlamellar space.

1.1 Smectites

Smectites are 2:1 or TOT clays, consisting of an octahedral $MgO_4(OH)_2$ or $AlO_4(OH)_2$ layer sandwiched between two SiO_4 tetrahedral layers. In the case of Al two out of three octahedral positions are occupied and the smectite is called dioctahedral; in the case of Mg all octahedral positions are occupied and the clay is called trioctahedral. The ideal structures are electrically neutral, but isomorphous substitution in octahedral or in the tetrahedral layer imposes a negative charge on the layers, which is neutralized by exchangeable cations in the interlamellar space. Typical smectites are given in Table 2.

Table 2
Typical smectites

Name	Structural formula
Montmorillonite	$(Si_4)^{IV}(Al_{2-x}Mg_x)^{VI}(OH)_2O_{10}xM^+ \; nH_2O$
Beiddelite	$(Si_{4-x}Al_x)^{IV}(Al_2)^{VI}(OH)_2O_{10}xM^+ \; nH_2O$
Hectorite	$(Si_4)^{IV}(Mg_{3-x}Li_x)^{VI}(OH)_2O_{10}xM^+ \; nH_2O$
Saponite	$(Si_{4-x}Al_x)^{IV}(Mg_3)^{VI}(OH)_2O_{10}xM^+ \; nH_2O$

IV, tetrahedral layer; VI, octahedral layer

The chemical compositions of Table 2 are ideal and x is in the range 0.2 - 0.6. Clay minerals are natural materials and invariably contain traces of Fe^{2+}, Fe^{3+} and Ti^{4+}, to name only the most important. There may also be some amorphous oxides associated with them. These and other contaminations are largely removed by taking only the size fraction $< 2\,\mu m$.

In water, the exchangeable cations in the interlamellar space attrack water and the clay swells. The degree of swelling, or the amount of water taken up in the interlamellar space is determined by the interplay between the cation-lattice and the cation-water interactions. Small, monovalent cations give the clay the greatest tendency to swell. The ultimate swollen case, individual clay platelets, is probably never attained. A schematic drawing of various types of aggregates is shown in Figure 2.

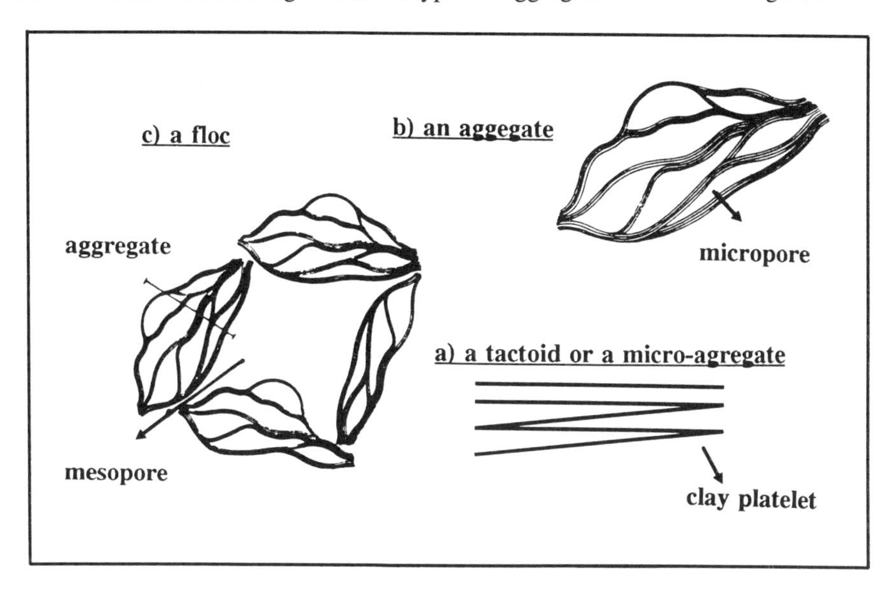

Figure 2. Various forms of aggregates of clay platelets.

Organophilic clays have organic cations in the interlamellar space, preferentially ammonium cations of the type $R_{4-x}H_xN^+$ with R = C_8-C_{18}. These cations take typical supramolecular arrangements, depending on the length of the alkyl chain and the charge density of the clay (Figure 3). They swell also, not in water but in slightly polar organics such as alcohols [10].

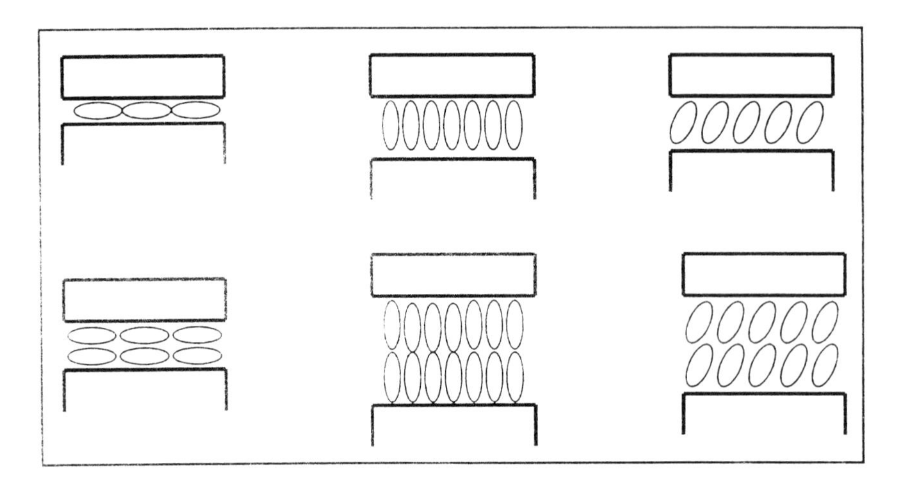

Figure 3. Idealized supramolecular arrangements of mono-alkyl-ammonium cations in the interlamellar space of clay minerals.

1.2 Zirconium phoshate

Figure 4 gives the layered structure of $Zr(HPO_4)_2$ and schematic drawings of an intercalate and of an organic derivative $Zr(ROPO_3)_2$. Intercalation of $Zr(HPO_4)_2$ is a difficult process because neighbouring layers are strongly held together via H-bonds, their number being about 6.3 meq/g. These protons are acidic and exchangeable if the incoming cation is capable of propping the layers apart [11]. If this is not the case, the intercalate may be obtained via direct synthesis. Organic derivatives can also be obtained via direct synthesis with R being any functionality with the desired properties for the supramolecular assembly [12].

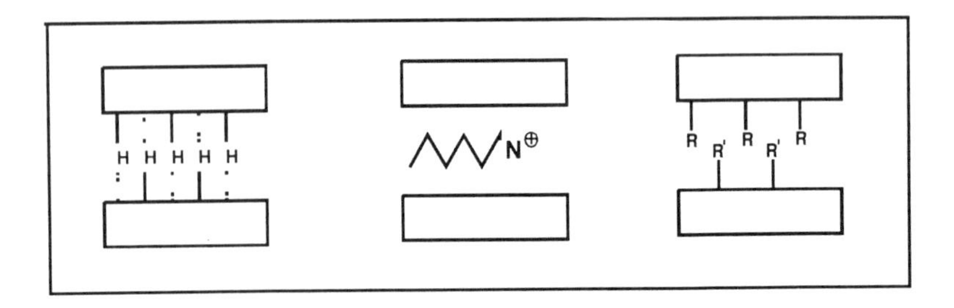

Figure 4. Schematic representations of layered phosphates.

2. AQUEOUS CLAY-DYE COMPEXES

2.1 Spectroscopic properties of dyes

There is a long list of dyes, which have been adsorbed on clay minerals in order to study (1) the surface chemistry of clays; (2) the distribution of molecules over the surface; and (3) the clay particle aggregation in aqueous suspension. Most of these dyes have a pronounced tendency to aggregate in aqueous solution, dimerization being the most commonly encountered process. The dimerization constant for the equilibrium 2 $D \rightleftharpoons D_2$ is K_d ($mole^{-1}$ dm^3) = $[D_2]/[D]^2$, where D represents a dye molecule. Values of K_d range between 500 and 5000. The dimer is a face-to-face aggregate of two monomers with their transition moment vectors parallel to each other, but pointing in opposite direction. Both exciton theory and perturbation theory predict that the excited state is split into two levels, one red-shifted and one blue-shifted with respect to the monomer excited state level [13,14]. Only the latter is allowed. Thus dimerization is evidenced by a new band at higher energy than the monomer absorption band. The low energy transition becomes partially allowed, when the transition moment vectors of the two monomers in the dimer are not parallel.

For higher aggregates the same reasoning holds. The equilibrium

$$n\,D \rightleftharpoons D_n \qquad (1)$$

is characterized by an equilibrium constant

$$K_n = \frac{[D_n]}{[D]^n} \qquad (2)$$

and the absorption of the n-mer is blue shifted with respect to that of the monomer. However, the resulting broad absorption band of the n-mer usually encompasses all aggregates with $n > 3$.

2.2 Clay-dye complexes in aqueous suspension.

A common feature of the adsorption of cationic dyes on clay minerals in aqueous suspensions via an ion exchange reaction, is their high affinity for the clay surface. The adsorption is quantitative and easily exceeds the cation exchange capacity (CEC) via incorporation of anions in the adsorption complex [15]. When it is however the intention to form stable dye-clay complexes in aqueous medium the loadings should be kept low, less than 25% of the CEC of the clays. The exact loading depends on the particle size and the type of co-exchanged cation. In any case micron-sized particles are obtained with optical properties which depend on the type of dye on the surface, the type of dye-surface interaction, the aggregation of the dye on the surface and the aggregation of the clay platelets in the aqueous supension. The measured optical properties are the average of those of the individual dye-clay aggregates in the suspension.

The complexity of the systems can be realized with reference to Figure 2. Because of the high affinity of the dyes for the clay surfaces, they do not distinguish between aggregates with different sizes or shapes, nor do they distinguish between diiferent types of sites. The dye molecules will adsorb on the first sites they encounter. The result is (1) a non-random distribution of the dye molecules over the available sites; (2) a non-equilibrium distribution of the molecules. One expects a slow evolution of the systems due to redistribution of dye molecules and to rearrangements of the clay aggregates. In any case, the dye molecules are concentrated in a small surface volume and this will promote aggregation. These ideas are illustrated below with specific examples.

2.3 Aggregation of dye molecules

There is a wealth of experimental information on dye aggregation at clay surfaces in aqueous suspension even at loadings, far below 1% of the CEC. The reported monomer and dimer absorption and emission bands are given in Table 3. For MB a trimer absorption has also been reported with an absorption maximum at 575 nm [16]. If the surface aggregation reaction is considered as a solution-like process, then equations (1) and (2) apply, and the aggregation constants can be derived from the decrease of the intensity of the monomer absorption with increasing loading. One obtains values, which are two to three orders of magnitude higher than the solution values [18]. Some authors proposed an effective volume around the clay particles in which the aggregation reactions take place [19, 20], as the surface area of the clay particles times 1 nm thickness. Then, the aggregation numbers and aggregation constants of Table 4 are obtained. The K_n values are now smaller than the solution values, but depend on the definition of the surface volume: the larger this volume is chosen, the higher is the resulting K_n value. Also, the average aggregation numbers are close to 2 and 3, showing that respectively dimerization and trimerization are the main phenomena.

Table 3.
Dyes and their spectroscopic characteristics in aqueous solution.

	monomer		dimer
Name	absorption/nm	emission/nm	absorption/nm
methylene blue,MB	664	-	605
proflavine, PF	445	506	430
methylviologen, MV	255	-	-
crystal violet, CV	588	-	540
rhodamine 6 G, RG	526	546	475
porphyrins, PP	421(Soret)	680(broad)	-
$Ru(bipy)_3^{2+}$	450	600	-
pyrene, PY	320(0-0)	480(excimer)	-

Table 4.
Aggregation constants and aggregation numbers in an effective surface volume around clay particles.

molecule	clay	n	Kn
MB	hectorite *	1.94	126
	barasym	2.82	772
PF	hectorite	1.95	266

* all clays are in their Na^+-forms

This treatment of the absorption spectra of aqueous suspensions only works for Na^+- and Ca^{2+}-exchanged hectorite and barasym, not for laponite, a synthetic analog of hectorite with very small particle sizes, and for K^+- and Cs^+-exchanged clays. In the latter cases, the solution-like behaviour is not obeyed, due to (1) restricted accessibility of the surface sites; (2) non-random distribution of the dyes over the surface. Solution-like behaviour is only garantueed on fully swollen clays, or when only the external surface of the aggregates is available for adorption (barasym). A fully swollen clay is a suspension without aggregates, or a material in which the individual clay platelets move independently of each other in the suspension. One can also envisage aggregates for which the interlamellar distance between the clay platelets in the aggregates are so large than free diffusion of the dyes in and out of the interlamellar space is possible. Such situations clearly only occur with Na^+- and Ca^{2+}-hectorite and with barasym.

The spectra fully reveal the situation, as illustrated for MB in Figure 5. One notices that the decrease of the monomer absorption in the region 650-670 nm with increasing loading is far more important for Na^+-hectorite than for Cs^+-hectorite. In the aggregates of the Cs^+-clay there is not enough space in the interlamellar space for dimerization to occur.

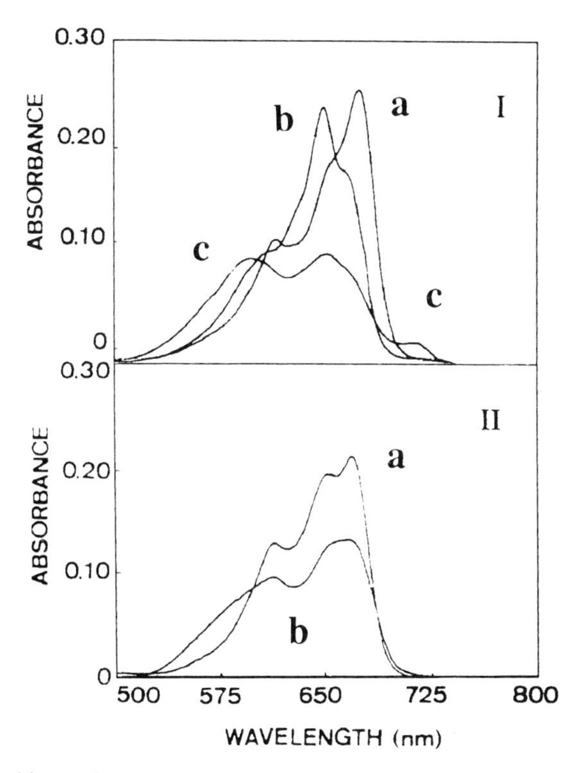

Figure 5. Absorption spectra of methylene blue-clay suspensions. I:a, K^+-laponite (a); Na^+-hectorite (b), both with a loading of 0.2% of the CEC; Na^+-laponite (c) with a loading of 20% of the CEC, II: Cs^+-laponite with a loading of 0.2% (a) and 23.5% (b) of the CEC.

Aggregation leads to fluorescence quenching [18,21], because of efficient energy transfer from the monomers to the dimers or aggregates. This follows Perrin's model of energy transfer. This means that (1) statistical distribution of monomers and aggregates is not garantueed; (2) the lifetime of the excited state of the monomer is shorter than the average time between two successive translational jumps of the molecules. These two consequences are indicative for non-solution-like behavior, even for completely swollen clays, such as Na^+-hectorite. Therefore, the physical meaning of the aggregation constants derived for these heterogeneous systems on the basis of absorption spectra is not at all clear.

The excimer emission of pyrene- or derivatized pyrene-clay complexes in aqueous suspensions is also an indication of dye aggregation on the surface [22-24]. Excimers are present at all loadings, showing that a random distribution of, in this case, P3N (= (3-(1-pyrenyl)propyl)trimethylammonium) does not occur at all in aqueous suspensions. Instead, the first incoming molecules are preferentially adsorbed on the external surface of aggregates, and slowly redistribute over the available surface. This redistribution is evidenced by a decrease of the I_e/I_m ratio with time. As shown in Figure 6, the redistribution is a process which occurs on the timescale of a few thousands of seconds, and it is influenced by the ionic strength of the medium. If the ionic strength increases e.g. by addition of $CaCl_2$, the I_e/I_m ratio is initially higher, but it decreases with a larger rate constant, than in the absence of ions in solution. This is because the clay particles tend to aggregate with increasing ionic strength.

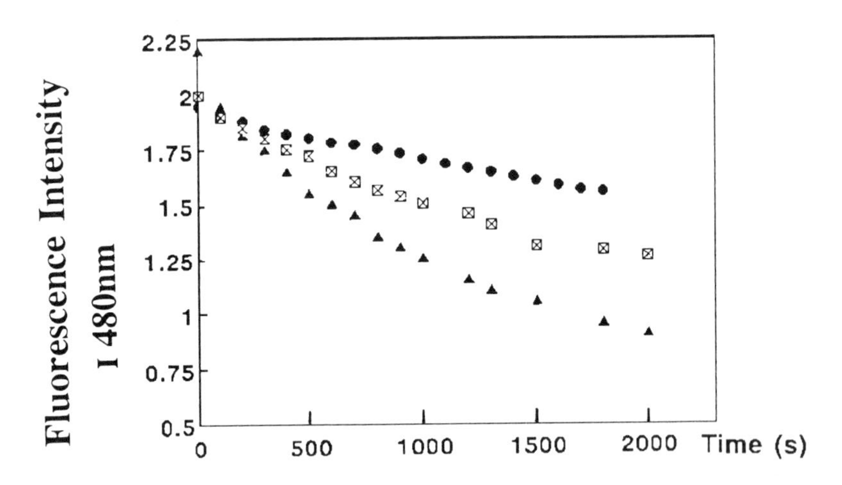

Figure 6. Decrease of the excimer emission of $P3N^+$-laponite suspentions with time: ●, $[Ca^{2+}]=0$; ⊠, $[Ca^{2+}]=10^{-4}M$; ▲, $[Ca^{2+}]=10^{-3}M$.

On the longer time scale (days) I_e/I_m increases with time and the viscosity of the medium decreases (Figure 7). This is due to aggregation of clay particles to larger aggregates, which settle out under gravity. In the long run there are less particles in the suspension and, as a consequence, the viscosity decreases. When particles, loaded with monomers of P3N, aggregate, excimers are formed and I_e/I_m increases, as observed.

Summarizing, there are at least two types of rearrangements: on the short timescale, redistribution of P3N molecules with a decrease of the I_e/I_m ratio; on the long timescale, aggregation of the clay particles and increase of the I_e/I_m ratio. Whether these two redistributions can be considered as totally independent of each other, as we assumed in the present discussion, is probably an oversimplification. In any case, time dependent fluorescence seems to be a promising technique to study the dynamics of clay suspensions.

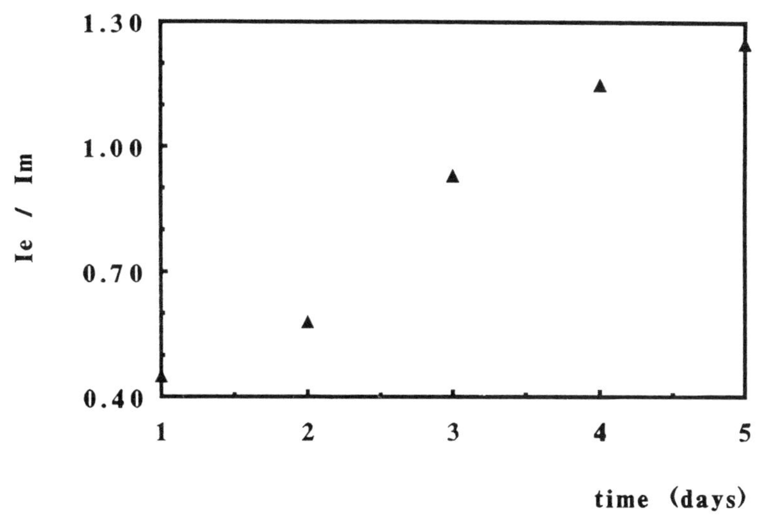

Figure 7. Time dependence of the excimer and monomer emission intensities ratio of P3N in a laponite suspension.

2.4 Site heterogeneity

The existence on the elementary clay particle of an edge surface and a planar surface (Figure 2) is a source of site heterogeneity. In addition, every site on the planar surface as well as on the edge surface has its own micro-environment, determined by the local chemical composition and structure. Thus, the absorption and emission bands of the adsorbed dyes are inhomogeneously broadened. Several attempts have been published on the distinction between edge sites and planar sites on the basis of spectroscopy of adsorbed molecules. Relevant data are gathered in Table 5 [16,18,25,26].

Table 5.
Spectra of monomeric dyes on edge and planar sites of clays.

Dye	Edge site λ/nm	Planar site λ/nm	type of spectroscopy
MV	254	278	absorption
		328(H)330(M)	fluorescence
PA	452	490	absorption(Soret)
R6G	544	544	emission
MB	676	656	absorption
PF	453	453	absorption

The data of Table 5 reveal (1) a red shift of all bands of adsorbed monomers with respect to the aqueous solution values, except for MB on the planar surface; (2) in the case of MV, PH and MB there are two monomer bands, interpreted as due to monomers on the external (edge) surface and on the interlamellar or planar surface. For R6G and PF no such two absorption bands were observed, although in the first case the shape of the absorption band changed with loading and two different emission maxima were observed.

In most cases, the shifts were interpreted as due to environmental effects. The local polarity being different on the interlamellar surface from that on the edge surface and from water. This is of course true for all the dyes and does not explain why some undego soetimes dramatic shifts and others no shoft at all. At least two other factors must be included: (1) perturbation of the conformation of the molecule by adsorption; (2) the nature of the absorption process, or the direction of the transition moment vector with respect to the surface. At present, there are no data in the literature which allow to resolve the relative contributions of each of these factors to the band shifts.

Site heterogeneity is also evidenced in the multi-exponential decays, which are invariably observed for adsorbed luminescent dyes [9,26]. The decays are usually fitted by two or three exponentials, but one could also invoke a Gaussian distribution of decay time constants, or use micellar kinetics to describe the decay. An additional problem with the interpretation of time dependence of the decay data is the quenching by Fe^{3+} impurities in the lattice or on the surface

3. CLAY-DYE MATERIALS

Clay-dye materials are defined as organized assemblies of elementary clay platelets and dye molecules. Organization must be achieved at two levels: at the level of the clay platelets and at the level of the dye molecules. If a two-dimensional molecule is considered with a transition moment vector in the plane of the molecule, several arrangements can be envisaged, some of them being illustrated in Figure 8. Three general situations can be envisaged: (1) perfect alignment of the clay platelets and of the transition moment vectors; (2) aligned transition moment vectors and randomly oriented clay platelets, and (3) both the transition moment vectors and the clay platelets are randomly oriented. If materials can be prepared as defined under (1), they must show characteristic optical properties, the optical anisotropy being the most obvious one.

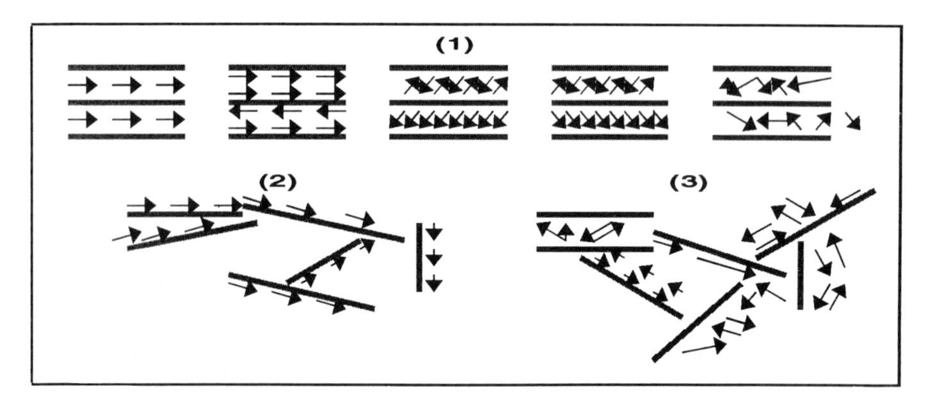

Figure 8. Organization of transition moment vectors and clay particles.

There are three ways to achieve one of the situations depicted in Figure 8: solid-liquid intercalation; (2) solid-solid intercalation, and (3) direct synthesis.

In solid-liquid intercalation the starting situation is one with a more or less random distribution of dye molecules and of clay particles in the suspension. Typical examples are the clay-dye aqueous suspensions, discussed above. The dye-clay complex settles under gravity. Water evaporates at room temperature and clay-dye films are formed with preferential face-to-face aggregation. This may lead to significant changes in the spectra of adsorbed dyes. Thus, whereas in aqueous suspension the clay-methylene blue complex is characterized by an absorption spectrum of aggregates, the monomer absorption band predominates in the film. Presumably, a significant rearrangement of both the clay platelets and the dye molecules has taken place during drying. The same observation has been made with crystal violet [27].

The particle size and crystallinity of the clay platelets, the type of exchangeable cation, the type of dye and the nature of the solvent are the most important parameters, which determine the type of material obtained from this solid-liquid intercalation. In the case of alkylammonium cations (Figure 3) quite regular arrangements have been achieved. The planar dyes will have a strong tendency to lay flat on the surface to maximize the interaction. This has been put into a rule [28]: two-dimensional surfaces prefer two-dimensional molecules.

An alternative procedure is to pre-organize the clay platelets in the suspension. It can be achieved by changing the type of exchangeable cation or by changing the polarity of the solvent. A third way is to pre- or co-adsorb organic molecules. The most popular molecules are alkylammonium cations and polymers such as polyvinylalcohols (PVA) and polyvinyl pyrrolidones (PVP) [29,30]. The dye molecules are then located in the interlamellar space of the pre-organized clay platelets in an organophilic environment. This decreases their tendency to aggregate.

In this way, the photochromic behavior of montmorillonite-viologen intercalation compounds has been studied, in which PVP acts as the acceptor of the electron from viologen [29]. The photochromism of spyropyrans in clay interlayers is another interesting example [31,32]. The kinetics of the isomerization involve at least two rate equations, possibly due to aggregation of part of the spyropyran molecules. Ogawa et al. [33] studied the photochemical hole burning of 1,4-dihydroxyanthraquinone intercalated in a saponite loaded with tetramethylammonium cations. The latter serve as spacers for the anthraquinone molecules.

On the two other intercalation methods proposed to achieve the organizations of Figure 8, only preliminary results are available in the open literature. Thus, Ogawa et al. [34] obtained intercalation compounds by solid-solid interaction of small neutral molecules such as acrylamide, methacrylamide and urea, with montmorillonite. Carrado et al. [35,36] used dyes as templates in clay synthesis, based on the early work of R. Barrer and the vast literature on templated zeolite synthesis [37,38].

The configurations adopted by the dye-clay systems (Figure 8) depend on both the type of clay and the type of dye. Clearly, the systems tend to an energetic minimum, which is determined by all the interactions, the most important being the ionic interaction between the negative lattice charge and the positive charge on the dyes. In addition, as all the dyes are rather bulky molecules, there will be an important contribution from van der Waals - type of interactions dye-clay and dye-dye, not to forget the co-adsorbed solvent.

The subtlety of these interaction energies is shown in the adsorption of $M(phen)_3^{2+}$ and $Ru(bipy)_3^{2+}$ complexes (M = Fe, Ru; phen = 1,10-phenanthroline; bipy = 2,2'-bipyridyl). In the former case, the racemic mixture is adsorbed to 2 times the CEC, and the enantiomer within the CEC. The reverse observation has been made with $Ru(bipy)_3^{2+}$. Sato et al. (39,40) have attempted to calculate the interaction energies of the phen complexes with the clay surface on the basis of ionic energies and a Lennard-Jones potential with Monte Carlo simulations. They showed that the racemic pair forms a more compact and energetically more stable pair than the enantiomeric pair. This is the basis for the difference in adsorbed amount. Why the reverse holds for the bipy complexes is unclear. These complexes are supposed to occupy three adjacent ditrigonal holes on the clay surface. (Figure 9). If this geometrical factor were determining, no distinction between the different complexes should be observable. The fact that there is, means that geometry, size and energetics are important in the geometrical arrangements of these complexes with important optical properties.

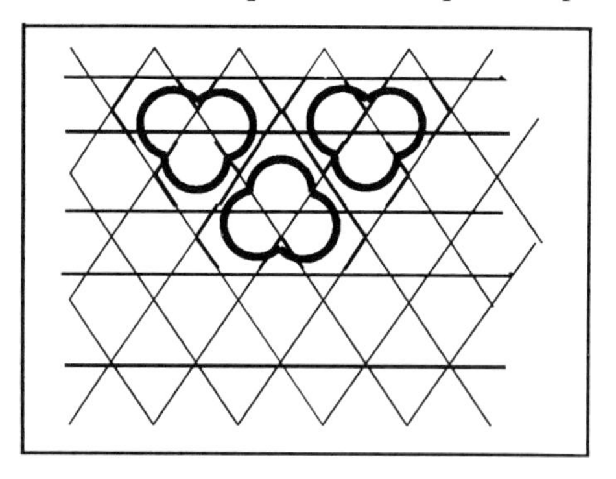

Figure 9. Theoretical arrangement of $M(bipy)^{n+}_3$ or $M(phen)^{n+}_3$ complexes on the planar clay surface.

4. $Zr(HPO_4)_2$ - DYE COMPLEXES

The principles governing the preparation and properties of $Zr(HPO_4)_2$ - dye complexes are the same as those of clay-dye complexes. The difference is that the CEC is about 6-10 times higher than that of a normal clay. Furthermore, the exchangeable cation is a proton (P-O-H), hydrogen-bonded to an oxygen of the next $Zr(HPO_4)_2$ layer (Figure 4). These two factors make the intercalation of dyes in $Zr(HPO_4)_2$ via the solid-liquid or solid-solid route a difficult task. It is possible first to delaminate $Zr(HPO_4)_2.H_2O$ by partial exchange with propylammonium cations, followed by adsorption of the dye e.g. $Ru(bipy)_3^{2+}$, via an ion exchange process, but the yield is low [41].

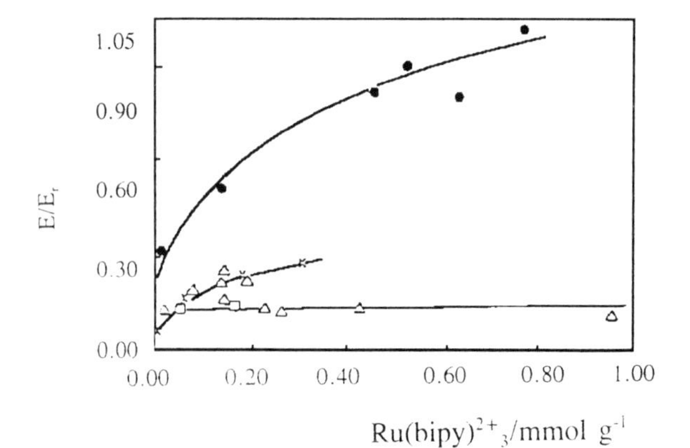

Figure 10. Dependence of the emission intensity of $Ru(bipy)^{2+}_3$ on the loading for crystalline intercalates (●), amorphous materials (△,□), and semi-crystalline zirconium phosphates (X), after ref.[42].

Direct synthesis is possible, if the dye is acid-resistant. This is the case for $Ru(bipy)_3^{2+}$. The advantage of the direct synthesis route is that a continuous series of materials can be prepared from amorphous to fully crystalline. The effect of the crystallinity is seen in Figure 10, wherein the emission intensity of intercalated $Ru(bipy)_3^{2+}$ is plotted against the loading [42]. This emission intensity is much more intense and increases with loading for the crystalline samples, whereas it is low and independent of the loading for the amorphous materials. This indicates that, in the former case, the complexes are intercalated in monolayers, whereas in the latter case they form non-intercalated clusters.

The most interesting feature of the layered phosphates is the direct synthesis of layered zirconium phosphonate compounds, the phosphonate carrying the required dye. Vermeulen and Thompson [43] synthesized layered viologen compounds (Figure 11), which form, upon exposure to solar radiation, a coloured, long-lived charge separated state, stable in air. The stability is ascribed to the crystallinity of the material and to the perfect molecular organization of the viologen moieties. Oxygen is unable to diffuse in the interlamellar space which is closely packed with viologen moieties.

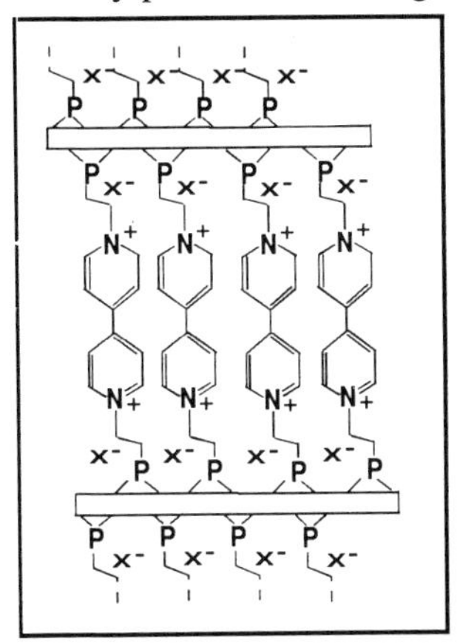

Figure 11. Zirconium phosphonate/viologen compound with X= Cl⁻ or Br⁻ after ref. [43].

Analogous componds are made by derivatization of clays:

clay-Na^+ + XR —> clay-R + NaX (3)

If R contains reactive groups further derivatization is possible for heterogeneisation of homogeneous catalysts [44] and for making clay-dye complexes with predetermined optical properties.

REFERENCES

1. G.A. Ozin and G. Gil, Chem. Rev., 89 (1989) 1749.
2. G.D. Stucky and J.E. Mac Dougall, Science, 247 (1990) 669.
3. Y. Wang and N. Herron, J. Phys. Chem., 95 (1991) 525.
4. G.A. Ozin, S. Özkar and R.A. Prokopowicz, Acc. Chem. Res., 25 (1992) 553.
5. G.A. Ozin, Adv. Mater., 4 (1992) 612.
6. G.A. Ozin and S. Özkar, Adv. Mater., 4 (1992) 11.
7. S.L. Suib, Chem. Rev., 93 (1993) 803.
8. J.K. Thomas, Acc. Chem. Res., 21 (1988) 275.
9. J.K. Thomas, Chem. Rev., 93 (1993) 301.
10. B.K.G. Theng, The Chemistry of Clay-Organic Reactions, A. Hilger, London, 1974.

11. A. Clearfield, Chem. Rev., 88 (1988) 125.
12. G. Cao, H.-G. Hong and T.E. Mallouk, Acc. Chem. Res., 25 (1992) 420.
13. M. Kasha, H.R. Rawls and M.A. El-Bayoumi, Pure & Applied Chem., 11 (1965) 371.
14. C.R. Cantor and P.R. Schimmel, Biophysical Chemistry, Part II, Freeman, San Francisco, 1981, pp. 846.
15. R.A. Schoonheydt, J. Pelgrims, Y. Heroes and J.B. Uytterhoeven, Clay Min., 13 (1978) 435.
16. J. Cenens and R.A. Schoonheydt, Clays and Clay Min., 36 (1988) 214.
17. J. Cenens, D.P. Vliers, R.A. Schoonheydt and F.C. De Schryver in Proc. Int. Clay Conf. Denver 1985 (L.G. Schultz, H. van Olphen and F.A. Mumpton, eds.), The Clay Minerals Society, Bloomington (1987), 352.
18. M.J. Tapia Estévez, F. Lopez Arbeloa, T. Lopez Arbeloa, I. Lopez Arbeloa and R.A. Schoonheydt, Clay Min., in press.
19. J. Cenens and R.A. Schoonheydt, Proc. 9th Int. Clay Conf. 1989 (V.C. Farmer and Y. Tardy, eds.) Sci. Géol. Mém., 85 (1990) 15.
20. J.K. Thomas, J. Phys. Chem., 88 (1984) 964.
21. J. Cenens, R.A. Schoonheydt and F.C. De Schryver in "Spectroscopic Characterization of Minerals and their Surfaces" (L.M. Coyne, S.W.S. McKeever and D.F. Blake, eds.), ACS Symp. Ser., 415 (1990) 378.
22. K. Viaene, J. Cagui, R.A. Schoonheydt and F.C. De Schryver, Langmuir, 3 (1987) 107.
23. K. Viaene, R.A. Schoonheydt, M. Crutzen, B. Kunyima and F.C. De Schryver, Langmuir, 4 (1988) 749.
24. B. Kunyima, K. Viaene, M.M. Hassan Khalil, R.A. Schoonheydt, M. Crutzen and F.C. De Schryver, Langmuir, 6 (1990) 482.
25. V.G. Kuykendahl and J.K. Thomas, Langmuir, 6 (1990) 1350.
26. G. Villemure, C. Detellier and A.G. Szabo, Langmuir, 7 (1991) 1215.
27. D. Pieters, engineering thesis, K.U.Leuven, 1993.
28. A. Maes, R.A. Schoonheydt, A. Cremers and J.B. Uytterhoeven, J. Phys. Chem., 84 (1980) 2795.
29. H. Miyata, Y. Sugihara, K. Kuroda and C. Kato, J. Chem. Soc., Faraday Trans. I, 83 (1987) 1851.
30. M. Ogawa, M. Inagaki, N. Kodama, K. Kuroda and C. Kato, J. Phys. Chem., 97 (1993) 3819.
31. K. Takagi, T. Kurematsu and Y. Sawaki, J. Chem. Soc. Perkin Trans. 2, (1991) 1517.
32. T. Saki and K. Ichimura, Macromolecules, 23 (1990) 31.
33. M. Ogawa, T. Handa, K. Kuroda, C. Kato and T. Tani, J. Phys. Chem., 96 (1992) 8116.
34. M. Ogawa, K. Kuroda and C. Kato, Chem. Lett., (1989) 1659.
35. K. Carrado, Ind. Eng. Chem. res., 31 (1992) 1654.
36. K. Carrado, Inorg. Chem., 30 (1991) 794.
37. R.M. Barrer and L.W.R. Dicks, J. Chem. Soc. A, Inorg. Phys. Theor., (1967) 1523.
38. P.A. Jacobs and J.A. Martens, Synthesis of high-silica aluminosilicate zeolites, Elsevier, Amsterdam, 1987.
39. H. Sato, A. Yamagishi and S. Kato, J. Phys. Chem., 96 (1992) 9377.
40. H. Sato, A. Yamagishi and S. Kato, J. Phys. Chem., 96 (1992) 9382.
41. D.P. Vliers, D. Collin, R.A. Schoonheydt and F.C. De Schryver, Langmuir, 2 (1986) 165.
42. D.P. Vliers, R.A. Schoonheydt and F.C. De Schryver, J. Chem. Soc. Faraday Trans. I, 81 (1985) 2009.
43. L.A. Vermeulen and M.E. Thompson, Nature, 358 (1992) 656.
44. B.M. Choudhary, K.R. Kumar, Z. Jamil and G. Thyagarajan, J. Chem. Soc. Chem. Comm., (1985) 931.

Part VI: Microelectrochemistry and Microphotoconversion

MICROCHEMISTRY
Spectroscopy and Chemistry in Small Domains
Edited by H. Masuhara et al.

Functionalized photoelectrochemistry

Kenichi Honda

Tokyo Institute of Polytechnics, 2-9-5 Honcho, Nakano-ku, Tokyo 164, Japan

Photoelectrochemistry is developing to show a variety of new functions. One of the important features of functionalized photoelectrochemistry is the trends towards microchemistry.

1. INTRODUCTION

Photoelectrochemistry deals with electron transfer reactions across the interface between solid electrode and the electrolyte solution where one of these two phases or the interfacial layer is excited electronically by light absorption as shown in Figure 1.

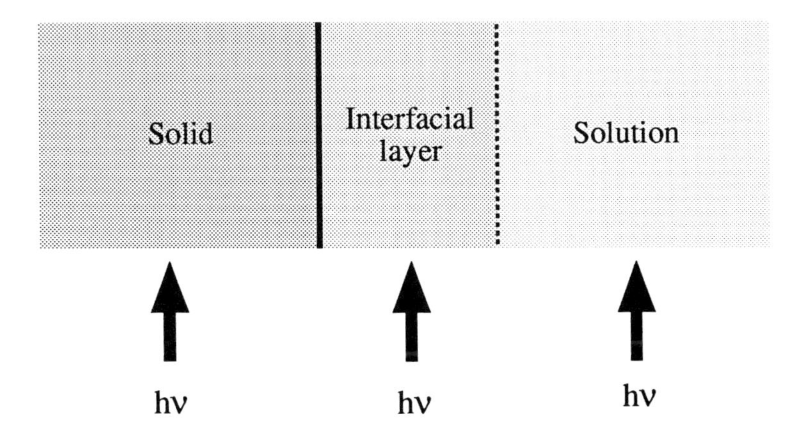

Figure 1. The principle of photoelectrochemistry.

Based on the above principle, the scope of modern photoelectrochemistry has been expanded and now covers the following areas.

(1) Electrochemical processes in the electronically excited state of each of the following phases
    solid electrode
    interface
    solution

(2) Electrochemical formation of electronically excited state
    luminescence

(3) Heterogeneous light-induced electron transfer
    surface
    membrane

(4) Hybrid or sequential systems
    photochemical + electrochemical
    photopolarography

(5) Correlation between the electrode potential and the molecular orbital

At the early stage of its evolution, the photoelectrochemistry had been developing with much concern with the light energy conversion. On the other hand, photochemistry itself has been expanding its area of application based on the evolution of new interdisciplines between conventional photochemistry and other disciplines such as electrochemistry, biochemistry, surface chemistry, polymer chemistry, solid state physics, imaging science and so on. Accordingly, the results of the growth of the above interdisciplines have given rise to a variety of new functions related to photoelectrochemistry.

In the course of the photoelectrochemical processes, the various types of energy and information conversion among the optical, electrical, chemical, thermal, mechanical, biological and other signal inputs take place, offering the novel functions such as sensor, display, imaging, recognition, memory and others. Some of these are expected to be applied to the metallurgy, the medical therapy, the environmental problem and the synthesis of new materials.

Research trend toward functionalized photoelectrochemistry reported in the literature is summarized.

2. GENERAL CONCEPT OF FUNCTIONALIZED PHOTOELECTRO-CHEMICAL SYSTEM

Figure 2 illustrates the concept of the functionality of the photoelectrochemical system. In the functional system both the input and the response can have any type of information or energy. In the case of the photoelectrochemical system, either the input or the response has the form of light and furthermore, the third input can be added to the system to control or to modulate the function.

Figure 3 shows only a few typical examples of applications from the functional photoelectrochemical system.

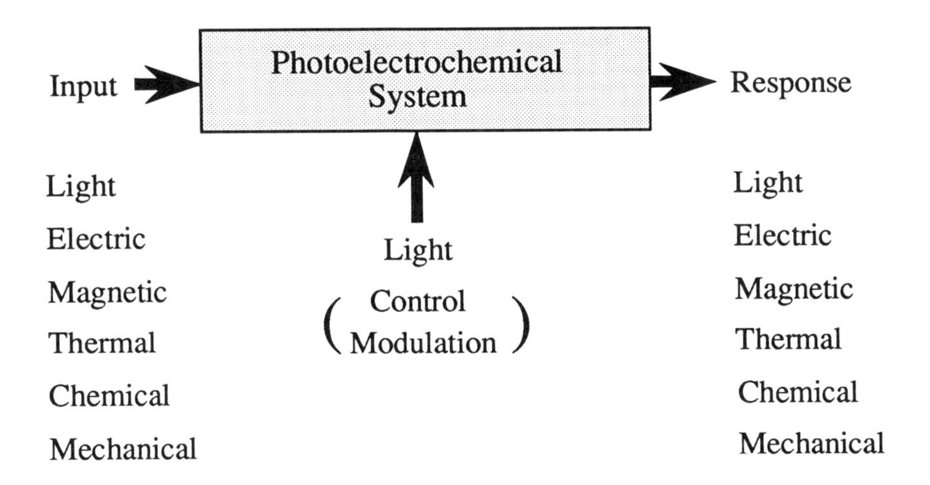

Figure 2. Functionality of photoelectrochemistry.

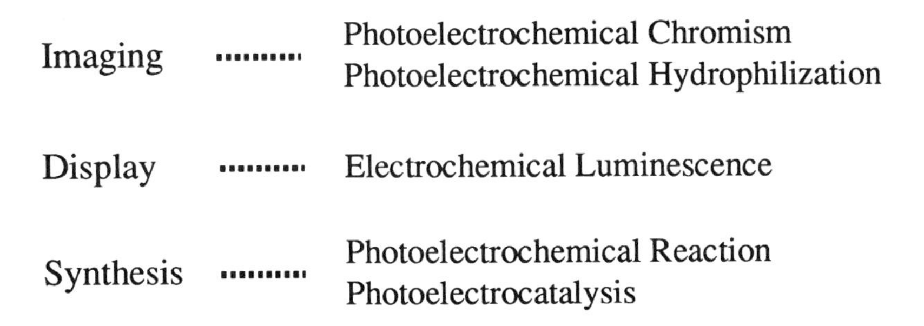

Figure 3. Typical examples of the application of the photoelectrochemical system.

3. PHOTOELECTROCHEMICAL IMAGING

One of the three examples of Figure 3, the photoelectrochemical imaging or patterning will be described. Four types of imaging methods are given in Figure 4.

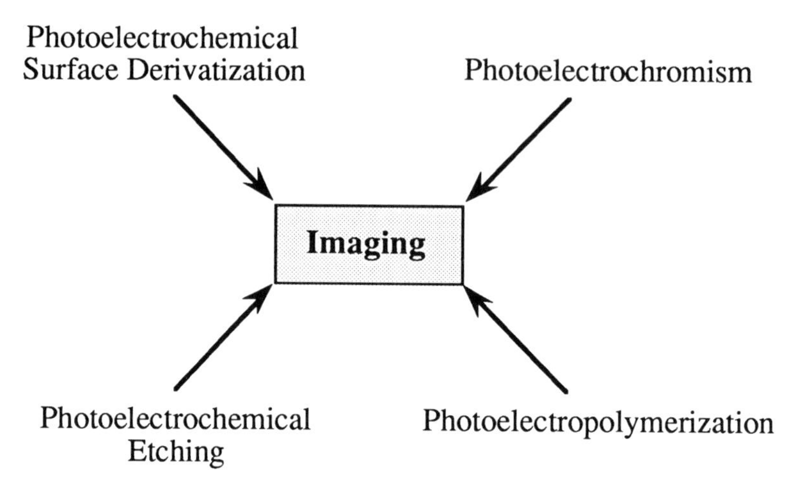

Figure 4. Four types of the photoelectrochemical imaging system.

3.1 Photoelectrochemical surface derivatization

This method is based on the photoelectrochemical hydrophilization of the semiconductor electrode which is made beforehand hydrophobic by the surface derivatization.

The photoelectrochemical oxidation destroys the hydrophobic derivatized layer on the semiconductor electrode substrate. Hence, the printing ink cannot be coated on the irradiated part of the semiconductor electrode (negative image).

Three procedures were proposed as shown in Figure 5 [1, 2].

Photoelectrochemical Hydrophilization

(Photo-sedimentation)

Hydrophilic-Hydrophobic Metamorphosis

1. Hydrophobic coating (Paraffin coating)
2. LB film
3. Surface Modification (Silanization)

Figure 5. Three types of photoelectrochemical hydrophilization.

(i) Hydrophobic coating with paraffin

The naked surface of TiO_2 semiconductor is normally hydrophilic because of the presence of OH^-. The surface is then coated with paraffin to make hydrophobic. Figure 6 shows the decrease of contact angle of a water drop on the paraffin coated surface as a function of the amount of the photoanodic current corresponding to the destroyed amount of hydrophobic material.

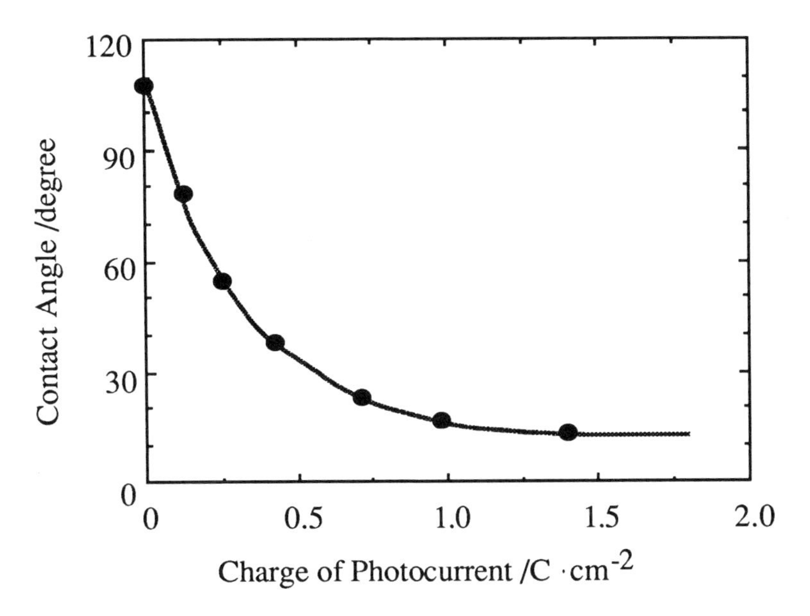

Figure 6. Photoanodic reaction of paraffin coated TiO_2 electrode.
Electrolyte: 0.2 mol dm^{-3} K_2SO_4 aqueous solution
Applied potential: 1 V vs. SCE

(ii) LB film coating

The semiconductor substrate plate becomes hydrophobic by the LB monolayer coating of amphiphilic molecules like stearic acid. As in the case of (i), the photoelectrochemical oxidation can give rise to the formation of the hydrophilic pattern on the substrate electrode.

(iii) Surface modification

Only one example of the surface modification by silanization will be referred here [2]. By treating TiO_2 surface with Cl_3SiCH_3, the following hydrophobic structure is formed.

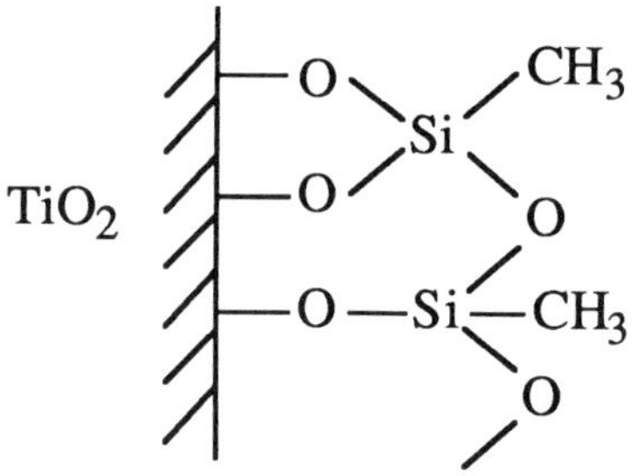

This modified layer is also subject to the photoelectrochemical oxidation.

3.2 Photoelectrochromism (Photoelectrochemical chromism)

The image formation by the color change brought about by the photoelectrochemical process is introduced for the following three types of color changes.

- Photoelectrochemical synthesis
- Semiconductor photocatalysis - assisted electrochromism
- Light - enhanced electrochromism

(i) Photoelectrochemical synthesis

The color change on the substrate can be given by the photoelectrochemical oxidation of the metallic ion to deposit the metal oxide on the semiconductor electrode [3]. The image formed by the color change in this way can be erased by the reverse electrochemical reduction. An example of the case of Tl^+ is given:

imaging,

$$2Tl^+ + 3H_2O + 4p^+ \rightarrow Tl_2O_3 + 6H^+$$

erasing,

$$Tl_2O_3 + 6H^+ + 4e \rightarrow 2Tl^+ + 3H_2O$$

With TiO_2 electrode, the color changes from gray to yellow with irradiation.

(ii) Semiconductor photocatalysis - assisted electrochromism

Yoneyama et al. studied a new type of the electronic imaging system using polyaniline incorporated with semiconductor particle [4]. Polyaniline is known to show the electrochromic property. It is photoelectrochemically reduced by TiO_2 particle under irradiation to show yellow color. Upon the electrochemical oxidation, polyaniline comes to show its original blue color. The scheme of the erasable image formation is given in Figure 7.

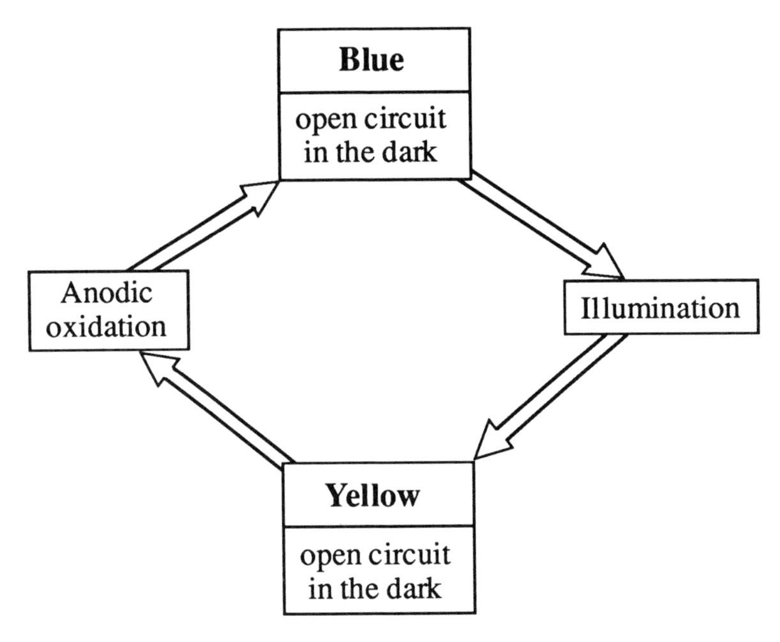

Figure 7. Light-induced electrochromic display system by Yoneyama et al. [4].

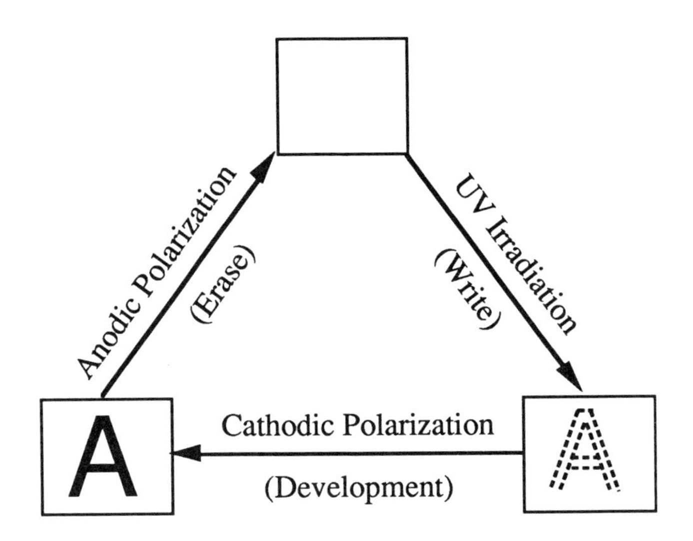

Figure 8. Schematic diagram of imaging using light-enhanced electrochromism [5].

(iii) Light-enhanced electrochromism

This is not the case of the authentic photoelectrochemistry. It can be classified to the hybrid system of photochemical and electrochemical processes as described in the Introduction.

MoO_3 is known to show the electrochromism. Fujishima et al. reported that the pre-irradiation of MoO_3 gives rise to the enhancement of the intensity of its electrochromism [5]. As a result, the pre-irradiation can be regarded as a kind of the latent image which can allow the image formation (development) by the electrochemical polarization. A schematic diagram of the process is shown in Figure 8 [5].

3.3 Photoelectropolymerization

It is known that many polymerization reactions can take place photoelectrochemically with semiconductor substrate electrode. Hence, the pattern formation is one of the interesting applications of the photoelectrochemistry. Only one paper dealing with the photoelectrochemical polymerization of pyrrole is referred here [6].

3.4 Photoelectrochemical etching

Many papers studied the anodic etching of n-type semiconductor such as GaAs, InP and others under irradiation with light. It is expected that this photoelectrochemical etching will serve for the patterning of IC fabrication. The resolution of the micropattern is being improved.

4. DESIGN OF THE PHOTOELECTROCHEMICAL SYSTEM FOR THE NEW FUNCTIONS

To expand and improve the various functions of the photoelectrochemical system, a variety of subtle design of the system has been proposed as shown in Figure 9.

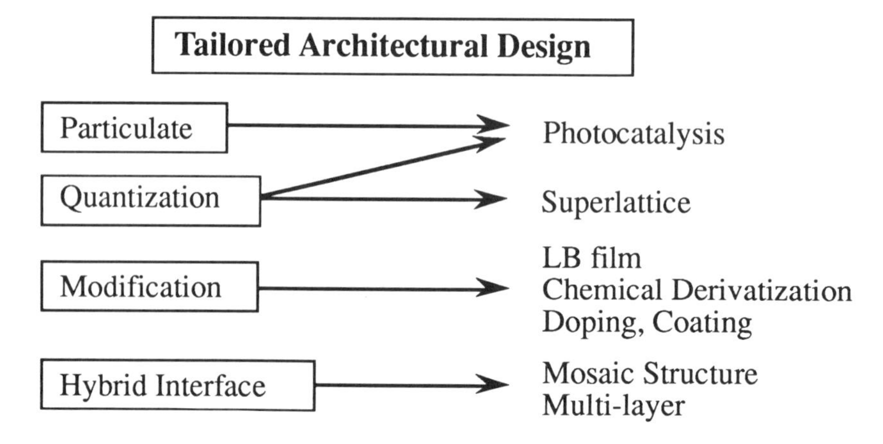

Figure 9. Design of the photoelectrochemical system.

In this paper, only the semiconductor particle system will shortly be described. It is well known as the photocatalytic system. Many new and important syntheses are expected from this system and a big number of papers have already been reported.

Among really a variety of syntheses, the following photoelectrocatalytic reactions should be pointed out as key reactions for the future of human beings.

(1) H_2O splitting [7]

(2) N_2 fixation [8] (Chemical evolution [9])

(3) CO_2 reduction [10] (Artificial photosynthesis)

(4) Halocarbon decomposition (Waste treatment)

(5) Water gas reaction [11]

5. TRENDS TOWARDS MICROPHOTOELECTROCHEMISTRY

One of the important features of functionalized photoelectrochemistry is the trends towards the microchemistry. In the imaging application, the resolution of the photoelectrochemical etching and the photoelectrochemical polymerization is being much improved and they are expected to be useful for the industrial IC fabrication.

In the photocatalytic application, the size of semiconductor particle becomes smaller and smaller, say less than 10 Åϕ. The quantization such as superlattice structure of photoelectrode will give a new aspect of the microchemistry.

REFERENCES

1. A. Fujishima, T. Kato, E. Maekawa, and K. Honda, Denki Kagaku, 54 (1986) 153.
2. T. Kato, A. Fujishima, E. Maekawa, and K. Honda, J. Chem. Soc. Japan, 1986, 8.
3. T. Inoue, A. Fujishima, and K. Honda, J. Electrochem. Soc., 127 (1980) 1582.
4. H. Yoneyama, Adv. Mater., 5 (1993) 394.
5. J. N. Yao, B. H. Loo, K. Hashimoto, and A. Fujishima, Ber. Bunsenges. Phys. Chem., 95 (1991) 537.
6. M. Okano, K. Ito, A. Fujishima, and K. Honda, J. Electrochem. Soc., 134 (1987) 837.
7. A. Fujishima and K. Honda, Nature, 238 (1972) 37.
8. N. Schrauzer and T. D. Guth, J. Am. Chem. Soc., 99 (1977) 7189.
9. H. Reiche and A. J. Bard, J. Am. Chem. Soc., 101 (1979) 3127.
10 T. Inoue, A. Fujishima, S. Konishi, and K. Honda, Nature, 277 (1979) 637.
11. T. Sakata and T. Kawai, Nature, 282 (1979) 283.

MICROCHEMISTRY
Spectroscopy and Chemistry in Small Domains
Edited by H. Masuhara et al.

Preparation and Characterization of Microelectrochemical Devices: Self-Assembly of Redox-Active Molecular Monolayers on Microelectrode Arrays

Mark S. Wrighton, C. Daniel Frisbie, Timothy J. Gardner, and Doris Kang

Department of Chemistry, Massachusetts Institute of Technology, Cambridge, Massachusetts 02139 USA

INTRODUCTION

In this article we wish to summarize recent work in our laboratory related to the preparation and characterization of chemically modified microelectrode arrays. In particular, we wish to present work related to the use of molecular self-assembly and photochemistry to prepare tailored surfaces of microelectrodes. In the context of this article, self-assembly refers to the spontaneous formation of a molecular monolayer onto particular surfaces upon exposure of the surface to a solution of the self-assembly reagent, L, as illustrated in Scheme I. Several reports of such

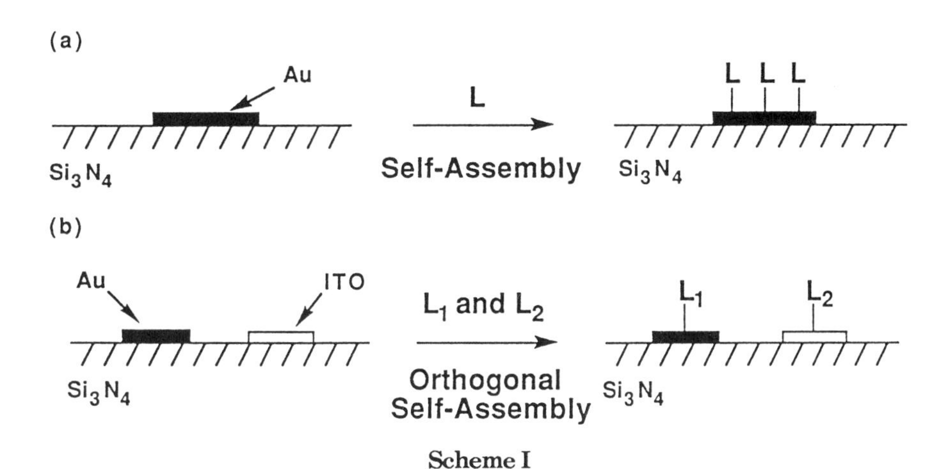

Scheme I

chemistry have already appeared from this laboratory [1-5]. An important feature of self-assembly is that L reacts selectively, covering the microelectrode surface but not the substrate, by virtue of a functional group of L that brings about selective attachment. For example, it is well-known that thiols react selectively with Au. However, it is also appreciated that structural features of L beyond the surface-ligating group are also important, since a small alkylthiol (e.g. CH_3SH) does not yield durable self-assembled monolayers. However, long chain alkylthiols (e.g. $CH_3(CH_2)_{17}SH$) do yield very durable monolayer-modified surfaces. Developing selective surface coordination chemistry for self-assembly is an important goal of our research.

Beyond the self-assembly represented in Scheme Ia, we seek to develop new systems for "orthogonal" self-assembly [6, 7], as illustrated in Scheme Ib. In this case, two different self-assembly reagents, L_1 and L_2, are simultaneously exposed to a microelectrode array consisting of two different kinds of microelectrode materials, M_1 and M_2, on a common substrate. The orthogonality refers to the fact that L_1 has an affinity for one surface and L_2 has an affinity for a second surface such that the self-assembly process yields a surface as represented in Scheme Ib where L_1 is on M_1 and L_2 is on M_2 and neither is on the substrate. Such orthogonal self-assembly was first demonstrated with M_1 = Au, L_1 = thiol; M_2 = Al/Al_2O_3, L_2 = carboxylic acid [6]. Unfortunately, the oxide coating on Al renders it useless as an electrode, since it is an insulator. Recently, research has demonstrated the viability of using M_2 = indium/tin oxide as a conducting oxide electrode [8]. Orthogonal self-assembly, in principle, can be extended in complexity to more than two reagents and surfaces. Work in this area is needed in order to develop rational methodology for preparation of complex structures of molecular materials on surfaces.

Work in the area of molecular modification of microelectrodes by spontaneous self-assembly has recently been extended to include self-assembly reagents which include photosensitive groups such as aryl azide and metal carbonyls [9-13]. Scheme II illustrates how such reagents can be used. In short, with the Mn carbonyl derivative, it is possible to introduce a wide range of two-electron donors on to the surface where photoexcitation occurs in the presence of a ligand. The aryl azide offers a route to functionalization of surfaces using primary and secondary amines. Thus, using focused light for "direct write" or by illumination through a mask, self-assembled monolayers can be photochemically patterned. Photochemical patterning of surfaces (macroscopic and microscopic) elaborates the methods available for functionalization of surfaces. Such photochemistry may be useful in fabrication processes for nano- and microfabricated devices.

Underlying all of molecular chemistry is a desire to relate structure and composition to function. In the case of modified surfaces it has become important to apply modern surface analytical methods to the characterization. It has been particularly important to be able to establish composition at high lateral resolution in characterizing modified microelectrode arrays. Advances in the use of SEM, SIMS, and Auger electron spectroscopy (AES) have proven useful in our work.

Photochemical Reactions of Self-Assembled Monolayers

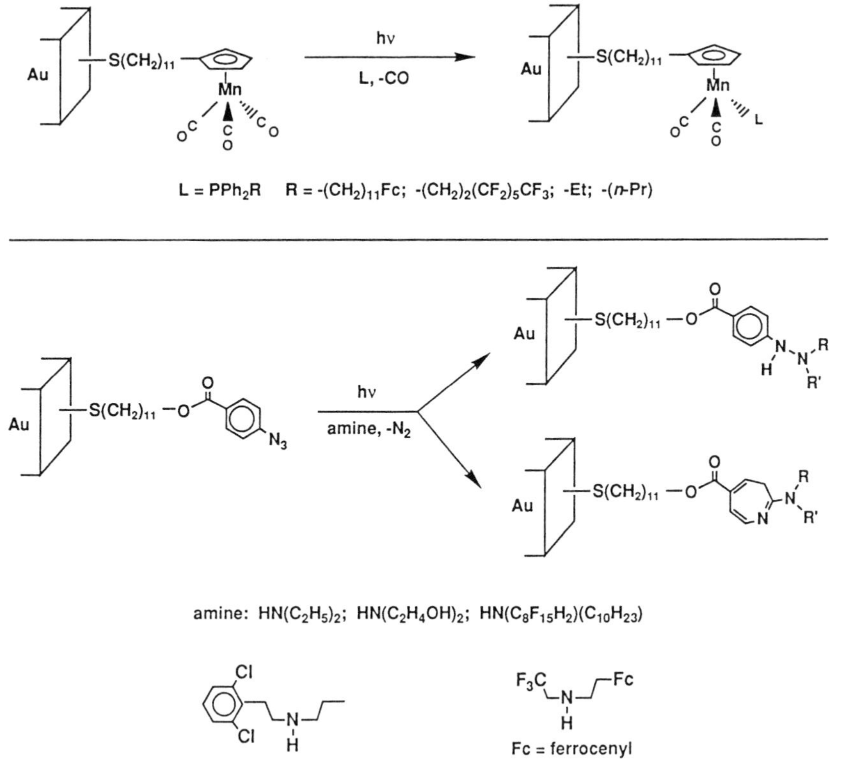

Au —S(CH$_2$)$_{11}$— Mn(CO)$_3$ —hν, L, -CO→ Au —S(CH$_2$)$_{11}$— Mn(CO)$_2$L

L = PPh$_2$R R = -(CH$_2$)$_{11}$Fc; -(CH$_2$)$_2$(CF$_2$)$_5$CF$_3$; -Et; -(*n*-Pr)

Au —S(CH$_2$)$_{11}$ —O— (C=O) — C$_6$H$_4$ — N$_3$ —hν, amine, -N$_2$→ Au —S(CH$_2$)$_{11}$ —O— (C=O) — C$_6$H$_4$ — NH—NRR' ; Au —S(CH$_2$)$_{11}$ —O— (C=O) — azepine — NRR'

amine: HN(C$_2$H$_5$)$_2$; HN(C$_2$H$_4$OH)$_2$; HN(C$_8$F$_{15}$H$_2$)(C$_{10}$H$_{23}$)

Fc = ferrocenyl

Scheme II

1. ORTHOGONAL SELF-ASSEMBLY

As illustrated in Scheme I, orthogonal self-assembly can be envisioned as a methodology for the functionalization of complex microfabricated structures. We have recently extended examples [6,7] of orthogonal self-assembly to include the useful electrode material indium/tin oxide (ITO) [8]. Molecules bearing carboxylate or phosphonate functionality, **I-III** appear to self-assemble selectively onto ITO in the presence of thiols such as **IV**. The work with the carboxylates follows from work done earlier on Al/Al_2O_3, and the use of phosphonates stem from work [14] with molecules such as **V** which binds to Au via the thiol end group and interacts strongly with oxides via the phosphonate.

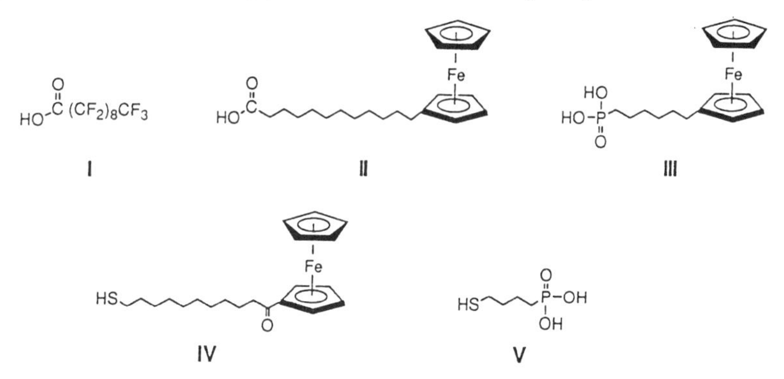

One of the key contributions underlying our recent work has been the development of a process for the preparation of ITO microelectrode (~2 x 100μm) arrays and arrays of interdigitated Au and ITO microelectrodes (~2μm spacings). The process for the preparation of ITO microelectrode arrays is summarized in Scheme III, and it is a relatively straightforward procedure to prepare the interdigitated microelectrodes needed for studies of orthogonal self-assembly [8]. A key to highly conductive ITO microelectrodes is the use of the multilayer arrangement incorporating Au to enhance the conductivity.

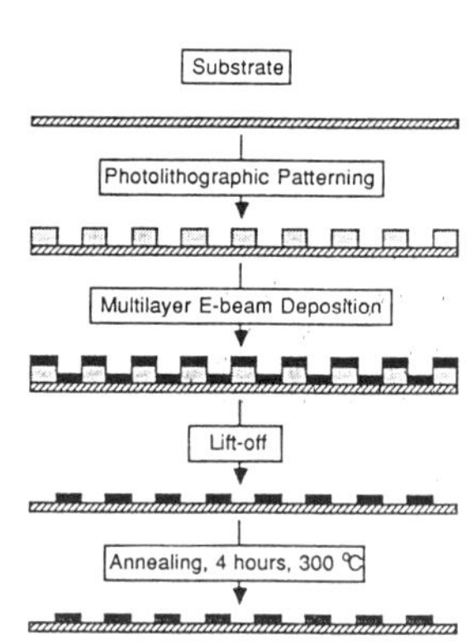

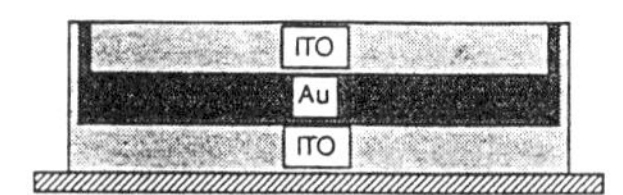

Electrode layer structure in cross section, as deposited, 1500 Å total thickness

Scheme III

Aside from the fundamental challenge associated with uncovering chemistry applicable to orthogonal self-assembly, our effort may prove useful in orienting molecules between two dissimilar metals in a fashion that yields diode functions or non-linear optical effects stemming from the orientation of the molecules. The sections that follow provide evidence for orthogonal self-assembly involving **I-IV** and Au and ITO as substrates.

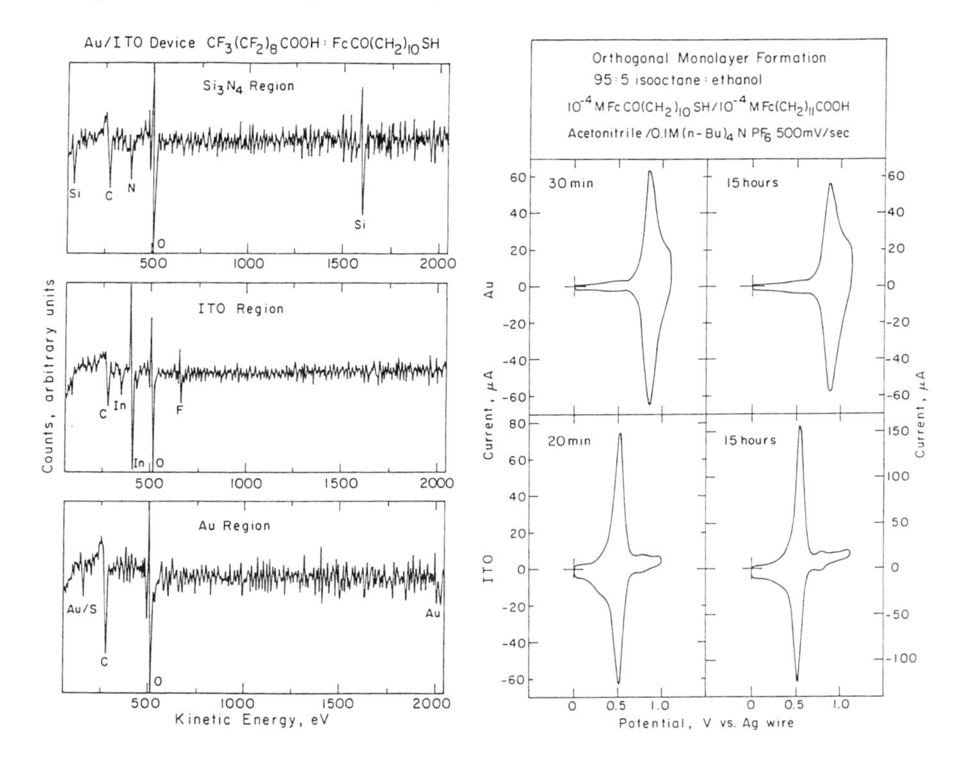

Figure 1. AES of Au and ITO surfaces.

Figure 2. Cyclic Voltammetry Showing Orthogonal Self-Assembly on Macroelectrodes.

1.1 Auger Electron Spectroscopy

AES has been used to establish the lateral position of molecular substances following self-assembly chemistry as represented in Scheme I. An advantage of AES is that the exciting beam is a beam of electrons that can be focused, providing high lateral resolution. A second advantage is that AES is very sensitive, even to substantially sub-monolayer coverages, depending on the element being assayed. However, a disadvantage is that AES does not provide much molecular specific information. AES is thus used as a method for assaying the presence or absence of particular elements over a surface. In assessing the surface distribution of self-assembly reagents using AES it is useful to introduce unique elements into the molecule, because most surfaces exposed to solvents and the atmosphere will show adventitious C, O, and N from impurities.

Figure 1 illustrates survey AES for an interdigitated array of Au and ITO microelectrodes withdrawn from a 1/1 ratio of **I** and **IV** in 5% EtOH in isooctane solution. The critical features are that the survey scan in the ITO region reveals a signal for F, but no S is detected on ITO, and the survey scan in the Au region shows a S signal, but no signal for F is found on Au. These findings are consistent with the conclusion that the carboxylate **I** (containing F) binds selectively to the ITO and that the thiol **IV** (containing S) binds selectively to the Au. The absence of both F and S on the Si_3N_4 substrate shows that neither molecule binds to the substrate. A complete map of the Auger electron intensities for the elements of interest shows that the molecules spontaneously assemble over the entire microelectrode array in the fashion shown in Scheme Ib, giving a uniform coating of F-containing material on the ITO, S-containing material on the Au, and no molecular material on the Si_3N_4. Thus, the orthogonal self-assembly as in Scheme Ib appears to obtain. However, the AES alone does not provide compelling evidence that self-assembly yields a monolayer of the self-assembly reagent. AES does not provide as quantitative a measure of surface coverage as is needed to make this determination.

1.2 Electrochemical Characterization

Cyclic voltammetry of microelectrodes modified by self-assembly can be a quantitative method for the measurement of surface coverage of molecules. Cyclic voltammetry can also provide quantitative measures of surface coverage for two or more reagents, since the formal potential of the redox substituents can vary sufficiently that the cyclic voltammetry waves do not overlap. The integral of the cyclic voltammetry wave is proportional to surface coverage, and in favorable cases all surface-bound reagents are electrochemically accessible. The ferrocene centers of the self-assembly reagents which prefer ITO, **II** and **III**, have a formal potential significantly negative of that for the reagent that prefers Au, **IV**. Thus, cyclic voltammetry of electrodes modified by self-assembly from a solution containing both **II** and **IV** or **III** and **IV** can be used to measure the surface coverage of both self-assembly reagents. Cyclic voltammetry, of course, does not provide information regarding the presence or absence of redox active molecules on the insulating Si_3N_4 substrate. Thus, the combination of AES and cyclic voltammetry can provide information regarding the distribution and surface concentration of molecules on the surface of a microelectrode array.

Cyclic voltammetry of ITO microelectrodes modified by self-assembly with **II** shows about 3×10^{-10} mol/cm^2 of ferrocene centers. Thus, the self-assembly technique yields about one molecular monolayer on the ITO surface. Figure 2 shows a comparison of the cyclic voltammetry for Au and ITO surfaces modified by self-assembly by immersion into a solution containing a 1/1 ratio of **II** and **IV**: the Au with the more positive cyclic voltammetry wave shows predominantly **IV** on its surface, while the ITO with the more negative cyclic voltammetry wave of the alkylferrocene centers shows predominantly **II** on its surface. Both Au and ITO show about one monolayer coverage from integration of the cyclic voltammetry waves. The combination of **III** and **IV** also demonstrates orthogonal self-assembly of a monolayer on ITO and Au, respectively, by immersion in a solution containing a 1/1 ratio of **III** and **IV**. Our current assessment is that the phosphonate/thiol combination represents a superior combination for orthogonal self-assembly, compared to the carboxylate/thiol, but quantitative comparisons need to be made.

2. CHARACTERIZATION OF PHOTOCHEMICALLY PATTERNED MONOLAYERS

Photochemistry has long been used in the preparation of microfabricated structures. Photochemical reactions of self-assembled monolayers represent a potentially practical way to pattern surfaces with diverse chemical functionality. Critical features of the photosensitive self-assembly reagents are that the molecules be thermally inert and capable of attaching a broad range of chemical groups. The aryl azide [15] and the Mn carbonyl [16] photosensitive groups fulfill both of these criteria, and we have recently reported [9-13] their use in self-assembly chemistry followed by photochemical patterning.

2.1 Aryl azide-based Systems

A self-assembled monolayer of the aryl azide, Scheme II, undergoes photochemical reaction in the presence of primary and secondary amines to attach the amine via the azepine or hydrazine product. The application of this technique to the functionalization of a microelectrode array is illustrated in Figure 3. What is shown is the cyclic voltammogram of each of the seven Au microelectrodes (~2 x 100 μm, spaced ~2 μm apart) of the array after self-assembly of the aryl azide onto all seven followed by masked irradiation of four of the electrodes through a thin film of a ferrocenylamine. Clearly, the irradiated electrodes each show a cyclic voltammogram consistent with the presence of an electroactive ferrocene, whereas the non-irradiated electrodes are essentially electrochemically silent. Integration of the waves for the irradiated electrodes shows about one monolayer of redox active material. The small electrochemical signal on the non-irradiated electrodes is likely a consequence of "light-piping" through the thin film of ferrocenylamine. In any event, it is evident that photochemical functionalization occurs. Support for the particular surface products shown comes from SIMS, electrochemistry, XPS, and FTIR studies [9-11].

Data presented in Figure 4 suggest that it should be possible to photochemically functionalize microfabricated structures with a variety of of amines. What is illustrated are SIMS maps for a Au microwire array on Si_3N_4 first functionalized with the aryl azide by self-assembly followed by masked irradiation through a thin film of a Cl-containing amine and then masked irradiation through a thin film of a F-containing amine. The SIMS maps show that the aryl azide self-assembles only onto the Au and not onto the Si_3N_4 substrate, because the Cl and F maps are in registration with the underlying Au and not with the exposed Si_3N_4 substrate. Further, the Cl map shows evidence that the Cl-containing amine is attached only where the microwires have been irradiated in the presence of the Cl-containing amine, and the F map confirms the expectation that the F-containing amine is found only where the microwires have been irradiated in the presence of the F-containing amine. The electrochemistry and SIMS data allow the conclusion that the aryl azide functional group provides a viable route to photochemically tailoring self-assembled monolayers. Primary and secondary amines provide a rich diversity of functional groups for tailoring surfaces by this method.

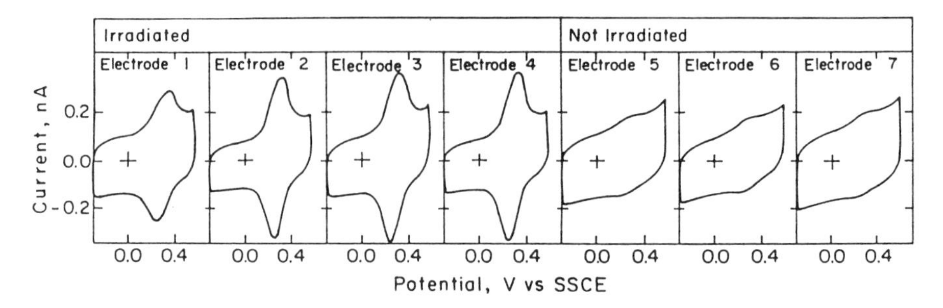

Figure 3. Cyclic Voltammetry of Photopatterned Microwire Array.

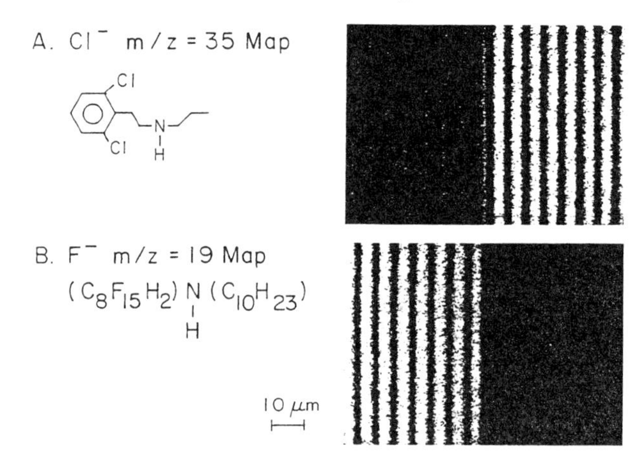

Figure 4 . SIMS Element Maps of a Photopatterned Microwire Array.

2.2 Mn Carbonyl-based Systems

It is well known that $(\eta^5\text{-}C_5H_5)Mn(CO)_3$ is extraordinarily thermally robust and sensitive to ultraviolet light [16]. Moreover, the dissociative loss of CO is the only detected photoprocess for the molecule. Thus, there is the reasonable prospect of functionalizing surfaces with a variety of ligands which will take up the vacant coordination site generated by the light-induced loss of CO from the Mn center. We have only investigated two-electron donor entering groups, e.g. phosphines, but oxidative addition substrates might also prove viable for modification of surfaces.

Self-assembly of the Mn carbonyl onto Au and subsequent photochemical reactions as represented in Scheme II have recently been

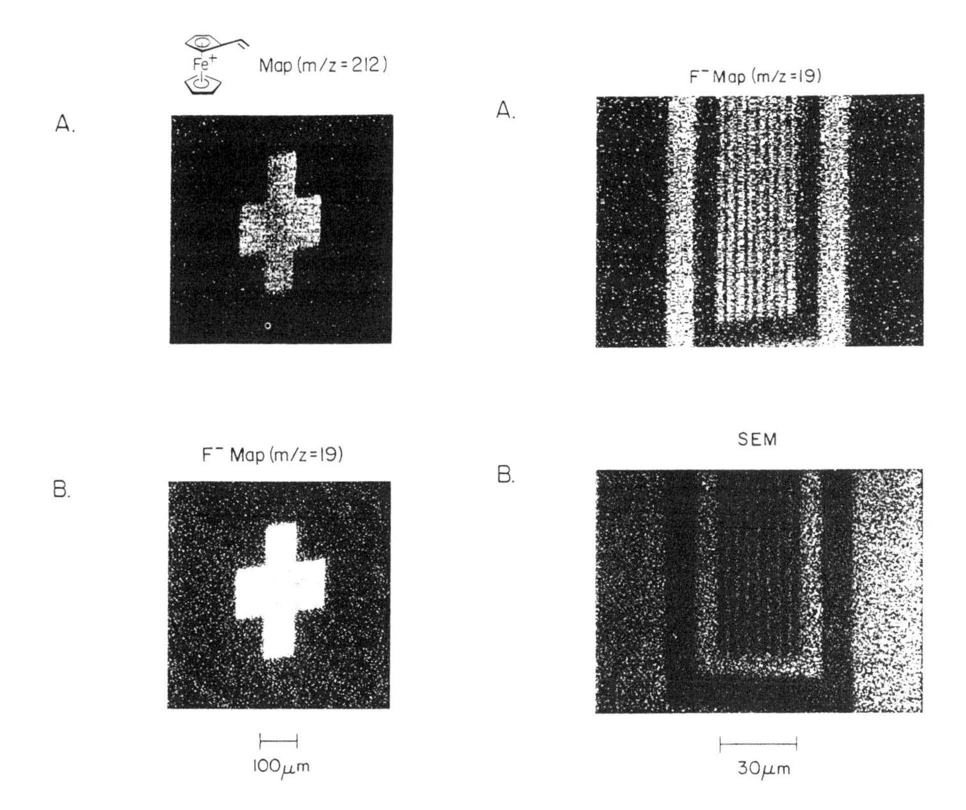

Figure 5. SIMS Maps of Photopatterned Monolayer.

Figure 6. SIMS Maps at High Lateral Resolution.

reported [12,13]. Electrochemical measurements of the amount of phosphine confined to the electrode surface using $PPh_2(CH_2)_{11}Fc$ (Fc = ferrocenyl) show that the photosubstitution yields about one monolayer of redox active material on the surface [12]. Interestingly, FTIR data show that the surface-bound molecule only yields photosubstitution of one CO, whereas the solution analogs give rapid photochemical formation of $(\eta^5$-$C_5H_5)Mn(CO)L_2$ species. Moreover, the ability to photochemically introduce the phosphines is more difficult for the surface-confined species, since higher concentrations of the entering phosphine are needed than for the solution photochemistry. It is possible that the photogenerated intermediate on the surface undergoes decomposition reactions with surface functional groups. The higher concentrations of entering groups are needed, in order to more rapidly scavenge the coordinatively unsaturated Mn center prior to such decomposition processes.

Despite the difficulties in solution vs. surface photoreactions of the Mn species, the surface-confined species has been shown to take up a variety of phosphines as listed in Scheme II. The functionalization of macroscopic Au surfaces with the Mn self-assembly reagent followed by masked irradiation yields photochemically patterned Au surfaces [13]. Similar photochemical patterning can be effected using the aryl azide self-assembly reagent on Au surfaces [9-11]. Evidence for the photochemical patterning of the macroscopic Au surfaces comes from SIMS and SEM, Figures 5 and 6. In Figure 5a the patterned (cross) region has been imaged by mapping the vinylferrocenium fragment from positive ion SIMS after masked irradiation of the self-assembled Mn carbonyl system in the presence of the $PPh_2(CH_2)_{11}Fc$. The bright region corresponds to high concentration of the indicated fragment on the surface. In Figure 5b the same mask was used in an irradiation in the presence of $PPh_2(CH_2)_2(CF_2)_5CF_3$. The negative ion SIMS map for F^- shows the replication of the mask on the surface on the irradiated monolayer. Thus, SIMS shows image transfer to the irradiated surface in that unique elements and fragments due to the entering ligand are found where light strikes the surface. The non-irradiated regions of the surface only show background signals for the entering group and the expected strong signals for the self-assembled Mn carbonyl.

Relatively high lateral resolution patterning of surfaces using either the aryl azide or the Mn carbonyl systems has been demonstrated. Figures 3 and 4 provide some evidence for this conclusion in that selective irradiation through a simple rectangular mask reveals that particular microwires of an array can be modified photochemically. Figure 6 illustrates that resolution in the micron regime is possible. Figure 6a shows the F^- SIMS map following masked irradiation of a self-assembled Mn carbonyl monolayer in the presence of the F-containing phosphine. The distribution of F^- replicates the mask which has 4 micron center-to-center spacings of the smallest features. The patterned surface can also be imaged by SEM, Figure 6b. The use of SEM to image patterned monolayers appears to be a viable method for rapidly assessing the fidelity of image transfer in such systems. It is likely that SEM will allow even higher lateral resolution imaging of photochemically (or electron beam) patterned self-assembled monolayers [10, 17].

CONCLUSIONS

Rational methods for modification of microfabricated structures based on self-assembly and photochemistry have been demonstrated for a wide range of molecular functional groups, including redox active molecules. The ability to rationally tailor surfaces with molecular functional groups opens potential applications in electrochemical devices, including sensors and analogs of transistors and diodes. The practical utility of these devices rests on their unique characteristics and cost. Development of the chemistry of surfaces contributes to the ability to design and fabricate chemical systems having desired functions. The ability to introduce a wide range of functional groups by relatively simple methods, as illustrated by the work outlined in this article, is essential to encouraging development of applications of microelectrochemical devices.

ACKNOWLEDGMENTS

We are grateful for support of the research summarized in this article. Sponsoring agencies have included the National Science Foundation, the United States Department of Energy, Office of Basic Energy Sciences, Division of Chemical Sciences, and the Advanced Research Projects Agency.

REFERENCES

1. J. J. Hickman, C. Zou, D. Ofer, P. D. Harvey, M. S. Wrighton, P. E. Laibinis, C. D. Bain, G. M. Whitesides, *J. Am. Chem. Soc.*, **111** (1989) 7271.

2. J. J. Hickman, D. Ofer, P. E. Laibinis, G. M. Whitesides, M. S.Wrighton, *Science*, **252** (1991) 688.

3. J. J. Hickman, C. Zou, M. S. Wrighton, P. E. Laibinis and G. M. Whitesides, *J. Am. Chem. Soc.*, **113** (1991) 1128.

4. C. D. Frisbie, I. Fritsch-Faules, E. W. Wollman and M. S. Wrighton, *Thin Solid Films*, **210/211** (1992) 341.

5. C. D. Frisbie, J. R. Martin, R. R. Duff and M. S. Wrighton, *J. Am. Chem. Soc.*, **114** (1992) 7142.

6. P. E. Laibinis, J. J. Hickman, M. S. Wrighton and G. M. Whitesides, *Science*, **245** (1989) 845.

7. J. J. Hickman, P. E. Laibinis, D. I.Auerbach, C. Zou, T. J. Gardner, G. M. Whitesides and M. S. Wrighton, *Langmuir*, **8** (1992) 357.

8. T. J. Gardner, C. D. Frisbie and M. S. Wrighton, submitted for publication.

9. C. D. Frisbie, E. W. Wollman, J. R. Martin and M. S. Wrighton, *J. Vac. Sci. & Technol. A*, **11**(4) (1993) 2368.

10. E. W. Wollman, C. D. Frisbie and M. S. Wrighton, *Langmuir*, **9** (1993) 1517.

11. E. W. Wollman, I. M. Lorkovic, C. D. Frisbie and M. S. Wrighton, submitted for publication to *J. Am. Chem. Soc.*

12. (a) D. Kang and M. S. Wrighton, *Langmuir*, **7** (1991) 2169. (b) D. Kang, E. W. Wollman and M. S. Wrighton in "Photosensitive Metal-Organic

Systems", C. Kutal and N. Serpone, (eds.), Advances in Chemistry Series, vol. 238, American Chemical Society, Washington, D. C., 1993, pp. 45-65.

13. D. Kang, C. D. Frisbie and M. S. Wrighton, submitted for publication to *J. Am. Chem. Soc.*

14. H.-G. Hong and T. E. Mallouk, *Langmuir*, **7** (1991) 2362.

15. E. F. V. Scriven, Ed., Azides and Nitrenes, Academic Press, San Diego, CA, 1984.

16. (a) G. L. Geoffroy and M. S. Wrighton, Organometallic Photochemistry, Academic Press, New York, 1979. (b) M. Wrighton, *Chem. Rev.*, **4** (1974) 401. (c) A. Cox, *Photochemistry*, **14** (1983) 158.

17. G. P. Lopez, H. A. Biebuyck and G. M. Whitesides, *Langmuir*, **9**, (1993) 1513.

MICROCHEMISTRY
Spectroscopy and Chemistry in Small Domains
Edited by H. Masuhara et al.

The Characterization and Fabrication of Small Domains by Scanning Electrochemical Microscopy

Allen J. Bard

Department of Chemistry and Biochemistry,
The University of Texas at Austin, Austin, Texas 78712 USA*

Following a brief discussion of the general principles, theory, and instrumentation of SECM, the application of this technique to imaging of surfaces and characterization of polymers and semiconductors is discussed. Fabrication by electrodeposition and etching is also described.

1. INTRODUCTION

Microelectrochemical methods are useful in the characterization and fabrication of small domains on surfaces and provide an alternative to the widely used spectroscopic and photochemical (e.g., lithographic) approaches. High-resolution electrochemistry generally uses ultramicroelectrodes (electrodes whose exposed area is in the nm—μm range) and, in the work described here, the scanning electrochemical microscope (SECM) [1-4]. In these methods, the size of this electrode (the tip) controls the resolution of the instrument and the minimum size of the features that can be produced. Thus the SECM, as other scanning probe techniques, is capable of very high resolution, easily in the sub-μm range, and is not limited by wave length or, so far, by quantum effects.

1.1. SECM Principles

SECM is based on the changes in the faradaic current that flows during a redox process at a tip as it moves above a substrate that is immersed in a solution containing an appropriate species in the oxidized (Ox) or reduced (Red) state. In addition to the tip electrode, auxiliary and reference electrodes are immersed in the solution to form an electrochemical cell, and the potential of the tip is controlled by a potentiostat [5]. For example, if the solution contains the species Ox (e.g., Fe^{3+}) and the tip potential is adjusted with respect to the reference electrode to reduce Ox at a mass-transfer-controlled rate, the reaction Ox + ne $\rightarrow$ Red occurs. If the ultramicroelectrode tip is far from any substrate, the steady-state current that flows, $i_{T,\infty}$, is given by equation (1) [6]

$$i_{T,\infty} = 4nFDca \qquad (1)$$

* The support of the Robert A. Welch Foundation and the National Science Foundation are gratefully acknowledged.

where F is the Faraday, c is the concentration of Ox, D is its diffusion coefficient, and a is the radius of the tip. This current represents the flux of Ox to the electrode through the essentially hemispherical diffusion layer around the tip. However, when the tip is close to the substrate, i.e., within a few tip radii, the current is perturbed by the presence of the substrate. If the reverse reaction Red $\rightarrow$ Ox + ne cannot occur on the substrate, e.g., if the substrate is an electrical insulator that does not react with Red, the current will be smaller than $i_{T,\infty}$, because the substrate simply blocks the diffusion of Ox to the tip. In SECM parlance, this effect is termed "negative feedback." However, for a conductive substrate, the oxidation of Red to Ox can occur. This provides an additional source of Ox for the tip, and the current observed is greater than $i_{T,\infty}$. Thus when the reverse reaction can occur at the substrate,"positive feedback" is observed. These principles are illustrated in Figure 1. In general in SECM the relative magnitude of the current compared to $i_{T,\infty}$ is a measure of the nature of the distance between tip and substrate, d, and depends upon the nature of the substrate.

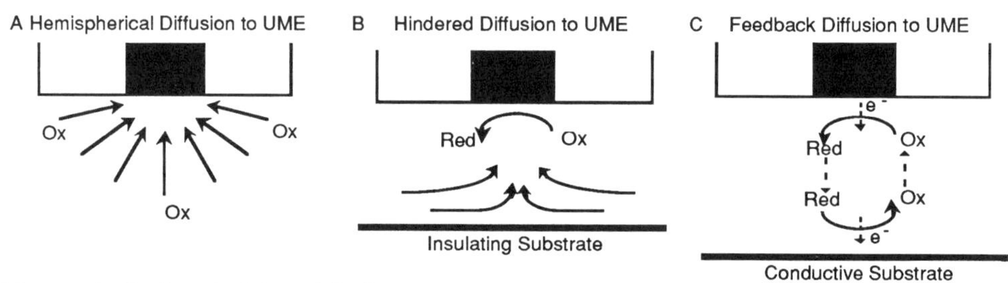

Figure 1. Basic principles of SECM. (A) With UME far from substrate, diffusion leads to a steady-state current $i_{T,\infty}$. (B) UME near an insulating substrate. Hindered diffusion leads to $i_T < i_{T,\infty}$. (C) UME near a conductive substrate. Positive feedback leads to $i_T > i_{T,\infty}$. (Reprinted from Ref. 2. Copyright 1991 American Association for the Advancement of Science.)

An advantage of SECM compared to other scanning probe microscopies, e.g., scanning tunneling microscopy (STM), is that the theoretical behavior is readily accessible through typical electrochemical diffusion-kinetic treatments [3,7-12]. Curves can be cast in the dimensionless form of $i_T/i_{T,\infty}$ vs. d/a, which are independent of solution concentration and diffusion coefficient. Curves showing the predicted behavior for an unreactive insulator (i.e., a material where the rate constant, k_f, of the Red $\rightarrow$ Ox reaction is 0) and for a conductive substrate (where $k_f \rightarrow \infty$) are shown in Figure 2. Thus by measuring the $i_T/i_{T,\infty}$ ratio, one can immediately estimate the tip-substrate distance, d, if the tip radius, a, is known. When the rate constant of the Red $\rightarrow$ Ox reaction is of an intermediate value, i.e., $0<k_f<\infty$, a family of curves is obtained that spans the behavior between the two limits illustrated in Figure 2 [10,13]. The dependence of $i_T/i_{T,\infty}$ on k_f allows one to carry out reaction-rate imaging of surfaces, as discussed below.

The apparatus used for SECM [1,4] basically combines electrochemical and STM instrumentation. Thus a potentiostat (or a bipotentiostat, if the substrate potential is controlled) is used to adjust the tip potential with respect to the reference electrode [5]. The tip is moved towards and away from the substrate (the z-direction) and across the surface (x- and y-directions) by means of piezoelectric scanners. Tip potential and position are adjusted via a digital computer and the associated A/D and D/A cards. This arrangement allows one to obtain "approach curves" of $i_T/i_{T,\infty}$ vs. d/a, and surface scans showing i_T at a given d as a function of x,y position. SECM scans with tip position modulation and constant current operation are also possible [14,15].

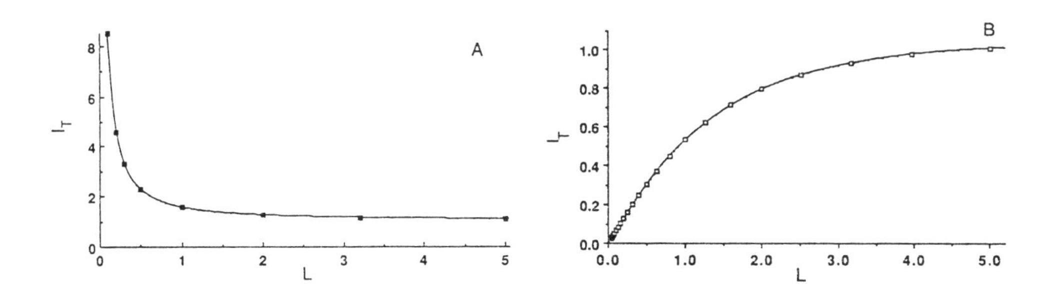

Figure 2. Diffusion-controlled steady state normalized tip current ($I_T=i_T/i_{T,\infty}$) as a function of dimensionless tip-substrate separation (L=da). (A) Substrate is a conductor where the reverse reaction is diffusion-controlled. (B) Substrate is an unreactive insulator. Data from Ref. 7. (Reprinted from Ref. 4. Copyright 1993 Marcel Dekker.)

1.2 Resolution and small domains

As mentioned above, resolution depends upon tip size, and the tip configuration is critical in SECM. Details about preparation of tips are given in several reviews [4,6]. This usually involves the electrochemical etching of a small wire, most frequently Pt, followed by coating with glass, wax or polymer to insulate all but the very end of the tip. The best method of characterizing small tips (a<1 μm) is probably in the SECM, by examining $i_{T,\infty}$ and the approach curves [16]. This allows one to estimate a and the shape of the tip, and warns of tips of improper geometry (e.g., with the tip recessed back in a space within the glass insulator) (see, e.g., [17-19]).

Let us consider the size of the small domains probed by the SECM, i.e., the effective volumes of solution near the tip, the areas on the surface, and the amount of material in these. We consider a tip of radius d/2 spaced a distance d away from the substrate. Roughly, in the time it takes for species Red generated at the tip to diffuse across the gap, approximately $(2Dt)^{1/2}$, it will also diffuse an equal distance in the x,y plane. The approximate volumes and areas addressed by the tip can be calculated as shown in Table 1.

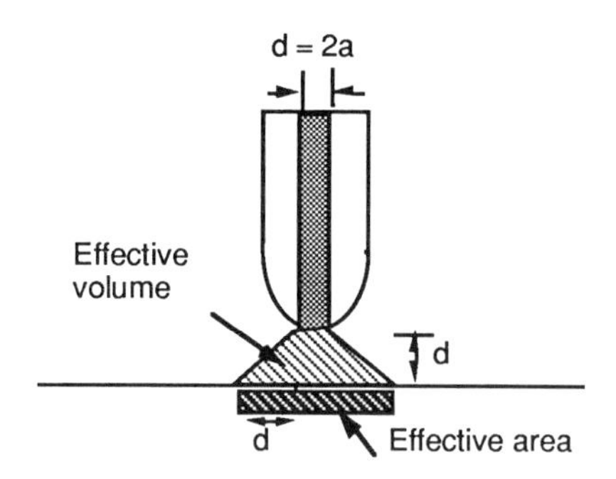

Table 1
Effective volumes (V) and areas (A) below tip, and approximate amount of species in those zones, assuming a concentration of 0.1 M and surface coverage of 10^{-10} mol cm^{-2}, for different values of d, with D=5 x 10^{-6} cm^2/s.

d	V (cm^3)	Contents (molecules)	A (cm^2)	Contents (molecules)	Transit Time
10 μm	3 x10^{-9}	2 x 10^{11}	7 x 10^{-6}	4 x 10^8	0.1 s
1 μm	10^{-12}	10^8	10^{-8}	10^6	1 ms
0.1 μm	10^{-15}	10^5	10^{-10}	10^4	10 μs
10 nm	10^{-18}	10^2	10^{-12}	10^2	100 ns

Adapted from Table 1, Ref. 4.

Clearly, the SECM is useful in addressing very small domains and potentially for measuring small amounts of materials in small volumes and surface areas. The ultimate limits depend upon the tip size and the measurement and control of small currents. Note, for example, that a current of 0.1 pA, which is roughly the lower limit now used in SECM, represents a faradaic process of 10^{-18} mol/s (n=1) or only about 600,000 electrons/s crossing the tip/solution interface.

2. SURFACE CHARACTERIZATION

2.1 Imaging

Many different types of surfaces have been imaged by SECM, including electrodes, minerals, semiconductors, membranes and biological specimens [2-4]. Most of these imaging studies have been carried out in the constant height mode, where the tip is rastered across the substrate at a constant reference plane above the sample surface and variations in the tip current are recorded. These are then used to produce topographic plots or gray-scale images of the sample surfaces. Fairly high resolution images of a polycarbonate membrane filter with a nominal pore size of 0.2 μm, obtained with a Pt/Ir tip (nominal exposed area about 50 nm) and a solution of 0.2 M $K_4Fe(CN)_6$ and 0.5 M Na_2SO_4 are shown in Figure 3.

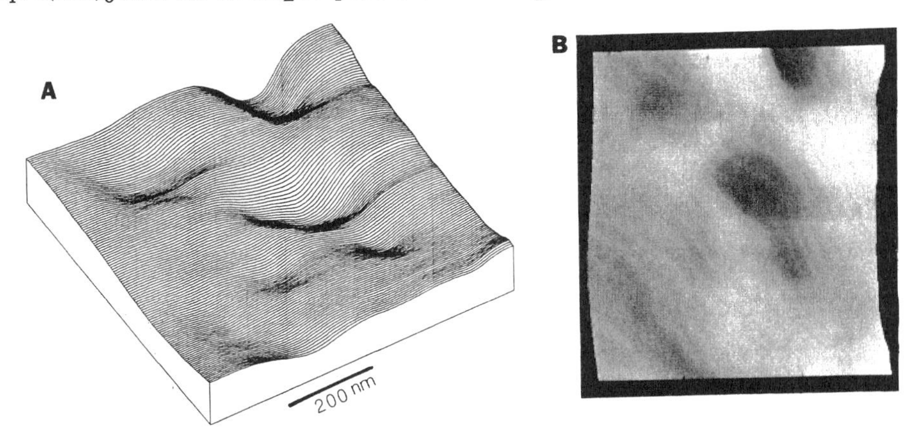

Figure 3. A 720 nm x 650 nm constant height SECM image of a Cyclopore polycarbonate membrane filter with nominal 0.2 μm pores. The solution contained 0.2 M $K_4Fe(CN)_6$ and 0.5 M $NaSO_4$. The tip potential was kept at 0.8 V vs. SCE. The rastering rate was 0.25 Hz. The current range is 0.15-0.45 nA. (A) Topographic view. The image has been inverted, so the lower parts of the image represent the pore sites. (B) Gray scale image. (Reprinted from A. J. Bard and F.-R. F. Fan, Faraday Discuss. 94 (1992) 1. Copyright 1992 Royal Society of Chemistry.)

It is also possible to obtain images in the constant current mode, as is frequently used in STM. This is especially important with small tips where constant height scanning can lead to tip crashes because of vibrations and substrate surface irregularities. As in STM, constant tip current imaging is carried out by changing the tip-sample spacing, d, with the z-piezo during the scan. However, since a decrease in d causes an increase in i_T with a conductive substrate, but a decrease in i_T with an insulator, a means must be available to identify the nature of the substrate and vary the sense of the piezo-movement. This can be accomplished by modulating the position of the tip and noting the phase of the modulated i_T with respect to the modulation [14,15]. Tip position modulation is also useful in improving the signal-to-noise ratio, especially with insulating substrates. A constant current image of a Kel-F/gold composite, containing both insulating (Kel-F) and conductive (Au) zones is given in Figure 4.

2.2 Reaction rate imaging and heterogeneous kinetics

Since the tip response depends upon the rate of the electron transfer reaction on the particular site being imaged on the substrate surface, by selecting the proper solution mediator, Ox/Red, and imaging conditions (e.g., the applied potential to the substrate) one can distinguish sites of different reactivity. The principles of this approach are illustrated in Figure 5. This was first demonstrated for a glassy carbon (GC) surface containing embedded Au sites, where Fe^{3+} was reduced at the tip, and the generated Fe^{2+} oxidized on the GC/Au substrate [13]. Not only is it possible to show qualitatively differences in the reaction rate at different locations on a surface, but, by quantitative measurements of i_T as a function of d and substrate potential, one can obtain values for heterogeneous electron transfer rate constants of surface reactions [10,20]. In addition to electron transfer reactions, other types of surface reactions, such as reactions of enzymes on surfaces [21,22] can be studied.

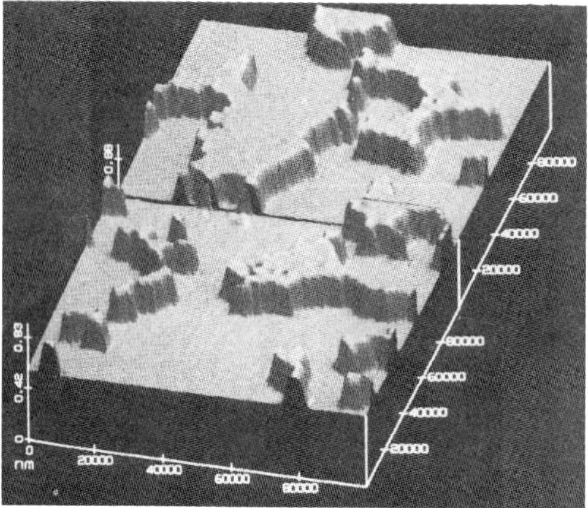

Figure 4. Surface plot of the tip current recorded during a constant-current image scan. The image shown is a composite of two consecutive scans. Scan size is 100 μm x 200 μm. Vertical axis is current in nA (0-0.83 nA) (Reprinted from Ref. 15. Copyright 1993 American Chemical Society.)

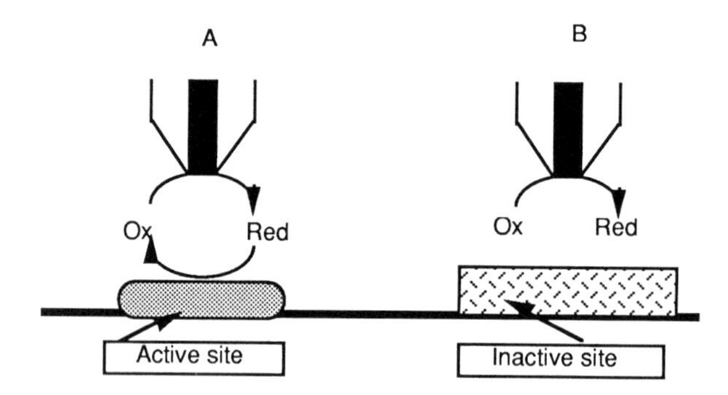

Figure 5. Principle of reaction rate imaging. A. An active site produces positive feedback. B. An inactive site produces negative feedback.

Chemical reactions of tip-generated species with a substrate can also be studied by a similar approach. For example, the rate of reaction of tip-generated $Ru(NH_3)_6^{2+}$ with a film of AgBr

$$Ru(NH_3)_6^{2+} + AgBr \rightarrow Ru(NH_3)_6^{3+} + Ag + Br^- \quad (2)$$

could be obtained from the SECM approach curve (Figure 6) [23] to yield a rate constant of 0.082 cm/s. Transient measurements at the Ag/AgBr electrode could be employed to determine the diffusion coefficient, D_{Br}, of Br^- in the AgBr film. In this case the tip is held in close proximity to the film surface and its potential is adjusted to a value (+0.9 V vs SCE) where Br^- is oxidized. When the substrate is stepped to a negative potential, the reaction $AgBr + e \rightarrow Ag + Br^-$ occurs at the Ag/AgBr interface, and Br^- diffuses through the AgBr film, and the small solution gap to the tip. By measuring the transit time, one can determine D_{Br} (5.6×10^{-7} cm^2/s). Earlier studies of a chemical reactions at interfaces by SECM involved protonation/deprotonation reactions of TiO_2 and Albite [24]. Note that in these types of studies the tip plays the role of a precise and controlled source of a species that can react at the substrate. These thus represent coulometric titrations in a confined zone with high accuracy and spatial resolution.

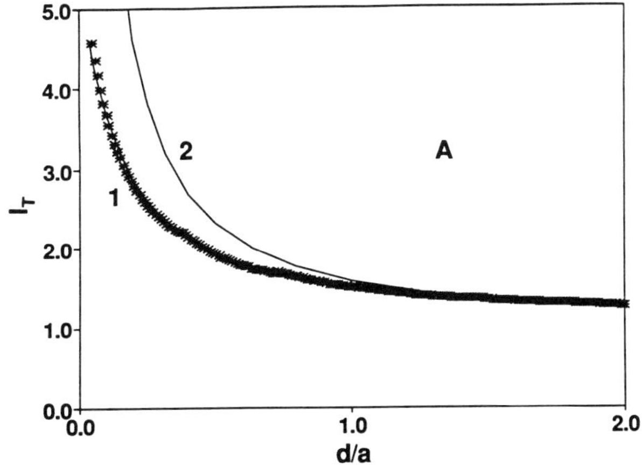

Figure 6. SECM approach curve showing positive feedback caused by the reaction of $Ru(NH_3)_6^{2+}$ with a 10µm thick AgBr film. The theoretical curve drawn through the experimental data (curve 1) was computed with a rate constant for the reaction, k, with $k/D=1.25 \times 10^4$. Curve 2 shows a diffusion controlled reaction. The solution contained 5 mM $Ru(NH_3)_6^{2+}$ and 0.5 M KNO_3 and the tip potential was +0.8 V vs SCE. (Reprinted from Ref. 23. Copyright 1993 American Chemical Society.)

Another way the SECM can be used to study surface reactions is by the use of potentiometric probes [25,26]. These tips are ion selective electrodes which respond logarithmically to the activity of species generated at a substrate surface (e.g., H^+, Cl^-).

2.3 Polymers and semiconductors

SECM has also been used to study thin polymer films and membranes [4]. As an example of an application where the SECM could be used to probe directly inside of a thin film, consider the case of a 2200 Å thick film of the ion-exchange polymer Nafion containing $Os(bpy)_3^{2+}$ [27]. In this experiment the film was immersed in an aqueous solution of 40 mM $NaClO_4$, containing no electroactive species, and was probed with a small conical tip (30 nm radius, 30 nm height). The experiment was started with the tip in the solution above the film and an approach curve was recorded as the tip moved into the film and ultimately contacted the indium tin oxide (ITO) conductor beneath the Nafion film (Figure 7). The tip was held at a potential (+0.80 V vs SCE) where the reaction $Os(bpy)_3^{2+} - e \rightarrow Os(bpy)_3^{3+}$ was diffusion-controlled and slowly (30 Å/s) moved towards and into the film. The curves in Figure 7 can be understood in terms of the different processes that occur during this tip movement. Curve 2, zone a, where essentially no current flows, represents the tip completely in the aqueous medium. The tip then enters the film (zone b) and is finally totally inside the film, where the current is essentially constant at about 1.3 pA (zone c). As the tip gets closer to the ITO the SECM positive feedback effect occurs and the current begins to increase (zone d). Finally the tip gets within tunneling distance of the ITO and a large increase in current is seen (curve 1). Measurements like these can provide a value for the film thickness, even when the film is immersed and swollen by solvent, by noting the difference between the initial and final current increases. It also allows determination of the diffusion coefficient of electroactive species in the film, via $i_{T,\infty}$. Finally one can do voltammetric measurements within the film and use these, as is typically done in electrochemistry [5], to determine standard potentials, rate constants, and mechanistic information.

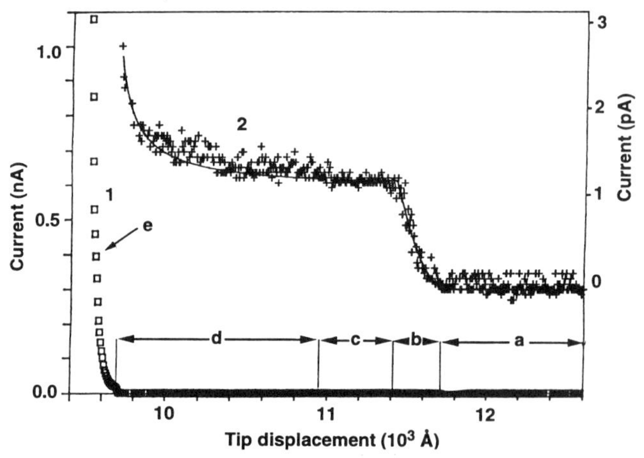

Figure 7. Approach curve for a small conical tip (30 nm base, 30 nm height) with a 2200 Å Nafion film on ITO substrate. The tip was biased at +0.80 V vs SCE and the ITO at +0.20 V vs SCE. The tip was moved in the z-direction at 30 Å/s. Letters correspond to different stages, as discussed in the text. (Reprinted from Ref. 27. Copyright 1992 American Association for the Advancement of Science.)

SECM can also be used to monitor fluxes through membranes, e.g., ones separating a compartment containing an electroactive species from one containing the tip and electrolyte. This approach was employed [28] to image the iontophoretic flux of Fe^{2+} through mouse skin, where single pores could be imaged. In principle this technique would allow quantitative flux measurements at a single, sub-μm size pore.

We have recently employed SECM to study processes on a semiconductor electrode surface [29]. By using the tip as a probe or collector of products generated on the semiconductor surface, one can determine the efficiency of generation of different species. These measurements can be used to estimate heterogeneous electron transfer rate constants at the semiconductor, even when they occur in parallel with corrosion processes. By performing reaction rate imaging, one can note how surface structure affects reaction rate. For example, in imaging a WSe_2 surface, one can see different rates at the basal plane and on edge sites [29]. Photoprocesses on the semiconductor surface can be studied in a similar way. Note that in these applications the SECM is somewhat similar to the rotating ring-disk electrode (RRDE), where a metal ring collects products from the semiconductor disk. However, with the SECM it is not necessary to construct an RRDE with each semiconductor sample; the effective mass transfer between tip and sample is higher than that attainable at an RRDE, and the SECM can provide high resolution information about the sample surface.

3. FABRICATION WITH THE SECM

3.1 Principles

Electrochemical methods are widely used in fabrication, for example in metal plating and electromachining. The SECM can be used in an analogous manner to produce microstructures on surfaces, either by deposition of materials or by etching the substrate. Representative examples of such processes are shown in Table 2. Two different approaches, the direct method [30], where the reaction occurs without the use of a mediator, and the feedback method [31], where a tip-generated species causes the desired reaction, have been used (Figure 8). Thus in the direct deposition mode the tip acts as the working electrode with the desired electrodeposition occurring at the end of the tip. In etching reactions, the tip is the counter electrode and etching occurs on the substrate immediately below the tip. In feedback methods a mediator couple is used, as in SECM characterization experiments, and it is the reaction of the tip-generated species on the substrate that brings about the desired deposition or etching reaction.

3.1 Metal deposition

The earliest approach to the direct deposition of a metal is shown schematically in Figure 8A [30,32,33]. A thin film of an ion exchange polymer, such as Nafion or poly(vinylpyridine) (PVP) was coated on the surface of the substrate, which was then immersed in a solution of the metal to be reduced in the appropriate charge state (e.g., Ag^+ with Nafion or $AuCl_4^-$ with PVP). The film was then removed from the solution, placed in the SECM and scanned in air, with the tip the cathode. In this case the tip only slightly penetrated the polymer film, so that the tip area was defined by the penetration depth, as in the experiment with the Nafion film discussed in 2.3 above. The extent of tip penetration, and the rate of metal deposition, was controlled by maintaining the current at the desired level. The smaller the current, the smaller the amount of tip penetration and the smaller the deposition rate and feature size. The result of deposition of Ag by this approach at a W tip moved in the desired pattern at 90 nm/s with a current of 0.5 nA is shown in Figure 9. Lines of Ag as narrow as 0.3 μm were produced by this procedure. By changing the polarity of the tip, metal deposition directly on the substrate is possible, with a counter reaction, such as the oxidation of water, occurring at the tip.

Table 2
Representative reactions for fabrication with the SECM

Process	Reactions	Examples
Metal deposition (Direct)	$M^{n+} + ne \rightarrow M$	Ag, Cu, Au
Metal deposition (Feedback)	$Ox + e \rightarrow Red$ (tip) $M^{n+} + nRed \rightarrow M + nOx$	Au, Pd
Metal or semiconductor etching (Direct)	$M - ne \rightarrow M^{n+}$	Cu, GaAs
Metal or semiconductor etching (Feedback)	$Red - e \rightarrow Ox$ (tip) $M + nOx \rightarrow M^{n+} + nRed$	Cu, GaAs
Polymer deposition[a]	$P - e \rightarrow P^+ \rightarrow P_n$	Polyaniline
Hydrous oxide precipitation[b]	$H_2O + 2e \rightarrow 2OH^- + H_2$ $M^{n+} + nOH^- \rightarrow M(OH)_n$	SnO_2

In these reactions Ox/Red represents a 1e redox mediator of appropriate potential. [a]P represents a suitable monomer. Polymer production could also be a reductive process. Indirect formation of polymer through a suitable tip-generated mediator should also be possible. [b]Other precipitation reactions with tip generated species (e.g., sulfide, halide) are also possible.

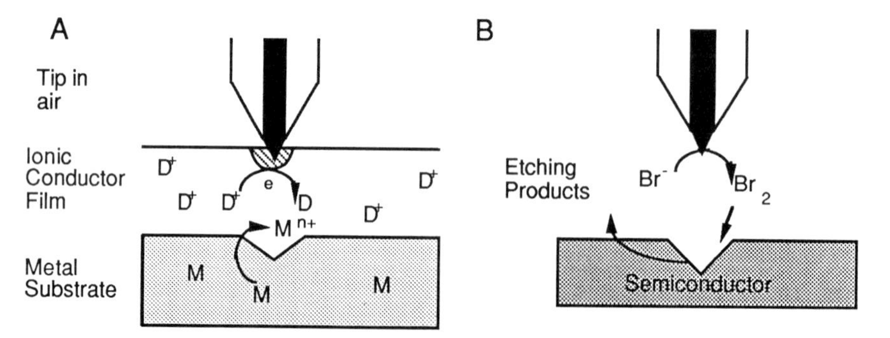

Figure 8. Schematic representations of the use of the SECM for fabrication of small features. A. Direct mode deposition (of D) and etching (of M) in a conductive film. B. Feedback mode etching.

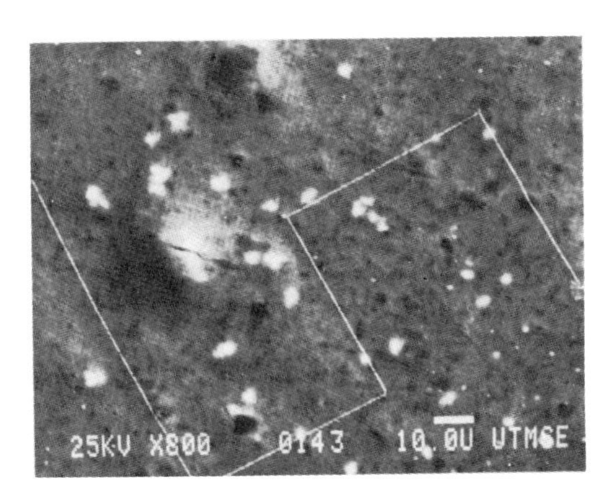

Figure 9. Scanning electron micrograph of silver lines deposited in a Nafion film by using the SECM in the direct mode. Tip material, tungsten; bias, 5 V, tip current, 0.5 nA; scan rate, 900 Å/s. (Reprinted from Ref. 33. Copyright 1989 The Electrochemical Society, Inc.)

In the feedback mode of metal deposition the same arrangement as in SECM imaging is used, with the tip reaction chosen to cause the desired reduction reaction at the substrate. For example deposition occurs when a reductant generated at the tip, e.g., $Ru(NH_3)_6^{2+}$, reacts with a metal ion, e.g., $AuCl_4^{2-}$, contained in a thin polymer (PVP) film on the substrate. The feedback mode has the advantage that the substrate need not be a conductor, as is necessary in the direct mode. However, the size of features deposited with a given tip will probably be larger than with the direct mode.

3.2 Etching

Direct etching of a metal substrate, like Cu, covered with a polymer film is accomplished by biasing the metal positive with respect to the tip, as suggested in Figure 8A [32]. It is useful to have a mediator in the film to prevent deposition of metal etched from the substrate on the tip. We have found that in the direct etching mode it is sometimes difficult to prevent etching of substrate at greater distances from the tip; even when the tip is in close proximity to the substrate, the electrical field distribution can lead to anodic processes in a wide area of substrate, depending on film resistivity. Note also that some processes carried out in air with the STM might actually be SECM reactions occurring in the thin film of liquid on the substrate surface. A typical example is the production of pits on highly oriented pyrolytic graphite (HOPG) when it is biased positive with respect to the tip in air [35]. Pits are not formed under similar conditions with the STM in high vacuum, suggesting the importance of a water layer on the substrate surface and an electrochemical route to the process.

Etching in the feedback mode involves generation of an appropriate oxidant (etchant) at the tip. For example in the etching of Cu, $Os(bpy)_3^{3+}$ is generated at the tip, diffuses to the Cu substrate, and oxidizes the Cu to Cu^{2+}, while it in turn is reduced to $Os(bpy)_3^{2+}$ which diffuses back to the tip to cause a positive feedback effect [31]. The observation of positive feedback is an indication that etching is occurring and allows one to position the tip precisely with respect to the substrate. As discussed in 2.2, approach curves determined during etching can be used to study the kinetics of the etching reaction. Another advantage of this approach is that there is no possibility of metal deposition on the tip (held at a positive potential) and less of a tendency to etch metal far from the tip position because of electric field effects.

The feedback method has also been used with semiconductors, like GaAs, by generating an appropriate oxidant, e.g., Br_2, at the tip [36]. The size and depth of the features produced in the semiconductor depended upon the length of time the biased tip was held above the particular site on the GaAs surface. The feedback etching mode is also useful in studying the mechanism of the semiconductor etching process [37]. For example by varying the mediator, pH, and doping of the GaAs (n- or p-) one could show that etching only occurred when the tip-generated oxidant was sufficiently energetic to inject a hole into the valence band of the semiconductor. Moreover, for most one-electron oxidants, p-GaAs was not etched, because holes injected beneath the tip were rapidly removed by reaction with electrons injected into the GaAs by the reduced form of the mediator at more remote sites of the semiconductor surface.

3.3 Other reactions

All kinds of other reactions that have been carried out electrochemically are potentially also available with the SECM. Thus the production of conducting polymers, frequently synthesized by electrochemical oxidation, or the formation of organic metals, like TTF-TCNQ, could also be carried out with high resolution by direct or feedback methods. For example, polyaniline has been deposited by direct oxidation at a Pt tip [38]. Polymerization reactions by reduction of an activated olefin should also be possible. Semiconductors have been deposited electrochemically, e.g., CdSe by reduction of solutions containing Cd^{2+} and selenous acid. Production of semiconductor features by SECM should similarly be possible. By generating pH gradients in the gap, precipitation of hydrous oxides or other salts of weak acids on a substrate might be carried out. However, we have found that scaling a reaction down to SECM dimensions often involves special considerations and that considerable experimentation is sometimes necessary to find the proper conditions for successful high resolution fabrication.

The resolution attainable by SECM depends upon a number of factors, e.g., tip size, tip current, and tip scan speed. Faster scans across the surface, to draw lines, tend to yield smaller features. Typical maximum scan speed used so far are of the order of 500 nm/s. Instrumental factors, such as vibration-damping, feedback response and temperature control can also be significant.

4. CONCLUSIONS

This brief review outlines some of the applications of SECM to microelectrochemical studies and microfabrication. A number of systems have been investigated, but much work remains to be done to fabricate actual devices and to push the resolution limits to even smaller levels. In closing, I would like to thank my many coworkers, whose names are given in the references, for their hard work and brilliant contributions to the development of SECM.

REFERENCES

1. A. J. Bard, F.-R. F. Fan, J. Kwak and O. Lev, Anal. Chem. 61 (1989) 132.
2. A. J. Bard, F.-R. F. Fan, D. T. Pierce, P. R. Unwin, D. O. Wipf and F. Zhou, Science 254 (1991) 68.
3. A. J. Bard, G. Denuault, C. Lee, D. Mandler and D. O. Wipf, Acc. Chem. Res. 23 (1990) 357.
4. A. J. Bard, F.-R. F. Fan and M. V. Mirkin, in Electroanalytical Chemistry, Vol. 18, (A. J. Bard, ed.), Marcel Dekker, New York 1993, in press.
5. A. J. Bard and L. R. Faulkner, Electrochemical Methods, Fundamentals and Applications, Wiley, New York, 1980.
6. R. M. Wightman and D. O. Wipf, in Electroanalytical Chemistry, Vol. 15, (A. J. Bard, ed.), Marcel Dekker, New York, 1988, p. 267.
7. J. Kwak and A. J. Bard, Anal. Chem. 61 (1989) 1221.
8. M. V. Mirkin and A. J. Bard, J. Electroanal. Chem. 323 (1992) 1.
9. M. V. Mirkin and A. J. Bard, J. Electroanal. Chem. 323 (1992) 29.
10. A. J. Bard, M. V. Mirkin, P. R. Unwin and D. O. Wipf, J. Phys. Chem. 96 (1992) 1861.
11. A. J. Bard, G. Denault, R. A. Friesner, B. C. Dornblaser and L. S. Tuckerman, Anal. Chem. 63 (1991) 1282.
12. P. R. Unwin and A. J. Bard, J. Phys. Chem. 95 (1991) 7814.
13. D. O. Wipf and A. J. Bard, J. Electrochem. Soc. 138 (1991) 469.
14. D. O. Wipf and A. J. Bard, Anal. Chem. 64 (1992)1362.
15. D. O. Wipf and A. J. Bard, Anal. Chem. 65 (1993) 1373.
16. M. V. Mirkin, F.-R. F. Fan and A. J. Bard, J Electroanal. Chem. 328 (1992) 47.
17. R. M. Penner, M. J. Heben, T. L. Longin and N. S. Lewis, Science 250 (1990) 1118.
18. A. S. Baranski, J. Electroanal. Chem. 307 (1991) 287.
19. K. B. Oldham, Anal. Chem. 64 (1992) 646.
20. M. V. Mirkin, T. C. Richards and A. J. Bard, J. Phys. Chem., in press.
21. D. T. Pierce, P. R. Unwin and A. J. Bard, Anal. Chem. 64 (1992) 1795.
22. D. T. Pierce and A. J. Bard, Anal. Chem., submitted.
23. M. V. Mirkin, M. Arca and A. J. Bard, J. Phys. Chem., in press.
24. P. R. Unwin and A. J. Bard, J. Phys. Chem. 96 (1992) 5035.
25. B. R. Horrocks, M. V. Mirkin, D. T. Pierce, A. J. Bard, G. Nagy and K. Toth, Anal. Chem. 65 (1993) 1213.
26. G. Denuault, M. H. Trise Frank and L.M. Peter, Faraday Discuss. 94 (1992).
27. M. V. Mirkin, F.-R. F. Fan and A.J. Bard, Science 257 (1992) 364.
28. E. R. Scott, H. S. White and J. B Phipps, Solid State Ionics 53 (1992) 176.
29. B. R. Horrocks, M. V. Mirkin and A. J. Bard, in preparation.

30. D. H. Craston, C. W. Lin and A. J. Bard, J. Electrochem. Soc. 135 (1988) 785.
31. D. Mandler and A. J. Bard, J. Electrochem. Soc. 136 (1989) 3143.
32. O. E. Hüsser, D. H. Craston and A. J. Bard, J. Vac. Sci. Technol. B 6 (1988) 1873.
33. O. E. Hüsser, D. H. Craston and A. J. Bard, J. Electrochem. Soc. 136 (1989) 3222.
34. D. Mandler and A. J. Bard, J. Electrochem. Soc. 137 (1990) 1079.
35. R. L. McCarley, S. A. Hendricks and A. J. Bard, J. Phys. Chem. 96 (1992) 10089.
36. D. Mandler and A. J. Bard, J. Electrochem. Soc. 137 (1990) 2468.
37. D. Mandler and A. J. Bard, Langmuir 6 (1990) 1489.
38. Y.-M. Wuu, F.-R. F. Fan and A. J. Bard, J. Electrochem. Soc. 136 (1989) 885.

MICROCHEMISTRY
Spectroscopy and Chemistry in Small Domains
Edited by H. Masuhara et al.

Electrochemical micromodification and imaging of ion conducting films by direct-mode scanning electrochemical microscopy

Hiroyuki Sugimura* and Noboru Kitamura**

Microphotoconversion Project, ERATO, Research Development Corporation of Japan, 15 Morimoto-cho, Shimogamo, Sakyo-ku, Kyoto, 606, Japan

Direct-mode scanning electrochemical microscopy (SECM) has been employed for microfabrication and observation of ionic conductive material surfaces. Ionic conductive polymer films was chemically modified in arbitrary sub-μm patterns through redox reaction at the SECM tip/film interface. A direct-mode SECM system with triple electrodes, by which an SECM tip potential can be defined electrochemically, was developed to elucidate mechanism of the pattern formation. Electrochemical imaging of Prussian blue surface in an electrolyte solution and air was also achieved by SECM with μm ~ nm resolution.

1. INTRODUCTION

Characterization of material surfaces in solution is of primary importance in elucidating chemical processes occurring at the material/solution interface, and is also indispensable to design and control of various functions on materials. Scanning probe microscopes such as a scanning tunneling microscope (STM) , an atomic force microscope (AFM), or a scanning electrochemical microscope (SECM) are powerful means for *in-situ* measurements of material surfaces in solution with high spatial resolutions from μm to an atomic scale, and hence the microscopes have been applied in a variety of research fields [1,2]. Among these, SECM developed by Bard and co-workers is essentially based on observation of rates of electrochemical reactions at a probe tip, and the technique has unique capabilities

* present address; Tsukuba Research Laboratory, Nikon Co., 5-9-1 Tohkodai, Tsukuba 300-26, Japan
** present address; Department of Chemistry, Faculty of Science, Hokkaido Univ., Sapporo 060, Japan

for investigating the chemical and physical properties of heterogeneous material surfaces in sub-μm dimensions [3,4]. Two types of SECM have been so far reported; *direct-mode SECM* and *feedback-mode SECM*. Feedback-mode SECM uses an electrochemical mediator(s) in the solution phase for electron transfer between a tip electrode and a sample to probe the sample surface in solution. A microdisk electrode surrounded by an insulator ring is biased at an appropriate potential and positioned in close proximity to the surface. The microelectrode is then scanned over the sample with a constant electrode-sample distance to probe a faradaic current as a function of the electrode position [5].

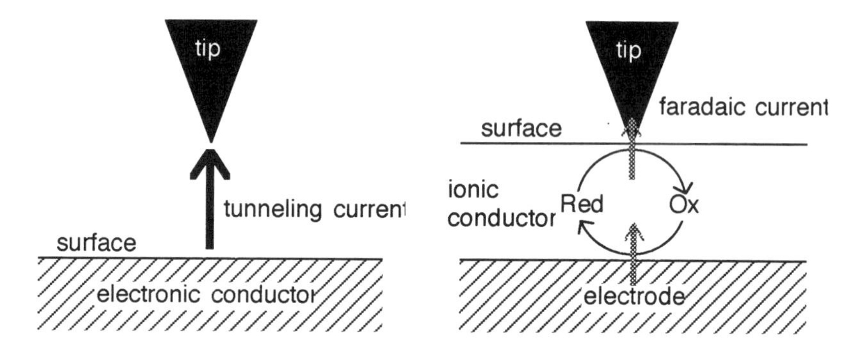

Figure 1. Schematic illustrations of STM and direct-mode SECM.

The principle of direct-mode SECM is schematically illustrated in Figure 1 together with that of STM. One of the advantages of direct-mode SECM is an applicability pt he method to *ionic conductive materials*. This is contrasting to STM, by which samples applied are limited to *electronic conductors* such as metals and semiconductors. In direct-mode SECM, a very sharp tip similar to a probe tip in STM can be used for a probing electrode as shown in Figure 1. The tip electrode is contacted with an ionic conductor film supported by a counter electrode to promote redox reactions at the tip/film and the film/electrode interfaces, and is then scanned across the film surface. During tip scanning, a faradaic current is kept a constant value through the control of the contacting area at the tip/film interface. By an appropriate choice of the probing faradaic current, the contact area can be held to be very small and, therefore, electrochemical reactions are induced only in a minute region. As an important application, SECM has been currently extended to electrochemical fabrication of materials as demonstrated for the deposition of metal lines on ionic conductive polymer films as well as for etching of metal films with sub-μm resolution [6,7].

In this paper, we review our recent studies on direct-mode SECM, demonstrating modification of ionic conductive polymer films doped with organic molecules [8], mechanistic investigation of electrochemical processes by triple electrodes SECM [9], and electrochemical imaging of ionic conductor film surfaces with resolution of μm ~ nm scales [10].

2. ELECTROCHEMICAL MODIFICATION BY SECM

Besides microfabrication of metals and semiconductors by SECM [3], a variety of materials will be modified with arbitrary chemical functions and spatial patterns, if various redox reactions are induced at the tip/materials interface. We thus explored micropatterning of ionic conductive polymer films through chemical reactions of *organic* compounds in the film by direct-mode SECM. Although metal micropatterns deposited or etched by SECM can be easily confirmed by a conventional microscope, detection of the spatial patterns produced by chemical reactions of organic molecules is very difficult since the patterns generally has no change in the morphology. Among various methods, fluorescence spectrometry is highly sensitive and widely applied to characterize polymer structures in minute dimensions, so that fluorescence microscopy is most suitable to detect chemically modified miropatterns.

In this section, we describe fabrication of perfluorosulfonated polymer films (Flemion, Asahi Glass Co.) doped with methylviologen (MV^{2+}), and detection of the modified patterns by fluorescent microscopy [8,9].

2.1. Fluorescent micropattern formation

Flemion (5 wt.% in ethanol) and rohdamine-6G (R-6G, 1.7×10^{-3} M in ethanol) were mixed with a volume ratio of 3:1, and the mixture was spin coated onto a platinum (Pt)-sputtered glass plate which works as a counter electrode in SECM. The film were then kept at 65 °C in an oven for 3 h to remove the solvent. The thickness of the film was estimated to be 0.2 μm by a step profiler. The R-6G/Flemion film was set in an SECM cell, and soaked in an aqueous solution of 50 mM MV^{2+} for 30 min to absorb MV^{2+} to the film. Before SECM scans on the sample, the MV^{2+} solution was removed, and the film was rinsed with pure water.

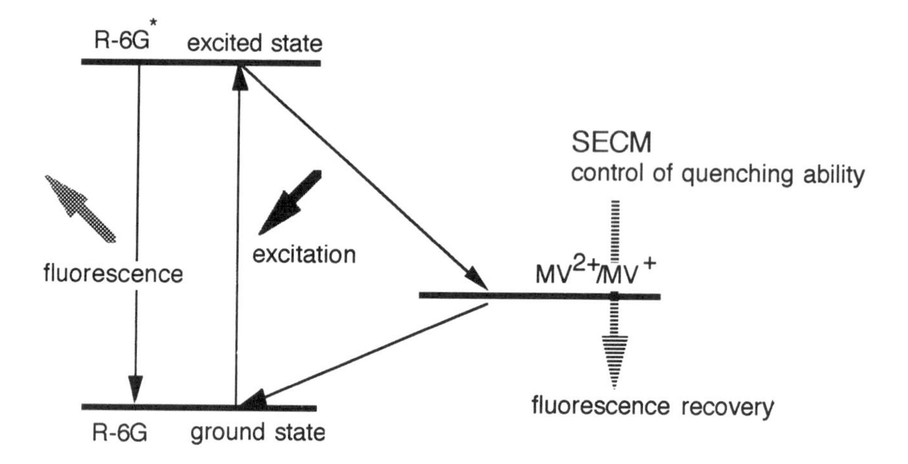

Figure 2. Principle of the fluorescent micropattern formation.

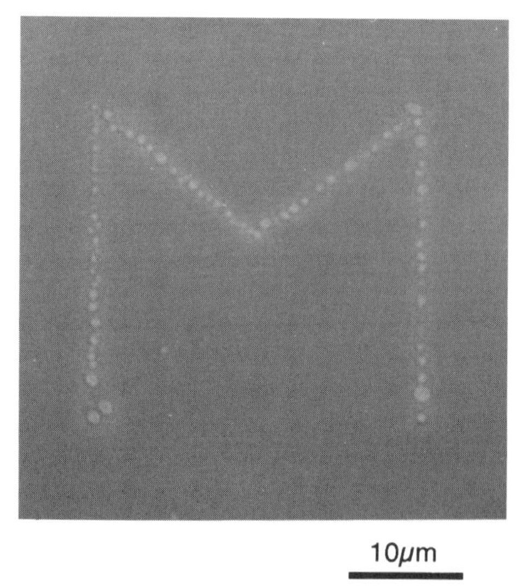

Figure 3. Fluorescent micropattern formation on the R-6G/MV^{2+}/Flemion film.

After an exposure to air for an appropriate period, the film was scanned by an SECM tip negatively biased versus the counter electrode on which the film was coated.

The principle of fluorescent micropattern formation is schematically illustrated in Figure 2. In the absence of MV^{2+}, R-6G in the Flemion film showed strong fluorescence with the maximum at 546 nm. Contrarily, the R-6G/MV^{2+} film was non-fluorescent owing to electron transfer quenching of the excited state of R-6G by MV^{2+} in the film. If the quenching ability of MV^{2+} is controlled electrochemically, R-6G fluorescence will be recovered and, therefore, fluorescent micropatterns will be formed on the R-6G/MV^{2+} film based on the SECM technique. Indeed, we succeeded in producing a fluorescent pattern on the Flemion film as shown in a florescence micrograph of Figure 3. The fluorescent pattern of the letter "M" was fabricated on the film along the trace of the SECM tip operated with a bias voltage of -4.0 V, a reference current of 0.1 nA, and a scan rate of 1 μm/s, while other regions where the tip was not scanned remained non-fluorescent. The fluorescent pattern was stable in air at least for several days. It is noteworthy that the fabricated pattern has no morphological change, so that observation of the pattern is impossible by optical microscope or an electron microscope. Fluorescence detection of the pattern is indispensable for the present experiment.

The spatial resolution of the fluorescent pattern is governed by generation and diffusion rates of chemical species produced under the tip. The amount of moisture in the film influences the ionic conductivity of the film as well as the diffusion coefficients of ions in the film, so that the spatial resolution of the pattern is strongly dependent on the water content in the film as reported previously in detail [8]. Also, the spatial resolution is influenced by the tip scan rate, Namely, the slower the scan rate, the worse the resolution owing to diffusion of the ionic species in the lateral direction. Therefore, the amount of moisture in the film and the tip scan rate should be optimized to obtain highly resolved fluorescent patterns. The best resolution of 0.5 μm was achieved at the present stage in the investigation.

2.2. Direct-mode SECM with triple electrodes

The direct mode SECM described above is operated with two electrodes as shown in Figure 4(A), so that the SECM tip is biased with respect to the counter electrode. Therefore, a tip potential is electrochemically undefined, and electrochemical reactions occurring at the tip/film interface during SECM scans can not be elucidated precisely. In order to establish the direct-mode SECM method, we have developed the SECM system with triple electrodes, in which a reference electrode (RE; Ag/AgCl) is assembled in the system to define the electrochemical potential of the SECM tip as illustrated in Figure 4(B).

The triple electrodes SECM system was employed to elucidate the mechanism of the fluorescent micropattern formation [9]. In order to know the electrochemical origin, affects of a tip potential on the micropattern formation should be studied in detail. When the potential of the Pt or tungsten (W) tip-electrode was biased more positively than -0.6 V vs. Ag/AgCl (the potential range I in Figure 5), the tip penetrated into the Flemion film and both the film and the tip were damaged. In this potential range I, electrochemical reactions scarcely occurred at the tip/film interface, so that no faradaic current flowed. Therefore, the tip went down into the film until a tunneling current flowed between the tip and the counter electrode. On the other hand, when the tip potential was biased more negatively than -0.6 V, an enough faradaic current began to flow and SECM could work well without serious damages of the tip and the film (II in Figure 5). When the tip potential was set at a bias more negative than a threshold value, a fluorescent pattern was fabricated (III in Figure 5). The threshold potentials for the Pt and W tips were -0.9 and -1.2 V vs. Ag/AgCl, respectively.

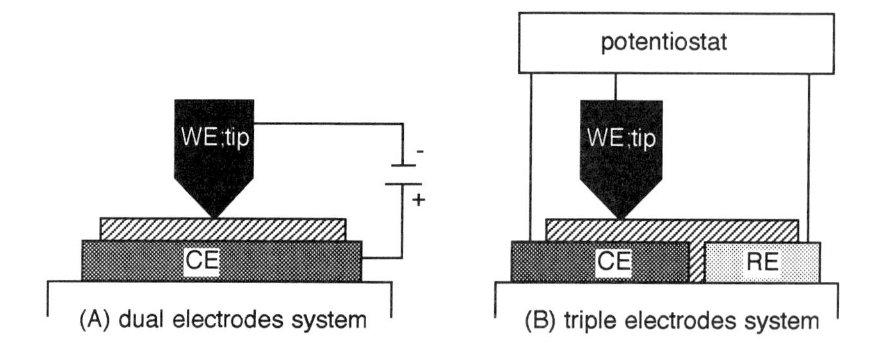

Figure 4. Dual (A) and triple (B) electrodes direct-mode SECM. WE; Working electrode (tip).CE; Counter electrode. RE; Reference electrode.

As reference experiments, voltammetry of R-6G/Flemion films coated on Pt and W macro electrodes was performed in an aqueous MV^{2+} solution. In both cases, e;electrochemical reduction of MV^{2+} to MV^{+}. took place at the potential more negative than -0.6 V vs. Ag/AgCl. Therefore, the faradaic current by SECM on the range II is ascribed to the reduction of MV^{2+}. When a potential of the Pt or W macro electrode was set at a further negative bias, electrolysis of water occurred at the potential of -0.9 or -1.2 V vs. Ag/AgCl, respectively. These potentials agree with the threshold potentials of the fluorescent pattern formation for Pt and W tips (region III), respectively. Therefore, we conclude that the origin of the fluorescent pattern formation by the SECM is reduction of water in the film. This can be confirmed the results in Figure 6, where a relationship between the recovery of the fluorescence intensity of R-6G at 570 nm and the W tip potential in the range III is summarized. In this experiment, fluorescent patterns of 20 × 20 μm were fabricated with a reference current of 0.1 nA. As clearly seen in Figure 6, the fluorescence intensity increased with shifting the potential more negatively in the range of -1.0 ~-2.0 V, in accordance with increased efficiency of water electrolysis. The leveling-off of the fluorescence intensity at < -2.0 V indicates that the ratio of the water reduction current to the SECM reference current becomes almost constant.

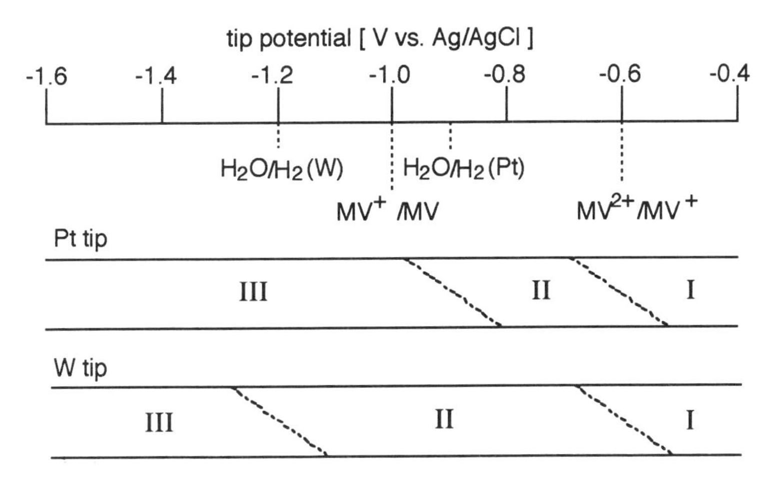

Figure 5. Relationships between the tip potential and SECM. See main text.

It has been reported that MV^{2+} is decomposed in alkaline solution [11]. Thus, as shown in Equation (1) and (2), OH^- ions produced by water reduction at the tip/film interface

$$2H_2O + 2e^- \longrightarrow H_2 + 2OH^- \qquad (1)$$

are expected to react with MV^{2+} in the Flemion film.

$$Me^{+}N\langle\bigcirc\rangle-\langle\bigcirc\rangle N^{+}Me + OH^{-} \longrightarrow Me^{+}N\langle\bigcirc\rangle-\langle\bigcirc\rangle N + MeOH \qquad (2)$$

Therefore, the concentration of MV^{2+}, namely, the quencher concentration for the excited state of R-6G, decreases in the scanned area by the SECM tip, resulting recovery of the fluorescence intensity of R-6G. This is the primary origin of the fluorescent micropattern formation in the Flemion/R-6G/MV^{2+} systems.

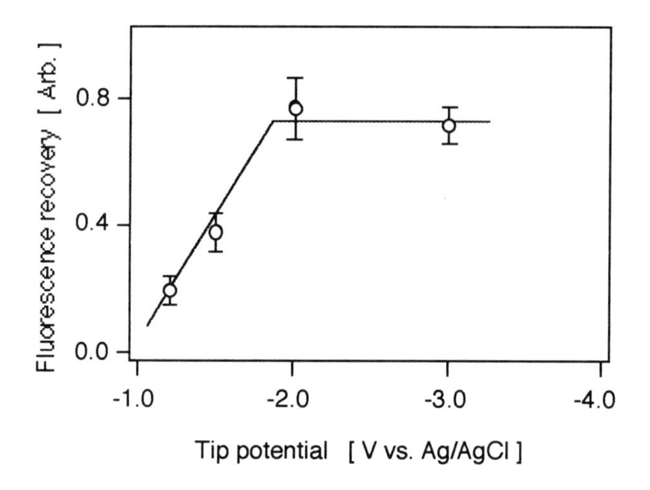

Figure 6. Effects of the tip potential on Fluorescence recovery vs. SECM tip potential.

3. ELECTROCHEMICAL IMAGING BY SECM

In principle, direct-mode SECM can be applied to topographical observation of ionic conductive material surfaces in solution or even in air. Nevertheless, the method has never been employed for surface imaging so far. In this section, we describe the first demonstration of topographical imaging of redox materials by direct-mode SECM in solution and air. Prussian blue (PB; $Fe_4^{III}[Fe^{II}(CN)_6]_3$) films were selected as a redox sample, since PB has been received much attentions for electrochromic devices, electrocatalysis and a representative mixed-valence compound [12]. PB films with 1 ~ 2 μm in thickness were prepared on Pt-sputtered glass plates as reported in the literature [13].

3.1. SECM in solution

For SECM experiments, the PB sample was placed at the bottom of an SECM cell (Figure 6), and immersed in an aqueous solution of KCl (20 mM, pH4.0). An insulated tip was used as a probe tip, and potentials of the PB sample and the tip were controlled independently by a bipotentiostat with respect to a reference electrode (Ag/AgCl).

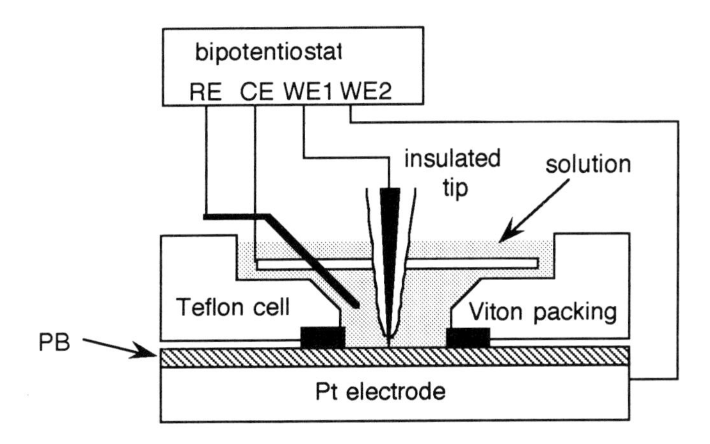

Figure 7. Electrochemical cell for SECM in solution.

PB is insoluble in water [12] and the solution contains no redox species, so that no faradaic current flows between the tip and the sample until the tip touches the PB surface. When appropriate potentials are applied independently to the tip and the substrate electrodes, the tip can be scanned over the PB surface with probing a faradaic current. In order to obtained stable SECM images, tip and substrate potentials, (abbreviate E_t and E_s, respectively) should be applied correctly with respect to the redox potential of PB (E_{PB} = +0.22 V vs. Ag/AgCl). Four cases of potential settings for E_t and E_s are summarized in Figure 7 together with the cyclic voltammogram (CV) of PB. The CV curve indicates the reversible redox reaction of PB (Equation (3)).

$$Fe_4^{III}[Fe^{II}(CN)_6]_3 + 4K^+ + 4e^- \rightleftharpoons K_4^+Fe_4^{II}[Fe^{II}(CN)_6]_3 \qquad (3)$$

According to CV, PB is reduced to colorless Prussian White (PW) at a potential more negative than E_{PB}, while PW is reoxidized to PB at a potential more positive than E_{PB}.

When E_t and E_S are set to more negative and more positive potentials relative to E_{PB} (Figure 8(A)), respectively, PB is reduced to PW at the tip/film interface. Thereby the faradaic current flows and the tip can trace the PB surface with probing the faradaic current. A typical SECM image observed with this potential setting is shown in Figure 9. Analogous SECM imaging of PW surface was also possible under the reversed potential setting with $E_t > E_{PB}$ and $E_S < E_{PB}$ (Figure 8(B)), as shown in Figure 10. In both SECM images of PB and PW, surface structures of the crystal grains could be observed even in solution with sub-μm resolution.

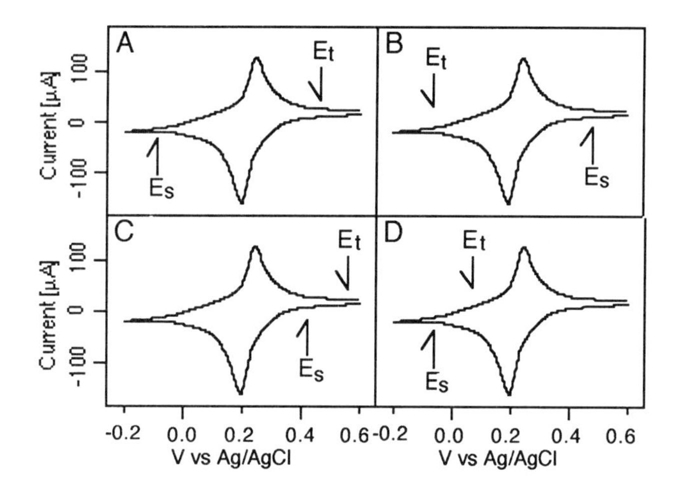

Figure 8. Cyclic voltammogram of PB and the potential setting (E_t and E_S) for SECM imaging. A and B: SECM images were obtained. C and D: No SECM image was obtained.

Flow of the faradaic current accompanied by the redox reactions of PB and PW is essential for SECM imaging. When both E_t and E_S were set at potentials more positive (Figure 8(C)) or more negative (Figure 8(D)) than E_{PB}, no SECM image was obtained. At the potentials of E_t = +0.3 and E_S = +0.6 V vs. Ag/AgCl (case C), a faradaic current does not flow even if the tip touches the sample surface, since no redox reaction occurs. Under the condition, the tip penetrated into the sample until a tunneling current flowed between the tip and Pt substrate electrodes, thus a serious damage which could be confirmed by optical microscopy was induced in the PB film. Analogous results were also obtained for a potential setting of E_t = -0.1 and E_S = -0.4 V vs. Ag/AgCl (case D). These results clearly indicate that the

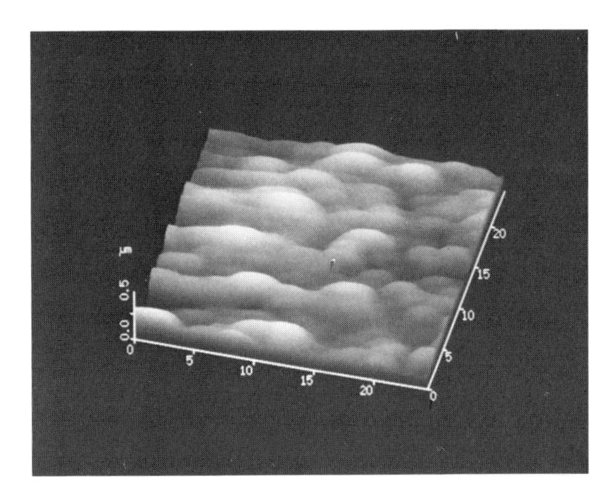

Figure 9. 25 × 25 µm SECM image of the crystal grains in a PB film in a 20 mM KCl solution (pH4.0) with E_S = +0.5, E_t = 0.1 V vs. Ag/AgCl, a tip scan rate = 10 µm/s and a reference current = 0.5 nA. Tip reaction, PB + $4e^-$ + $4K^+$ → PW. Substrate reaction, PW → PB + $4e^-$ + $4K^+$.

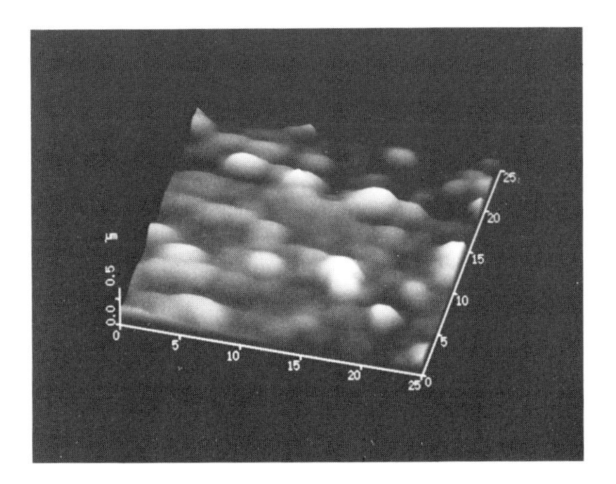

Figure 10. 25 × 25 µm SECM image of the crystal grains in a PW film in a 20 mM KCl solution (pH4.0) with E_S = -0.2, E_t = +0.3 V vs. Ag/AgCl, a tip scan rate = 10 µm/s and a reference current = 0.5 nA. Tip reaction, PW → PB + $4e^-$ + $4K^+$. Substrate reaction, PB + $4e^-$ + $4K^+$ → PW.

topographical imaging in Figures 9 and 10 is essentially based on the faradaic current accompanied by the redox reaction of PB or PW (SECM mode), but not on a tunneling current between the tip and the sample (STM mode).

In addition to the potential setting of E_t and E_s, reversibility of redox reactions in a sample is very important for SECM. When a PB film was poorly adhered on a Pt substrate, the SECM image changed with every scan since inefficient redox cycles between PB and PW resulted in damage of the sample surface. SECM can therefore detect both electrochemical reactivities and topography of a sample.

Figure 11. 3.8 × 3.8 μm SECM image of the crystal grains in a PB film in air. Tip reaction, Fe(III/II) → Fe(II/II): substrate reaction, Fe(III/II) → Fe(III/III).

3.2. SECM in air

Prussian blue is known to contain a considerable amount of both coordinate and interstitial water, and to show redox reactions in air without the presence of any electrolyte [14]. Thus, we expected to observe SECM images of PB even in air. The PB samples stored under a water-saturated atmosphere were exposed in air for 30 min prior to SECM experiments to remove excess water in the PB surface. We succeeded in SECM imaging of PB in air with the resolution better than 100 nm as demonstrated in Figure 11. In this experiment, a very sharp tungsten tip was used and a bias voltage was applied to the tip electrode with respect to the Pt substrate coated with PB (in two electrodes mode). Figure 11 shows an SECM image of PB obtained at E_t = +4.0 V (vs. Pt substrate), where the scan rate and the reference current were set at 2 μm/sec and 0.1 nA, respectively. Since the experiments were performed in air, a relatively high bias voltage could be applied. Electrochemical reactions of PB occurring in air will be oxidation of a low-spin iron (Fe(III/II) → Fe(III/III)) at the anode and reduction of a high-spin iron (Fe(III/II) → Fe(II/II)) at the cathode [14]. Electrolysis of coordinated and interstitial water molecules will also play a role for SECM imaging. The SECM observation of PB in air was successful both at positive and negative tip voltages. To the best of our knowledge, this is the first demonstration for SECM observation of material surface in air.

4. SUMMARY

Fluorescent micropatterns were fabricated on the R-6G/MV^{2+}/Flemion films by direct-mode SECM, and the electrochemical origin of the pattern formation was elucidated by the triple-electrodes system developed in this study. The detailed studies revealed that the primary key step of the pattern formation was reduction of water at the tip/film interface and subsequent decomposition of MV^{2+} by OH^- ions generated by the water reduction.

Electrochemical imaging of PB by the SECM both in solution and air has been also demonstrated. Images were obtained only when the tip and the substrate electrodes were biased at more positive (or negative) and negative (or positive) potentials, respectively, relative to the redox potential of PB. Such results indicate that the images are obviously based on probing of the faradaic current accompanied by the redox reaction of PB.

Since SECM has been shown to be applicable both to electrochemical imaging and micromodification, characterization and fabrication of various ionic conductive materials will be greatly advanced. Furthermore, direct-mode SECM can be performed in the absence of an electrochemical mediator in solution, so that the method enables direct electrochemical characterization of material surfaces in very minute dimensions.

REFERENCES

1. T. R. I. Cataldi, I. G. Blackham, G. Andrew, D. Briggs, J. B. Pethica and H. A. O. Hill, J. Electroanal. Chem., 290 (1990) 1.
2. S. R. Snyder and H. S. White, Anal. Chem. 64 (1992) 116R.
3. A. J. Bard, G. Denualt, C. Lee, D. Mandeler and D. O. Wipf, Acc. Chem. Res., 23 (1990) 357.
4. A. J. Bard, F.-R. F. Fan, D. T. Pierce, R. R. Unwin, D. O. Wipf and F. Zhou, Science 254 (1991) 68.
5. A. J. Bard, F.-R. F. Fan, J. Kwak and O. Lev., Anal. Chem., 61 (1989) 132.
6. O. E. Hüsser, D. H. Craston, and A. J. Bard, J. Vac. Sci. & Technol., B6 (1988) 1873
7. O. E. Hüsser, D. H. Craston, and A. J. Bard, J. Electrochem. Soc., 136 (1989) 3222
8. H. Sugimura, T. Uchida, N. Shimo, N. Kitamura and H. Masuhara, Ultramicroscopy, 42-44 (1992) 468.
9. H. Sugimura, N. Kitamura, N. Shimo and H. Masuhara, J. Electroanal. Chem. 346 (1993) 147.
10. H. Sugimura, T. Uchida, N. Kitamura, N. Shimo and H. Masuhara, J. Electroanal. Chem., (in press)
11. J. A. Farrington, A. Ledwith, and M. F. Stam, J. Chem. Soc., Chem. Commun., (1969) 259.
12. K. Itaya, I. Uchida and V. D. Neff, Acc. Chem. Res., 19 (1986) 162.
13. K. Itaya, K. Shibayama, H. Akahoshi, and S. Thoshima, J. Appl. Phys., 53 (1982) 804.
14. B. J. Feldman and R. W. Murray, Inorg. Chem., 26 (1987) 1702.

MICROCHEMISTRY
Spectroscopy and Chemistry in Small Domains
Edited by H. Masuhara et al.

Photoelectrochemical characteristics of semiconductor-based microelectrode arrays

T. Uchida,§ A. Sekiguchi,‡ K. Pásztor,# and N. Kitamura*

Microphotoconversion Project,† ERATO Program, Research Development Corporation of Japan, 15 Morimoto-cho, Shimogamo, Sakyo-ku, Kyoto 606, Japan

Titanium dioxide/platinum (TiO_2/Pt) and SnO_2 microelectrode arrays were fabricated to elucidate characteristic features of photoelectrochemistry in micrometer dimension. In the TiO_2/Pt microelectrode systems, a photocurrent at TiO_2 accompanied by oxygen evolution from water was shown to be controlled by an applied potential to the neighboring Pt microelectrode as well as by the interelectrode distance between TiO_2 and Pt. Photochemical generation and electrochemical collection of redox active species by a pair of microelectrodes were demonstrated for the first time in the SnO_2 microelectrode array - tris(2,2'-bipyridine)ruthenium(II) complex systems. An approach to design and construct microchemical systems is also discussed.

1. INTRODUCTION

Recent advances in microfabrication techniques have enabled one to arrange micrometer reaction sites arbitrarily in space. As an example, the techniques have been successfully applied to fabricate various microelectrodes [1]. Among these, microelectrode arrays have been so far studied during these years in reference to development of chemical microtransistors and sensors as well as to high performance measurements of redox intermediates based on a platinum, silver, or gold micro-

---

* To whom correspondence should be sent to the present address at Department of Chemistry, Faculty of Science, Hokkaido University, Sapporo 060, Japan.
§ Present address ; Electrochemiscopy Project, ERATO, Research Development Corporation of Japan, 2-1-1 Yagiyama-minami, Taihaku-ku, Sendai 982, Japan.
‡ Present address ; Anelva Co. Ltd., 5-8-1 Yotsuya, Fuchu-shi, Tokyo 183, Japan.
# Present address ; Department of Electron Devices, Technical University of Budapest, Goldmann Gy. ter 3, H-1521 Budapest, Hungary.
† Five-year term project ; October 1988 ~ September 1993.

electrode array [2 - 4]. Although such metal-based microelectrode arrays are certainly useful as electrochemical devices, chemical modification of individual microelectrode surfaces will lead to further advances in the microelectrode chemistry, and also provide a future possibility for development of microchemical systems where cascade reactions proceed sequentially along a series of micrometer reaction sites. A possible candidate of the electrodes for such studies is semiconductor-based microelectrode arrays, by which photochemical and electrochemical control of reactions along the microelectrodes is expected based on photoelectrochemical reactivities of the materials and, such microelectrode arrays will be served as a prototype of the microchemical systems. We therefore fabricated semiconductor-based (TiO_2 and SnO_2) microelectrode arrays and explored to elucidate characteristic features of the chemistry in micrometer dimension, with focusing our attention on effects of the spatial arrangement of the electrodes on photoelectrochemical response of the semiconductors.

The choice of TiO_2 and SnO_2 as semiconductor materials is as follows. TiO_2 electrodes or particles are of primary importance in wide areas of chemistry because of their excellent photocatalytic activities towards various redox reactions. In particular, photocatalytic systems based on platinized TiO_2 (TiO_2/Pt) have been extensively studied over the past ten years in special reference to conversion of solar energy to chemical energy [5, 6]. In order to fabricate microchemical systems, therefore, TiO_2/Pt microelectrode arrays are one of the promising candidates to conduct efficient chemical reactions in micrometer dimension. On the other hand, SnO_2 is known as a transparent semiconducting material in the visible region, so that various photochemical reactions will be induced at the electrode/solution interface when the solution phase is irradiated from the electrode side. Such circumstances are very favorable for photochemical and electrochemical control of reactions along the microelectrodes in the array. In the present article, we describe recent results on photoelectrochemistry of the TiO_2/Pt microelectrode array as well as on photo-sensitized reactions in the SnO_2 microelectrode array - $Ru(bpy)_3^{2+}$ systems.

2. PHOTOELECTROCHEMISTRY OF TiO_2/Pt MICROELECTRODE ARRAY

2.1. Fabrication and Characterization of TiO_2 Microelectrode

Semiconductor-based microelectrode arrays were fabricated by conventional photolithography. As a typical example, a photograph of a TiO_2/Pt microelectrode array used in this study is shown in Figure 1. A titanium film prepared on a glass substrate by ion plating was patterned in micrometer to produce eight Ti microelectrodes (100 μm long x 10 μm wide x 0.3 μm thick, 10 μm spacing) in an array. The surface of the array was protected with a Si_3N_4 film (0.6 μm thick) by plasma-enhanced chemical vapor deposition except for those of the reaction window and the eight bonding pads. One Ti electrode (**#1** in Figure 1) was then anodized to TiO_2 by anodic potential sweep (1 mV/s) between ~ -0.55 V (vs. SCE) and an appropriate potential in an aqueous H_2SO_4 solution (0.05 M) [7]. The thickness of the TiO_2 film on Ti was estimated to be ~ 40 nm. The surfaces of other seven Ti

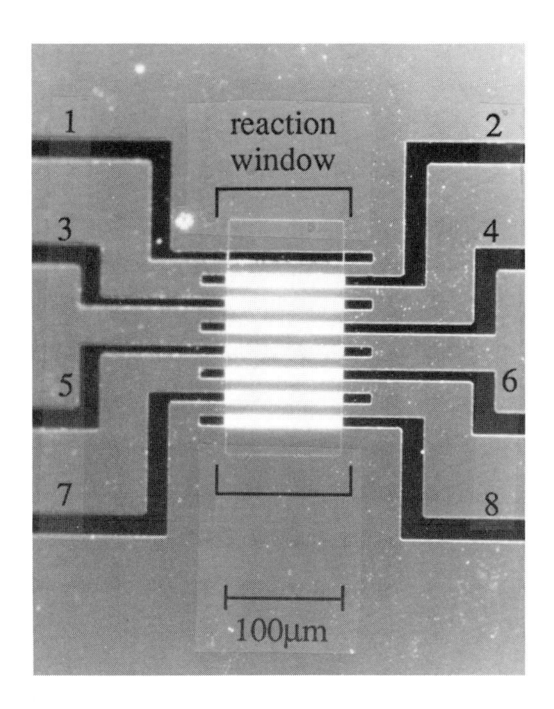

Figure 1. A photograph of TiO_2/Pt microelectrode array.

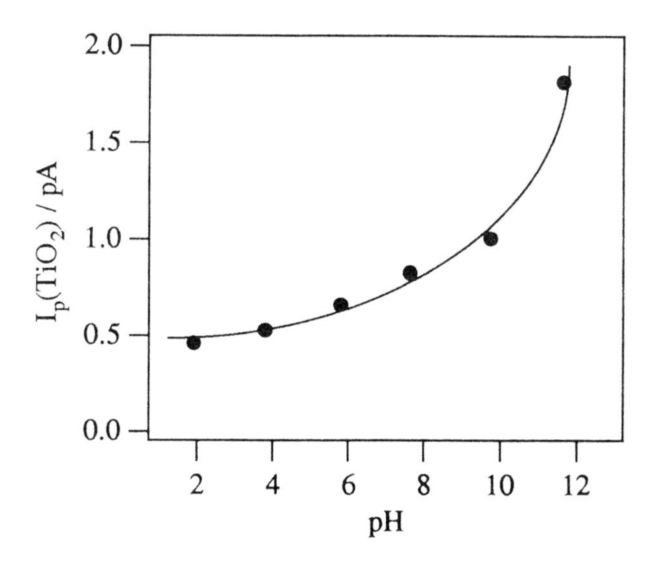

Figure 2. pH dependence of photocurrent at TiO_2 microelectrode.

microelectrodes, **#2~8**, were platinized electrochemically from an electroplating solution [8]. A cyclic voltammogram of $Ru(NH_3)_6Cl_2$ (5 mM) in an aqueous 0.1 M KCl solution obtained at each Pt electrode showed a sigmoidal curve characteristic of a microelectrode and the observed redox potential of the complex proved the correct response of the Pt electrode. The **#1** and one of the Pt electrodes were used as working electrodes, and a potential of each electrode was set with respect to a saturated calomel electrode (SCE).

It is well known that photoillumination of TiO_2 in an aqueous solution leads to generation of a photocurrent through oxygen evolution from water. In an aqueous Na_2SO_4 solution (0.5 M), indeed, we observed a photocurrent (I_p) when the **#1** TiO_2 microelectrode was selectively irradiated at 365 nm under an optical microscope. The action spectrum of I_p observed at TiO_2 was very similar to that reported by Leitner et al. for anodized TiO_2 films [9]. Detailed studies on the I_p - V (voltage) relationships indicated that TiO_2 was best characterized as polycrystalline. As a characteristic feature, furthermore, I_p at the TiO_2 electrode was dependent on the pH of the solution as shown in Figure 2. It is clearly seen that I_p increases with increasing the solution pH. Further details on the pH dependence of the photocurrent including I_p - V characteristics of the TiO_2 microelectrode are reported in elsewhere [7].

2.2. Spatial Control of Photocurrent in Micrometer Dimension

In the TiO_2 - Pt microelectrode array system, OH^- can be generated arbitrarily at one of the seven Pt microelectrodes through electrolysis of water, so that effects of OH^- generation (i.e., local solution pH) at each Pt electrode on the photocurrent at TiO_2 can be studied in detail. Diffusion processes of OH^- in the solution phase is also expected to be revealed on the basis of photocurrent measurements at TiO_2 under OH^- generation at one of the Pt electrodes. Actually, the photocurrent at the **#1** TiO_2 microelectrode accompanied by oxygen evolution from water (eqs. (1) and (2)) was strongly dependent on a bias voltage (V_g) applied to the adjacent Pt electrode, **#2** [8]. As shown in Figure 3, I_p at TiO_2 increased sharply with shifting V_g more negative than -0.7 V and almost saturated around $V_g < -1.0$ V (vs. SCE). The results indicate that I_p can be enhanced by a factor of ~ 2.5 through applying V_g (< -0.7 V) to the **#2** Pt microelectrode. The Pt electrode potential of -0.7 V coincides with that for electrolysis of water (eq.(3)). Therefore, the increase in I_p with V_g is primarily responsible for OH^- ions generated at the Pt electrode and subsequent change of the solution pH in the vicinity of the TiO_2 electrode through diffusion of OH^-. The more negative V_g than -0.7 V, the higher the amount of OH^- generated at Pt, so that I_p increases with V_g as demonstrated in Figure 3.

$$\text{at } TiO_2: \quad TiO_2 + h\nu \rightarrow e^- + p^+ + TiO_2 \qquad (1)$$

$$p^+ + 1/2H_2O \rightarrow 1/4O_2 + H^+ \qquad (2)$$

$$\text{at Pt}: \quad H_2O + e^- \rightarrow 1/2H_2 + OH^- \qquad (3)$$

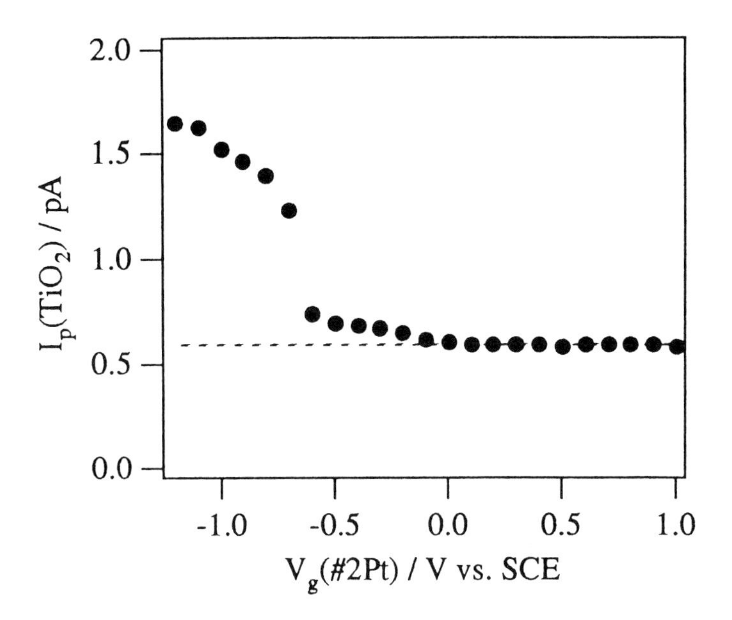

Figure 3. Effects of applied potential to **#2** Pt microelectrode on photocurrent at TiO_2.

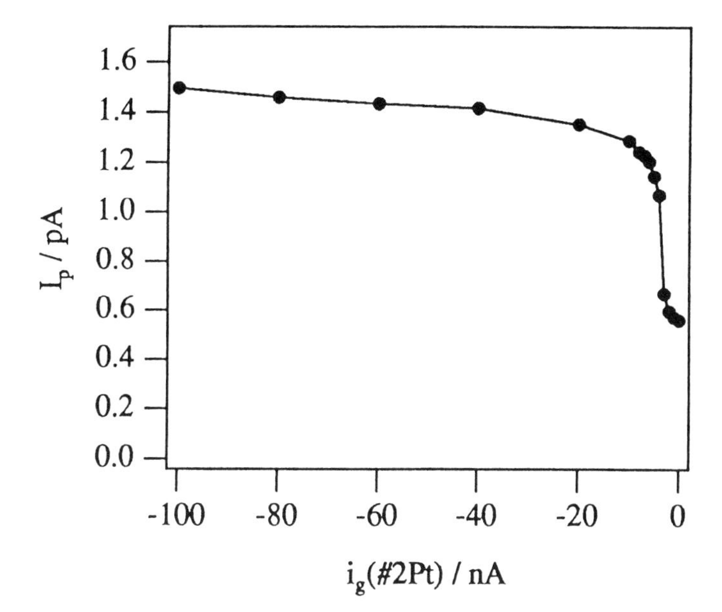

Figure 4. Effects of constant current electrolysis of water at **#2** Pt microelectrode on photocurrent at TiO_2.

In order to confirm the above discussion, we performed constant current (i_g) electrolysis of water at the **#2** Pt electrode and, measured I_p at TiO_2 as a function of i_g. The relationship between I_p and i_g thus obtained is summarized in Figure 4. The amount of OH^- generated at Pt is proportional to i_g, so that I_p should increase linearly with an increase in i_g if the OH^- concentration near TiO_2 directly governs I_p. However, the results did not agree with such the expectation, but increased with i_g in a logarithmic manner. This logarithmic dependence of I_p on i_g clearly indicates that the photocurrent at TiO_2 is governed by the solution pH.

Generation of OH^- ions was also conducted at other Pt microelectrode. Since the photocurrent is governed by the pH near the TiO_2 electrode through generation and diffusion of OH^- as mentioned above, I_p will be strongly dependent on the distance between the TiO_2 and Pt microelectrodes. Indeed, the photocurrent sharply decreases with increasing the interelectrode distance (d) as demonstrated in Figure 5. When OH^- generation was performed at **#2** Pt (i.e., d = 20 μm) with i_g = - 10 nA, I_p was enhanced by a factor of 2.3 as compared with that at i_g = 0. For d = 140 μm (**#8** Pt microelectrode), on the other hand, electrolysis of water had almost no effect on I_p, since the interelectrode distance was large long to change the solution pH near TiO_2 via OH^- diffusion.

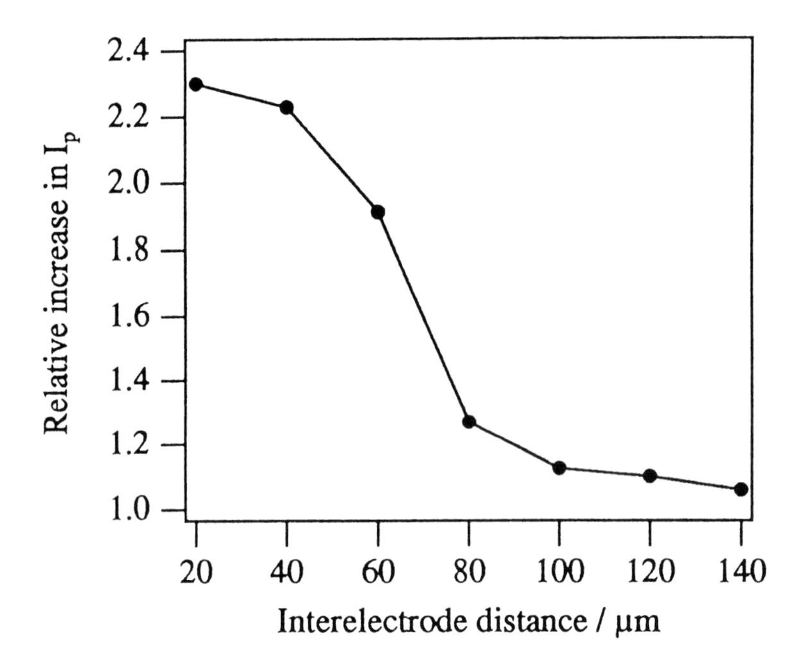

Figure 5. TiO_2 - Pt interelectrode distance dependence of photocurrent at TiO_2 microelectrode. Constant current electrolysis was performed at each Pt electrode with i_g = - 10 nA.

The interelectrode distance between TiO_2 and Pt also influences the time response of the photocurrent. When electrolysis of water was performed at the **#2** Pt electrode (d = 20 μm), the photocurrent at TiO_2 showed a very fast time response with the time constant faster than 0.2 s (Figure 6), whereas the rise time of I_p was ~ 1 or ~ 4 s for the electrolysis of water at the **#6** (d = 100 μm) or **#8** Pt microelectrode (d = 140 μm), respectively. It is noteworthy that these time constants agrees with those estimated by Fick's law and the diffusion coefficient of OH^- in water (D = 5.27 x 10^{-9} $m^2 s^{-1}$ [10]). All the results indicate that I_p is essentially determined by the solution pH in the vicinity of the TiO_2 microelectrode. According to the relation between I_p and pH described in the previous section (Figure 2), the solution pH was estimated to increase from 6.2 to > 12 upon electrolysis of water at the Pt microelectrode. Since the local pH can be increased by ~ 6 pH units within 0.2 s, the method will be applied to control a solution pH or pH jump in micrometer dimension as well.

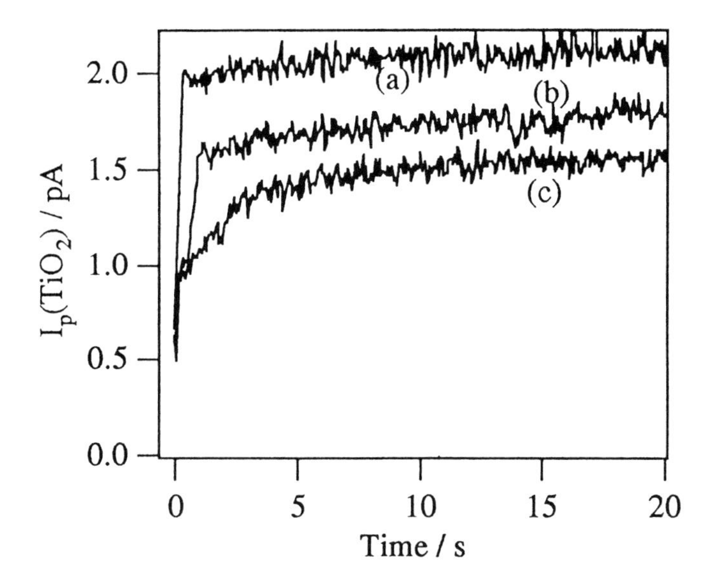

Figure 6. Time response of photocurrent at TiO_2 upon constant current electrolysis of water (i_g = - 2 μA) at a) **#2**, b) **#6**, or c) **#8** Pt microelectrode.

The I_p - V characteristics (Figure 3) and the TiO_2 - Pt distance dependence of I_p (Figures 5 and 6) are not expected for platinum- or gold-based microelectrode arrays, and are responsible for the semiconducting properties of TiO_2. The present results clearly indicate that chemical reactions in micrometer dimension can be controlled electrochemically and photochemically by the use of semiconductor-based micro-electrode arrays. Microchemical systems will be designed and constructed through such approaches.

3. PHOTOSENSITIZED REACTIONS OF SnO_2 MICROELECTRODE ARRAY

3.1. $Ru(bpy)_3^{2+}$ Sensitized Photocurrent at SnO_2 Microelectrode

SnO_2 microelectrode arrays are also very suitable for photoelectrochemical control of reactions in micrometer dimension, since SnO_2 shows semiconducting properties and does not possess strong absorption in the visible region. Therefore, we fabricated SnO_2 microelectrode arrays through photolithography of SnO_2/sapphire substrates. The size and the structure of the array consisting of eight SnO_2 microelectrodes were the same with those in Figure 1. The use of a sapphire plate as a substrate allowed one to irradiate the array from the substrate side and to facilitate photoreactions at the electrode/solution interface or in the solution phase near the electrode.

When the SnO_2 array is placed in a 0.05 M aqueous H_2SO_4 solution of tris(2,2'-bipyridine)ruthenium (II) complex ($Ru(bpy)_3^{2+}$; 7.1 mM) and the solution phase near the electrode (biased + 1.0 V vs. Ag/AgCl) is irradiated in the visible region, we can observe a cathodic photocurrent at the electrode (I_p). The action spectrum of the photocurrent above 400 nm agreed satisfactorily with the absorption spectrum of the metal-to-ligand charge transfer band of the Ru(II) complex (λ_{max} ~ 450 nm). Since SnO_2 does not absorb strongly above 400 nm, I_p at SnO_2 is essentially responsible for the excited state of $Ru(bpy)_3^{2+}$ in the solution phase. Under such conditions, following reactions are expected to proceed at the SnO_2 electrode (Scheme I). Firstly, oxidation of the excited state of Ru(II) (*Ru(II)) to Ru(III) will take place through electron transfer from the highest-occupied molecular orbital of *Ru(II) to the conduction band (CB) of SnO_2, which leads to an anodic current at SnO_2 (path **a** in Scheme I). Secondly, reduction of *Ru(II) to Ru(I) is also expected to occur through electron tunneling from CB of SnO_2 to the lowest-unoccupied molecular orbital of *Ru(II) (path **b**). If Ru(III) exists in the reaction system, reduction of Ru(III) to Ru(II) will take place at the electrode (path **c**). It is noteworthy that the latter two processes (path **b** and **c**) lead to a cathodic photocurrent at SnO_2. Our experimental observation of the cathodic photocurrent at SnO_2 coincides very well with those reported by several research groups for the $Ru(bpy)_3^{2+}$ - bulk SnO_2 electrode (high doping) systems in air [11, 12]. Since we could not observe an anodic photocurrent, the major reactions proceeded in the present system were concluded to be those in path **b** and **c**.

Besides these reactions at the electrode/solution interface, reactive species generated in the bulk solution phase will also influence the photocurrent at SnO_2 through diffusion of the species to the electrode surface. Ru(III) produced by electron transfer between *Ru(II) and an electron acceptor (**A**) and/or *Ru(II) generated in the bulk phase will play a role to some extent for the photocurrent at SnO_2. Although separation of the surface and bulk processes mentioned above is in general difficult, microelectrode array systems can provide an inside look at the surface and bulk processes. Namely, since the lifetime and diffusion length of Ru(III) is much longer than those of *Ru(II), the contributions of the electrode reactions of Ru(III) and the excited state of Ru(II) to I_p will be separated when photocurrent measurements are performed at the **#1** SnO_2 electrode under photoirradiation to

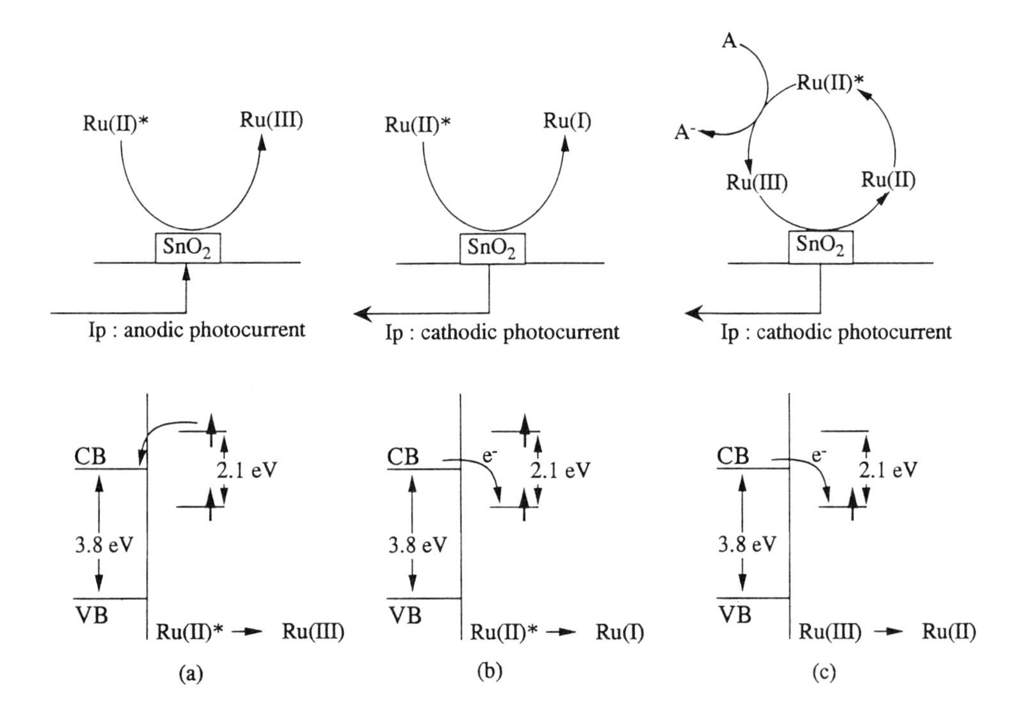

Scheme I. Electrochemical reactions at SnO_2 microelectrode.

various positions in the array. The contribution of the reduction of Ru(III) to I_p will become more important with increasing the distance between the **#1** electrode and irradiated position (*d*).

The curve **a** in Figure 7 shows the time response of the photocurrent at **#1** SnO_2 when the solution phase near the electrode ($d = \sim 10$ μm) is irradiated. It is clear that I_p increases sharply with time ($t < \sim 0.2$ s) and then gradually increases in the t range of 2 ~ 15 s. The fast component will be ascribed to the reduction of *Ru(II) and/or Ru(III) produced in the close vicinity of the SnO_2 electrode (path **b** and/or **c**, respectively, in Scheme I). On the other hand, the slow component showing the rise time of ~ 10 s is clearly too slow to ascribe to the reduction of *Ru(II) at **#1** SnO_2, since the excited state lifetime of the complex is about 0.5 μs. Therefore, the slow component in Figure 7 is primarily responsible for the reduction of Ru(III) generated in the bulk solution phase. Since Ru(III) should diffuse from the bulk phase to the **#1** SnO_2 electrode surface, the photocurrent gradually increases with t (~ 10 s). This was also confirmed from the d dependence of the photocurrent as shown by the curves **b** - **d** in Figure 7. The time response of I_p becomes slower with increasing d from 10 to 130 μm, and the fast component cannot be observed for $d = 130$ μm. The slow time response is ascribed to diffusion of Ru(III) from the bulk phase to the **#1** SnO_2 electrode and subsequent reduction to Ru(II).

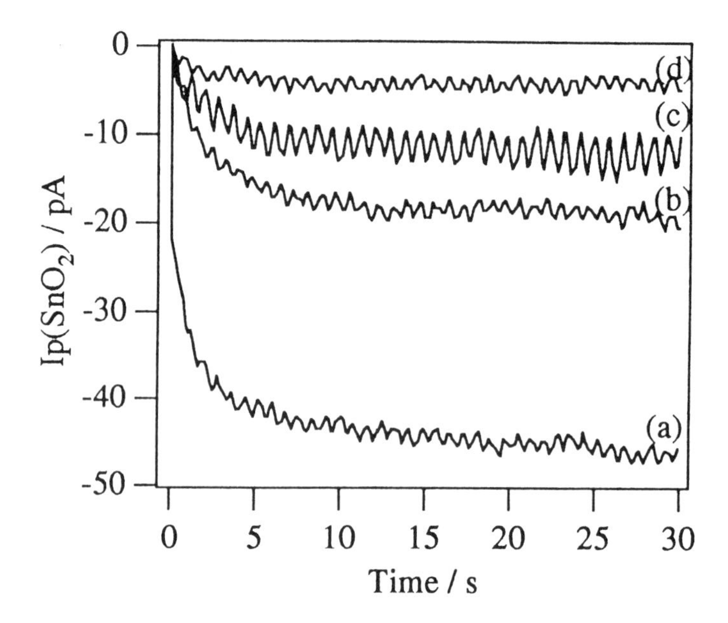

Figure 7. Time response of photocurrent at **#1** SnO_2 microelectrode upon photoirradiation at d = 10 (a), 30 (b), 70 (c), or 130 μm (d).

Since the surface and bulk processes are separated, the latter chemical reactions are now discussed briefly. In the present system, we did not add **A** (electron acceptor) for the excited state of Ru(II) and, therefore, O_2 should act as **A**. Although it has been reported that the excitation energy transfer from the excited state of Ru(II) to O_2 prefers to electron transfer [13], we conclude the oxidation of *Ru(II) by O_2 producing Ru(III) takes place in the present system as confirmed by the cathodic photocurrent at SnO_2. On the other hand, Ru(III) has been known to react with H_2O [14]. When the solution phase near the **#8** electrode is irradiated, Ru(III) should diffuse over the distance of 130 μm for I_p to be observed at **#1** SnO_2 (Figure 7). Assuming the diffusion length (Δ = 130 μm) and the diffusion coefficient of Ru(II) (D ~ 5 x 10^{-6} cm^2 s^{-1}), the relation of $(D/k)^{1/2} = \Delta$ gives the value of k ~ 3 x 10^{-2} s^{-1}, which corresponds to the rate constant for the reaction between Ru(III) and H_2O. Previously, k was reported to be 1.4 x 10^{-4} s^{-1} (pH = 3 ~ 4.8) or 9 x 10^{2} s^{-1} (pH = 1) by Creutz and Sutin [14] or Memming et al. [11], respectively. Although discussion on k is not the main issue of the study, the observed value is comparable to that reported by Creutz and Sutin. Solution phase redox reactions can be also studied based on the microelectrode array system.

3.2. Photochemical Generation and Electrochemical Collection of Redox Active Species by SnO_2 Microelectrode Array

An another important feature of microelectrode arrays is generation and collection of redox active species by a pair of microelectrodes as has been demonstrated by Wrighton and his co-workers [4, 15]. For example, Ru(III) produced by oxidation of Ru(II) at one microelectrode (generator electrode ; GE) can be regenerated to Ru(II) at the adjacent microelectrode (collector electrode ; CE) if appropriate potentials are applied to these electrodes. Redox cycles between Ru(II) and Ru(III) can be actually driven by the SnO_2 microelectrodes as the relevant cyclic voltammograms are shown in Figure 8. The figure proves that Ru(III) is generated by one SnO_2 electrode and collected by the other electrode. In this case, the collection efficiency, defined as the ratio of the current at CE (i_c) to that at GE (i_g), was i_c/i_g ~ 30 %, which almost agreed with the value expected from the separation of two microelectrodes (d = 20 μm) [15]. It is very important to note, furthermore, that Gleria and Memming [16] reported that the bulk SnO_2 electrode - Ru(II) experiments indicated strong adsorption of the complex onto the electrode surface and such adsorption resulted in characteristic electrochemical response of the electrode. Nonetheless, the collection efficiency of ~ 30 % obtained by the present CV experiments indicates that the Ru(II) complex does not adsorb on **#1** SnO_2 microelectrode, but Ru(III) diffuses to the **#2** SnO_2 electrode. Although we cannot explain the discrepancy between these two experiments, the microelectrode array can provide a mechanistic information on the reactions as well.

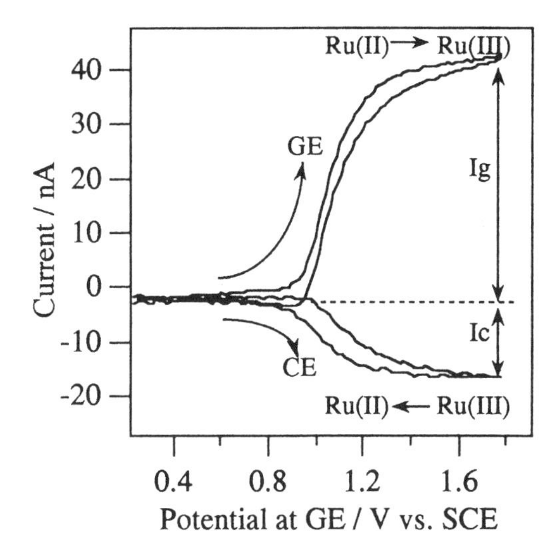

Figure 8. Cyclic voltammograms of $Ru(bpy)_3^{2+}$ by a pair of SnO_2 microelectrodes.

In the present Ru(II)-sensitized reactions, photochemical generation and electrochemical collection of redox active species can be attained as illustrated in Scheme II. Namely, Ru(III) produced by electron transfer between *Ru(II) and **A** in the bulk phase can be reduced to Ru(II) by the **#2** SnO_2 electrode (i.e., CE) when **#2** electrode is biased at an appropriate potential (left figure in Scheme II). Alternatively, the anion radical of **A** produced by the electron transfer can be also collected and oxidized to A at the **#2** SnO_2 (right figure in Scheme II). This implies that an applied potential to the neighboring **#2** microelectrode will suppress consumption of both Ru(II) and **A**, which is expected to lead to enhancement of the photocurrent at **#1** SnO_2. Indeed, an applied potential of 0.3 V (vs. Ag/AgCl) to the **#2** SnO_2 electrode

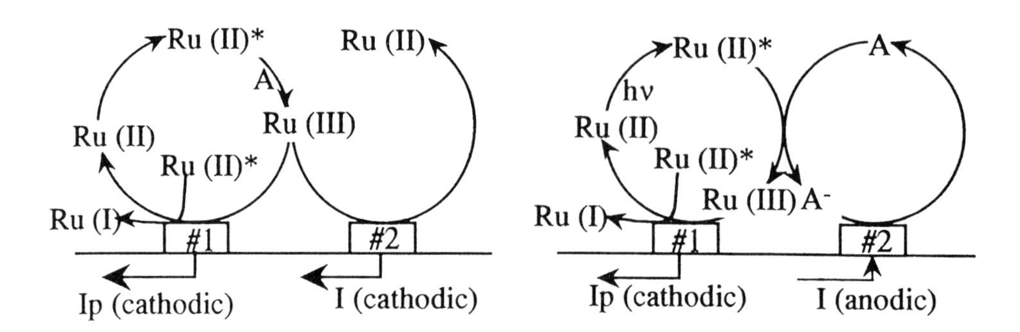

Scheme II. Redox cycles of Ru(II) and **A** by a pair of microelectrode.

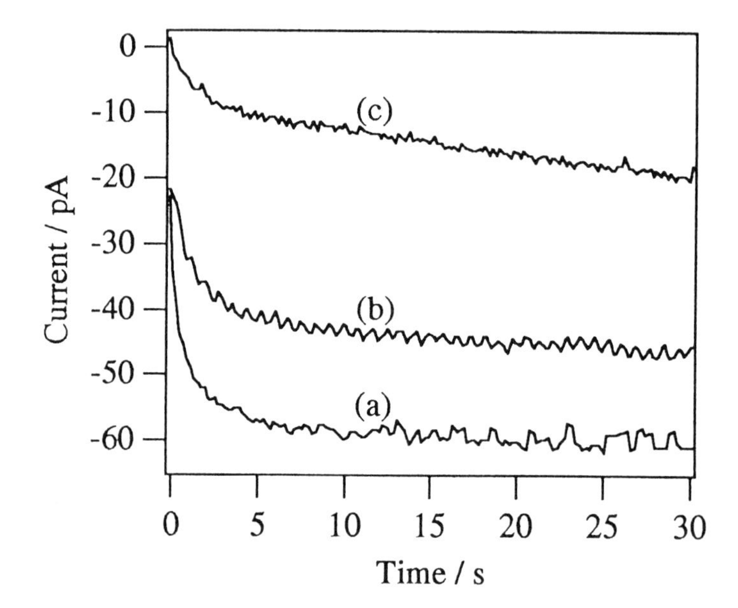

Figure 9. Time response of photocurrent at **#1** SnO_2 microelectrode with (**a**) and without (**b**) applied potential to **#2** SnO_2 electrode. Curve **c** represents the current at **#2** SnO_2 electrode.

resulted in the increase in the cathodic photocurrent at **#1** SnO_2 through collection of Ru(III) and the acceptor anion radical as demonstrated by the curve **a** in Figure 9, in which I_p was shown to be larger than that without an applied potential to **#2** SnO_2 (curve **b**). The curve **c** represents the current at **#2** SnO_2 microelectrode (not photocurrent). Close inspection of the curve **c** indicates that the current is composed of both anodic and cathodic currents and, therefore, reduction of Ru(III) and oxidation of the acceptor anion radical corresponding to the cathodic and anodic currents, respectively, take place at the **#2** SnO_2 electrode. The increase in the photocurrent is ascribed to photochemical generation and electrochemical collection of redox active species in the SnO_2 microelectrode system.

In the SnO_2 - $Ru(bpy)_3{}^{2+}$ system, we showed that the surface and bulk photochemical reactions could be separated by the use of the microelectrode array. Also, redox active species generated photochemically in the solution phase can be collected by the microelectrode, leading to enhancement of the photocurrent at other microelectrode. Photochemical generation and electrochemical collection of redox active species are certainly very fruitful to conduct efficient chemical reactions in micrometer dimension.

4. TOWARDS MICROPHOTOCONVERSION SYSTEMS

Microfabrication and micromodification techniques have realized to prepare semiconductor-based microelectrode arrays, by which photochemical and electrochemical control of reactions along the electrodes can be attained. Photoelectrochemical control of reactions is one of the important characteristics of semiconductor-based microelectrode arrays, in particular, the photochemical generation - electrochemical collection mode as demonstrated for the first time in the SnO_2 - $Ru(bpy)_3{}^{2+}$ system. On the other hand, recent advances in laser techniques are successful to conduct three-dimensional manipulation of various microparticles undergoing Brownian motion in solution by a focused laser beam under an optical microscope [17, 18]. Photochemical/physical and electrochemical processes occurring in individual laser-manipulated particles [19 - 21] or at the solution/microdroplet interface [22 - 24] can be also studied on the basis of a laser manipulation - spectroscopy - electrochemistry technique in a nanosecond ~ picosecond time region and a current resolution of picoampere. We expect that microphotoconversion systems where efficient and selective materials conversion proceeds along a series of microchemical reaction sites [25, 26] will be realized in near future through potential applications of the laser manipulation - spectroscopy - electrochemistry techniques combined and spatially-arranged semiconductor-based microelectrode arrays.

REFERENCES

1. M. Fleischmann, S. Pons, D. R. Rolison and P. P. Schmidt (eds.), Ultramicroelectrodes, Datatech Systems, North Carolina, 1987.
2. H. S. White, G. P. Kittlesen and M. S. Wrighton, J. Am. Chem. Soc., 106 (1984) 5357.
3. C. E. D. Chidsey and R. W. Murray, Science, 231 (1986) 25.
4. S. Licht, V. Cammarata and M. S. Wrighton, Science, 243 (1989) 1176.
5. A. J. Bard, J. Photochem., 10 (1979) 59.

6. D. Duonghong, E. Borgarello and M. Grätzel, J. Am. Chem. Soc., 103 (1981) 4685.
7. T. Uchida, H. Sugimura, K. Pasztor, A. Sekiguchi, N. Shimo, N. Kitamura and H. Masuhara, in preparation.
8. T. Uchida, H. Sugimura, A. Sekiguchi, N. Kitamura, N. Shimo and H. Masuhara, J. Electroanal. Chem., 351 (1993) 343.
9. K. Leitner, J. W. Schultze and U. Stimming, J. Electrochem. Soc., 133 (1986) 1561.
10. W. Olthuis, Anal. Chim. Acta, 229 (1990) 71.
11. R. Memming, F. Schroppel and U. Bringmann, J. Electroanal. Chem., 100 (1979) 307.
12. A. K. Mesmaeker, M. Rochus-Dewitt and J. Nasielski, J. Phys. Chem., 90 (1986) 6657.
13. R. Bensasson, C. Salet and V. Balzani, J. Am.Chem.Soc., 98 (1976) 3722.
14. C. Creutz and N. Sutin, Proc. Natl. Acad. Sci., USA, 72 (1975) 2858.
15. S. Licht, V. Cammarata, and M. S. Wrighton, J. Phys. Chem., 94 (1990) 6133.
16. M. Gleria and R. M. Memming, Z. Phys. Chem., 98 (1975) 303.
17. H. Misawa, M. Koshioka, K. Sasaki, N. Kitamura and H. Masuhara, J. Appl. Phys., 70 (1991) 3829.
18. K. Sasaki and H. Misawa, this volume.
19. M. Koshioka, H. Misawa, K. Sasaki, N. Kitamura and H. Masuhara, J. Phys. Chem., 96 (1992) 2909.
20. K. Nakatani, H. Misawa, K. Sasaki, N. Kitamura and H. Masuhara, J. Phys. Chem., 97 (1993) 1701.
21. K. Sasaki and M. Koshioka, this volume.
22. K. Nakatani, T. Uchida, H. Misawa, N. Kitamura and H. Masuhara, J. Phys. Chem., 97 (1993) 5197.
23. K. Nakatani, T. Uchida, H. Misawa, N. Kitamura and H. Masuhara, J. Electroanal. Chem., in press.
24. K. Nakatani, T. Uchida and N. Kitamura, this volume.
25. H. Masuhara, Pure Appl. Chem., 64 (1992) 1278.
26 H. Masuhara, Photochemical Processes in Organized Molecular Systems, K. Honda, N. Kitamura, H. Masuhara, T. Ikeda, M. Sisido and M. A. Winnik (eds.), p.509, Elsevier, Amsterdam, 1991.

MICROCHEMISTRY
Spectroscopy and Chemistry in Small Domains
Edited by H. Masuhara et al.

Chemical reaction control in small domains: laser trapping-electrochemistry-photochemistry of a single microdroplet

K. Nakatani*,#, T. Uchida$ and N. Kitamura#

Microphotoconversion Project†, ERATO, Research Development Corporation of Japan, 15 Morimoto-cho, Shimogamo, Sakyo-ku, Kyoto 606, Japan

Photochemistry and electrochemistry of a single oil droplet dispersed in water have been studied to understand and control chemical reactions in micrometer small domains. The experiments were performed by a new technique combined with laser trapping, fluorescence/absorption microspectroscopy, and a microelectrochemical method (laser trapping-spectroscopy-electrochemistry technique) and we demonstrated direct observations of mass transfer and chemical reactions between a single oil droplet and the surrounding water phase. A potential means of the technique and characteristic features of microchemistry are discussed.

1. INTRODUCTION

Oil-in-water or water-in-oil emulsions composed of microdroplets in solution are particularly interested as microchemical reaction systems, in which various chemical reactions and mass transfer of solutes proceed across the droplet/solution interface. As an example, a hydrophobic solute in an oil droplet reacts with a solute in the surrounding water phase at the droplet/water interface and the product(s) is extracted into the droplet or water phase via a distribution equilibrium [1-3]. In such the case, the reactions and/or mass transfer are expected to depend on the diameter of the droplet (d), since a surface area/volume ratio increases with decreasing d. Furthermore, an interfacial potential and a surface tention at the droplet/solution boundary will vary with compositions and concentrations of solutes dissolved in both phases [4]. Diffusion of solutes in a droplet and the surrounding solution phase is also expected to play a key role for chemical reactions in emulsions. Although chemical reactions and/or phenomena in microdroplets or emulsions are supposed to relate complicatedly to various factors as mentioned above, a suitable

* To whom correspondence should be addressed.
# Present address; Department of Chemistry, Faculty of Science, Hokkaido University, Sapporo 060, Japan.
$ Present address; Electrochemiscopy Project, ERATO, Research Development Corporation of Japan, Sendai 982, Japan.
† Five-year term project : October 1988 ~ September 1993.

method to resolve these factors governing the chemistry in individual microdroplets has not been so far available. Development of a method to study and control chemical reactions in a single microdroplet is therefore certainly necessary for advances in chemistry of emulsions as well as for application of emulsions to microchemical systems.

In order to study and control reactions and mass transfer in emulsions, microelectrochemical and spectroscopic techniques are quite useful. Namely, microelectrodes or their arrays with various spatial arrangements can be prepared by potential applications of microfabrication techniques, so that redox reactions will be easily induced, controlled, and monitored in μm-dimension by a microelectrochemical method. Mass transfer processes are known to be followed by microelectrode voltammetry as well [5,6]. Quite recently, furthermore, we developed a laser trapping-spectroscopy method, by which individual microparticles in solution could be freely manipulated in three-dimensional space and chemical reactions occurring in the particle could be simultaneously measured by absorption and fluorescence spectroscopy in ps ~ ns time regime [7-9]. A combination of the microelectrochemical method with laser trapping-spectroscopy of individual microdroplets (laser trapping-spectroscopy-electrochemistry) is therefore highly fruitful to elucidate chemical reactions and mass transfer proceeding in emulsions. Microchemical systems based on microdroplets will be also constructed through the laser trapping-spectroscopy-electrochemistry system and an appropriate choice of a reaction system.

In this review, we describe electrochemistry of a single microdroplet in water, with the discussion being focused on electron transfer across the droplet/electrode interface and mass transfer in and across the droplet and water phases [10,11]. Control of photoinduced electron transfer and dye formation reactions in individual microdroplets by mass transfer of a solute across the droplet/water interface and the distance between the electrode and a droplet in μm-dimension is also described in detail [10,12].

2. LASER TRAPPING-SPECTROSCOPY-ELECTROCHEMISTRY [10,12]

Electrochemical studies of redox active species in a single biological cell or microcapsule have been so far conducted by inserting an ultramicroelectrode into the cell or capsule [13-15]. However, immobilization of a microparticle undergoing vigorous Brownian motion in solution is generally difficult, so that the number of the work on electrochemical measurements on individual microparticles is quite limited. However, a laser trapping-spectroscopy-electrochemistry system has now provided a powerful means to study electron transfer and mass transfer across the microdroplet/water interface.

A block diagram of the system is shown in Figure 1. A 1064 nm laser beam from a continuous-wave Nd^{3+} : YAG laser (Spectron, SL-903U) was introduced to an optical microscope (Nikon, Optiphoto XF) and focused (spot size ~ 1 μm) on a droplet through an objective lens (x 100, NA = 0.75). For fluorescence

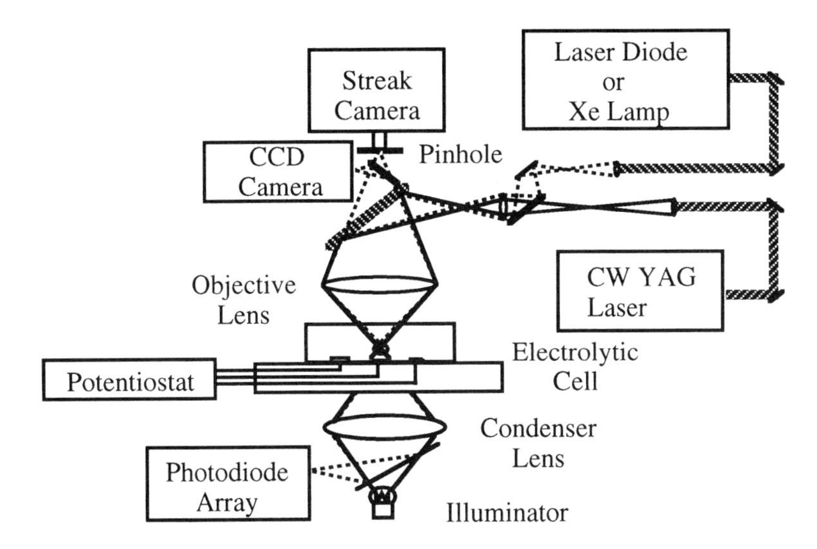

Figure 1. A block diagram of a laser trapping-spectroscopy-electrochemistry system.

measurements, 391.5 nm pulses from a picosecond diode laser (Hamamatsu Photonics, PLP-02, ~30 ps, 1 MHz) as an excitation light source was led to the microscope coaxially with the 1064 nm laser beam and focused (< 5 μm) on the droplet. Fluorescence from the droplet was collected by the same objective lens and detected by a streak camera (Hamamatsu Photonics, C4334, time resolution ~15 ps). For absorption measurements, xenon flash pulses (Tokyo Instruments, XF80-60, pulse duration ~ 70 μs, 3 Hz) as a monitor light beam were led to the microscope coaxially with the 1064 nm beam, and irradiated (spot size ~ 6 μm) to the droplet. The monitor light beam was detected by a photodiode array (Princeton Instruments, DSIDA). Electrochemical behaviour was measured with an electrochemical analyzer (BAS, BAS100B - low current module). An Au microelectrode (8 μm wide x 33 μm long x ~ 0.3 μm thick) fabricated by microlithgraphy or Au wire (d = 10 μm, length ~ 50 μm) as a working electrode was set on a slide glass, and a sample solution was placed between the slide and cover glasses. An Au or Pt wire was used as a counter electrode. All potentials were controlled relative to an Ag/AgCl electrode.

3. ELECTROCHEMISTRY OF FERROCENE IN A SINGLE DROPLET [10,11]

As an example of electrochemical measurements on individual microdroplets, we studied direct electrolysis of ferrocene (FeCp ; 2.0 x 10^{-2} M) in a single nitrobenzene

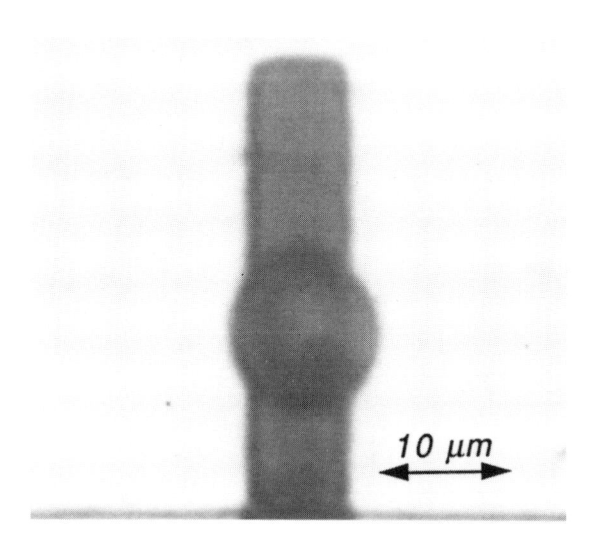

Figure 2. Single nitrobenzene droplet on an Au microelectrode.

(NB) droplet dispersed in water ($MgSO_4$; 0.10 M). As a common cation, tetra-n-butylammonium tetraphenylborate (TBA^+TPB^- ; 0.10 M) and tetra-n-butylammonium chloride (TBA^+Cl^- ; 1.0×10^{-3} M) were added to the NB and the water phases, respectively. Under the conditions, the potential difference at the NB/water interface was calculated to be -131 mV on the basis of the concentration of TBA^+ in both phases [16]. An individual NB droplet was freely manipulated by the 1064 nm laser beam similar to the previous results [7] and contacted with an Au microelectrode, as shown in Figure 2. After contact of the droplet with the electrode, the Brownian motion of the droplet was completely frozen without laser trapping, so that the following electrochemical measurements were performed without irradiating the 1064 nm laser beam to the droplet.

Figure 3 shows current (I) - potential (E) curves of FeCp in a single droplet with d ~ 8 μm. Since NB is not oxidized in the E range of 0 - 0.7 V, I is ascribed to oxidation of FeCp in the droplet. The I-E curve of FeCp in the droplet was distorted from a symmetrical parabolic shape showing an anodic peak (E_{pa}) around ~400 mV. Althuogh E_{pa} was highly depended on a potential sweep rate (v), the total electron charge (Q) during a forward potential sweep calculated by integration of the I-E curve was ~400 pC irrespective of v (50, 100 or 200 mV/s). The Q value agrees with the concentration of FeCp in the droplet ([FeCp(NB)]) with d = 7.3 μm. The result indicates that oxidation of FeCp in the droplet is completed during the first potential sweep.

A cathodic current corresponding to reduction of ferrocenium cation ($FeCp^+$) to FeCp was not observed in the v range of 25 ~ 500 mV/s. If $FeCp^+$ remains in the electrolyzed droplet, the cathodic current should be observed even at a slow scan rate. Although a distribution coefficient of $FeCp^+$ ($[FeCp^+(NB)]$ / $[FeCp^+(W)]$, where $[FeCp^+(NB)]$ and $[FeCp^+(W)]$ are the concentrations of $FeCp^+$ in NB and water, respectively) is estimated to be ~20 from the Gibbs energy of transfer of $FeCp^+$ [17], a volume ratio of the droplet (~2 x 10^{-10} ml for d ~ 8 μm) to the water phase (~2 ml; solution volume) is very small. Therefore, $FeCp^+$ is expected to exit from the droplet owing to the distribution equilibrium of $FeCp^+$ between the droplet and water phases during the forward potential sweep.

Figure 4 shows a time dependence of Q in a single droplet (d ~ 8 μm) after electrolysis of FeCp. Since oxidation of FeCp to $FeCp^+$ is a one-electron transfer step, the Q value should agree with the concentration of FeCp in the droplet

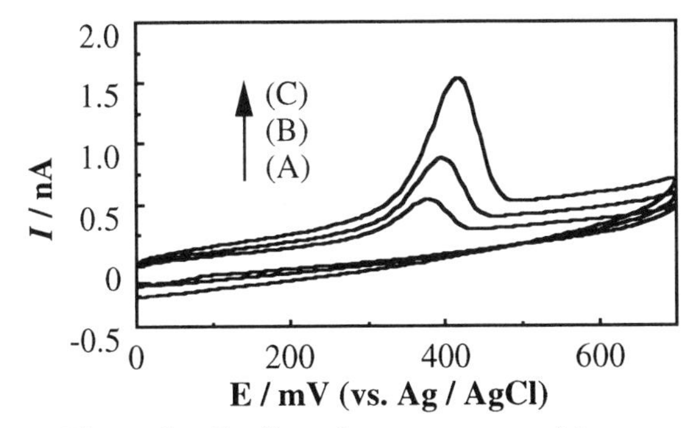

Figure 3. Cyclic voltammograms of ferrocene (v = 50 mV/s (A), 100 mV/s (B), and 200 mV/s (C)) in a single nitrobenzene droplet.

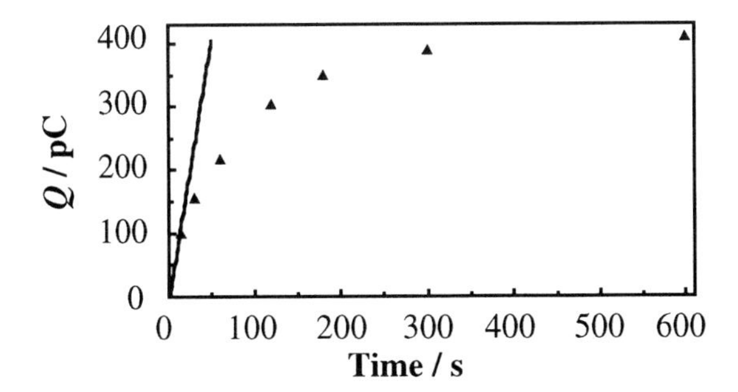

Figure 4. Time dependences of observed (▲) and calculated (–) Q in a single nitrobenzene droplet after electrolysis of ferrocene.

([FeCp(NB)]) and, therefore, the time dependence of Q corresponds to that of [FeCp(NB)]. Recovery of Q with time in Figure 4 suggests that the decrease in [FeCp(NB)] upon electrolysis is compensated by redistribution of FeCp to the droplet either from the water phase or other nonelectrolyzed droplets. The concentration of FeCp in water, [FeCp(W)], was determined to be 3 x 10^{-6} M by a distribution coefficient of FeCp ; [FeCp(NB)] / [FeCp(W)] = 7 x 10^{3}. The total mole number of FeCp ([FeCp(NB)] = 2.0 x 10^{-2} M) in the single droplet with $d \sim 8$ μm is 10^{6} times smaller than that in the water phase with a solution volume of ~ 2 ml, so that the mole number of FeCp solubilized in the water phase is enough to account for recovery of FeCp in the single electrolyzed droplet. This implies that the recovery rate is solely determined by the distribution reequilibrium rate of FeCp between the droplet and water phases. The entrance process from the water phase into the droplet is expected to be governed by mass transfer of FeCp through diffusion in water, permeation across the NB/water interface, or diffusion in the droplet. In order to analyze the redistribution process, the time dependence of Q was simulated by a diffusion-limited model for the collision between the droplet and FeCp in water [18];

$$Q = \int 4\pi r^{2} F D\, [\mathrm{FeCp(W)}]\, [1 / (\pi D\, t)^{1/2} + 1 / r]\; dt \qquad (1)$$

where r, F, and D are the radius of a droplet (4 μm), the Faraday constant, and the diffusion coefficient of FeCp in water, respectively. For the calculation, D was assumed to be equal to that of (ferrocenylmethyl)trimethylammonium cation in water (6 x 10^{-6} $cm^{2}s^{-1}$). The calculated curve agreed very well with the observed one in the initial time region as demonstrated in Figure 4, indicating that a rate-determining step of the entrance process is diffusion of FeCp in water. As distribution of FeCp from the water phase into the droplet proceeds, the contribution of the exit process of FeCp from the droplet into the water phase increases to establish the distribution reequilibrium. In the actual experiments, therefore, the Q value saturates in the time region later than ~300 s. When I-E curves of the same droplet were measured again at ~300 s after electrolysis, therefore, the analogous results with those in Figure 3 were satisfactorily reproduced and I-E curves at various v were obtained for a single droplet. For smaller size of the droplet ($d < \sim 8$ μm), the Q value saturated in a shorter time scale (< ~300 s) as expected from eq.(1) [19].

With increasing v from 50 to 100 or 200 mV/s (Figure 3), the I-E curve was shifted to the positive direction. FeCp in the droplet is completely oxidized as discussed above, so that the anodic peak current (I_{pa}) should be directly proportional to v, similar to the results of thin-layer electrolysis. Besides I_{pa}, we found that E_{pa} was proportionally related to ln$v$ (v = 25 ~ 500 mV/s). Figure 5 shows v dependences of I_{pa} and E_{pa} in a single droplet with $d \sim 8$ μm. The observed I values were in the range of ~nA so that ohmic drop owing to resistance (R) of the NB droplet was very small and, therefore, the shift in E_{pa} with v is not ascribed to IR drop. As an another possibility, electrolysis of FeCp in a droplet is expected to accompany migration of TBA^{+} across the NB droplet/water interface to compensate a difference in the cation concentration between the two phases. According to separate experiments, however, we concluded that such the effect did not explain the present data. The resluts imply that the E_{pa} shift is ascribed to electron transfer of FeCp.

The v dependence of the I-E curve was further analyzed by the relationships known for electrolysis in thin-layer systems. For a totally irreversible electron transfer reaction, I_{pa} and E_{pa} are given by eqs.(2) and (3) [18,20], respectively,

$$I_{pa} = n(1-\alpha)n_0F^2V^0v\,[\mathrm{FeCp(NB)}] / (2.718)RT \tag{2}$$

$$E_{pa} = E^0 + \{RT / (1-\alpha)n_0F\}\ \ln[(1-\alpha)n_0FvV^0/ ARTk^0] \tag{3}$$

where n and n_0 are the numbers of electron transferred and involved in a rate-determining step, respectively, which are assumed to be unity in the present system. α is a transfer coefficient at the transition state of the reaction. V^0 and A correspond to the volume of a droplet and the contacting area between the droplet and the Au microelectrode, respectively. R and T are ususal meanings. E^0 and k^0 are a standard potential including the contribution from the NB/water interfacial potential and the relevant rate constant, respectively. According to eqs.(2) and (3), α and V^0 can be estimated from the slopes of the E_{pa} - ln$v$ and I_{pa} - v curves. The calculated V^0 value was consistent with that of the droplet measured, while α was determined to be ~0.1. α is related to a transition probability from the reactant to product potential energy surfaces along the reaction coordinate, and is known to be ~0.5 when the potential energy surfaces are symmetrical. The calculated value of α

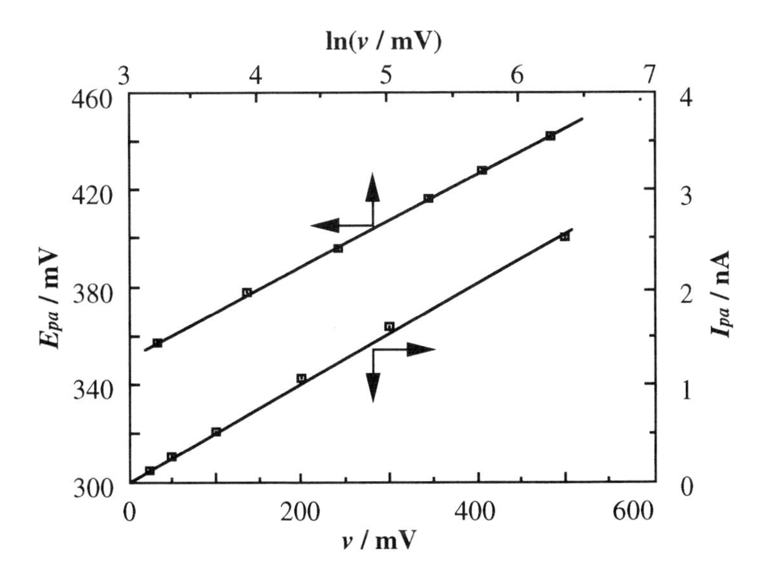

Figure 5. v dependences of I_{pa} and E_{pa} in a single nitrobenzene droplet.

~ 0.1 is clearly too small for electron transfer of FeCp across the NB/electrode interface. For potential sweep measurements on FeCp in bulk systems, a change in v does not lead to an E_{pa} shift because of fast electron transfer in solution and the highly reversible redox reaction. However, the present result of $\alpha \sim 0.1$ indicates that oxidation of FeCp in a NB droplet by an Au microelectrode is irreversible in marked contrast to the results in the bulk system. A possible origin of unusual α will be to assume a decrease in the transition probability by the presence of an adsorption layer between the droplet and the electrode or to assume fast exit of $FeCp^+$ from the droplet to the water phase upon potential sweep. Further detailed studies are necessary to understand the primary reason for $\alpha \sim 0.1$.

We succeeded in quantitative measurements of electrochemical behaviour of FeCp in a single μm-sized droplet and directly observed mass transfer of FeCp/$FeCp^+$ across the droplet/water interface for the first time. Such studies are only possible based on the laser trapping-electrochemistry technique, and we think that the approach is very fruitful to understand both the mechanisms and the factors controlling electron transfer and mass transfer proceeding in individual microdroplets dispersed in solution.

4. CONTROL OF REACTIONS IN A SINGLE DROPLET

4.1. Control of a photoinduced electron transfer reaction by mass transfer across the droplet/water interface [10]

As described in the preceding section, the distribution equilibrium of FeCp between the oil droplet and the water phase is easily perturbed by electrolysis of FeCp in the droplet and, therefore, the concentration of FeCp in a single droplet can be controlled electrochemically. We extended the method to control a photoinduced electron transfer reaction of 9,10-diphenylanthracene (DPA) in a droplet by mass transfer of FeCp across the droplet/water interface. Tri-n-butyl phosphate (TBP) microdroplets containing FeCp (0.10 M), DPA (1.0×10^{-2} M), and TBA^+TPB^- (1.0×10^{-2} M) dispersed in an aqueous TBA^+Cl^- (1.0×10^{-2} M) solution were used as a sample emulsion. The photoinduced electron transfer reaction in the DPA/FeCp droplet is assumed to proceed via the following model (Scheme I);

$$\begin{array}{ccc} DPA^* + FeCp & \xrightarrow{k_{ET}} & DPA^- + FeCp^+ \\ \downarrow k_0 & & \downarrow \\ DPA & & DPA + FeCp \end{array} \qquad \text{Scheme I}$$

$$[DPA^*(t)] = [DPA^*(t = 0)] \exp[-(k_{ET}[FeCp(TBP)] + k_0)\, t] \tag{4}$$

$$k_0 = k_F + k_G$$

where [DPA*(t)] gives a time response of DPA fluorescence, and k_{ET}, k_F and k_G are electron transfer, radiative, and non-radiative rate constants of DPA*, respectively.

In a bulk TBP solution, $1/k_0$ and $1/(k_{ET}[FeCp] + k_0)$ were determined to be ~ 9 and ~ 1 ns, respectively, by separate experiments ([FeCp] = 0.10 M).

Figure 6 shows fluorescence decay curves of DPA in a droplet with d ~ 20 μm under several conditions. Before electrolysis, the decay time of DPA* was determined to be ~1 ns (a), indicating that electron transfer from DPA* to FeCp took place efficiently in the droplet. Also, the value agreed very well with that in the bulk DPA/FeCp solution. On the other hand, the fluorescence lifetime during electrolysis was as long as 7 ~ 8 ns (b), similar to that in the absence of FeCp, while the lifetime at about 10 min after electrolysis of the droplet was ~1 ns (c). The results demonstrate that photoinduced electron transfer between DPA* and FeCp is controlled by mass transfer of FeCp in the single droplet system.

Figure 7. shows a schematic illustration of reaction control by mass transfer

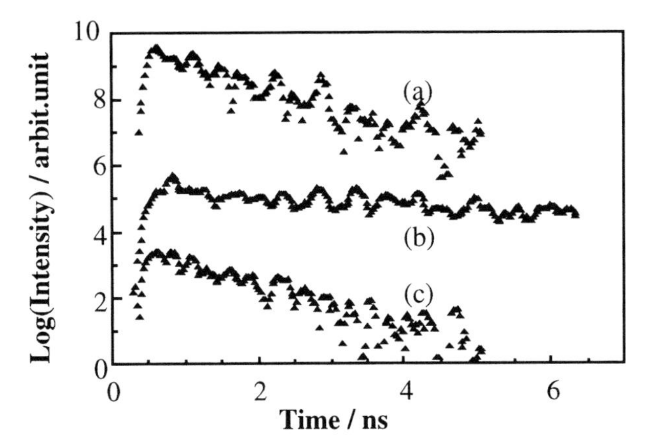

Figure 6. Fluorescence decay curves of DPA before (a), during (b), and after (c) electrolysis of FeCp in a single TBP droplet.

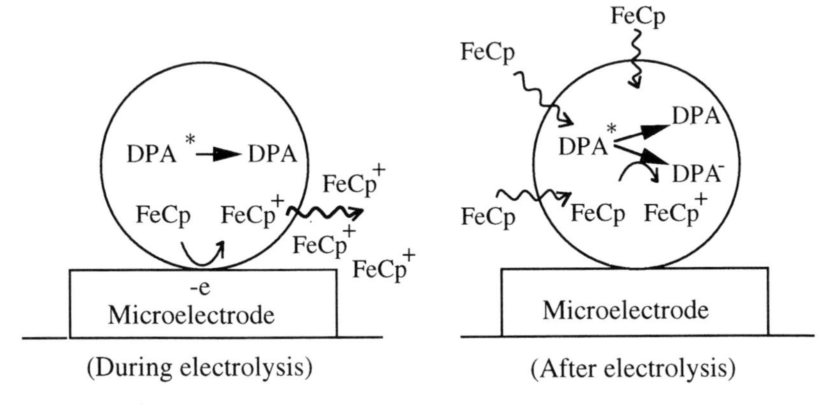

Figure 7. Schematic illustration of control of reactions by mass transfer across droplet/water interface.

across the droplet/water interface. In the presence of FeCp in the droplet before and sufficiently after electrolysis (Figure 6a and 6c, respectively), DPA* is completely quenched owing to the enough amount of FeCp in the droplet ([FeCp] = 0.1 M). On the other hand, if electrolysis of FeCp to $FeCp^+$ in the droplet takes a long time, DPA* will react with unelectrolyzed FeCp still exists in the droplet. If such the case, the fluorescence lifetime of DPA* will be much shorter than 9 ns irrespective the experimental conditions ; Figure 6a, b, and c. Therefore, an electrolytic time (t) of FeCp in a single droplet is a very important factor for controlling photoinduced electron transfer in the droplet. Actually, t can be easily estimated by eq.(5),

$$\Delta = (2Dt)^{1/2} \tag{5}$$

where Δ is the thickness of the diffusion layer, and we assume Δ to be equal to d. D of FeCp was determined to be 2×10^{-6} cm^2s^{-1} in a bulk TBP solution. In the case of $d = \Delta = 20$ μm as a typical example of a μm-sized droplet, t was calculated to be ~ 1 s while that was ~ 2500 s for a mm-sized droplet (d = 1 mm). The calculation clearly indicates that control of photoinduced electron transfer between FeCp and DPA* is very difficult for mm-sized droplets. For μm-sized droplets, on the other hand, FeCp is electrolyzed rapidly, so that mass transfer of $FeCp/FeCp^+$ can be controlled directly through the distribution reequilibria of the solutes across the droplet/water interface. This is one of the important characteristics of chemistry in individual microdroplets. Various chemical reactions in a single droplet competing with electrode reactions will be also controlled by taking mass transfer processes of redox species across the interface into account for designing microchemical systems.

4.2. Control of a dye formation reaction by the droplet-electrode distance [12]

Besides reaction control through mass transfer across the droplet/water interface, we have succeeded in controlling a dye formation reaction in a single microdroplet

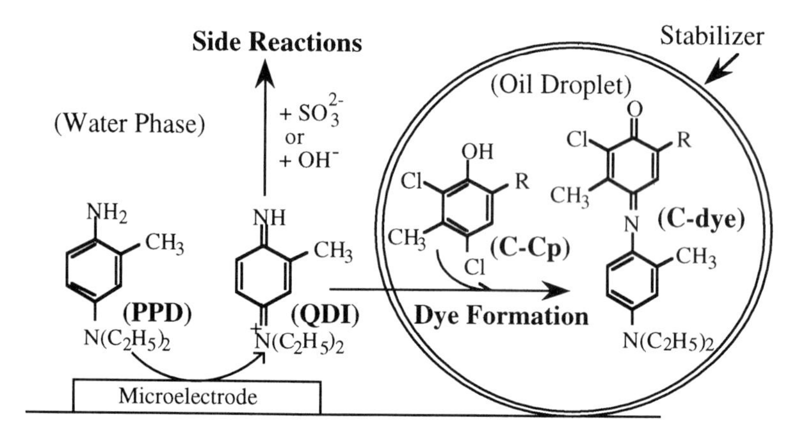

Scheme II. Dye formation reaction at the droplet/water interface.

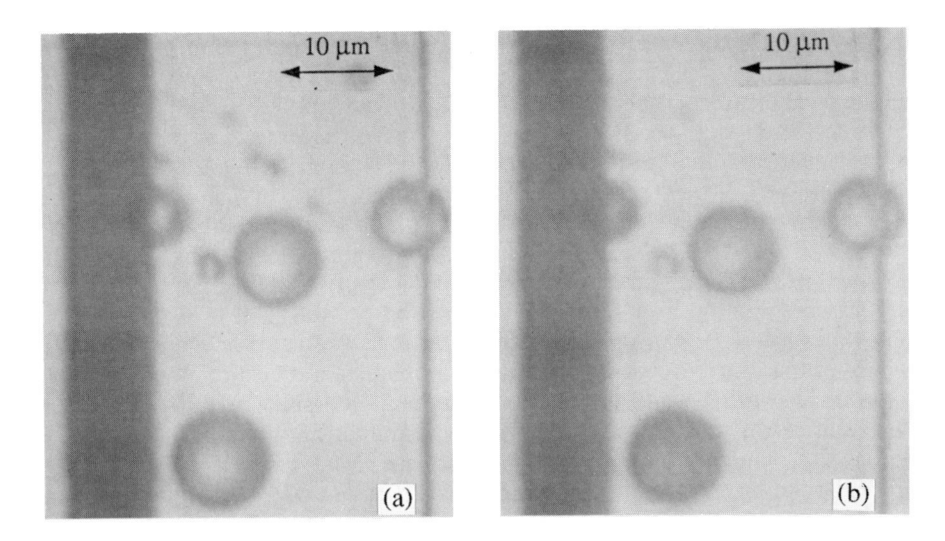

Figure 8. Photographs of laser-trapped di-n-butyl phthalate droplets before (a) and after (b) 15 s electrolysis of PPD in water at the microelectrode.

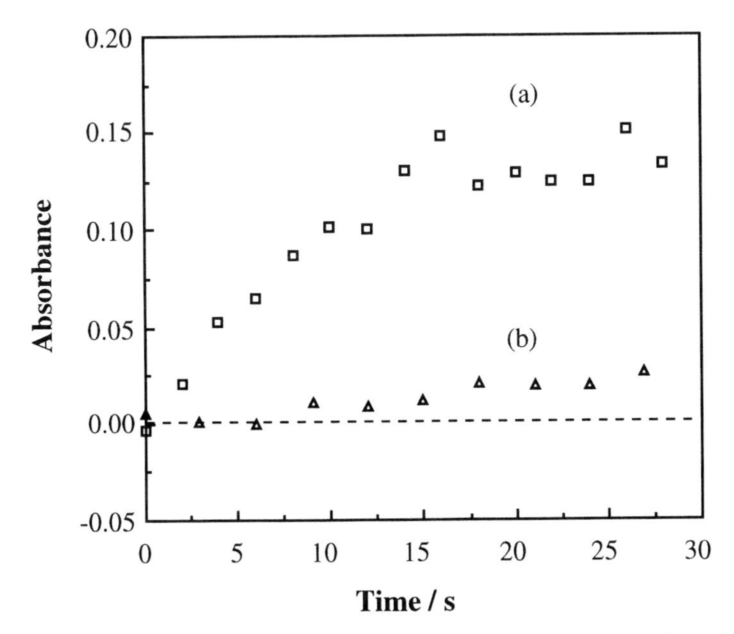

Figure 9. Time dependence on peak absorbance of the C-dye in a single, laser-trapped di-n-butyl phthalate droplet upon electrolysis of PPD in water; (a) $\delta \sim 0\ \mu m$, (b) $\delta \sim 9\ \mu m$.

by the droplet-microelectrode distance (δ). As an oil-in-water emulsion, di-n-butyl phthalate containing a phenol derivative (C-Cp ; 0.13 M) was vigorously mixed with an aqueous solution of gelatin and sodium dodecylsulfate. The emulsion was dispersed in a pH = 10 buffer solution containing 4-N,N-diethyl-2-methylphenylenediamine (PPD ; 5.0 x 10^{-3} M) and sodium sulfite (2.0 x 10^{-2} M).

The cyan-dye (C-dye) formation is assumed to proceed via Scheme II as reported previously for color developing processes in photographic emulsions [3]. In the presence of OH^- and SO_3^{2-} in the water phase, 4-N,N-diethyl-2-methylquinonediimine (QDI) produced by oxidation of PPD at the microelectrode undergoes deamination by OH^- and/or sulfonation by SO_3^{2-}. If QDI diffuses to the droplet/water interface and reacts with C-Cp, the C-dye is produced in the droplet. This implies that the C-dye formation in the droplet should be strongly dependent on the lifetime (τ) of QDI, since the dye formation competes with the deamination and/or sulfonation reactions in water. The deamination and sulfonation rates of N,N-diethylquinonediimine are 0.2 s^{-1} and 2 x 10^4 $M^{-1}s^{-1}$ at pH = 10, respectively [21]. Under the present condition of $[SO_3^{2-}]$ = 20 mM, τ of QDI is mainly governed by the sulfonation reaction and is estimated to be ~3 ms, which corresponds to Δ of QDI to be ~2 μm as calculated by eq.(5), where D = 5 x 10^{-6} cm^2s^{-1} of PPD was used for the calculation and t was assumed to be equal to τ. The Δ value and above discussion indicate that the dye formation can be controlled by δ in μm-dimension.

Photographs of laser-trapped droplets are shown in Figure 8. Knowing the redox potential of PPD to be ~ 30 mV, PPD in water was oxidized at 100 mV by a potential step method. The smaller the δ value, the more efficient the C-dye formation. For δ > ~10 μm, the dye formation was scarcely observed as expected from $\delta >> \Delta$ (~2 μm). Figure 9 shows time dependences of peak absorbance of the C-dye in the droplets (d ~ 10 μm) with different δ. For δ ~ 0 μm, absorbance was saturated at ~20 s after the potential step, while that for δ ~ 9 μm was only 20 % of the value for the droplet at δ ~ 0 μm even after ~30 s electrolysis. The results in Figure 8 were proved by absorption spectroscopy on the individual droplets. In photographic emulsions, it has been reported that C-Cp dissociates into a C-Cp anion and a proton at the droplet/water interface in an alkaline solution, and a rate-determining step of the C-dye formation is coupling between QDI and the C-Cp anion at the oil/water interface for the droplets with d < ~0.25 μm [3]. For di-n-

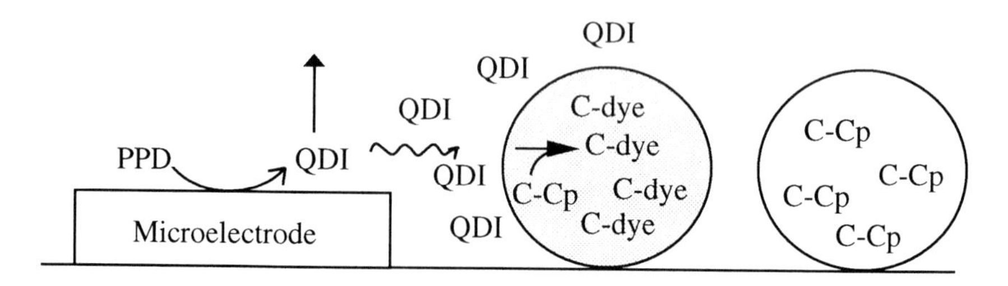

Figure 10. Schematic illustration of control of reacions by droplet-electrode distance

butyl phthalate droplets with $d \sim 10$ μm, the viscosity is so high (~ 20 cP) that diffusion of C-Cp and the C-dye to the droplet surface and interior, respectively, will be limited. Therefore, the C-dye formation is expected to proceed efficiently at the droplet/water interfacial layer in the initial stage of electrolysis.

We demonstrated that chemical reactions in individual droplets coupled with electrode reactions could be controlled by the distance between the droplet and the electrode (Figure 10). Diffusion of QDI from the microelectrode surface is in a cyrindrical manner and, therefore, [QDI] at $\delta \sim 10$ μm is half of that at $\delta \sim 0$ μm even if no side reactions of QDI proceed, as expected from a collection efficiency of redox species by a pair of microelectrodes [4]. This is the primary reason for the success in controlling the dye formation reaction in μm-dimension. It is worth noting that the result is in contrast to that in mm-dimension, where δ or d is too large to diffuse QDI from the electrode surface to a droplet. μm-dimension is certainly unique for controlling various chemical reactions.

5. CONCLUSION

We revealed the characteristic features of electrochemistry proceeding in a single μm-sized droplet, and extended the works to control photoinduced electron transfer and dye formation reactions in μm-dimension. The present results clearly show a potential means of the laser trapping-spectroscopy-electrochemistry technique to induce, monitor, and control the reactions in minute volume. Our ultimate goal of the researches is to propose and construct microchemical reaction systems having high selectivity and efficiency, named Microphotoconversion systems [22], similar to reactions proceeding in biological cell systems. Manipulation and assembling of several microparticles [7,23,24] and fabrication of functionalized microelectrode arrays [25] have been realized, so that successive reactions in individual laser-trapped particles in solution along addressable microelectrode arrays will be conducted and controlled arbitrarily by the sophisticated method of laser trapping-spectroscopy-electrochemistry. Microphotoconversion systems will be designed and constructed through such approaches.

REFERENCES

1. I. Willner, W. E. Ford, J. W. Otvos and M. Calvin, Nature, 280 (1979) 823.
2. J. Kiwi and M. Gratzel, J. Am. Chem. Soc., 27 (1978) 6314.
3. T. H. James, The Theory of the Photographic Process, Macmillian, New York, 1977.
4. J. Koryta and P. Vanysek, Advances in Electrochemistry and Electrochemical Engineering, H. Gerischer and C. W. Tobias (eds.), John Wiley, New York, 12 (1981) 113.
5. Ultramicroelectrodes, M. Fleischmann, S. Ponds, D. Roison and P. P. Schmidt (eds.), Datatech Science, Morgaton, N. C., 1987.

6. A. J. Bard, J. A. Crayston, G. P. Kittlesen, T. V. Shea and M. S. Wrighton, Anal. Chem., 58 (1986) 2321.
7. H. Misawa, N. Kitamura and H. Masuhara, J. Am. Chem. Soc., 113 (1991) 7859.
8. S. Funakura, K. Nakatani, H. Misawa, N. Kitamura and H. Masuhara, J. Phys. Chem., submitted.
9. M. Koshioka, H. Misawa, K. Sasaki, N. Kitamura and H. Masuhara, J. Phys. Chem., 96 (1992) 2909.
10. K. Nakatani, T. Uchida, H. Misawa, N. Kitamura and H. Masuhara, J. Phys. Chem., 97 (1993) 5197.
11. K. Nakatani, T. Uchida, H. Misawa, N. Kitamura and H. Masuhara, J. Electroanal. Chem., in press.
12. K. Nakatani, T. Uchida, S. Funakura, A. Sekiguchi, H. Misawa, N. Kitamura and H. Masuhara, Chem. Lett., (1993) 717.
13. T. Abe, T. Matsue and I. Uchida, Chem. Lett., (1989) 301.
14. T. Matsue, S. Koike, T. Abe, T. Itabashi and I. Uchida, Biochim. Biophys. Acta, 1101 (1992) 69.
15. C. Lee, J. Kwak and A. J. Bard, J. Proc. Natl. Acad. Sci. USA, 87 (1990) 1740.
16. T. Kakiuchi and M. Senda, Bull. Chem. Soc. Jpn., 60 (1987) 3099.
17. Z. Samec, V. Marecek and J. Weber, J. Electroanal. Chem., 103 (1979) 11.
18. A. J. Bard and L. R. Faulkner, Electrochemical Methods. Fundamentals and Applications, John Wiley, New York, 1980.
19. K. Nakatani, T. Uchida, N. Kitamura and H. Masuhara, J. Electroanal. Chem., submitted.
20. A. T. Hubbard, J. Electroanal. Chem., 22 (1969) 165.
21. H. Kobayashi, K. Yoshida, H. Takano, T. Ohno and S. Mizusawa, J. Imaging Sci., 32 (1988) 90.
22 H. Masuhara, Photochemical Processes in Organized Molecular Systems, K. Honda et al. (eds.), Elsevier, Amsterdam, 491 (1991).
23. K. Sasaki, M. Koshioka, H. Misawa, N. Kitamura and H. Masuhara, Jpn. J. Appl. Phys., 30 (1991) L907.
24. H. Misawa, K. Sasaki, M. Koshioka, N. Kitamura and H. Masuhara, Macromolecules, 26 (1993) 282.
25. C. D. Frisbie, I. Fritsch-Faules, E.W. Wollman, M. S. Wrighton, Thin Solid Films, 210-211 (1992) 341.

INDEX

Subject Index